D1069007

Die Grundlehren der mathematischen Wissenschaften

in Einzeldarstellungen
mit besonderer Berücksichtigung
der Anwendungsgebiete

Band 52

Formulas and Theorems for the Special Functions of Mathematical Physics

Dr. Wilhelm Magnus

Professor at the New York University
Courant Institute of Mathematical Sciences

Dr. Fritz Oberhettinger

Professor at the Oregon State University
Department of Mathematics

Dr. Raj Pal Soni

Mathematician
International Business Machines Corporation

Third enlarged Edition

Springer-Verlag New York Inc. 1966

Geschäftsführende Herausgeber:

Library of Congress Catalog Card Number 66-28437
Printed in the United States of America

Title No. 5035

Preface

This is a new and enlarged English edition of the book which, under the title "Formeln und Sätze für die Speziellen Funktionen der mathematischen Physik" appeared in German in 1946. Much of the material (part of it unpublished) did not appear in the earlier editions.

We hope that these additions will be useful and yet not too numerous for the purpose of locating with ease any particular result. Compared to the first two (German) editions a change has taken place as far as the list of references is concerned. They are generally restricted to books and monographs and accomodated at the end of each individual chapter. Occasional references to papers follow those results to which they apply. The authors felt a certain justification for this change. At the time of the appearance of the previous edition nearly twenty years ago much of the material was scattered over a number of single contributions. Since then most of it has been included in books and monographs with quite exhaustive bibliographies. For information about numerical tables the reader is referred to "Mathematics of Computation", a periodical published by the American Mathematical Society; "Handbook of Mathematical Functions" with formulas, graphs and mathematical tables National Bureau of Standards Applied Mathematics Series, 55, 1964, 1046 pp., Government Printing Office, Washington, D.C., and FLETCHER, MILLER, ROSENHEAD, Index of Mathematical Tables, Addison-Wesley, Reading, Mass.).

There is a list of symbols and abbreviations at the end of the book. The formulas within each section are arranged in order of increasing complexity (a concept which, cannot of course, be sharply defined).

We gratefully acknowledge the assistance of many people who contributed towards the improvement of the book. Finally, the authors wish to thank the staff of Springer-Verlag for their patience and splendid cooperation.

New York, N.Y. — WILHELM MAGNUS
Corvallis, Oregon — FRITZ OBERHETTINGER
Endicott, N.Y., July 1966 — RAJ PAL SONI

Contents

Chapter I

The gamma function and related functions

1.1 The gamma function

The function $\Gamma(z)$ is a meromorphic function of z with simple poles at $z = -n$, $(n = 0, 1, 2, \ldots)$ with the respective residue $\frac{(-1)^n}{n!}$.

Definitions by an infinite product

$$\Gamma(z) = \lim_{n\to\infty} \frac{n!\, n^z}{z(z+1)\,(z+2)\,\ldots\,(z+n)} = z^{-1} \prod_{n=1}^{\infty} \left(1+\frac{1}{n}\right)^z \left(1+\frac{z}{n}\right)^{-1},$$

$$\frac{1}{\Gamma(z)} = z\, e^{\gamma z} \prod_{n=1}^{\infty} \left(1+\frac{z}{n}\right) e^{-\frac{z}{n}},$$

$$\gamma = \lim_{m\to\infty} \left(\sum_{l=1}^{m} \frac{1}{l} - \log m\right) = 0.577215\ldots$$

Integral representations

$$\Gamma(z) = \int_0^\infty e^{-t} t^{z-1}\, dt = \int_0^1 \left[\log\left(\frac{1}{t}\right)\right]^{z-1} dt, \quad Re\, z > 0,$$

$$\Gamma(z) = \int_0^\infty \left[e^{-t} - 1 + t - \frac{t^2}{2!} + \cdots + (-1)^{n+1} \frac{t^u}{n!}\right] t^{z-1}\, dt,$$

$$-(n+1) < Re\, z < -n,$$

$$\Gamma(z) = \sigma^z \int_0^{\infty e^{i\delta}} e^{-t\sigma} t^{z-1}\, dt, \quad -\frac{1}{2}\pi - \delta < \arg\sigma < \frac{1}{2}\pi - \delta, \quad Re\, z > 0,$$

$$\Gamma(z) = \sigma^z (e^{2\pi i z} - 1)^{-1} \int_{\infty e^{i\delta}}^{(0+)} e^{-t\sigma} t^{z-1}\, dt$$

$$-\frac{1}{2}\pi - \delta < \arg\sigma < \frac{1}{2}\pi - \delta, \ \delta \leq \arg t \leq 2\pi + \delta,$$

$$\sigma \neq 0, \ \pm 1, \ \pm 2, \ldots,$$

$$\frac{1}{\Gamma(z)} = \frac{(\sigma e^{-i\pi})^{1-z}}{2\pi i} \int_{\infty e^{i\delta}}^{(0+)} t^{-z} e^{-t\sigma}\, dt$$

$$-\frac{1}{2}\pi - \delta < \arg\sigma < \frac{1}{2}\pi - \delta, \quad \delta \leq \arg t \leq 2\pi + \delta.$$

The symbol $\int_{\infty e^{i\delta}}^{(0+)}$ denotes an integral taken over a contour loop coming from the point $\infty e^{i\delta}$, encircling the origin counter-clockwise and returning to its starting point

Series expressions for $\Gamma(z)$

$$\Gamma(z) = \sum_{n=0}^{\infty} \frac{(-1)^n}{n!\,(z+n)} + \int_1^{\infty} e^{-t} t^{z-1}\, dt,$$

$$\Gamma(z) = \frac{(-1)^n}{n!} \left\{ \frac{1}{z+n} + \psi(n+1) + \frac{1}{2}(z+n)\left[\frac{1}{3}\pi^2 + \psi^2(n+1) - \psi'(n+1)\right] + O[(z+n)^2] \right\}$$

(series expansion in the neighborhood of a singular point $z = -n$, $(n = 0, 1, 2, \ldots)$

$$\frac{1}{\Gamma(1+z)} = 1 + \gamma z + \left(\gamma^2 - \frac{\pi^2}{6}\right)\frac{z^2}{2} + \cdots$$

Functional equations

$$\Gamma(1+z) = z\Gamma(z),$$

$$\Gamma(n+z) = z(z+1)\cdots(z+n-1)\,\Gamma(z),$$

$$\Gamma(z-n) = (-1)^n \Gamma(z)\,\frac{\Gamma(1-z)}{\Gamma(n+1-z)} = \frac{(-1)^n \pi}{\sin(\pi z)\,\Gamma(n+1-z)},$$

$$\Gamma(z)\,\Gamma(1-z) = \frac{\pi}{\sin(\pi z)},$$

$$\Gamma(z)\,\Gamma(-z) = \frac{-\pi}{z\sin(\pi z)},$$

$$\Gamma\left(\frac{1}{2}+z\right)\Gamma\left(\frac{1}{2}-z\right) = \frac{\pi}{\cos(\pi z)},$$

$$\frac{\Gamma(n+z)\,\Gamma(n-z)}{[(n-1)!]^2} = \frac{\pi z}{\sin(\pi z)} \prod_{m=1}^{n-1}\left(1 - \frac{z^2}{m^2}\right),$$

$$n = 2, 3, 4, \ldots,$$

$$\Gamma\left(n+\frac{1}{2}+z\right)\Gamma\left(n+\frac{1}{2}-z\right) = \frac{\pi}{\cos(\pi z)} \prod_{m=1}^{n}\left[\left(m-\frac{1}{2}\right)^2 - z^2\right],$$

$$n = 1, 2, 3, \ldots,$$

$$\lim_{z\to -m} \frac{\Gamma(n+z)}{\Gamma(z)} = (-1)^m \frac{m!}{(n-m)!}, \quad n \le m$$
$$= \quad 0, \qquad n > m,$$

m, n positive integers.

Multiplication theorems

$$\prod_{l=0}^{m-1} \Gamma\left(z + \frac{l}{m}\right) = (2\pi)^{\frac{1}{2}m - \frac{1}{2}} m^{\frac{1}{2} - mz} \Gamma(mz),$$
$$m = 2, 3, 4, \ldots,$$

$$\prod_{l=0}^{m-1} \Gamma\left(kz + \frac{kl}{m}\right) = (2\pi)^{\frac{1}{2}m - \frac{1}{2}k} \left(\frac{k}{m}\right)^{mkz + \frac{1}{2}(mk - m - k)} \times \prod_{n=0}^{k-1} \Gamma\left(mz + \frac{mn}{k}\right),$$
$$k, m = 1, 2, 3, \ldots$$

Particularly for $m = 2$ and $m = 3$

$$\Gamma(2z) = \pi^{-\frac{1}{2}} 2^{2z-1} \Gamma(z) \Gamma\left(\frac{1}{2} + z\right),$$
$$\Gamma(3z) = (2\pi)^{-1} 3^{3z - \frac{1}{2}} \Gamma(z) \Gamma\left(\frac{1}{3} + z\right) \Gamma\left(\frac{2}{3} + z\right).$$

Special values

$$\Gamma\left(\frac{1}{2}\right) = \pi^{\frac{1}{2}}, \quad \Gamma\left(\frac{3}{2}\right) = \frac{1}{2}\pi^{\frac{1}{2}}, \quad \Gamma\left(-\frac{1}{2}\right) = -2\pi^{\frac{1}{2}},$$
$$\left[\Gamma\left(\frac{1}{4}\right)\right]^2 = 16\pi^2 \left[\frac{3^2}{3^2-1} \cdot \frac{5^2-1}{5^2} \cdot \frac{7^2}{7^2-1} \cdot \frac{9^2-1}{9^2} \cdots\right],$$
$$\Gamma\left(\frac{1}{2} + n\right) = \pi^{\frac{1}{2}} 2^{-2n} \frac{(2n)!}{n!},$$
$$\Gamma\left(\frac{1}{2} - n\right) = (-1)^n \pi^{\frac{1}{2}} 2^{2n} \frac{n!}{(2n)!}.$$

Some symbols expressed in terms of gamma functions

Hankel's symbol

$$(\alpha, l) = \frac{2^{-2l}}{l!} \{(4\alpha^2 - 1)(4\alpha^2 - 3^2) \cdots [4\alpha^2 - (2l-1)^2]\}$$
$$= \frac{\Gamma\left(\frac{1}{2} + \alpha + l\right)}{l!\,\Gamma\left(\frac{1}{2} + \alpha - l\right)}, \quad l = 1, 2, 3, \ldots$$
$$= (\pi l!)^{-1} (-1)^l \cos(\pi\alpha)\, \Gamma\left(\frac{1}{2} + \alpha + l\right) \Gamma\left(\frac{1}{2} - \alpha + l\right).$$

Pochhammer's symbol

$$(\alpha)_n = \alpha(\alpha+1) \cdots (\alpha + n - 1) = \frac{\Gamma(\alpha+n)}{\Gamma(\alpha)}.$$

Binomial coefficient

$$\binom{\alpha}{m} = \alpha(\alpha-1)\cdots\frac{(\alpha-m+1)}{m!}$$

$$= (-1)^m \frac{\Gamma(m-\alpha)}{m!\,\Gamma(-\alpha)} = \frac{\Gamma(1+\alpha)}{m!\,\Gamma(1+\alpha-m)},$$

$$m = 1, 2, 3, \ldots$$

Some infinite products expressed in terms of the gamma function

$$\prod_{n=1}^{\infty}\left(1-\frac{z^2}{n^2}\right) = \frac{\sin(\pi z)}{\pi z} = \frac{1}{\Gamma(1+z)\,\Gamma(1-z)},$$

$$\prod_{n=1}^{\infty}\left(1-\frac{z^4}{n^4}\right) = (\pi z)^{-2}\sin(\pi z)\sinh(\pi z),$$

$$\prod_{n=1}^{\infty}\left[1+(-1)^n\frac{z}{n}\right] = 2\pi^{\frac{1}{2}}\left[z\Gamma\left(\frac{1}{2}z\right)\Gamma\left(\frac{1}{2}-\frac{1}{2}z\right)\right]^{-1},$$

$$\prod_{n=1}^{\infty}\left[1+\frac{\zeta}{z+n}\right]e^{-\frac{\zeta}{n+1}} = e^{-\gamma\zeta}\frac{\Gamma(z)}{\Gamma(z+\zeta)},$$

$$\prod_{n=0}^{\infty}\left(1+\frac{x}{y+n}\right)\left(1-\frac{x}{z+n}\right) = \frac{\Gamma(y)\,\Gamma(z)}{\Gamma(x+y)\,\Gamma(z-x)},$$

$$\prod_{n=0}^{\infty}\left(1+\frac{y}{n+x}\right)e^{-\frac{y}{n+x}} = \frac{\Gamma(x)}{\Gamma(x+y)}e^{y\psi(x)},$$

$$\prod_{n=1}^{\infty}\left(1+\frac{iy}{n+x}\right)^{-1}e^{i\frac{y}{n}} = \left(1+\frac{iy}{x}\right)\frac{\Gamma(x+iy)}{\Gamma(x)}e^{i\gamma y},$$

$$\prod_{n=1}^{\infty}\left[\pi^{-\frac{1}{2}}\Gamma\left(\frac{1}{2}+2^{-n}z\right)\right] = 2^{-2z}\Gamma(1+z),$$

$$\prod_{n=0}^{\infty}\left[1+\left(\frac{y}{x+n}\right)^2\right] = \frac{[\Gamma(x)]^2}{|\Gamma(x+iy)|^2},$$

$$\prod_{n=0}^{\infty}\left[1+\left(\frac{y}{x+n}\right)^2\right]^{-1} = \Gamma(x+iy)\,\Gamma(x-iy)\,[\Gamma(x)]^{-2},$$

$$\prod_{n=1}^{\infty}\left[1-\left(\frac{z}{n}\right)^m\right] = -\frac{z^{-m}}{\Gamma(-\alpha_1 z)\,\Gamma(-\alpha_2 z)\cdots\Gamma(-\alpha_m z)},$$

where

$$\alpha_r = \exp\left(i\frac{2\pi r}{m}\right),$$

$$\prod_{n=1}^{\infty}\exp\left[\frac{xy}{n(n+x)}\right] = \exp\left[y\left(\psi(x)+\gamma+\frac{1}{x}\right)\right],$$

$$\prod_{n=1}^{\infty}\left[1+\left(\frac{z}{n}\right)^m\right] = \frac{z^{-m}}{\Gamma(-\beta_1 z)\,\Gamma(-\beta_2 z)\cdots\Gamma(-\beta_m z)},$$

where

$$\beta_r = \exp\left[i\,\frac{(2r+1)\,\pi}{m}\right],$$

$$\prod_{n=1}^{\infty}\left[1-\left(\frac{z}{n}\right)^{2m}\right] = -\frac{z^{-2m}}{\Gamma(-\alpha_1 z)\,\Gamma(-\beta_1 z)\cdots\Gamma(-\alpha_m z)\,\Gamma(-\beta_m z)},$$

$$\prod_{n=1}^{\infty}\left[\frac{n^m - z^m}{n^m + z^m}\right] = -\frac{\Gamma(-\beta_1 z)\;\Gamma(-\beta_2 z)\;\cdots\;\Gamma(-\beta_m z)}{\Gamma(-\alpha_1 z)\;\Gamma(-\alpha_2 z)\;\cdots\;\Gamma(-\alpha_m z)}.$$

α_r and β_r as before

Some infinite series expressed in terms of the gamma function [Dougall's formula].

$$\sum_{n=-\infty}^{\infty}\frac{\Gamma(a+n)\,\Gamma(b+n)}{\Gamma(c+n)\,\Gamma(d+n)} = \frac{\pi^2}{\sin(\pi a)\,\sin(\pi b)} \times \frac{\Gamma(c+d-a-b-1)}{\Gamma(c-a)\,\Gamma(d-a)\,\Gamma(c-b)\,\Gamma(d-b)},$$

$Re\,(a+b-c-d) < 1;\ a, b$ not integers.

$$\sum_{n=0}^{\infty}(-1)^n\binom{y-1}{n}(x+n)^{-1} = \frac{\Gamma(x)\,\Gamma(y)}{\Gamma(x+y)},$$

$$\sum_{n=0}^{\infty}\frac{(2n)!}{(2^n\,n!)^2\,(z+n)} = \pi^{\frac{1}{2}}\,\frac{\Gamma(z)}{\Gamma\left(\frac{1}{2}+z\right)},$$

$$\sum_{n=0}^{\infty}(-1)^n\binom{a}{n+1}\frac{n+1}{z+n} = \frac{\Gamma(z)\,\Gamma(1+a)}{\Gamma(a+z)},$$

$$\sum_{n=0}^{\infty}\frac{(\alpha)_n\,(\beta)_n}{(1-\beta+\alpha)_n}\,\frac{1}{n!}\left[\frac{1}{\frac{1}{2}\alpha+n-z} - \frac{1}{\frac{1}{2}\alpha+n+z}\right]$$

$$= \frac{\Gamma\left(\frac{1}{2}\alpha - z\right)\Gamma\left(\frac{1}{2}\alpha + z\right)\Gamma(1-\beta+\alpha)\,\Gamma(1-\beta)}{\Gamma\left(1-\beta+\frac{1}{2}\alpha-z\right)\Gamma\left(1-\beta+\frac{1}{2}\alpha+z\right)\Gamma(\alpha)},$$

$Re\,\beta < 1,$

$$\frac{\Gamma\left(\frac{1}{2}\alpha+z\right)\Gamma\left(\frac{1}{2}\alpha-z\right)}{\Gamma\left(\gamma-\frac{1}{2}\alpha+z\right)\Gamma\left(1-\beta+\frac{1}{2}\alpha-z\right)} = \frac{\Gamma(\alpha)}{\Gamma(\gamma)\,\Gamma(1-\beta)}$$

$$\times\sum_{n=0}^{\infty}\frac{(\alpha)_n\,(\beta)_n}{n!\,(\gamma)_n}\,\frac{1}{\left(\frac{1}{2}\alpha+n-z\right)} + \frac{\Gamma(\alpha)}{\Gamma(1+\alpha-\beta)\,\Gamma(\gamma-\alpha)}$$

$$\times\sum_{n=0}^{\infty}\frac{(\alpha)_n\,(1-\gamma+\alpha)_n}{(1-\beta+\alpha)_n\,n!}\,\frac{1}{\frac{1}{2}\alpha+n+z},$$

$Re\,(\alpha+\beta-\gamma) < 1.$

Some definite integrals expressed in terms of the gamma function

$$\int_0^1 t^{x-1}(1-t)^{y-1}(t+p)^{-x-y}\,dt = \frac{\Gamma(x)\,\Gamma(y)}{\Gamma(x+y)}(1+p)^{-x}p^{-y},$$

$Re\,x > 0,\ Re\,y > 0$; p real and positive.

$$\int_0^\infty \frac{t^x\,dt}{(1+t)^{1+y}} = \frac{\Gamma(x+1)\,\Gamma(y-x)}{\Gamma(1+y)},$$

$Re\,y > Re\,x > -1.$

$$\int_0^\infty \frac{t^x\,dt}{(1+t^z)^{1+y}} = \frac{1}{z}\,\frac{\Gamma\left(\frac{x+1}{z}\right)\Gamma\left(y-\frac{x-z+1}{z}\right)}{\Gamma(1+y)},$$

$Re\,z > 0$; $Re\,y > Re\,\frac{x-z+1}{z}$; $Re\,x > -1$; $Re\,y > -1.$

$$\int_0^\infty e^{-at^x}\,t^y\,dt = (1+y)^{-1}\,\Gamma\left(\frac{x+y+1}{x}\right)a^{-\frac{(y+1)}{x}},$$

$Re\,x > 0$; $Re\,y > -1$; $Re\,a > 0.$

$$\int_0^1 (1-t^x)^{-\frac{1}{y}}\,dt = \frac{\Gamma\left(1+\frac{1}{x}\right)\Gamma\left(1-\frac{1}{y}\right)}{\Gamma\left(1+\frac{1}{x}-\frac{1}{y}\right)},$$

$x > 0$; $y > 1.$

$$\int_0^1 \frac{dt}{\sqrt{1-t^3}} = \frac{1}{2\pi}\,2^{-\frac{1}{3}}\,3^{-\frac{1}{2}}\left[\Gamma\left(\frac{1}{3}\right)\right]^2,$$

$$\int_0^1 \frac{t\,dt}{\sqrt{1-t^3}} = \frac{1}{2\pi}\,2^{\frac{1}{3}}\,3^{\frac{1}{2}}\left[\Gamma\left(\frac{2}{3}\right)\right]^2,$$

$$\int_0^1 \frac{dt}{\sqrt{1-t^4}} = \frac{1}{4\sqrt{2\pi}}\left[\Gamma\left(\frac{1}{4}\right)\right]^2.$$

Integrals with gamma function in the integrand occur mainly in the Mellin transforms (see chap. XI).

$$\int_{-\infty}^\infty \frac{e^{itz}\,dt}{\Gamma(\mu+t)\,\Gamma(\nu-t)} = \begin{cases} \dfrac{\left(2\cos\frac{z}{2}\right)^{\mu+\nu-2}}{\Gamma(\mu+\nu-1)}\exp\left[i\,\frac{z}{2}(\nu-\mu)\right], & |z| < \pi \\ 0 & |z| > \pi, \end{cases}$$

z real and $Re\,(\mu+\nu)>1$.

$$\int_0^1 t^{x-1}(1-t)^{y-1}\,dt=\frac{\Gamma(x)\,\Gamma(y)}{\Gamma(x+y)}=B(x,y).$$

The function $B(x,y)$ is called the beta function and is defined by the integral above. The function $B(x,y)$ satisfies the following functional equations:

$$B(x,y)=B(y,x),$$

$$B(x,y+1)=\frac{y}{x}\,B(x+1,y)=\frac{y}{x+y}\,B(x,y),$$

$$B(x,y)\,B(x+y,z)=B(y,z)\,B(y+z,x)$$

$$=B(z,x)\,B(z+x,y)=\frac{\Gamma(x)\,\Gamma(y)\,\Gamma(z)}{\Gamma(x+y+z)}.$$

More generally

$$B(x_1,x_2)\,B(x_1+x_2,x_3)\cdots B(x_1+x_2+\cdots+x_{n-1},x_n)$$

$$=\frac{\Gamma(x_1)\,\Gamma(x_2)\cdots\Gamma(x_n)}{\Gamma(x_1+x_2+\cdots+x_n)},$$

$$[B(n,m)]^{-1}=m\binom{n+m-1}{n-1}=n\binom{n+m-1}{m-1},$$

$$n=1,2,3,\ldots,$$

$$m=1,2,3,\ldots$$

Integral representations of the function $B(x,y)$

$$B(x,y)=\int_0^\infty t^{x-1}(1+t)^{-x-y}\,dt,\quad Re\,x>0,\ Re\,y>0,$$

$$=\int_0^1 (1+t)^{-x-y}\,[t^{x-1}+t^{y-1}]\,dt,$$

$$Re\,x>0,\quad Re\,y>0,$$

$$B(z+\alpha,\beta-\alpha)=\int_0^\infty e^{-t(z+\alpha)}\,[1-e^{-t}]^{\beta-\alpha-1}\,dt,$$

$$Re\,(z+\alpha)>0,\quad Re\,(\beta-\alpha)>0,$$

$$B(x,y)=2^{-x-y+1}\int_0^1 [(1+t)^{x-1}(1-t)^{y-1}+(1+t)^{y-1}(1-t)^{x-1}]\,dt,$$

$$B(x,y)=(1+b)^x\int_0^1 t^{x-1}(1-t)^{y-1}(1+bt)^{-x-y}\,dt,$$

$$Re\,x>0,\ Re\,y>0,\ b>-1,$$

$$B\left(\frac{x}{z}, y\right) = z \int_0^1 t^{x-1} (1 - t^z)^{y-1}\, dt, \quad z > 0.$$

$$B(x, y) = a\, 2^{-x-y+2} \int_{-a}^{a} (a + t)^{2x-1} (a - t)^{2y-1} (a^2 + t^2)^{-x-y}\, dt.$$

$$a > 0.$$

$$B(x, y) = (a - b)^{-x-y+1} \int_b^a (t - b)^{x-1} (a - t)^{y-1}\, dt,$$

$$= \frac{(a - c)^x (b - c)^y}{(a - b)^{x+y-1}} \int_b^a (t - b)^{x-1} (a - t)^{y-1} (t - c)^{-x-y}\, dt.$$

$$a > b > c$$

$$= \frac{(c - a)^x (c - b)^y}{(a - b)^{x+y-1}} \int_b^a (t - b)^{x-1} (a - t)^{y-1} (c - t)^{-x-y}\, dt$$

$$c > a > b$$

$$= b^x \int_0^\infty t^{x-1} (1 + bt)^{-x-y}\, dt, \quad b > 0.$$

$$b^{-\frac{1+x}{z}} B\left(\frac{1 + x}{2}, y - \frac{1 + x}{2}\right) = z \int_0^\infty t^{x} (1 + bt^z)^{-y}\, dt,$$

$$z > 0,\ b > 0,\ 0 < Re\,\frac{x + 1}{2} < y.$$

Integrals with trigonometric integrands

$$\int_0^{\frac{\pi}{2}} (\sin t)^{2x-1} (\cos t)^{2y-1}\, dt = \frac{1}{2} B(x, y),$$

$$Re\, x > 0, \ Re\, y > 0.$$

$$\int_0^{\frac{\pi}{2}} (1 + a \sin^2 t)^{-x-y} (\sin t)^{2x-1} (\cos t)^{2y-1}\, dt = \frac{1}{2} (1 + a)^{-x} B(x, y),$$

$$a > -1,$$

$$\int_0^{\frac{\pi}{2}} (\cos^2 t + a \sin^2 t)^{-x-y} (\sin t)^{2x-1} (\cos t)^{2y-1}\, dt = \frac{a^{-x}}{2} B(x, y),$$

$$a > 0,$$

$$\int_0^{\pi} (\sin t)^{\alpha} \exp(i\beta t)\, dt = \frac{2^{-\alpha} \pi\, \Gamma(1 + \alpha) \exp\left(i \frac{\pi}{2} \beta\right)}{\Gamma\left(\frac{1}{2}\alpha + \frac{1}{2}\beta + 1\right) \Gamma\left(\frac{1}{2}\alpha - \frac{1}{2}\beta + 1\right)},$$

$$Re\, \alpha > -1,$$

$$\int_0^{\frac{\pi}{2}} (\cos t)^\alpha \cos(\beta t)\, dt = \frac{\pi\,\Gamma(1+\alpha)\, 2^{-\alpha-1}}{\Gamma\left(1+\frac{1}{2}\alpha+\frac{1}{2}\beta\right)\Gamma\left(1+\frac{1}{2}\alpha-\frac{1}{2}\beta\right)},$$

$Re\,\alpha > -1$,

$$\int_0^{\frac{\pi}{2}} (\cos t)^x \cos(yt)\, dt = \frac{\pi}{2^{x+1}} \frac{\Gamma(x+1)}{\Gamma\left(\frac{x+y}{2}+1\right)\Gamma\left(\frac{x-y}{2}+1\right)},$$

$Re\,x > -1$,

$$\int_0^{\frac{\pi}{2}} \sin^{\frac{3}{2}} t\, dt = \int_0^{\frac{\pi}{2}} \cos^{\frac{3}{2}} t\, dt = \frac{1}{6\sqrt{2\pi}}\,\Gamma^2\left(\frac{1}{4}\right),$$

$$\int_0^\infty \cos(xt)\, t^{-y}\, dt = \frac{\Gamma(1-y)}{x^{1-y}} \sin\left(\frac{\pi y}{2}\right),$$

x, y real and positive; $0 < y < 1$,

$$\int_0^\infty \sin(xt)\, t^{-y} dt = \frac{\Gamma(1-y)}{x^{1-y}} \cos\left(\frac{\pi y}{2}\right),$$

x, y real and positive; $0 < y < 2$.

$$\int_0^{\frac{\pi}{2}} (\sin t)^{\alpha-1} (\cos t)^{\beta-1} \exp[i(\alpha+\beta)t]\, dt = B(\alpha,\beta) \exp\left(i\frac{\pi}{2}\alpha\right),$$

$$\int_0^\infty t^{z-1} \cos(at^x)\, dt = \frac{\Gamma\left(\frac{z}{x}\right)\cos\left(\frac{\pi}{2x}z\right)}{x\, a^{\frac{z}{x}}},$$

$x > Re\,z > 0$.

$$\int_0^\infty t^{z-1} \sin(at^x)\, dt = \frac{\Gamma\left(\frac{z}{x}\right)\sin\left(\frac{\pi}{2x}z\right)}{x\, a^{\frac{z}{x}}},$$

$-x < Re\,z < x$.

Integrals with exponential and hyperbolic functions in the integrand

$$\int_0^\infty t^y e^{-\alpha t^x}\, dt = \frac{1}{x}\Gamma\left(\frac{1+y}{x}\right)\alpha^{-\frac{1+y}{x}},$$

$Re\,\alpha > 0$, $Re\,x > 0$, $Re\,y > -1$.

$$\int_0^\infty [1-\exp(-tz)]^{y-1} \exp(-xt)\, dt = \frac{1}{z} B\left(\frac{x}{z}, y\right),$$

$Re\,\frac{x}{z} > 0$, $Re\,z > 0$, $Re\,y > 0$,

$$\int_0^\infty \frac{\cosh(2yt)}{(\cosh t)^{2x}}\,dt = 2^{2x-2}\,\frac{\Gamma(x+y)\,\Gamma(x-y)}{\Gamma(2x)},$$

$(Re\,x > |Re\,y|)$,

$$\int_0^\infty (\sinh t)^\alpha (\cosh t)^{-\beta}\,dt = \frac{1}{2}\,B\left(\frac{1+\alpha}{2}, \frac{\beta-\alpha}{2}\right),$$

$Re(\beta-\alpha) > 0,\ Re\,\alpha > -1$

$$\int_0^\infty \cosh(2\alpha t)\,[\cosh(at)]^{-2\beta}\,dt = \frac{4^{\beta-1}}{a}\,B\left(\beta+\frac{\alpha}{a}, \beta-\frac{\alpha}{a}\right),$$

$a > 0,\ Re\left(\beta \pm \frac{\alpha}{a}\right) > 0.$

The following two results are special cases of the integral above

$$\int_0^\infty \frac{\cosh(2\alpha t)}{\cosh(at)}\,dt = \frac{\pi}{2a}\sec\left(\frac{\pi\alpha}{a}\right),$$

$Re\,a > 2\,Re\,\alpha$,

$$\int_0^\infty \frac{\cosh(2\alpha t)}{\cosh^2(at)}\,dt = \frac{\pi\alpha}{a^2}\operatorname{cosec}\left(\frac{\pi\alpha}{a}\right),$$

$Re\,a > Re\,\alpha$,

$$\int_0^\infty t^{\alpha-1} e^{-pt} \exp(iqt)\,dt = \frac{\Gamma(\alpha)}{(p^2+q^2)^{\frac{\alpha}{2}}}\,e^{i\alpha\arctan\left(\frac{q}{p}\right)},$$

$p > 0,\ Re\,\alpha > 0$, however if $p = 0$.

α must satisfy the condition $0 < Re\,\alpha < 1$,

$$\int_0^\infty t^{\alpha-1} e^{-pt}\cos(qt)\,dt = \frac{\Gamma(\alpha)}{(p^2+q^2)^{\frac{\alpha}{2}}}\cos\left[\alpha\arctan\left(\frac{q}{p}\right)\right],$$

$$\int_0^\infty t^{\alpha-1} e^{-pt}\sin(qt)\,dt = \frac{\Gamma(\alpha)}{(p^2+q^2)^{\frac{\alpha}{2}}}\sin\left[\alpha\arctan\left(\frac{q}{p}\right)\right],$$

$$\int_0^\infty t^{\alpha-1}\cos(\beta t)\,dt = \beta^{-\alpha}\,\Gamma(\alpha)\cos\left(\frac{\pi}{2}\alpha\right),$$

$0 < Re\,\alpha < 1$,

$$\int_0^\infty t^{\alpha-1}\sin(\beta t)\,dt = \beta^{-\alpha}\,\Gamma(\alpha)\sin\left(\frac{\pi}{2}\alpha\right),$$

$-1 < Re\,\alpha < 1.$

Expressions for $\log \Gamma(z)$

$$\log \Gamma(z) = \int_0^\infty \left[z - 1 - \frac{1 - \exp\{-(z-1)t\}}{1 - \exp(-t)}\right] \frac{e^{-t}}{t}\, dt,$$

$Re\, z > 0$ [Malmstén's formula]

$$= \left(z - \frac{1}{2}\right) \log z - z + \frac{1}{2} \log(2\pi) + \int_0^\infty \left(\frac{1}{e^t - 1} - \frac{1}{t} + \frac{1}{2}\right) \frac{e^{-tz}}{t}\, dt,$$

$Re\, z > 0$ [BINET],

$$= \left(z - \frac{1}{2}\right) \log z - z + \frac{1}{2} \log(2\pi) + 2 \int_0^\infty \frac{\arctan\left(\frac{t}{z}\right)}{e^{2\pi t} - 1}\, dt$$

[BINET].

$$\log \Gamma(z) = \log\left(\frac{\pi}{\sin \pi z}\right) - \int_0^\infty \left[\frac{e^{zt} - 1}{1 - e^{-t}} - z\right] \frac{e^{-t}}{t}\, dt,$$

$Re\, z < 1$,

$$= \frac{1}{2} \log \pi - \frac{1}{2} \log \sin(\pi z).$$

$$+ \frac{1}{2} \int_0^\infty \left[\frac{\sinh\left(\frac{1}{2} - z\right) t}{\sinh\left(\frac{t}{2}\right)} - (1 - 2z)\, e^{-t}\right] \frac{dt}{t},$$

$0 < Re\, z < 1$,

$$\log \Gamma(z) = \frac{1}{2} \log(2\pi) + \sum_{n=1}^\infty \left\{\frac{\cos(2\pi n z)}{2n} + [\gamma + \log(2\pi n)] \frac{\sin(2\pi n z)}{\pi n}\right\}$$

$$= \left(\frac{1}{2} - z\right)(\gamma + \log 2) + (1 - z) \log \pi$$

$$- \frac{1}{2} \log(\sin \pi z) + \sum_{n=1}^\infty \frac{\log n}{\pi n} \cdot \sin(2\pi n z)$$

[KUMMER].

The following are power series representations of the function $\log \Gamma(z)$.

$$\log \Gamma(z + 1) = -\gamma z - \frac{1}{2} \log\left(\frac{\sin \pi z}{\pi z}\right) - \sum_{n=1}^\infty \zeta(2n + 1) \frac{z^{2n+1}}{2n + 1},$$

$|z| < 1$,

where $\zeta(z)$ is the Riemann zeta function defined in section 1.3.

$$\log \Gamma(1+z) = (1-\gamma)\, z + \frac{1}{2}\log\left(\frac{\pi z}{\sin \pi z}\right) - \frac{1}{2}\log\left(\frac{1+z}{1-z}\right) + \sum_{n=1}^{\infty} \frac{1-\zeta(2n+1)}{2n+1}\, z^{2n+1},$$

$$|z| < 1.$$

Binet's formula

$$\log \Gamma(z) = \left(z-\frac{1}{2}\right)\log z - z + \frac{1}{2}\log(2\pi) + \frac{1}{2}\sum_{n=1}^{\infty} \frac{n\,\zeta(n+1, z+1)}{(n+1)(n+2)},$$

where $\zeta(z, \alpha)$ is the generalized zeta function defined in section 1.4.

Burnside's formula

$$\log \Gamma(z) = \left(z-\frac{1}{2}\right)\log\left(z-\frac{1}{2}\right) - z - \frac{1}{2} + \frac{1}{2}\log(2\pi) - \sum_{n=1}^{\infty} \frac{\zeta(2n, z)}{2n(2n+1)}\, 2^{-2n},$$

$$Re\, z \geq -\frac{1}{2}.$$

Asymptotic expansion for $\Gamma(z)$ *and* $\log \Gamma(z)$

For large values of $|z|$ and $|\arg z| < \pi$

$$\log \Gamma(z) = \left(z-\frac{1}{2}\right)\log z - z + \frac{1}{2}\log(2\pi) + \sum_{m=1}^{n} \frac{B_{2m}}{2m(2m-1)}\, z^{-2m+1} + O(z^{-2n-1})$$

[Stirling's formula],

where B_{2m} are the Bernoulli numbers defined in section 1.5.1

$$\Gamma(z) = z^{-\frac{1}{2}}\, e^{z(\log z - 1)}\, (2\pi)^{\frac{1}{2}} \left\{1 + \frac{1}{12z} + \frac{1}{288z^2} - \frac{139}{51840z^3} + O(z^{-4})\right\},$$

$$\log \Gamma(z+\alpha) = \left(z+\alpha-\frac{1}{2}\right)\log z - z + \frac{1}{2}\log(2\pi) + O(z^{-1})$$

for $|z| \ll |\alpha|$ and $|\arg z| < \pi$.

$$\frac{\Gamma(z+\alpha)}{\Gamma(z+\beta)} \approx z^{\alpha-\beta}\left[1 + \frac{1}{2z}(\alpha-\beta)(\alpha+\beta-1) + O(z^{-2})\right], |\arg z| < \pi,$$

$$\lim_{z\to\infty} e^{-\alpha \log z}\, \frac{\Gamma(z+\alpha)}{\Gamma(z)} = 1, \; |\arg z| < \pi,$$

$$\Gamma(k+ix)\,\Gamma(k-ix) = O(x^{2k-1}\, e^{-\pi x}) \text{ for large } x.$$

$$\lim_{|y|\to\infty} |\Gamma(x+iy)|\, e^{\frac{\pi}{2}|y|}\, |y|^{\frac{1}{2}-x} = (2\pi)^{\frac{1}{2}} \quad x, y \text{ real}.$$

$$\log \Gamma(z+\alpha) = \left(z+\alpha-\frac{1}{2}\right)\log z - z + \frac{1}{2}\log(2\pi)$$
$$+ \sum_{m=1}^{n} (-1)^{m+1} \frac{B_{m+1}(\alpha)}{m(m+1)} z^{-m} + O(z^{-n-1}),$$
$$|\arg z| < \pi, \quad n = 1, 2, 3, \ldots$$

where $B_n(\alpha)$ are the Bernoulli polynomials defined in section 1.5.1.

1.2 The function $\psi(z)$

The function $\psi(z)$ is defined as the logarithmic derivative of the function $\Gamma(z)$.

$$\psi(z) = \frac{d}{dz}\log\Gamma(z) = \frac{\Gamma'(z)}{\Gamma(z)}.$$

The function $\psi(z)$ is meromorphic with simple poles at $z = 0, -1, -2, \ldots$

Series representations for $\psi(z)$.

$$\psi(z) = -\gamma + \sum_{n=0}^{\infty}\left(\frac{1}{n+1} - \frac{1}{z+n}\right)$$
$$= \log z - \sum_{n=0}^{\infty}\left[\frac{1}{z+n} - \log\left(1+\frac{1}{z+n}\right)\right]$$
$$= -\gamma + \sum_{m=2}^{\infty}(-1)^m \zeta(m)\, z^{m-1}$$
$$= \frac{1}{2z} - \gamma - \frac{\pi}{2}\cot(\pi z) - \sum_{m=1}^{\infty}\zeta(2m+1)\, z^{2m}.$$

The function $\psi(z)$ is also given by the following limit

$$\psi(z) = \lim_{n\to\infty}\left[\log n - \frac{1}{z} - \frac{1}{z+1} - \cdots - \frac{1}{z+n}\right].$$

Lerch's series

$$\psi(z)\sin(\pi x) = -\frac{\pi}{2}\cos(\pi x) - (\gamma + \log 2\pi)\sin(\pi x)$$
$$+ \sum_{n=1}^{\infty}\log\left(\frac{n}{n+1}\right)\sin[(2n+1)\pi x]$$
$$0 < x < 1.$$

Near $z = -m$, $(m = 0, 1, 2, \ldots)$ the Laurent expansion for $\psi(z)$ is given by

$$\psi(z) = -\frac{1}{z+m} + \psi(m+1) + \sum_{n=2}^{\infty} A_n (z+m)^{n-1},$$

where

$$A_n = (-1)^n \zeta(n) + \sum_{l=1}^{m} l^{-n}.$$

The expansion (LAURENT) shows that the function $\psi(z)$ has simple poles at $z = -m$, $(m = 0, 1, 2, \ldots)$ with residue -1.

$$\psi'(z) = \sum_{n=0}^{\infty} (z+n)^{-2} = \zeta(2, z),$$

$$\frac{d^n}{dz^n}\psi(z) = \psi^{(n)}(z) = (-1)^{n+1} \int_0^{\infty} \frac{t^n}{1-e^{-t}} e^{-zt}\, dt, \qquad Re\, z > 0,$$

$$n = 1, 2, 3, \ldots$$

Functional equations

$$\psi(z+n) = \sum_{m=0}^{n-1} \frac{1}{z+m} + \psi(z), \quad n = 1, 2, 3, \ldots,$$

$$\psi(z-n) = -\sum_{m=1}^{n} \frac{1}{z-m} + \psi(z), \quad n = 1, 2, 3, \ldots,$$

$$\psi(1+n) = \sum_{m=1}^{n} \frac{1}{m} - \gamma, \quad n = 1, 2, 3, \ldots,$$

$$\psi(z+1) = \psi(z) + \frac{1}{z},$$

$$\psi(z) = \psi(1-z) - \pi \cot(\pi z),$$

$$\psi(z) = \psi(-z) - \frac{1}{z} - \pi \cot(\pi z),$$

$$\psi(z+1) = \psi(1-z) + \frac{1}{z} - \pi \cot(\pi z),$$

$$\psi\left(\frac{1}{2} + z\right) = \psi\left(\frac{1}{2} - z\right) + \pi \tan(\pi z),$$

$$\psi\left(\frac{1}{2} + n\right) = \psi\left(\frac{1}{2} - n\right) = -\gamma + 2\left(\sum_{m=0}^{n-1} \frac{1}{2m+1} - \log 2\right),$$

$$n = 1, 2, 3, \ldots,$$

$$\psi^{(n)}(z+1) = \psi^{(n)}(z) + (-1)^n n!\, z^{-n-1},$$

$$\psi^{(n)}(1-z) = (-1)^n \left[\psi^{(n)}(z) + \pi \frac{d^n}{dz^n} \cot(\pi z)\right].$$

Multiplication formulas

$$\psi(mz) = \frac{1}{m} \sum_{l=0}^{m-1} \psi\left(z + \frac{l}{m}\right) + \log m, \quad m = 2, 3, 4, \ldots,$$

$$\psi^{(n)}(mz) = \frac{1}{m^{n+1}} \sum_{l=0}^{m-1} \psi^{n}\left(z + \frac{l}{m}\right).$$

Special values of the argument

$$\psi(1) = -\gamma, \ \psi(n) = -\gamma + \sum_{l=1}^{n-1} \frac{1}{l},$$

$$\psi\left(\frac{1}{2}\right) = -\gamma - 2\log 2, \ \psi\left(-\frac{1}{2}\right) = -\gamma - 2\log 2 + 2.$$

$$\psi\left(n + \frac{1}{2}\right) = -\gamma - 2\log 2 + 2\sum_{l=0}^{n-1} \frac{1}{2l+1},$$

$$\operatorname{Im}\psi(iy) = \frac{1}{2y} + \frac{\pi}{2}\coth(\pi y),$$

$$\operatorname{Im}\psi\left(\frac{1}{2} + iy\right) = \frac{\pi}{2}\tanh(\pi y),$$

$$\operatorname{Im}\psi(n + iy) = \frac{\pi}{2}\coth(\pi y) - y\left[\frac{y^{-2}}{2} + \sum_{l=1}^{n-1} \frac{1}{y^2 + l^2}\right],$$

$$n = 1, 2, 3, \ldots,$$

$$\operatorname{Im}\psi\left(n + \frac{1}{2} + iy\right) = \frac{\pi}{2}\tanh(\pi y) + y\sum_{l=0}^{n} \frac{1}{\left(l - \frac{1}{2}\right)^2 + y^2},$$

$$n = 0, 1, 2, \ldots$$

Gauss' formula for rational argument

$$\psi\left(\frac{p}{q}\right) = -\gamma - \log q - \frac{\pi}{2}\cot\left(\frac{\pi p}{q}\right),$$

$$+ \sum_{n=1}^{\leq \frac{1}{2}q}{}' \cos\left(2\pi n \frac{p}{q}\right)\log\left[2 - 2\cos\frac{2\pi n}{q}\right],$$

where $0 < p < q$ and p, q are positive integers. The prime $'$ indicates that if q is even, only one half of the last term is to be taken in the sum.

$$Re\,\psi(1 + iy) = -\gamma + 1 - \frac{1}{1 + y^2} + \sum_{n=1}^{\infty} (-1)^{n+1}[\zeta(2n+1) - 1]\,y^{2n},$$

$$|y| < 2$$

$$= -\gamma + y^2 \sum_{n=1}^{\infty} [n(n^2 + y^2)]^{-1},$$

$$|y| < \infty,$$

$$\psi^{(n)}(1) = (-1)^{n+1}\, n!\, \zeta(n+1), \quad n = 1, 2, 3, \ldots,$$

$$\psi^{(n)}(m+1) = (-1)^n\, n!\left[1 + \sum_{r=2}^{m} r^{-n-1} - \zeta(n+1)\right].$$

$$\psi^{n}\left(\frac{1}{2}\right) = (-1)^{n+1}\, n!\,(2^{n+1} - 1)\,\zeta(n+1), \quad n = 1, 2, \ldots,$$

$$\psi'\left(n + \frac{1}{2}\right) = \frac{1}{2}\pi^2 - 4\sum_{r=1}^{n} (2r-1)^{-2}.$$

Integral representations for $\psi(z)$

$$\psi(z) = -\gamma + \int_0^1 (1-t)^{-1}(1-t^{z-1})\,dt, \quad Re\, z > 0,$$

$$\psi(z) = -\gamma - \pi\cot(\pi z) + \int_0^1 (1-t)^{-1}(1-t^{-z})\,dt, \quad Re\, z < 1,$$

$$\psi(z) = \int_0^\infty \left[\frac{e^{-t}}{t} - \frac{e^{-zt}}{1-e^{-t}}\right] dt, \quad Re\, z > 0,$$

[GAUSS].

$$\psi(z) = \int_0^\infty [e^{-t} - (1+t)^{-z}]\frac{dt}{t}, \quad Re\, z > 0,$$

$$= -\gamma + \int_0^\infty [(1+t)^{-1} - (1+t)^{-z}]\frac{dt}{t}, \quad Re\, z > 0.$$

[Dirichlet's formulas].

$$\psi(z) = \log z + \int_0^\infty \left[\frac{1}{t} - \frac{1}{1-e^{-t}}\right] e^{-zt}\,dt, \quad Re\, z > 0,$$

$$= \log z - \frac{1}{2z} - \int_0^\infty \left[\frac{1}{1-e^{-t}} - \frac{1}{t} - \frac{1}{2}\right] e^{-zt}\,dt, \quad Re\, z > 0,$$

$$\psi(z) = \log z + \int_0^\infty \left[\frac{1}{1-e^{t}} + \frac{1}{t} - 1\right] e^{-zt}\,dt, \quad Re\, z > 0,$$

$$= \log z - \frac{1}{2z} - \int_0^\infty \left[(e^t-1)^{-1} - t^{-1} + \frac{1}{2}\right] e^{-zt}\,dt, \quad Re\, z > 0,$$

$$= \log z - \frac{1}{2z} - 2\int_0^\infty t\,[(t^2+z^2)(e^{2\pi t}-1)]^{-1}\,dt, \quad Re\, z > 0,$$

$$= \log z - 2\int_0^\infty t(t^2+z^2)^{-1}[(e^{2\pi t}-1)^{-1} - (2\pi t)^{-1}]\,dt,$$

$Re\, z > 0,$

$$= -\gamma + \int_0^\infty \frac{e^{-t} - e^{-zt}}{1-e^{-t}}\,dt, \quad Re\, z > 0,$$

$$= -\gamma + 2\int_0^\infty e^{-zt}\frac{\sinh[(z-1)t]}{\sinh t}\,dt, \quad Re\, z > 0,$$

$$\psi\left(\frac{1}{2}+z\right) = \log z + 2\int_0^\infty \frac{t\,dt}{(1+e^{\pi t})(t^2+4z^2)}, \quad Re\, z > 0,$$

$$= \log z + \frac{\pi}{4}\int_0^\infty \log(t^2+4z^2)\left[\cosh\left(\frac{\pi}{2}t\right)\right]^{-2} dt, \quad Re\, z > 0.$$

$$\psi(z) = -\gamma - \frac{1}{2z} - \frac{\pi}{2}\cot(\pi z)$$

$$+ \frac{\pi}{2}\int_0^1 \tan\left(\frac{\pi}{2}t\right)\left[\frac{\sin(\pi z t)}{\sin(\pi z)} - t\right] dt, \quad z \neq \pm 1, \pm 2, \ldots$$

Some series expressed in terms of the ψ-function

$$2\sum_{n=0}^\infty \frac{(-1)^n}{z+n} = \psi\left(\frac{1+z}{2}\right) - \psi\left(\frac{1}{2}z\right) = G(z)$$

$$= \frac{2}{z}\,{}_2F_1(1, z; 1+z; -1).$$

$$\sum_{n=0}^\infty [z^2 - (a+nd)^2]^{-1} = (2zd)^{-1}\left[\psi\left(\frac{a-z}{d}\right) - \psi\left(\frac{a+z}{d}\right)\right],$$

$$\sum_{n=1}^\infty \frac{(-1)^{n-1}}{a+nb} = \frac{1}{2b}\left[\psi\left(1+\frac{a}{2b}\right) - \psi\left(\frac{1}{2}+\frac{a}{2b}\right)\right],$$

$$\sum_{n=0}^m (a+nb)^{-1} = \frac{1}{b}\left[\psi\left(1+m+\frac{a}{b}\right) - \psi\left(\frac{a}{b}\right)\right],$$

$$\sum_{n=1}^{2m} \frac{(-1)^{n+1}}{a+nb} = \frac{1}{2b}\left[\psi\left(m+\frac{1}{2}+\frac{a}{2b}\right) - \psi\left(\frac{1}{2}+\frac{a}{2b}\right)\right.$$

$$\left. - \psi\left(m+1+\frac{a}{2b}\right) + \psi\left(1+\frac{a}{2b}\right)\right],$$

$$\frac{z}{a} - \frac{1}{2}\frac{z(z-1)}{a(a+1)} + \frac{1}{3}\frac{z(z-1)(z-2)}{a(a+1)(a+2)} + \cdots + = \psi(a+z) - \psi(a),$$

$$Re\,(a+z) > 0,\ a \neq 0, -1, -2, \ldots$$

For further properties of the function $G(z)$ see ERDELYI (1953).

Some definite integrals connected with the ψ-function

$$\int_0^1 \frac{t^{z-1}}{t+1}\,dt = \frac{1}{2}\left[\psi\left(\frac{1+z}{2}\right) - \psi\left(\frac{1}{2}z\right)\right]$$

$$= \frac{1}{2}G(z), \quad Re\, z > 0.$$

[All zeros of the function $G(z)$ are complex and have real part $< -\frac{1}{2}$. For each $n = 1, 2, 3, \ldots$ the strip $-\left(2n-\frac{1}{2}\right) < Re\, z < -\left(2n-\frac{3}{2}\right)$

contains exactly two zeros of the G-function]

$$\int_0^1 (1-t)^{-1} t^{x-1}(1-t^y)\,dt = \psi(x+y) - \psi(x),$$

$Re\,(x+y) > 0,\ Re\,x > 0.$

$$\int_0^1 (t^x - t^y)\,[t - t^2]^{-1}\,dt = -\psi(x) + \psi(y),$$

$Re\,x > 0,\ Re\,y > 0.$

$$\int_0^\infty e^{-\alpha t}\frac{\sinh(\beta t)}{\sinh(\gamma t)}\,dt = \frac{1}{2\gamma}\left[\psi\left(\frac{\alpha}{2\gamma}+\frac{1}{2}+\frac{\beta}{2\gamma}\right) - \psi\left(\frac{\alpha}{2\gamma}+\frac{1}{2}-\frac{\beta}{2\gamma}\right)\right].$$

$Re\,(\alpha+\gamma\pm\beta) > 0.$

$$\int_0^\infty e^{-\alpha t}\frac{\cosh(\beta t)}{\cosh(\lambda t)}\,dt = \frac{1}{4\lambda}\left[\psi\left(\frac{3}{4}+\frac{\alpha+\beta}{4\lambda}\right) + \psi\left(\frac{3}{4}+\frac{\alpha-\beta}{4\lambda}\right) - \psi\left(\frac{1}{4}+\frac{\alpha+\beta}{4\lambda}\right) - \psi\left(\frac{1}{4}+\frac{\alpha-\beta}{4\lambda}\right)\right].$$

$Re\,(\alpha+\lambda\pm\beta) > 0.$

$$\int_0^\infty e^{-\alpha t}\frac{\sinh(\beta t)}{\cosh(\lambda t)}\,dt = \frac{1}{4\lambda}\left[\psi\left(\frac{1}{4}+\frac{\alpha+\beta}{4\lambda}\right) + \psi\left(\frac{3}{4}+\frac{\alpha-\beta}{4\lambda}\right) - \psi\left(\frac{1}{4}+\frac{\alpha-\beta}{4\lambda}\right) - \psi\left(\frac{3}{4}+\frac{\alpha+\beta}{4\lambda}\right)\right],$$

$Re\,(\alpha+\lambda\pm\beta) > 0.$

$$\int_0^\infty \sin(xy)\,[e^{\alpha x}-1]^{-1}\,dx = \frac{\pi}{\alpha}\left[\left(e^{2\pi\frac{y}{\alpha}}-1\right)^{-1} + \frac{1}{2} - \frac{\alpha}{2\pi y}\right].$$

$$\int_0^\infty \frac{\sin(xy)}{e^{\alpha x}+1}\,dx = \frac{1}{2y} - \frac{\pi}{2\alpha\sinh\left(\pi\frac{y}{\alpha}\right)}.$$

In the last two integrals $|Im\,y| < Re\,\alpha$, $Re\,\alpha > 0$. *Asymptotic expansions*

For $z\to\infty$ and $|\arg z| < \pi$

$$\psi(z) = \log z - \frac{1}{2z} - \sum_{n=1}^{m}\frac{B_{2n}}{2n}\frac{1}{z^{2n}} + O(z^{-2m-2})$$
$$= \log z - \frac{1}{2z} - \frac{1}{12z^2} + \frac{1}{120z^4} - \frac{1}{252z^6} + O(z^{-8}).$$

For $y\to\infty$,

$$Re\,\psi(1+iy) \approx \log y + \sum_{n=1}^{\infty}\frac{(-1)^{n-1}}{2n}B_{2n}\,y^{-2n},$$

$$\psi^{(n)}(z) \approx (-1)^{n-1}\left[\frac{(n-1)!}{z^n} + \frac{n!}{2}z^{-n-1} + \sum_{l=1}^{\infty}B_{2l}\frac{(n+2l-1)!}{(2l)!}z^{-n-2l}\right]$$

as $|z|\to\infty$ and $|\arg z| < \pi$.

1.3 The Riemann zeta function $\zeta(z)$

The function $\zeta(z)$ is a meromorphic function of z and has its only singularity a simple pole, with residue 1, at the point $z = 1$.

Series representations

$$\zeta(z) = \sum_{n=1}^{\infty} n^{-z}, \quad Re\, z > 1$$

$$= [1 - 2^{(1-z)}]^{-1} \sum_{n=1}^{\infty} (-1)^{n-1} n^{-z}, \quad Re\, z > 0$$

$$= [1 - 2^{-z}]^{-1} \sum_{n=0}^{\infty} (2n + 1)^{-z}, \quad Re\, z > 1.$$

As an infinite product, the function $\zeta(z)$ can be represented as

$$\zeta(z) = \prod_{p} (1 - p^{-z})^{-1},$$

where the infinite product is taken over all primes p.

Functional equations

$$\zeta(z) = 2(2\pi)^{z-1} \sin\left(\frac{\pi}{2} z\right) \Gamma(1 - z)\, \zeta(1 - z),$$

$$\zeta(1 - z) = 2(2\pi)^{-z} \Gamma(z) \cos\left(\frac{\pi}{2} z\right) \zeta(z).$$

Laurent expansion

In the neighborhood of the point $z = 1$

$$\zeta(z) = \frac{1}{z-1} + \gamma + \sum_{n=1}^{\infty} a_n (z - 1)^n,$$

where γ is Euler's constant and

$$a_n = \lim_{m\to\infty} \left[\sum_{l=1}^{m} l^{-1} (\log l)^n - \frac{(\log m)^{n+1}}{n+1} \right].$$

Special values of the argument

$$\zeta(0) = -\frac{1}{2},$$

$$\zeta(-m) = -\frac{B_{m+1}}{m+1}, \quad m = 1, 2, 3, \ldots,$$

$$\zeta(-2m) = 0, \; m = 1, 2, 3, \ldots,$$

$$\zeta(-2m + 1) = -\frac{B_{2m}}{2m}, \; m = 1, 2, 3, \ldots,$$

$$\zeta(2m) = (-1)^{m+1} \frac{(2\pi)^{2m}}{2(2m)!} B_{2m}, \; m = 1, 2, 3, \ldots,$$

$$= \frac{(2\pi)^{2m}}{2(2m)!} |B_{2m}|, \; m = 1, 2, 3, \ldots$$

$$\zeta(2m+1) = \frac{(-1)^{m+1}(2\pi)^{2m+1}}{2(2m+1)!} \int_0^1 B_{2m+1}(t) \cot(\pi t)\, dt, \quad m = 1, 2, 3, \ldots,$$

$$\zeta(2) = \sum_{n=1}^{\infty} \frac{1}{n^2} = \frac{\pi^2}{6},$$

$$\zeta(4) = \sum_{n=1}^{\infty} \frac{1}{n^4} = \frac{\pi^4}{90}.$$

Repesentation in terms of definite integrals

$$\zeta(z) = \frac{1}{\Gamma(z)} \int_0^{\infty} [e^t - 1]^{-1} t^{z-1}\, dt, \quad Re\, z > 1$$

$$= \frac{1}{\Gamma(z+1)} \int_0^{\infty} e^t t^z [e^t - 1]^{-2}\, dt, \quad Re\, z > 1$$

$$= \frac{[1 - 2^{1-z}]^{-1}}{\Gamma(z)} \int_0^{\infty} t^{z-1} [e^t + 1]^{-1}\, dt, \quad Re\, z > 0$$

$$= \frac{[1 - 2^{-z}]^{-1}}{2\Gamma(z)} \int_0^{\infty} t^{z-1} [\sinh t]^{-1}\, dt, \quad Re\, z > 1$$

$$= \frac{1}{2} + \frac{1}{z-1} + \frac{1}{\Gamma(z)} \int_0^{\infty} e^{-t} \left[\frac{1}{e^t - 1} - \frac{1}{t} + \frac{1}{2}\right] t^{z-1}\, dt, \; Re\, z > -1$$

$$= \frac{[1 - 2^{1-z}]^{-1}}{\Gamma(z+1)} \int_0^{\infty} t^z e^t [e^t + 1]^{-2}\, dt, \quad Re\, z > 0$$

$$= \frac{1}{z-1} + \pi^{-1} \sin(\pi z) \int_0^{\infty} [\log(1+t) - \psi(1+t)]\, t^{-z}\, dt,$$

$$0 < Re\, z < 1$$

$$= \frac{1}{2} + \frac{1}{z-1} + \sum_{m=1}^{n} \frac{B_{2m}}{(2m)!} \frac{\Gamma(z + 2m - 1)}{\Gamma(z)}$$

$$+ \frac{1}{\Gamma(z)} \int_0^{\infty} t^{z-1} e^{-t} \left[\frac{1}{e^t - 1} - \frac{1}{t} + \frac{1}{2} - \sum_{m=1}^{n} \frac{B_{2m}}{(2m)!} t^{2m-1}\right] dt,$$

$$n = 1, 2, 3, \ldots, \; Re\, z > -(2n+1),$$

$$\zeta(1+z) = \frac{\sin(\pi z)}{\pi z} \int_0^{\infty} \psi'(1+t)\, t^{-z}\, dt, \quad 0 < Re\, z < 1,$$

$$= \frac{\sin(\pi z)}{\pi} \int_0^{\infty} [\psi(t+1) + \gamma]\, t^{-z-1}\, dt, \quad 0 < Re\, z < 1,$$

$$\zeta(m+z) = \frac{(-1)^{m-1}}{\pi} \frac{\Gamma(z)}{\Gamma(m+z)} \sin(\pi z) \int_0^{\infty} \psi^{(m)}(1+t)\, t^{-z}\, dt,$$

$$0 < Re\, z < 1,$$

$$\zeta(z) = \frac{1}{z-1} - \frac{\sin(\pi z)}{\pi(z-1)} \int_0^\infty \left[\psi'(1+t) - \frac{1}{1+t}\right] t^{1-z}\, dt,$$
$$0 < Re\, z < 1$$
$$= -\frac{\sin(\pi z)}{\pi} \int_0^\infty [\psi(1+t) - \log t]\, t^{-z}\, dt,$$
$$0 < Re\, z < 1$$
$$= \frac{2^{z-1}}{\Gamma(z+1)} \int_0^\infty t^z [\sinh t]^{-2}\, dt, \quad Re\, z > -1$$
$$= 2^{z-1} \frac{[1-2^{1-z}]^{-1}}{\Gamma(1+z)} \int_0^\infty t^{-z} [\cosh t]^{-2}\, dt, \quad Re\, z > -1,$$
$$\zeta(z) = \frac{\Gamma(1-z)}{2\pi} \int_{-\pi}^{\pi} [\log(1+e^{i\varphi})]^{z-1}\, d\varphi.$$

The following integral representations of $\zeta(z)$ are valid in the entire z-plane except at the point $z = 1$

$$\zeta(z) = 2^{z-1}(z-1)^{-1} - 2^z \int_0^\infty (1+t^2)^{-\frac{z}{2}} \sin(z \arctan t)\, [e^{\pi t}+1]^{-1}\, dt,$$
$$= 2^{z-1}[1-2^{1-z}]^{-1} \int_0^\infty (1+t^2)^{-\frac{z}{2}} \cos(z \arctan t)\, \operatorname{sech}\left(\frac{\pi}{2} t\right) dt,$$
$$= \frac{1}{2} + \frac{1}{z-1} + 2 \int_0^\infty (1+t^2)^{-\frac{z}{2}} \sin(z \arctan t)\, (e^{2\pi t}-1)^{-1}\, dt,$$
$$= \frac{\pi^{\frac{z}{2}}}{z(z-1)\,\Gamma\left(\frac{1}{2}z\right)} + \frac{\pi^{\frac{z}{2}}}{\Gamma\left(\frac{1}{2}z\right)} \int_1^\infty \left[t^{\frac{(1-z)}{2}} + t^{\frac{z}{2}}\right] \frac{\omega(t)}{t}\, dt,$$

where

$$\omega(t) = \sum_{n=1}^\infty \exp(-n^2 \pi t) = \frac{1}{2}[\vartheta_3(0, it) - 1],$$

ϑ_3 being the elliptic theta function.

Contour integral representations of $\zeta(z)$

$$\zeta(z) = -\frac{\Gamma(1-z)}{2\pi i} \int_\infty^{(0+)} (-t)^{z-1} [e^t - 1]^{-1}\, dt,$$
$$= -\frac{\Gamma(1-z)\,[1-2^{1-z}]^{-1}}{2\pi i} \int_\infty^{(0+)} [e^t+1]^{-1} (-t)^{z-1}\, dt,$$

where

$$|\arg(-t)| < \pi$$

and

$$z \neq 1, 2, 3, \ldots$$

Some expansions in terms of $\zeta(z)$

$$\zeta(z) - \frac{1}{z-1} = 1 - \frac{z}{2}[\zeta(z+1) - 1] - \frac{z(z+1)}{2 \cdot 3}[\zeta(z+2) - 1] + \cdots,$$

$$\Gamma(z-1) = \sum_{r=0}^{\infty} \frac{1}{(r+1)!} \Gamma(z+r)\,[\zeta(z+r) - 1],$$

$$(1 - 2^{1-z})\,\zeta(z) = z\,\frac{\zeta(z+1)}{2^{z+1}} + \frac{z(z+1)}{2!}\,\frac{\zeta(z+2)}{2^{z+2}} + \cdots,$$

$$(1 - 2^{-z})\,\Gamma(z)\,\zeta(z) = \sum_{r=0}^{\infty} \frac{\Gamma(z+r)\,\zeta(z+r)}{r!\;2^{z+r}},$$

$$\sum_{d/n} \frac{f(d)}{d^z} = \Gamma(z)\,\zeta(z) + \sum_{r=0}^{\infty} \frac{f_r(n)\,\Gamma(z+r)}{r!\;n^{z+r}}\,\zeta(z+r),$$

where

$$f_r(n) = \sum_{m=0}^{n-1} \sum_{d/(m,n)} f(d)\,m^r, \quad r = 1, 2, \ldots$$

and

$$f_0(n) = \sum_{m=0}^{n-1} \sum_{d/n} f(d).$$

1.4 The generalized zeta function $\zeta(z, \alpha)$

The function $\zeta(z, \alpha)$ is generally defined by

$$\zeta(z, \alpha) = \sum_{n=0}^{\infty} (n + \alpha)^{-z}, \quad \alpha \neq 0, -1, -2, \ldots$$

It is a meromorphic function everywhere except for a simple pole at $z = 1$ with residue 1. The function $\zeta(z, \alpha)$ reduces to $\zeta(z)$ for $\alpha = 1$.

Functional equations and series representation of $\zeta(z, \alpha)$

$$\zeta(z, \alpha) = \zeta(z, m + \alpha) + \sum_{n=0}^{m-1} (\alpha + n)^{-z}, \quad m = 1, 2, 3, \ldots,$$

$$\zeta\left(z, \frac{1}{2}\alpha\right) - \zeta\left(z, \frac{\alpha+1}{2}\right) = 2^z \sum_{n=0}^{\infty} (-1)^n (\alpha + n)^{-z},$$

$$\zeta(z, \alpha) = 2(2\pi)^{z-1}\,\Gamma(1-z) \sum_{n=1}^{\infty} n^{z-1} \sin\left(2n\pi\alpha + \frac{\pi}{2}z\right),$$

$$0 < \alpha \leq 1, \quad Re\, z < 0.$$

Special values of the argument and the parameter

$$\zeta(z, 1) = \zeta(z) = [2^z - 1]^{-1} \zeta\left(z, \frac{1}{2}\right),$$

$$\zeta(0, \alpha) = \frac{1}{2} - \alpha,$$

$$\zeta(z) = \lim_{\alpha \to 0} [\zeta(z, \alpha) - \alpha^{-z}].$$

$$\lim_{z \to 1} [\zeta(z, \alpha) - (z - 1)^{-1}] = -\psi(\alpha),$$

$$\zeta(-m, \alpha) = -\frac{B_{m+1}(\alpha)}{m+1}, \quad m = 0, 1, 2, \ldots,$$

where $B_n(x)$ are Bernoulli polynomials defined in (1.5.1).

Rademacher's formula

$$\zeta\left(z, \frac{p}{q}\right) = 2\Gamma(1-z)(2\pi q)^{z-1}\left[\sin\left(\frac{\pi}{2} z\right) \sum_{n=1}^{q} \cos\left(\frac{2\pi p}{q} n\right) \zeta\left(1 - z, \frac{n}{q}\right) + \cos\left(\frac{\pi}{2} z\right) \sum_{n=1}^{q} \sin\left(\frac{2\pi p}{q} n\right) \zeta\left(1 - z, \frac{n}{q}\right)\right].$$

p, q positive integers.

Derivatives of $\zeta(z, \alpha)$

$$\frac{\partial}{\partial \alpha} \zeta(z, \alpha) = -z \zeta(z + 1, \alpha),$$

$$\frac{\partial}{\partial z} \zeta(z, \alpha) \Big|_{z=0} = \log \Gamma(\alpha) - \frac{1}{2} \log(2\pi).$$

Integral representations

$$\zeta(z, \alpha) = \frac{1}{\Gamma(z)} \int_0^\infty [1 - e^{-t}]^{-1} t^{z-1} e^{-\alpha t}\, dt, \quad Re\, z > 1,\ Re\, \alpha > 0.$$

$$\zeta(z, \alpha) = \frac{1}{\Gamma(z)} \int_0^\infty \frac{t^{z-1} \exp[-(\alpha - 1) t]}{\exp(t) - 1}\, dt, \quad Re\, z > 1,\ Re\, \alpha > 0,$$

$$\zeta(z, \alpha) = \frac{1}{2} \alpha^{-z} - \frac{\alpha^{1-z}}{1 - z} + \frac{1}{\Gamma(z)} \int_0^\infty \left[\frac{1}{e^t - 1} - \frac{1}{t} + \frac{1}{2}\right] e^{-\alpha t} t^{z-1}\, dt,$$

$$Re\, z > -1, \quad Re\, \alpha > 0.$$

$$\zeta(z, \alpha) = \frac{1}{2} \alpha^{-z} + \frac{\alpha^{1-z}}{z - 1} + 2 \int_0^\infty \frac{(\alpha^2 + t^2)^{-\frac{z}{2}}}{e^{2\pi t} - 1} \sin\left[z \arctan\left(\frac{t}{\alpha}\right)\right] dt,$$

$$z \neq 1, \quad Re\, \alpha > 0.$$

$$\zeta(z, \alpha) = \frac{\pi\, 2^{z-2}}{z - 1} \int_0^\infty [t^2 + (2\alpha - 1)^2]^{\frac{1-z}{2}} \frac{\cos\left[(z - 1) \arctan\left(\frac{t}{(2\alpha - 1)}\right)\right]}{\cosh^2\left(\frac{\pi}{2} t\right)}\, dt,$$

$$Re\, \alpha > \frac{1}{2}.$$

$$\zeta(z,\alpha)=\cos\left(\frac{\pi}{2}z\right)\sin(2\pi\alpha)\int_0^\infty t^{-z}[\cosh(2\pi t)-\cosh(2\pi\alpha)]^{-1}dt$$

$$+\sin\left(\frac{\pi}{2}z\right)\int_0^\infty t^{-z}[\cosh(2\pi\alpha)-e^{-2\pi t}][\cosh(2\pi t)-\cosh(2\pi\alpha)]^{-1}dt.$$

$Re\, z<1,\ 0<Re\,\alpha<1;$ for $\alpha=1,\ Re\, z<0.$

$$\zeta(z,\alpha)=\alpha^{-z}+\sum_{n=0}^{m}\frac{\Gamma(n+z-1)}{\Gamma(z)}\frac{B_n}{n!}\alpha^{-n-z+1}$$

$$+\frac{1}{\Gamma(z)}\int_0^\infty t^{z-1}e^{-\alpha t}\left[\frac{1}{e^t-1}-\sum_{n=0}^{m}\frac{B_n}{n!}t^{n-1}\right]dt.$$

$Re\, z>-(2m+1).$

The following is a contour integral representation of $\zeta(z,\alpha)$.

$$\zeta(z,\alpha)=-\frac{\Gamma(1-z)}{2\pi i}\int_\infty^{(0+)}[1-e^{-t}]^{-1}(-t)^{z-1}e^{-\alpha t}\,dt,$$

$Re\,\alpha>0,\ |\arg(-t)|<\pi,\ z\neq 1,2,3,\ldots$

Some integrals associated with $\zeta(z,\alpha)$

$$\zeta\left(z,\frac{1}{2}\alpha\right)-\zeta\left(z,\frac{1+\alpha}{2}\right)=\frac{2^z}{\Gamma(z)}\int_0^\infty[1+e^{-t}]^{-1}t^{z-1}e^{-\alpha t}\,dt,$$

$Re\, z>0,\ Re\,\alpha>0;$ for $Re\,\alpha=0$ (but $Im\,\alpha\neq 0$) $0<Re\, z<1,$

$$\zeta\left(z,\frac{1+\alpha}{4}\right)-\zeta\left(z,\frac{3+\alpha}{4}\right)=2^{z-1}\int_0^\infty(t^2+\alpha^2)^{-\frac{z}{2}}\frac{\cosh\left[z\arctan\frac{t}{\alpha}\right]}{\cosh\left(\frac{\pi}{2}t\right)}dt$$

$$=\frac{2^{2z-1}}{\Gamma(z)}\int_0^\infty\frac{t^{z-1}}{\cosh t}e^{-\alpha t}\,dt.$$

$Re\, z>0,\ Re\,\alpha>-1.$

$$\int_0^1\frac{(\log t)^2}{\sqrt{1-t^2}}dt=\frac{\pi}{2}(\log 2)^2+\frac{\pi}{8}\left[\zeta\left(2,\frac{1}{2}\right)-\zeta(2)\right].$$

$$\int_0^1\frac{(\log t)^3}{\sqrt{1-t^2}}dt=-\frac{\pi}{8}\left[\pi^2\log 2+4(\log 2)^3+\zeta\left(3,\frac{1}{2}\right)-\zeta(3)\right].$$

$$\int_0^1\frac{(\log t)^{n+1}}{\sqrt{1-t^2}}dt=-\log 2\int_0^1\frac{(\log t)^n}{\sqrt{1-t^2}}dt$$

$$+\sum_{l=1}^{n}(-1)^{l+1}\frac{l!}{2^{l+1}}\binom{n}{l}\left[\zeta\left(l+1,\frac{1}{2}\right)-\zeta(l+1)\right]\int_0^1\frac{(\log t)^{n-l}}{\sqrt{1-t^2}}dt.$$

$$\zeta^*(z,\alpha) = \sum_{n=0}^{\infty} (-1)^n (\alpha+n)^{-z}$$

$$= 2^{-z}\left[\zeta\left(z,\frac{1}{2}\alpha\right) - \zeta\left(z,\frac{1+\alpha}{2}\right)\right]$$

$$= 2\Gamma(1-z)\,\pi^{z-1} \sum_{n=1}^{\infty} (2n-1)^{z-1} \sin\left[(2n-1)\,\pi\alpha + \frac{\pi}{2} z\right].$$

$Re\, z < 1,\ 0 < \alpha \leq 1.$

$$\zeta(z,\alpha) = \sum_{n=0}^{\infty} (-1)^n \frac{(\alpha-1)^n}{n!} \Gamma(n+z)\,\zeta(z+n),$$

$|\alpha - 1| < 1.$

Asymptotic expansion

For large $|\alpha|$ and $|\arg \alpha| < \pi$

$$\zeta(z,\alpha) = \frac{1}{\Gamma(z)}\left[\alpha^{1-z}\Gamma(z-1) + \frac{1}{2}\Gamma(z)\,\alpha^{-z}\right.$$

$$\left. + \sum_{n=1}^{m-1} \frac{B_{2n}}{(2n)!} \Gamma(z+2n-1)\,\alpha^{-2n-z+1}\right] + O(\alpha^{-2m-z-1}).$$

If $z = -l$, $(l = 1, 2, 3, \ldots)$ the right hand side has only a finite number of terms.

1.5 Bernoulli and Euler polynomials

1.5.1 Bernoulli numbers and polynomials

The Bernoulli numbers B_n and the Bernoulli polynomials $B_n(x)$ are defined by:

$$\frac{z}{\exp(z) - 1} = \sum_{n=0}^{\infty} \frac{B_n}{n!} z^n, \quad |z| < 2\pi,$$

$$\frac{z e^{xz}}{\exp(z) - 1} = \sum_{n=0}^{\infty} \frac{B_n(x)}{n!} z^n, \quad |z| < 2\pi.$$

More explicitly

$$B_n(x) = \sum_{l=0}^{n} \binom{n}{l} B_l x^{n-l}.$$

The following expansion is valid for the Bernoulli polynomials and gives the above result as a special case.

$$B_n(x+y) = \sum_{m=0}^{n} \binom{n}{m} B_m(x)\, y^{n-m}.$$

Special values

$$B_n(0) = (-1)^n B_n(1) = B_n, \quad n = 0, 1, 2, \ldots,$$

$$B_{2n+1}(0) = 0, \quad n = 1, 2, 3, \ldots,$$

$$B_n\left(\frac{1}{2}\right) = -(1 - 2^{1-n}) B_n, \quad n = 0, 1, 2, \ldots,$$

$$B_n\left(\frac{1}{4}\right) = (-1)^n B_n\left(\frac{3}{4}\right)$$

$$= -2^{-n}(1 - 2^{1-n}) B_n - n 4^{-n} E_{n-1},$$

where E_n are Euler's numbers (see 1.5.2)

$$B_{2n}\left(\frac{1}{6}\right) = B_{2n}\left(\frac{5}{6}\right)$$

$$= \frac{1}{2}(1 - 2^{1-2n})(1 - 3^{1-2n}) B_{2n}.$$

More generally the following symmetry relations are valid:

$$B_n(1 - x) = (-1)^n B_n(x), \quad n = 0, 1, 2, \ldots,$$

$$B_n(-x) = (-1)^n \{B_n(x) + n x^{n-1}\}. \quad n = 0, 1, 2, \ldots,$$

$$B_0(x) = 1,$$

$$B_1(x) = x - \frac{1}{2},$$

$$B_2(x) = x^2 - x + \frac{1}{6},$$

$$B_3(x) = x^3 - \frac{3}{2} x^2 + \frac{1}{2} x,$$

$$B_4(x) = x^4 - 2x^3 + x^2 - \frac{1}{30},$$

$$\sum_{m=1}^{N} m^n = \frac{1}{n+1}\left(B_{n+1}(N + 1) - B_{n+1}\right).$$

Functional equations and differentiation formulas

$$\frac{d}{dx} B_n(x) = B_n'(x) = n B_{n-1}(x), \quad n = 1, 2, 3, \ldots,$$

$$B_n(x + 1) - B_n(x) = n x^{n-1}, \quad n = 0, 1, 2, \ldots,$$

$$\sum_{m=0}^{n-1} \binom{n}{m} B_m(x) = n x^{n-1}, \quad n = 2, 3, 4, \ldots,$$

$$B_n(mx) = m^{n-1} \sum_{l=0}^{m-1} B_n\left(x + \frac{l}{m}\right), \quad \begin{array}{l} n = 0, 1, 2, \ldots, \\ m = 1, 2, \ldots \end{array}$$

Representations by trigonometric series

$$B_1(x) = x - \frac{1}{2} = -\sum_{m=1}^{\infty} \frac{\sin(2\pi m x)}{\pi m}, \quad 0 < x < 1,$$

$$B_n(x) = -2 \cdot \frac{n!}{(2\pi)^n} \sum_{m=1}^{\infty} \frac{1}{m^n} \cos\left[2\pi m x - n\frac{\pi}{2}\right],$$

$$B_{2n}(x) = 2(-1)^{n+1}(2n)! \sum_{m=1}^{\infty} (2\pi m)^{-2n} \cos(2\pi m x), \quad n = 1, 2, \ldots,$$
$$0 \leq x \leq 1$$

$$B_{2n+1}(x) = 2(-1)^{n+1}(2n+1)! \sum_{m=1}^{\infty} (2\pi m)^{-2n-1} \sin(2\pi m x).$$
$$0 \leq x \leq 1$$

Integral representations

$$B_{2n}(x) = (-1)^{n+1}(2n) \int_0^{\infty} \frac{\cos(2\pi x) - e^{-2\pi t}}{\cosh(2\pi t) - \cos(2\pi x)} t^{2n-1}\, dt,$$

$0 < Re\, x < 1, \quad n = 1, 2, 3, \ldots,$

$$B_{2n+1}(x) = (-1)^{n+1}(2n+1) \int_0^{\infty} \frac{\sin(2\pi x)}{\cosh(2\pi t) - \cos(2\pi x)} t^{2n}\, dt,$$

$0 < Re\, x < 1, \; n = 0, 1, 2, \ldots$

From the integral representations it follows that $i^n B_n\left(\frac{1}{2} + iy\right)$ is real for y real and $n = 0, 1, 2, \ldots$

Some integrals connected with Bernoulli polynomials

$$\int_a^x B_n(t)\, dt = \frac{1}{n+1}\{B_{n+1}(x) - B_{n+1}(a)\},$$

$$\int_0^1 B_n(t)\, B_m(t)\, dt = (-1)^{n-1} \frac{m!\, n!}{(m+n)!} B_{m+n},$$

$n = 1, 2, 3, \ldots; \quad m = 1, 2, 3, \ldots$

For relations between the polynomials of BERNOULLI and EULER see next section (1.5.2).

Bernoulli numbers

$$B_n = B_n(0) = B_n(1), \quad n = 2, 3, 4, \ldots,$$

$$B_{2n+1} = 0, \quad n = 1, 2, 3, \ldots,$$

$$B_{2n} = 2(-1)^{n+1}(2n)! \sum_{m=1}^{\infty} (2\pi m)^{-2n}, \quad n = 1, 2, \ldots,$$

$$B_0 = 1, \; B_1 = -\frac{1}{2}, \; B_2 = \frac{1}{6},$$

$$B_4 = -\frac{1}{30}, \; B_6 = \frac{1}{42}, \; B_8 = -\frac{1}{30}, \ldots$$

Other representations of B_n

$$B_{2n} = 2(-1)^{n+1}(2\pi)^{-2n}(2n)!\,\zeta(2n), \; n = 0, 1, 2, \ldots$$

$$= -2n\,\zeta(1-2n), \; n = 1, 2, 3, \ldots$$

$$= (-1)^{n+1}\,4n \int_0^\infty t^{2n-1}(e^{2\pi t}-1)^{-1}\,dt,$$

$$B_{2n} = (-1)^{n+1}\,4n(1-2^{1-2n})^{-1} \int_0^\infty t^{2n-1}(1+e^{2\pi t})^{-1}\,dt$$

$$= (-1)^{n+1}\,2n(2^{2n}-1)^{-1} \int_0^\infty t^{2n-1}(\sinh \pi t)^{-1}\,dt$$

$$= (-1)^{n+1}\,\pi \int_0^\infty t^{2n}(\sinh \pi t)^{-2}\,dt$$

$$= \pi(-1)^{n+1}(1-2^{1-2n})^{-1} \int_0^\infty t^{2n}[\cosh(\pi t)]^{-2}\,dt,$$

$$n = 1, 2, 3, \ldots$$

Recurrence formula

$$\sum_{m=0}^{n-1} \binom{n}{m} B_m = 0, \quad n = 2, 3, 4, \ldots$$

Generalized Bernoulli numbers and polynomials

The generalized Bernoulli numbers $B_n^{(m)}$ and the generalized Bernoulli polynomials $B_n^{(m)}(x)$ are defined by: ($m = 0, 1, 2, \ldots$)

$$\left(\frac{z}{e^z-1}\right)^m = \sum_{n=0}^\infty \frac{B_n^{(m)}}{n!} z^n, \quad |z| < 2\pi,$$

$$e^{xz}\left(\frac{z}{e^z-1}\right)^m = \sum_{n=0}^\infty \frac{B_n^{(m)}(x)}{n!} z^n, \quad |z| < 2\pi.$$

Recurrence formulas

$$(n-m)\,B_n^{(m)}(x) = n x\,B_{n-1}^{(m)}(x) + n\left(\frac{x}{m}-1\right) B_{n-1}^{(m)}(x),$$

$$B_n^{(m+1)}(x) = \left(1-\frac{n}{m}\right) B_n^{(m)}(x) + n\left(\frac{x}{m}-1\right) B_{n-1}^{(m)}(x),$$

$$n\,B_{n-1}^{(m)}(x) = \sum_{l=1}^{n} \binom{n}{l} (-1)^{l-1}(l-1)!\,B_{n-l}^{(m-l)}(x).$$

Some inequalities involving Bernoulli numbers and Bernoulli polynomials

$$|B_{2n}(x)| < |B_{2n}|, \quad n = 1, 2, 3, \ldots \quad 0 < x < 1,$$

$$\frac{2(2n)!}{(2\pi)^{2n}}(1 - 2^{1-2n})^{-1} > (-1)^{n+1} B_{2n} > \frac{2(2n)!}{(2\pi)^{2n}},$$

$$n = 1, 2, 3, \ldots,$$

$$\frac{2(2n+1)!}{(2\pi)^{2n+1}}(1 - 2^{-2n})^{-1} > (-1)^{n+1} B_{2n+1}(x) > 0,$$

$$n = 1, 2, 3, \ldots; \quad 0 < x < \frac{1}{2}.$$

1.5.2 Euler's numbers and polynomials

The Euler numbers E_n and the Euler polynomials $E_n(x)$ are defined by

$$\frac{1}{\cosh z} = \frac{2e^z}{e^{2z}+1} = \sum_{n=0}^{\infty} \frac{E_n}{n!} z^n, \quad |z| < \frac{\pi}{2},$$

$$\frac{2e^{xz}}{e^z+1} = \sum_{n=0}^{\infty} \frac{E_n(x)}{n!} z^n, \quad |z| < \pi.$$

Relations between Bernoulli and Euler polynomials

$$B_n(x) = 2^{-n} \sum_{m=0}^{n} \binom{n}{m} B_{n-m} E_m(2x), \quad n = 0, 1, 2, \ldots,$$

$$E_n(x) = \frac{2^{n+1}}{n+1}\left[B_{n+1}\left(\frac{x+1}{2}\right) - B_{n+1}\left(\frac{x}{2}\right)\right]$$

$$= \frac{2}{n+1}\left[B_{n+1}(x) - 2^{n+1} B_{n+1}\left(\frac{x}{2}\right)\right]$$

$$= 2\binom{n+1}{2}^{-1} \sum_{m=0}^{n} \binom{n+2}{m} (2^{n-m+2} - 1)\, B_m(x)\, B_{n-m+2},$$

$$n = 0, 1, 2, \ldots$$

The following expansions are valid for Euler polynomials:

$$E_n(x+y) = \sum_{m=0}^{n} \binom{n}{m} E_m(x)\, y^{n-m},$$

$$E_n(x) = \sum_{m=0}^{n} \binom{n}{m} 2^{-m} \left(x - \frac{1}{2}\right)^{n-m} E_m.$$

Special values

$$E_{2n+1} = 0,$$

$$E_n(0) = (-1)^n E_n(1)$$

$$= 2(n+1)^{-1}(1 - 2^{n+1})\, B_{n+1}, \quad n = 1, 2, \ldots,$$

$$E_n\left(\frac{1}{2}\right) = 2^{-n} E_n, \quad n = 0, 1, 2, \ldots,$$

$$E_{2n-1}\left(\frac{1}{3}\right) = -E_{2n-1}\left(\frac{2}{3}\right) = -\frac{1}{2n}(1 - 3^{1-2n})(2^{2n} - 1) B_{2n}.$$

More generally the following symmetry relations are valid:

$$E_n(1-x) = (-1)^n E_n(x), \quad n = 0, 1, 2, \ldots,$$

$$E_n(-x) = (-1)^{n+1}\{E_n(x) - 2x^n\}, \quad n = 0, 1, 2, \ldots,$$

$$\sum_{m=1}^{N} (-1)^{N-m} m^n = \frac{1}{2}\{E_n(N+1) + (-1)^N E_n(0)\},$$

$$n = 1, 2, \ldots; m = 1, 2, \ldots$$

Functional equations and differentiation formulas

$$\frac{d}{dx} E_n(x) = E_n'(x) = n E_{n-1}(x),$$

$$E_n(x+1) + E_n(x) = 2x^n,$$

$$E_n(x+1) = \sum_{m=0}^{n} \binom{n}{m} E_m(x),$$

$$\sum_{m=0}^{n} \binom{n}{m} E_m(x) + E_n(x) = 2x^n,$$

$$E_n(mx) = m^n \sum_{l=0}^{m-1} (-1)^l E_n\left(x + \frac{l}{m}\right),$$

$$n = 0, 1, 2, \ldots; \quad m \text{ odd},$$

$$E_n(mx) = -\frac{2m^n}{n+1} \sum_{l=0}^{m-1} (-1)^l B_{n+1}\left(x + \frac{l}{m}\right),$$

$$n = 0, 1, 2, \ldots; \quad m \text{ even}$$

$$\sum_{n=0}^{\infty} [E_n(x+1) + E_n(x)] \frac{z^n}{n!} = 2e^{xz}.$$

Representations by trigonometric series

$$E_1(x) = -4 \sum_{n=0}^{\infty} [(2n+1)\pi]^{-2} \cos[(2n+1)\pi x],$$

$$0 \leq x \leq 1.$$

$$E_n(x) = 4\frac{n!}{\pi^{n+1}} \sum_{m=0}^{\infty} \frac{\sin\left[(2m+1)\pi x - \frac{n\pi}{2}\right]}{(2m+1)^{n+1}},$$

$$n > 0,\ 0 \leq x \leq 1; \quad n = 0,\ 0 < x < 1,$$

$$E_{2n}(x) = 4(-1)^n (2n)! \sum_{m=0}^{\infty} [(2m+1)\pi]^{-2n-1} \sin[(2m+1)\pi x],$$

$$n > 0,\ 0 \leq x \leq 1; \quad n = 0,\ 0 < x < 1,$$

$$E_{2n+1}(x) = 4(-1)^{n+1} (2n+1)! \sum_{m=0}^{\infty} [(2m+1)\pi]^{-2n-2} \cos[(2m+1)\pi x],$$

$$n = 0, 1, 2, \ldots;\ 0 \leq x \leq 1.$$

Integral representations

$$E_{2n}(x) = 4(-1)^n \int_0^{\infty} \frac{\sin(\pi x)\cosh(\pi t)}{\cosh(2\pi t) - \cos(2\pi x)} t^{2n}\, dt,$$

$$n = 0. 1, 2, \ldots;\ 0 < x < 1,$$

$$E_{2n+1}(x) = 4(-1)^{n+1} \int_0^{\infty} \frac{\cos(\pi x)\sinh(\pi t)}{\cosh(2\pi t) - \cos(2\pi x)} t^{2n+1}\, dt,$$

$$n = 0, 1, 2, \ldots;\ 0 < x < 1.$$

Some integrals involving $E_n(x)$

$$\int_a^x E_n(t)\, dt = \frac{E_{n+1}(x) - E_{n+1}(a)}{n+1},$$

$$\int_0^1 E_n(t)\, E_m(t)\, dt = 4(-1)^n (2^{m+n+2} - 1) \frac{m!\, n!}{(m+n+2)!} B_{m+n+2},$$

$$m = 0, 1, 2, \ldots;\ n = 0, 1, 2, \ldots$$

Euler's numbers

$$E_{2n+1} = 0, \quad n = 1, 2, 3, \ldots,$$

$$E_n = 2^n E_n\left(\frac{1}{2}\right).$$

$$E_{2n} = (-1)^n\, 2(2n)! \sum_{m=0}^{\infty} (-1)^m \left[(2m+1)\frac{\pi}{2}\right]^{-2n-1},$$

$$n = 1, 2, 3, \ldots,$$

$$E_0 = 1,\ E_2 = -1,$$

$$E_4 = 5,\ E_6 = -61,\ E_8 = 1385, \ldots$$

Other representations of E_n

$$E_{2n} = (-1)^n\, 2^{2n+1} \int_0^{\infty} \operatorname{sech}(\pi t)\, t^{2n}\, dt.$$

Recurrence relation

$$\sum_{m=0}^{n} \binom{2n}{2m} E_{2m} = 0.$$

Generalized Euler numbers and polynomials

The generalized Euler numbers $E_n^{(m)}$ and the generalized Euler polynomials $E_n^{(m)}(x)$ are defined by

$$\left(\frac{2e^z}{e^{2z}+1}\right)^m = \sum_{n=0}^{\infty} \frac{E_n^{(m)}}{n!} z^n, \quad |z| < \frac{\pi}{2},$$

$$e^{xz}\left(\frac{2}{e^z+1}\right)^m = \sum_{n=0}^{\infty} \frac{E_n^{(m)}(x)}{n!} z^n, \quad |z| < \pi.$$

Recurrence formulas

$$m\, E_n^{m+1}(x) = 2E_{n+1}^{m}(x) + 2(m-x)\, E_n^{(m)}(x),$$

$$E_n^{(m)} = 2^n E_n^{(m)}\left(\frac{1}{2} m\right).$$

For more details see Erdelyi (1953).

Some inequalities involving Euler numbers and polynomials

$$|E_{2n}| > (-4)^n E_{2n}(x) > 0$$

$$0 < x < \frac{1}{2}, n = 1, 2, \ldots,$$

$$4\frac{\Gamma(2n)}{\pi^{2n}}\left(1 + \frac{1}{2^{2n}-2}\right) > (-1)^n E_{2n-1}(x) > 0,$$

$$0 < x < \frac{1}{2}, \quad n = 1, 2, 3, \ldots,$$

$$4^{n+1}\frac{\Gamma(2n+1)}{\pi^{2n+1}} > (-1)^n E_{2n} > 4^{n+1}\frac{\Gamma(2n+1)}{\pi^{2n+1}}\left(\frac{1}{1+3^{-2n-1}}\right),$$

$$n = 0, 1, 2, \ldots$$

1.6 Lerch's transcendent $\Phi(z, s, \alpha)$

The function $\Phi(z, s, \alpha)$ defined by

$$\Phi(z, s, \alpha) = \sum_{n=0}^{\infty} (\alpha + n)^{-s} z^n, \quad |z| < 1$$

satisfies the equation

$$\Phi(z, s, \alpha) = z^m \Phi(z, s, m + \alpha) + \sum_{n=0}^{m-1} (\alpha + n)^{-s} z^n,$$

$$m = 1, 2, 3, \ldots, \quad \alpha \neq 0, -1, -2, \ldots$$

$\Phi(z, s, \alpha)$ is an analytic function of z in the cut z-plane, the cut being from 1 to ∞ along the positive real z-axis.

In the limiting case $z \to 1$

$$\lim_{z\to1} [\Phi(z, s, \alpha) - \Gamma(1-s)\,[-\log z]^{s-1} z^{-\alpha}] = \zeta(s, \alpha).$$

Special cases of $\Phi(z, s, \alpha)$

Riemann zeta function

$$\zeta(s) = \sum_{n=1}^{\infty} n^{-s} = \Phi(1, s, 1).$$

Generalized zeta function

$$\zeta(s, \alpha) = \sum_{n=0}^{\infty} (\alpha + n)^{-s} = \Phi(1, s, \alpha).$$

Dilogrithm function

$$\sum_{n=1}^{\infty} \frac{z^n}{n^2} = z\,\Phi(z, 2, 1).$$

Fermi-Dirac function

$$F_k(\eta) = e^{\eta} \int_0^{\infty} \frac{x^k}{e^x + e^{\eta}}\,dx = e^h\,\Phi(e^{\eta}, k+1, 1),$$

$$F_k(\eta) = \Gamma(k+1) \sum_{n=1}^{\infty} (-1)^{n-1} \frac{e^{n\eta}}{n^{k+1}}, \quad \eta \leq 0.$$

Furthermore

$$-F_k(\eta) + e^{i\pi k} F_k(-\eta) = (2\pi i)^{k+1} \zeta\left(-k, \frac{1}{2} + \frac{\eta}{2\pi i}\right),$$

$$-F_m(\eta) + (-1)^m F_m(-\eta) = -\,(2\pi)^{m+1} e^{i\frac{\pi}{2}(m+1)} \frac{B_{m+1}\left(\frac{1}{2} + \frac{\eta}{2\pi i}\right)}{m+1}.$$

Jonquière's function

$$F(z, s) = \sum_{n=1}^{\infty} \frac{z^n}{n^s} = z\,\Phi(z, s, 1).$$

The function $F(z, s)$ satisfies the following equation

$$F(z, s) + e^{i\pi s} F(z^{-1}, s) = \frac{(2\pi)^s}{\Gamma(s)}\, e^{i\frac{\pi}{2}s}\, \zeta\left(1 - s, \frac{\log z}{2\pi i}\right).$$

In the particular case $s = m$, $(m = 1, 2, 3, \ldots)$

$$F(z, m) + (-1)^m F(z^{-1}, m) = -\frac{(2\pi i)^m}{m!} B_m\left(\frac{\log z}{2\pi i}\right).$$

Finally,

$$\Phi(z, 1, \alpha) = \sum_{n=0}^{\infty} \frac{z^n}{n+\alpha}, \qquad |z| < 1, \quad \alpha \neq 0, -1, -2, \ldots$$

$$= z^{-\alpha}\left[-\gamma - \psi(\alpha) - \log\left(\log\frac{1}{z}\right) - \sum_{m=0}^{\infty} \frac{B_{m+1}(\alpha)}{\Gamma(m+2)} (\log z)^{m+1}\right].$$

Functional equation by Lerch

$$\Phi(e^{-v}, s, \alpha) - \Gamma(1-s)\, e^{v\alpha}\, v^{s-1}$$

$$= -i(2\pi)^{s-1}\,\Gamma(1-s)\, e^{v\alpha}\left[e^{i\left(2\pi\alpha+\frac{\pi}{2}s\right)}\Phi\left(e^{i2\pi\alpha}, 1-s, 1+\frac{v}{2\pi i}\right)\right.$$

$$\left. - e^{-i\left(2\pi\alpha+\frac{\pi}{2}s\right)}\Phi\left(e^{-i2\pi\alpha}, 1-s, 1-\frac{v}{2\pi i}\right)\right].$$

Also written in the form

$$\Phi(z, s, \alpha) = i\, z^{-\alpha}\,\Gamma(1-s)\,(2\pi)^{s-1}.$$

$$\times\left[e^{-i\pi\frac{s}{2}}\Phi\left(e^{-2\pi i\alpha}, 1-s, \frac{\log z}{2\pi i}\right) - \right.$$

$$\left. - e^{i\pi\left(\frac{s}{2}+2\alpha\right)}\Phi\left(e^{i2\pi\alpha}, 1-s, 1-\frac{\log z}{2\pi i}\right)\right]$$

Integral representations

$$\Gamma(s)\,\Phi(z, s, \alpha) = \int_0^\infty t^{s-1}\, e^{-\alpha t}[1 - z e^{-t}]^{-1}\, dt$$

$$= \int_0^\infty t^{s-1}\, e^{-(\alpha-1)t}[e^t - z]^{-1}\, dt.$$

$Re\,\alpha > 0$, z not on the real axis between 1 and ∞, $Re\,s > 0$

$$\Phi(z, s, \alpha) = \frac{1}{2}\alpha^{-s} + \int_0^\infty (\alpha+t)^{-s} z^t\, dt$$

$$- 2\int_0^\infty \frac{\sin\left[t\log z - s\arctan\left(\frac{t}{\alpha}\right)\right]}{(t^2+\alpha^2)^{\frac{s}{2}}\,(e^{2\pi t}-1)}\, dt,$$

$Re\,\alpha > 0.$

Contour integral representation

$$\Phi(z, s, \alpha) = -\frac{\Gamma(1-s)}{2\pi i}\int_\infty^{(0+)} (-t)^{s-1}\, e^{-\alpha t}[1 - z e^{-t}]^{-1}\, dt,$$

$Re\,\alpha > 0$, $|\arg(-t)| < \pi$.

Series representations

$$\Phi(z, s, \alpha) = \Gamma(1-s)\, z^{-\alpha}\sum_{n=-\infty}^{\infty} e^{i2\pi n\alpha}\left[\log\frac{1}{z} + 2\pi i n\right]^{s-1},$$

$|\arg(-\log z + 2\pi i n)| < \pi$, $Re\,s < 1$, $0 < \alpha < 1$,

$$\Phi(z, s, \alpha) = \Gamma(1-s)\, z^{-\alpha}\left(\log\frac{1}{z}\right)^{s-1} + \sum_{n=0}^{\infty}\zeta(s-n, \alpha)\frac{(\log z)^n}{n!},$$

$$s \neq 1, 2, 3, \ldots, \quad \alpha \neq 0, -1, -2, \ldots, \quad |\log z| < 2\pi,$$

$$\Phi(z, m, \alpha) = z^{-\alpha} \left[\sum_{n=0}^{\infty}{}' \zeta(m-n, \alpha) \frac{(\log z)^n}{n!} + \frac{(\log z)^{m-1}}{(m-1)!} \left\{ \psi(m) - \psi(\alpha) - \log\left(\log \frac{1}{z}\right) \right\} \right],$$

$$m = 2, 3, 4, \ldots, \quad \alpha \neq 0, -1, -2, \ldots, \quad |\log z| < 2\pi.$$

The prime indicates the omission of the term for which $n = m - 1$.

1.7 Miscellaneous results

Euler's constant γ can be represented by the following integrals and infinite series

$$\gamma = -\int_0^\infty e^{-t} \log t \, dt$$

$$= -\int_0^1 \log\left(\log \frac{1}{t}\right) dt$$

$$= \int_0^1 [(\log t)^{-1} + (1-t)^{-1}] \, dt$$

$$= -\int_0^\infty [\cos t - (1+t)^{-1}] \frac{dt}{t}$$

$$= -\int_0^\infty [e^{-t} - (1+t)^{-1}] \frac{dt}{t}$$

$$= -\int_0^\infty [e^{-t} - (1+t^2)^{-1}] \frac{dt}{t}$$

$$= 1 - \int_0^\infty \left[\frac{\sin t}{t} - \frac{1}{1+t}\right] \frac{dt}{t}$$

$$= \sum_{n=2}^\infty \frac{(-1)^n}{n} \zeta(n)$$

$$= 1 - 2\left[\frac{\zeta(3)}{3 \cdot 4} + \frac{\zeta(5)}{5 \cdot 6} + \frac{\zeta(7)}{7 \cdot 8} + \cdots\right]$$

$$= \frac{5}{6} - 12\left[\frac{\zeta(3)}{3 \cdot 5 \cdot 6} + \frac{\zeta(5)}{5 \cdot 7 \cdot 8} + \frac{\zeta(7)}{7 \cdot 9 \cdot 10} + \cdots\right]$$

$$= \frac{47}{60} - 120\left[\frac{\zeta(3)}{3 \cdot 6 \cdot 7 \cdot 8} + \frac{\zeta(5)}{5 \cdot 8 \cdot 9 \cdot 10} + \frac{\zeta(7)}{7 \cdot 10 \cdot 11 \cdot 12} + \cdots\right],$$

$$\gamma = \frac{319}{420} - 1680\left[\frac{\zeta(3)}{3 \cdot 7 \cdot 8 \cdot 9 \cdot 10} + \frac{\zeta(5)}{5 \cdot 9 \cdot 10 \cdot 11 \cdot 12} + \frac{\zeta(7)}{7 \cdot 11 \cdot 12 \cdot 13 \cdot 14} + \cdots\right].$$

Power series expansions of some trigonometric and hyperbolic functions

$$\frac{z}{\sin z} = 2 \sum_{n=0}^{\infty} (-1)^n (1 - 2^{2n-1}) \frac{B_{2n}}{(2n)!} z^{2n}, \quad |z| < \pi,$$

$$z \cot z = \sum_{n=0}^{\infty} (-1)^n 2^{2n} \frac{B_{2n}}{(2n)!} z^{2n}$$

$$= -2 \sum_{n=0}^{\infty} \frac{\zeta(2n)}{\pi^{2n}} z^{2n}, \quad |z| < \pi,$$

$$\tan z = \sum_{n=1}^{\infty} (-1)^{n+1} 2^{2n} (2^{2n} - 1) \frac{B_{2n}}{(2n)!} z^{2n-1}$$

$$= 2 \sum_{n=1}^{\infty} \frac{(2^{2n} - 1)}{\pi^{2n}} \zeta(2n) z^{2n-1}, \quad |z| < \frac{\pi}{2},$$

$$\log \cos z = \sum_{n=1}^{\infty} (-1)^n 2^{2n-1} (2^{2n} - 1) \frac{B_{2n}}{n(2n)!} \cdot z^{2n}, \quad |z| < \frac{\pi}{2},$$

$$\tanh z = 2 \coth(2z) - \coth z$$

$$= \sum_{n=1}^{\infty} 2^{2n} (2^{2n} - 1) \frac{B_{2n}}{(2n)!} z^{2n-1}$$

$$= 2 \sum_{n=1}^{\infty} (-1)^{n+1} \frac{(2^{2n} - 1)}{\pi^{2n}} \zeta(2n) z^{2n-1}, \quad |z| < \frac{\pi}{2},$$

$$z \coth z = z + \frac{2z}{e^{2z} - 1}$$

$$= \sum_{n=0}^{\infty} 2^{2n} \frac{B_{2n}}{(2n)!} z^{2n}$$

$$= 2 \sum_{n=0}^{\infty} \frac{(-1)^{n+1}}{\pi^{2n}} \zeta(2n) z^{2n}.$$

$$\log \left[\frac{2 \sinh\left(\frac{z}{2}\right)}{z} \right] = \frac{1}{2} z + \log\left(\frac{1 - e^{-z}}{z}\right)$$

$$= \sum_{n=1}^{\infty} \frac{B_{2n}}{2n(2n)!} z^{2n}, \quad |z| < 2\pi.$$

Literature

BROMWICH, T. J. I'a.: An introduction to the theory of infinite series. London: Macmillan 1947.

ERDÉLYI, A.: Higher transcendental functions, Vol. 1. New York: McGraw-Hill 1953.

FORT, T.: Finite differences. Oxford 1948.

KNOPP, K.: Theory and Application of infinite series. London: Blackie and sons 1951.

LEWIN, L.: Dilogaritms and associated functions. London: MacDonald 1958.
MILNE-THOMPSON, L. M.: The Calculus of finite differences. London: Macmillan 1951.
NIELSEN, N.: Handbuch der Theorie der Gammafunktion. Leipzig: Teubner 1906.
NÖRLUND, N. E.: Vorlesungen über Differenzenrechnung. Berlin: Springer 1924.
TITCHMARSH, E. C.: The Zeta function of Riemann. Cambridge (England) 1930.
WHITTAKER, E. T., and G. N. WATSON: A course of modern analysis. Cambridge (England) 1927.

Chapter II

The hypergeometric function

2.1 Definitions and elementary relations

The function represented by the infinite series $\sum_{n=0}^{\infty} \frac{(a)_n (b)_n}{(c)_n} \frac{z^n}{n!}$ within its circle of convergence and all the analytic continuations is called the hypergeometric function ${}_2F_1(a, b; c; z)$.* The symbol $(a)_n$ is defined as

$$(a)_n = a(a+1)(a+2)\cdots(a+n-1)$$
$$= \frac{\Gamma(a+n)}{\Gamma(a)}.$$

Thus

$${}_2F_1(a, b; c; z) = \frac{\Gamma(c)}{\Gamma(b)\,\Gamma(a)} \sum_{n=0}^{\infty} \frac{\Gamma(a+n)\,\Gamma(b+n)}{\Gamma(c+n)} \frac{z^n}{n!}.$$

The series converges in the unit circle $|z| < 1$. The behavior of the series on its circle of convergence $|z| = 1$ is given by

a) divergent for $Re\,(a+b-c) \geq 1$,

b) absolutely convergent for $Re\,(a+b-c) < 0$,

c) conditionally convergent for $0 \leq Re(a+b-c) < 1$, the point $z = 1$ being excluded.

The series reduces to a polynomial of degree n in z when a or b is a negative integer $-n$, $(n = 0, 1, 2, 3, \ldots)$. The series is undefined when $c = -m$, $(m = 0, 1, 2, \ldots)$ unless a or b is a negative integer $-k$ such that $k < m$, however, $\lim_{c \to -m} \frac{1}{\Gamma(c)}\, {}_2F_1(a, b; c; z)$ exists and is given by

* The symbol $F(a, b; c; z)$ is also used instead of ${}_2F_1(a, b; c; z)$.

the following result:

$$\lim_{c \to -m} \frac{1}{\Gamma(c)} {}_2F_1(a, b; c; z)$$
$$= \frac{(a)_{m+1} (b)_{m+1}}{(m+1)!} z^{m+1} {}_2F_1(a+m+1, b+m+1; m+2; z).$$

The function ${}_2F_1(a, b; c; z)$ is a single-valued analytic function of z in the whole z-plane with a branch cut along the positive real axis from one to infinity.

Some elementary cases of the hypergeometric function

$$(1+z)^a = {}_2F_1(-a, b; b; -z),$$

$$(1+z)(1-z)^{-2a-1} = {}_2F_1(2a, a+1; a; z),$$

$$\log(1 \pm z) = z\, {}_2F_1(1, 1; 2; \mp z),$$

$$\log\left(\frac{1+z}{1-z}\right) = 2z\, {}_2F_1\left(\frac{1}{2}, 1; \frac{3}{2}; z^2\right),$$

$$\text{arc sin } z = z\, {}_2F_1\left(\frac{1}{2}, \frac{1}{2}; \frac{3}{2}; z^2\right)$$

$$= z(1-z^2)^{\frac{1}{2}} {}_2F_1\left(1, 1; \frac{3}{2}; z^2\right),$$

$$\arctan z = z\, {}_2F_1\left(\frac{1}{2}, 1; \frac{3}{2}; -z^2\right),$$

$$\sin \alpha z = \alpha \sin z\, {}_2F_1\left(\frac{1+\alpha}{2}, \frac{1-\alpha}{2}; \frac{3}{2}; \sin^2 z\right),$$

$$\sin[2(\alpha-1)z] = (\alpha-1)\sin(2z)\, {}_2F_1\left(\alpha, 2-\alpha; \frac{3}{2}; \sin^2 z\right),$$

$$\cos \alpha z = {}_2F_1\left(\frac{\alpha}{2}, -\frac{\alpha}{2}; \frac{1}{2}; \sin^2 z\right)$$

$$= \cos z\, {}_2F_1\left(\frac{1+\alpha}{2}, \frac{1-\alpha}{2}; \frac{1}{2}; \sin^2 z\right)$$

$$= (\cos z)^\alpha\, {}_2F_1\left(-\frac{\alpha}{2}, \frac{1-\alpha}{2}; \frac{1}{2}; -\tan^2 z\right),$$

$$\sin[\alpha \arcsin z] = \alpha z\, {}_2F_1\left(\frac{1+\alpha}{2}, \frac{1-\alpha}{2}; \frac{3}{2}; z^2\right),$$

$$= \alpha z \sqrt{1-z^2}\, {}_2F_1\left(1+\frac{\alpha}{2}, 1-\frac{\alpha}{2}; \frac{3}{2}; z^2\right),$$

$$\cos[\alpha \arcsin z] = {}_2F_1\left(\frac{\alpha}{2}, -\frac{\alpha}{2}; \frac{1}{2}; z^2\right)$$

$$= \sqrt{1-z^2}\, {}_2F_1\left(\frac{1+\alpha}{2}, \frac{1-\alpha}{2}; \frac{1}{2}; z^2\right),$$

$$\left(1+\sqrt{1-z^2}\right)^{-2a} = 2^{-2a}\, {}_2F_1\left(a, a+\frac{1}{2}; 2a+1; z^2\right),$$

$$(1-z^2)^{-\frac{1}{2}}\left(1+\sqrt{1-z^2}\right)^{-2a}=2^{-2a}\,{}_2F_1\left(a+1,a+\frac{1}{2};2a+1;z^2\right),$$

$$(1-z)^{-\frac{1}{2}}\left(1+\sqrt{1-z}\right)^{1-2a}=2^{1-2a}\,{}_2F_1\left(a,a+\frac{1}{2};2a;z\right),$$

$$\log\left(z+\sqrt{1+z^2}\right)=z\,{}_2F_1\left(\frac{1}{2},\frac{1}{2};\frac{3}{2};-z^2\right)$$

$$=z(1+z^2)^{\frac{1}{2}}\,{}_2F_1\left(1,1;\frac{3}{2};-z^2\right),$$

$$(1+z)^{-2a}+(1-z)^{-2a}=2\,{}_2F_1\left(a,a+\frac{1}{2};\frac{1}{2};z^2\right),$$

$$(1+z)^{1-2a}-(1-z)^{1-2a}=2z(1-2a)\,{}_2F_1\left(a,a+\frac{1}{2};\frac{3}{2};z^2\right),$$

$$\left(z+\sqrt{1+z^2}\right)^{2a}+\left(\sqrt{1+z^2}-z\right)^{2a}=2\,{}_2F_1\left(-a,a;\frac{1}{2};-z^2\right),$$

$$(1+z^2)^{-\frac{1}{2}}\left[\left(\sqrt{1+z^2}+z\right)^{2a-1}+\left(\sqrt{1+z^2}-z\right)^{2a-1}\right]$$

$$=2\,{}_2F_1\left(a,1-a;\frac{1}{2};-z^2\right),$$

$${}_2F_1\left(a+1,a+\frac{1}{2};2a+1;\operatorname{sech}^2 z\right)=e^{-2az}\coth z\,(2\cosh z)^{2a}.$$

Higher transcendental functions as special cases of ${}_2F_1(a,b;c;z)$

When a or b is equal to a negative integer $-m$,

$${}_2F_1(-m,b;c;z)=\sum_{n=0}^{m}\frac{(-m)_n\,(b)_n}{(c)_n}\frac{z^n}{n!}$$

provided that c is not a negative integer $-k$ such that $k<m$. If c is not a negative integer and a or b is a negative integer $-m$, the hypergeometric function reduces to a polynomial. The result can be extended to include the case

$$a=-m,\ c=-m-l;\ l=0,-1,-2,\ldots,\ m=0,-1,-2,\ldots$$

In this case

$${}_2F_1(-m,b;-m-l;z)=\sum_{n=0}^{m}\frac{(-m)_n\,(b)_n}{(-m-l)_n}\frac{z^n}{n!}.$$

Some special cases of this result are:

$$T_n(1-2x)={}_2F_1\left(-n,n;\frac{1}{2};x\right),$$

$$P_n(1-2x)={}_2F_1(-n,n+1;1;x),$$

$$C_n^{(\alpha)}(1-2x)=\frac{(2\alpha)_n}{n!}\,{}_2F_1\left(-n,n+2\alpha;\frac{1}{2}+\alpha;x\right),$$

$$P_n^{(\alpha,\beta)}(1-2x)=\frac{(1+\alpha)_n}{n!}\,{}_2F_1(-n,n+\alpha+\beta+1;1+\alpha;x),$$

where

$$T_n(x),\ P_n(x),\ C_n^{(\alpha)}(x) \quad \text{and} \quad P_n^{(\alpha,\beta)}(x)$$

denote the polynomials of CHEBYSHEV, LEGENDRE, GEGENBAUER and JACOBI respectively. For more details about these polynomials see chap. V. For the relation between ${}_2F_1(a, b; c; z)$ and the Legendre functions see chap. IV and also 2.4.3.

Special values of the argument and parameters

$${}_2F_1\left(\frac{1}{2}, \frac{1}{2}; 1; k^2\right) = \frac{2}{\pi} K,$$

$${}_2F_1\left(\frac{1}{2}, -\frac{1}{2}; 1; k^2\right) = \frac{2}{\pi} E,$$

where K, E are the complete elliptic integrals of the first and second kind respectively.

$${}_2F_1(a, b; c; 1) = \frac{\Gamma(c)\,\Gamma(c-a-b)}{\Gamma(c-a)\,\Gamma(c-b)},$$

$$Re\,(a+b-c) < 0,\ c \neq 0, -1, -2, \ldots,$$

$${}_2F_1(a, b; a-b+1; -1) = 2^{-a}\,\pi^{\frac{1}{2}} \frac{\Gamma(1+a-b)}{\Gamma\left(1+\frac{1}{2}a-b\right)\Gamma\left(\frac{1+a}{2}\right)},$$

$$1+a-b \neq 0, -1, -2, \ldots,$$

$${}_2F_1(a, b; a-b+2; -1) = 2^{-a}\,\pi^{\frac{1}{2}}\,(b-1)^{-1}\cdot\Gamma(a-b+2)$$

$$\times\left[\frac{1}{\Gamma\left(\frac{1}{2}a\right)\Gamma\left(\frac{3}{2}+\frac{1}{2}a-b\right)} - \frac{1}{\Gamma\left(\frac{1+a}{2}\right)\Gamma\left(1+\frac{a}{2}-b\right)}\right],$$

$$a-b+2 \neq 0, -1, -2, \ldots,$$

$${}_2F_1(1, a; a+1; -1) = \frac{a}{2}\left[\psi\left(\frac{1+a}{2}\right) - \psi\left(\frac{a}{2}\right)\right],$$

where $\psi(z) = \dfrac{\Gamma'(z)}{\Gamma(z)}$ is the function discussed in chap. I.

$${}_2F_1\left(a, b; \frac{a+b+1}{2}; \frac{1}{2}\right) = \pi^{\frac{1}{2}} \frac{\Gamma\left(\frac{a+b+1}{2}\right)}{\Gamma\left(\frac{a+1}{2}\right)\Gamma\left(\frac{b+1}{2}\right)},$$

$$a+b+1 \neq 0, -2, -4, -6, \ldots,$$

$${}_2F_1\left(a, b; \frac{a+b+2}{2}; \frac{1}{2}\right) = 2\pi^{\frac{1}{2}}\,(a-b)^{-1}\,\Gamma\left(\frac{a+b}{2}+1\right)$$

$$\times\left[\frac{1}{\Gamma\left(\frac{a}{2}\right)\Gamma\left(\frac{b+1}{2}\right)} - \frac{1}{\Gamma\left(\frac{b}{2}\right)\Gamma\left(\frac{a+1}{2}\right)}\right],$$

$$a+b \neq -2, -4, -6, \ldots,$$

$${}_2F_1\left(a, 1-a; b; \frac{1}{2}\right) = 2^{1-b}\pi^{\frac{1}{2}}\,\Gamma(b)\left[\Gamma\left(\frac{a+b}{2}\right)\Gamma\left(\frac{1+b-a}{2}\right)\right]^{-1},$$

$$b \neq 0, -1, -2, \ldots,$$

$${}_2F_1\left(1, 1; c+1; \frac{1}{2}\right) = c\left[\psi\left(\frac{c+1}{2}\right) - \psi\left(\frac{c}{2}\right)\right],$$

$$c \neq -1, -2, \ldots,$$

$${}_2F_1\left(a, a; a+1; \frac{1}{2}\right) = 2^{a-1}a\left[\psi\left(\frac{a+1}{2}\right) - \psi\left(\frac{a}{2}\right)\right],$$

$$a \neq -1, -2, -3, \ldots,$$

$${}_2F_1\left(a, a+\frac{1}{2}; \frac{3}{2} - 2a; -\frac{1}{3}\right) = \left(\frac{8}{9}\right)^{-2a}\frac{\Gamma\left(\frac{4}{3}\right)\Gamma\left(\frac{3}{2}-2a\right)}{\Gamma\left(\frac{3}{2}\right)\Gamma\left(\frac{4}{3}-2a\right)},$$

$$\frac{3}{2} - 2a \neq 0, -1, -2, \ldots$$

$${}_2F_1\left(a, a+\frac{1}{2}; a+\frac{5}{6}; \frac{1}{9}\right) = \left(\frac{3}{4}\right)^a \pi^{\frac{1}{2}}\frac{\Gamma\left(\frac{5}{6}+\frac{2}{3}a\right)}{\Gamma\left(\frac{1}{2}+\frac{1}{3}a\right)\Gamma\left(\frac{5}{6}+\frac{a}{3}\right)},$$

$$\frac{2}{3}a + \frac{5}{6} \neq 0, -1, -2, -3, \ldots$$

$${}_2F_1\left(a, \frac{a+1}{3}; \frac{2}{3}(a+1); e^{i\frac{\pi}{3}}\right) = 2^{\frac{2}{3}a+\frac{2}{3}}\pi^{\frac{1}{2}}3^{-\frac{1}{2}(a+1)}$$

$$\times\, e^{i\frac{\pi}{6}a}\frac{\Gamma\left(\frac{1}{3}a+\frac{5}{6}\right)}{\Gamma\left(\frac{2}{3}\right)\Gamma\left(\frac{a+2}{3}\right)},$$

$$\frac{5}{6} + \frac{a}{3} \neq 0, -1, -2, -3 \ldots$$

Differentiation formulas

$$\frac{d}{dz}\,{}_2F_1(a, b; c; z) = \frac{ab}{c}\,{}_2F_1(a+1, b+1; c+1; z).$$

More generally

$$\frac{d^n}{dz^n}\,{}_2F_1(a, b; c; z) = \frac{(a)_n\,(b)_n}{(c)_n}\,{}_2F_1(a+n, b+n; c+n; z),$$

$$\frac{d^n}{dz^n}\left[z^{a+n-1}\,{}_2F_1(a, b; c; z)\right] = (a)_n\,z^{a-1}\,{}_2F_1(a+n, b; c; z),$$

$$\frac{d^n}{dz^n}\left[z^{c-1}\,{}_2F_1(a, b; c; z)\right] = (c-n)_n\,z^{c-n-1}\,{}_2F_1(a, b; c-n; z),$$

$$\frac{d^n}{dz^n}\left[z^{c-a+n-1}(1-z)^{a+b-c}\,{}_2F_1(a, b; c; z)\right]$$

$$= (c-a)_n\,z^{c-a-1}(1-z)^{a+b-c-n}\,{}_2F_1(a-n, b; c; z),$$

$$\frac{d^n}{dz^n}\left[(1-z)^{a+b-c}\,{}_2F_1(a,b;c;z)\right] = \frac{(c-a)_n\,(c-b)_n}{(c)_n}(1-z)^{a+b-c-n}\,{}_2F_1(a,b;c+n;z),$$

$$\frac{d^n}{dz^n}\left[(1-z)^{a+n-1}\,{}_2F_1(a,b;c;z)\right] = (-1)^n\frac{(a)_n\,(c-b)_n}{(c)_n} \times (1-z)^{a-1}\,{}_2F_1(a+n,b;c+n;z),$$

$$\frac{d^n}{dz^n}\left[z^{c-1}(1-z)^{b-c+n}\,{}_2F_1(a,b;c;z)\right] = (c-n)_n\,z^{c-n-1} \times (1-z)^{b-c}\,{}_2F_1(a-n,b;c-n;z),$$

$$\frac{d^n}{dz^n}\left[z^{c-1}(1-z)^{a+b-c}\,{}_2F_1(a,b;c;z)\right] = (c-n)_n\cdot z^{c-n-1} \times (1-z)^{a+b-c-n}\,{}_2F_1(a-n,b-n;c-n;z).$$

2.2 The hypergeometric differential equation

The differential equation

$$\text{(I)}\qquad z(1-z)\frac{d^2w}{dz^2}+[c-(a+b+1)z]\frac{dw}{dz}-abw=0.$$

Called "the hypergeometric differential equation", has three singular points $z=0,1,\infty$. These are regular singular points of the differential equation and the pairs of exponents at these points are

$$\varrho^{(0)}=0,1-c;\ \varrho^{(1)}=0,c-a-b;$$

and

$$\varrho^{(\infty)}=a,b$$

respectively.
From the general theory of the differential equations of the Fuchsian type, one distinguishes the following cases.

1. None of the numbers c, $a-b$, $c-a-b$, is equal to an integer

In this case a system of two linearly independent solutions of (I) in the vicinity of the singular points $z=0,1,\infty$ are given by

$$w_1^{(0)}(z)={}_2F_1(a,b;c;z) = (1-z)^{c-a-b}\,{}_2F_1(c-a,c-b;c;z),$$

$$w_2^{(0)}(z)=z^{1-c}\,{}_2F_1(a-c+1,b-c+1;2-c;z) = z^{1-c}(1-z)^{c-a-b}\,{}_2F_1(1-a,1-b;2-c;z),$$

$$w_1^{(1)}(z)={}_2F_1(a,b;a+b-c+1;1-z) = z^{1-c}\,{}_2F_1(1+b-c,a-c+1;a+b-c+1;1-z),$$

$$w_2^{(1)}(z) = (1-z)^{c-a-b}\,{}_2F_1(c-b, c-a; c-a-b+1; 1-z)$$
$$= z^{1-c}(1-z)^{c-a-b}\,{}_2F_1(1-a, 1-b; c-a-b+1; 1-z),$$
$$w_1^{(\infty)}(z) = z^{-a}\,{}_2F_1\left(a, a-c+1; a-b+1; \frac{1}{z}\right)$$
$$= z^{b-c}(z-1)^{c-a-b}\,{}_2F_1\left(1-b, c-b; a-b+1; \frac{1}{z}\right),$$
$$w_2^{(\infty)}(z) = z^{-b}\,{}_2F_1\left(b, b-c+1; b-a+1; \frac{1}{z}\right)$$
$$= z^{a-c}(-1+z)^{c-a-b}\,{}_2F_1\left(1-a, c-a; b-a+1; \frac{1}{z}\right).$$

Other representations of the solutions are:

$$w_1^{(0)}(z) = (1-z)^{-a}\,{}_2F_1\left(a, c-b; c; \frac{z}{z-1}\right)$$
$$= (1-z)^{-b}\,{}_2F_1\left(c-a, b; c; \frac{z}{z-1}\right),$$
$$w_2^{(0)}(z) = z^{1-c}(1-z)^{c-a-1}\,{}_2F_1\left(a-c+1, 1-b; 2-c; \frac{z}{z-1}\right)$$
$$= z^{1-c}(1-z)^{c-b-1}\,{}_2F_1\left(1-a, b-c+1; 2-c; \frac{z}{z-1}\right),$$
$$w_1^{(1)}(z) = z^{-a}\,{}_2F_1\left(a, a-c+1; a+b-c+1; 1-\frac{1}{z}\right)$$
$$= z^{-b}\,{}_2F_1\left(b, b-c+1; a+b-c+1; 1-\frac{1}{z}\right),$$
$$w_2^{(1)}(z) = z^{a-c}(1-z)^{c-a-b}\,{}_2F_1\left(c-a, 1-a; c-a-b+1; 1-\frac{1}{z}\right)$$
$$= z^{b-c}(1-z)^{c-a-b}\,{}_2F_1\left(c-b, 1-b; c-a-b+1; 1-\frac{1}{z}\right),$$
$$w_1^{(\infty)}(z) = (z-1)^{-a}\,{}_2F_1\left(a, c-b; -b+a+1; \frac{1}{1-z}\right)$$
$$= (z-1)^{-b}\,{}_2F_1\left(b, c-a; b-a+1; \frac{1}{1-z}\right),$$
$$w_2^{(\infty)}(z) = z^{1-c}(z-1)^{c-a-1}\,{}_2F_1\left(a-c+1, 1-b; a-b+1; \frac{1}{1-z}\right)$$
$$= z^{1-c}(z-1)^{c-b-1}\,{}_2F_1\left(1-a, b-c+1; b-a+1; \frac{1}{1-z}\right).$$

These are the 24 solutions of the hypergeometric differential equation due to KUMMER.

2. One of the numbers $a, b, c-a, c-b$, is an integer

In this case one of the hypergeometric series terminates and the corresponding solution is of the form

$$w(z) = z^{\lambda}(1-z)^{\mu}\,P_n(z),$$

where $P_n(z)$ is an nth degree polynomial in z. This is the degenerate case of the hypergeometric differential equation and its solutions are given elsewhere (see chap. V).

3. $c - a - b$ is an integer and c is not an integer

In this case $c = a + b + m$ where m is an integer. The expressions

$$ {}_2F_1(a, b; a+b; z) = \frac{\Gamma(a+b)}{\Gamma(a)\,\Gamma(b)} \sum_{n=0}^{\infty} \frac{(a)_n\,(b)_n}{(n!)^2} $$

$$ \times [2\psi(n+1) - \psi(a+n) - \psi(b+n) - \log(1-z)]\,(1-z)^n, $$

$$ |\arg(1-z)| < \pi, \; |1-z| < 1, $$

$$ {}_2F_1(a, b; a+b+m; z) = \frac{\Gamma(a+b+m)\,\Gamma(m)}{\Gamma(a+m)\,\Gamma(b+m)} \sum_{n=0}^{m-1} \frac{(a)_n\,(b)_n}{(1-m)_n\,n!}(1-z)^n $$

$$ - \frac{\Gamma(a+b+m)}{\Gamma(a)\,\Gamma(b)}(z-1)^m \sum_{n=0}^{\infty} \frac{(a+m)_n\,(b+m)_n}{n!\,(n+m)!} $$

$$ \times [\log(1-z) + \psi(a+n+m) + \psi(b+n+m) $$

$$ - \psi(n+1) - \psi(n+m+1)]\,(1-z)^n, $$

$$ m = 1, 2, 3, \ldots, \quad |\arg(1-z)| < \pi, \; |1-z| < 1, $$

$$ {}_2F_1(a, b; a+b-m; z) = \frac{\Gamma(a+b-m)\,\Gamma(m)}{\Gamma(a)\,\Gamma(b)}(1-z)^{-m} $$

$$ \times \sum_{n=0}^{m-1} \frac{(a-m)_n\,(b-m)_n}{n!\,(1-m)_n}(1-z)^n - \frac{(-1)^m\,\Gamma(a+b-m)}{\Gamma(a-m)\,\Gamma(b-m)} $$

$$ \times \sum_{n=0}^{\infty} \frac{(a)_n\,(b)_n}{n!\,(n+m)!} [\log(1-z) + \psi(a+n) + \psi(b+n) $$

$$ - \psi(n+1) - \psi(n+m+1)]\,(1-z)^n, $$

$$ m = 1, 2, 3, \ldots, \quad |\arg(1-z)| < \pi, \; |1-z| < 1 $$

give the analytic continuation of $w_1^{(0)}(z)$ and $w_2^{(0)}(z)$ into the neighborhood of $z = 1$. Similarly the expressions

$$ {}_2F_1(a, a; c; z) = \frac{\Gamma(c)}{\Gamma(a)\,\Gamma(c-a)} \sum_{n=0}^{\infty} \frac{(a)_n\,(a-c+1)_n}{(n!)^2} $$

$$ \times [\log(-z) - \psi(a+n) - \psi(c-a-n) + 2\psi(n+1)]\,z^{-n}, $$

$$ |\arg(-z)| < \pi, \; |z| > 1, \; c - a \neq 0, \pm 1, \pm 2, \ldots $$

and

$$
\begin{aligned}
{}_2F_1(a, a+m; c; z) &= {}_2F_1(a+m, a; c; z) \\
&= \frac{\Gamma(c)\,(-z)^{-a-m}}{\Gamma(c-a)\,\Gamma(a+m)} \sum_{n=0}^{\infty} \frac{(a)_{n+m}\,(a-c+1)_{n+m}}{(n+m)!\,n!} \cdot z^{-n} \\
&\quad \times [\log(-z) - \psi(a+m+n) - \psi(b+m+n) \\
&\quad + \psi(1+n) + \psi(m+n+1)] \\
&\quad + \frac{\Gamma(c)}{\Gamma(a+m)} (-z)^{-a} \sum_{n=0}^{m-1} \frac{(a)_n}{n!} \frac{\Gamma(m-n)}{\Gamma(c-a-n)} z^{-n},
\end{aligned}
$$

$$m = 1, 2, 3, \ldots; \; |\arg(-z)| < \pi, \; |z| > 1; \; c - a \neq 0, \pm 1, \pm 2, \ldots$$

give the analytic continuation of $w_1^{(0)}(z)$ and $w_2^{(0)}(z)$ into the neighborhood of $z = \infty$.

4. $c = 1$

The fundamental system is given by

$$
\begin{aligned}
w_1^{(0)}(z) &= {}_2F_1(a, b; 1; z), \\
w_2^{(0)}(z) &= {}_2F_1(a, b; 1; z) \log z + \sum_{n=1}^{\infty} \frac{(a)_n\,(b)_n}{(n!)^2} z^n \\
&\times [\psi(a+n) + \psi(b+n) - \psi(a) - \psi(b) - 2\psi(n+1) + 2\psi(1)], \\
&|z| < 1.
\end{aligned}
$$

5. $c = m + 1, \; m = 1, 2, 3, \ldots$

A fundamental system of solution is

$$
\begin{aligned}
w_1^{(0)}(z) &= {}_2F_1(a, b; m+1; z), \\
w_2^{(0)}(z) &= {}_2F_1(a, b; m+1; z) \log z - \sum_{n=1}^{m} \frac{(n-1)!\,(-m)_n}{(1-a)_n\,(1-b)_n} z^{-n} \\
&+ \sum_{n=1}^{\infty} \frac{(a)_n\,(b)_n}{n!\,(1+m)_n} z^n [\psi(a+n) + \psi(b+n) - \psi(a) - \psi(b) \\
&+ \psi(1) - \psi(n+1) + \psi(m+1) - \psi(n+m+1)], \\
&|z| < 1, \quad a, b \neq 0, 1, 2, \ldots, m-1.
\end{aligned}
$$

6. $c = 1 - m, \; m = 1, 2, 3, \ldots$

A fundamental system of solution is

$$w_1^{(0)}(z) = z^m\, {}_2F_1(a+m, b+m; 1+m; z),$$

$$w_2^{(0)}(z) = z^m \, {}_2F_1(a+m, b+m; m+1; z) \log z - \sum_{n=1}^{m} \frac{(n-1)!\,(-m)_n}{(1-a-m)_n}$$
$$\times \frac{z^{m-n}}{(1-b-m)_n} + z^m \sum_{n=1}^{\infty} \frac{(a+m)_n\,(b+m)_n}{n!\,(1+m)_n}$$
$$\times [\psi(1) - \psi(n+1) + \psi(m+1) - \psi(a+m) - \psi(b+m)$$
$$- \psi(m+n+1) + \psi(a+m+n) + \psi(b+m+n)]\, z^n,$$
$$|z| < 1, \quad a, b \neq 0, -1, -2, \ldots, \quad -m+1.$$

2.3 Gauss' contiguous relations

The six functions ${}_2F_1(a \pm 1, b; c; z)$, ${}_2F_1(a, b \pm 1; c; z)$, ${}_2F_1(a, b; c \pm 1; z)$ associated with ${}_2F_1(a, b; c; z)$ are called contiguous to it. A relation between ${}_2F_1(a, b; c; z)$ and any two contiguous functions is called a contiguous relation. These relations were given by GAUSS. By repeated application of these relations it is possible to express the function ${}_2F_1(a+l, b+m; c+n; z)$ as a linear combination of ${}_2F_1(a, b; c; z)$ and one of its contiguous functions such that the coefficients are rational functions of a, b, c, z provided that l, m, n are integers and $c + n \neq 0, -1, -2, -3, \ldots$

$$(c-a)\, {}_2F_1(a-1, b; c; z) + [2a - c - (a-b)\, z]\, {}_2F_1(a, b; c; z)$$
$$+ a(z-1)\, {}_2F_1(a+1, b; c; z) = 0,$$

$$(c-b)\, {}_2F_1(a, b-1; c; z) + [2b - c - (b-a)\, z]\, {}_2F_1(a, b; c; z)$$
$$+ b(z-1)\, {}_2F_1(a, b+1; c; z) = 0,$$

$$c(c-1)(z-1)\, {}_2F_1(a, b; c-1; z)$$
$$+ c[c - 1 - (2c - a - b - 1)\, z]\, {}_2F_1(a, b; c; z)$$
$$+ (c-a)(c-b)\, z\, {}_2F_1(a, b; c+1; z) = 0,$$

$$(b-a)\, {}_2F_1(a, b; c; z) + a\, {}_2F_1(a+1, b; c; z)$$
$$- b\, {}_2F_1(a, b+1; c; z) = 0,$$

$$(b-c)\, {}_2F_1(a, b-1; c; z) + (c-a-b)\, {}_2F_1(a, b; c; z)$$
$$+ a(1-z)\, {}_2F_1(a+1, b; c; z) = 0,$$

$$c[a - (c-b)\, z]\, {}_2F_1(a, b; c; z) - ac(1-z)\, {}_2F_1(a+1, b; c; z)$$
$$+ (c-a)(c-b)\, z\, {}_2F_1(a, b; c+1; z) = 0,$$

$$(1-c)\, {}_2F_1(a, b; c-1; z) + (c-a-1)\, {}_2F_1(a, b; c; z)$$
$$+ a\, {}_2F_1(a+1, b; c; z) = 0,$$

$$(a-c)\,{}_2F_1(a-1,b;c;z)+(c-a-b)\,{}_2F_1(a,b;c;z)$$
$$+\,b(1-z)\,{}_2F_1(a,b+1;c;z)=0,$$

$$(a-c)\,{}_2F_1(a-1,b;c;z)+(c-b)\,{}_2F_1(a,b-1;c;z)$$
$$+\,(b-a)(1-z)\,{}_2F_1(a,b;c;z)=0,$$

$$(-c)\,{}_2F_1(a-1,b;c;z)+c(1-z)\,{}_2F_1(a,b;c;z)$$
$$+\,(c-b)\,z\,{}_2F_1(a,b;c+1;z)=0,$$

$$(c-a)\,{}_2F_1(a-1,b;c;z)-(c-1)(1-z)\,{}_2F_1(a,b;c-1;z)$$
$$+\,[a-1-(c-b-1)\,z]\,{}_2F_1(a,b;c;z)=0,$$

$$c\,[b-(c-a)\,z]\,{}_2F_1(a,b;c;z)-bc(1-z)\,{}_2F_1(a,b+1;c;z)$$
$$+\,(c-a)(c-b)\,z\,{}_2F_1(a,b;c+1;z)=0,$$

$$(1-c)\,{}_2F_1(a,b;c-1;z)+(c-b-1)\,{}_2F_1(a,b;c;z)$$
$$+\,b\,{}_2F_1(a,b+1;c;z)=0,$$

$$(-c)\,{}_2F_1(a,b-1;c;z)+c(1-z)\,{}_2F_1(a,b;c;z)$$
$$+\,(c-a)\,z\,{}_2F_1(a,b;c+1;z)=0,$$

$$(c-b)\,{}_2F_1(a,b-1;c;z)-(c-1)(1-z)\,{}_2F_1(a,b;c-1;z)$$
$$+\,[b-1-(c-a-1)\,z]\,{}_2F_1(a,b;c;z)=0,$$

$$c(1-c)(1-z)\,{}_2F_1(a,b;c-1;z)$$
$$+\,c\,[c-1-(2c-a-b-1)\,z]\,{}_2F_1(a,b;c;z)$$
$$+\,(c-a)(c-b)\,z\,{}_2F_1(a,b;c+1;z)=0.$$

2.4 Linear and higher order transformations

2.4.1 Linear transformations

$${}_2F_1(a,b;c;z)=(1-z)^{c-a-b}\,{}_2F_1(c-a,c-b;c;z),$$

$${}_2F_1(a,b;c;z)=(1-z)^{-a}\,{}_2F_1\left(a,c-b;c;\frac{z}{z-1}\right),$$

$${}_2F_1(a,b;c;z)=(1-z)^{-b}\,{}_2F_1\left(b,c-a;c;\frac{z}{z-1}\right),$$

$${}_2F_1(a,b;c;z)=\frac{\Gamma(c)\,\Gamma(c-a-b)}{\Gamma(c-b)\,\Gamma(c-a)}\,{}_2F_1(a,b;a+b-c+1;1-z)$$
$$+\,(1-z)^{c-a-b}\,\frac{\Gamma(c)\,\Gamma(a+b-c)}{\Gamma(a)\,\Gamma(b)}\,{}_2F_1(c-a,c-b;c-a-b+1;1-z),$$
$$|\arg(1-z)|<\pi,\quad c-a-b\neq\pm m,\quad m=0,1,2,3,\ldots,$$

$$ {}_2F_1(a, b; c; z) = \frac{\Gamma(c)\,\Gamma(b-a)}{\Gamma(b)\,\Gamma(c-a)} (-z)^{-a}\, {}_2F_1\left(a, a-c+1; a-b+1; \frac{1}{z}\right) $$

$$ + \frac{\Gamma(c)\,\Gamma(a-b)}{\Gamma(a)\,\Gamma(c-b)} (-z)^{-b}\, {}_2F_1\left(b, b-c+1; b-a+1; \frac{1}{z}\right), $$

$$ |\arg(-z)| < \pi, \; a - b \neq \pm m, \; m = 0, 1, 2, 3, \ldots, $$

$$ {}_2F_1(a, b; c; z) = (1-z)^{-a} \frac{\Gamma(c)\,\Gamma(b-a)}{\Gamma(b)\,\Gamma(c-a)}\, {}_2F_1\left(a, c-b; a-b+1; \frac{1}{1-z}\right) $$

$$ + (1-z)^{-b} \frac{\Gamma(c)\,\Gamma(a-b)}{\Gamma(a)\,\Gamma(c-b)}\, {}_2F_1\left(b, c-a; b-a+1; \frac{1}{1-z}\right), $$

$$ |\arg(1-z)| < \pi, \; a - b \neq \pm m, \; m = 0, 1, 2, 3, \ldots, $$

$$ {}_2F_1(a, b; c; z) = \frac{\Gamma(c)\,\Gamma(c-a-b)}{\Gamma(c-a)\,\Gamma(c-b)} $$

$$ \times z^{-a}\, {}_2F_1\left(a, a-c+1; a+b-c+1; 1-\frac{1}{z}\right) $$

$$ + \frac{\Gamma(c)\,\Gamma(a+b-c)}{\Gamma(a)\,\Gamma(b)} (1-z)^{c-a-b} $$

$$ \times z^{a-c}\, {}_2F_1\left(c-a, 1-a; c-a-b+1; 1-\frac{1}{z}\right), $$

$$ |\arg z| < \pi, \quad |\arg(1-z)| < \pi, $$

$$ c - a - b \neq \pm m, \; m = 0, 1, 2, 3, \ldots, $$

$$ {}_2F_1(a, a; c; z) = \frac{\Gamma(c)}{\Gamma(a)\,\Gamma(c-a)} (-z)^{-a} \sum_{n=0}^{\infty} \frac{(a)_n\,(a-c+1)_n}{(n!)^2} $$

$$ \times [\log(-z) - \psi(a+n) - \psi(c-a-n) + 2\psi(n+1)]\, z^{-n}, $$

$$ |\arg(-z)| < \pi, \; |z| > 1, \; c - a \neq 0, \pm 1, \pm 2, \ldots, $$

$$ {}_2F_1(a, a+m; c; z) = \frac{\Gamma(c)}{\Gamma(c-a)\,\Gamma(a+m)} (-z)^{-a-m} \sum_{n=0}^{\infty} \frac{(a)_{n+m}\,(a-c+1)_{m+n}}{n!\,(n+m)!} $$

$$ \times z^{-n} [\log(-z) + \psi(1+n) - \psi(a+m+n) $$

$$ - \psi(c-a-m-n) + \psi(1+m+n)] $$

$$ + \frac{\Gamma(c)}{\Gamma(a+m)} (-z)^{-a} \sum_{n=0}^{m-1} \frac{(a)_n\,\Gamma(m-n)}{n!\,\Gamma(c-a-n)} z^{-n}, $$

$$ |\arg(-z)| < \pi, \; |z| > 1, \quad c - a \neq 0, \pm 1, \pm 2, \ldots $$

In the last two results the case $c - a = 0, -1, -2, \ldots,$ becomes elementary and may be obtained directly from the differentiation formulas. The case $c - a = 1, 2, 3, \ldots$ is obtained from the formula

$$ \frac{d^n}{dz^n} [(1-z)^{a+b-c}\, {}_2F_1(a, b; c; z)] $$

$$ = \frac{(c-a)_n\,(c-b)_n}{(c)_n} (1-z)^{a+b-c-n}\, {}_2F_1(a, b; c+n; z) $$

by taking the limit as $c \to a$.

$$
{}_2F_1(a, b; a+b; z) = \frac{\Gamma(a+b)}{\Gamma(a)\,\Gamma(b)} \sum_{n=0}^{\infty} \frac{(a)_n\,(b)_n}{(n!)^2}
$$
$$
\times [2\psi(n+1) - \psi(a+n) - \psi(b+n) - \log(1-z)]\,(1-z)^n,
$$
$$
|\arg(1-z)| < \pi, \quad |1-z| < 1,
$$

$$
{}_2F_1(a, b; a+b+m; z)
$$
$$
= \frac{\Gamma(m)\,\Gamma(a+b+m)}{\Gamma(a+m)\,\Gamma(b+m)} \sum_{n=0}^{m-1} \frac{(a)_n\,(b)_n}{n!\,(1-m)_n} (1-z)^n
$$
$$
- \frac{\Gamma(a+b+m)}{\Gamma(a)\,\Gamma(b)} (z-1)^m \sum_{n=0}^{\infty} \frac{(a+m)_n\,(b+m)_n}{n!\,(n+m)!} (1-z)^n
$$
$$
\times [\log(1-z) - \psi(n+1) + \psi(a+m+n)
$$
$$
+ \psi(b+m+n) - \psi(1+m+n)],
$$
$$
|\arg(1-z)| < \pi, \quad |1-z| < 1, \quad m = 1, 2, 3, \ldots,
$$

$$
{}_2F_1(a, b; a+b-m; z)
$$
$$
= \frac{\Gamma(m)\,\Gamma(a+b-m)}{\Gamma(a)\,\Gamma(b)} (1-z)^{-m} \sum_{n=0}^{m-1} \frac{(a-m)_n\,(b-m)_n}{n!\,(1-m)_n} (1-z)^n
$$
$$
- \frac{(-1)^m\,\Gamma(a+b-m)}{\Gamma(a-m)\,\Gamma(b-m)} \sum_{n=0}^{\infty} \frac{(a)_n\,(b)_n}{n!\,(n+m)!} (1-z)^n
$$
$$
\times [\log(1-z) + \psi(a+n) + \psi(b+n) - \psi(1+n) - \psi(1+m+n)],
$$
$$
|\arg(1-z)| < \pi, \quad |1-z| < \pi, \quad m = 1, 2, 3, \ldots
$$

2.4.2 Quadratic transformations

There exists a quadratic transformation if and only if the numbers $\pm(1-c), \pm(a-b), \pm(c-a-b)$ are such that

(i) Two of these numbers are equal.

(ii) One of them is equal to $\frac{1}{2}$.

The basic formulas given below are due to KUMMER

$$
{}_2F_1\left(a, b; a+b+\frac{1}{2}; \sin^2\theta\right) = {}_2F_1\left(2a, 2b; a+b+\frac{1}{2}; \sin^2\frac{\theta}{2}\right),
$$

$$
{}_2F_1\left(a, b; a+b+\frac{1}{2}; z\right) = {}_2F_1\left(2a, 2b; a+b+\frac{1}{2}; \frac{1}{2} - \frac{1}{2}\sqrt{1-z}\right)
$$
$$
= \left(\frac{1+\sqrt{1-z}}{2}\right)^{-2a} {}_2F_1\left(2a, a-b+\frac{1}{2}; a+b+\frac{1}{2}; \frac{\sqrt{1-z}-1}{\sqrt{1-z}+1}\right),
$$

$$ {}_2F_1\left(a, b; a+b-\frac{1}{2}; z\right) = \frac{1}{\sqrt{1-z}} $$

$$ \times {}_2F_1\left(2a-1, 2b-1; a+b-\frac{1}{2}; \frac{1-\sqrt{1-z}}{2}\right) $$

$$ = \frac{(1+\sqrt{1-z})^{1-2a}}{2^{-2a+1}\sqrt{1-z}} {}_2F_1\left(2a-1, a-b+\frac{1}{2}; a+b-\frac{1}{2}; \frac{\sqrt{1-z}-1}{\sqrt{1-z}+1}\right), $$

$$ {}_2F_1\left(a, b; \frac{a+b+1}{2}; z\right) $$

$$ = (1-2z)^{-a} {}_2F_1\left(\frac{a}{2}, \frac{a+1}{2}; \frac{a+b+1}{2}; \frac{4z^2-4z}{(1-2z)^2}\right) $$

$$ = {}_2F_1\left(\frac{a}{2}, \frac{b}{2}; \frac{a+b+1}{2}; 4z-4z^2\right), $$

$$ {}_2F_1\left(a, a-b+\frac{1}{2}; b+\frac{1}{2}; z^2\right) = (1+z)^{-2a} {}_2F_1\left(a, b; 2b; \frac{4z}{(1+z)^2}\right), $$

$$ {}_2F_1(2a, 2a-c+1; c; z) = (1+z)^{-2a} {}_2F_1\left(a, a+\frac{1}{2}; c; \frac{4z}{(1+z)^2}\right), $$

$$ {}_2F_1(a, 1-a; c; z) $$

$$ = (1-z)^{c-1} {}_2F_1\left(\frac{c-a}{2}, \frac{c+a-1}{2}; c; 4z-4z^2\right) $$

$$ = (1-z)^{c-1}(1-2z)^{a-c} {}_2F_1\left(\frac{c-a}{2}, \frac{c-a+1}{2}; c; \frac{4z^2-4z}{(1-2z)^2}\right), $$

$$ {}_2F_1(a, b; a-b+1; z) $$

$$ = (1+z)^{-a} {}_2F_1\left(\frac{a}{2}, \frac{a+1}{2}; a-b+1; \frac{4z}{(1+z)^2}\right) $$

$$ = (1-z)^{-a} {}_2F_1\left(\frac{a}{2}, \frac{a+1}{2}-b; a-b+1; \frac{-4z}{(1-z)^2}\right) $$

$$ = \left(1 \pm \sqrt{z}\right)^{-2a} {}_2F_1\left(a, a-b+\frac{1}{2}; 2a-2b+1; \pm\frac{4\sqrt{z}}{(1\pm\sqrt{z})^2}\right), $$

$$ {}_2F_1\left(a, a+\frac{1}{2}; c; z\right) $$

$$ = (1-z)^{-a} {}_2F_1\left(2a, 2c-2a-1; c; \frac{\sqrt{1-z}-1}{2\sqrt{1-z}}\right) $$

$$ = \left(1 \pm \sqrt{z}\right)^{-2a} {}_2F_1\left(2a, c-\frac{1}{2}; 2c-1; \pm\frac{2\sqrt{z}}{1\pm\sqrt{z}}\right) $$

$$ = \left(\frac{1+\sqrt{1-z}}{2}\right)^{-2a} {}_2F_1\left(2a, 2a-c+1; c; \frac{1-\sqrt{1-z}}{1+\sqrt{1-z}}\right), $$

$$
\begin{aligned}
&{}_2F_1(a, b; 2b; z)\\
&= (1-z)^{-\frac{a}{2}}\,{}_2F_1\left(\frac{a}{2}, b-\frac{a}{2}; b+\frac{1}{2}; \frac{z^2}{4z-4}\right)\\
&= (1-z)^{-\frac{a}{2}}\,{}_2F_1\left(a, 2b-a; b+\frac{1}{2}; -\frac{(1-\sqrt{1-z})^2}{4\sqrt{1-z}}\right)\\
&= \left(1-\frac{z}{2}\right)^{-a}{}_2F_1\left(\frac{a}{2}, \frac{a+1}{2}; b+\frac{1}{2}; \frac{z^2}{(2-z)^2}\right)\\
&= \left(\frac{1+\sqrt{1-z}}{2}\right)^{-2a}{}_2F_1\left(a, a-b+\frac{1}{2}; b+\frac{1}{2}; \left(\frac{1-\sqrt{1-z}}{1+\sqrt{1-z}}\right)^2\right).
\end{aligned}
$$

For a complete list of quadratic transformations and the results on cubic transformations see ERDÉLYI (1953). In the results given above, the square root is defined in such a way that its value is positive and real when $0 \leq z \leq 1$. All these results are valid in the vicinity of the origin i.e. $z = 0$.

2.4.3 Legendre functions and the hypergeometric function

The Legendre functions are related to those special cases of the hypergeometric function for which there exists a quadratic transformation. If $\mathfrak{P}_\nu^\mu(z)$ denotes the Legendre function of the first kind for argument z in the complex plane cut along the real axis from 1 to $-\infty$ and $P_\nu^\mu(x)$, the Legendre function of the first kind for real values of x for which $-1 < x < 1$, then

$$
\begin{aligned}
{}_2F_1(a, b; 2b; z) &= \Gamma\left(b+\frac{1}{2}\right) 2^{2b-1} z^{\frac{1}{2}-b} (1-z)^{\frac{1}{2}\left(b-a-\frac{1}{2}\right)}\\
&\quad\times \mathfrak{P}^{\frac{1}{2}-b}_{a-b-\frac{1}{2}}\left[\frac{\left(1-\frac{z}{2}\right)}{\sqrt{1-z}}\right]\\
&= \frac{\Gamma\left(b+\frac{1}{2}\right)}{\Gamma(2b-a)}\pi^{-\frac{1}{2}}\left(\frac{z}{4}\right)^{-b}(1-z)^{\frac{1}{2}(b-a)} e^{i\pi(a-b)}\\
&\quad\times \mathfrak{Q}^{b-a}_{b-1}\left(\frac{2}{z}-1\right),
\end{aligned}
$$

$$
\begin{aligned}
{}_2F_1(a, b; 2b; -z) &= \frac{\Gamma\left(b+\frac{1}{2}\right)}{\Gamma(a)}\frac{\pi^{-\frac{1}{2}}}{\left(\frac{z}{4}\right)^b}(1+z)^{\frac{1}{2}(b-a)} e^{-i\pi(a-b)}\\
&\quad\times \mathfrak{Q}^{a-b}_{b-1}\left(1+\frac{2}{z}\right),
\end{aligned}
$$

$$|\arg z| < \pi, \quad |\arg (1 \pm z)| < \pi,$$

$${}_2F_1\left(a, a+\frac{1}{2}; c; z\right) = 2^{c-1}\Gamma(c)\, z^{\frac{1}{2}-\frac{1}{2}c}(1-z)^{\frac{1}{2}c-\frac{1}{2}-a}\,\mathfrak{P}^{1-c}_{2a-c}\left(\frac{1}{\sqrt{1-z}}\right),$$

$|\arg z| < \pi$, $|\arg(1-z)| < \pi$; z not a negative real number,

$${}_2F_1\left(a, b; a+b+\frac{1}{2}; z\right) = 2^{a+b-\frac{1}{2}}\Gamma\left(a+b+\frac{1}{2}\right)(-z)^{\frac{1}{4}-\frac{1}{2}a-\frac{1}{2}b}$$
$$\times\,\mathfrak{P}^{\frac{1}{2}-a-b}_{a-b-\frac{1}{2}}\left(\sqrt{1-z}\right),$$

$|\arg(-z)| < \pi$; z not between 0 and 1.

$${}_2F_1(a, 1-a; c; z) = \Gamma(c)\,(-z)^{\frac{1}{2}-\frac{1}{2}c}(1-z)^{\frac{1}{2}c-\frac{1}{2}}\,\mathfrak{P}^{1-c}_{-a}(1-2z),$$

$|\arg(-z)| < \pi$, $|\arg(1-z)| < \pi$; z not between 0 and 1,

$${}_2F_1(a, b; a-b+1; z) = \Gamma(a-b+1)\, z^{\frac{1}{2}b-\frac{1}{2}a}(1-z)^{-b}\,\mathfrak{P}^{b-a}_{-b}\left(\frac{1+z}{1-z}\right),$$

$|\arg(1-z)| < \pi$; z not between 0 and $-\infty$.

$${}_2F_1\left(a, b; a+b-\frac{1}{2}; z\right)$$
$$= 2^{a+b-\frac{3}{2}}\Gamma\left(a+b-\frac{1}{2}\right)(-z)^{\frac{3}{4}-\frac{1}{2}a-\frac{1}{2}b}(1-z)^{-\frac{1}{2}}\,\mathfrak{P}^{\frac{3}{2}-a-b}_{b-a-\frac{1}{2}}\left(\sqrt{1-z}\right),$$

$|\arg(-z)| < \pi$, $|\arg(1-z)| < \pi$; $Re\sqrt{1-z} > 0$; z not between 0 and 1,

$${}_2F_1\left(a, b; \frac{a}{2}+\frac{b}{2}+\frac{1}{2}; z\right)$$
$$= \Gamma\left(\frac{a+b+1}{2}\right)[z^2-z]^{\frac{1}{4}(1-a-b)}\,\mathfrak{P}^{\frac{1}{2}-\frac{1}{2}a-\frac{1}{2}b}_{\frac{1}{2}a-\frac{1}{2}b-\frac{1}{2}}(1-2z),$$

$|\arg z| < \pi$, $|\arg(z-1)| < \pi$; z not between 0 and 1,

$${}_2F_1\left(a, b; \frac{1}{2}; z\right) = \frac{2^{a+b-\frac{3}{2}}}{\sqrt{\pi}}\Gamma\left(a+\frac{1}{2}\right)\Gamma\left(b+\frac{1}{2}\right)(z-1)^{\frac{1}{4}-\frac{a+b}{2}}$$
$$\times\left\{\mathfrak{P}^{\frac{1}{2}-a-b}_{a-b-\frac{1}{2}}\left(\sqrt{z}\right) + \mathfrak{P}^{\frac{1}{2}-a-b}_{a-b-\frac{1}{2}}\left(-\sqrt{z}\right)\right\},$$

$|\arg z| < \pi$, $|\arg(z-1)| < \pi$; z not between 0 and 1,

$${}_2F_1\left(a, b; \frac{1}{2}; -z\right) = \frac{2^{a-b-1}}{\sqrt{\pi}}\Gamma\left(a+\frac{1}{2}\right)\Gamma(1-b)\,(z+1)^{-\frac{a+b}{2}}$$
$$\times\, e^{\pm i\frac{\pi}{2}(b-a)}\left\{\mathfrak{P}^{b-a}_{a+b-1}\left[\frac{\sqrt{z}}{\sqrt{1+z}}\right] + \mathfrak{P}^{b-a}_{a+b-1}\left[\frac{-\sqrt{z}}{\sqrt{1+z}}\right]\right\},$$

where z is not between 0 and $-\infty$; the sign is $\pm$ according as $\operatorname{Im} z \gtrless 0$.

$$ {}_2F_1(a, 1-a; c; x) = \Gamma(c)\, x^{\frac{1}{2}-\frac{1}{2}c} (1-x)^{\frac{1}{2}c-\frac{1}{2}} P_{-a}^{1-c}(1-2x) $$

$0 < x < 1,$

$$ {}_2F_1\left(a, a+\frac{1}{2}; c; x\right) = 2^{c-1}\, \Gamma(c)\, (-x)^{\frac{1}{2}-\frac{1}{2}c} (1-x)^{\frac{1}{2}c-\frac{1}{2}-a} \times P_{2a-c}^{1-c}\left(\frac{1}{\sqrt{1-x}}\right), $$

$-\infty < x < 0,$

$$ {}_2F_1(a, b; a-b+1; x) = \Gamma(a-b+1)\, (-x)^{\frac{1}{2}b-\frac{1}{2}a} (1-x)^{-b} \times P_{-b}^{b-a}\left(\frac{1+x}{1-x}\right), $$

$-\infty < x < 0,$

$$ {}_2F_1\left(a, b; a+b+\frac{1}{2}; x\right) = 2^{a+b-\frac{1}{2}}\, \Gamma\left(a+b+\frac{1}{2}\right) x^{\frac{1}{2}\left(\frac{1}{2}-a-b\right)} \times P_{a-b-\frac{1}{2}}^{\frac{1}{2}-a-b}\left[\sqrt{1-x}\right], $$

$0 < x < 1,$

$$ {}_2F_1\left(a, b; a+b-\frac{1}{2}; x\right) = 2^{a+b-\frac{3}{2}}\, \Gamma\left(a+b-\frac{1}{2}\right) x^{\frac{1}{2}\left(\frac{3}{2}-a-b\right)} \times (1-x)^{-\frac{1}{2}} P_{b-a-\frac{1}{2}}^{\frac{3}{2}-a-b}\left(\sqrt{1-x}\right), $$

$0 < x < 1,$

$$ {}_2F_1\left(a, b; \frac{a+b+1}{2}; x\right) = \Gamma\left(\frac{a+b+1}{2}\right) (x-x^2)^{\frac{1}{4}-\frac{a}{4}-\frac{b}{4}} \times P_{\frac{a}{2}-\frac{b+1}{2}}^{\frac{1}{2}-\frac{a+b}{2}}(1-2x), $$

$0 < x < 1,$

$$ {}_2F_1\left(a, b; \frac{1}{2}; x\right) = \frac{\Gamma\left(a+\frac{1}{2}\right) \Gamma\left(b+\frac{1}{2}\right) 2^{a+b-\frac{3}{2}}}{\sqrt{\pi}} (1-x)^{\frac{1}{2}\left(\frac{1}{2}-a-b\right)} \times \left[P_{a-b-\frac{1}{2}}^{\frac{1}{2}-a-b}\left(\sqrt{x}\right) + P_{a-b-\frac{1}{2}}^{\frac{1}{2}-a-b}\left(-\sqrt{x}\right)\right], $$

$0 < x < 1,$

$$ {}_2F_1\left(a, b; \frac{1}{2}; -x\right) = \frac{2^{a-b-1}}{\sqrt{\pi}} \Gamma\left(a + \frac{1}{2}\right) \Gamma(1 - b) (1 + x)^{-\frac{a+b}{2}} $$

$$ \times \left[P_{a+b-1}^{b-a}\left(\frac{\sqrt{x}}{\sqrt{1+x}}\right) + P\left(-\frac{\sqrt{x}}{\sqrt{1+x}}\right)\right], $$

$0 < x < \infty$,

$$ {}_2F_1\left(a, b; \frac{3}{2}; x\right) = -\frac{\Gamma\left(a - \frac{1}{2}\right) \Gamma\left(b - \frac{1}{2}\right) 2^{a+b-\frac{7}{2}}}{\sqrt{\pi x}} (1 - x)^{\frac{1}{2}\left(\frac{3}{2} - a - b\right)} $$

$$ \times \left[P_{a-b-\frac{1}{2}}^{\frac{3}{2}-a-b}\left(\sqrt{x}\right) - P_{a-b-\frac{1}{2}}^{\frac{3}{2}-a-b}\left(-\sqrt{x}\right)\right], $$

$0 < x < 1$.

2.5 Integral representations

$$ {}_2F_1(a, b; c; z) = \frac{\Gamma(c)}{\Gamma(b)\,\Gamma(c-b)} \int_0^1 t^{b-1} (1 - t)^{c-b-1} (1 - tz)^{-a}\, dt, $$

$Re\, c > Re\, b > 0$, $|\arg(1 - z)| < \pi$.

The integral on the right hand side represents a single-valued analytic function of z in the z-plane cut along the positive real axis from 1 to ∞ and hence gives the analytic continuation of ${}_2F_1(a, b; c; z)$. Mellin-Barnes integral,

$$ {}_2F_1(a, b; c; z) $$

$$ = \frac{\Gamma(c)}{2\pi i\, \Gamma(a)\, \Gamma(b)} \int_{-i\infty}^{i\infty} \frac{\Gamma(a+t)\, \Gamma(b+t)\, \Gamma(-t)}{\Gamma(c+t)} (-z)^t\, dt $$

$$ = \frac{i}{2} \frac{\Gamma(c)}{\Gamma(a)\, \Gamma(b)} \int_{-i\infty}^{i\infty} \frac{\Gamma(a+t)\, \Gamma(b+t)}{\Gamma(1+t)\, \Gamma(c+t)} \operatorname{cosec}(\pi t)\, (-z)^t\, dt, $$

$-\pi < \arg(-z) < \pi$.

The path of integration is chosen such that the poles of $\Gamma(a + t)$ and $\Gamma(b + t)$ are to the left and the poles of $\Gamma(-t)$ or $\operatorname{cosec}(\pi t)$ are to the right of the contour. The cases in which $-a$, $-b$, or $-c$ are non-negative integers or $(a - b)$ an integer are excluded. Finally,

$$ {}_2F_1(a, b; c; z) $$

$$ = \frac{[\Gamma(c)]^2}{\Gamma(a)\, \Gamma(b)\, \Gamma(c-a)\, \Gamma(c-b)} $$

$$ \times \int_0^1 \int_0^1 t^{b-1} \tau^{a-1} (1 - t)^{c-b-1} (1 - \tau)^{c-a-1} (1 - t\tau z)^{-c}\, dt\, d\tau. $$

2.5.1 Some integrals associated with ${}_2F_1(a, b; c; z)$

$$ {}_2F_1\left(-n, -\frac{n}{2} - \frac{\alpha}{2}; 1 - \frac{n}{2} - \frac{\alpha}{2}; -1\right) = (-2)^n \frac{(n+\alpha)}{\sin(\pi\alpha)} \int_0^\pi \cos^n t \cos(\alpha t)\, dt, $$

$$ n = 0, 1, 2, \ldots; \quad \alpha \neq 0, \pm 1, \pm 2, \ldots, $$

$$ {}_2F_1(\alpha, \alpha + n; 1 + n; z^2) = \frac{1}{2\pi} z^{-n} \frac{\Gamma(\alpha)\, n!}{\Gamma(\alpha + n)} \int_0^{2\pi} \frac{\cos n\varphi\, d\varphi}{(1 - 2z\cos\varphi + z^2)^\alpha}, $$

$$ n = 0, 1, 2, \ldots; \alpha \neq 0, -1, -2, \ldots; |z| < 1, $$

$$ {}_2F_1\left(a, a - b + \frac{1}{2}; b + \frac{1}{2}; z^2\right) = \frac{\Gamma\left(b + \frac{1}{2}\right)}{\sqrt{\pi}\, \Gamma(b)} \int_0^\pi \frac{(\sin t)^{2b-1}}{(1 + 2z\cos t + z^2)^a}, $$

$$ Re\, b > 0, |z| < 1, $$

$$ {}_2F_1(a, b; c; z) = \frac{\Gamma(c)}{\Gamma(\lambda)\, \Gamma(c - \lambda)} \int_0^1 x^{\lambda-1}(1 - x)^{c-\lambda-1}\, {}_2F_1(a, b; \lambda; xz)\, dx, $$

$$ Re\, c > Re\, \lambda > 0; \ |\arg(1 - z)| < \pi; \ z \neq 1 $$

$$ = \frac{\Gamma(c)}{\Gamma(\lambda)\, \Gamma(c - \lambda)} \int_0^1 x^{\lambda-1}(1 - x)^{c-\lambda-1} (1 - xz)^{-a'} $$

$$ \times {}_2F_1(a - a', b; \lambda; xz)\, {}_2F_1\left(a', b - \lambda; c - \lambda; \frac{z(1 - x)}{1 - xz}\right) dx, $$

$$ Re\, c > Re\, \lambda > 0; \ |\arg(1 - z)| < \pi; \ z \neq 1, $$

$$ \int_0^{\frac{\pi}{2}} (\cos t)^\lambda (\sin t)^\mu e^{it(\lambda+\mu+2)}\, dt = e^{i\frac{\pi}{2}(1+\mu)} \frac{\Gamma(\lambda + 1)\, \Gamma(1 + \mu)}{\Gamma(\lambda + \mu + 2)}, $$

$$ Re\, \lambda > -1, \ Re\, \mu > -1, $$

$$ \int_0^\infty {}_2F_1(a, b; c; -t)\, t^{-z-1}\, dt = \frac{\Gamma(c)\, \Gamma(-z)\, \Gamma(a + z)\, \Gamma(b + z)}{\Gamma(a)\, \Gamma(b)\, \Gamma(c + z)}, $$

$$ Re\, z > 0, \ Re\,(a + z) > 0, \ Re\,(b + z) > 0; \ c \neq 0, -1, -2, \ldots, $$

$$ \int_0^\pi (\cos t)^{2\mu} (\sin t)^{2\nu} e^{i2\lambda t}\, dt = \frac{\pi\, 4^{-\mu-\nu} e^{i\pi(\lambda-\mu)}\, \Gamma(1 + 2\nu)}{\Gamma(1 + \mu + \nu - \lambda)\, \Gamma(1 - \mu + \nu + \lambda)} $$

$$ \times {}_2F_1(-2\mu, \lambda - \mu - \nu; 1 - \mu + \nu + \lambda; -1), $$

where

$$ Re\, \mu > -\frac{1}{2}, \quad Re\, \nu > -\frac{1}{2} $$

and

$$(\cos t)^{2\mu} = e^{-i2\pi\mu}\left[\sin\left(t-\frac{\pi}{2}\right)\right]^{2\mu}, \quad \frac{\pi}{2} \leq t \leq \pi,$$

$$\int_0^{\frac{\pi}{2}} (\cos t)^\lambda (\sin t)^\mu e^{it(\lambda+\mu+2)}\, dt = 2^{-\lambda-\mu-1} e^{i\frac{\pi}{2}(1+\mu)}$$

$$\times\left[\frac{1}{1+\mu}\,{}_2F_1(-\lambda, 1; 2+\mu; -1) + \frac{1}{1+\lambda}\,{}_2F_1(-\mu, 1; 2+\lambda; -1)\right],$$

$$Re\,\lambda > -1, \quad Re\,\mu > -1,$$

$$\int_{-\frac{\pi}{2}}^{\frac{\pi}{2}} e^{i\alpha t} (\cos t)^\beta (\alpha^2 e^{it} + \lambda^2 e^{-it})^\mu\, dt = \frac{\pi\, 2^{-\beta}\,\Gamma(1+\beta)}{\Gamma\left(1+\frac{\alpha+\beta-\mu}{2}\right)}$$

$$\times \frac{1}{\Gamma\left(1+\frac{\beta-\alpha+\mu}{2}\right)}\,{}_2F_1\left(-\mu, \frac{\alpha-\beta-\mu}{2}; 1+\frac{\alpha+\beta-\mu}{2}; \frac{\alpha^2}{\lambda^2}\right),$$

$$Re\,\beta > -1, \quad |\alpha| < |\lambda|.$$

2.6 Asymptotic expansions

a) For large $|z|$, the behavior of ${}_2F_1(a, b; c; z)$ can be obtained from the transformation formulas given in sections (2.4.1) and (2.4.2).

b) For a, b, z fixed and $|c|$ large, one has [MacRobert]

$${}_2F_1(a, b; c; z) = \sum_{m=0}^{n} \frac{(a)_m (b)_m}{(c)_m} \frac{z^m}{m!} + O(|c|^{-n-1}).$$

c) For a, c, z fixed ($c \neq 0, -1, -2, \ldots$), $0 < |z| < 1$ and $|b|$ large, one has [H. T. F. (vol 1)].

c 1) $${}_2F_1(a, b; c; z) = e^{-i\pi a} \frac{\Gamma(c)}{\Gamma(c-a)} (bz)^{-a}\,[1 + O(|bz|^{-1})]$$
$$+ \frac{\Gamma(c)}{\Gamma(a)} e^{bz} (bz)^{a-c}\,[1 + O(|bz|^{-1})],$$
$$-3\frac{\pi}{2} < \arg(bz) < \frac{\pi}{2}.$$

c 2) $${}_2F_1(a, b; c; z) = e^{i\pi a} \frac{\Gamma(c)}{\Gamma(c-a)} (bz)^{-a}\,[1 + O(|bz|^{-1})]$$
$$+ \frac{\Gamma(c)}{\Gamma(a)} e^{bz} (bz)^{a-c}\,[1 + O(|bz|^{-1})].$$
$$-\frac{\pi}{2} < \arg z < \frac{3\pi}{2}$$

For the cases when more than one of the parameters is large see Erdélyi (1953).

2.7 The Riemann differential equation

The hypergeometric differential equation is a special case of Riemann's differential equation

$$\frac{d^2w}{dz^2} + \left[\frac{1-\alpha-\alpha'}{z-a} + \frac{1-\beta-\beta'}{z-b} + \frac{1-\gamma-\gamma'}{z-c}\right]\frac{dw}{dz}$$
$$+ \left[\frac{\alpha\alpha'(a-b)(a-c)}{z-a} + \frac{\beta\beta'(b-c)(b-a)}{z-b} + \frac{\gamma\gamma'(c-a)(c-b)}{z-c}\right]$$
$$\times \frac{w}{(z-a)(z-b)(z-c)} = 0.$$

Riemann's differential equation has three regular singular points $z = a, b, c$ and becomes the hypergeometric differential equation for the singular points $0, 1, \infty$.

$\alpha, \alpha'; \beta, \beta'; \gamma, \gamma'$ are called the exponents with respect to the singular points a, b, c respectively. The exponents must satisfy the additional condition

$$\alpha + \alpha' + \beta + \beta' + \gamma + \gamma' = 1.$$

The complete set of solutions of Riemann's differential equation is denoted by the symbol.

$$w = P\left\{\begin{matrix} a & b & c & \\ \alpha & \beta & \gamma & z \\ \alpha' & \beta' & \gamma' & \end{matrix}\right\}.$$

This symbol is also used to denote the differential equation for w.

Special cases of Riemann's differential equation:

The hypergeometric differential equation

$$z(1-z)\frac{d^2w}{dz^2} + [c - (a+b+1)z]\frac{dw}{dz} - abw = 0$$

can be written as

$$w = P\left\{\begin{matrix} 0 & \infty & 1 & \\ 0 & a & 0 & z \\ 1-c & b & c-a-b & \end{matrix}\right\}.$$

The generalized hypergeometric differential equation

$$\frac{d^2w}{dz^2} + \left[\frac{1-\alpha-\alpha'}{z} + \frac{1-\gamma-\gamma'}{z-1}\right]\frac{dw}{dz}$$
$$+ \left[\frac{-\alpha\alpha'}{z} + \beta\beta' + \frac{\gamma\gamma'}{z-1}\right]\frac{w}{z(z-1)} = 0$$

is represented by

$$w = P\left\{\begin{matrix} 0 & \infty & 1 & \\ \alpha & \beta & \gamma & z \\ \alpha' & \beta' & \gamma' & \end{matrix}\right\}.$$

The Legendre differential equation

$$(1-z^2)\frac{d^2w}{dz^2}-2z\frac{dw}{dz}+\left[\frac{-\mu^2}{1-z^2}+\nu(\nu+1)\right]w=0$$

is written

$$w=P\left\{\begin{matrix}0 & \infty & 1 & \\ -\frac{\nu}{2} & \frac{\mu}{2} & 0 & (1-z^2)^{-1} \\ \frac{\nu+1}{2} & -\frac{\mu}{2} & \frac{1}{2} & \end{matrix}\right\}.$$

The Gegenbauer polynomials $C_n^{(\lambda)}(z)$ which are related to the Legendre functions $\mathfrak{P}_\nu^\mu(z)$ and are given by

$$(1-2hz+h^2)^{-\nu}=\sum_{n=0}^{\infty}C_n^{(\nu)}(z)\,h^n$$

satisfy the differential equation

$$w=P\left\{\begin{matrix}-1 & \infty & 1 & \\ \frac{1}{2}-\lambda & n+2\lambda & \frac{1}{2}-\lambda & z \\ 0 & -n & 0 & \end{matrix}\right\}.$$

The differential equation of the (confluent) hypergeometric function

$$\frac{d^2w}{dz^2}+\frac{dw}{dz}+\left(\frac{\frac{1}{4}-\mu^2}{z^2}+\frac{\varkappa}{z}\right)w=0$$

is written as

$$w=P\left\{\begin{matrix}0 & \infty & c & \\ \frac{1}{2}+\mu & -c & c-\varkappa & z \\ \frac{1}{2}-\mu & 0 & \varkappa & \end{matrix}\right\}\quad\text{with }\lim c\to\infty.$$

2.8 Transformation formulas for Riemann's *P*-function

The following two transformation formulas are valid for the *P*-function

(i)
$$\left(\frac{z-a}{z-b}\right)^k\left(\frac{z-c}{z-b}\right)^l P\left\{\begin{matrix}a & b & c & \\ \alpha & \beta & \gamma & z \\ \alpha' & \beta' & \gamma' & \end{matrix}\right\}$$
$$=P\left\{\begin{matrix}a & b & c & \\ \alpha+k & \beta-k-l & \gamma+l & z \\ \alpha'+k & \beta'-k-l & \gamma'+l & \end{matrix}\right\};$$

(ii)
$$P\left\{\begin{matrix}a & b & c & \\ \alpha & \beta & \gamma & z \\ \alpha' & \beta' & \gamma' & \end{matrix}\right\}=P\left\{\begin{matrix}a_1 & b_1 & c_1 & \\ \alpha & \beta & \gamma & z_1 \\ \alpha' & \beta' & \gamma' & \end{matrix}\right\},$$

where the quantities a_1, b_1, c_1, z_1 are obtained from the following fractional linear transformations

$$a = \frac{A a_1 + B}{C a_1 + D}, \quad b = \frac{A b_1 + B}{C b_1 + D}, \quad c = \frac{A c_1 + B}{C c_1 + D}, \quad z = \frac{A z_1 + B}{C z_1 + D},$$

where $BC - AD \neq 0$, otherwise A, B, C, D are arbitrary constants.

From section 2.7, if

$$w = P \begin{Bmatrix} a & b & c & \\ \alpha & \beta & \gamma & z \\ \alpha' & \beta' & \gamma' & \end{Bmatrix},$$

then $w_1 = \left(\frac{z-a}{z-b}\right)^k \left(\frac{z-c}{z-b}\right)^l \cdot w$ satisfies a differential equation of second order with the singular points a, b, c and exponents $\alpha + k$, $\alpha' + k$; $\beta - l - k$, $\beta' - l - k$; $\gamma + l$, $\gamma' + l$.

The second transformation formula (ii) takes a differential equation with singularities a, b, c, the exponents α, α'; β, β'; γ, γ' and variable z into one with the same exponents (α, α'; β, β'; γ, γ'), the singularities a_1, b_1, c_1 and the variable z_1, where z_1 is defined by $z = \frac{A z_1 + B}{C z_1 + D}$. The constants a_1, b_1, c_1 are obtained from the constants a, b, c by the same linear transformation.

Solution of Riemann's differential equation in terms of the hypergeometric function:

By successive application of the two transformation formulas, the Riemann differential equation may be reduced to the hypergeometric differential equation

for $$k = -\alpha, \quad l = -\gamma, \quad z_1 = \frac{(z-a)(c-b)}{(z-b)(c-a)}$$

we have

$$W = P \begin{Bmatrix} a & b & c & \\ \alpha & \beta & \gamma & z \\ \alpha' & \beta' & \gamma' & \end{Bmatrix}$$

$$= \left\{\frac{z-a}{z-b}\right\}^\alpha \left(\frac{z-c}{z-b}\right)^\gamma P \begin{Bmatrix} a & b & c & \\ 0 & \beta + \alpha + \gamma & 0 & z \\ \alpha' - \alpha & \beta' + \alpha + \gamma & \gamma' - \gamma & \end{Bmatrix}$$

$$= \left(\frac{z-a}{z-b}\right)^\alpha \left(\frac{z-c}{z-b}\right)^\gamma P \begin{Bmatrix} 0 & \infty & 1 & \\ 0 & \beta + \alpha + \gamma & 0 & \frac{(z-a)(c-b)}{(z-b)(c-a)} \\ \alpha' - \alpha & \beta' + \alpha + \gamma & \gamma' - \gamma & \end{Bmatrix}.$$

A solution of the Riemann differential equation may therefore be expressed by the hypergeometric series:

$$W = \left(\frac{z-a}{z-b}\right)^{\alpha} \left(\frac{z-c}{z-b}\right)^{\gamma} \times {}_2F_1\left(\alpha+\beta+\gamma, \alpha+\beta'+\gamma; 1+\alpha-\alpha'; \frac{(z-a)(c-b)}{(z-b)(c-a)}\right).$$

Since Riemann's equation remains unchanged when the constants $a, b, c;\ \alpha, \alpha';\ \beta, \beta';\ \gamma, \gamma'$ are interchanged in a suitable manner, one gets 24 solutions of the differential equation, which under the assumption that the values $\alpha-\alpha'$, $\beta-\beta'$, $\gamma-\gamma'$ are neither zero nor integers take the following form:

$$W_1 = \left(\frac{z-a}{z-b}\right)^{\alpha} \left(\frac{z-c}{z-b}\right)^{\gamma} \times {}_2F_1\left(\alpha+\beta+\gamma, \alpha+\beta'+\gamma; 1+\alpha-\alpha'; \frac{(z-a)(c-b)}{(z-b)(c-a)}\right),$$

$$W_2 = \left(\frac{z-a}{z-b}\right)^{\alpha'} \left(\frac{z-c}{z-b}\right)^{\gamma} \times {}_2F_1\left(\alpha'+\beta+\gamma, \alpha'+\beta'+\gamma; 1+\alpha'-\alpha; \frac{(z-a)(c-b)}{(z-b)(c-a)}\right),$$

$$W_3 = \left(\frac{z-a}{z-b}\right)^{\alpha} \left(\frac{z-c}{z-b}\right)^{\gamma'} \times {}_2F_1\left(\alpha+\beta+\gamma', \alpha+\beta'+\gamma'; 1+\alpha-\alpha'; \frac{(z-a)(c-b)}{(z-b)(c-a)}\right),$$

$$W_4 = \left(\frac{z-a}{z-b}\right)^{\alpha'} \left(\frac{z-c}{z-b}\right)^{\gamma'} \times {}_2F_1\left(\alpha'+\beta+\gamma', \alpha'+\beta'+\gamma'; 1+\alpha'-\alpha; \frac{(z-a)(c-b)}{(z-b)(c-a)}\right),$$

$$W_5 = \left(\frac{z-b}{z-c}\right)^{\beta} \left(\frac{z-a}{z-c}\right)^{\alpha} \times {}_2F_1\left(\beta+\gamma+\alpha, \beta+\gamma'+\alpha; 1+\beta-\beta'; \frac{(z-b)(a-c)}{(z-c)(a-b)}\right),$$

$$W_6 = \left(\frac{z-b}{z-c}\right)^{\beta'} \left(\frac{z-a}{z-c}\right)^{\alpha} \times {}_2F_1\left(\beta'+\gamma+\alpha, \beta'+\gamma'+\alpha; 1+\beta'-\beta; \frac{(z-b)(a-c)}{(z-c)(a-b)}\right),$$

$$W_7 = \left(\frac{z-b}{z-c}\right)^{\beta} \left(\frac{z-a}{z-c}\right)^{\alpha'} \times {}_2F_1\left(\beta+\gamma+\alpha', \beta+\gamma'+\alpha'; 1+\beta-\beta'; \frac{(z-b)(a-c)}{(z-c)(a-b)}\right),$$

$$W_8 = \left(\frac{z-b}{z-c}\right)^{\beta'} \left(\frac{z-a}{z-c}\right)^{\alpha'} \times {}_2F_1\left(\beta'+\gamma+\alpha', \beta'+\gamma'+\alpha'; 1+\beta'-\beta; \frac{(z-b)(a-c)}{(z-c)(a-b)}\right),$$

$$W_9 = \left(\frac{z-c}{z-a}\right)^{\gamma} \left(\frac{z-b}{z-a}\right)^{\beta}$$

$$\times {}_2F_1\left(\gamma+\alpha+\beta, \gamma+\alpha'+\beta; 1+\gamma-\gamma'; \frac{(z-c)\,(b-a)}{(z-a)\,(b-c)}\right),$$

$$W_{10} = \left(\frac{z-c}{z-a}\right)^{\gamma'} \left(\frac{z-b}{z-a}\right)^{\beta}$$

$$\times {}_2F_1\left(\gamma'+\alpha+\beta, \gamma'+\alpha'+\beta; 1+\gamma'-\gamma; \frac{(z-c)\,(b-a)}{(z-a)\,(b-c)}\right),$$

$$W_{11} = \left(\frac{z-c}{z-a}\right)^{\gamma} \left(\frac{z-b}{z-a}\right)^{\beta'}$$

$$\times {}_2F_1\left(\gamma+\alpha+\beta'\ \gamma+\alpha'+\beta'; 1+\gamma-\gamma'; \frac{(z-c)\,(b-a)}{(z-a)\,(b-c)}\right),$$

$$W_{12} = \left(\frac{z-c}{z-a}\right)^{\gamma'} \left(\frac{z-b}{z-a}\right)^{\beta'}$$

$$\times {}_2F_1\left(\gamma'+\alpha+\beta', \gamma'+\alpha'+\beta'; 1+\gamma'-\gamma; \frac{(z-c)\,(b-a)}{(z-a)\,(b-c)}\right),$$

$$W_{13} = \left(\frac{z-a}{z-c}\right)^{\alpha} \left(\frac{z-b}{z-c}\right)^{\beta}$$

$$\times {}_2F_1\left(\alpha+\gamma+\beta, \alpha+\gamma'+\beta; 1+\alpha-\alpha'; \frac{(z-a)\,(b-c)}{(z-c)\,(b-a)}\right),$$

$$W_{14} = \left(\frac{z-a}{z-c}\right)^{\alpha'} \left(\frac{z-b}{z-c}\right)^{\beta}$$

$$\times {}_2F_1\left(\alpha'+\gamma+\beta, \alpha'+\gamma'+\beta; 1+\alpha'-\alpha; \frac{(z-a)\,(b-c)}{(z-c)\,(b-a)}\right),$$

$$W_{15} = \left(\frac{z-a}{z-c}\right)^{\alpha} \left(\frac{z-b}{z-c}\right)^{\beta'}$$

$$\times {}_2F_1\left(\alpha+\gamma+\beta', \alpha+\gamma'+\beta'; 1+\alpha-\alpha'; \frac{(z-a)\,(b-c)}{(z-c)\,(b-a)}\right),$$

$$W_{16} = \left(\frac{z-a}{z-c}\right)^{\alpha'} \left(\frac{z-b}{z-c}\right)^{\beta'}$$

$$\times {}_2F_1\left(\alpha'+\gamma+\beta', \alpha'+\gamma'+\beta'; 1+\alpha'-\alpha; \frac{(z-a)\,(b-c)}{(z-c)\,(b-a)}\right),$$

$$W_{17} = \left(\frac{z-c}{z-b}\right)^{\gamma} \left(\frac{z-a}{z-b}\right)^{\alpha}$$

$$\times {}_2F_1\left(\gamma+\beta+\alpha, \gamma+\beta'+\alpha; 1+\gamma-\gamma'; \frac{(z-c)\,(a-b)}{(z-b)\,(a-c)}\right),$$

$$W_{18} = \left(\frac{z-c}{z-b}\right)^{\gamma'} \left(\frac{z-a}{z-b}\right)^{\alpha}$$

$$\times {}_2F_1\left(\gamma'+\beta+\alpha, \gamma'+\beta'+\alpha; 1+\gamma'-\gamma; \frac{(z-c)\,(a-b)}{(z-b)\,(a-c)}\right),$$

$$W_{19} = \left(\frac{z-c}{z-b}\right)^{\gamma} \left(\frac{z-a}{z-b}\right)^{\alpha'}$$

$$\times {}_2F_1\left(\gamma+\beta+\alpha', \gamma+\beta'+\alpha'; 1+\gamma-\gamma'; \frac{(z-c)\,(a-b)}{(z-b)\,(a-c)}\right),$$

$$W_{20} = \left(\frac{z-c}{z-b}\right)^{\gamma'} \left(\frac{z-a}{z-b}\right)^{\alpha'}$$
$$\times {}_2F_1\left(\gamma' + \beta + \alpha', \gamma' + \beta' + \alpha'; 1 + \gamma' - \gamma; \frac{(z-c)\,(a-b)}{(z-b)\,(a-c)}\right),$$

$$W_{21} = \left(\frac{z-b}{z-a}\right)^{\beta} \left(\frac{z-c}{z-a}\right)^{\gamma}$$
$$\times {}_2F_1\left(\beta + \alpha + \gamma, \beta + \alpha' + \gamma; 1 + \beta - \beta'; \frac{(z-b)\,(c-a)}{(z-a)\,(c-b)}\right),$$

$$W_{22} = \left(\frac{z-b}{z-a}\right)^{\beta'} \left(\frac{z-c}{z-a}\right)^{\gamma}$$
$$\times {}_2F_1\left(\beta' + \alpha + \gamma, \beta' + \alpha' + \gamma; 1 + \beta' - \beta; \frac{(z-b)\,(c-a)}{(z-a)\,(c-b)}\right),$$

$$W_{23} = \left(\frac{z-b}{z-a}\right)^{\beta} \left(\frac{z-c}{z-a}\right)^{\gamma'}$$
$$\times {}_2F_1\left(\beta + \alpha + \gamma', \beta + \alpha' + \gamma'; 1 + \beta - \beta'; \frac{(z-b)\,(c-a)}{(z-a)\,(c-b)}\right),$$

$$W_{24} = \left(\frac{z-b}{z-a}\right)^{\beta'} \left(\frac{z-c}{z-a}\right)^{\gamma'}$$
$$\times {}_2F_1\left(\beta' + \alpha + \gamma', \beta' + \alpha' + \gamma'; 1 + \beta' - \beta; \frac{(z-b)\,(c-a)}{(z-a)\,(c-b)}\right).$$

2.9 The generalized hypergeometric series

$${}_pF_q(\alpha_1, \alpha_2, \ldots, \alpha_p; \beta_1, \beta_2, \ldots, \beta_q; z) = \sum_{n=0}^{\infty} \frac{(\alpha_1)_n \cdots (\alpha_p)_n}{(\beta_1)_n \cdots (\beta_q)_n} \frac{z^n}{n!},$$

where

$$(\alpha)_n = \alpha(\alpha+1)\cdots(\alpha+n-1) = \frac{\Gamma(\alpha+n)}{\Gamma(\alpha)},$$
$$(\alpha)_0 = 1.$$

The values of the q parameters $\beta_1, \ldots, \beta_q$ are always different from $0, -1, -2, \ldots$ In this notation we get $F(a, b; c; z) = {}_2F_1(a, b; c; z)$. ${}_pF_q$ converges for all finite z if $p \leq q$, converges for $|z| < 1$ if $p = q + 1$, and diverges for all $z \neq 0$ if $p > q + 1$. This excludes certain integer values of the parameters for which the series terminates or fails to make sense. If one or more of the parameters $\alpha_1, \alpha_2, \ldots, \alpha_p$ is a non-positive integer, the series terminates; the cases in which one of the parameters $\beta_1, \beta_2, \ldots, \beta_q$ is a non-positive integer are excluded.

If $p = q + 1$, $s = Re\,(\alpha_1 + \alpha_2 + \cdots + \alpha_p - \beta_1 - \beta_2 \cdots - \beta_q)$ then the series converges for all $|z| = 1$ if $s < 0$; it converges for all $|z| = 1$, $z \neq 1$ if $0 \leq s < 1$ and diverges if $s \geq 1$.

Let $\theta \equiv z\dfrac{d}{dz}$, then $w = {}_pF_q$ satisfies the differential equation

$$\{\theta(\theta + \beta_1 - 1) \cdots (\theta + \beta_q - 1) - z(\theta + \alpha_1) \cdots (\theta + \alpha_p)\}\, w = 0.$$

Recurrence relations and identities

$$\left[{}_2F_1\left(a, b; a+b+\frac{1}{2}; z\right)\right]^2 = {}_3F_2\left(2a, a+b, 2b; a+b+\frac{1}{2}, a+2b; z\right),$$

$${}_0F_1(\alpha; z)\, {}_0F_1(\beta; z) = {}_2F_3\left(\frac{\alpha+\beta}{2}, \frac{\alpha+\beta-1}{2}; \alpha, \beta, \alpha+\beta-1; 4z\right),$$

$${}_0F_1(\alpha; z)\, {}_0F_1(\alpha; -z) = {}_0F_3\left(\alpha, \frac{1}{2}\alpha, \frac{1}{2}\alpha+\frac{1}{2}; -\frac{z^2}{4}\right),$$

$${}_2F_0(\alpha, \beta; z)\, {}_2F_0(\alpha, \beta; -z) = {}_4F_1\left(\alpha, \beta, \frac{\alpha+\beta}{2}, \frac{\alpha+\beta+1}{2}; \alpha+\beta; 4z^2\right),$$

$${}_1F_1(\alpha; \beta; z)\, {}_1F_1(\alpha; \beta; -z) = {}_2F_3\left(\alpha, \beta-\alpha; \beta, \frac{1}{2}\beta, \frac{\beta+1}{2}; \frac{1}{4}z^2\right),$$

$${}_1F_1(\alpha; 2\alpha; z)\, {}_1F_1(\beta; 2\beta; -z)$$
$$= {}_2F_3\left(\frac{\alpha+\beta}{2}, \frac{\alpha+\beta+1}{2}; \alpha+\frac{1}{2}, \beta+\frac{1}{2}, \alpha+\beta; \frac{1}{4}z^2\right),$$

$${}_2F_1\left(\alpha, \beta; \alpha+\beta-\frac{1}{2}; z\right) {}_2F_1\left(\alpha, \beta; \alpha+\beta+\frac{1}{2}; z\right)$$
$$= {}_3F_2\left(2\alpha, 2\beta, \alpha+\beta; 2\alpha+2\beta-1, \alpha+\beta+\frac{1}{2}; z\right),$$

$${}_2F_1\left(\alpha, \beta; \alpha+\beta-\frac{1}{2}; z\right) {}_2F_1\left(\alpha-1, \beta; \alpha+\beta-\frac{1}{2}; z\right)$$
$$= {}_3F_2\left(2\alpha-1, 2\beta, \alpha+\beta-1; 2(\alpha+\beta-1), \alpha+\beta-\frac{1}{2}; z\right).$$

The formulas above are valid at least for $|z| < 1$.

Some integrals associated with ${}_pF_q(\alpha_1, \alpha_2, \ldots, \alpha_p; \beta_1, \ldots, \beta_q; z)$

$$\int_0^\infty t^{\lambda-1} e^{-zt}\, {}_pF_q(\alpha_1, \alpha_2, \ldots, \alpha_p; \beta_1, \beta_2, \ldots, \beta_q; kt)\, dt$$
$$= \Gamma(\lambda)\, z^{-\lambda}\, {}_{p+1}F_q\left(\alpha_1, \ldots, \alpha_p, \lambda; \beta_1, \ldots, \beta_q; \frac{k}{z}\right),$$
$$p < q,\ Re\,\lambda > 0,\ Re\,z > 0,$$

or

$$p = q,\ Re\,\lambda > 0,\ Re\,z > Re\,k,$$

$$\int_0^\infty e^{-zt}\left\{\int_0^t \tau^{\lambda-1}\, {}_pF_q(\alpha_1, \ldots, \alpha_p; \beta_1, \ldots, \beta_q; k\tau)\, (t-\tau)^{\mu-1}\right.$$
$$\left.\times\, {}_{p'}F_{q'}(\alpha'_1, \ldots, \alpha'_{p'}; \beta'_1, \ldots, \beta'_{q'}; k'(t-\tau))\, d\tau\right\} dt$$
$$= \Gamma(\lambda)\,\Gamma(\mu)\, z^{-\lambda-\mu}\, {}_{p+1}F_q\left(\alpha_1, \ldots, \alpha_p, \lambda; \beta_1, \ldots, \beta_q; \frac{k}{z}\right)$$
$$\times\, {}_{p'+1}F_{q'}\left(\alpha'_1, \ldots, \alpha'_{p'}, \mu; \beta'_1, \ldots, \beta'_{q'}; \frac{k'}{z}\right),$$
$$p < q,\ p' < q',\ Re\,\lambda > 0,\ Re\,\mu > 0,\ Re\,z > 0$$

or

$$p=q,\ p'=q',\ Re\,\lambda>0,\ Re\,\mu>0,\ Re\,z>Re\,k,$$

$$\int_0^\infty e^{-zt}\,t^{\alpha-1}\,{}_pF_q(\alpha_1,\ldots,\alpha_p;\beta_1,\ldots,\beta_q;kt)$$
$$\times\,{}_{p'}F_{q'}(\alpha_1',\ldots,\alpha_{p'}';\beta_1',\ldots,\beta_{q'}';k't)\,dt$$
$$=z^{-\alpha}\,\Gamma(\alpha)\sum_{n=0}^\infty\frac{(\alpha_1)_n\cdots(\alpha_p)_n\,(\alpha)_n\,k^n}{(\beta_1)_n\cdots(\beta_q)_n\,z^n\,n!}$$
$$\times\,{}_{p'+1}F_{q'}\left(\alpha_1',\ldots\alpha_{p'}',\alpha+n;\beta_1',\ldots,\beta_{q'}';\frac{k'}{z}\right),$$

$$\frac{t^{\lambda-1}}{\Gamma(\lambda)}\,{}_pF_{q+1}(\alpha_1,\ldots,\alpha_p;\beta_1,\ldots,\beta_q,\lambda;kt)$$
$$=\frac{1}{2\pi i}\int_{c-i\infty}^{c+i\infty}e^{zt}\,z^{-\lambda}\,{}_pF_q\left(\alpha_1,\ldots,\alpha_p;\beta_1,\ldots,\beta_q;\frac{k}{z}\right)dz,$$

$$p<q+1,\ Re\,z>0,\ Re\,\lambda>0,$$

or

$$p=q+1,\ Re\,z>Re\,k,\ Re\,\lambda>0.$$

2.10 Miscellaneous results

The generalized hypergeometric function reduces to a polynomial in the cases below

$$F_n(z)={}_3F_2\left(-n,n+1,\frac{1}{2}+\frac{1}{2}z;1,1;1\right),\qquad n=0,1,2,\ldots,$$

$$Z_n(z)={}_2F_2(-n,n+1;1,1;z),\qquad n=0,1,2,\ldots,$$

$$z^{-u}\,J_n^{u,v}(z)=\frac{\Gamma\left(v+n+1+\frac{1}{2}u\right)}{n!\,\Gamma(u+1)\,\Gamma\left(v+1+\frac{1}{2}u\right)}$$
$$\times\,{}_1F_2\left(-n;u+1,v+\frac{1}{2}u+1;z^2\right),$$

$$H_n(\xi,\alpha,z)={}_3F_2(-n,n+1,\xi;1,\alpha;z),$$

where $n=0,1,2,\ldots$ and ξ,α,z are complex variables but $\alpha\neq-n-1,-n-2,\ldots$

$$H_n(\xi,\alpha,z)=\frac{\Gamma(\alpha)}{\Gamma(\xi)\,\Gamma(\alpha-\xi)}\int_0^1 t^{\xi-1}(1-t)^{\alpha-\xi-1}\,P_n(1-2zt)\,dt,$$

$$Re\,\alpha>Re\,\xi>0,\ P_n(z)={}_2F_1\left(-n,n+1;1;\frac{1}{2}-\frac{1}{2}z\right).$$

The generating function of H_n is

$$(1-t)^{-1}\,{}_2F_1\left(\xi,\frac{1}{2};\alpha;-4zt(1-t)^{-2}\right)=\sum_{n=0}^\infty t^n\,H_n(\xi,\alpha,z).$$

An asymptotic expression for $H_n(\xi, \alpha, 1)$ as $n \to \infty$ is

$$H_n(\xi, \alpha, 1) \sim \frac{\Gamma(\alpha)\, n^{-2\xi}}{\Gamma(\alpha - \xi)\, \Gamma(1 - \xi)} + (-1)^n \frac{\Gamma(\alpha)\, n^{2\xi - 2\alpha}}{\Gamma(\xi - \alpha + 1)\, \Gamma(\xi)}.$$

For $z = 1$, one has

$${}_4F_3\left[\begin{matrix} a, & \frac{1}{2}a + 1, & b, & -n; & \\ & \frac{1}{2}a, & 1 + a - b, & 1 + a + n; & 1 \end{matrix}\right] = \frac{(1 + a)_n \left(\frac{1}{2} + \frac{1}{2}a - b\right)_n}{\left(\frac{1}{2} + \frac{1}{2}a\right)_n (1 + a - b)_n}.$$

For more results on the generalized hypergeometric function see ERDÉLYI (1953).

Literature

ERDÉLYI, A.: Higher transcendental functions, Vol. 1. New York: McGraw-Hill 1953.

KAMPÉ DE FÉRIET, J.: La fonction hypergeometrique. Paris: Gauthiers- Villars 1937.

KLEIN, F.: Vorlesungen über die hypergeometrische Funktion. Berlin: Teubner 1933.

MACROBERT, T. M.: Proc. Edinburgh Math. Soc. **42** (1923) 84—88.

— Functions of a complex variable. London: Macmillan 1954.

Chapter III

Bessel functions

3.1 Solutions of the Bessel and the modified Bessel differential equation

Bessel functions are solutions of Bessel's differential equation

$$z^2 \frac{d^2w}{dz^2} + z \frac{dw}{dz} + (z^2 - \nu^2)\, w = 0,$$

ν, z can be arbitrarily complex.

Special solutions of this equation are the Bessel, Neumann and Hankel functions $J_\nu(z)$, $Y_\nu(z)$, $H_\nu^{(1)}(z)$, $H_\nu^{(2)}(z)$ respectively. The latter three are linear combinations of the first. They are defined by

$$J_\nu(z) = \sum_{m=0}^{\infty} \frac{(-1)^m \left(\frac{z}{2}\right)^{\nu+2m}}{m!\, \Gamma(\nu + m + 1)} = \frac{\left(\frac{1}{2}z\right)^\nu}{\Gamma(\nu + 1)}\, {}_0F_1\left(\nu + 1; -\frac{1}{4}z^2\right).$$

[For fixed z, considered as function of ν, $J_\nu(z)$ represents an entire function. While

$$\left(\frac{z}{2}\right)^{-\nu} J_\nu(z) = \sum_{m=0}^{\infty} \frac{(-1)^m \left(\frac{z}{2}\right)^{2m}}{m!\,\Gamma(\nu+m+1)}$$

is an entire function of z.]

$$Y_\nu(z) = [\sin(\pi\nu)]^{-1} [J_\nu(z)\cos(\pi\nu) - J_{-\nu}(z)],$$

$$H_\nu^{(1)}(z) = J_\nu(z) + i\,Y_\nu(z) = (i\sin\pi\nu)^{-1} [J_{-\nu}(z) - J_\nu(z)\,e^{-i\pi\nu}],$$

$$H_\nu^{(2)}(z) = J_\nu(z) - i\,Y_\nu(z) = (i\sin\pi\nu)^{-1} [J_\nu(z)\,e^{i\pi\nu} - J_{-\nu}(z)].$$

(In case ν is an integer or zero the right hand sides of the equations above become indeterminate for this case see 3.2.)

These definitions lead to

$$J_{-\nu}(z) = J_\nu(z)\cos(\pi\nu) - Y_\nu(z)\sin(\pi\nu),$$

$$Y_{-\nu}(z) = J_\nu(z)\sin(\pi\nu) + Y_\nu(z)\cos(\pi\nu),$$

$$H_{-\nu}^{(1)}(z) = e^{i\pi\nu} H_\nu^{(1)}(z), \qquad H_{-\nu}^{(2)}(z) = e^{-i\pi\nu} H_\nu^{(2)}(z),$$

$$\overline{J_\nu}(z) = J_{\bar\nu}(\bar z), \qquad \overline{Y_\nu}(z) = Y_{\bar\nu}(\bar z),$$

$$\overline{H}_\nu^{(1)}(z) = H_{\bar\nu}^{(2)}(\bar z), \qquad \overline{H}_\nu^{(2)}(z) = H_{\bar\nu}^{(1)}(\bar z),$$

($\bar\zeta$ means the quantity conjugate complex to ζ).

Modified Bessel functions. The modified Bessel functions are solutions of the modified Bessel differential equation

$$z^2\frac{d^2w}{dz^2} + z\frac{dw}{dz} - (z^2+\nu^2)\,w = 0.$$

(The modified Bessel differential equation is obtained replacing z by iz in Bessel's differential equation.)

Special solutions are

$$I_\nu(z) = e^{-i\frac{\pi}{2}\nu} J_\nu\left(z e^{i\frac{\pi}{2}}\right) = \sum_{m=0}^{\infty} \frac{\left(\frac{1}{2}z\right)^{\nu+2m}}{m!\,\Gamma(\nu+m+1)}$$

$$= \frac{\left(\frac{1}{2}z\right)^\nu}{\Gamma(1+\nu)}\, {}_0F_1\left(\nu+1; \frac{1}{4}z^2\right)$$

and

$$K_\nu(z) = \frac{1}{2}\pi\,[\sin(\pi\nu)]^{-1} [I_{-\nu}(z) - I_\nu(z)]$$

or

$$K_\nu(z) = \frac{1}{2}\pi\,[\sin(\pi\nu)]^{-1} \left[e^{i\frac{\pi}{2}\nu} J_{-\nu}\left(z e^{i\frac{\pi}{2}}\right) - e^{-i\frac{\pi}{2}\nu} J_\nu\left(z e^{i\frac{\pi}{2}}\right)\right]$$

or, expressed in terms of Hankel's functions

$$K_\nu(z) = \frac{1}{2} i\pi e^{i\frac{\pi}{2}\nu} H_\nu^{(1)}\left(z e^{i\frac{\pi}{2}}\right) = -\frac{1}{2} i\pi e^{-i\frac{\pi}{2}\nu} H_\nu^{(2)}\left(z e^{-i\frac{\pi}{2}}\right),$$

$$K_{-\nu}(z) = K_\nu(z),$$

$$Y_\nu\left(z e^{i\frac{\pi}{2}}\right) = i e^{i\nu\frac{\pi}{2}} I_\nu(z) - \frac{2}{\pi} e^{-i\nu\frac{\pi}{2}} K_\nu(z),$$

$$K_\nu\left(z e^{i\frac{\pi}{2}}\right) = \frac{1}{2} i\pi e^{i\frac{\pi}{2}\nu} H_\nu^{(1)}(z e^{i\pi}) = -\frac{1}{2} i\pi e^{-i\frac{\pi}{2}\nu} H_\nu^{(2)}(z),$$

$$K_\nu\left(z e^{-i\frac{\pi}{2}}\right) = \frac{1}{2} i\pi e^{i\frac{\pi}{2}\nu} H_\nu^{(1)}(z).$$

3.1.1 Recurrence relations, differential formulas and Wronskians

Denote by $C_\nu(z)$ any of the functions $J_\nu(z)$, $Y_\nu(z)$, $H_\nu^{(1)}(z)$, $H_\nu^{(2)}(z)$ then

$$C_{\nu-1}(z) - C_{\nu+1}(z) = 2C_\nu'(z),$$

$$C_{\nu-1}(z) + C_{\nu+1}(z) = 2\nu z^{-1} C_\nu(z),$$

$$C_{\nu+1}(z) = \frac{\nu}{z} C_\nu(z) - C_\nu'(z),$$

$$C_\nu'(z) = C_{\nu-1}(z) - \frac{\nu}{z} C_\nu(z),$$

$$\left(\frac{d}{z\,dz}\right)^m [z^\nu C_\nu(z)] = z^{\nu-m} C_{\nu-m}(z),$$

$$\left(\frac{d}{z\,dz}\right)^m [z^{-\nu} C_\nu(z)] = (-1)^m z^{-\nu-m} C_{\nu+m}(z),$$

$$(m = 1, 2, 3, \ldots).$$

For the modified Bessel functions one has

$$I_{\nu-1}(z) - I_{\nu+1}(z) = 2\frac{\nu}{z} I_\nu(z), \qquad I_{\nu-1}(z) + I_{\nu+1}(z) = 2I_\nu'(z),$$

$$K_{\nu-1}(z) - K_{\nu+1}(z) = -2\frac{\nu}{z} K_\nu(z); \qquad K_{\nu-1}(z) + K_{\nu+1}(z) = -2K_\nu'(z),$$

$$\left(\frac{d}{z\,dz}\right)^m [z^\nu I_\nu(z)] = z^{\nu-m} I_{\nu-m}(z),$$

$$\left(\frac{d}{z\,dz}\right)^m [z^{-\nu} I_\nu(z)] = z^{-\nu-m} I_{\nu+m}(z),$$

$$\left(\frac{d}{z\,dz}\right)^m [z^\nu K_\nu(z)] = (-1)^m z^{\nu-m} K_{\nu-m}(z),$$

$$\left(\frac{d}{z\,dz}\right)^m [z^{-\nu} K_\nu(z)] = (-1)^m z^{-\nu-m} K_{\nu+m}(z),$$

$$(m = 1, 2, 3, \ldots).$$

Wronskian determinants and related formulas. Define by $W(w_1, w_2) = w_1 w_2' - w_2 w_1'$.

Then

$$W(J_\nu, J_{-\nu}) = -2(\pi z)^{-1} \sin(\pi\nu),$$
$$W(J_\nu, Y_\nu) = 2(\pi z)^{-1},$$
$$W(H_\nu^{(1)}, H_\nu^{(2)}) = -4i(\pi z)^{-1},$$
$$W(J_\nu, H_\nu^{(1),(2)}) = \pm 2i(\pi z)^{-1},$$
$$W(I_\nu, I_{-\nu}) = -2(\pi z)^{-1} \sin(\pi\nu),$$
$$W(I_\nu, K_\nu) = -z^{-1}.$$

It follows from these Wronskians that $J_\nu(z)$ and $J_{-\nu}(z)$ are not linearly independent solutions of Bessel's differential equation when $\nu = 0, \pm 1, \pm 2, \ldots$

$$J_\nu(z) J_{-\nu+1}(z) + J_{-\nu}(z) J_{\nu-1}(z) = 2(\pi z)^{-1} \sin(\pi\nu),$$
$$H_\nu^{(1)}(z) H_{\nu-1}^{(2)}(z) - H_{\nu-1}^{(1)}(z) H_\nu^{(2)}(z) = -4i(\pi z)^{-1},$$
$$J_\nu(z) Y_{\nu-1}(z) - Y_\nu(z) J_{\nu-1}(z) = 2(\pi z)^{-1},$$
$$J_{\nu-1}(z) H_\nu^{(1)}(z) - J_\nu(z) H_{\nu-1}^{(1)}(z) = 2(\pi i z)^{-1},$$
$$J_\nu(z) H_{\nu-1}^{(1),(2)}(z) - J_{\nu-1}(z) H_\nu^{(1),(2)}(z) = \pm 2(\pi i z)^{-1},$$

(+ for the 1st, − for the 2d Hankel function)

$$I_\nu(z) I_{-\nu+1}(z) - I_{-\nu}(z) I_{\nu-1}(z) = -2(\pi z)^{-1} \sin(\pi\nu),$$
$$K_{\nu+1}(z) I_\nu(z) + K_\nu(z) I_{\nu+1}(z) = z^{-1}.$$

3.1.2 Reduction to the principal branch

The Bessel and modified Bessel functions are generally many valued. They are one valued for all points z of the principal branch $-\pi < \arg z < \pi$. But the values for these functions for points z not on the principal branch can be reduced to the principal value by means of the relations

$$J_\nu(z e^{im\pi}) = e^{im\pi\nu} J_\nu(z),$$
$$Y_\nu(z e^{im\pi}) = e^{-m\pi i\nu} Y_\nu(z) + 2i \sin(m\pi\nu) \cot(\pi\nu) J_\nu(z),$$
$$H_\nu^{(1)}(z e^{im\pi}) = -\frac{\sin[(m-1)\pi\nu]}{\sin(\pi\nu)} H_\nu^{(1)}(z) - e^{-i\pi\nu} \frac{\sin(\pi\nu m)}{\sin(\pi\nu)} H_\nu^{(2)}(z),$$
$$H_\nu^{(2)}(z e^{im\pi}) = \frac{\sin[(m+1)\pi\nu]}{\sin(\pi\nu)} H_\nu^{(2)}(z) + e^{i\pi\nu} \frac{\sin(\pi\nu m)}{\sin(\pi\nu)} H_\nu^{(1)}(z),$$

$$I_\nu(z e^{im\pi}) = e^{im\pi\nu} I_\nu(z),$$

$$K_\nu(z e^{im\pi}) = e^{-im\pi\nu} K_\nu(z) - i\pi \frac{\sin(m\pi\nu)}{\sin(\pi\nu)} I_\nu(z),$$

$$m = \pm 1, \pm 2, \pm 3, \ldots$$

3.1.3 Differentiation with respect to the order

$$\frac{\partial J_\nu(z)}{\partial \nu} = J_\nu(z)\log\left(\frac{1}{2}z\right) - \sum_{m=0}^{\infty}(-1)^m\left(\frac{1}{2}z\right)^{\nu+2m}\frac{\psi(\nu+m+1)}{m!\,\Gamma(\nu+m+1)}$$

$$\frac{\partial J_{-\nu}(z)}{\partial \nu} = -J_{-\nu}(z)\log\left(\frac{1}{2}z\right) + \sum_{m=0}^{\infty}(-1)^m\left(\frac{1}{2}z\right)^{-\nu+2m}\frac{\psi(-\nu+m+1)}{m!\,\Gamma(-\nu+m+1)},$$

$$\frac{\partial Y_\nu(z)}{\partial \nu} = \cot(\pi\nu)\frac{\partial J_\nu(z)}{\partial \nu} - \operatorname{cosec}(\pi\nu)\frac{\partial J_{-\nu}(z)}{\partial \nu} - \pi\operatorname{cosec}(\pi\nu)\,Y_{-\nu}(z),$$

$$\frac{\partial I_\nu(z)}{\partial \nu} = I_\nu(z)\log\left(\frac{1}{2}z\right) - \sum_{m=0}^{\infty}\left(\frac{1}{2}z\right)^{\nu+2m}\frac{\psi(\nu+m+1)}{m!\,\Gamma(\nu+m+1)},$$

$$\frac{\partial K_\nu(z)}{\partial \nu} = -\pi\cot(\pi\nu)\,K_\nu(z) + \frac{1}{2}\pi\operatorname{cosec}(\pi\nu)\left[\frac{\partial I_{-\nu}(z)}{\partial \nu} - \frac{\partial I_\nu(z)}{\partial \nu}\right].$$

For the cases $\nu = n$ or $\nu = \pm\left(n + \frac{1}{2}\right)$ see pp. 74.

3.2 Bessel functions of integer order

For $\nu = n$, $(n = 0, 1, 2, \ldots)$

$$Y_n(z) = \frac{2}{\pi}\left[\gamma + \log\left(\frac{1}{2}z\right)\right]J_n(z) - \frac{1}{\pi}\sum_{m=0}^{n-1}\frac{\left(\frac{1}{2}z\right)^{2m-n}}{m!}(n-m-1)!$$

$$-\frac{1}{\pi}\sum_{m=0}^{\infty}(-1)^m\frac{\left(\frac{1}{2}z\right)^{n+2m}}{m!(n+m)!}(k_{m+n}+k_m),$$

$$k_l = 1 + \frac{1}{2} + \frac{1}{3} + \cdots + \frac{1}{l}, \quad l = 1, 2, 3, \ldots; \quad k_0 = 0,$$

$$K_n(z) = (-1)^{n+1}\left[\gamma + \log\left(\frac{1}{2}z\right)\right]I_n(z)$$

$$+\frac{1}{2}\sum_{m=0}^{n-1}(-1)^m\frac{\left(\frac{1}{2}z\right)^{2m-n}}{m!}(n-m-1)!$$

$$+\frac{1}{2}(-1)^n\sum_{m=0}^{\infty}\frac{\left(\frac{1}{2}z\right)^{n+2m}}{m!\,(n+m)!}(k_{m+n}+k_m).$$

For $n = 0$; $\sum_{m=0}^{n-1}() = 0$.

Also

$$Y_0(z) = \frac{2}{\pi}\left[\gamma + \log\left(\frac{1}{2}z\right)\right] J_0(z) - \frac{4}{\pi}\sum_{n=1}^{\infty}\frac{(-1)^n}{n} J_{2n}(z),$$

$$K_0(z) = -\left[\gamma + \log\left(\frac{1}{2}z\right)\right] I_0(z) + 2\sum_{n=1}^{\infty}\frac{1}{n} I_{2n}(z).$$

And more general.

$$Y_n(z) = -n!\,\frac{\left(\frac{1}{2}z\right)^{-n}}{\pi}\sum_{k=0}^{n-1}\frac{\left(\frac{1}{2}z\right)^k J_k(z)}{(n-k)\,k!}$$

$$+\frac{2}{\pi}\left[\log\left(\frac{1}{2}z\right) - \psi(n+1)\right] J_n(z)$$

$$-\frac{2}{\pi}\sum_{k=1}^{\infty}\frac{(-1)^k (n+2k)\,J_{n+2k}(z)}{k(n+k)},$$

$$K_n(z) = (-1)^{n-1}\left[\log\left(\frac{1}{2}z\right) - \psi(n+1)\right] I_n(z)$$

$$+\frac{1}{2}n!\left(\frac{1}{2}z\right)^{-n}\sum_{k=0}^{n-1}(-1)^k\frac{\left(\frac{1}{2}z\right)^k I_k(z)}{(n-k)\,k!}$$

$$+(-1)^n\sum_{k-1}^{\infty}\frac{(n+2k)\,I_{n+2k}(z)}{k(n+k)}.$$

For negative integer order

$$J_{-n}(z) = (-1)^n J_n(z); \quad Y_{-n}(z) = (-1)^n Y_n(z); \quad I_{-n}(z) = I_n(z).$$

3.2.1 Generating functions

$$e^{\frac{1}{2}z\left(t-\frac{\alpha^2}{t}\right)} = \sum_{n=-\infty}^{\infty}\left(\frac{t}{\alpha}\right)^n J_n(\alpha z),$$

$$e^{iz\cos\alpha} = \sum_{n=-\infty}^{\infty} i^n e^{in\alpha} J_n(z) = \sum_{n=0}^{\infty}\varepsilon_n(-1)^n J_{2n}(z)\cos(2n\alpha)$$

$$+ 2i\sum_{n=0}^{\infty}(-1)^n J_{2n+1}(z)\cos[(2n+1)\alpha],$$

$$e^{iz\sin\alpha} = \sum_{n=-\infty}^{\infty} e^{in\alpha} J_n(z) = \sum_{n=0}^{\infty}\varepsilon_n J_{2n}(z)\cos(2n\alpha)$$

$$+ 2i\sum_{n=0}^{\infty} J_{2n+1}(z)\sin[(2n+1)\alpha].$$

3.2.2 Finite series involving Bessel functions of integer order

$$J_{2n}(z) = (-1)^n n \sum_{l=0}^{n} \frac{(n+l-1)!}{(n-l)!} (-1)^l \frac{\left(\frac{z}{2}\right)^{-l}}{l!} J_l(z),$$

$$Y_{2n}(z) = (-1)^n n \sum_{l=0}^{n} \frac{(n+l-1)!}{(n-l)!} (-1)^l \frac{\left(\frac{z}{2}\right)^{-l}}{l!} Y_l(z),$$

$$I_{2n}(z) = n \sum_{l=0}^{n} \frac{(n+l-1)!}{(n-l)!} (-1)^l \frac{\left(\frac{z}{2}\right)^{-l}}{l!} I_l(z),$$

$$K_{2n}(z) = n \sum_{l=0}^{n} \frac{(n+l-1)!}{(n-l)!} \frac{\left(\frac{z}{2}\right)^{-l}}{l!} K_l(z),$$

$$\left(\frac{1}{2}z\right)^{-n} J_n(z) = n! \sum_{l=0}^{n} \varepsilon_l [(n+l)!\,(n-l)!]^{-1} J_{2l}(z),$$

$$\left(\frac{1}{2}z\right)^{-n} Y_n(z) = n! \sum_{l=0}^{n} \varepsilon_l [(n+l)!\,(n-l)!]^{-1} Y_{2l}(z),$$

$$\left(\frac{1}{2}z\right)^{-n} I_n(z) = n! \sum_{l=0}^{n} (-1)^l \varepsilon_l [(n+l)!\,(n-l)!]^{-1} I_{2l}(z),$$

$$\left(\frac{1}{2}z\right)^{-n} K_n(z) = (-1)^n n! \sum_{l=0}^{n} (-1)^l \varepsilon_l [(n+l)!\,(n-l)!]^{-1} K_{2l}(z).$$

3.2.3 Differentiation with respect to the order

$$\left[\frac{\partial J_\nu(z)}{\partial \nu}\right]_{\nu=\pm n} = \frac{1}{2}\pi (\pm 1)^n Y_n(z) \pm (\pm 1)^n \frac{1}{2} n! \sum_{l=0}^{n-1} \frac{\left(\frac{1}{2}z\right)^{l-n} J_l(z)}{l!\,(n-l)},$$

$$\left[\frac{\partial Y_\nu(z)}{\partial \nu}\right]_{\nu=\pm n} = -\frac{1}{2}\pi (\pm 1)^n J_n(z) \pm (\pm 1)^n \frac{1}{2} n! \sum_{l=0}^{n-1} \frac{\left(\frac{1}{2}z\right)^{l-n} Y_l(z)}{l!\,(n-l)},$$

$$\left[\frac{\partial I_\nu(z)}{\partial \nu}\right]_{\nu=\pm n} = (-1)^{n+1} K_n(z) \pm (-1)^n \frac{1}{2} n! \sum_{l=0}^{n-1} \frac{(-1)^l \left(\frac{1}{2}z\right)^{l-n} I_l(z)}{l!\,(n-l)},$$

$$\left[\frac{\partial K_\nu(z)}{\partial \nu}\right]_{\nu=\pm n} = \pm \frac{1}{2} n! \sum_{l=0}^{n-1} \frac{\left(\frac{1}{2}z\right)^{l-n} K_l(z)}{l!\,(n-l)}.$$

Special cases of the above formulas are

$$\left[\frac{\partial J_\nu(z)}{\partial \nu}\right]_{\nu=0} = \frac{1}{2}\pi Y_0(z); \quad \left[\frac{\partial Y_\nu(z)}{\partial \nu}\right]_{\nu=0} = -\frac{1}{2}\pi J_0(z),$$

$$\left[\frac{\partial I_\nu(z)}{\partial \nu}\right]_{\nu=0} = -K_0(z); \quad \left[\frac{\partial K_\nu(z)}{\partial z}\right]_{\nu=0} = 0$$

3.3 Half odd integer order

The Bessel functions of the order $\nu = n + \frac{1}{2}$ $(n = 0, \pm 1, \pm 2, \ldots)$ are elementary functions. For these we have the representations for $n = 0, 1, 2, \ldots$

$$J_{n+\frac{1}{2}}(z) = \left(\frac{1}{2}\pi z\right)^{-\frac{1}{2}} \left[\sin\left(z - \frac{1}{2} n\pi\right) \sum_{m=0}^{\leq \frac{1}{2}n} (-1)^m \left(n + \frac{1}{2}, 2m\right)(2z)^{-2m}\right.$$

$$\left. + \cos\left(z - \frac{1}{2} n\pi\right) \sum_{m=0}^{\leq \frac{1}{2}n - \frac{1}{2}} (-1)^m \left(n + \frac{1}{2}, 2m + 1\right)(2z)^{-2m-1}\right],$$

$$Y_{n+\frac{1}{2}}(z) = \left(\frac{1}{2}\pi z\right)^{-\frac{1}{2}} \left[\sin\left(z - \frac{1}{2} n\pi\right) \sum_{m=0}^{\leq \frac{1}{2}n - \frac{1}{2}} (-1)^m\right.$$

$$\times \left(n + \frac{1}{2}, 2m + 1\right)(2z)^{-2m-1} - \cos\left(z - \frac{1}{2} n\pi\right) \sum_{m=0}^{\leq \frac{1}{2}n} (-1)^m$$

$$\left. \times \left(n + \frac{1}{2}, 2m\right)(2z)^{-2m}\right],$$

$$H^{(1,2)}_{n+\frac{1}{2}}(z) = \left(\frac{1}{2}\pi z\right)^{-\frac{1}{2}} e^{\mp i\frac{\pi}{2}(n+1) \pm iz} \sum_{m=0}^{n} (\pm i)^m \left(n + \frac{1}{2}, m\right)(2z)^{-m},$$

$$K_{n+\frac{1}{2}}(z) = \left(\frac{\frac{1}{2}\pi}{z}\right) e^{-z} \sum_{m=0}^{n} \left(n + \frac{1}{2}, m\right)(2z)^{-m}.$$

Furthermore for $n = 0, 1, 2, \ldots$

$$J_{-n-\frac{1}{2}}(z) = (-1)^{n+1} Y_{n+\frac{1}{2}}(z), \qquad Y_{-n-\frac{1}{2}}(z) = (-1)^n J_{n+\frac{1}{2}}(z),$$

$$H^{(1)}_{-n-\frac{1}{2}}(z) = i(-1)^n H^{(1)}_{n+\frac{1}{2}}(z), \qquad H^{(2)}_{-n-\frac{1}{2}}(z) = -i(-1)^n H^{(2)}_{n+\frac{1}{2}}(z),$$

$$I_{-n-\frac{1}{2}}(z) = (-1)^n \frac{2}{\pi} K_{n+\frac{1}{2}}(z) + I_{n+\frac{1}{2}}(z).$$

One can also write

$$J_{n+\frac{1}{2}}(z) = (-1)^n \left(\frac{1}{2}\pi z\right)^{-\frac{1}{2}} z^{n+1} \left(\frac{d}{z\,dz}\right)^n \frac{\sin z}{z},$$

$$Y_{n+\frac{1}{2}}(z) = -(-1)^n \left(\frac{1}{2}\pi z\right)^{-\frac{1}{2}} z^{n+1} \left(\frac{d}{z\,dz}\right)^n \frac{\cos z}{z},$$

$$H^{(1,2)}_{n+\frac{1}{2}}(z) = \mp i(-1)^n \left(\frac{1}{2}\pi z\right)^{-\frac{1}{2}} z^{n+1} \left(\frac{d}{z\,dz}\right)^n \frac{e^{\pm iz}}{z},$$

$$I_{n+\frac{1}{2}}(z) = \left(\frac{1}{2}\pi z\right)^{-\frac{1}{2}} z^{n+1} \left(\frac{d}{z\,dz}\right)^n \frac{\sinh z}{z},$$

$$K_{n+\frac{1}{2}}(z) = (-1)^n \left(\frac{\frac{1}{2}\pi}{z}\right)^{\frac{1}{2}} z^{n+1} \left(\frac{d}{z\,dz}\right)^n \frac{e^{-z}}{z}.$$

The so called spherical Bessel functions $\psi_n(z)$ and $\zeta_n(z)$ are defined by

$$\psi_n(z) = \left(\frac{\frac{1}{2}\pi}{z}\right)^{\frac{1}{2}} J_{n+\frac{1}{2}}(z),$$

$$\zeta_n^{(1,2)}(z) = \left(\frac{\frac{1}{2}\pi}{z}\right)^{\frac{1}{2}} H^{(1,2)}_{n+\frac{1}{2}}(z).$$

Special cases are

$$J_{\frac{1}{2}}(z) = Y_{-\frac{1}{2}}(z) = \left(\frac{1}{2}\pi z\right)^{-\frac{1}{2}} \sin z,$$

$$Y_{\frac{1}{2}}(z) = -J_{-\frac{1}{2}}(z) = -\left(\frac{1}{2}\pi z\right)^{-\frac{1}{2}} \cos z,$$

$$H^{(1)}_{\frac{1}{2}}(z) = -i H^{(1)}_{-\frac{1}{2}}(z) = -i\left(\frac{1}{2}\pi z\right)^{-\frac{1}{2}} e^{iz},$$

$$H^{(2)}_{\frac{1}{2}}(z) = i H^{(2)}_{-\frac{1}{2}}(z) = i\left(\frac{1}{2}\pi z\right)^{-\frac{1}{2}} e^{-iz},$$

$$I_{\frac{1}{2}}(z) = \left(\frac{1}{2}\pi z\right)^{-\frac{1}{2}} \sinh z, \qquad I_{-\frac{1}{2}}(z) = \left(\frac{1}{2}\pi z\right)^{-\frac{1}{2}} \cosh z,$$

$$K_{\frac{1}{2}}(z) = K_{-\frac{1}{2}}(z) = \left(\frac{\frac{1}{2}\pi}{z}\right)^{\frac{1}{2}} e^{-z}.$$

3.3.1 Generating functions

$$e^{z\cos\alpha}\,\mathrm{Erfc}\left[(2z)^{\frac{1}{2}} \cos\left(\frac{1}{2}\alpha\right)\right] = \sum_{n=0}^{\infty} (-1)^n \varepsilon_n I_{\frac{n}{2}}(z) \cos\left(\frac{n}{2}\alpha\right),$$

$$e^{z\cos\alpha}\,\mathrm{Erf}\left[(2z)^{\frac{1}{2}} \cos\left(\frac{1}{2}\alpha\right)\right] = 2\sum_{n=0}^{\infty} I_{n+\frac{1}{2}}(z) \cos\left[\left(n+\frac{1}{2}\right)\alpha\right],$$

$$\left(\frac{i}{\pi}\right)^{\frac{1}{2}} e^{iz\cos 2v} \int_{-\infty}^{u} e^{-it^2}\,dt = \frac{1}{2}\sum_{n=0}^{\infty} \varepsilon_n\, e^{in\frac{\pi}{4}} J_{\frac{n}{2}}(z)\cos(nv),$$

$$u = (2z)^{\frac{1}{2}}\cos v,$$

$$z^{-1}\sinh\left[(z^2+2izt)^{\frac{1}{2}}\right] = \sum_{n=0}^{\infty} \frac{(-it)^n}{n!}\left(\frac{1}{2}\pi z\right)^{\frac{1}{2}} I_{-n+\frac{1}{2}}(z),$$

$$z^{-1}\cosh\left[(z^2+2izt)^{\frac{1}{2}}\right] = \sum_{n=0}^{\infty} \frac{(it)^n}{n!}\left(\frac{1}{2}\pi z\right)^{\frac{1}{2}} I_{n-\frac{1}{2}}(z).$$

3.3.2 Duplication formulas

$$J_{n+\frac{1}{2}}(2z) = -\pi^{\frac{1}{2}}\, n!\, z^{n+\frac{1}{2}} \sum_{k=0}^{n} \frac{(2n-2k+1)}{k!\,(2n-k+1)!} J_{n-k+\frac{1}{2}}(z)\, Y_{n-k+\frac{1}{2}}(z)$$

$$Y_{n+\frac{1}{2}}(2z) = \frac{1}{2}\pi^{\frac{1}{2}}\, n!\, z^{n+\frac{1}{2}} \sum_{k=0}^{n} \frac{(2n-2k+1)}{k!\,(2n-k+1)!} \times\left\{\left[J_{n-k+\frac{1}{2}}(z)\right]^2 - \left[Y_{n-k+\frac{1}{2}}(z)\right]^2\right\},$$

$$K_{n+\frac{1}{2}}(2z) = n!\,\pi^{-\frac{1}{2}} z^{n+\frac{1}{2}} \sum_{k=0}^{n} \frac{(-1)^k (2n-2k+1)}{k!\,(2n-k+1)!}\left[K_{n-k+\frac{1}{2}}(z)\right]^2,$$

$$I_{n+\frac{1}{2}}(2z) = n!\,\pi^{\frac{1}{2}} z^{n+\frac{1}{2}} \sum_{k=0}^{n} (-1)^{n-k} \frac{(2n-2k+1)}{k!\,(2n-k+1)} \times I_{n-k+\frac{1}{2}}(z)\, I_{-\left(n-k+\frac{1}{2}\right)}(z).$$

3.3.3 Differentiation with respect to the order

$$\left[\frac{\partial J_\nu(z)}{\partial\nu}\right]_{\nu=\frac{1}{2}} = \left(\frac{1}{2}\pi x\right)^{-\frac{1}{2}} [\sin x\, Ci(2x) - \cos x\, Si(2x)],$$

$$\left[\frac{\partial J_\nu(z)}{\partial\nu}\right]_{\nu=-\frac{1}{2}} = \left(\frac{1}{2}\pi x\right)^{-\frac{1}{2}} [\cos x\, Ci(2x) + \sin x\, Si(2x)],$$

$$\left[\frac{\partial Y_\nu(z)}{\partial\nu}\right]_{\nu=\frac{1}{2}} = \left(\frac{1}{2}\pi x\right)^{-\frac{1}{2}} \{\cos x\, Ci(2x) + \sin x\,[Si(2x) - \pi]\},$$

$$\left[\frac{\partial Y_\nu(z)}{\partial\nu}\right]_{\nu=-\frac{1}{2}} = -\left(\frac{1}{2}\pi x\right)^{-\frac{1}{2}} \{\sin x\, Ci(2x) - \cos x\,[Si(2x) - \pi]\},$$

$$\left[\frac{\partial I_\nu(z)}{\partial \nu}\right]_{\nu=\pm\frac{1}{2}} = (2\pi x)^{-\frac{1}{2}} [e^x Ei(-2x) \mp e^{-x} \overline{Ei}(2x)],$$

$$\left[\frac{\partial K_\nu(z)}{\partial \nu}\right]_{\nu=\pm\frac{1}{2}} = \mp \left(\frac{\frac{1}{2}\pi}{x}\right)^{\frac{1}{2}} e^x Ei(-2x).$$

3.4 The Airy functions and related functions

The Airy functions are linear combinations of Bessel functions of order $\pm\frac{1}{3}$. We define

$$Ai(z) = \frac{1}{\pi}\left(\frac{z}{3}\right)^{\frac{1}{2}} K_{\frac{1}{3}}\left(\frac{2}{3} z^{\frac{3}{2}}\right),$$

$$Bi(z) = \left(\frac{z}{3}\right)^{\frac{1}{2}} \left[I_{-\frac{1}{3}}\left(\frac{2}{3} z^{\frac{3}{2}}\right) + I_{\frac{1}{3}}\left(\frac{2}{3} z^{\frac{3}{2}}\right)\right].$$

They are entire functions of z.

Also

$$Ai(-z) = \frac{1}{3} z^{\frac{1}{2}} \left[J_{\frac{1}{3}}\left(\frac{2}{3} z^{\frac{3}{2}}\right) + J_{-\frac{1}{3}}\left(\frac{2}{3} z^{\frac{3}{2}}\right)\right],$$

$$Bi(-z) = \left(\frac{z}{3}\right)^{\frac{1}{2}} \left[J_{-\frac{1}{3}}\left(\frac{2}{3} z^{\frac{3}{2}}\right) - J_{\frac{1}{3}}\left(\frac{2}{3} z^{\frac{3}{2}}\right)\right].$$

A number of integrals involving trigonometric or (and) exponential functions can be expressed in terms of Airy's or related functions. For instance:

$$\int_0^\infty \cos(a^3 t^3 \pm xt)\, dt = \frac{1}{3a}\left(\frac{x}{a}\right)^{\frac{1}{2}} K_{\frac{1}{3}}\left[2\left(\frac{x}{3a}\right)^{\frac{3}{2}}\right] = 3^{-\frac{1}{3}} \frac{\pi}{a} Ai\left(\pm 3^{-\frac{1}{3}} \frac{x}{a}\right),$$

$$\int_0^\infty [e^{-a^3t^3 \pm zt} + \sin(a^3 t^3 \pm zt)]\, dt = 3^{-\frac{1}{3}} \frac{\pi}{a} Bi\left(\pm 3^{-\frac{1}{3}} \frac{z}{a}\right),$$

a and x are positive numbers and z is arbitrary complex.

Also

$$\int_0^\infty \sin(a^3 t^3 - xt)\, dt = \frac{\pi}{9a}\left(\frac{x}{a}\right)^{\frac{1}{2}} \left[I_{\frac{1}{3}}(v) + I_{-\frac{1}{3}}(v) + 2i J_{\frac{1}{3}}(iv) - 2i J_{-\frac{1}{3}}(iv)\right],$$

$$\int_0^\infty \sin(a^3t^3 + xt)\,dt = \frac{\pi}{9a}\left(\frac{x}{a}\right)^{\frac{1}{2}}\Big[J_{-\frac{1}{3}}(v) - J_{\frac{1}{3}}(v) - 2\boldsymbol{J}_{\frac{1}{3}}(v) + 2\boldsymbol{J}_{-\frac{1}{3}}(v)\Big],$$

$$\int_0^\infty e^{-a^3t^3 - zt}\,dt = z^{-1} v\, S_{0,\frac{1}{3}}(v) \quad \text{with} \quad v = 2\left(\frac{z}{3a}\right)^{\frac{3}{2}}.$$

Again, $a, x > 0$, z arbitrary complex. The Airy functions $Ai(z)$ and $Bi(z)$ are linearly independent solutions of the differential equation

$$\frac{d^2w}{dz^2} - zw = 0.$$

So are

$$Ai(z),\ Ai\left(ze^{\pm i2\frac{\pi}{3}}\right) \text{ with the Wronskians}$$

$$W[Ai(z), Bi(z)] = \pi^{-1},$$

$$W\left[Ai(z), Ai\left(ze^{\pm i2\frac{\pi}{3}}\right)\right] = \frac{1}{2}\pi^{-1}e^{\mp i\frac{\pi}{6}},$$

$$W\left[Ai\left(ze^{i2\frac{\pi}{3}}\right), Ai\left(ze^{-i2\frac{\pi}{3}}\right)\right] = \frac{1}{2}i\pi^{-1}.$$

Also

$$Bi(z) = e^{i\frac{\pi}{6}}Ai\left(ze^{i2\frac{\pi}{3}}\right) + e^{-i\frac{\pi}{6}}Ai\left(ze^{-i2\frac{\pi}{3}}\right),$$

$$Ai(z) + e^{i2\frac{\pi}{3}}Ai\left(ze^{i2\frac{\pi}{3}}\right) + e^{-i2\frac{\pi}{3}}Ai\left(ze^{-i2\frac{\pi}{3}}\right) = 0,$$

$$Ai\left(ze^{\pm i2\frac{\pi}{3}}\right) = \frac{1}{2}e^{\pm i\frac{\pi}{3}}[Ai(z) \mp i\,Bi(z)].$$

The asymptotic expansions of these functions for large $|z|$ are readily available from the general expressions [3.14.1]. $Ai(z)$, $Ai'(z)$ have zeros on the negative real axis only. $Bi(z)$, $Bi'(z)$ have zeros on the negative real axis and in the sector $\frac{1}{3}\pi < |\arg z| < \frac{1}{2}\pi$.

Products of Airy functions

The functions $Ai^2(z)$, $Ai(z)\,Bi(z)$, $Bi^2(z)$ are linearly independent solutions of

$$\frac{d^3w}{dz^3} - 4z\frac{dw}{dz} - 2w = 0$$

with the Wronskian

$$W[Ai^2(z), Ai(z)\,Bi(z), Bi^2(z)] = 2\pi^{-3}.$$

Generalizations

Denote by

$$T_n(t, x) = t^n {}_2F_1\left(-\frac{1}{2}n, \frac{1}{2} - \frac{1}{2}n; 1-n; -4xt^{-2}\right)$$

or

$$T_2(t, x) = t^2 + 2x,$$
$$T_3(t, x) = t^3 + 3tx,$$
$$T_4(t, x) = t^4 + 4t^2x + 2x^2.$$

Then for $x > 0$, $n = 1, 2, 3, \ldots$; $m = 1, 2, 3, \ldots$

$$\int_0^\infty \cos\left[T_n(t, -x)\right] dt = \frac{\pi}{2n} x^{\frac{1}{2}} \operatorname{cosec}\left(\frac{\pi}{2n}\right)\left[J_{\frac{1}{n}}\left(2x^{\frac{1}{2}n}\right) + J_{-\frac{1}{n}}\left(2x^{\frac{1}{2}n}\right)\right],$$

$$\int_0^\infty \cos\left[T_{2m}(t, x)\right] dt = \frac{\pi}{4m} x^{\frac{1}{2}} \operatorname{cosec}\left(\frac{\pi}{4m}\right)\left[J_{-\frac{1}{2m}}(2x^m) - J_{\frac{1}{2m}}(2x^m)\right],$$

$$\int_0^\infty \cos\left[T_{2m+1}(t, x)\right] dt = \left(m + \frac{1}{2}\right)^{-1}$$
$$\times\, x^{\frac{1}{2}} \cos\left[\frac{\pi}{2}(2m+1)^{-1}\right] K_{\frac{1}{2m+1}}\left(2x^{m+\frac{1}{2}}\right),$$
$$m = 1, 2, \ldots$$

The right hand side of the equations above represent again entire functions of x.

3.5 Differential equations and a power series expansion for the product of two Bessel functions

Let Z_ν be an arbitrary solution of Bessel's differential equation; then a solution of

$$z^2 \frac{d^2w}{dz^2} + (1 - 2\alpha)\, z \frac{dw}{dz} + \left[(\beta\gamma z^\gamma)^2 + \alpha^2 - \nu^2\gamma^2\right] w = 0$$

is

$$w = z^\alpha Z_\nu(\beta z^\gamma).$$

Special cases of this equation are

$$\frac{d^2w}{dz^2} + c^2 z^{2q-2} w = 0, \text{ solution } w = z^{\frac{1}{2}} Z_{\frac{1}{2q}}\left(\frac{c}{q} z^q\right),$$

$$\frac{d^2w}{dz^2} + (e^{2z} - \nu^2)\, w = 0, \text{ solution } w = Z_\nu(e^z),$$

$$\frac{d^2w}{dz^2} + z^{-4}\left(e^{\frac{2}{z}} - \nu^2\right) w = 0, \text{ solution } w = z Z_\nu\left(e^{\frac{1}{z}}\right).$$

The functions $J_\nu(z)$, $Y_\nu(z)$, $H_\nu^{(1)}(z)$, $H_\nu^{(2)}(z)$, $I_\nu(z)$, $K_\nu(z)$ are solutions of the differential equation of the fourth order

$$\frac{d^4 w}{dz^4} + \frac{2}{z}\frac{d^3 w}{dz^3} - \frac{(2\nu^2+1)}{z^2}\frac{d^2 w}{dz^2} + \frac{(2\nu^2+1)}{z^3}\frac{dw}{dz} + \left(\frac{\nu^4 - 4\nu^2}{z^4} - 1\right) w = 0.$$

The functions $J'_\nu(z)$ and $b J_\nu(z) + a z J'_\nu(z)$ satisfy the following differential equations respectively

$$z^2(z^2-\nu^2)\frac{d^2 w}{dz^2} + z(z^2 - 3\nu^2)\frac{dw}{dz} + [(z^2-\nu^2)^2 - (z^2+\nu^2)]\, w = 0,$$

$$z^2[a^2(z^2-\nu^2)+b^2]\frac{d^2 w}{dz^2} - z[a^2(z^2+\nu^2) - b^2]\frac{dw}{dz}$$

$$+ [a^2(z^2-\nu^2)^2 + 2abz^2 + b^2(z^2-\nu^2)]\, w = 0.$$

A general solution of the inhomogeneous Bessel differential equation

$$z^2\frac{d^2 w}{dz^2} + z\frac{dw}{dz} + (z^2-\nu^2)\, w = f(z)$$

has the form

$$w = A w_1(z) + B w_2(z) + u(z),$$

where w_1 and w_2 are linearly independent solutions of the homogeneous equation and

$$W\, u(z) = -w_1(z)\int_{z_0}^{z} t^{-1} w_2(t)\, f(t)\, dt + w_2(z)\int_{z_0}^{z} t^{-1} w_1(t)\, f(t)\, dt$$

and W is the Wronskian of w_1 and w_2.

The Ricatti equation

$$\frac{dw}{dz} + a w^2 = b z^m$$

has the solution

$$w = u^{-1}\frac{du}{dz} \quad \text{where} \quad u = z^{\frac{1}{2}}\, Z_{\frac{1}{m+2}}\left[\frac{2i(ab)^{\frac{1}{2}}}{m+2} z^{1+\frac{1}{2}m}\right].$$

Power series expansion of the product of two Bessel functions

$$\Gamma(\nu+1)\, J_\nu(\beta z)\, J_\mu(\alpha z) = \left(\frac{1}{2}\alpha z\right)^\mu \left(\frac{1}{2}\beta z\right)^\nu$$

$$\times \sum_{m=0}^{\infty} \frac{(-1)^m \left(\frac{1}{2}\alpha z\right)^{2m}}{m!\,\Gamma(\mu+m+1)}\, {}_2F_1\left(-m, -\mu-m; \nu+1; \frac{\beta^2}{\alpha^2}\right).$$

Special cases are

$$\Gamma(\nu+1)\,\Gamma(\mu+1)\,J_\nu(z)\,J_\mu(z)$$

$$=\left(\frac{1}{2}z\right)^{\nu+\mu}{}_2F_3\left(\frac{1}{2}+\frac{1}{2}\nu+\frac{1}{2}\mu,\,1+\frac{1}{2}\nu+\frac{1}{2}\mu;\right.$$

$$\left.1+\nu,\,1+\mu,\,1+\nu+\mu;\,-z^2\right)$$

and

$$e^{\pm iz}J_\nu(z)=\pi^{-\frac{1}{2}}(2z)^\nu\sum_{n=0}^{\infty}\frac{\Gamma\left(\nu+n+\frac{1}{2}\right)(\pm 2iz)^n}{n!\,\Gamma(2\nu+n+1)}.$$

3.6 Integral representations for Bessel, Neumann and Hankel functions

3.6.1 Integer order

$$J_n(z)=\frac{1}{2\pi}\int_{-\pi}^{\pi}e^{iz\cos t}\,e^{in\left(t-\frac{1}{2}\pi\right)}\,dt=\frac{i^{-n}}{\pi}\int_0^\pi e^{iz\cos t}\cos(nt)\,dt$$

$$=\frac{1}{\pi}\int_0^\pi\cos(z\sin t-nt)\,dt,$$

$$J_{2n}(z)=\frac{2}{\pi}\int_0^{\frac{1}{2}\pi}\cos(z\sin t)\cos(2nt)\,dt$$

$$=\frac{2}{\pi}(-1)^n\int_0^{\frac{1}{2}\pi}\cos(z\cos t)\cos(2nt)\,dt,$$

$$J_{2n+1}(z)=\frac{2}{\pi}\int_0^{\frac{1}{2}\pi}\sin(z\sin t)\sin[(2n+1)t]\,dt.$$

(In these formulas, $n=0,1,2,\ldots$).

3.6.2 Formulas of the Poisson type, Mehler — Sonine, Schlaefli's and Heine's formulas

$$\Gamma\left(\frac{1}{2}+\nu\right)J_\nu(z)=2\pi^{-\frac{1}{2}}\left(\frac{1}{2}z\right)^\nu\int_0^{\frac{1}{2}\pi}\cos(z\cos t)\sin^{2\nu}t\,dt$$

$$=2\pi^{-\frac{1}{2}}\left(\frac{1}{2}z\right)^\nu\int_0^1(1-t^2)^{\nu-\frac{1}{2}}\cos(zt)\,dt,$$

$$Re\,\nu>-\frac{1}{2},$$

$$\Gamma\left(\frac{1}{2}+\nu\right) J_{-\nu}(z) = 2\pi^{-\frac{1}{2}}\left(\frac{1}{2}z\right)^{\nu}\left[\int_0^1 (1-t^2)\cos(zt+\nu\pi)\,dt\right.$$

$$\left.+\sin(\pi\nu)\int_0^\infty (1+t^2)^{\nu-\frac{1}{2}} e^{-zt}\,dt\right],$$

$$Re\,\nu > -\frac{1}{2}, \quad Re\,z > 0,$$

Generalization of Poisson's formula

$$z^{-\nu} J_{\nu+n}(z) = 2^{-\nu}\pi^{-\frac{1}{2}}(-i)^n\, n!\,\Gamma(2\nu)\left[\Gamma\left(\frac{1}{2}+\nu\right)\Gamma(2\nu+n)\right]^{-1}$$

$$\times\int_0^\pi e^{iz\cos t}\, C_n^\nu(\cos t)\sin^{2\nu} t\,dt,$$

$$Re\,\nu > -\frac{1}{2}, \quad n = 0, 1, 2, \ldots,$$

$$\Gamma\left(\frac{1}{2}+\nu\right) Y_\nu(z) = 2\pi^{-\frac{1}{2}}\left(\frac{1}{2}z\right)^{\nu}\left[\int_0^1 (1-t^2)^{\nu-\frac{1}{2}}\sin(zt)\,dt\right.$$

$$\left.-\int_0^\infty e^{-zt}(1+t^2)^{\nu-\frac{1}{2}}\,dt\right],$$

$$Re\,\nu > -\frac{1}{2}, \quad Re\,z > 0,$$

$$\Gamma\left(\frac{1}{2}+\nu\right) H_\nu^{(1)}(z) = \left(\frac{1}{2}\pi z\right)^{-\frac{1}{2}} e^{i\left(z-\frac{1}{2}\nu\pi-\frac{1}{4}\pi\right)}$$

$$\times\int_0^{\infty e^{i\delta}} e^{-t}\, t^{\nu-\frac{1}{2}}\left(1+i\frac{t}{2z}\right)^{\nu-\frac{1}{2}} dt,$$

$$Re\,\nu > -\frac{1}{2}, \quad |\delta| < \frac{1}{2}\pi, \quad \delta-\frac{1}{2}\pi < \arg z < \delta+\frac{3}{2}\pi,$$

$$\Gamma\left(\frac{1}{2}+\nu\right) H_\nu^{(2)}(z) = \left(\frac{1}{2}\pi z\right)^{-\frac{1}{2}} e^{-i\left(z-\frac{1}{2}\nu\pi-\frac{1}{4}\pi\right)}$$

$$\times\int_0^{\infty e^{i\delta}} e^{-t}\, t^{\nu-\frac{1}{2}}\left(1-i\frac{t}{2z}\right)^{\nu-\frac{1}{2}} dt,$$

$$Re\,\nu > -\frac{1}{2}, \quad |\delta| < \frac{1}{2}\pi, \quad -3\frac{\pi}{2}+\delta < \arg z < \frac{1}{2}\pi+\delta.$$

Mehler-Sonine formulas

$$\Gamma\left(\frac{1}{2}-\nu\right) J_\nu(x) = 2\pi^{-\frac{1}{2}}\left(\frac{1}{2}x\right)^{-\nu}\int_1^\infty (t^2-1)^{-\nu-\frac{1}{2}}\sin(xt)\,dt,$$

$$\Gamma\left(\frac{1}{2}-\nu\right) Y_\nu(x) = -2\pi^{-\frac{1}{2}}\left(\frac{1}{2}x\right)^{-\nu}\int_1^\infty (t^2-1)^{-\nu-\frac{1}{2}}\cos(xt)\,dt.$$

(In both formulas $x > 0$ (real), $-\frac{1}{2} < Re\,\nu < \frac{1}{2}$)

$$J_\nu(x) = \frac{2}{\pi} \int_0^\infty \sin\left(x \cosh t - \frac{1}{2}\pi\nu\right) \cosh(\nu t)\, dt,$$

$$Y_\nu(x) = -\frac{2}{\pi} \int_0^\infty \cos\left(x \cosh t - \frac{1}{2}\pi\nu\right) \cosh(\nu t)\, dt.$$

(In both formulas $x > 0$, (real) and $-1 < Re\,\nu < 1$).

Schlaefli's formulas

$$\pi J_\nu(z) = \int_0^\pi \cos(z \sin t - \nu t)\, dt - \sin(\pi\nu) \int_0^\infty e^{-z\sinh t} e^{-\nu t}\, dt,$$

$$\pi Y_\nu(z) = \int_0^\pi \sin(z \sin t - \nu t) dt - \int_0^\infty e^{-z\sinh t} [e^{\nu t} + \cos(\pi\nu) e^{-\nu t}]\, dt.$$

(In both formulas $Re\,z > 0$).

Heine's formulas

$$\pi J_\nu(z) = e^{i\frac{\pi}{2}\nu}\left[\int_0^\pi e^{-iz\cos t} \cos(\nu t)\, dt - \sin(\pi\nu) \int_0^\infty e^{iz\cosh t - \nu t}\, dt\right],$$
$$0 < \arg z < \pi,$$

$$\pi J_\nu(z) = e^{-i\frac{\pi}{2}\nu}\left[\int_0^\pi e^{iz\cos t} \cos(\nu t)\, dt - \sin(\pi\nu) \int_0^\infty e^{-iz\cosh t - \nu t}\, dt\right],$$
$$-\pi < \arg z < 0,$$

$$\pi H_\nu^{(1)}(z) = -2ie^{-i\frac{\pi}{2}\nu} \int_0^\infty e^{iz\cosh t} \cosh(\nu t) d t,$$
$$0 < \arg z < \pi,$$

$$\pi H_\nu^{(2)}(z) = 2ie^{i\frac{\pi}{2}\nu} \int_0^\infty e^{-iz\cosh t} \cosh(\nu t)\, dt,$$
$$-\pi < \arg z < 0.$$

Generalized "Schlaefli" formulas

$$\pi\left(\frac{z-\zeta}{z+\zeta}\right)^{-\frac{1}{2}\nu} J_\nu\left[(z^2 - \zeta^2)^{\frac{1}{2}}\right]$$
$$= \int_0^\pi e^{\zeta\cos t} \cos(z \sin t - \nu t)\, dt - \sin(\pi\nu) \int_0^\infty e^{-z\sinh t - \zeta\cosh t - \nu t}\, dt,$$
$$Re\,(z + \zeta) > 0,$$

$$\pi\left(\frac{z-\zeta}{z+\zeta}\right)^{-\frac{1}{2}\nu} Y_\nu\left[(z^2 - \zeta^2)^{\frac{1}{2}}\right] = \int_0^\pi e^{\zeta\cos t} \sin(z \sin t - \nu t)\, dt$$
$$- \int_0^\infty [e^{\nu t + \zeta\cosh t} + e^{-\nu t - \zeta\cosh t} \cos(\pi\nu)]\, e^{-z\sinh t}\, dt,$$
$$Re\,(z \pm \zeta) > 0.$$

Generalized "Heine" formulas

$$\pi\left(\frac{z+\zeta}{z-\zeta}\right)^{\frac{1}{2}\nu} H_\nu^{(1)}\left[(z^2-\zeta^2)^{\frac{1}{2}}\right]$$

$$= -i\, e^{-i\frac{1}{2}\pi\nu} \int_{-\infty}^{\infty} e^{iz\cosh t + i\zeta\sinh t - \nu t}\, dt,$$

$$\operatorname{Im}(z \pm \zeta) > 0,$$

$$\pi\left(\frac{z+\zeta}{z-\zeta}\right)^{\frac{1}{2}\nu} H_\nu^{(2)}\left[(z^2-\zeta^2)^{\frac{1}{2}}\right]$$

$$= i\, e^{i\frac{1}{2}\pi\nu} \int_{-\infty}^{\infty} e^{-iz\cosh t - i\zeta\sinh t - \nu t}\, dt,$$

$$\operatorname{Im}(z \pm \zeta) < 0.$$

3.6.3 Mellin-Barnes integrals

$$4\pi J_\nu(x) = \int_{c-i\infty}^{c+i\infty} \left(\frac{1}{2}x\right)^{-s} \Gamma\left(\frac{1}{2}\nu + \frac{1}{2}s\right)\left[\Gamma\left(1+\frac{1}{2}\nu-\frac{1}{2}s\right)\right]^{-1} ds,$$

$$x > 0, \quad -Re\,\nu < c < 1,$$

$$2\pi i J_\nu(x) = \int_{-i\infty}^{i\infty} \Gamma(-s)\, [\Gamma(\nu+s+1)]^{-1} \left(\frac{1}{2}x\right)^{\nu+2s} ds,$$

$$x > 0, \quad Re\,\nu > 0,$$

$$2\pi^2 H_\nu^{(1)}(z) = -\, e^{-i\frac{1}{2}\nu\pi} \int_{-c-i\infty}^{-c+i\infty} \Gamma(-s)\, \Gamma(-\nu-s) \left(-\frac{1}{2}iz\right)^{\nu+2s} ds,$$

$$-\frac{\pi}{2} < \arg(-iz) < \frac{\pi}{2},$$

$$2\pi^2 H_\nu^{(2)}(z) = e^{i\frac{1}{2}\nu\pi} \int_{-c-i\infty}^{-c+i\infty} \Gamma(-s)\, \Gamma(-\nu-s) \left(\frac{1}{2}iz\right)^{\nu+2s} ds,$$

$$-\frac{\pi}{2} < \arg(iz) < \frac{\pi}{2}.$$

In both formulas c is any positive number exceeding $Re\,\nu$

$$H_\nu^{(1)}(z) = -\,\pi^{-\frac{5}{2}} e^{i(z-\nu\pi)} \cos(\pi\nu)\, (2z)^\nu \int_{-i\infty}^{i\infty} \Gamma(-s)\, \Gamma(-2\nu-s)$$

$$\times\, \Gamma\left(\nu+s+\frac{1}{2}\right)(-2iz)^s\, ds,$$

$$-\frac{3\pi}{2} < \arg(-iz) < \frac{3\pi}{2}, \quad 2\nu \text{ not an odd integer},$$

$$H_\nu^{(2)}(z) = \pi^{-\frac{5}{2}} e^{-i(z-\nu\pi)} \cos(\pi\nu)\,(2z)^\nu \int_{-i\infty}^{i\infty} \Gamma(-s)\,\Gamma(-2\nu - s)$$
$$\times\, \Gamma\left(\nu + s + \frac{1}{2}\right)(2iz)^s\, ds,$$
$$-\frac{3\pi}{2} < \arg(iz) < \frac{3\pi}{2}, \quad 2\nu \text{ not an odd integer}.$$

3.6.4 Loop integrals and Sommerfeld's representations

$$2\pi J_\nu(\alpha z) = -iz^\nu \int_{-\infty}^{(0+)} e^{\frac{1}{2}\alpha\left(t-\frac{z^2}{t}\right)} t^{-\nu-1}\, dt,$$
$$Re\,\alpha > 0,\ |\arg t| \leq \pi.$$

Loop integral of the Poisson type

$$2\pi J_\nu(z) = \pi^{-\frac{1}{2}} \Gamma\left(\frac{1}{2} + \nu\right) e^{i3\pi\nu} \left(\frac{1}{2} z\right)^{-\nu} \int_{\infty e^{i\delta}}^{(-1+,1+)} e^{izt} (t^2 - 1)^{-\nu-\frac{1}{2}}\, dt,$$
$$\delta \leq \arg t \leq 2\pi + \delta,\ -\delta < \arg z < \pi - \delta.$$

Sommerfeld's integrals

$$\pi H_\nu^{(1)}(z) = \int_{C_1} e^{iz\cos t}\, e^{i\nu\left(t - \frac{1}{2}\pi\right)}\, dt,$$
$$\pi H_\nu^{(2)}(z) = \int_{C_2} e^{iz\cos t}\, e^{i\nu\left(t - \frac{1}{2}\pi\right)}\, dt,$$
$$2\pi J_\nu(z) = \int_{C_3} e^{iz\cos t}\, e^{i\nu\left(t - \frac{1}{2}\pi\right)}\, dt.$$

The contours of integration C_1, C_2, C_3 in the complex t plane are defined as follows (η is any number between 0 and π)

C_1: From $-\eta + i\infty$ to $\eta - i\infty$.

C_2: From $\eta - i\infty$ to $2\pi - \eta + i\infty$.

C_3: From $-\eta + i\infty$ to $2\pi - \eta + i\infty$

and the above formulas are valid in the domain $-\eta < \arg z < \pi - \eta$

$$\pi H_\nu^{(1)}(z) = -i \int_{-\infty}^{\infty + i\pi} e^{z\sinh t - \nu t}\, dt,$$
$$\pi H_\nu^{(2)}(z) = i \int_{-\infty}^{\infty - i\pi} e^{z\sinh t - \nu t}\, dt,$$
$$2\pi J_\nu(z) = -i \int_{\infty - i\pi}^{\infty + i\pi} e^{z\sinh t - \nu t}\, dt.$$

3.6.5 Miscellaneous representations

$$H_\nu^{(1)}(az) = -2ie^{-i\nu\pi}\pi^{-\frac{1}{2}} \frac{\left(\frac{\frac{1}{2}z}{a}\right)^\nu}{\Gamma\left(\frac{1}{2}+\nu\right)} \int_0^\infty t^{2\nu}(t^2+a^2)^{-\frac{1}{2}} e^{iz(t^2+a^2)^{\frac{1}{2}}}\,dt,$$

$$\operatorname{Im} z > 0,\ Re\,\nu > -\frac{1}{2},$$

$$H_\nu^{(2)}(az) = 2ie^{i\nu\pi}\pi^{-\frac{1}{2}} \frac{\left(\frac{\frac{1}{2}z}{a}\right)^\nu}{\Gamma\left(\frac{1}{2}+\nu\right)} \int_0^\infty t^{2\nu}(t^2+a^2)^{-\frac{1}{2}} e^{-iz(t^2+a^2)^{\frac{1}{2}}}\,dt,$$

$$\operatorname{Im} z < 0,\ Re\,\nu > -\frac{1}{2},$$

$$H_\nu^{(1)}(az) = -i2\pi^{-\frac{1}{2}} \frac{\left(\frac{\frac{1}{2}z}{a}\right)^{-\nu}}{\Gamma\left(\frac{1}{2}-\nu\right)} \int_0^\infty t^{-2\nu}(t^2+a^2)^{-\frac{1}{2}} e^{iz(t^2+a^2)^{\frac{1}{2}}}\,dt,$$

$$\operatorname{Im} z > 0,\ Re\,\nu < \frac{1}{2},$$

$$H_\nu^{(2)}(az) = i2\pi^{-\frac{1}{2}} \frac{\left(\frac{\frac{1}{2}z}{a}\right)^{-\nu}}{\Gamma\left(\frac{1}{2}-\nu\right)} \int_0^\infty t^{-2\nu}(t^2+a^2)^{-\frac{1}{2}} e^{-iz(t^2+a^2)^{\frac{1}{2}}}\,dt,$$

$$\operatorname{Im} z < 0,\ Re\,\nu < \frac{1}{2},$$

$$H_\nu^{(1)}\left(z\sqrt{\alpha^2}\right) = -\frac{i}{\pi} e^{-i\frac{1}{2}\nu\pi} (\alpha^2)^{\frac{1}{2}\nu} \int_0^\infty e^{i\frac{1}{2}z\left(t+\frac{\alpha^2}{t}\right)} t^{-\nu-1}\,dt,$$

$$\operatorname{Im} z > 0,\ \operatorname{Im}(\alpha^2 z) > 0.$$

3.7 Integral representations for the modified Bessel functions

$$\Gamma\left(\frac{1}{2}+\nu\right) I_\nu(z) = \pi^{-\frac{1}{2}}\left(\frac{1}{2}z\right)^\nu \int_{-1}^{1} e^{-zt}(1-t^2)^{\nu-\frac{1}{2}}\,dt,$$

$$Re\,\nu > -\frac{1}{2},$$

$$\Gamma\left(\frac{1}{2}+\nu\right) I_{-\nu}(z) = 2\pi^{-\frac{1}{2}}\left(\frac{1}{2}z\right)^\nu \left[\frac{1}{2}\int_{-1}^{1} e^{zt}(1-t^2)^{\nu-\frac{1}{2}}\,dt + \sin(\pi\nu)\int_1^\infty e^{-zt}(t^2-1)^{\nu-\frac{1}{2}}\,dt\right],$$

$$Re\,\nu > -\frac{1}{2}, \quad Re\,z > 0,$$

$$\Gamma\left(\frac{1}{2}+\nu\right) K_\nu(z) = \pi^{\frac{1}{2}} \left(\frac{1}{2}z\right)^\nu \int_0^\infty e^{-z\cosh t} (\sinh t)^{2\nu}\, dt,$$

$$Re\,z > 0, \quad Re\,\nu > -\frac{1}{2},$$

$$K_\nu(z) = \int_0^\infty e^{-z\cosh t} \cosh(\nu t)\, dt,$$

$$Re\,z > 0,$$

$$z^\alpha K_\nu(z) = \left(\frac{1}{2}\pi\right)^{\frac{1}{2}} \int_1^\infty (t^2-1)^{-\frac{1}{2}\alpha-\frac{1}{4}} P_{\nu-\frac{1}{2}}^{\alpha+\frac{1}{2}}(t)\, e^{-zt}\, dt,$$

$$Re\,\alpha < \frac{1}{2}, \quad Re\,z > 0,$$

$$K_\nu(x) = \pi^{-\frac{1}{2}} \left(\frac{1}{2}x\right)^{-\nu} \Gamma\left(\frac{1}{2}+\nu\right) \int_0^\infty (\cosh t)^{-2\nu} \cos(x \sinh t)\, dt,$$

$$x > 0, \quad Re\,\nu > -\frac{1}{2},$$

$$2\left(\frac{z}{\zeta}\right)^{\frac{1}{2}\nu} K_\nu\left[2(z\zeta)^{\frac{1}{2}}\right] = \int_0^\infty e^{-zt}\, e^{-\frac{\zeta}{t}}\, t^{-\nu-1}\, dt,$$

$$Re\,z > 0, \quad Re\,\zeta > 0,$$

$$K_\nu\left(z\sqrt{\alpha^2}\right) = \frac{1}{2} e^{i\frac{1}{2}\nu\pi} (\alpha^2)^{\frac{1}{2}\nu} \int_0^\infty e^{i\frac{1}{2}z\left(t-\frac{\alpha^2}{t}\right)} t^{-\nu-1}\, dt,$$

$$\operatorname{Im} z > 0, \quad \operatorname{Im}(\alpha^2 z) < 0,$$

$$K_\nu(x) \cos\left(\frac{1}{2}\pi\nu\right) = \int_0^\infty \cos(x\sinh t) \cosh(\nu t)\, dt,$$

$$x > 0, \quad -1 < Re\,\nu < 1,$$

$$K_\nu(x) \sin\left(\frac{1}{2}\pi\nu\right) = \int_0^\infty \sin(x\sinh t) \sinh(\nu t)\, dt,$$

$$x > 0, \quad -1 < Re\,\nu < 1,$$

$$K_\nu(az) = \pi^{-\frac{1}{2}} \left(\frac{2z}{a}\right)^\nu \Gamma\left(\frac{1}{2}+\nu\right) \int_0^\infty (t^2+z^2)^{-\nu-\frac{1}{2}} \cos(at)\, dt,$$

$$Re\,\nu > -\frac{1}{2}, \quad Re\,z > 0, \; a > 0,$$

$$\left(\frac{z-\zeta}{z+\zeta}\right)^{\frac{1}{2}\nu} K_\nu\left[a(z^2-\zeta^2)^{\frac{1}{2}}\right] = \frac{1}{2}\int_{-\infty}^{\infty} e^{-z\cosh t-\zeta\sinh t+\nu t}\,dt,$$

$$Re\,(z \pm \zeta) > 0,$$

$$K_\nu\left[(z^2+\zeta^2)^{\frac{1}{2}}\right]\cos\left(\nu\arctan\frac{\zeta}{z}\right) = \int_0^\infty e^{-z\cosh t}\cos(\zeta\sinh t)\cosh(\nu t)\,dt,$$

$$K_\nu\left[(z^2+\zeta^2)^{\frac{1}{2}}\right]\sin\left(\nu\arctan\frac{\zeta}{z}\right) = \int_0^\infty e^{-z\cosh t}\sin(\zeta\sinh t)\sinh(\nu t)\,dt,$$

$Re(z \pm \zeta i) > 0$ in both formulas.

Mellin-Barnes type

$$4\pi^2 K_\nu(z) = -i\left(\frac{\frac{1}{2}\pi}{z}\right)^{\frac{1}{2}} e^{-z}\cos(\pi\nu)\int_{-i\infty}^{i\infty}\Gamma(s)\,\Gamma\left(\frac{1}{2}-s-\nu\right)$$

$$\times\,\Gamma\left(\frac{1}{2}-s+\nu\right)(2z)^s\,ds,$$

$$-\frac{3\pi}{2} < \arg z < \frac{3\pi}{2},\quad 2\nu \text{ not an odd integer.}$$

Sommerfeld type

$$K_\nu(z) = \frac{1}{2}i\int_{C_1} e^{-z\cos t}\,e^{i\nu t}\,dt,$$

$$-\frac{1}{2}\pi-\eta < \arg z < \frac{1}{2}\pi-\eta$$

$$= -\frac{1}{2}i\int_{C_2} e^{z\cos t}\,e^{i\nu(t-\pi)}\,dt,$$

$$\frac{\pi}{2}-\eta < \arg z < \frac{3\pi}{2}-\eta.$$

(For the description of the contours C_1 and C_2 see section before.)

3.8 Integrals involving Bessel functions

3.8.1 Indefinite Integrals

$$\int z^{\nu+1}J_\nu(z)\,dz = z^{\nu+1}J_{\nu+1}(z),$$

$$\int z^{-\nu+1}J_\nu(z)\,dz = -z^{-\nu+1}J_{\nu-1}(z),$$

$$\int J_\nu(z)\,dz = 2\sum_{m=0}^{n-1} J_{\nu+2m+1}(z) + \int J_{\nu+2n}(z)\,dz.$$

The same formulas hold for the Neumann and for the Hankel functions.

$$\int J_\nu(z)\,dz = 2\sum_{m=0}^{\infty} J_{\nu+2m+1}(z),$$

$$\int z^\nu J_\nu(z)\,dz = 2^{\nu-1}\pi^{\frac{1}{2}}\,\Gamma\left(\frac{1}{2}+\nu\right) z\,[J_\nu(z)\,\boldsymbol{H}_{\nu-1}(z) - \boldsymbol{H}_\nu(z)\,J_{\nu-1}(z)],$$

$$\int z^\mu J_\nu(z)\,dz = (\mu+\nu-1)\,z J_\nu(z)\,S_{\mu-1,\nu-1}(z) - z J_{\nu-1}(z)\,S_{\mu,\nu}(z),$$

$$\int z^\nu Y_\nu(z)\,dz = 2^{\nu-1}\pi^{\frac{1}{2}}\,\Gamma\left(\frac{1}{2}+\nu\right) z\,[Y_\nu(z)\,\boldsymbol{H}_{\nu-1}(z) - \boldsymbol{H}_\nu(z)\,Y_{\nu-1}(z)],$$

$$\int z^\mu Y_\nu(z)\,dz = (\mu+\nu-1)\,z Y_\nu(z)\,S_{\mu-1,\nu-1}(z) - z Y_{\nu-1}(z)\,S_{\mu,\nu}(z),$$

$$\int z^{\nu+1} I_\nu(z)\,dz = z^{\nu+1} I_{\nu+1}(z),$$

$$\int z^{1-\nu} I_\nu(z)\,dz = z^{-\nu+1} I_{\nu-1}(z),$$

$$\int z^\nu I_\nu(z)\,dz = 2^{\nu-1}\pi^{\frac{1}{2}}\,\Gamma\left(\frac{1}{2}+\nu\right) z\,[I_\nu(z)\,\boldsymbol{L}_{\nu-1}(z) - \boldsymbol{L}_\nu(z)\,I_{\nu-1}(z)],$$

$$\int z^\mu I_\nu(z)\,dz = e^{-i\frac{\pi}{2}\mu}\,[(\mu+\nu-1)\,z I_\nu(z)\,S_{\mu-1,\nu-1}(iz) + iz I_{\nu-1}(z)\,S_{\mu,\nu}(iz)],$$

$$\int z^{\nu+1} K_\nu(z)\,dz = -z^{\nu+1} K_{\nu+1}(z),$$

$$\int z^{-\nu+1} K_\nu(z)\,dz = -z^{-\nu+1} K_{\nu-1}(z),$$

$$\int z^\nu K_\nu(z)\,dz = 2^{\nu-1}\pi^{\frac{1}{2}}\,\Gamma\left(\frac{1}{2}+\nu\right) z\,[K_\nu(z)\,\boldsymbol{L}_{\nu-1}(z) + \boldsymbol{L}_\nu(z)\,K_{\nu-1}(z)],$$

$$\int z^\mu K_\nu(z)\,dz = e^{-i\frac{\pi}{2}\mu}\,[(\mu+\nu-1)\,z K_\nu(z)\,S_{\mu-1,\nu-1}(iz) - iz K_{\nu-1}(z)\,S_{\mu,\nu}(iz)].$$

Let $w_\nu(z)$ be an arbitrary solution of the Bessel differential equation with parameter (order) ν and similarly $W_\mu(z)$ a solution of Bessel's differential equation with parameter μ.

Then

$$\int\left[[(\beta^2-\alpha^2)\,z + \frac{(\nu^2-\mu^2)}{z}\right] w_\nu(\alpha z)\,W_\mu(\beta z)\,dz$$
$$= z\,[\alpha W_\mu(\beta z)\,w'_\nu(\alpha z) - \beta w_\nu(\alpha z)\,W'_\mu(\beta z)]$$
$$= \alpha z W_\mu(\beta z)\,w_{\nu-1}(\alpha z) - \beta z W_{\mu-1}(\beta z)\,w_\nu(\alpha z) + (\mu-\nu)\,W_\mu(\beta z)\,w_\nu(\alpha z),$$

$$\int z w_\nu(\alpha z)\,W_\nu(\beta z)\,dz = z(\beta^2-\alpha^2)^{-1} \times [\beta W_{\nu+1}(\beta z)\,w_\nu(\alpha z) - \alpha W_\nu(\beta z)\,w_{\nu+1}(\alpha z)],$$

$$\int z w_\nu(\alpha z)\, W_\nu(\alpha z)\, dz$$
$$= \frac{1}{4} z^2 \left[2 w_\nu(\alpha z)\, W_\nu(\alpha z) - w_{\nu+1}(\alpha z)\, W_{\nu-1}(\alpha z) - w_{\nu-1}(\alpha z)\, W_{\nu+1}(\alpha z)\right],$$

$$\int z^{-1} w_\nu(\alpha z)\, W_\nu(\alpha z)\, dz = \frac{1}{2} \nu^{-1} w_\nu(\alpha z)\, W_\nu(\alpha z)$$
$$+ \frac{1}{2} \nu^{-1} \alpha z \left[w_{\nu+1}(\alpha z) \frac{\partial W_\nu(\alpha z)}{\partial \nu} - w_\nu(\alpha z) \frac{\partial W_{\nu+1}(\alpha z)}{\partial \nu} \right].$$

For any two solution $v_\nu(z)$, $V_\mu(z)$ of the modified Bessel differential equation with the parameter (order ν and μ respectively) one has.

$$\int \left[(\beta^2 - \alpha^2)\, z + \frac{(\mu^2 - \nu^2)}{z} \right] v_\nu(\alpha z)\, V_\mu(\beta z)\, dz$$
$$= z\left[-\alpha V_\mu(\beta z)\, v'_\nu(\alpha z) + \beta v_\nu(\alpha z)\, V'_\mu(\beta z)\right],$$

$$\int z [v_\nu(\alpha z)]^2\, dz = -\frac{1}{2} z^2 \{[v'_\nu(\alpha z)]^2 - [v_\nu(\alpha z)]^2$$
$$\times [1 + \nu^2 (\alpha^2 z^2)^{-1}]\}.$$

3.8.2 Finite integrals

$$\int_0^{\frac{1}{2}\pi} J_\mu(\alpha \sin t) \sin^{\mu+1} t \cos^{2\nu+1} t\, dt = 2^\nu \Gamma(1+\nu)\, \alpha^{-1-\nu} J_{\nu+\mu+1}(\alpha),$$
$$Re\, \nu > -1, \quad Re\, \mu > -1,$$

$$\int_0^{\frac{1}{2}\pi} J_\mu(\alpha \sin t)\, J_\nu(\beta \cos t) \sin^{\mu+1} t \cos^{\nu+1} t\, dt$$
$$= \alpha^\mu \beta^\nu (\alpha^2 + \beta^2)^{-\frac{1}{2}(\nu+\mu+1)} J_{\nu+\mu+1}\left[(\alpha^2+\beta^2)^{\frac{1}{2}}\right],$$
$$Re\, \nu > -1, \quad Re\, \mu > -1,$$

$$\int_0^{\frac{1}{2}\pi} J_\mu(z \sin^2 t)\, J_\nu(z \cos^2 t)\, (\sin t \cos t)^{-1}\, dt = \frac{1}{2}(\nu^{-1} + \mu^{-1})\, z^{-1} J_{\nu+\mu}(z),$$
$$Re\, \nu > 0, \quad Re\, \mu > 0,$$

$$\int_0^{\frac{1}{2}\pi} J_\mu(z \sin^2 t)\, J_\nu(z \cos^2 t) \sin^{2\mu+1} t \cos^{2\nu+1} t\, dt$$
$$= \frac{1}{2} [\Gamma(\nu+\mu+1)]^{-1}\, \Gamma\left(\frac{1}{2}+\nu\right) \Gamma\left(\frac{1}{2}+\mu\right) (2\pi z)^{-\frac{1}{2}} J_{\nu+\mu+\frac{1}{2}}(z),$$

$$\int_0^{\frac{1}{2}\pi} J_\mu(z \sin^2 t)\, J_\nu(z \cos^2 t) \cot t\, dt = \frac{1}{2} \mu^{-1} J_{\nu+\mu}(z),$$
$$Re\, \nu > -1, \quad Re\, \mu > 0,$$

$$\int_0^{\frac{1}{2}\pi} J_\mu(z\sin^2 t)\, J_\nu(z\cos^2 t)\sin t\cos t\, dt = z^{-1}\sum_{m=0}^{\infty}(-1)^m J_{\nu+\mu+2m+1}(z),$$

$$Re\,\nu > -1,\ Re\,\mu > -1,$$

$$\int_0^{\frac{1}{2}\pi} J_\nu^2(z\sin t)\sin t\, dt = z^{-1}\sum_{m=0}^{\infty} J_{2\nu+2m+1}(z),$$

$$Re\,\nu > -1,$$

$$\int_0^{\frac{1}{2}\pi} J_\nu(2z\sin t)\sin^\nu t\cos^{2\nu} t\, dt = \frac{1}{2}\pi^{\frac{1}{2}} z^{-\nu}\,\Gamma\left(\frac{1}{2}+\nu\right) J_\nu^2(z).$$

$$Re\,\nu > -\frac{1}{2},$$

$$\int_0^{\frac{1}{2}\pi} J_{\nu+\mu}(2z\cos t)\cos[(\mu-\nu)t]\, dt = \frac{1}{2}\pi J_\nu(z)\, J_\mu(z),$$

$$Re\,(\nu+\mu) > -1,$$

$$\int_0^{\frac{1}{2}\pi} J_{2\nu}\left[2(z\zeta)^{\frac{1}{2}}\sin t\right]\cos[(z-\zeta)\cos t]\, dt = \frac{1}{2}\pi J_\nu(z)\, J_\nu(\zeta),$$

$$Re\,\nu > -\frac{1}{2},$$

$$\int_0^{\frac{1}{2}\pi} I_{2\nu}\left[2(z\zeta)^{\frac{1}{2}}\sin t\right]\cos[(z+\zeta)\cos t]\, dt = \frac{1}{2}\pi J_\nu(z)\, J_\nu(\zeta),$$

$$Re\,\nu > -\frac{1}{2},$$

$$\int_{-\frac{1}{2}\pi}^{\frac{1}{2}\pi} e^{it(\mu-\nu)}\cos^{\nu+\mu} t\,(\lambda z)^{-\nu-\mu} J_{\nu+\mu}(\lambda z)\, dt$$

$$= \pi(2\alpha z)^{-\mu}(2\beta z)^{-\nu} J_\mu(\alpha z)\, J_\nu(\beta z),$$

$$\lambda = [2\cos t(\alpha^2 e^{it} + \beta^2 e^{-it})]^{\frac{1}{2}},\quad Re\,(\nu+\mu) > -1,$$

$$\int_0^{\frac{1}{2}\pi} J_\nu(z\sin t)(\sin t)^{1-\nu}\, dt = \left(\frac{2z}{\pi}\right)^{-\frac{1}{2}} \boldsymbol{H}_{\nu-\frac{1}{2}}(z),$$

$$\int_0^{\frac{1}{2}\pi} J_\nu(z\sin t)(\cos t)^{2\mu+1}(\sin t)^{1-\nu}\, dt = 2^{1-\nu}[\Gamma(\nu)]^{-1} z^{-\mu-1} s_{\mu+\nu,\mu-\nu+1}(z),$$

$$Re\,\mu > -1.$$

Let $Z_\nu(z)$ be any solution of Bessel's differential equation, then

$$\int_0^\pi (a^2+b^2-2ab\cos t)^{-\frac{1}{2}\nu} Z_\nu\left[(a^2+b^2-2ab\cos t)^{\frac{1}{2}}\right] \sin^{2\nu} t\, dt$$

$$=\pi^{\frac{1}{2}}\Gamma\left(\frac{1}{2}+\nu\right)\left(\frac{1}{2}ab\right)^{-\nu} Z_\nu(b)\, J_\nu(a),$$

$$Re\,\nu > -\frac{1}{2},\quad a<b \text{ (for } a>b \text{ interchange } b \text{ and } a),$$

$$\int_0^\pi (a^2+b^2-2ab\cos t)^{-\frac{1}{2}\nu} Z_\nu\left[(a^2+b^2-2ab\cos t)^{\frac{1}{2}}\right] C_m^\nu(\cos t)\sin^{2\nu} t\, dt$$

$$=2\pi\Gamma(m+2\nu)\,[m!\,\Gamma(\nu)]^{-1}(2ab)^{-\nu} Z_{\nu+m}(b)\, J_{\nu+m}(a),$$

$$Re\,\nu > -\frac{1}{2},\quad a<b \text{ (for } a>b \text{ as above)},$$

$$m=0,1,2,\ldots$$

For the modified Bessel functions

$$\int_0^\pi (a^2+b^2-2ab\cos t)^{-\frac{1}{2}\nu} I_\nu\left[(a^2+b^2-2ab\cos t)^{\frac{1}{2}}\right] \sin^{2\nu} t\, dt$$

$$=\pi^{\frac{1}{2}}\Gamma\left(\frac{1}{2}+\nu\right)\left(\frac{1}{2}ab\right)^{-\nu} I_\nu(a)\, I_\nu(b),$$

$$Re\,\nu > -\frac{1}{2},$$

$$\int_0^\pi (a^2+b^2-2ab\cos t)^{-\frac{1}{2}\nu} K_\nu\left[(a^2+b^2-2ab\cos t)^{\frac{1}{2}}\right] \sin^{2\nu} t\, dt$$

$$=\pi^{\frac{1}{2}}\Gamma\left(\frac{1}{2}+\nu\right)\left(\frac{1}{2}ab\right)^{-\nu} I_\nu(a)\, K_\nu(b),$$

$$Re\,\nu > -\frac{1}{2},\quad a<b,$$

$$\int_0^\pi (a^2+b^2-2ab\cos t)^{-\frac{1}{2}\nu} I_\nu\left[(a^2+b^2-2ab\cos t)^{\frac{1}{2}}\right] C_m^\nu(\cos t)\sin^{2\nu} t\, dt$$

$$=(-1)^m\, 2\pi\Gamma(m+2\nu)\,[m!\,\Gamma(\nu)]^{-1}(2ab)^{-\nu} I_{\nu+m}(b)\, I_{\nu+m}(a),$$

$$Re\,\nu > -\frac{1}{2},\quad m=0,1,2,\ldots,$$

$$\int_0^\pi (a^2+b^2-2ab\cos t)^{-\frac{1}{2}\nu} K_\nu\left[(a^2+b^2-2ab\cos t)^{\frac{1}{2}}\right] C_m^\nu(\cos t)\sin^{2\nu} t\, dt$$

$$=2\pi\Gamma(m+2\nu)\,[m!\,\Gamma(\nu)]^{-1}(2ab)^{-\nu} K_{\nu+m}(b)\, I_{\nu+m}(a),$$

$$Re\,\nu > -\frac{1}{2},\quad a<b,\quad m=0,1,2,\ldots$$

3.8.3 Infinite Integrals, the integrand involving Bessel functions, powers of the variable and exponential functions

$$\int_0^\infty t^{\mu-1} J_\nu(at)\,dt = 2^{\mu-1} a^{-\mu} \frac{\Gamma\left(\frac{1}{2}\nu + \frac{1}{2}\mu\right)}{\Gamma\left(1 + \frac{\nu}{2} - \frac{1}{2}\mu\right)},$$

$$-\,Re\,\nu < Re\,\mu < \frac{3}{2},$$

$$\int_0^\infty t^{\mu-1} Y_\nu(at)\,dt = -2^{\mu-1}\pi^{-1} a^{-\mu} \cos\left[\frac{\pi}{2}(\mu-\nu)\right]$$

$$\times\,\Gamma\left(\frac{1}{2}\nu - \frac{1}{2}\mu\right)\Gamma\left(\frac{1}{2}\nu + \frac{1}{2}\mu\right),$$

$$Re\,(\mu \pm \nu) > 0, \quad Re\,\mu < \frac{3}{2},$$

$$\int_0^\infty t^{\mu-1} K_\nu(at)\,dt = 2^{\mu-2} a^{-\mu}\,\Gamma\left(\frac{1}{2}\mu + \frac{1}{2}\nu\right)\Gamma\left(\frac{1}{2}\mu - \frac{1}{2}\nu\right),$$

$$Re\,(\mu \pm \nu) > 0,$$

$$\int_0^\infty e^{-\beta t} J_\nu(\alpha t)\,dt = \alpha^{-\nu}(\beta^2+\alpha^2)^{-\frac{1}{2}}\left[(\beta^2+\alpha^2)^{\frac{1}{2}} - \beta\right]^\nu,$$

$$Re\,\nu > -1, \quad Re\,(\beta \pm i\alpha) > 0,$$

$$\int_0^\infty t^\mu e^{-\beta t} J_\nu(\alpha t)\,dt = \Gamma(\nu+\mu+1)\,(\beta^2+\alpha^2)^{-\frac{1}{2}\mu-\frac{1}{2}}$$

$$\times\,P_\mu^{-\nu}\left[\beta(\beta^2+\alpha^2)^{-\frac{1}{2}}\right],$$

$$Re\,(\nu+\mu) > -1, \quad Re\,\beta > |\mathrm{Im}\,\alpha|,$$

$$\int_0^\infty e^{-\beta t} t^\nu J_\nu(\alpha t)\,dt = \pi^{-\frac{1}{2}}(2\alpha)^\nu\,\Gamma\left(\frac{1}{2}+\nu\right)(\beta^2+\alpha^2)^{-\nu-\frac{1}{2}},$$

$$Re\,\nu > -\frac{1}{2}, \quad Re\,(\beta \pm i\alpha) > 0,$$

$$\int_0^\infty e^{-\beta t} t^{\nu+1} J_\nu(\alpha t)\,dt = 2\beta\pi^{-\frac{1}{2}}(2\alpha)^\nu\,\Gamma\left(\frac{3}{2}+\nu\right)(\beta^2+\alpha^2)^{-\nu-\frac{3}{2}},$$

$$Re\,\nu > -1, \quad Re\,\beta > |\mathrm{Im}\,\alpha|,$$

$$\int_0^\infty e^{-\beta t} t^{-1} J_\nu(\alpha t)\,dt = \nu^{-1}\alpha^{-\nu}\left[(\alpha^2+\beta^2)^{\frac{1}{2}} - \beta\right]^\nu,$$

$$Re\,\nu > 0, \; Re\,\beta > |\mathrm{Im}\,\alpha|,$$

$$\int_0^\infty e^{-\beta t} J_\nu(\alpha t)\,dt = (\beta^2+\alpha^2)^{-\frac{1}{2}}\alpha^\nu\left[\beta + (\beta^2+\alpha^2)^{\frac{1}{2}}\right]^{-\nu},$$

$$Re\,\nu > -1, \; Re\,\beta > |\mathrm{Im}\,\alpha|,$$

$$\int_0^\infty t^\nu e^{-\beta t} I_\nu(\alpha t)\, dt = \pi^{-\frac{1}{2}} (2\alpha)^\nu \Gamma\left(\frac{1}{2}+\nu\right) (\beta^2 - \alpha^2)^{-\nu-\frac{1}{2}},$$

$$Re\,\nu > -\frac{1}{2}, \quad Re\,(\beta \pm \alpha) > 0,$$

$$\int_0^\infty t^\mu e^{-\beta t} I_\nu(\alpha t)\, dt = \Gamma(\nu+\mu+1)\,(\beta^2-\alpha^2)^{-\frac{1}{2}\mu-\frac{1}{2}} \times \mathfrak{P}_\mu^{-\nu}\left[\beta(\beta^2-\alpha^2)^{-\frac{1}{2}}\right],$$

$$Re\,(\mu+\nu) > -1, \quad Re\,\beta > |Re\,\alpha|,$$

$$\int_0^\infty e^{-\beta t} K_\nu(\alpha t)\, dt = \frac{1}{2}\pi \operatorname{cosec}(\pi\nu)\, \alpha^{-\nu} (\beta^2-\alpha^2)^{-\frac{1}{2}} \times \left\{\left[\beta + (\beta^2-\alpha^2)^{\frac{1}{2}}\right]^\nu - \left[\beta - (\beta^2-\alpha^2)^{\frac{1}{2}}\right]^\nu\right\},$$

$$-1 < Re\,\nu < 1, \quad Re\,(\beta+\alpha) > 0,$$

$$\int_0^\infty t^\mu e^{-\beta t} K_\nu(\alpha t)\, dt = \left(\frac{\pi}{2a}\right)^{\frac{1}{2}} \Gamma(\mu+1-\nu)\,\Gamma(\mu+1+\nu) \times (\alpha^2-\beta^2)^{-\frac{1}{2}\mu-\frac{1}{4}} P_{\nu-\frac{1}{2}}^{-\mu-\frac{1}{2}}\left(\frac{\beta}{\alpha}\right),$$

$$Re\,(\mu+1\pm\nu) > 0, \quad \beta, \alpha \text{ real}, -\alpha < \beta < \alpha$$

$$\int_0^\infty t^\mu e^{-\beta t} K_\nu(\alpha t)\, dt = \left(\frac{\pi}{2\alpha}\right)^{\frac{1}{2}} \Gamma(\mu+1-\nu)\,\Gamma(\mu+1+\nu) \times (\beta^2-\alpha^2)^{-\frac{1}{2}\mu-\frac{1}{4}} \mathfrak{P}_{\nu-\frac{1}{2}}^{-\mu-\frac{1}{2}}\left(\frac{\beta}{\alpha}\right),$$

$$Re\,(\mu+1\pm\nu) > 0, \; Re\,\beta > -Re\,\alpha,$$

$$\int_0^\infty e^{-\beta t^2} J_\nu(\alpha t)\, dt = \frac{1}{2}\left(\frac{\pi}{\beta}\right)^{\frac{1}{2}} e^{-\frac{\alpha^2}{8\beta}} I_{\frac{1}{2}\nu}\left(\frac{\alpha^2}{8\beta}\right),$$

$$Re\,\beta > 0, \quad Re\,\nu > -1,$$

$$\int_0^\infty e^{-\beta t^2} I_\nu(\alpha t)\, dt = \frac{1}{2}\left(\frac{\pi}{\beta}\right)^{\frac{1}{2}} e^{-\frac{\alpha^2}{8\beta}} I_{\frac{1}{2}\nu}\left(\frac{\alpha^2}{8\beta}\right),$$

$$Re\,\beta > 0, \quad Re\,\nu > -1,$$

$$\int_0^\infty e^{-\beta t^2} K_\nu(\alpha t)\, dt = \frac{1}{4}\left(\frac{\pi}{\beta}\right)^{\frac{1}{2}} \sec\left(\frac{1}{2}\pi\nu\right) e^{\frac{\alpha^2}{8\beta}} K_{\frac{1}{2}\nu}\left(\frac{\alpha^2}{8\beta}\right),$$

$$-1 < Re\,\nu < 1, \quad Re\,\beta > 0,$$

$$\int_0^\infty t^{\nu+1} e^{-\beta t^2} J_\nu(\alpha t)\, dt = \alpha^\nu (2\beta)^{-\nu-1} e^{-\frac{\alpha^2}{4\beta}},$$

$$Re\,\nu > -1, \quad Re\,\beta > 0,$$

$$\int_0^\infty t^{\nu+1} e^{-\beta t^2} I_\nu(\alpha t)\, dt = \alpha^\nu (2\beta)^{-\nu-1} e^{\frac{\alpha^2}{4\beta}},$$

$$Re\,\nu > -1, \; Re\,\beta > 0,$$

$$\int_0^\infty J_\nu(\alpha t)\, J_\nu(\beta t)\, e^{-\gamma t^2}\, t\, dt = \frac{1}{2}\gamma^{-1} e^{\frac{-(\alpha^2+\beta^2)}{(4\gamma)}} I_\nu\left(\frac{\frac{1}{2}\alpha\beta}{\gamma}\right),$$

$$Re\,\nu > -1, \quad Re\,\gamma > 0,$$

$$\int_0^\infty I_\nu(\alpha t)\, I_\nu(\beta t)\, e^{-\gamma t^2}\, t\, dt = \frac{1}{2}\gamma^{-1} e^{\frac{(\alpha^2+\beta^2)}{(4\gamma)}} J_\nu\left(\frac{\frac{1}{2}\alpha\beta}{\gamma}\right),$$

$$Re\,\nu > -1, \quad Re\,\gamma > 0.$$

3.8.4 Infinite integrals representing products of two Bessel functions

$$J_\nu^2(z) + Y_\nu^2(z) = 8\pi^{-2} \int_0^\infty K_0(2z \sinh t) \cosh(2\nu t)\, dt,$$

$$Re\,z > 0,$$

$$J_\mu(z)\, J_\nu(z) + Y_\mu(z)\, Y_\nu(z)$$

$$= 4\pi^{-2} \int_0^\infty K_{\nu+\mu}(2z \sinh t)\, [e^{(\mu-\nu)t} \cos(\pi\nu) + e^{-(\mu-\nu)t} \cos(\pi\mu)]\, dt,$$

$$Re\,z > 0, \quad -1 < Re\,(\nu+\mu) < 1,$$

$$J_\mu(x)\, J_\nu(x) - Y_\mu(x)\, Y_\nu(x) = 4\pi^{-1} \int_0^\infty Y_{\nu+\mu}(2x \cosh t) \cosh[(\mu-\nu)t]\, dt,$$

$$x > 0, \quad -\frac{3}{2} < Re\,(\mu-\nu) < \frac{3}{2},$$

$$J_\mu(z)\, Y_\nu(z) - J_\nu(z)\, Y_\mu(z)$$

$$= 4\pi^{-2} \operatorname{cosec}[\pi(\mu-\nu)] \int_0^\infty K_{\nu-\mu}(2z \sinh t)\, e^{-(\nu+\mu)t}\, dt,$$

$$Re\,z > 0, \quad -1 < Re\,(\nu-\mu) < 1,$$

$$J_\nu(x)\, Y_\mu(y) + J_\mu(y)\, Y_\nu(x)$$

$$= -2\pi^{-1} \int_{-\infty}^\infty e^{-(\mu-\nu)t} \left(\frac{ye^t + xe^{-t}}{ye^{-t} + xe^t}\right)^{\frac{1}{2}(\nu+\mu)} J_{\nu+\mu}\left[(x^2+y^2+2xy\cosh t)^{\frac{1}{2}}\right] dt,$$

$$-\frac{3}{2} < Re\,(\nu-\mu) < \frac{3}{2},$$

$$J_\nu(x)\,J_\mu(y) - Y_\nu(x)\,Y_\mu(y)$$

$$= 2\pi^{-1} \int_{-\infty}^{\infty} e^{-(\mu-\nu)t} \left(\frac{ye^t + xe^{-t}}{ye^{-t} + xe^t}\right)^{\frac{1}{2}(\nu+\mu)} Y_{\nu+\mu}\left[(x^2 + y^2 + 2xy\cosh t)^{\frac{1}{2}}\right] dt,$$

$$-\frac{3}{2} < Re\,(\nu - \mu) < \frac{3}{2},$$

$$J_\nu^2\left[(2\alpha z)^{\frac{1}{2}}\right] + Y_\nu^2\left[(2\alpha z)^{\frac{1}{2}}\right] = 2\pi^{-2}\int_0^\infty t^{-1} e^{-\frac{\alpha}{t}} e^{zt} K_\nu(zt)\,dt,$$

$$J_\mu(x)\,Y_\nu(x) + J_\nu(x)\,Y_\mu(x) = -4\pi^{-1}\int_0^\infty J_{\nu+\mu}(2x\cosh t)\cosh[(\mu-\nu)t]\,dt,$$

$$x > 0, \quad -\frac{3}{2} < Re\,(\nu - \mu) < \frac{3}{2},$$

$$J_\mu(z)\,Y_\nu(z) - J_\nu(z)\,Y_\mu(z)$$

$$= 4\pi^{-2}\int_0^\infty K_{\nu+\mu}(2z\sinh t)\,[e^{(\nu-\mu)t}\sin(\pi\mu) - e^{-(\nu-\mu)t}\sin(\pi\nu)]\,dt,$$

$$Re\,z > 0, \quad -1 < Re\,(\nu + \mu) < 1,$$

$$J_\nu(z)\,H_{-\nu}^{(2)}(\zeta) = 2i\pi^{-1}\int_0^\infty J_{2\nu}\left[(2z\zeta)^{\frac{1}{2}}\sinh t\right] e^{-i(z+\zeta)\cosh t}\,dt,$$

$$Re\,\nu > -\frac{1}{2}, \quad \mathrm{Im}\left(z^{\frac{1}{2}} + \zeta^{\frac{1}{2}}\right) < 0,$$

$$H_\nu^{(2)}(z)\,H_\nu^{(2)}(\zeta) = 4i\pi^{-1}\int_0^\infty H_{2\nu}^{(2)}\left[(2z\zeta)^{\frac{1}{2}}\sinh t\right] e^{-i(z+\zeta)\cosh t}\,dt,$$

$$-\frac{1}{2} < Re\,\nu < \frac{1}{2}, \quad \mathrm{Im}\left(z^{\frac{1}{2}} + \zeta^{\frac{1}{2}}\right) < 0,$$

$$J_\nu(x)\,J_{-\nu}(y) = 2\pi^{-1}\int_0^\infty J_{2\nu}\left[2(xy)^{\frac{1}{2}}\sinh t\right]\sin[(x+y)\cosh t]\,dt,$$

$$Re\,\nu > -\frac{1}{2},$$

$$J_\nu(x)\,Y_{-\nu}(y) = -2\pi^{-1}\int_0^\infty J_{2\nu}\left[2(xy)^{\frac{1}{2}}\sinh t\right]\cos[(x+y)\cosh t]\,dt,$$

$$Re\,\nu > -\frac{1}{2},$$

$$J_\nu^2(z) + Y_\nu^2(z) = 8\pi^{-2}\int_0^\infty K_{2\nu}(2z\cosh t)\cosh(2z\sinh t)\,dt,$$

$$J_\nu(az)\,J_\nu(bz) + Y_\nu(az)\,Y_\nu(bz)$$

$$= 8\pi^{-2}\cos(\pi\nu)\int_0^\infty K_{2\nu}\left[2z(ab)^{\frac{1}{2}}\sinh t\right]\cos[z(a-b)\cosh t]\,dt,$$

$$-\frac{1}{2} < Re\,\nu < \frac{1}{2},$$

$$J_\nu(bz)\,Y_\nu(az) - J_\nu(az)\,Y_\nu(bz)$$

$$= 8\pi^{-2}\cos(\pi\nu)\int_0^\infty K_{2\nu}\left[2z(ab)^{\frac{1}{2}}\sinh t\right]\sin\left[z(a-b)\cosh t\right]dt,$$

$$-\frac{1}{2} < Re\,\nu < \frac{1}{2},$$

$$H_\nu^{(2)}(x)\,H_{-\nu}^{(2)}(y) = 2i\pi^{-1}\int_0^\infty H_0^{(2)}\left[(x^2+y^2+2xy\cosh t)^{\frac{1}{2}}\right]\cosh(\nu t)\,dt,$$

$$-\frac{3}{4} < Re\,\nu < \frac{3}{4},$$

$$J_\nu(ax)\,J_\nu(ay) = \pi^{-1}\int_0^\pi \cos(\nu t)\,J_0\left[a(x^2+y^2-2xy\cos t)^{\frac{1}{2}}\right]dt$$

$$-\pi^{-1}\sin(\pi\nu)\int_0^\infty e^{-\nu t}\,J_0\left[a(x^2+y^2+2xy\cosh t)^{\frac{1}{2}}\right]dt,$$

$$Re\,\nu < \frac{3}{4},$$

$$J_\nu(ay)\,Y_\nu(ax) = \pi^{-1}\int_0^\pi \cos(\nu t)\,Y_0\left[a(x^2+y^2-2xy\cos t)^{\frac{1}{2}}\right]dt$$

$$-\pi^{-1}\sin(\pi\nu)\int_0^\infty e^{-\nu t}\,Y_0\left[a(x^2+y^2+2xy\cosh t)^{\frac{1}{2}}\right]dt,$$

$$x > y, \quad Re\,\nu < \frac{3}{4},$$

$$\int_0^\infty t^{\varrho-1}(t^2+\zeta^2)^{-\frac{1}{2}\mu}\,J_\mu\left[b(t^2+\zeta^2)^{\frac{1}{2}}\right](t^2+z^2)^{-1}$$

$$\times\left\{\cos\left[\frac{1}{2}\pi(\varrho-\nu)\right]J_\nu(at)+\sin\left[\frac{1}{2}\pi(\varrho-\nu)\right]Y_\nu(at)\right\}dt$$

$$= -z^{\varrho-2}(\zeta^2-z^2)^{-\frac{1}{2}\mu}\,J_\mu\left[b(\zeta^2-z^2)^{\frac{1}{2}}\right]K_\nu(az),$$

$$a \geq b, \quad Re(\pm\nu) < Re\,\varrho < 4 + Re\,\mu, \quad Re\,z > 0,$$

$$\int_0^\infty t^{\varrho-1}(t^2+\zeta^2)^{-\frac{1}{2}\mu}\,J_\mu\left[b(t^2+\zeta^2)^{\frac{1}{2}}\right](t^2+z^2)^{-m-1}$$

$$\times\left\{\cos\left[\frac{1}{2}\pi(\varrho-\nu)\right]J_\nu(at)+\sin\left[\frac{1}{2}\pi(\varrho-\nu)\right]Y_\nu(at)\right\}dt$$

$$= (-1)^{m+1}\frac{2^{-m}}{m!}\left(\frac{d}{z\,dz}\right)^m\left\{z^{\varrho-2}(\zeta^2-z^2)^{-\frac{1}{2}\mu}\,J_\mu\left[b(\zeta^2-z^2)^{\frac{1}{2}}\right]K_\nu(az)\right\},$$

$$a \geq b, \quad Re(\pm\nu) < Re\,\varrho < 2m + 4 + Re\,\mu; \quad Re\,z > 0, \quad m = 0, 1, 2, \ldots,$$

$$K_\mu(ax)\,K_\mu(bx) = \pi^{\frac{1}{2}}\,2^{-\frac{3}{2}}\,\Gamma(\nu+\mu+1)\,\Gamma(\nu-\mu+1)$$

$$\times (ab)^{-\nu-1}\,x^{-\nu}\int_0^\infty (z^2-1)^{-\frac{1}{2}\nu-\frac{1}{4}}\,P_{\mu-\frac{1}{2}}^{-\frac{1}{2}\nu-\frac{1}{4}}(z)\,t^{\nu+1}\,J_\nu(tx)\,dt,$$

$$z = \frac{1}{2}(ab)^{-1}(t^2+a^2+b^2), \qquad -\frac{3}{4}-\frac{1}{2}\nu < Re\,\mu < \frac{3}{4}+\frac{1}{2}\nu,$$

$$K_\nu(az)\,K_\nu(bz) = \left(\frac{1}{2}\pi\right)^{\frac{1}{2}}(abz)^{-\alpha}$$

$$\times \int_1^\infty (x^2-1)^{-\frac{1}{2}\alpha-\frac{1}{4}}\,(a^2+b^2+2abx)^{\frac{1}{2}\alpha}\,K_\alpha\left[z\,(a^2+b^2+2abx)^{\frac{1}{2}}\right]$$

$$\times\,\mathfrak{P}_{\nu-\frac{1}{2}}^{\alpha+\frac{1}{2}}(x)\,dx, \qquad Re\,\alpha < \frac{1}{2}, \quad Re\,z > 0,$$

$$z^{\nu-\mu}\,I_\mu(bz)\,K_\nu(az) = \int_0^\infty t^{\nu-\mu+1}\,J_\mu(bt)\,J_\nu(at)\,(t^2+z^2)^{-1}\,dt,$$

$$a \geq b, \quad Re\,\nu > -1, \quad Re\,(\nu-\mu) < 2, \quad Re\,z > 0,$$

$$\frac{2}{\pi}\sin(\pi\nu)\,K_\nu(az)\,K_\nu(bz) = \int_0^\infty t\,(t^2+z^2)^{-1}\,[J_{-\nu}(at)\,J_{-\nu}(bt) - J_\nu(at)\,J_\nu(bt)]\,dt,$$

$$Re\,z > 0, \quad -1 < Re\,\nu < 1,$$

$$(-1)^{l+1}\,2\pi^{-1}z^{\nu+\mu+2l}K_\nu(az)\,K_\mu(bz)$$

$$= \int_0^\infty t^{\nu+\mu+1+2l}(z^2+t^2)^{-1}\,[J_\nu(at)\,Y_\mu(bt) + Y_\nu(at)\,J_\mu(bt)]\,dt,$$

$$Re\,(\nu+l) > -1, \quad Re\,(\mu+l) > -1, \quad l-1 < Re\,(\nu+\mu+2l) < l,$$

$$Re\,z > 0, \quad l = 0, \pm 1, \pm 2, \ldots,$$

$$(-1)^{l+1}\,2\pi^{-1}z^{\nu+\mu+2l-1}K_\nu(az)\,K_\mu(bz)$$

$$= \int_0^\infty t^{\nu+\mu+2l}(z^2+t^2)^{-1}\,[J_\nu(at)\,J_\mu(bt) - Y_\nu(at)\,Y_\mu(bt)]\,dt,$$

$$Re\,(\nu+l) > -\frac{1}{2}, \quad Re\,(\mu+l) > -\frac{1}{2}, \quad l-\frac{1}{2} < Re\,(\nu+\mu+2l) < l,$$

$$l = 0, \pm 1, \pm 2, \ldots, \quad Re\,z > 0,$$

$$K_\nu(z)\,K_\mu(z) = 2\int_0^\infty K_{\nu\pm\mu}(2z\cosh t)\cosh[(\mu\mp\nu)\,t]\,dt,$$

$$Re\,z > 0,$$

$$K_\nu(x)\,I_\mu(x) = \int_0^\infty J_{\nu+\mu}(2x\sinh t)\,e^{(\nu-\mu)t}\,dt,$$

$$x > 0, \quad Re\,(\nu + \mu) > -1, \quad Re\,(\nu - \mu) < \frac{3}{2},$$

$$J_\nu(x)\, K_\nu(x) = \frac{1}{2} \int_0^\infty J_0\left[x\,(2 \sinh t)^{\frac{1}{2}}\right] e^{-\nu t}\, dt,$$

$$x > 0, \quad Re\,\nu > -\frac{3}{4},$$

$$K_\nu^2(x) = \pi \operatorname{cosec}(\pi\nu) \int_0^\infty J_0(2x \sinh t) \sinh(2\nu t)\, dt,$$

$$x > 0, \quad -\frac{3}{4} < Re\,\nu < \frac{3}{4},$$

$$I_\nu(x)\, K_\mu(x) + I_\mu(x)\, K_\nu(x)$$

$$= 2 \int_0^\infty J_{\nu+\mu}(2x \sinh t) \cosh\,[(\mu - \nu)\, t]\, dt,$$

$$Re\,(\nu + \mu) > -1, \quad -\frac{3}{2} < Re\,(\mu - \nu) < \frac{3}{2}, \quad x > 0,$$

$$I_\nu(x)\, K_\mu(x) - I_\mu(x)\, K_\nu(x) = 2 \int_0^\infty J_{\nu+\mu}(2x \sinh t) \sinh\,[(\mu - \nu)\, t]\, dt,$$

$$Re\,(\nu + \mu) > -2, \quad -\frac{3}{2} < Re\,(\mu - \nu) < \frac{3}{2}, \quad x > 0,$$

$$K_\nu^2(x) = -\pi \sec(\pi\nu) \int_0^\infty Y_0(2x \sinh t) \cosh(2\nu t)\, dt,$$

$$-\frac{3}{4} < Re\,\nu < \frac{3}{4}, \quad x > 0,$$

$$K_\nu(x)\, K_\nu(y) = \int_0^\infty \cosh(\nu t)\, K_0\left[(x^2 + y^2 + 2xy \cosh t)^{\frac{1}{2}}\right] dt,$$

$$K_\nu(x)\, I_\nu(y) = \pi^{-1} \int_0^\pi K_0\left[(x^2 + y^2 - 2xy \cos t)^{\frac{1}{2}}\right] \cos(\nu t)\, dt$$

$$-\pi^{-1} \sin(\pi\nu) \int_0^\infty e^{-\nu t} K_0\left[(x^2 + y^2 + 2xy \cosh t)^{\frac{1}{2}}\right] dt,$$

$$\pi^{-\frac{1}{2}} \left(\frac{1}{2} xy\right)^{-\nu} \Gamma\left(\frac{1}{2} + \nu\right) K_\nu(x)\, K_\nu(y)$$

$$= \int_0^\infty (\sinh t)^{2\nu} (x^2 + y^2 + 2xy \cosh t)^{-\frac{1}{2}\nu} K_\nu\left[(x^2 + y^2 + 2xy \cosh t)^{\frac{1}{2}}\right] dt,$$

$$Re\,\nu > -\frac{1}{2},$$

$$K_\nu(z)\, K_\mu(\zeta) = \int_{-\infty}^\infty e^{-(\mu-\nu)t} \left(\frac{\zeta e^t + z e^{-t}}{\zeta e^{-t} + z e^t}\right)^{\frac{1}{2}(\nu+\mu)}$$

$$\times K_{\nu+\mu}\left\{[z^2 + \zeta^2 + 2z\zeta \cosh(2t)]^{\frac{1}{2}}\right\} dt,$$

$$Re\,z > 0, \quad Re\,\zeta > 0,$$

$$2\pi K_\mu(x)\, I_\nu(y) = \int_{-\pi}^{\pi} e^{-i\nu t}\left(\frac{x - ye^{it}}{x - ye^{-it}}\right)^{\frac{1}{2}(\nu+\mu)}$$

$$\times K_{\nu+\mu}\left[(x^2 + y^2 - 2xy\cos t)^{\frac{1}{2}}\right] dt$$

$$- 2\sin(\pi\nu)\int_0^\infty e^{-\nu t}\left(\frac{x + ye^{-t}}{x + ye^{t}}\right)^{\frac{1}{2}(\nu+\mu)} K_{\nu+\mu}\left[(x^2 + y^2 + 2xy\cosh t)^{\frac{1}{2}}\right] dt,$$

$$x > y$$

$$2I_\nu\left(2az^{\frac{1}{2}}\right) K_\nu\left(2bz^{\frac{1}{2}}\right) = \int_0^\infty t^{-1} e^{-zt} e^{-(a^2+b^2)t^{-1}} I_\nu(2abt^{-1})\, dt,$$

$$a < b, \quad Re\, z > 0,$$

$$2K_\nu\left(2az^{\frac{1}{2}}\right) K_\nu\left(2bz^{\frac{1}{2}}\right) = \int_0^\infty t^{-1} e^{-zt} e^{-(a^2+b^2)t^{-1}} K_\nu(2abt^{-1})\, dt,$$

$$Re\, z > 0, \quad Re\,(a + b)^2 > 0,$$

$$I_\nu(z)\, K_\nu(\zeta) = \int_0^\infty J_{2\nu}\left[2(z\zeta)^{\frac{1}{2}}\sinh t\right] e^{-(\zeta - z)\cosh t}\, dt,$$

$$Re\, v > -\frac{1}{2}, \quad Re\,(\zeta - z) > 0,$$

$$K_\nu(z)\, K_\nu(\zeta) = 2\cos(\pi\nu)\int_0^\infty K_{2\nu}\left[2(z\zeta)^{\frac{1}{2}}\sinh t\right] e^{-(\zeta+z)\cosh t}\, dt,$$

$$-\frac{1}{2} < Re\,\nu < \frac{1}{2}, \quad Re\left(z^{\frac{1}{2}} + \zeta^{\frac{1}{2}}\right)^2 > 0,$$

$$K_\nu(z)\, K_\nu(\zeta) = 2\int_0^\infty K_{2\nu}\left[2(z\zeta)^{\frac{1}{2}}\cosh t\right] \cos[(\zeta - z)\sinh t]\, dt,$$

$$I_\nu(az)\, K_\nu(bz) = \int_0^\infty I_{2\nu}\left[2(ab)^{\frac{1}{2}} z\sinh t\right] e^{-z(a+b)\cosh t}\, dt,$$

$$Re\, z > 0, \quad a < b, \quad Re\,\nu > -\frac{1}{2},$$

$$\cos(\pi\nu)\, I_\nu(x)\, K_\nu(y) = \int_0^\infty J_{2\nu}\left[2(xy)^{\frac{1}{2}}\cosh t\right] \cos[(x + y)\sinh t]\, dt,$$

$$x < y,$$

$$I_\nu(z)\, I_{-\nu}(z) = 4\pi^{-2}\cos(\pi z)\int_0^\infty K_{2\nu}(2z\sinh t)\sinh(2z\cosh t)\, dt,$$

$$-\frac{1}{2} < Re\,\nu < \frac{1}{2}, \quad Re\, z > 0.$$

For further representations of products of cylindrical functions see also 4.7.3.

3.8.5 Weber-Schafheitlin discontinuous integrals and special cases

$$\int_0^\infty t^{-\varrho} J_\mu(at)\, J_\nu(bt)\, dt = \frac{a^\mu 2^{-\varrho} b^{\varrho-\mu-1} \Gamma\left(\frac{1}{2} + \frac{1}{2}\nu + \frac{1}{2}\mu - \frac{1}{2}\varrho\right)}{\Gamma\left(\frac{1}{2} + \frac{1}{2}\nu + \frac{1}{2}\varrho - \frac{1}{2}\mu\right) \Gamma(1+\mu)}$$

$$\times\, {}_2F_1\left(\frac{1}{2} + \frac{1}{2}\nu + \frac{1}{2}\mu - \frac{1}{2}\varrho, \frac{1}{2} + \frac{1}{2}\mu - \frac{1}{2}\nu - \frac{1}{2}\varrho; \mu+1; \frac{a^2}{b^2}\right),$$

$$Re(\nu + \mu - \varrho + 1) > 0, \quad Re\,\varrho > -1, \quad 0 < a < b,$$

$$\int_0^\infty t^{-\varrho} J_\mu(at)\, J_\nu(bt)\, dt = \frac{b^{-\nu} 2^{-\varrho} a^{\varrho-\nu-1} \Gamma\left(\frac{1}{2} + \frac{1}{2}\nu + \frac{1}{2}\mu - \frac{1}{2}\varrho\right)}{\Gamma\left(\frac{1}{2} + \frac{1}{2}\mu + \frac{1}{2}\varrho - \frac{1}{2}\nu\right) \Gamma(1+\nu)}$$

$$\times\, {}_2F_1\left(\frac{1}{2} + \frac{1}{2}\nu + \frac{1}{2}\mu - \frac{1}{2}\varrho, \frac{1}{2} + \frac{1}{2}\nu - \frac{1}{2}\mu - \frac{1}{2}\varrho; \nu+1; \frac{b^2}{a^2}\right),$$

$$Re\,(\nu + \mu - \varrho + 1) > 0, \quad Re\,\varrho > -1, \quad a > b > 0,$$

$$\int_0^\infty J_\mu(at)\, J_\nu(at)\, t^{-\varrho}\, dt = \left(\frac{1}{2}a\right)^{\varrho-1} \Gamma(\varrho)\, \Gamma\left(\frac{1}{2}\nu + \frac{1}{2}\mu + \frac{1}{2} - \frac{1}{2}\varrho\right)$$

$$\times \left[2\Gamma\left(\frac{1}{2} + \frac{1}{2}\nu - \frac{1}{2}\mu + \frac{1}{2}\varrho\right)\Gamma\left(\frac{1}{2} + \frac{1}{2}\nu + \frac{1}{2}\mu + \frac{1}{2}\varrho\right)\right.$$

$$\left.\times\, \Gamma\left(\frac{1}{2} + \frac{1}{2}\mu - \frac{1}{2}\nu + \frac{1}{2}\varrho\right)\right]^{-1},$$

$$Re\,(\nu + \mu + 1) > Re\,\varrho > 0.$$

Special cases of these formulas are

$$\int_0^\infty t^{-1} J_\mu(at) \sin(bt)\, dt = \mu^{-1} \sin\left[\mu \arcsin\left(\frac{b}{a}\right)\right], \quad b < a$$

$$= a^\mu \mu^{-1} \sin\left(\frac{1}{2}\pi\mu\right)\left[b + (b^2 - a^2)^{\frac{1}{2}}\right]^{-\mu}, \quad b > a,$$

$$Re\,\mu > -1,$$

$$\int_0^\infty t^{-1} J_\mu(at) \cos(bt)\, dt = \mu^{-1} \cos\left[\mu \arcsin\left(\frac{b}{a}\right)\right], \quad b < a$$

$$= \mu^{-1} a^\mu \cos\left(\frac{1}{2}\pi\mu\right)\left[b + (b^2 - a^2)^{\frac{1}{2}}\right]^{-\mu}, \quad b > a,$$

$$Re\,\mu > 0,$$

$$\int_0^\infty J_\mu(at) \cos(bt)\, dt = (a^2 - b^2)^{-\frac{1}{2}} \cos\left[\mu \arcsin\left(\frac{b}{a}\right)\right], \quad b < a$$

$$= -\,a^\mu \sin\left(\frac{1}{2}\pi\mu\right)(b^2 - a^2)^{-\frac{1}{2}}\left[b + (b^2 - a^2)^{\frac{1}{2}}\right]^{-\mu}, \quad b > a,$$

$$Re\,\mu > -1,$$

$$\int_0^\infty t^{-1} J_\mu(at)\, J_\nu(at)\, dt = 2\pi^{-1}(\nu^2 - \mu^2)^{-1} \sin\left[\frac{1}{2}\pi(\nu-\mu)\right],$$

$$Re\,(\nu+\mu) > 0,$$

$$\int_0^\infty J_\mu(at)\, J_\nu(at)\, t^{-\nu-\mu}\, dt$$

$$= \pi^{\frac{1}{2}} a^{-1}\left(\frac{1}{2}a\right)^{\nu+\mu} \Gamma(\nu+\mu)\left[\Gamma\left(\frac{1}{2}+\nu+\mu\right)\Gamma\left(\frac{1}{2}+\mu\right)\Gamma\left(\frac{1}{2}+\nu\right)\right]^{-1}$$

$$Re\,(\nu+\mu) > 0,$$

$$\int_0^\infty J_\mu(at)\, J_\nu(bt)\, t^{\mu-\nu+1}\, dt$$

$$= 2^{\mu-\nu+1} a^\mu b^{-\nu}(b^2 - a^2)^{\nu-\mu-1}\,[\Gamma(\nu-\mu)]^{-1} \qquad b > a,$$

$$= 0, \qquad b < a,$$

$$Re\,\nu > Re\,\mu > -1,$$

$$\int_0^\infty t^{-1} J_\mu(at)\, J_\mu(bt)\, dt = \begin{cases} \frac{1}{2}\mu^{-1}\left(\frac{b}{a}\right)^\mu, & a > b \\ \frac{1}{2}\mu^{-1}\left(\frac{a}{b}\right)^\mu, & a < b, \end{cases}$$

$$Re\,\mu > 0,$$

$$\int_0^\infty J_\nu(at)\, J_{\nu+1}(bt)\, dt = \begin{cases} a^\nu b^{-\nu-1} & 0 < a < b \\ \frac{1}{2a} & a = b \\ 0 & a > b, \end{cases}$$

$$\int_0^\infty \prod_{n=1}^m J_\nu(a_n t)\, t^{2\nu+1-\nu m}\, dt = 0,$$

$$a_1 > a_2 > a_3 \cdots > a_m > 0,\ a_1 > a_2 + a_3 + \cdots + a_m,$$

$$Re\,\nu > -1,$$

$$\int_0^\infty J_\nu(\beta t)\, K_\mu(\alpha t)\, t^{-\varrho}\, dt = 2^{-1-\varrho}\alpha^{\varrho-\nu-1}\beta^\nu\, \Gamma\left(\frac{1}{2}+\frac{1}{2}\nu-\frac{1}{2}\varrho+\frac{1}{2}\mu\right)$$

$$\times\, [\Gamma(1+\nu)]^{-1}\, \Gamma\left(\frac{1}{2}+\frac{1}{2}\nu-\frac{1}{2}\varrho-\frac{1}{2}\mu\right)$$

$$\times\, {}_2F_1\left(\frac{1}{2}+\frac{1}{2}\nu-\frac{1}{2}\varrho+\frac{1}{2}\mu,\ \frac{1}{2}+\frac{1}{2}\nu-\frac{1}{2}\varrho-\frac{1}{2}\mu;\ 1+\nu;\ -\frac{\beta^2}{\alpha^2}\right),$$

$$Re\,(\alpha \pm i\beta) > 0, \quad Re\,(\nu - \varrho + 1 \pm \mu) > 0,$$

$$\int_0^\infty K_\mu(\alpha t)\, J_\nu(\beta t)\, t^{\mu+\nu+1}\, dt = 2^{\nu+\mu}\alpha^\mu\beta^\nu\,\Gamma(\nu+\mu+1)\,(\alpha^2+\beta^2)^{-\mu-\nu-1},$$

$$Re\,(\nu+\mu) > |Re\,\mu|, \quad Re\,\alpha > |\mathrm{Im}\,\beta|.$$

(For further cases when the hypergeometric function reduces to a simpler expression see 2.1)

$$\int_0^\infty t^{-\varrho} K_\mu(\alpha t)\, K_\nu(\beta t)\, dt = 2^{-\varrho-2} \alpha^{\varrho-\nu-1} \beta^\nu [\Gamma(1-\varrho)]^{-1}$$

$$\times \Gamma\left(\frac{1}{2} + \frac{1}{2}\nu + \frac{1}{2}\mu - \frac{1}{2}\varrho\right) \Gamma\left(\frac{1}{2} + \frac{1}{2}\nu - \frac{1}{2}\mu - \frac{1}{2}\varrho\right)$$

$$\times \Gamma\left(\frac{1}{2} - \frac{1}{2}\nu + \frac{1}{2}\mu - \frac{1}{2}\varrho\right) \Gamma\left(\frac{1}{2} - \frac{1}{2}\nu - \frac{1}{2}\mu - \frac{1}{2}\varrho\right)$$

$$\times {}_2F_1\left(\frac{1}{2} + \frac{1}{2}\nu + \frac{1}{2}\mu - \frac{1}{2}\varrho, \frac{1}{2} + \frac{1}{2}\nu - \frac{1}{2}\mu - \frac{1}{2}\varrho; 1-\varrho; 1 - \frac{\beta^2}{\alpha^2}\right)$$

$$Re\,(\alpha + \beta) > 0, \quad Re\,(1 \pm \nu \pm \mu - \varrho) > 0.$$

Again, special cases of the above formula can be obtained by the aid of the formulas in 2.1. Some special cases are

$$\int_0^\infty x^{-\mu} J_\mu(ax)\, J_\nu(bx)\, dx = 2^{1-\mu} b^{-1} (a^2 - b^2)^{\frac{1}{2}\mu}$$

$$\times \left[\Gamma\left(\frac{1}{2} + \mu + \frac{1}{2}\nu\right) \Gamma\left(\frac{1}{2} + \mu - \frac{1}{2}\nu\right)\right]^{-1}$$

$$\times e^{-i\pi\mu} \mathfrak{Q}^\mu_{\frac{1}{2}\nu - \frac{1}{2}} (2a^2 b^{-2} - 1), \quad b < a,$$

$$= 2^{-\mu} b^{-1} (b^2 - a^2)^{\frac{1}{2}\mu} P^{-\mu}_{\frac{1}{2}\nu - \frac{1}{2}} (1 - 2a^2 b^{-2}), \quad b > a,$$

$$Re\,(\nu, \mu) > -1,$$

$$\int_0^\infty x^\mu J_\mu(ax)\, J_\nu(bx)\, dx = 2^{\mu+1} \pi^{-1} \cos\left(\frac{1}{2}\pi\nu\right) b^{-1} (a^2 - b^2)^{-\frac{1}{2}\mu}$$

$$\times e^{-i\pi\mu} \mathfrak{Q}^\mu_{\frac{1}{2}\nu - \frac{1}{2}} (2a^2 b^{-2} - 1), \quad b < a$$

$$= 2^\mu \Gamma\left(\frac{1}{2} + \frac{1}{2}\nu + \mu\right) \left[\Gamma\left(\frac{1}{2} + \frac{1}{2}\nu - \mu\right) b\right]^{-1}$$

$$\times (b^2 - a^2)^{-\frac{1}{2}\mu} P^{-\mu}_{\frac{1}{2}\nu - \frac{1}{2}} (1 - 2a^2 b^{-2}), \quad b > a,$$

$$Re\,(2\mu + \nu) > -1, \quad Re\,\mu < 1,$$

$$\int_0^\infty x^\mu K_\mu(ax)\, K_\nu(bx)\, dx$$

$$= 2^{\mu-2}\pi \sec\left(\frac{1}{2}\pi\nu\right) b^{-1}\Gamma\left(\frac{1}{2}+\mu+\frac{1}{2}\nu\right)\Gamma\left(\frac{1}{2}+\mu-\frac{1}{2}\nu\right)$$

$$\times \begin{cases} (a^2-b^2)^{-\frac{1}{2}\mu}\, \mathfrak{P}^{-\mu}_{-\frac{1}{2}-\frac{1}{2}\nu}(2a^2b^{-2}-1) & b<a \\ (b^2-a^2)^{-\frac{1}{2}\mu}\, P^{-\mu}_{-\frac{1}{2}-\frac{1}{2}\nu}(2a^2b^{-2}-1) & b>a \end{cases}$$

$$-1 < Re\,\nu < 1, \quad Re\left(\frac{1}{2}+\mu\pm\frac{1}{2}\nu\right) > 0,$$

$$\int_0^\infty x^\mu K_\mu(ax)\, J_\nu(bx)\, dx = 2^\mu b^{-1}(a^2+b^2)^{-\frac{1}{2}\mu} e^{-i\pi\mu}\mathfrak{Q}^\mu_{\frac{1}{2}\nu-\frac{1}{2}}(1+2a^2b^{-2}),$$

$$Re\,(\nu+\mu+1) > |Re\,\mu|,$$

$$\int_0^\infty x^\mu J_\mu(ax)\, K_\nu(bx)\, dx = 2^{\mu-1}\Gamma\left(\mu+\frac{1}{2}+\frac{1}{2}\nu\right)\Gamma\left(\mu+\frac{1}{2}-\frac{1}{2}\nu\right)\cdot b^{-1}$$

$$\times (a^2+b^2)^{-\frac{1}{2}\mu}\, \mathfrak{P}^{-\mu}_{\frac{1}{2}\nu-\frac{1}{2}}(1+2a^2b^{-2}),$$

$$Re\,(2\mu\pm\nu) > -1,$$

$$\int_0^\infty x^{-\mu} J_\mu(ax)\, K_\nu(bx)\, dx = 2^{-\mu-1}\pi \sec\left(\frac{1}{2}\pi\nu\right) b^{-1}$$

$$\times (a^2+b^2)^{\frac{1}{2}\mu}\, \mathfrak{P}^{-\mu}_{\frac{1}{2}\nu-\frac{1}{2}}(1+2a^2b^{-2}),$$

$$-1 < Re\,\nu < 1.$$

Finally, Hardy's integral. Denote by

$$I_{\mu,\nu} = \int_{-\infty}^\infty (x-a)^{-\mu} J_\mu[c(x-a)]\,(x-b)^{-\nu} J_\nu[c(x-b)]\, dx$$

then

$$I_{\mu,\nu} = \left(\frac{2\pi}{c}\right)^{\frac{1}{2}} \frac{\Gamma(\mu+\nu)}{\Gamma\left(\mu+\frac{1}{2}\right)\Gamma\left(\nu+\frac{1}{2}\right)}(a-b)^{\frac{1}{2}-\mu-\nu} J_{\mu+\nu-\frac{1}{2}}[c(a-b)],$$

$$Re\,(\mu+\nu) > 0,\ a, b \text{ real},\ c > 0.$$

3.8.6 Some integrals involving products of three Bessel functions

$$\int_0^\infty [J_\mu(ax)]^2 J_\nu(bx)\,dx$$

$$= b^{-1}\Gamma\left(\frac{1}{2}+\frac{1}{2}\nu+\mu\right)\left[\Gamma\left(\frac{1}{2}+\frac{1}{2}\nu-\mu\right)\right]^{-1}$$

$$\times \left\{P^{-\mu}_{-\frac{1}{2}+\frac{1}{2}\nu}\left[(1-4a^2b^{-2})^{\frac{1}{2}}\right]\right\}^2,$$

$$Re\,(\nu+2\mu)>-1, \quad b>2a,$$

$$\int_0^\infty J_\mu(ax)\,J_{-\mu}(ax)\,J_\nu(bx)\,dx$$

$$= b^{-1}P^{\mu}_{\frac{1}{2}\nu-\frac{1}{2}}\left[(1-4a^2b^{-2})^{\frac{1}{2}}\right]P^{-\mu}_{\frac{1}{2}\nu-\frac{1}{2}}\left[(1-4a^2b^{-2})^{\frac{1}{2}}\right],$$

$$Re\,\nu>-1, \quad b>2a,$$

$$\int_0^\infty [K_\mu(ax)]^2 J_\nu(bx)\,dx = \Gamma\left(\frac{1}{2}+\frac{1}{2}\nu+\mu\right)\left[\Gamma\left(\frac{1}{2}+\frac{1}{2}\nu-\mu\right)\right]^{-1}$$

$$\times e^{i2\pi\mu}\left\{\mathfrak{Q}^{-\mu}_{\frac{1}{2}\nu-\frac{1}{2}}\left[(1+4a^2b^{-2})^{\frac{1}{2}}\right]\right\}^2,$$

$$Re\left(\frac{1}{2}\nu\pm\mu\right)>-\frac{1}{2},$$

$$\int_0^\infty x^{\nu+1}K_\mu(ax)\,K(_\mu bx)\,J_\nu(cx)\,dx$$

$$= \frac{1}{2}\left(\frac{1}{2}\pi\right)^{\frac{1}{2}}\left(\frac{ab}{c}\right)^{-\nu-1}\Gamma(\nu+\mu+1)\,\Gamma(\nu-\mu+1)$$

$$\times (z^2-1)^{-\frac{1}{2}\nu-\frac{1}{4}}\,\mathfrak{P}^{-\nu-\frac{1}{2}}_{\mu-\frac{1}{2}}(z),$$

$$z=\frac{1}{2}\left[\frac{(a^2+b^2+c^2)}{(ab)}\right]$$

$$Re\,(\nu\pm\mu)>1, \quad Re\,\nu>-1,$$

$$\int_0^\infty x^{\mu+1}K_\mu(ax)\,J_\nu(bx)\,J_\nu(cx)\,dx$$

$$= (2\pi)^{-\frac{1}{2}}a^\mu(bc)^{-\mu-1}e^{-i\pi\left(\frac{1}{2}+\mu\right)}(z^2-1)^{-\frac{1}{2}\mu-\frac{1}{4}}\,\mathfrak{Q}^{\mu+\frac{1}{2}}_{\nu-\frac{1}{2}}(z),$$

$$z=\frac{1}{2}\left[\frac{(a^2+b^2+c^2)}{(ab)}\right]$$

$$Re\,\nu>-1, \quad Re\,(\mu+\nu)>-1.$$

3.8.7 The integrals of Sonine and Gegenbauer and related integrals

$$\int_0^\infty J_\mu(bt)\, J_\nu\left[a(t^2+z^2)^{\frac{1}{2}}\right](t^2+z^2)^{-\frac{1}{2}\nu}\, t^{\mu+1}\, dt$$

$$= 0, \qquad a < b$$

$$= b^\mu a^{-\nu} z^{1+\mu-\nu}(a^2-b^2)^{\frac{1}{2}\nu-\frac{1}{2}\mu-\frac{1}{2}} J_{\nu-\mu-1}\left[z(a^2-b^2)^{\frac{1}{2}}\right], \qquad a > b,$$

$$Re\,\nu > Re\,\mu > -1,$$

$$\int_0^\infty J_\mu(bt)\, K_\nu\left[a(t^2+z^2)^{\frac{1}{2}}\right](t^2+z^2)^{-\frac{1}{2}\nu}\, t^{\mu+1}\, dt$$

$$= b^\mu a^{-\nu} z^{\mu-\nu+1}(a^2+b^2)^{\frac{1}{2}\nu-\frac{1}{2}\mu-\frac{1}{2}} K_{\nu-\mu-1}\left[z(a^2+b^2)^{\frac{1}{2}}\right],$$

$$Re\,\mu > -1, \quad Re\,z > 0,$$

$$\int_0^\infty J_\mu(bt)\, K_\nu\left[a(t^2-y^2)^{\frac{1}{2}}\right](t^2-y^2)^{-\frac{1}{2}\nu}\, t^{\mu+1}\, dt$$

$$= \frac{1}{2}\pi e^{-i\pi\left(\nu-\mu-\frac{1}{2}\right)} b^\mu a^{-\nu} y^{1+\mu-\nu}(a^2+b^2)^{\frac{1}{2}(\nu-\mu-1)} H^{(2)}_{\nu-\mu-1}\left[y(a^2+b^2)^{\frac{1}{2}}\right],$$

$$Re\,\nu < 1, \quad Re\,\mu > -1, \quad \arg\left[(t^2-y^2)^{\frac{1}{2}}\right] = 0 \text{ for } t > y,$$

$$\arg\left[(t^2-y^2)^\sigma\right] = \pi\sigma \text{ for } t < y \text{ with } \sigma = \frac{1}{2} \text{ or } \sigma = -\frac{1}{2}\nu,$$

$$\int_0^\infty J_\mu(bt)\, H^{(2)}_\nu\left[a(t^2+x^2)^{\frac{1}{2}}\right](t^2+x^2)^{-\frac{1}{2}\nu}\, t^{\mu+1}\, dt$$

$$= a^{-\nu} b^\mu x^{1+\mu-\nu}(a^2-b^2)^{\frac{1}{2}(\nu-\mu-1)} H^{(2)}_{\nu-\mu-1}\left[x(a^2-b^2)^{\frac{1}{2}}\right], \qquad a > b,$$

$$= 2i\pi^{-1} b^\mu a^{-\nu} x^{1+\mu-\nu}(b^2-a^2)^{\frac{1}{2}(\nu-\mu-1)} K_{\nu-\mu-1}\left[x(b^2-a^2)^{\frac{1}{2}}\right], \qquad b > a,$$

$$Re\,\nu > Re\,\mu > -1.$$

Special cases of these formulas are

$$\int_0^\infty J_\nu\left[a(t^2+z^2)^{\frac{1}{2}}\right](t^2+z^2)^{-\frac{1}{2}\nu}\, t^{2\mu+1}\, dt$$

$$= 2^\mu a^{-\mu-1} z^{1+\mu-\nu}\Gamma(1+\mu)\, J_{\nu-\mu-1}(az),$$

$$Re\left(\frac{1}{2}\nu-\frac{1}{4}\right) > Re\,\mu > -1,$$

$$\int_0^\infty K_\nu\left[a(t^2+z^2)^{\frac{1}{2}}\right](t^2+z^2)^{-\frac{1}{2}\nu}\, t^{2\mu+1}\, dt$$

$$= 2^\mu a^{-\mu-1} z^{1+\mu-\nu}\Gamma(1+\mu)\, K_{\nu-\mu-1}(az),$$

$$Re\,\mu > -1,$$

$$\int_0^\infty J_\mu(bt)\,(t^2+z^2)^{-\nu}\,t^{\mu+1}\,dt$$

$$=\left(\frac{1}{2}b\right)^{\nu-1} z^{1+\mu-\nu}[\Gamma(\nu)]^{-1}\,K_{\nu-\mu-1}(bz),$$

$$Re\left(2\nu-\frac{1}{2}\right) > Re\,\mu > -1,\quad Re\,z > 0,$$

$$\int_0^\infty (t^2-y^2)^{-\frac{1}{2}}\,e^{-a(t^2-y^2)^{\frac{1}{2}}}\,t\,J_0(bt)\,dt = (a^2+b^2)^{-\frac{1}{2}}\,e^{-iy(a^2+b^2)^{\frac{1}{2}}},$$

$$\arg\left[(t^2-y^2)^{\frac{1}{2}}\right] = \frac{1}{2}\pi \text{ when } t < y,$$

$$\int_0^\infty K_0\left[a(t^2-y^2)^{\frac{1}{2}}\right]\cos(bt) = -\frac{1}{2}i\pi\int_0^\infty H_0^{(2)}\left[a(y^2-t^2)^{\frac{1}{2}}\right]\cos(bt)\,dt$$

$$=\frac{1}{2}\pi(a^2+b^2)^{-\frac{1}{2}}\,e^{-iy(a^2+b^2)^{\frac{1}{2}}},$$

$$\arg\left[(t^2-y^2)^{\frac{1}{2}}\right] = \frac{1}{2}\pi \text{ when } t < y,$$

$$\int_0^\infty t^{\nu+1}(t^2+z^2)^{-1}\,J_\nu(at)\,dt = z^\nu K_\nu(az),$$

$$Re\,z > 0,\quad -1 < Re\,\nu < \frac{3}{2},$$

$$\int_0^\infty t^{\nu+1}(t^2+z^2)^{-\mu-1}\,J_\nu(at)\,dt = \left(\frac{1}{2}a\right)^\mu z^{\nu-\mu}[\Gamma(1+\mu)]^{-1}\,K_{\nu-\mu}(az),$$

$$Re\,z > 0,\quad -1 < Re\,\nu < 2\,Re\,\mu + \frac{3}{2}.$$

3.8.8 Integrals with respect to the order

Ramanujan's formula

$$\int_{-\infty}^\infty a^{-\mu-x}J_{\mu+x}(a)\,b^{-\nu+x}J_{\nu-x}(b)\,e^{ixy}\,dx$$

$$=\left(2\cos\frac{1}{2}y\right)^{\frac{1}{2}(\nu+\mu)}\left(a^2e^{-i\frac{1}{2}y}+b^2e^{i\frac{1}{2}y}\right)^{-\frac{1}{2}\nu-\frac{1}{2}\mu} e^{i\frac{1}{2}y(\nu-\mu)}$$

$$\times J_{\nu+\mu}\left\{\left[2\cos\left(\frac{1}{2}y\right)\left(a^2e^{-i\frac{1}{2}y}+b^2e^{i\frac{1}{2}y}\right)\right]^{\frac{1}{2}}\right\},\qquad |y|<\pi$$

$$=0,\qquad |y|>\pi,$$

$$y \text{ real},\ a, b > 0,\quad Re\,(\nu+\mu) > 1.$$

Special case $y = 0$, $b = a$

$$\int_{-\infty}^\infty J_{\mu+x}(a)\,J_{\nu-x}(a)\,dx = J_{\mu+\nu}(2a),$$

$$a > 0,\quad Re\,(\mu+\nu) > 1,$$

$$\int_{-\infty}^{\infty} K_{i(\varepsilon+x)}(a)\, K_{i(\tau+x)}(b)\, e^{(\pi-\gamma)x}\, dx = K_{i(\varepsilon-\tau)}(c)\, e^{-\varepsilon\beta-\tau\alpha},$$

where α, β, γ are the angles of a triangle with sidelengths a, b, c.

Integrals of the addition theorem type

Gegenbauer's and Graf's addition theorem for the modified Hankel function [see 3.9] can also be written in the form of an integral

$$(a^2+b^2-2ab\cos\varphi)^{-\frac{1}{2}\nu} K_\nu\left[(a^2+b^2-2ab\cos\varphi)^{\frac{1}{2}}\right]$$
$$= 2^{\frac{1}{2}}\pi^{-\frac{3}{2}}(\sin\varphi)^{\frac{1}{2}-\nu}(ab)^{-\nu}$$
$$\int_0^\infty x\sinh(\pi x)\,\Gamma(\nu+ix)\,\Gamma(\nu-ix)\,P^{\frac{1}{2}-\nu}_{-\frac{1}{2}+ix}(-\cos\varphi)\,K_{ix}(a)\,K_{ix}(b)\,dx,$$
$$0 \leqq \varphi \leqq 2\pi$$

$$\left(\frac{a-be^{-i\varphi}}{a-be^{i\varphi}}\right)^{\frac{1}{2}\nu} K_\nu\left[(a^2+b^2-2ab\cos\varphi)^{\frac{1}{2}}\right]$$
$$= \pi^{-1}\int_{-\infty}^{\infty} K_{\frac{1}{2}\nu+ix}(a)\, K_{\frac{1}{2}\nu-ix}(b)\, e^{i(\pi-\varphi)\left(\frac{1}{2}\nu+ix\right)}\,dx,$$
$$0 \leqq \varphi \leqq 2\pi,$$

The first integral can also be written as

$$(a^2+b^2-2ab\cos\varphi)^{-\frac{1}{2}\nu} K_\nu\left[(a^2+b^2-2ab\cos\varphi)^{\frac{1}{2}}\right]$$
$$= \frac{1}{2}\Gamma(\nu)\left(\frac{1}{2}ab\right)^{-\nu}\int_{-\infty}^{\infty}\operatorname{sech}(\pi x)\left(\nu-\frac{1}{2}+ix\right)K_{\nu-\frac{1}{2}+ix}(a)\,I_{\nu-\frac{1}{2}+ix}(b)$$
$$\times\, C^\nu_{-\frac{1}{2}+ix}(-\cos\varphi)\,dx,$$
$$0 \leqq \varphi \leqq 2\pi.$$

The special case $\nu = 0$ yields

$$K_0\left[(a^2+b^2-2ab\cos\varphi)^{\frac{1}{2}}\right] = 2\pi^{-1}\int_0^\infty K_{ix}(a)\,K_{ix}(b)\cosh[x(\pi-|\varphi|)]\,dx,$$
$$0 \leqq \varphi \leqq 2\pi.$$

3.9 Addition theorems

Let

$$w = (\varrho^2+r^2-2r\varrho\cos\varphi)^{\frac{1}{2}},$$
$$0 < \varrho < r.$$

Denote by $Z_\nu(z)$ any solution of the Bessel differential equation (Bessel function, Neumann function, Hankel function or any linear combination). Then

$$w^{-\nu} Z_\nu(\gamma w) = 2^\nu \gamma^{-\nu} \Gamma(\nu)\,(r\varrho)^{-\nu} \times \sum_{m=0}^{\infty} (\nu + m)\, C_m^\nu(\cos\varphi)\, J_{\nu+m}(\gamma\varrho)\, Z_{\nu+m}(\gamma r),$$

$\nu \neq 0, -1, -2, \ldots, \gamma$ arbitrary complex. For the $C_m^\nu(x)$ see 5.3.

Especially for $\nu = 0$

$$Z_0(\gamma w) = J_0(\gamma\varrho)\, Z_0(\gamma r) + 2 \sum_{m=1}^{\infty} J_m(\gamma\varrho)\, Z_m(\gamma r) \cos(m\varphi).$$

For the modified Bessel functions we have under the same conditions

$$w^{-\nu} I_\nu(\gamma w) = 2^\nu \gamma^{-\nu} \Gamma(\nu)\,(r\varrho)^{-\nu} \times \sum_{m=0}^{\infty} (-1)^m (\nu + m)\, C_m^\nu(\cos\varphi)\, I_{\nu+m}(\gamma\varrho)\, I_{\nu+m}(\gamma r),$$

$$w^{-\nu} K_\nu(\gamma w) = 2^\nu \gamma^{-\nu} \Gamma(\nu)\,(r\varrho)^{-\nu} \times \sum_{m=0}^{\infty} (\nu + m)\, C_m^\nu(\cos\varphi)\, I_{\nu+m}(\gamma\varrho)\, K_{\nu+m}(\gamma r).$$

Especially for $\nu = 0$

$$I_0(\gamma w) = I_0(\gamma\varrho)\, I_0(\gamma r) + 2 \sum_{m=1}^{\infty} (-1)^m I_m(\gamma\varrho)\, I_m(\gamma r) \cos(m\varphi),$$

$$K_0(\gamma w) = I_0(\gamma\varrho)\, K_0(\gamma r) + 2 \sum_{m=1}^{\infty} I_m(\gamma\varrho)\, K_m(\gamma r) \cos(m\varphi).$$

These are the addition theorems of the Gegenbauer type.

Also under the same conditions as before

$$w^{-\nu} J_{-\nu}(\gamma w) = 2^\nu \gamma^{-\nu} \Gamma(\nu)\,(r\varrho)^{-\nu} \times \sum_{m=0}^{\infty} (-1)^m (\nu + m)\, C_m^\nu(\cos\varphi)\, J_{-\nu-m}(\gamma r)\, J_{\nu+m}(\gamma\varrho),$$

$$w^{-\nu} I_{-\nu}(\gamma w) = 2^\nu \gamma^{-\nu} \Gamma(\nu)\,(r\varrho)^{-\nu} \times \sum_{m=0}^{\infty} (-1)^m (\nu + m)\, C_m^\nu(\cos\varphi)\, I_{-\nu-m}(\gamma r)\, J_{\nu+m}(\gamma\varrho).$$

Special case $\nu = \frac{1}{2}$

$$\frac{e^{\pm i\gamma w}}{w} = \pm \frac{1}{2} i\pi (r\varrho)^{-\frac{1}{2}} \sum_{m=0}^{\infty} (2m+1)\, P_m(\cos\varphi)\, J_{m+\frac{1}{2}}(\gamma\varrho) \cdot H^{(1,2)}_{m+\frac{1}{2}}(\gamma r),$$

$$\frac{e^{-\gamma w}}{w} = (r\varrho)^{-\frac{1}{2}} \sum_{m=0}^{\infty} (2m+1)\, P_m(\cos\varphi)\, I_{m+\frac{1}{2}}(\gamma\varrho)\, K_{m+\frac{1}{2}}(\gamma r)$$

and for the special case $r \to \infty$

$$e^{\pm i\gamma\varrho\cos\varphi} = 2^\nu \Gamma(\nu) (\gamma\varrho)^{-\nu} \sum_{m=0}^{\infty} (\nu + m) (\pm i)^m J_{m+\nu}(\gamma\varrho) C_m^\nu(\cos\varphi),$$

$$e^{\gamma\varrho\cos\varphi} = 2^\nu \Gamma(\nu) (\gamma\varrho)^{-\nu} \sum_{m=0}^{\infty} (\nu + m) I_{\nu+m}(\gamma\varrho) C_m^\nu(\cos\varphi)$$

also for $\nu = \frac{1}{2}$

$$e^{\pm i\gamma\varrho\cos\varphi} = \left(\frac{2\gamma\varrho}{\pi}\right)^{-\frac{1}{2}} \sum_{m=0}^{\infty} (\pm i)^m (2m+1) J_{m+\frac{1}{2}}(\gamma\varrho) P_m(\cos\varphi),$$

$$e^{\gamma\varrho\cos\varphi} = \left(\frac{2\gamma\varrho}{\pi}\right)^{-\frac{1}{2}} \sum_{m=0}^{\infty} (2m+1) I_{m+\frac{1}{2}}(\gamma\varrho) P_m(\cos\varphi).$$

Another form of the addition theorem is due to GRAF.

Let

$$w = (z_1^2 + z_2^2 - 2z_1 z_2 \cos\varphi)^{\frac{1}{2}} = [(z_1 - z_2 e^{-i\varphi})(z_1 - z_2 e^{i\varphi})]^{\frac{1}{2}}$$

and let $Z_\nu(z)$ be any solution of Bessel's differential equation, then

$$\left(\frac{z_1 - z_2 e^{-i\varphi}}{z_1 - z_2 e^{i\varphi}}\right)^{\frac{1}{2}\nu} Z_\nu(w) = \sum_{n=-\infty}^{\infty} J_n(z_2) Z_{\nu+n}(z_1) e^{in\varphi}$$

and for the modified Bessel functions

$$\left(\frac{z_1 - z_2 e^{-i\varphi}}{z_1 - z_2 e^{i\varphi}}\right)^{\frac{1}{2}\nu} I_\nu(w) = \sum_{n=-\infty}^{\infty} (-1)^n I_n(z_2) I_{\nu+n}(z_1) e^{in\varphi}$$

$$\left(\frac{z_1 - z_2 e^{-i\varphi}}{z_1 - z_2 e^{i\varphi}}\right)^{\frac{1}{2}\nu} K_\nu(w) = \sum_{n=-\infty}^{\infty} I_n(z_2) K_{\nu+n}(z_1) e^{in\varphi}.$$

For representations of the above formulas in the form of integral expressions see 3.8.8.

3.10 Functions related to Bessel functions

3.10.1 Lommel's functions

A solution of the inhomogeneous Bessel differential equation

$$z^2 \frac{d^2 w}{dz^2} + z \frac{dw}{dz} + (z^2 - \nu^2) w = z^{\mu+1}$$

(μ, ν unrestricted constants) is

$$s_{\mu,\nu}(z) = \frac{z^{\mu+1}}{(\mu + \nu + 1)(\mu - \nu + 1)} {}_1F_2\left(1; \frac{\mu - \nu + 3}{2}, \frac{\mu + \nu + 3}{2}; -\frac{1}{4} z^2\right)$$

provided $\mu \pm \nu \neq -1, -2, -3, \ldots$

For arbitrary μ, ν a solution of the above differential equation is

$$S_{\mu,\nu}(z) = s_{\mu,\nu}(z) + 2^{\mu-1}\Gamma\left(\frac{\mu-\nu+1}{2}\right)\Gamma\left(\frac{\mu+\nu+1}{2}\right)$$

$$\times \operatorname{cosec}(\pi\nu)\left[\cos\left(\frac{1}{2}\pi\mu - \frac{1}{2}\pi\nu\right)J_{-\nu}(z) - \cos\left(\frac{1}{2}\pi\mu + \frac{1}{2}\pi\nu\right)J_{\nu}(z)\right]$$

$$= s_{\mu,\nu}(z) + 2^{\mu-1}\Gamma\left(\frac{\mu-\nu+1}{2}\right)\Gamma\left(\frac{\mu+\nu+1}{2}\right)$$

$$\times\left[\sin\left(\frac{1}{2}\pi\mu - \frac{1}{2}\pi\nu\right)J_{\nu}(z) - \cos\left(\frac{1}{2}\pi\mu - \frac{1}{2}\pi\nu\right)Y_{\nu}(z)\right],$$

$$s_{\mu,-\nu}(z) = s_{\mu,\nu}(z), \quad S_{\mu,-\nu}(z) = S_{\mu,\nu}(z).$$

When either of the numbers $\mu \pm \nu$ is an odd positive integer, $S_{\mu,\nu}(z)$ can be represented by the following terminating series in descending powers of z

$$S_{\mu,\nu}(z) = z^{\mu-1}\sum_{n=0}^{\infty}(-1)^n\left(\frac{1}{2} - \frac{1}{2}\mu + \frac{1}{2}\nu\right)_n\left(\frac{1}{2} - \frac{1}{2}\mu - \frac{1}{2}\nu\right)_n\left(\frac{z}{2}\right)^{-2n}.$$

The r. h. s. of the formula above gives in case of arbitrary μ, ν an asymptotic expansion for $S_{\mu,\nu}(z)$ for large z valid in $-\pi < \arg z < \pi$.

Recurrence relations

$$s_{\mu+2,\nu}(z) = z^{\mu+1} - [(\mu+1)^2 - \nu^2]\, s_{\mu,\nu}(z),$$

$$s'_{\mu,\nu}(z) + \left(\frac{\nu}{z}\right)s_{\mu,\nu}(z) = (\mu+\nu-1)\, s_{\mu-1,\nu-1}(z),$$

$$\left(\frac{2\nu}{z}\right)s_{\mu,\nu}(z) = (\mu+\nu-1)\, s_{\mu-1,\nu-1}(z) - (\mu-\nu-1)\, s_{\mu-1,\nu+1}(z),$$

$$2s'_{\mu,\nu}(z) = (\mu+\nu-1)\, s_{\mu-1,\nu-1}(z) + (\mu-\nu-1)\, s_{\mu-1,\nu-1}(z).$$

The same recurrence relations are valid when the $s_{\mu,\nu}(z)$ are replaced by $S_{\mu,\nu}(z)$

Integral representations for Lommel's functions

$$S_{\mu,\nu}(z) = z^{\mu}\int_0^{\infty} e^{-zt}\,{}_2F_1\left(\frac{1-\mu+\nu}{2}, \frac{1-\mu-\nu}{2}; \frac{1}{2}; -t^2\right)dt,$$

$$Re\, z > 0,$$

$$S_{\nu,\mu}(z) = z^{\mu+1}\int_0^{\infty} te^{-zt}\,{}_2F_1\left(\frac{1-\mu+\nu}{2}, \frac{1-\mu-\nu}{2}; \frac{3}{2}; -t^2\right)dt,$$

$$Re\, z > 0$$

equivalent with this is

$$2^{\nu}\pi^{-\frac{1}{2}}\frac{\Gamma\left(\frac{\mu+\nu}{2}\right)}{\Gamma\left(\frac{1+\mu-\nu}{2}\right)}z^{-\mu}S_{\mu,\nu}(z)$$

$$= \int_0^{\infty} e^{-zt}(1+t^2)^{\frac{1}{2}\mu-\frac{1}{2}}\left\{\sin\left(\frac{1}{2}\pi\mu + \frac{1}{2}\pi\nu\right)P^{\nu}_{\mu-1}\left[t(1+t^2)^{-\frac{1}{2}}\right]\right.$$

$$\left. + 2\pi^{-1}\cos\left(\frac{1}{2}\pi\mu + \frac{1}{2}\pi\nu\right)Q^{\nu}_{\mu-1}\left[t(1+t^2)^{-\frac{1}{2}}\right]\right\}dt,$$

$$Re\, z > 0,$$

$$2^{\nu}\pi^{-\frac{1}{2}}\frac{\Gamma\left(\frac{1+\nu-\mu}{2}\right)}{\Gamma\left(1-\frac{\nu+\mu}{2}\right)}z^{-\mu}S_{\mu,\nu}(z)$$

$$=\int_0^\infty e^{-zt}(1+t^2)^{\frac{1}{2}\mu-\frac{1}{2}}\left\{\cos\left(\frac{1}{2}\pi\mu-\frac{1}{2}\pi\nu\right)P^{\nu}_{-\mu}\left[t(1+t^2)^{-\frac{1}{2}}\right]\right.$$

$$\left.+\,2\pi^{-1}\sin\left(\frac{1}{2}\pi\mu-\frac{1}{2}\pi\nu\right)Q^{\nu}_{-\mu}\left[t(1+t^2)^{-\frac{1}{2}}\right]\right\}dt,$$

$$Re\, z>0,$$

$$2^{\frac{3}{2}+\mu}\pi^{\frac{1}{2}}\left[\Gamma\left(\frac{-\nu-\mu}{2}\right)\Gamma\left(\frac{\nu-\mu}{2}\right)\right]^{-1}z^{-\mu-1}S_{\mu,\nu}(z)$$

$$=\int_0^1(1-t^2)^{\frac{1}{2}\mu+\frac{1}{4}}P^{\mu+\frac{1}{2}}_{\nu-\frac{1}{2}}(t)\cos(zt)\,dt$$

$$+\int_1^\infty(t^2-1)^{\frac{1}{2}\mu+\frac{1}{4}}\mathfrak{P}^{\mu+\frac{1}{2}}_{\nu-\frac{1}{2}}(t)\cos(zt)\,dt,$$

$$2^{\frac{1}{2}+\mu}\pi^{\frac{1}{2}}\left[\Gamma\left(1-\frac{\mu+\nu}{2}\right)\Gamma\left(1-\frac{\mu-\nu}{2}\right)\right]^{-1}z^{-\mu}S_{\mu,\nu}(z)$$

$$=\int_0^1(1-t^2)^{\frac{1}{2}\mu-\frac{1}{4}}P^{\mu-\frac{1}{2}}_{\nu-\frac{1}{2}}(t)\sin(zt)\,dt$$

$$+\int_1^\infty(t^2-1)^{\frac{1}{2}\mu-\frac{1}{4}}\mathfrak{P}^{\mu-\frac{1}{2}}_{\nu-\frac{1}{2}}(t)\sin(zt)\,dt,$$

$$2^{-\mu-\frac{3}{2}}\pi^{\frac{1}{2}}\left[\Gamma\left(\frac{3+\mu+\nu}{2}\right)\Gamma\left(\frac{3+\mu-\nu}{2}\right)\right]^{-1}$$

$$\times(\nu-\mu-1)(-\nu-\mu-1)\,z^{-\mu-1}s_{\mu,\nu}(z)$$

$$=\int_0^1(1-t^2)^{\frac{1}{2}\mu+\frac{1}{4}}P^{-\mu-\frac{1}{2}}_{\nu-\frac{1}{2}}(t)\cos(zt)\,dt,$$

$$2^{-\mu-\frac{3}{2}}\pi^{\frac{1}{2}}\left[\Gamma\left(\frac{3+\mu+\nu}{2}\right)\Gamma\left(\frac{3+\mu-\nu}{2}\right)\right]^{-1}$$

$$\times(1+\mu+\nu)(1+\mu-\nu)\,z^{-\mu}s_{\mu,\nu}(z)$$

$$=\int_0^1(1-t^2)^{\frac{1}{2}\mu-\frac{1}{4}}P^{\frac{1}{2}-\mu}_{\nu-\frac{1}{2}}(t)\sin(zt)\,dt,$$

$$\Gamma\left(\frac{1-\mu-\nu}{2}\right)\Gamma\left(\frac{1-\mu+\nu}{2}\right)S_{\mu,\nu}(az)=\int_0^\infty t^{-\mu}(a^2+t^2)^{-1}K_\nu(zt)\,dt,$$

$$Re\,(\mu\pm\nu)<1,$$

$$2^{\frac{1}{2}(\nu-\mu-1)}\Gamma\Big(\frac{1+\nu-\mu}{2}\Big)a^{\nu}z^{-\frac{1}{2}(\nu+\mu+1)}S_{\mu,\nu}(az)$$
$$=\int_0^\infty t^{\frac{1}{2}(\nu-\mu+1)}(a^2+t^2)^{\frac{1}{2}(\nu+\mu-1)}K_{\frac{1}{2}(\nu-\mu-1)}(zt)\,dt,$$
$$Re\,(\mu-\nu)<1,$$
$$2^{\frac{1}{2}(\nu-\mu-1)}\Big[\Gamma\Big(\frac{\mu-\nu+1}{2}\Big)\Big]^{-1}a^{\nu}z^{-\frac{1}{2}(\nu+\mu+1)}s_{\mu,\nu}(az)$$
$$=\int_0^a t^{\frac{1}{2}(\nu-\mu+1)}(a^2-t^2)^{\frac{1}{2}(\nu+\mu-1)}J_{\frac{1}{2}(\mu-\nu+1)}(zt)\,dt,$$
$$Re\,(\nu+\mu)>-1,$$
$$z^{-\varrho-1}S_{\mu,\nu}(az)=2\Gamma\Big(\frac{3}{2}+\varrho-\frac{1}{2}\mu-\frac{1}{2}\nu\Big)$$
$$\times\Big[\Gamma(1+\varrho)\,\Gamma\Big(\frac{1}{2}-\frac{1}{2}\mu-\frac{1}{2}\nu\Big)\Big]^{-1}a^{-\nu}$$
$$\times\int_a^\infty t^{\nu-\varrho}(t^2-a^2)^{\varrho}S_{\mu-\varrho-1,\nu-\varrho-1}(tz)\,dt,$$
$$Re\,\varrho>-1,\quad Re\,(\mu+\nu)<1,$$
$$S_{0,\nu}(z)=\int_0^\infty e^{-z\sinh t}\cosh(\nu t)\,dt,$$
$$\nu S_{0,\nu}(z)=z\int_0^\infty e^{-z\sinh t}\sinh(\nu t)\cosh t\,dt,$$
$$S_{1,\nu}(z)=z\int_0^\infty e^{-z\sinh t}\cosh(\nu t)\cosh t\,dt,$$
$$Re\,z>0.$$

Differentiation with respect to a parameter

$$\Big[\frac{\partial S_{\mu,\nu}(z)}{\partial \nu}\Big]_{\nu=n}=\frac{1}{2}n!\,\frac{\Gamma\Big(\frac{1}{2}-\frac{1}{2}\mu-\frac{1}{2}n\Big)}{\Gamma\Big(\frac{1}{2}-\frac{1}{2}\mu+\frac{1}{2}n\Big)}$$
$$\times\sum_{l=0}^{n-1}\frac{z^{l-n}\Big(\frac{1}{2}-\frac{1}{2}\mu-\frac{1}{2}n\Big)_l}{l!\,(n-l)}S_{\mu+n-l,l}(z)$$
$$-\frac{1}{2}\Big[\psi\Big(\frac{1}{2}-\frac{1}{2}\mu+\frac{1}{2}n\Big)-\psi\Big(\frac{1}{2}-\frac{1}{2}\mu-\frac{1}{2}n\Big)\Big]S_{\mu,n}(z).$$

For instance

$$\Big[\frac{dS_{0,\nu}(z)}{d\nu}\Big]_{\nu=1}=2S_{-2,1}(z).$$

3.10.2 Special cases of Lommel's functions

$$s_{\nu,\nu}(z)=\pi^{\frac{1}{2}}2^{\nu-1}\Gamma\Big(\frac{1}{2}+\nu\Big)\boldsymbol{H}_{\nu}(z),$$

$$S_{\nu,\nu}(z) = \pi^{\frac{1}{2}} 2^{\nu-1} \Gamma\left(\frac{1}{2}+\nu\right) [\boldsymbol{H}_\nu(z) - Y_\nu(z)],$$

$$s_{\nu,\nu}(z) - S_{\nu,\nu}(z) = \pi^{\frac{1}{2}} 2^{\nu-1} \Gamma\left(\frac{1}{2}+\nu\right) Y_\nu(z),$$

$$s_{0,\nu}(z) = \frac{1}{2}\pi \operatorname{cosec}(\pi\nu) [\boldsymbol{J}_\nu(z) - \boldsymbol{J}_{-\nu}(z)],$$

$$S_{0,\nu}(z) = \frac{1}{2}\pi \operatorname{cosec}(\pi\nu) [\boldsymbol{J}_\nu(z) - \boldsymbol{J}_{-\nu}(z) - J_\nu(z) + J_{-\nu}(z)],$$

$$s_{-1,\nu}(z) = -\frac{1}{2}\pi\nu^{-1} \operatorname{cosec}(\pi\nu) [\boldsymbol{J}_\nu(z) + \boldsymbol{J}_{-\nu}(z)],$$

$$S_{-1,\nu}(z) = \frac{1}{2}\pi\nu^{-1} \operatorname{cosec}(\pi\nu) [J_\nu(z) + J_{-\nu}(z) - \boldsymbol{J}_\nu(z) - \boldsymbol{J}_{-\nu}(z)],$$

$$s_{1,\nu}(z) = 1 + \nu^2 s_{-1,\nu}(z); \quad S_{1,\nu}(z) = 1 + \nu^2 S_{-1,\nu}(z),$$

$$s_{0,\frac{1}{2}}(z) = \left(\frac{2\pi}{z}\right)^{\frac{1}{2}} [\sin z C(z) - \cos z S(z)],$$

$$S_{0,\frac{1}{2}}(z) = \left(\frac{2\pi}{z}\right)^{\frac{1}{2}} \left\{\cos z \left[\frac{1}{2} - S(z)\right] - \sin z \left[\frac{1}{2} - C(z)\right]\right\},$$

$$s_{-1,\frac{1}{2}}(z) = 2\left(\frac{2\pi}{z}\right)^{\frac{1}{2}} [\sin z\, S(z) + \cos z\, C(z)],$$

$$S_{-1,\frac{1}{2}}(z) = 2\left(\frac{2\pi}{z}\right)^{\frac{1}{2}} \left\{\cos z \left[\frac{1}{2} - C(z)\right] + \sin z \left[\frac{1}{2} - S(z)\right]\right\},$$

$$S_{\frac{1}{2},\frac{1}{2}}(z) = z^{-\frac{1}{2}}, \quad S_{\frac{3}{2},\frac{1}{2}}(z) = z^{\frac{1}{2}}, \quad S_{0,-1}(z) = z^{-1},$$

$$S_{-\frac{1}{2},\frac{1}{2}}(z) = z^{-\frac{1}{2}} [\sin z\, Ci(z) - \cos z\, si(z)],$$

$$S_{-\frac{3}{2},\frac{1}{2}}(z) = -z^{-\frac{1}{2}} [\sin z\, si(z) + \cos z\, Ci(z)],$$

$$\lim_{\mu=\nu} [\Gamma(\nu-\mu)]^{-1} s_{\mu-1,\nu}(z) = -2^{\nu-1}\Gamma(\nu) J_\nu(z),$$

$$S_{-1,0}(z) = \frac{1}{2}\sum_{m=0}^{\infty} \frac{(-1)^m \left(\frac{1}{2}z\right)^{2m}}{(m!)^2} \times \left\{\left[\log\left(\frac{1}{2}z\right) - \psi(m+1)\right]^2 - \frac{1}{2}\psi'(m+1) + \frac{\pi^2}{4}\right\},$$

$$S_{\nu-1,\nu}(z) = \frac{1}{4} z^\nu \Gamma(\nu) \sum_{m=0}^{\infty} \frac{(-1)^m \left(\frac{1}{2}z\right)^{2m}}{m!\,\Gamma(\nu+m+1)} \times \left[2\log\left(\frac{1}{2}z\right) - \psi(\nu+m+1) - \psi(m+1)\right] - 2^{\nu-2}\pi\,\Gamma(\nu) Y_\nu(z),$$

$$S_{\nu-2n-1,\nu}(z) = \sum_{m=0}^{n-1} \frac{(-1)^m z^{\nu-2n+2m}}{2^{2m+2}(-n)_{m+1}(\nu-n)_{m+1}} + \frac{(-1)^n S_{\nu-1,\nu}(z)}{2^{2n}\, n!\,(1-\nu)_n}.$$

Some series expansions

$$\sum_{n=1}^{\infty} (-1)^n\, n(n^2-z^2)^{-1}\, n^{-\mu} s_{\mu,\nu}(an) = -\frac{1}{2}\pi z^{-\mu} \operatorname{cosec}(\pi z)\, s_{\mu,\nu}(az),$$

$$0 < a < \pi, \quad Re\,\mu > -\frac{3}{2},$$

$$\sum_{n=0}^{\infty} (-1)^n\, \varepsilon_n (n^2-z^2)^{-1}\, n^{-\mu-1} s_{\mu,\nu}(an) = -\pi \operatorname{cosec}(\pi z)\, z^{-\mu-2} s_{\mu,\nu}(az),$$

$$0 < a < \pi, \quad Re\,\mu > -\frac{7}{2},$$

$$\sum_{n=1}^{\infty} (-1)^n\, n^{-\mu-3} s_{\mu,\nu}(an) = \frac{1}{2}\, \frac{a^{\mu+1}}{(\mu+1)^2-\nu^2} \left[\frac{a^2}{(\mu+3)^2-\nu^2} - \frac{\pi^2}{6}\right],$$

$$0 < a < \pi, \quad Re\,\mu > -\frac{7}{2}.$$

3.10.3 Struve's functions

The Struve function $w = \boldsymbol{H}_\nu(z)$ is a solution of the differential equation (inhomogeneous Bessel equation)

$$z^2 w'' + z w' + (z^2-\nu^2)\, w = \pi^{-\frac{1}{2}} \frac{4\left(\frac{1}{2}z\right)^{\nu+1}}{\Gamma\left(\nu+\frac{1}{2}\right)}.$$

Representations in the form of power series or integrals are

$$\boldsymbol{H}_\nu(z) = \sum_{m=0}^{\infty} (-1)^m \left(\frac{1}{2}z\right)^{\nu+2m+1} \left[\Gamma\left(m+\frac{3}{2}\right)\Gamma\left(\nu+m+\frac{3}{2}\right)\right]^{-1}$$

$$= 2\pi^{-\frac{1}{2}} \left(\frac{1}{2}z\right)^{\nu+1} \frac{{}_1F_2\left(1; \frac{3}{2}+\nu, \frac{3}{2}; -\frac{1}{4}z^2\right)}{\Gamma\left(\frac{3}{2}+\nu\right)}$$

$z^{-\nu}\boldsymbol{H}_\nu(z)$ is an entire function of ν and z.

$$\boldsymbol{H}_\nu(z)\, \Gamma\left(\frac{1}{2}+\nu\right) = 2\pi^{-\frac{1}{2}} \left(\frac{1}{2}z\right)^{\nu} \int_0^1 (1-t^2)^{\nu-\frac{1}{2}} \sin(zt)\, dt$$

$$= 2\pi^{-\frac{1}{2}} \left(\frac{1}{2}z\right)^{\nu} \int_0^{\frac{1}{2}\pi} \sin(z\cos t)\, (\sin t)^{2\nu}\, dt,$$

$$Re\,\nu > -\frac{1}{2},$$

$$\Gamma\left(\frac{1}{2}+\nu\right) [\boldsymbol{H}_\nu(z) - Y_\nu(z)] = 2\pi^{-\frac{1}{2}} \left(\frac{1}{2}z\right)^{\nu} \int_0^{\infty} (1+t^2)^{\nu-\frac{1}{2}} e^{-zt}\, dt,$$

$$Re\,z > 0,$$

$$H_\nu(z) = \left(\frac{2z}{\pi}\right)^{\frac{1}{2}} \int_0^{\frac{\pi}{2}} J_{\nu+\frac{1}{2}}(z \sin t)\,(\sin t)^{\frac{1}{2}-\nu}\,dt.$$

The modified Struve function is defined by

$$L_\nu(z) = -ie^{-i\frac{1}{2}\pi\nu} H_\nu\left(ze^{i\frac{1}{2}\pi}\right)$$

$$L_\nu(z) = \sum_{m=0}^{\infty} \left(\frac{1}{2}z\right)^{\nu+2m+1} \left[\Gamma\left(m+\frac{3}{2}\right)\Gamma\left(\nu+m+\frac{3}{2}\right)\right]^{-1}$$

$$= 2\pi^{-\frac{1}{2}} \left(\frac{1}{2}z\right)^{\nu+1} \frac{{}_1F_2\left(1;\frac{3}{2}+\nu,\frac{3}{2};\frac{1}{4}z^2\right)}{\Gamma\left(\frac{3}{2}+\nu\right)},$$

$$\Gamma\left(\frac{1}{2}+\nu\right) L_\nu(z) = 2\pi^{-\frac{1}{2}} \left(\frac{1}{2}z\right)^\nu \int_0^{\frac{1}{2}\pi} \sinh(z\cos t)\,(\sin t)^{2\nu}\,dt,$$

$$Re\,\nu > -\frac{1}{2},$$

$$\Gamma\left(\frac{1}{2}+\nu\right)[I_\nu(z) - L_\nu(z]] = 2\pi^{-\frac{1}{2}} \left(\frac{1}{2}z\right)^\nu \int_0^1 e^{-zt}(1-t^2)^{\nu-\frac{1}{2}}\,dt,$$

$$Re\,\nu > -\frac{1}{2},$$

$$\Gamma\left(\frac{1}{2}+\nu\right)[I_\nu(z - L_\nu(z)] = 2\pi^{-\frac{1}{2}} \left(\frac{1}{2}z\right)^\nu \int_0^1 e^{zt}(1-t^2)^{\nu-\frac{1}{2}}\,dt,$$

$$Re\,\nu > -\frac{1}{2},$$

$$\Gamma\left(\frac{1}{2}+\nu\right)[I_{-\nu}(x) - L_\nu(x)] = 2\pi^{-\frac{1}{2}} \left(\frac{1}{2}x\right)^\nu \int_0^\infty (1+t^2)^{\nu-\frac{1}{2}} \sin(xt)\,dt,$$

$$x > 0, \quad Re\,\nu < \frac{1}{2},$$

$$L_\nu(z) = \left(\frac{2z}{\pi}\right)^{\frac{1}{2}} \int_0^{\frac{\pi}{2}} I_{\nu+\frac{1}{2}}(z\sin t)\,(\sin t)^{\frac{1}{2}-\nu}\,dt.$$

For further representation use 3.7 and the fact that

$$s_{\nu,\nu}(z) = \pi^{\frac{1}{2}}\, 2^{\nu-1}\, \Gamma\left(\frac{1}{2}+\nu\right) H_\nu(z),$$

$$S_{\nu,\nu}(z) = \pi^{\frac{1}{2}}\, 2^{\nu-1}\, \Gamma\left(\frac{1}{2}+\nu\right)[H_\nu(z) - Y_\nu(z)].$$

Functional relations

$$\frac{d}{dz}[z^\nu H_\nu(z)] = z^\nu H_{\nu-1}(z),$$

$$\frac{d}{dz}[z^{-\nu} H_\nu(z)] = \frac{2^{-\nu}\pi^{-\frac{1}{2}}}{\Gamma\left(\frac{3}{2}+\nu\right)} - z^\nu H_{\nu+1}(z),$$

$$\boldsymbol{H}_\nu(z e^{im\pi}) = e^{i\pi m(\nu+1)} \boldsymbol{H}_\nu(z),$$

$$\boldsymbol{L}_\nu(z e^{im\pi}) = e^{i\pi m(\nu+1)} \boldsymbol{L}_\nu(z),$$

$$m = \pm 1, \pm 2, \ldots$$

Half an odd integer order

$$\boldsymbol{H}_{n+\frac{1}{2}}(z) = Y_{n+\frac{1}{2}}(z) + \pi^{-\frac{1}{2}} \sum_{m=0}^{n} \left(\frac{1}{2}z\right)^{-2m+n-\frac{1}{2}} \frac{(2m)!\, 2^{-2m}}{m!\,(n-m)!},$$

$$\boldsymbol{L}_{n+\frac{1}{2}}(z) = I_{n+\frac{1}{2}}(z) + 2\pi^{-1}(-1)^n K_{n+\frac{1}{2}}(z)$$

$$- \pi^{-\frac{1}{2}} \sum_{m=0}^{n} (-1)^m \left(\frac{1}{2}z\right)^{-2m+n-\frac{1}{2}} \frac{(2m)!\, 2^{-2m}}{m!\,(n-m)!},$$

$$\boldsymbol{H}_{-n-\frac{1}{2}}(z) = (-1)^n J_{n+\frac{1}{2}}(z); \qquad \boldsymbol{L}_{-n-\frac{1}{2}}(z) = I_{n+\frac{1}{2}}(z).$$

In these formulas $n = 0, 1, 2, \ldots$

$$\boldsymbol{H}_{\frac{1}{2}}(z) = \left(\frac{1}{2}\pi z\right)^{-\frac{1}{2}} (1 - \cos z); \quad \boldsymbol{L}_{-\frac{1}{2}}(z) = \left(\frac{1}{2}\pi z\right)^{-\frac{1}{2}} (1 - \cosh z)$$

integer order $n = 1, 2, 3, \ldots$

$$\boldsymbol{H}_n(z) = 4\pi^{-1} \sum_{m=0}^{\leq \frac{1}{2}n} \frac{(2m)!\,(n-m)!}{m!\,(2n-2m)!} (2z)^{n-2m-1} - \boldsymbol{E}_n(z),$$

$$\boldsymbol{H}_{-n}(z) = 4\pi^{-1} \sum_{m=0}^{\leq \frac{1}{2}n} \frac{(2n-2m-2)!\,(m+1)!}{(n-m-1)!\,(2m+2)!} (2z)^{2m-n+1} - \boldsymbol{E}_{-n}(z).$$

Asymptotic expansions

For large argument z and fixed order ν there exist the following asymptotic expansions

$$\boldsymbol{H}_\nu(z) = Y_\nu(z) + \frac{\left(\frac{1}{2}z\right)^{\nu-1} \pi^{-\frac{1}{2}}}{\Gamma\left(\frac{1}{2}+\nu\right)} \left[\sum_{m=0}^{p-1}{}' \frac{(-1)^m \left(\frac{1}{2}-\nu\right)_m (2m)!}{m!\, z^{2m}} + O(z^{-2p}) \right],$$

$$-\pi < \arg z < \pi,$$

$$\boldsymbol{L}_\nu(z) = I_\nu(z) - \frac{\left(\frac{1}{2}z\right)^{\nu-1} \pi^{-\frac{1}{2}}}{\Gamma\left(\frac{1}{2}+\nu\right)} \left[\sum_{m=0}^{p-1} \frac{\left(\frac{1}{2}-\nu\right)_m (2m)!}{m!\, z^{2m}} + O(z^{-2p}) \right],$$

$$-\frac{\pi}{2} < \arg z < \frac{\pi}{2}.$$

Some integrals involving Struve functions

$$\int_0^\infty t^{\varrho-1} \boldsymbol{H}_\nu(at)\,dt = \frac{1}{2}\left(\frac{1}{2}a\right)^{-\varrho} \tan\left[\frac{1}{2}\pi(\nu+\varrho)\right] \frac{\Gamma\left(\frac{1}{2}\nu+\frac{1}{2}\varrho\right)}{\Gamma\left(1+\frac{1}{2}\nu-\frac{1}{2}\varrho\right)},$$

$$Re\,\varrho < \frac{3}{2}, \quad -1 < Re\,(\varrho+\nu) < 1,$$

$$\int_0^\infty \boldsymbol{H}_\nu(t)\,\boldsymbol{H}_\mu(t)\,t^{-\mu-\nu}\,dt = 2^{-\nu-\mu}\pi^{\frac{1}{2}}\,\Gamma(\mu+\nu)$$

$$\times\left[\Gamma\left(\frac{1}{2}+\nu\right)\Gamma\left(\frac{1}{2}+\mu\right)\Gamma\left(\frac{1}{2}+\nu+\mu\right)\right]^{-1},$$

$$0 < Re\,(\mu+\nu) < 1,$$

$$\int_0^a t^{1-\nu}(a^2-t^2)^\mu\,\boldsymbol{H}_\nu(ty)\,dt$$

$$= 2^{-\nu}\Gamma(1+\mu)\left[\Gamma\left(\frac{3}{2}+\mu\right)\Gamma\left(\frac{1}{2}+\nu\right)\right]^{-1}$$

$$\times\, a^{\mu-\nu+1} y^{-\mu-1} s_{\mu+\nu+1,\mu-\nu+1}(ay),$$

$$Re\,\mu > -1,$$

$$\int_a^\infty t^{1+\nu}(t^2-a^2)^\mu\,[\boldsymbol{H}_\nu(ty) - Y_\nu(ty)]\,dt$$

$$= -\frac{1}{2}a^{\nu+\mu+1}\Gamma(1+\mu)\cos(\pi\nu)\sec[\pi(\mu+\nu)]\,z^{-\mu-1}$$

$$\times\,[\boldsymbol{H}_{\mu+1+\nu}(ay) - Y_{\mu+1+\nu}(ay)],$$

$$Re\,\mu > -1, \qquad Re\,(\mu+\nu) < -\frac{1}{2},$$

$$\int_1^\infty \boldsymbol{H}_\nu(2zt)\,(t^2-1)^{-\nu-\frac{1}{2}}\,t^{-\nu}\,dt = \frac{1}{2}\pi^{\frac{1}{2}}\,\Gamma\left(\frac{1}{2}-\nu\right)z^\nu\,[J_\nu(z)]^2,$$

$$z > 0, \quad -\frac{1}{2} < Re\,\nu < \frac{1}{2}.$$

Some series expansions

$$\Gamma\left(\frac{3}{2}+\nu\right)\sum_{m=1}^\infty\left(\frac{1}{2}mx\right)^{-\nu-1}\cos(mt)\,\boldsymbol{H}_\nu(mx) = -\left(\frac{1}{2}+\nu\right)\pi^{-\frac{1}{2}}$$

$$= -\pi^{-\frac{1}{2}} + \pi^{\frac{1}{2}}x^{-1}\left(1-\frac{t^2}{x^2}\right)^{\nu+\frac{1}{2}}{}_2F_1\left(\frac{1}{2}+\nu, \frac{1}{2}; \frac{3}{2}+\nu; 1-\frac{t^2}{x^2}\right),$$

$$Re\,\nu > -1.$$

The first result is valid in $0 < x < t < \pi$, the second in $0 < t < x < \pi$

$$\pi J_\nu(x) = 2^{3-\nu}\sum_{m=1}^\infty m^{1-\nu}(4m^2-1)^{-1}\,\boldsymbol{H}_\nu(2mx),$$

$$Re\,\nu > -\frac{3}{2}, \quad 0 \leq x \leq \pi, \quad \nu \geq -\frac{1}{2},$$

$$\sum_{n=1}^{\infty} n(n^2 - z^2)^{-1} n^{-\nu} \boldsymbol{H}_\nu(nx) = \frac{1}{2}\pi z^{-\nu}[J_\nu(zx) - \cot(\pi z)\boldsymbol{H}_\nu(zx)],$$
$$0 < x < 2\pi,$$

$$\sum_{n=1}^{\infty} (-1)^n n(n^2 - z^2)^{-1} n^{-\nu} \boldsymbol{H}_\nu(nx) = -\frac{1}{2}\pi z^{-\nu} \operatorname{cosec}(\pi z)\boldsymbol{H}_\nu(zx),$$
$$Re\,\nu > -\frac{3}{2}, \quad 0 < x < \pi,$$

$$\sum_{n=0}^{\infty} (-1)^n \varepsilon_n (n^2 - z^2)^{-1} n^{-\nu-1} \boldsymbol{H}_\nu(nx) = -\pi \operatorname{cosec}(\pi z)\, z^{-\nu-2} \boldsymbol{H}_\nu(zx),$$
$$Re\,\nu > -\frac{7}{2}, \quad 0 < x < \pi.$$

3.10.4 The functions of Anger and Weber

Anger's function $\boldsymbol{J}_\nu(z)$ and Weber's function $\boldsymbol{E}_\nu(z)$ are defined by integrals of the Bessel type

$$\boldsymbol{J}_\nu(z) \pm i\,\boldsymbol{E}_\nu(z) = \pi^{-1}\int_0^{\pi} e^{\pm i(\nu t - z\sin t)}\,dt,$$

also

$$\boldsymbol{J}_\nu(z) = J_\nu(z) + \pi^{-1}\sin(\nu\pi)\int_0^{\infty} e^{-z\sinh t - \nu t}\,dt,$$
$$Re\,z > 0,$$

$$\boldsymbol{E}_\nu(z) = -Y_\nu(z) - \pi^{-1}\int_0^{\infty} (e^{\nu t} + e^{-\nu t}\cos\pi\nu)\, e^{-z\sinh t}\,dt,$$
$$Re\,z > 0.$$

or, expressed as power series expansions

$$\boldsymbol{J}_\nu(z) = \cos\left(\frac{1}{2}\pi\nu\right)\sum_{m=0}^{\infty}(-1)^m\left(\frac{1}{2}z\right)^{2m}$$
$$\times\left[\Gamma\left(m+1+\frac{1}{2}\nu\right)\Gamma\left(m+1-\frac{1}{2}\nu\right)\right]^{-1}$$
$$+\sin\left(\frac{1}{2}\pi\nu\right)\sum_{m=0}^{\infty}(-1)^m\left(\frac{1}{2}z\right)^{2m+1}$$
$$\times\left[\Gamma\left(m+\frac{3}{2}+\frac{1}{2}\nu\right)\Gamma\left(m+\frac{3}{2}-\frac{1}{2}\nu\right)\right]^{-1},$$

$$\boldsymbol{E}_\nu(z) = \sin\left(\frac{1}{2}\pi\nu\right)\sum_{m=0}^{\infty}(-1)^m\left(\frac{1}{2}z\right)^{2m}$$
$$\times\left[\Gamma\left(m+1+\frac{1}{2}\nu\right)\Gamma\left(m+1-\frac{1}{2}\nu\right)\right]^{-1}$$
$$-\cos\left(\frac{1}{2}\pi\nu\right)\sum_{m=0}^{\infty}(-1)^m\left(\frac{1}{2}z\right)^{2m+1}$$
$$\times\left[\Gamma\left(m+\frac{3}{2}+\frac{1}{2}\nu\right)\Gamma\left(m+\frac{3}{2}-\frac{1}{2}\nu\right)\right]^{-1},$$

$$\mathbf{J}_n(z) = J_n(z); \qquad n = 0, 1, 2, \ldots,$$

$$\mathbf{E}_0(z) = -\mathbf{H}_0(z),$$

$$\left[\frac{\partial}{\partial\nu}\mathbf{J}_\nu(z)\right]_{\nu=0} = \frac{1}{2}\pi\,\mathbf{H}_0(z); \quad \left[\frac{\partial}{\partial\nu}\mathbf{E}_\nu(z)\right]_{\nu=0} = \frac{1}{2}\pi J_0(z).$$

Integrals expressed in terms of $\mathbf{J}_\nu(z)$ and $\mathbf{E}_\nu(z)$

$$\int_0^{\frac{1}{2}\pi} \cos(z\cos t)\cos(\nu t)\,dt = \frac{1}{4}\pi\sec\left(\frac{1}{2}\pi\nu\right)[\mathbf{J}_\nu(z) + \mathbf{J}_{-\nu}(z)]$$

$$= \frac{1}{4}\pi\operatorname{cosec}\left(\frac{1}{2}\pi\nu\right)[\mathbf{E}_\nu(z) - \mathbf{E}_{-\nu}(z)],$$

$$\int_0^{\frac{1}{2}\pi} \sin(z\cos t)\cos(\nu t)\,dt = \frac{1}{4}\pi\operatorname{cosec}\left(\frac{1}{2}\pi\nu\right)[\mathbf{J}_\nu(z) - \mathbf{J}_{-\nu}(z)]$$

$$= -\frac{1}{4}\pi\sec\left(\frac{1}{2}\pi\nu\right)[\mathbf{E}_\nu(z) + \mathbf{E}_{-\nu}(z)].$$

Functional relations

$$\sin(\pi\nu)\,\mathbf{J}_\nu(z) = \cos(\pi\nu)\,\mathbf{E}_\nu(z) - \mathbf{E}_{-\nu}(z),$$

$$\sin(\pi\nu)\,\mathbf{E}_\nu(z) = \mathbf{J}_{-\nu}(z) - \cos(\pi\nu)\,\mathbf{J}_\nu(z),$$

$$\mathbf{J}_{\nu-1}(z) + \mathbf{J}_{\nu+1}(z) = 2\nu z^{-\nu}\,\mathbf{J}_\nu(z) - 2(\pi z)^{-1}\sin(\pi\nu),$$

$$\mathbf{E}_{\nu-1}(z) + \mathbf{E}_{\nu+1}(z) = 2\nu z^{-1}\,\mathbf{E}_\nu(z) - 2(\pi z)^{-1}(1 - \cos\pi\nu),$$

$$\mathbf{J}_{\nu-1}(z) - \mathbf{J}_{\nu+1}(z) = 2\,\mathbf{J}'_\nu(z),$$

$$\mathbf{E}_{\nu-1}(z) - \mathbf{E}_{\nu+1}(z) = 2\,\mathbf{E}'_\nu(z).$$

The functions $\mathbf{J}_\nu(z)$ and $\mathbf{E}_\nu(z)$ satisfy the respective differential equations

$$w'' + z^{-1}w' + (1 - \nu^2 z^{-2})\,w = \pi^{-1}z^{-2}(z - \nu)\sin(\pi\nu)$$

and

$$w'' + z^{-1}w' + (1 - \nu^2 z^{-2})\,w = -\pi^{-1}z^{-2}[z + \nu + (z - \nu)\cos(\pi\nu)].$$

Asymptotic expansions

One has for fixed order ν and large argument z the following in the domain $-\pi < \arg z < +\pi$ valid asymptotic expansions

$$\mathbf{J}_\nu(z) = J_\nu(z) + (\pi z)^{-1}\sin(\pi\nu)\left\{\sum_{m=0}^{M-1}(-1)^m\,2^{2m}\left(\frac{1}{2} + \frac{1}{2}\nu\right)_m\left(\frac{1}{2} - \frac{1}{2}\nu\right)_m\right.$$

$$\times z^{-2m} + O(z^{-2M}) + \nu\sum_{m=0}^{M-1}(-1)^m\,2^{2m}\left(1 + \frac{1}{2}\nu\right)_m\left(1 - \frac{1}{2}\nu\right)_m$$

$$\left.\times z^{-2m-1} + \nu O(|z|^{-2M-1})\right\},$$

$$\boldsymbol{E}_\nu(z) = -Y_\nu(z) - (\pi z)^{-1}(1 + \cos\pi\nu)$$
$$\times\left\{\sum_{m=0}^{M-1}(-1)^m 2^{2m}\left(\frac{1}{2}+\frac{1}{2}\nu\right)_m\left(\frac{1}{2}-\frac{1}{2}\nu\right)_m z^{-2m} + O(z^{-2M})\right\}$$
$$-\nu(\pi z)^{-1}(1-\cos\pi\nu)$$
$$\times\left\{\sum_{m=0}^{M-1}(-1)^m 2^{2m}\left(1+\frac{1}{2}\nu\right)_m\left(1-\frac{1}{2}\nu\right)_m z^{-2m-1} + O(z^{-2M-1})\right\}.$$

Some special cases

$$\boldsymbol{J}_{-\frac{1}{2}}(z) = \boldsymbol{E}_{\frac{1}{2}}(z) = \left(\frac{1}{2}\pi z\right)^{-\frac{1}{2}}\{\cos z\,[C(z)+S(z)] - \sin z\,[C(z)-S(z)]\},$$

$$\boldsymbol{J}_{\frac{1}{2}}(z) = -\boldsymbol{E}_{-\frac{1}{2}}(z) = \left(\frac{1}{2}\pi z\right)^{-\frac{1}{2}}\{\cos z\,[C(z)-S(z)] + \sin z\,[C(z)+S(z)]\}.$$

Some series expansions

$$\sum_{n=0}^{\infty}\varepsilon_n(n^2-z^2)^{-1}[\boldsymbol{J}_\nu(an)+\boldsymbol{J}_{-\nu}(an)]$$
$$= -\pi z^{-1}\left\{\cot(\pi z)\,[\boldsymbol{J}_\nu(az)+\boldsymbol{J}_{-\nu}(az)] + \cot\left(\frac{1}{2}\pi\nu\right)\boldsymbol{J}_\nu(az) - \boldsymbol{J}_{-\nu}(az)]\right\},$$
$$0 < a < 2\pi,$$

$$\sum_{n=0}^{\infty}(-1)^n\varepsilon_n(n^2-z^2)^{-1}[\boldsymbol{J}_\nu(an)+\boldsymbol{J}_{-\nu}(an)]$$
$$= -\pi z^{-1}\operatorname{cosec}(\pi z)\,[\boldsymbol{J}_\nu(az)+\boldsymbol{J}_{-\nu}(az)],$$
$$0 < a < \pi,$$

$$\sum_{n=0}^{\infty}\varepsilon_n(n^2-z^2)^{-1}[\boldsymbol{J}_{cn}(b)+J_{-cn}(b)]$$
$$= -z^{-1}\pi\,[\boldsymbol{J}_{cz}(b)+\boldsymbol{J}_{-cz}(b)]\cdot\left[\cot(\pi z)+\tan\left(\frac{1}{2}\pi z\right)\right],$$
$$0 < c < 2,$$

$$\sum_{n=0}^{\infty}(-1)^n\varepsilon_n(n^2-z^2)^{-1}[\boldsymbol{J}_{cn}(b)+\boldsymbol{J}_{-cn}(b)]$$
$$= -\pi z^{-1}\operatorname{cosec}(\pi z)\,[\boldsymbol{J}_{cz}(b)+\boldsymbol{J}_{-cz}(b)],$$

$$\sum_{n=1}^{\infty}(-1)^n n(n^2-z^2)^{-1}[\boldsymbol{J}_\nu(na)+\boldsymbol{J}_{-\nu}(na)]$$
$$= -\frac{1}{2}\pi\operatorname{cosec}(\pi z)\,[\boldsymbol{J}_\nu(az)-\boldsymbol{J}_{-\nu}(az)].$$

3.11 Polynomials related to Bessel functions

3.11.1 Lommel's polynomials

The Lommel polynomials are defined by

$$R_{m,\nu}(z) = (\nu)_m \left(\frac{1}{2} z\right)^{-m} {}_2F_3\left(\frac{1}{2} - \frac{1}{2} m, -\frac{1}{2} m; \nu, -m, 1-\nu-m; -z^2\right),$$

$m = 0, 1, 2, \ldots$, ν arbitrary. They are polynomials in z^{-1} (degree m) and they connect three Bessel functions of order ν, $\nu + m$ and $\nu + m - 1$ such that

$$J_{\nu+m}(z) = J_\nu(z)\, R_{m,\nu}(z) - J_{\nu-1}(z)\, R_{m-1,\nu+1}(z),$$

$$(-1)^m J_{-\nu-m}(z) = J_{-\nu}(z)\, R_{m,\nu}(z) + J_{-\nu+1}(z)\, R_{m-1,\nu+1}(z),$$

$$m = 1, 2, 3, \ldots$$

The $R_{m,\nu}(z)$ can be expressed in terms of Bessel functions.

$$2(\pi z)^{-1} \sin(\pi\nu)\, R_{m,\nu}(z) = J_{\nu+m}(z)\, J_{-\nu+1}(z) + (-1)^m J_{-\nu-m}(z)\, J_{\nu-1}(z).$$

$$m = 1, 2, 3, \ldots$$

One has also explicitly

$$R_{m,\nu}(z) = \sum_{n=0}^{\leq \frac{1}{2} m} \frac{(-1)^n (m-n)!\, \Gamma(\nu+m-n)}{n!\,(m-2n)!\,\Gamma(\nu+n)} \left(\frac{1}{2} z\right)^{-m+2n}$$

When ν is a negative integer replace the quotient in the above series

$$\frac{\Gamma(\nu+m-n)}{\Gamma(\nu+n)} \text{ by } (-1)^m \frac{\Gamma(-\nu-n+1)}{\Gamma(-\nu-m+n+1)}$$

in part of the series.

The $R_{m,\nu}(z)$ are solutions of the differential equation.

$$[(\vartheta+m)(\vartheta+2\nu+m-2)(\vartheta-2\nu-m)(\vartheta-m-2)]\, y$$
$$+ 4z^2\vartheta(\vartheta+1)\, y = 0,$$

with $\vartheta = z\left(\frac{d}{dz}\right)$.

Differential formulas

$$R'_{m,\nu}(z) = -m z^{-1} R_{m,\nu}(z) + R_{m-1,\nu}(z) - R_{m-1,\nu+1}(z)$$
$$= (2\nu+m)\, z^{-1} R_{m,\nu}(z) - R_{m-1,\nu+1}(z) - R_{m+1,\nu}(z)$$
$$= -(2\nu+m-2)\, z^{-1} R_{m,\nu}(z) + R_{m+1,\nu-1}(z) + R_{m-1,\nu}(z).$$

Especially

$$R_{0,\nu} = 1, \quad R_{1,\nu} = 2\nu z^{-1}, \quad R_{2,\nu} = 4\nu(\nu+1)\, z^{-2} - 1,$$
$$R_{-1,\nu}(z) = 0, \quad R_{-2,\nu}(z) = -1,$$

$$J^2_{m+\frac{1}{2}}(z) + J^2_{-m-\frac{1}{2}}(z) = 2(\pi z)^{-1} \sum_{n=0}^{m} \frac{(2z)^{2n-2m}\,(2m-n)!\,(2m-2n)!}{[(m-n)!]^2\, n!}.$$

Recurrence relations

$$R_{m,\nu}(z)\, R_{m-n+1,\nu+n}(z) - R_{m+1,\nu}(z)\, R_{m-n,\nu+n}(z) = R_{n-1,\nu}(z),$$

$$R_{n,\nu}(z)\, R_{p-m-1,\nu+m+1}(z) + R_{p,\nu}(z)\, R_{m-n-1,\nu+n+1}(z)$$
$$+ R_{m,\nu}(z)\, R_{n-p-1,\nu+p+1}(z) = 0.$$

Hurwitz's limit of a Lommel polynomial

$$\lim_{m\to\infty} \frac{\left(\frac{1}{2}z\right)^{\nu+m} R_{m,\nu+1}(z)}{\Gamma(\nu+m+1)} = J_\nu(z).$$

3.11.2 Neumann's polynomials

Neumann's polynomials $O_n(z)$ are defined by the equation

$$(z-y)^{-1} = \sum_{n=0}^{\infty} \varepsilon_n J_n(y)\, O_n(z),$$
$$|y| < |z|.$$

They are of importance in the theory of the expansion of an analytic function $f(z)$ into a series.

$$f(z) = \sum_{n=0}^{\infty} a_n J_n(z).$$

(Neumann series; see 3.12.1).

Explicitly

$$O_{2n}(z) = \frac{1}{2} n \sum_{m=0}^{n} [(n-m)!]^{-1} (n+m-1)! \left(\frac{1}{2}z\right)^{-2m-1},$$

$$O_{2n+1}(z) = \frac{1}{2}\left(n+\frac{1}{2}\right) \sum_{m=0}^{n} [(n-m)!]^{-1} (n+m)! \left(\frac{1}{2}z\right)^{-2m-2},$$

$$O_n(z) = \frac{1}{4} \sum_{m=0}^{\leqq \frac{1}{2}n} n(n-m-1)! \left(\frac{1}{2}z\right)^{-2m-n-1} (m!)^{-1}.$$

In particular

$$O_0(z) = z^{-1}, \quad O_1(z) = z^{-2}, \quad O_2(z) = z^{-1} + 4z^{-3}.$$

Obviously $O_n(z)$ is a polynomial in z^{-1} of degree $n+1$.

There exists the inequality

$$|O_n(z)| \leq 2^{n-1} n!\, |z|^{-n-1}\, e^{\frac{1}{4}|z|^2}.$$

The series $\sum_{n=0}^{\infty} a_n J_n(y)\, O_n(z)$ is absolutely convergent whenever the series $\sum_{0}^{\infty} a_n \left(\frac{y}{z}\right)^n$ is absolutely convergent.

Recurrenc relaticns and differential formulas

$$O'(z) = -O_1(z) = -z^{-2},$$

$$2O'_n(z) = O_{n-1}(z) - O_{n+1}(z),$$

$$(n-1)\,O_{n+1}(z) + (n+1)\,O_{n-1}(z) - 2z^{-1}(n^2-1)\,O_n(z)$$
$$= 2\pi z^{-1}\left(\sin\frac{1}{2}n\pi\right)^2,$$

$$nzO_{n-1}(z) - (n^2-1)\,O_n(z) = (n-1)\,zO'_n(z) + n\left(\sin\frac{1}{2}n\pi\right)^2,$$

$$nzO_{n+1}(z) - (n^2-1)\,O_n(z) = -(n+1)\,zO'_n(z) + n\left(\sin\frac{1}{2}n\pi\right)^2,$$

$v = O_n(z)$ is a solution of the differential equation

$$z^2\frac{d^2v}{dz^2} + 3z\frac{dv}{dz} + (z^2+1-n^2)\,v = z\left(\cos\frac{1}{2}n\pi\right)^2 + n\left(\sin\frac{1}{2}n\pi\right)^2.$$

If C denotes a simple closed contour around $z = 0$ then

$$\int_C O_m(z)\,O_n(z)\,dz = 0, \qquad m, n = 0, 1, 2, \ldots,$$

$$\int_C J_m(z)\,O_n(z)\,dz = 0, \qquad m \neq n,$$
$$= \pi i, \qquad m = n \geq 1.$$

Consider the following expansion

$$(z^2-y^2)^{-1} = \sum_{n=0}^{\infty} \varepsilon_n \Omega_n(z)\,[J_n(y)]^2, \qquad |y| < |z|.$$

Then the $\Omega_n(z)$ are likewise polynomials in z^{-1}. The $\Omega_n(z)$ are likewise called Neumann polynomials. A generalisation of Neumann's polynomials are given by the following defining expansions.

$$y^\nu(z-y)^{-1} = \sum_{n=0}^{\infty} A_{n,\nu}(z)\,J_{\nu+n}(y),$$

$$y^{\nu+\mu}(z-y)^{-1} = \sum_{n=0}^{\infty} B_{n;\mu,\nu}(z)\,J_{\mu+\frac{1}{2}n}(y)\,J_{\nu+\frac{1}{2}n}(y),$$
$$|y| < |z|,$$

$$A_{n,\nu}(z) = z^{-n-1}\,2^{\nu+n}(\nu+n)\sum_{m=0}^{\leq\frac{1}{2}n}(m!)^{-1}\,\Gamma(\nu+n-m)\left(\frac{1}{2}z\right)^{2m}.$$

For further results on $A_{n,\nu}(z)$ and $B_{n;\mu,\nu}(z)$ see G. N. Watson, Theory of Bessel functions.

3.11.3 Schlaefli's polynomials

They are given by

$$S_0(z) = 0, \quad S_n(z) = \sum_{m=0}^{\leq \frac{1}{2} n} (n - m - 1)! \left(\frac{1}{2} z\right)^{-n+2m} (m!)^{-1}.$$

They are connected with the Neumann polynomials by

$$n S_n(z) = 2z O_n(z) - 2\left(\cos \frac{1}{2} \pi n\right)^2,$$

$$n = 1, 2, 3, \ldots .$$

They satisfy the relations

$$S_{-n}(z) = (-1)^{n+1} S_n(z),$$

$$n = 1, 2, 3, \ldots,$$

$$S_n(z - y) = \sum_{m=-\infty}^{\infty} S_{n+m}(z) J_m(y),$$

$$\left[z^2 \frac{d^2}{dz^2} + z \frac{d}{dz} + (z^2 - n^2)\right] S_n(z)$$

$$= 2z \sin^2\left(\frac{1}{2} \pi n\right) + 2n \cos^2\left(\frac{1}{2} \pi n\right).$$

3.12 Series of arbitrary functions in terms of Bessel functions

3.12.1 Neumann series

A Neumann series is a series of the type

$$\sum_0^{\infty} a_n J_{\nu+n}(z).$$

Its domain of convergence is a circle. Its radius of convergence is the same as the one of the power series

$$\sum_{n=0}^{\infty} a_n \left(\frac{1}{2} z\right)^{\nu+n} [\Gamma(\nu + n + 1)]^{-1}.$$

The special case $\nu = 0$

$$f(z) = \sum_0^{\infty} a_n J_n(z),$$

$$|z| < c$$

is an analytic function and c is the distance of the nearest singularity of $f(z)$ from $z = 0$. Vice versa

$$2\pi i a_n = \int_{|z|<c'} f(t) O_n(t)\, dt,$$

$$0 < c' < c.$$

If

$$f(z) = \sum_{l=0}^{\infty} b_l z^l,$$

then $f(z)$ can be expressed as

$$f(z) = z^{-\nu} \sum_{n=0}^{\infty} a_n J_{\nu+n}(z)$$

with

$$a_n = 2^{\nu+n}(\nu + n) \sum_{s=0}^{\leq \frac{1}{2} n} 2^{-2s} \Gamma(\nu + n - s)\, b_{n-2s}/s!$$

and conversely

$$b_l \Gamma(\nu + l + 1) = 2^{-l-\nu} \sum_{m=0}^{\leq \frac{1}{2} l} (-1)^m \binom{\nu + l}{m} a_{l-2m}.$$

Simple examples are

$$\left(\frac{1}{2} z\right)^{\nu} = \sum_{n=0}^{\infty} (\nu + 2n)\, \Gamma(\nu + n)\, J_{\nu+2n}(z)\, (n!)^{-1},$$

$$\nu \neq -1, -2, -3, \ldots$$

Gegenbauer's formula

$$z^{\nu} e^{i\gamma z} = 2^{\nu} \Gamma(\nu) \sum_{n=0}^{\infty} i^n (\nu + n)\, C_n^{\nu}(\gamma)\, J_{\nu+n}(z),$$

$$\nu \neq 0, -1, -2, \ldots,$$

$$\Gamma(\nu + 1) \left(\frac{1}{2} a z\right)^{\mu-\nu} J_{\nu}(a z)$$

$$= \sum_{n=0}^{\infty} {}_2F_1(-n, n + \mu; \nu + 1; a^2)\, \Gamma(\mu + n)\, (\mu + 2n)\, J_{\mu+2n}(z)\, (n!)^{-1},$$

$$s_{\mu,\nu}(z) = 2^{\mu+1} \sum_{n=0}^{\infty} \frac{(\mu + 1 + 2n)\, \Gamma(\mu + 1 + n)}{n!\, [(2n + 1 + \mu)^2 - \nu^2]} J_{\mu+1+2n}(z).$$

For more results see 4.13.1.

A series of the type

$$\sum_{n=0}^{\infty} a_n J_{\mu+\frac{1}{2} n}(z)\, J_{\nu+\frac{1}{2} n}(z)$$

is called a Neumann series of the second type. If

$$f(z) = \sum_{l=0}^{\infty} b_l z^l,$$

then $f(z)$ can be expressed as

$$f(z) = z^{-\nu-\mu} \sum_{n=0}^{\infty} a_n J_{\mu+\frac{1}{2} n}(z)\, J_{\nu+\frac{1}{2} n}(z),$$

where

$$\Gamma\left(\nu+1+\frac{1}{2}n\right)\Gamma\left(\mu+1+\frac{1}{2}n\right)b_l$$

$$= 2^{-l-\nu-\mu} \sum_{m=1}^{\leq \frac{1}{2}l} (-1)^m \binom{l+\nu+\mu}{m} a_{l-2m}$$

and conversely

$$a_n = 2^{\nu+\mu+n}(\nu+\mu+n)$$

$$\times \sum_{s=0}^{\leq \frac{1}{2}n} 2^{-2s}\, b_{n-2s} \frac{\Gamma(\nu+\mu+n-s)\Gamma\left(\nu+1-s+\frac{1}{2}n\right)\Gamma\left(\mu+1-s+\frac{1}{2}n\right)}{s!\,\Gamma(\nu+\mu+n-2s+1)}$$

$$\mu, \nu, \mu+\nu \neq -1, -2, -3, \ldots$$

A simple example is:

$$\left(\frac{1}{2}z\right)^{\mu+\nu} = \Gamma(\nu+1)\Gamma(\mu+1) \sum_{n=0}^{\infty} \frac{\nu+\mu+2n}{\nu+\mu+n} \binom{\nu+\mu+n}{n}$$

$$\times J_{\nu+n}(z)\, J_{\mu+n}(z).$$

A modified form of Neumann's series is

$$\sum_0^{\infty} a_n z^n J_{\nu+n}(z).$$

A series of the form $\sum_{l=0}^{\infty} b_l z^{2l}$ can be converted into

$$\sum_{l=0}^{\infty} b_l z^{2l} = z^{-\nu} \sum_{n=0}^{\infty} a_n z^n J_{\nu+n}(z),$$

where

$$a_n = \sum_{s=0}^{n} \frac{\Gamma(\nu+s+1)}{(n-s)!} 2^{2s-n+\nu} b_s$$

and conversely

$$\Gamma(\nu+\mu+1)\, b_n = \sum_{s=0}^{n} (-1)^s\, 2^{-\nu-n-s} a_{n-s} (s!)^{-1}.$$

Examples are

$$J_\nu(\lambda z) = \lambda^\nu \sum_{n=0}^{\infty} \left[\frac{1}{2} z (1-\lambda^2)\right]^n J_{\nu+n}(z)\, (n!)^{-1},$$

$$\left(\frac{1}{2}z\right)^\nu = \Gamma(\nu+1) \sum_{n=0}^{\infty} \left(\frac{1}{2}z\right)^n J_{\nu+n}(z)\, (n!)^{-1}.$$

For further examples see 4.13.1. The formal expansion of a function $f(x)$ into a Neumann series is

$$f(x) = \sum_{n=0}^{\infty} (2\nu + 2 + 4n)\, J_{\nu+2n+1}(x) \int_0^{\infty} t^{-1} f(t)\, J_{\nu+2n+1}(t)\, dt.$$

For the conditions with respect to $f(t)$ see J. E. WILKINS: Bull. Amer. Math. Soc. 54, (1948) 232—234; Trans. Amer. Math. Soc. 64, (1948) 359—385; 69, (1950) 55—65; Amer. J. Math. 15, (1950) 187—191.

3.12.2 Kapteyn series

A Kapteyn series is of the form

$$\sum_{n=0}^{\infty} a_n J_{\nu+n}[(\nu + n)z]$$

it converges throughout a domain in which

$$\sum_{n=0}^{\infty} a_n [w(z)]^n; \quad w(z) = z e^{(1-z^2)^{\frac{1}{2}}} \left[1 + (1 - z^2)^{\frac{1}{2}}\right]^{-1}$$

is absolutely convergent. Expansion of a power of z into a Kapteyn series

$$\left(\frac{1}{2} z\right)^l = \left(\frac{1}{2} z\right)^{-\nu} (\nu + l)^2$$

$$\times \sum_{n=0}^{\infty} \Gamma(\nu + l + n)\, (\nu + l + 2n)^{-\nu-l-1}\, J_{\nu+l+2n}[(\nu + l + 2n)\, z]\, (n!)^{-1},$$

$$l = 0, 1, 2, \ldots$$

If

$$f(z) = \sum_{l=0}^{\infty} b_l z^l,$$

then also

$$f(z) = z^{-\nu} \sum_{n=0}^{\infty} a_n J_{\nu+n}[(\nu + n)z],$$

$$\nu \neq -1, -2, \ldots,$$

where

$$a_n = \frac{1}{2} \sum_{s=0}^{\leq \frac{1}{2} n} (\nu + n - 2s)^2\, \Gamma(\nu + n - s) \left(\frac{1}{2}\nu + \frac{1}{2} n\right)^{2s-n-\nu-1}.$$

This series is absolutely convergent when $|w(z)| < 1$ and $|w(z)| < |w(\varrho)|$. ϱ is the radius of convergence of the power series.

A Kapteyn series of the second kind is a series of the type

$$\sum_{n=0}^{\infty} a_n J_{\frac{1}{2}\nu+\frac{1}{2}n}\left[\left(\frac{1}{2}\nu+\frac{1}{2}\varrho+n\right)z\right] J_{\frac{1}{2}\varrho+\frac{1}{2}n}\left[\left(\frac{1}{2}\nu+\frac{1}{2}\varrho+n\right)z\right].$$

Further expansions

$$y^{\nu}(z-y)^{-1}=\sum_{n=0}^{\infty} A_{n,\nu}(z)\, J_{\nu+n}[(\nu+n)\, y],$$

$$0<|w(y)|<\operatorname{Min}(1,|w(z)|),$$

where

$$A_{n,\nu}(z)=\frac{1}{2}\sum_{m=0}^{\leqq\frac{1}{2}n}\left(\frac{1}{2}\nu+\frac{1}{2}n\right)^{2m-1-\nu-n} z^{2m-n-1}(m!)^{-1}$$

$$\times(\nu+n-2m)^2\,\Gamma(\nu+n-m).$$

3.12.3 Schloemilch series

A Schloemilch series is of the type

$$f(x)=\sum_{m=0}^{\infty}\left[a_m J_\nu(mx)+b_m \boldsymbol{H}_\nu(mx)\right]\left(\frac{1}{2}mx\right)^{-\nu}.$$

If one chooses a_m and b_m such that

$$\Gamma\left(\frac{1}{2}-\nu\right)a_m=m\pi^{-\frac{1}{2}}\int_{-\pi}^{\pi} t\sin(mt)$$

$$\times\left[\int_0^{\frac{1}{2}\pi} f(t\sin\vartheta)(\sin\vartheta)^{2\nu+1}(\cos\vartheta)^{-2\nu}\,d\vartheta\right]dt,$$

$$\Gamma\left(\frac{1}{2}-\nu\right)b_m=-m\pi^{-\frac{1}{2}}\int_{-\pi}^{\pi} t\cos(tm)$$

$$\times\left[\int_0^{\frac{1}{2}\pi} f(t\sin\vartheta)(\sin\vartheta)^{2\nu+1}(\cos\vartheta)^{-2\nu}\,d\vartheta\right]dt,$$

then the series above is called the Schloemilch series of $f(x)$. The theory of this expansion is given by R. G. Cooke, Proc. London Math. Soc. 28, (1928), 201—247. Simple examples are

$$\pi(ax)^{-\nu}\boldsymbol{H}_\nu(ax)=-2^{-\nu+1}\sin(a\pi)\sum_{m=1}^{\infty}$$

$$\times(-1)^m m(m^2-a^2)^{-1}\left(\frac{1}{2}mx\right)^{-\nu}\boldsymbol{H}_\nu(mx),$$

$$-\pi < x < \pi, \ Re\,\nu > -\frac{3}{2},$$

$$\pi (a x)^{-\nu} J_\nu(a x) = -2^{1-\nu} a^{-1} \sin (a\pi)$$

$$\times \sum_{m=1}^{\infty} (-1)^m m^2 (m^2 - a^2)^{-1} \left(\frac{1}{2} m x\right)^{-\nu} J_\nu(m x),$$

$$0 < x < \pi, \ Re\,\nu > -\frac{1}{2}.$$

For further examples see 4.13.3.

3.12.4 Fourier-Bessel and Dini series

Let ν be a real number and denote by $\gamma_{\nu,m}$ and $\lambda_{\nu,m}$ the mth positive root (with respect to x) of $J_\nu(x) = 0$ and $x J_\nu'(x) + a J_\nu(x) = 0$ respectively arranged in ascending order of magnitude (see 3.15). Then the series

$$f(x) = \sum_{m=1}^{\infty} a_m J_\nu(\gamma_{\nu,m} x)$$

with

$$a_m = 2 [J_{\nu+1}(\gamma_{\nu,m})]^{-2} \int_0^1 t f(t) J_\nu(\gamma_{\nu,m} t)\, dt$$

is called the Fourier-Bessel series of $f(x)$. Similarly the series

$$f(x) = \sum_{m=1}^{\infty} b_m J_\nu(\lambda_{\nu,m} x)$$

with

$$b_m = 2\lambda_{\nu,m}^2 \{\lambda_{\nu,m}^2 [J_\nu'(\lambda_{\nu,m})]^2 + (\lambda_{\nu,m}^2 - \nu^2) [J_\nu(\lambda_{\nu,m})]^2\}^{-1} \int_0^1 t J_\nu(\lambda_{\nu,m} t) f(t)\, dt$$

is called the Dini series of $f(x)$. Simple examples are

$$x^\nu = \sum_{m=1}^{\infty} 2 J_\nu(\gamma_{\nu,m} x) [\gamma_{\nu,m} J_{\nu+1}(\gamma_{\nu,m})]^{-1},$$

$$0 \le x < 1,$$

$$x^\nu = \sum_{m=1}^{\infty} 2\lambda_{\nu,m} J_\nu(\lambda_{\nu,m} x) J_{\nu+1}(\lambda_{\nu,m}) (\lambda_{\nu,m} - \nu^2)^{-1}$$

$$\times \{[J_\nu(\lambda_{\nu,m})]^2 + \lambda_{\nu,m}^2 [J_\nu'(\lambda_{\nu,m})]^2\}^{-1},$$

$$0 \le x \le 1.$$

For further examples see 3.13.4.

3.13 A list of series involving Bessel functions

3.13.1 Series of the Neumann type

$$z^\nu e^{\gamma z} = 2^\nu \Gamma(\nu) \sum_{n=0}^{\infty} (\nu + n)\, C_n^\nu(\gamma)\, I_{\nu+n}(z),$$

$$\left(\frac{1}{2} z\right)^{\mu-\nu} J_\nu(z) = \sum_{n=0}^{\infty} \frac{\Gamma(\mu + n)\, \Gamma(\nu + 1 - \mu)\, (\mu + 2n)}{n!\, \Gamma(\nu + 1 - \mu - n)\, \Gamma(\nu + n + 1)} J_{\mu+2n}(z),$$

For $\nu - \mu = 0, 1, 2, \ldots$ the series terminates

$$J_\nu(z \sin\vartheta) = \left(\frac{1}{2}\pi z\right)^{-\frac{1}{2}} (\sin\vartheta)^\nu$$

$$\times \sum_{n=0}^{\infty} \frac{\left(\nu + \frac{1}{2} + 2n\right) \Gamma\left(\frac{1}{2} + n\right)}{\Gamma(n + \nu + 1)} \Gamma\left(\frac{1}{2} + \nu\right) C_{2n}^{\frac{1}{2}+\nu}(\cos\vartheta)\, J_{\nu+\frac{1}{2}+2n}(z),$$

$$\boldsymbol{H}_\nu(z)\, \Gamma\left(\frac{1}{2} + \nu\right) = 4\pi^{-\frac{1}{2}} \sum_{n=0}^{\infty} \frac{(\nu + 1 + 2n)\, \Gamma(\nu + 1 + n)}{n!\, (2n + 2\nu + 1)\, (2n + 1)} J_{\nu+1+2n}(z),$$

$$z^\nu = 2^\nu \Gamma\left(1 + \frac{1}{2}\nu\right) \sum_{n=0}^{\infty} \left(\frac{1}{2} z\right)^{\frac{1}{2}\nu+n} J_{\frac{1}{2}\nu+n}(z)\, (n!)^{-1},$$

$$\Gamma(\nu - \mu)\, J_\nu(z) = \Gamma(1 + \mu) \sum_{n=0}^{\infty} \frac{\Gamma(\nu - \mu + n)}{n!\, \Gamma(\nu + n + 1)} \left(\frac{1}{2} z\right)^{\nu-\mu+n} J_{\mu+n}(z),$$

$$\nu \neq 0, \quad \mu \neq -1, -2, \ldots,$$

$$J_\nu(z \cos\vartheta)\, J_\nu(z \sin\vartheta) = \sum_{n=0}^{\infty} \frac{\left(\frac{1}{2} z \sin 2\vartheta\right)^{\nu+2n}}{n!\, \Gamma(\nu + n + 1)} J_{\nu+2n}(z),$$

$$\nu \neq -1, -2, -3, \ldots,$$

$$(z + h)^{\pm\frac{1}{2}\nu} J_\nu\left[(z + h)^{\frac{1}{2}}\right] = \sum_{n=0}^{\infty} (n!)^{-1} \left(\pm \frac{1}{2} h\right)^n z^{\pm\frac{1}{2}\nu-\frac{1}{2}n} J_{\nu\mp n}\left(z^{\frac{1}{2}}\right),$$

$$(z + h)^{\pm\frac{1}{2}\nu} Y_\nu\left[(z + h)^{\frac{1}{2}}\right] = \sum_{n=0}^{\infty} (n!)^{-1} \left(\pm \frac{1}{2} h\right)^n z^{\pm\frac{1}{2}\nu-\frac{1}{2}n} Y_{\nu\mp n}\left(z^{\frac{1}{2}}\right),$$

$$|h| < |z|,$$

$$H_0^{(1)}\left[z(1 - a)^{\frac{1}{2}}\right] = \sum_{n=-\infty}^{\infty} [\Gamma(\mu - \nu + 1)]^{-1} \left(\frac{1}{2} a z\right)^{n-\nu} H_{n-\nu}^{(1)}(z),$$

$$H_1^{(1)}\left[z(1 - a)^{\frac{1}{2}}\right] = (1 - a)^{\frac{1}{2}} \sum_{n=-\infty}^{\infty} [\Gamma(n - \nu + 1)]^{-1}$$

$$\times \left(\frac{1}{2} a z\right)^{n-\nu} H_{n-\nu+1}^{(1)}(z),$$

$$\left(\frac{1}{2}\pi z\right)^{-\frac{1}{2}}\cos\left[(z^2-2zt)^{\frac{1}{2}}\right]=\sum_{n=-\infty}^{\infty}[\Gamma(n-\nu+1)]^{-1}t^{n-\nu}J_{n-\nu-\frac{1}{2}}(z)$$

$$\left(\frac{1}{2}\pi z\right)^{-\frac{1}{2}}\sin\left[(z^2+2zt)^{\frac{1}{2}}\right]$$

$$=\sum_{n=-\infty}^{\infty}[\Gamma(n-\nu+1)]^{-1}t^{n-\nu}J_{-\left(n-\nu-\frac{1}{2}\right)}(z),$$

$$(s^2-\tau^2)^{-\frac{1}{2}\nu}H_\nu^{(1)}\left[z(s^2-\tau^2)^{\frac{1}{2}}\right]=\sum_{n=0}^{\infty}(n!)^{-1}\left(\frac{1}{2}z\tau^2\right)^n s^{-\nu-n}H_{\nu+n}^{(1)}(zs),$$

$$(s^2-\tau^2)^{-\frac{1}{2}\nu}K_\nu\left[z(s^2-\tau^2)^{\frac{1}{2}}\right]=\sum_{n=0}^{\infty}(n!)^{-1}\left(\frac{1}{2}z\tau^2\right)^n s^{-\nu-n}K_{\nu+n}(zs),$$

$$(\pi\nu)^{-1}\sin(\pi\nu)=\sum_{n=0}^{\infty}\varepsilon_n J_{n+\nu}(z)\,J_{n-\nu}(z),$$

$$[J_\nu(z)]^2=2\sum_{n=1}^{\infty}(-1)^{n-1}J_{\nu+n}(z)\,J_{\nu-n}(z),$$

$$Re\, z>0,$$

$$J_{2\nu}(2z)=\frac{1}{2}\pi z^{\frac{1}{2}}\sum_{n=0}^{\infty}(-1)^n\left[n!\,\Gamma\left(\frac{3}{2}-n\right)\right]^{-1}J_{\nu+n}(z)\,J_{\nu+n-\frac{1}{2}}(z),$$

$$\sum_{n=-\infty}^{\infty}e^{inc}J_{\nu-\frac{a}{\pi}(y+nd)}(b)\,J_{\nu+\frac{a}{\pi}(y+nd)}(b)$$

$$=\pi(ad)^{-1}\sum_{m_1}^{m_2}\lambda(m)\,J_{2\nu}\left[2b\cos\left(\frac{\frac{1}{2}\pi t}{a}\right)\right],$$

$$Re\,\nu>-\frac{1}{2}\text{ and } Re\,\nu>0 \text{ if } c=ad.$$

For λ_m see 3.13.3

3.13.2 Series of the Kapteyn type

$$(1-z)^{-1}=\sum_{0}^{\infty}\varepsilon_n J_n(nz),$$

$$\frac{1}{2}z^2(1-z^2)^{-1}=\sum_{1}^{\infty}J_{2m}(2mz),$$

provided

$$\left|ze^{(1-z^2)^{\frac{1}{2}}}\left[1+(1-z^2)^{\frac{1}{2}}\right]^{-1}\right|<1,$$

$$\frac{1}{2}z(1+z)^{-1} = \sum_1^\infty (-1)^{n-1} J_n(nz),$$

$$\frac{1}{2}z^2 = \sum_1^\infty n^{-2} J_{2n}(2nz),$$

$$\frac{1}{2}z = \sum_1^\infty (2n-1)^2 J_{2n-1}[(2n-1)z].$$

3.13.3 Series of the Schloemilch type

$$\sum_0^\infty \varepsilon_n J_0(nx) = 2x^{-1} + 4\sum_{k=1}^{r} (x^2 - 4k^2\pi^2)^{-\frac{1}{2}},$$

$$2r\pi < x < 2(r+1)\pi,$$

$$r = 0, 1, 2, \ldots,$$

$$\sum_0^\infty (-1)^n \varepsilon_n J_0(nx) = 0, \qquad 0 < x < \pi$$

$$= 4\sum_{k=1}^{r+1} [x^2 - (2k-1)^2\pi^2]^{-\frac{1}{2}},$$

$$(2r+1)\pi < x < (2r+3)\pi,$$

$$r = 0, 1, 2, \ldots,$$

$$\sum_0^\infty e^{-nz} J_0\left[n(x^2+y^2)^{\frac{1}{2}}\right]$$

$$= r^{-1} + \frac{1}{2} + \sum_{n=1}^\infty [(2n)!]^{-1} (-1)^{n-1} B_{2n} r^{2n-1} P_{2n-1}\left(\frac{z}{r}\right),$$

$$0 < r < 2\pi,$$

$$r = (x^2+y^2+z^2)^{\frac{1}{2}}.\ B_{2n}$$

are the Bernoulli numbers.

The following formulas hold for $x > 0$ and $0 \leq t < 1$, $m_1, m_2 = 0, 1, 2, \ldots$

$$\sum_1^\infty J_0(nx)\cos(nxt) = -\frac{1}{2} + \sum_{m=1}^{m_1} [x^2 - (2\pi m + tx)^2]^{-\frac{1}{2}}$$

$$+ x^{-1}(1-t^2)^{-\frac{1}{2}} + \sum_{m=1}^{m_2} [x^2 - (2\pi m - tx)^2]^{-\frac{1}{2}}.$$

Here

$$2\pi m_1 < x(1-t) < 2(m_1+1)\pi,$$
$$2\pi m_2 < x(1+t) < 2(m_2+1)\pi,$$

$$\sum_{n=1}^{\infty} (-1)^n J_0(nx)\cos(nxt) = -\frac{1}{2} + \sum_{m=1}^{m_1} \{x^2 - [(2m-1)\pi + tx]^2\}^{-\frac{1}{2}} + \sum_{m=1}^{\infty} \{x^2 - [(2m-1)\pi - tx]^2\}^{-\frac{1}{2}}.$$

Here

$$(2m_1-1)\pi < x(1-t) < (2m_1+1)\pi,$$
$$(2m_2-1)\pi < x(1+t) < (2m_2+1)\pi.$$

For $x > 0$ and $t > 1$ we have

$$\sum_1^{\infty} J_0(nx)\cos(nxt) = -\frac{1}{2} + \sum_{m=m_1+1}^{m_2} [x^2 - (2\pi m - tx)^2]^{-\frac{1}{2}},$$

with

$$2\pi m_1 < x(t-1) < 2(m_1+1)\pi,$$
$$2\pi m_2 < x(t+1) < 2(m_2+1)\pi,$$

$$\sum_1^{\infty} (-1)^n J_0(nx)\cos(nxt) = -\frac{1}{2} + \sum_{m=m_1+1}^{m_2} \{x^2 - [(2m_2-1)\pi - tx]^2\}^{-\frac{1}{2}},$$

with

$$(2m_1-1)\pi < x(t-1) < (2m_1+1)\pi,$$
$$(2m_2-1)\pi < x(t+1) < (2m_2+1)\pi.$$

Sums whose upper limit is smaller than its lower limit have to be replaced by zero

$$\sum_1^{\infty} (-1)^m m^{-\nu} J_\nu\left(\frac{m\pi x}{a}\right) = -\frac{1}{2}[\Gamma(\nu+1)]^{-1}\left(\frac{\frac{1}{2}\pi x}{a}\right)^{\nu},$$
$$0 < x < a, \quad \nu \geqq 0,$$

$$\sum_1^{\infty} \cos(mt)\left(\frac{1}{2}mx\right)^{-\nu} J_\nu(mx) = -\frac{1}{2}[\Gamma(1+\nu)]^{-1}, \qquad 0 < x < t \leqq \pi$$
$$= [\Gamma(\nu+1)]^{-1}\left[\frac{1}{2} + \pi^{\frac{1}{2}} x^{-1}\left(1 - \frac{t^2}{x^2}\right)^{\nu-\frac{1}{2}}\right], \quad 0 < t < x < \pi,$$
$$Re\,\nu > -\frac{1}{2}.$$

The following parameters are explained as follows.

$$\lambda(m) = e^{-iy\frac{(2\pi m + c)}{d}}, \quad m_1 = -\left[\frac{ad+c}{2\pi}\right], \quad m_2 = \left[\frac{ad-c}{2\pi}\right],$$

$$t = \frac{(2\pi m + c)}{d},$$

$$\sum_{n=-\infty}^{\infty} e^{inc}(y+nd)^{-\nu} J_\nu[a(y+nd)]$$

$$= \pi^{\frac{1}{2}} 2^{1-\nu} a^{-\nu} \left[d\Gamma\left(\frac{1}{2}+\nu\right)\right]^{-1} \sum_{m_1}^{m_2} \lambda(m)\,(a^2-t^2)^{\nu-\frac{1}{2}},$$

$$Re\,\nu > -\frac{1}{2}.\ Re\,\nu > \frac{1}{2} \text{ if } c = ad,$$

$$\sum_{n=-\infty}^{\infty} e^{inc}(y+nd)^{-\nu} J_{\nu+2l}[a(y+nd)]$$

$$= (-1)^l \left(\frac{1}{2}a\right)^{-\nu} \Gamma(\nu)\,(2l)!\,[d\Gamma(2l+2\nu)]^{-1} \sum_{m_1}^{m_2} \lambda(m)\,(a^2-t^2)^{\nu-\frac{1}{2}} C_{2l}^{\nu}\left(\frac{t}{a}\right),$$

$$l = 0, 1, 2, \ldots,\quad Re\,\nu > -\frac{1}{2}.\ Re\,\nu > \frac{1}{2} \text{ if } c = ad,$$

$$\sum_{n=-\infty}^{\infty} e^{inc} J_{2l}[a(y+nd)] = (-1)^l d^{-1}\, 2 \sum_{m_1}^{m_2} \lambda(m)\,(a^2-t^2)^{-\frac{1}{2}} T_{2l}\left(\frac{t}{a}\right),$$

$$l = 1, 2, 3, \ldots,$$

$$\sum_{n=-\infty}^{\infty} e^{inc} J_\nu\left[\frac{1}{2}a(y+nd)\right] J_{-\nu}\left[\frac{1}{2}a(y+nd)\right]$$

$$= 2\,(ad)^{-1} \sum_{m_1}^{m_2} \lambda(m)\, P_{\nu-\frac{1}{2}}(2a^{-2}t^2-1),$$

$$\sum_{n=-\infty}^{\infty} e^{inc} J_0\left\{a\left[(y+nd)^2+b^2\right]^{\frac{1}{2}}\right\}$$

$$= 2d^{-1} \sum_{m_1}^{m_2} \lambda(m)\,(a^2-t^2)^{-\frac{1}{2}} \cos\left\{b(a^2-t^2)^{\frac{1}{2}}\right\},$$

$$c \neq ad,$$

$$\sum_{n=-\infty}^{\infty} e^{inc} J_0\left\{a\left[(y+nd)^2-b^2\right]^{\frac{1}{2}}\right\}$$

$$= 2d^{-1} \sum_{m_1}^{m_2} \lambda(m)\,(a^2-t^2)^{-\frac{1}{2}} \cosh\left[b(a^2-t^2)^{\frac{1}{2}}\right],$$

$$c \neq ad,$$

$$\sum_{n=-\infty}^{\infty} e^{inc}\left[b^2+(y+nd)^2\right]^{-\frac{1}{2}\nu} J_\nu\left\{a\left[b^2+(y+nd)^2\right]^{\frac{1}{2}}\right\}$$

$$= (2\pi b)^{\frac{1}{2}} d^{-1}(ab)^{-\nu} \sum_{m_1}^{m_2} \lambda(m)\,(a^2-t^2)^{\frac{1}{2}\nu-\frac{1}{4}} J_{\nu-\frac{1}{2}}\left[b(a^2-t^2)^{\frac{1}{2}}\right],$$

$$Re\,\nu > -\frac{1}{2}. \quad Re\,\nu > \frac{1}{2} \text{ if } c = a\,d,$$

$$\sum_{n=-\infty}^{\infty} e^{inc} J_\nu\left\{\frac{1}{2} a\left([b^2 + (y + nd)^2]^{\frac{1}{2}} + y + nd\right)\right\}$$

$$\times J_\nu\left\{\frac{1}{2} a\left([b^2 + (y + nd)^2]^{\frac{1}{2}} - y - nd\right)\right\}$$

$$= 2d^{-1} \sum_{m_1}^{m_2} \lambda(m)\,(a^2 - t^2)^{-\frac{1}{2}} J_{2\nu}\left[b\,(a^2 - t^2)^{\frac{1}{2}}\right],$$

$$Re\,\nu > -\frac{1}{2}. \quad Re\,\nu > \frac{1}{2} \text{ if } c = a\,d.$$

3.13.4 Series of the Fourier-Bessel type

$$\frac{1}{2} \log X = - \sum_{n=1}^{\infty} J_0(x\gamma_{0,n})\, J_0(X\gamma_{0,n})\, [\gamma_{0,n} J_1(\gamma_{0,n})]^{-1},$$

$$0 \leq x \leq X \leq 1,$$

$$\frac{J_\nu(xz)}{J_\nu(z)} = 2 \sum_{n=1}^{\infty} \gamma_{\nu,n} J_\nu(x\gamma_{\nu,n})\, [(\gamma_{\nu,n}^2 - z^2)\, J_{\nu+1}(\gamma_{\nu,n})]^{-1}$$

$$= x^\nu + 2z^2 \sum_{n=1}^{\infty} J_\nu(x\gamma_{\nu,n})\, [\gamma_{\nu,n}(\gamma_{\nu,n}^2 - z^2)\, J_{\nu+1}(\gamma_{\nu,n})]^{-1},$$

$$0 \leq x < 1,$$

$$\frac{J_{\nu+1}(xz)}{J_\nu(z)} = 2z \sum_{n=1}^{\infty} J_{\nu+1}(\gamma_{\nu,n} x)\, [(\gamma_{\nu,n}^2 - z^2)\, J_{\nu+1}(\gamma_{\nu,n})]^{-1},$$

$$\frac{\pi J_\nu(xz)}{4\, J_\nu(z)} [J_\nu(z)\, Y_\nu(Xz) - J_\nu(Xz)\, Y_\nu(z)]$$

$$= \sum_{n=1}^{\infty} J_\nu(x\gamma_{\nu,n})\, J_\nu(X\gamma_{\nu,n})\, [J_{\nu+1}(\gamma_{\nu,n})]^{-2} (z^2 - \gamma_{\nu,n}^2)^{-1},$$

$$0 \leq x \leq X \leq 1,$$

$$\frac{1}{2} \frac{I_\nu(xz)}{I_\nu(z)} [I_\nu(z)\, K_\nu(Xz) - K_\nu(z)\, I_\nu(Xz)]$$

$$= \sum_{n=1}^{\infty} J_\nu(x\gamma_{\nu,n})\, J_\nu(X\gamma_{\nu,n})\, [J_{\nu+1}(\gamma_{\nu,n})]^{-2} (z^2 + \gamma_{\nu,n}^2)^{-1},$$

$$0 \leq x \leq X \leq 1.$$

In the preceeding formulas ν and z are arbitrary, but $\nu \neq -1, -2, -3, \ldots$ The zeros of $z^{-\nu} J_\nu(z)$ arranged in ascending magnitudes of $Re(\gamma_{\nu,n}) > 0$ are $\pm\gamma_{\nu,n}$ ($n = 1, 2, 3, \ldots$).

3.13.5 Series of the addition theorem type

$$z^{\nu} J_{\nu}(2z \sin \vartheta) = \Gamma(\nu)\,(4 \sin \theta)^{\nu} \sum_{n=0}^{\infty} (\nu + n)\, C_n^{\nu}(\cos 2\vartheta)\, [J_{\nu+n}(z)]^2,$$

$$z^{\nu} J_{-\nu}(2z \sin \vartheta) = \Gamma(\nu)\,(4 \sin \vartheta)^{\nu} \sum_{n=0}^{\infty} (-1)^n (\nu + n)\, C_n^{\nu}(\cos 2\vartheta) \times J_{-\nu-n}(z)\, J_{\nu+n}(z),$$

$$z^{\nu} H_{\nu}^{(1),(2)}(2z \sin \vartheta) = \Gamma(\nu)\,(4 \sin \vartheta)^{\nu} \sum_{n=0}^{\infty} (\nu + n)\, C_n^{\nu}(\cos 2\vartheta) \times J_{\nu+n}(z)\, H_{\nu+n}^{(1),(2)}(z),$$

$$z^{\nu} K_{\nu}(2z \sin \vartheta) = \Gamma(\nu)\,(4 \sin \vartheta)^{\nu} \sum_{n=0}^{\infty} (\nu + n)\, C_n^{\nu}(\cos 2\vartheta)\, I_{\nu+n}(z)\, K_{\nu+n}(z),$$

$$J_{2\nu}(2z \sin \vartheta) = \sum_{n=0}^{\infty} (-1)^n \varepsilon_n J_{\nu-n}(z)\, J_{\nu+n}(z) \cos(2n\vartheta), \quad 0 < \vartheta < \frac{1}{2}\pi, \quad Re\,\nu > -\frac{1}{2},$$

$$J_{2\nu}(2z \sin \vartheta) = 2 \sum_{n=1}^{\infty} (-1)^n J_{\nu-n-\frac{1}{2}}(z)\, J_{\nu+n+\frac{1}{2}}(z) \sin[(2n+1)\vartheta], \quad 0 < \vartheta < \frac{1}{2}\pi, \quad Re\,\nu > -\frac{1}{2},$$

$$K_{2\nu}(2z \sin \vartheta) = \frac{1}{2} \sec(\pi\nu) \sum_{n=0}^{\infty} \varepsilon_n \cos(2n\vartheta) \times [I_{n+\nu}(z)\, K_{n-\nu}(z) + I_{n-\nu}(z)\, K_{n+\nu}(z)], \quad -\frac{1}{2} < Re\,\nu < \frac{1}{2}, \quad 0 < \vartheta < \frac{1}{2}\pi,$$

$$\boldsymbol{H}_0(2z \cos \vartheta) = 2 \sum_{n=0}^{\infty} (-1)^n \cos[(2n+1)\vartheta] \left[J_{n+\frac{1}{2}}(z)\right]^2,$$

$$I_0(2z \cos \vartheta) - \boldsymbol{L}_0(2z \cos \vartheta) = 4\pi^{-1} \sum_{n=0}^{\infty} (-1)^n \cos[(2n+1)\vartheta] \times I_{n+\frac{1}{2}}(z)\, K_{n+\frac{1}{2}}(z),$$

$$Y_0(2z \cos \vartheta) = \sum_{n=0}^{\infty} (-1)^n \varepsilon_n J_n(z)\, Y_n(z) \cos(2n\vartheta),$$

$$t^{\nu} J_{\nu}[z(t + t^{-1})] = \sum_{n=-\infty}^{\infty} t^{2n} J_{\nu-n}(z)\, J_n(z), \quad |t| < 1 \text{ for } \nu \neq 0, \pm 1, \pm 2, \ldots,$$

$$t^\nu I_\nu[z(t^{-1}-t)] = \sum_{n=-\infty}^{\infty} (-1)^n t^{2n} J_{\nu-n}(z) J_n(z),$$

$$|t| < 1 \text{ for } \nu \neq 0, \pm 1, \pm 2, \ldots,$$

$$\cos(z\cos\vartheta) = 2^\nu \Gamma(\nu) \sum_{n=0}^{\infty} (-1)^n (\nu+2n) z^{-\nu} J_{\nu+2n}(z) C_{2n}^\nu(\cos\vartheta),$$

$$\sin(z\cos\vartheta) = 2^\nu \Gamma(\nu) \sum_{n=0}^{\infty} (-1)^n (\nu+2n+1) z^{-\nu} J_{\nu+2n+1}(z) C_{2n+1}^\nu(\cos\vartheta),$$

$$(\sin\alpha\sin\beta)^{\frac{1}{2}-\nu} J_{\nu-\frac{1}{2}}(z\sin\alpha\sin\beta)\, e^{iz\cos\alpha\cos\beta}$$

$$= 2^{2\nu}(2\pi z)^{-\frac{1}{2}} [\Gamma(\nu)]^2 \sum_{n=0}^{\infty} \frac{i^n\, n!\,(\nu+n)}{\Gamma(2\nu+n)} J_{\nu+n}(z) C_n^\nu(\cos\alpha) C_n^\nu(\cos\beta).$$

3.13.6 Cardinal type series

A cardinal series of a function $f(z)$ is of the form

$$f(z) = \frac{\sin(\pi z)}{\pi} \sum_{n=-\infty}^{\infty} (-1)^n \frac{f(n)}{z-n}$$

The series listed here represent meromorphic functions with poles of the first order at $\pm 1, \pm 2, \pm 3, \ldots$ and a pole of the second order at zero. Some of these series listed here are simultaneously of the Schloemilch type

$$\sum_{n=0}^{\infty} \varepsilon_n (n^2 - z^2)^{-1} n^{-\nu} J_\nu(an) = -\pi z^{-\nu-1}[\cot(\pi z) J_\nu(az) + \mathbf{H}_\nu(az)],$$

$$Re\,\nu > -\frac{5}{2}, \quad 0 < a < 2\pi,$$

$$\sum_{n=0}^{\infty} (-1)^n \varepsilon_n (n^2 - z^2)^{-1} n^{-\nu} J_{\nu+2m}(an) = -\pi \operatorname{cosec}(\pi z) z^{-\nu-1} J_{\nu+2m}(az),$$

$$Re\,\nu > -\frac{5}{2}, \quad 0 < a < \pi, \quad m = 0, 1, 2, \ldots,$$

$$\sum_{n=1}^{\infty} (-1)^n n^{2-\nu} (n^2 - z^2)^{-1} J_\nu(an) = -\frac{1}{2}\pi \operatorname{cosec}(\pi z) z^{1-\nu} J_\nu(az),$$

$$Re\,\nu > -\frac{1}{2}, \quad 0 < a < \pi,$$

$$\sum_{n=1}^{\infty} (-1)^n n^{1-\nu} (n^2 - z^2)^{-1} J_{\nu+2m+1}(na) = -\frac{1}{2}\pi \operatorname{cosec}(\pi z) z^{-\nu} J_{\nu+2m+1}(az),$$

$$Re\,\nu > -\frac{1}{2}, \quad 0 < a < \pi, \quad m = 0, 1, 2, \ldots,$$

$$\sum_{n=0}^{\infty} (-1)^n \varepsilon_n (b^2+n^2)^{-\frac{1}{2}\nu} J_\nu\left[a(b^2+n^2)^{\frac{1}{2}}\right](n^2-z^2)^{-1}$$
$$= -\pi z^{-1} \operatorname{cosec}(\pi z)\,(b^2+z^2)^{-\frac{1}{2}\nu} J_\nu\left[a(b^2+z^2)^{\frac{1}{2}}\right],$$
$$Re\,\nu > -\frac{5}{2}, \quad 0 < a < 1,$$

$$\sum_{n=1}^{\infty} (-1)^n n (n^2-z^2)^{-1}\left[J_{m+\frac{1}{2}}(an)\right]^2$$
$$= -\frac{1}{2}\pi \operatorname{cosec}(\pi z)\left[J_{m+\frac{1}{2}}(az)\right]^2,$$
$$0 < a < \frac{1}{2}\pi, \quad m = 0, 1, 2, \ldots,$$

$$\sum_{n=0}^{\infty} (-1)^n \varepsilon_n (n^2-z^2)^{-1} J_\nu(an)\,J_{-\nu}(an) = -\pi z^{-1}\operatorname{cosec}(\pi z)\,J_\nu(az)\,J_{-\nu}(az),$$
$$0 < a < \frac{1}{2}\pi,$$

$$\sum_{n=0}^{\infty} (-1)^n \varepsilon_n (n^2-z^2)^{-1}\, T_{2m}\left[n(b^2+n^2)^{-\frac{1}{2}}\right] J_{2m}\left[a(b^2+n^2)^{\frac{1}{2}}\right]$$
$$= -\pi z^{-1}\operatorname{cosec}(\pi z)\, T_{2m}\left[z(b^2+z^2)^{-\frac{1}{2}}\right] J_{2m}\left[a(b^2+z^2)^{\frac{1}{2}}\right],$$
$$0 < a < \pi, \quad m = 0, 1, 2, \ldots,$$

$$\sum_{n=0}^{\infty} (-1)^n \varepsilon_n (n^2-z^2)^{-1}\, T_{2m}\left[b(b^2+n^2)^{-\frac{1}{2}}\right] J_{2m}\left[a(b^2+n^2)^{\frac{1}{2}}\right]$$
$$= -\pi z^{-1}\operatorname{cosec}(z\pi)\, T_{2m}\left[b(b^2+z^2)^{-\frac{1}{2}}\right] J_{2m}\left[a(b^2+z^2)^{\frac{1}{2}}\right],$$
$$0 < a < \pi, \quad m = 0, 1, 2, \ldots,$$

$$\sum_{n=0}^{\infty} (-1)^n \varepsilon_n (n^2-z^2)^{-1} J_\nu\left\{\frac{1}{2}a\left[(b^2+n^2)^{\frac{1}{2}}+n\right]\right\}$$
$$\times J_\nu\left\{\frac{1}{2}a\left[(b^2+n^2)^{\frac{1}{2}}-n\right]\right\}$$
$$= -\pi \operatorname{cosec}(\pi z)\, J_\nu\left\{\frac{1}{2}a\left[(b^2+z^2)^{\frac{1}{2}}+z\right]\right\} J_\nu\left\{\frac{1}{2}a\left[(b^2+z^2)^{\frac{1}{2}}-z\right]\right\},$$
$$0 < a < \pi, \quad Re\,\nu > -\frac{3}{2},$$

$$\sum_{n=0}^{\infty} (-1)^n \varepsilon_n (n^2-z^2)^{-1} J_{\nu-an}(b)\, J_{\nu+an}(b)$$
$$= -\pi z^{-1}\operatorname{cosec}(\pi z)\, J_{\nu-az}(b)\, J_{\nu+az}(b),$$

$$0 < a < 1,$$

$$\sum_{n=1}^{\infty} (-1)^n (n^2 - z^2)^{-1} n^{-2\nu} J_{\nu+\mu}(an) J_{\nu-\mu}(an)$$

$$= -\frac{1}{2} \operatorname{cosec}(\pi z) z^{-1-2\nu} J_{\nu+\mu}(az) J_{\nu-\mu}(az)$$

$$+ 2^{-2\nu-1} a^{2\nu} z^{-2} [\Gamma(1+\nu+\mu)\Gamma(1+\nu-\mu)]^{-1},$$

$$0 < a < \frac{1}{2}\pi, \; Re\,\nu > -1,$$

$$\sum_{n=1}^{\infty} (-1)^n (n^2 - z^2)^{-1} n^{-\nu-\mu} J_\mu(na) J_\nu(na)$$

$$= -\frac{1}{2} \pi z^{-\nu-\mu-1} \operatorname{cosec}(\pi z) J_\mu(az) J_\nu(az)$$

$$+ \frac{1}{2} z^{-2} \left(\frac{1}{2} a\right)^{\nu+\mu} [\Gamma(1+\mu)\Gamma(1+\nu)]^{-1},$$

$$Re(\nu+\mu) > -2, \qquad 0 < a < \frac{1}{2}\pi,$$

$$\sum_{n=1}^{\infty} (-1)^n (n^2 - z^2)^{-1} n^{-\nu} J_\nu\left\{a\left[(b^2+n^2)^{\frac{1}{2}} + b\right]\right\} J_\nu\left\{a\left[(b^2+n^2)^{\frac{1}{2}} - b\right]\right\}$$

$$= -\frac{1}{2} \pi z^{-\nu-1} \operatorname{cosec}(\pi z) J_\nu\left\{a\left[(b^2+z^2)^{\frac{1}{2}} + z\right]\right\} J_\nu\left\{a\left[(b^2+z^2)^{\frac{1}{2}} - z\right]\right\}$$

$$+ \frac{1}{2} z^{-2} \left(\frac{\frac{1}{4} a}{b}\right)^{\nu} [\Gamma(1+\nu)]^{-1} J_\nu(2ab),$$

$$0 < a < \frac{1}{2}\pi, \qquad Re\,\nu > -1.$$

3.14 Asymptotic representations

Asymptotic expressions for the cylindrical functions can be roughly divided into the following types:

1. Expansions of the "Hankel" type for large argument and fixed order.

2. Expansions of the "Debye" type for large argument x and large order ν both real and positive such that $\frac{\nu}{x}$ is fixed and simultaneously $|x - \nu| = O\left(x^{\frac{1}{3}}\right)$.

3. Expressions of the "Nicholson" type for the transitional region $\frac{x}{\nu}$ nearly equal to unity while $|x - \nu|$ is large.

4. Expressions of the Langer type which represent uniform asymptotic expansions.

3.14.1 Series of the Hankel type

$$J_\nu(z) = \left(\frac{1}{2}\pi z\right)^{-\frac{1}{2}}$$

$$\times \left\{\cos\left(z - \frac{1}{2}\pi\nu - \frac{1}{4}\pi\right)\left[\sum_{m=0}^{M-1} (-1)^m (\nu, 2m) (2z)^{-2m} + O(|z|^{-2M})\right]\right.$$

$$\left. - \sin\left(z - \frac{1}{2}\pi\nu - \frac{1}{4}\pi\right)\left[\sum_{m=0}^{M-1} (-1)^m (\nu, 2m+1) (2z)^{-2m-1} + O(|z|^{-2M-1})\right]\right\},$$

$-\pi < \arg z < \pi,$

$$Y_\nu(z) = \left(\frac{1}{2}\pi z\right)^{-\frac{1}{2}}$$

$$\times \left\{\sin\left(z - \frac{1}{2}\pi\nu - \frac{1}{4}\pi\right)\left[\sum_{m=0}^{M-1} (-1)^m (\nu, 2m) (2z)^{-2m} + O(|z|^{-2M})\right]\right.$$

$$+ \cos\left(z - \frac{1}{2}\pi\nu - \frac{1}{4}\pi\right)$$

$$\left. \times \left[\sum_{m=0}^{M-1} (-1)^m (\nu, 2m+1) (2z)^{-2m-1} + O(|z|^{-2M-1})\right]\right\},$$

$-\pi < \arg z < \pi,$

$$H_\nu^{(1),(2)}(z) = \left(\frac{1}{2}\pi z\right)^{-\frac{1}{2}} e^{\pm i\left(z - \frac{1}{2}\pi\nu - \frac{1}{4}\pi\right)} \left[\sum_{m=0}^{M-1} (\nu, m) (\mp 2iz)^{-m} + O(|z|^{-M})\right].$$

For the first Hankel function the upper sign, valid in $-\pi < \arg z < 2\pi$. For the second Hankel function the lower sign, valid in $-2\pi < \arg z < \pi$

$$K_\nu(z) = \left(\frac{\frac{1}{2}\pi}{z}\right)^{\frac{1}{2}} e^{-z} \left[\sum_{m=0}^{M-1} (\nu, m) (2z)^{-m} + O(|z|^{-M})\right],$$

$$-\frac{3\pi}{2} < \arg z < \frac{3\pi}{2},$$

$$I_\nu(z) = (2\pi z)^{-\frac{1}{2}} \left\{e^z \left[\sum_{m=0}^{M-1} (-1)^m (\nu, m) (2z)^{-m} + O(|z|^{-M})\right]\right.$$

$$\left. + i e^{-z + i\nu\pi} \left[\sum_{m=0}^{M-1} (\nu, m) (2z)^{-m} + O(|z|^{-M})\right]\right\},$$

$$-\frac{1}{2}\pi < \arg z < \frac{3}{2}\pi,$$

$$I_\nu(z) = \pi^{-1} (2\pi z)^{-\frac{1}{2}} \cos(\pi\nu) \left\{\sum_{m=0}^{M-1} [e^z - i(-1)^m e^{-i\pi\nu - z}]\right.$$

$$\left. \times \Gamma\left(m + \frac{1}{2} - \nu\right) \Gamma\left(m + \frac{1}{2} + \nu\right) \frac{(2z)^{-m}}{m!} + e^z O(|z|^{-M})\right\},$$

$$-\frac{3}{2}\pi < \arg z < \frac{1}{2}\pi.$$

3.14.2 Expansions of the Debye type

In these formulas both ν and x are large and real positive numbers such that $\frac{\nu}{x}$ is fixed as $\nu, x \to \infty$. The following abbreviations will be used:

$$a_0 = 1,\quad a_1 = -\frac{1}{8} + \frac{5}{24}\left(1 + \frac{x^2}{\nu^2}\right)^{-1},\quad a_2 = \frac{3}{128} - \frac{77}{576}\left(1 + \frac{x^2}{\nu^2}\right)^{-1} + \frac{385}{3456}\left(1 + \frac{x^2}{\nu^2}\right)^{-2}, \ldots,$$

$$b_0 = 1,\quad b_1 = \frac{1}{8} - \frac{5}{24}\left(1 - \frac{x^2}{\nu^2}\right)^{-1},\quad b_2 = \frac{3}{128} - \frac{77}{576}\left(1 - \frac{x^2}{\nu^2}\right)^{-1} + \frac{385}{3456}\left(1 - \frac{x^2}{\nu^2}\right)^{-2}, \ldots,$$

$$B_0(\varepsilon x) = 1,\quad B_1(\varepsilon x) = \varepsilon x,\quad B_2(\varepsilon x) = \frac{1}{2}(\varepsilon x)^2 - \frac{1}{20},$$

$$B_3(\varepsilon x) = \frac{1}{6}(\varepsilon x)^3 - \frac{1}{15}\varepsilon x, \ldots,$$

$$C_0(\varepsilon x) = 1,\quad C_1(\varepsilon x) = \varepsilon x,\quad C_2(\varepsilon x) = \frac{1}{2}(\varepsilon x)^2 + \frac{1}{20},$$

$$C_3(\varepsilon x) = \frac{1}{6}(\varepsilon x)^3 + \frac{1}{15}\varepsilon x, \ldots,$$

$$2\pi I_\nu(x) = 2^{\frac{1}{2}}(\nu^2 + x^2)^{-\frac{1}{4}} \exp\left[(\nu^2 + x^2)^{\frac{1}{2}} - \nu \sinh^{-1}\left(\frac{\nu}{x}\right)\right] \times \left[\sum_{m=0}^{M-1} (-2)^m a_m \Gamma\left(\frac{1}{2} + m\right)(\nu^2 + x^2)^{-\frac{1}{2}m} + O(x^{-M})\right],$$

$$K_\nu(x) = 2^{-\frac{1}{2}}(\nu^2 + x^2)^{-\frac{1}{4}} \exp\left[-(\nu^2 + x^2)^{\frac{1}{2}} + \nu \sinh^{-1}\left(\frac{\nu}{x}\right)\right] \times \left[\sum_{m=0}^{M-1} 2^m a_m \Gamma\left(\frac{1}{2} + m\right)(\nu^2 + x^2)^{-\frac{1}{2}m} + O(x^{-M})\right],$$

$$2\pi J_{i\nu}(x) = 2^{\frac{1}{2}}(\nu^2 + x^2)^{-\frac{1}{4}} \exp\left[i(\nu^2 + x^2)^{\frac{1}{2}} - i\nu \sinh^{-1}\left(\frac{\nu}{x}\right) - \frac{1}{4}i\pi\right] \times e^{\frac{1}{2}\nu\pi}\left[\sum_{m=0}^{M-1} (2i)^m a_m \Gamma\left(\frac{1}{2} + m\right)(\nu^2 + x^2)^{-\frac{1}{2}m} + O(x^{-M})\right],$$

$$\pi H_{i\nu}^{(1)}(x) = 2^{\frac{1}{2}}(\nu^2 + x^2)^{-\frac{1}{4}} \exp\left[i(\nu^2 + x^2)^{\frac{1}{2}} - i\nu \sinh^{-1}\left(\frac{\nu}{x}\right) - i\frac{1}{4}\pi\right] \times e^{\frac{1}{2}\pi\nu}\left[\sum_{m=0}^{M-1} (-2i)^m b_m \Gamma\left(\frac{1}{2} + m\right)(\nu^2 + x^2)^{-\frac{1}{2}m} + O(x^{-M})\right],$$

$$\pi H_{i\nu}^{(2)}(x) = \pi e^{-\pi\nu}\overline{H_{i\nu}^{(1)}}(x).$$

The above formulas are valid for both cases $x > \nu$ or $x < \nu$. In the following formulas we have to distinguish the cases $x > \nu$ and $x < \nu$, while $\frac{\nu}{x}$ is fixed as $\nu, x \to \infty$.

$$\pi H_\nu^{(1)}(x) = 2^{\frac{1}{2}} (x^2 - \nu^2)^{-\frac{1}{2}} \exp\left[i (x^2 - \nu^2)^{\frac{1}{2}} + i\nu \arcsin\left(\frac{\nu}{x}\right)\right]$$

$$\times e^{-i\frac{1}{2}\pi\left(\frac{1}{2}+\nu\right)} \left[\sum_{m=0}^{M-1} 2^m b_m \Gamma\left(\frac{1}{2}+m\right)(-i)^m (x^2-\nu^2)^{-\frac{1}{2}m} + O(x^{-M})\right],$$

$$x > \nu,$$

$$\pi H_\nu^{(1)}(x) = -i\, 2^{\frac{1}{2}} (\nu^2 - x^2)^{-\frac{1}{4}} \exp\left[-(\nu^2 - x^2)^{\frac{1}{2}} + \nu \cosh^{-1}\left(\frac{\nu}{x}\right)\right]$$

$$\times \left[\sum_{m=0}^{M-1} (-1)^m 2^m b_m \Gamma\left(\frac{1}{2}+m\right)(\nu^2-x^2)^{-\frac{1}{2}m} + O(x^{-M})\right].$$

$$\nu > x.$$

If ν, x large while $\frac{\nu}{x}$ is fixed we have

$$\pi H_\nu^{(1)}(x) \sim -\frac{2}{3\pi} \sum_{m=0}^{\infty} e^{\frac{2}{3}\pi i(m+1)} B_m(\varepsilon x) \sin\left(m\frac{\pi}{3}+\frac{1}{3}\right)$$

$$\times \Gamma\left(\frac{1}{3}+\frac{1}{3}m\right)\left(\frac{x}{6}\right)^{-\frac{1}{3}(m+1)},$$

$$\nu \approx x, \quad \varepsilon = 1 - \frac{\nu}{x}, \quad \varepsilon = 0\left(x^{-\frac{2}{3}}\right).$$

The formulas for the second Hankel functions are obtained by the relation (since ν and x real > 0) $H_\nu^{(2)} = \overline{H_\nu^{(1)}}$. The functions $J_\nu(x)$ and $Y_\nu(x)$ can then be obtained by combining the above formulas. This, however, would in the case $\nu > x$ lead to the result $J_\nu(x) = 0$. This is due to the fact that $J_\nu(x)$ tends to zero with a higher magnitude than $H_\nu^{(1)}(x)$. One has for this case

$$2\pi J_\nu(x) = 2^{\frac{1}{2}} (\nu^2 - x^2)^{-\frac{1}{4}} \exp\left[(\nu^2 - x^2)^{\frac{1}{2}} - \nu \sinh^{-1}\left(\frac{\nu}{x}\right)\right]$$

$$\times \left[\sum_{m=0}^{M-1} 2^m b_m \Gamma\left(\frac{1}{2}+m\right)(\nu^2-x^2)^{-\frac{1}{2}m} + O(x^{-M})\right],$$

$$\nu > x,$$

$$K_{i\nu}(x) = 2^{-\frac{1}{2}} (x^2 - \nu^2)^{-\frac{1}{4}} \exp\left[-(x^2 - \nu^2)^{\frac{1}{2}} - \nu \arcsin\left(\frac{\nu}{x}\right)\right]$$

$$\times \left[\sum_{m=0}^{M-1} (-1)^m 2^m b_m \Gamma\left(\frac{1}{2}+m\right)(x^2-\nu^2)^{-\frac{1}{2}m} + O(x^{-M})\right],$$

$$x > \nu,$$

$$K_{i\nu}(x) = 2^{\frac{1}{2}} (\nu^2 - x^2)^{-\frac{1}{4}} e^{-\frac{1}{2}\pi\nu} \cdot \left\{ \sum_{m=0}^{M-1} 2^m b_m \Gamma\left(\frac{1}{2} + m\right) (\nu^2 - x^2)^{-\frac{1}{2}m} \right.$$

$$\left. \times \sin\left[\frac{1}{2}\pi m + \nu \cosh^{-1}\left(\frac{\nu}{x}\right) - (\nu^2 - x^2)^{\frac{1}{2}} + \frac{1}{4}\pi\right] + O(x^{-M}) \right\},$$

$$\nu > x,$$

$$K_{i\nu}(x) \sim \frac{i}{3} e^{-\frac{\pi}{2}\nu} \sum_{m=0}^{\infty} (-1)^m \sin\left[2\frac{\pi}{3}(m+1)\right] B_m(-i\varepsilon x)$$

$$\times \Gamma\left(\frac{1}{3} + \frac{1}{3}m\right) \left(\frac{x}{6} e^{2\pi i}\right)^{-\frac{1}{3}(m+1)},$$

$$\nu \approx x, \quad \varepsilon = 1 - \frac{\nu}{x}, \quad \varepsilon = o\left(x^{-\frac{2}{3}}\right).$$

3.14.3 Formulas of the Nicholson type

Here x, ν are positive numbers and n is a positive integer

$$J_n(x) \sim \left(\frac{1}{2}x\right)^{-\frac{1}{3}} Ai\left[\left(\frac{1}{2}x\right)^{-\frac{1}{3}} (n - x)\right],$$

$$Y_n(x) \sim -\left(\frac{1}{2}x\right)^{-\frac{1}{3}} Bi\left[\left(\frac{1}{2}x\right)^{-\frac{1}{3}} (n - x)\right],$$

$$K_{i\nu}(x) \sim \pi e^{-\frac{1}{2}\pi\nu} \left(\frac{1}{2}x\right)^{-\frac{1}{3}} Ai\left[-\left(\frac{1}{2}x\right)^{-\frac{1}{3}} (\nu - x)\right].$$

The functions $Ai(x)$ and $Bi(x)$ are the Airy integrals (see 3.4). Expressed in terms of Bessel functions these formulas read:

$$J_n(x) \sim \pi^{-1} \left[\frac{2(n-x)}{3x}\right]^{\frac{1}{2}} K_{\frac{1}{3}}\left[\frac{2^{\frac{3}{2}} (n-x)^{\frac{3}{2}}}{3x^{\frac{1}{2}}}\right],$$

$$Y_n(x) \sim -\left[\frac{2(n-x)}{3x}\right]^{\frac{1}{2}} \left[I_{\frac{1}{3}} + I_{-\frac{1}{3}}\right],$$

$$K_{i\nu}(x) \sim \frac{1}{3}\pi e^{-\frac{1}{2}\pi\nu} \left[\frac{2(\nu - x)}{x}\right]^{\frac{1}{2}} \left[J_{\frac{1}{3}} + J_{-\frac{1}{3}}\right]$$

valid for $n > x$, $\nu > x$ and the arguments in the Bessel functions are the same.

$$J_n(x) \sim \frac{1}{3} \left[\frac{2(x-n)}{x}\right]^{\frac{1}{2}} \left[J_{-\frac{1}{3}} + J_{\frac{1}{3}}\right],$$

$$Y_n(x) \sim -\left[\frac{2(x-n)}{3x}\right]^{\frac{1}{2}} \left[J_{-\frac{1}{3}} - J_{\frac{1}{3}}\right],$$

$$K_{i\nu}(x) \sim \left[\frac{2(x-\nu)}{3x}\right]^{\frac{1}{2}} e^{-\frac{1}{2}\pi\nu} K_{\frac{1}{3}},$$

valid for $x > n$, $x > \nu$ and the argument of all Bessel functions is equal to

$$\frac{1}{3}\,[2(x-n)]^{\frac{3}{2}} \cdot x^{-\frac{1}{2}}.$$

Extension of Nicholson's formulas

$$J_\nu(z) \sim \sum_{n=0}^{\infty} (-1)^n \left(\frac{1}{2}z\right)^{-\frac{1}{3}-\frac{2}{3}n} [L_n(\xi)\, Ai(\xi) + M_n(\xi)\, Ai'(\xi)]$$

$$-Y_\nu(z) \sim \sum_{n=0}^{\infty} (-1)^n \left(\frac{1}{2}z\right)^{-\frac{1}{3}-2n} [L_n(\xi)\, Bi(\xi) + M_n(\xi)\, Bi'(\xi)],$$

with

$$\xi = (\nu - z)\left(\frac{1}{2}z\right)^{-\frac{1}{3}}, \quad |\arg z| \leq \pi - \varepsilon, \quad z - \nu = O\left(z^{\frac{1}{3}}\right).$$

Here $L_n(\xi)$ and $M_n(\xi)$ denote the following polynomials:
[$Ai(\xi)$ and $Bi(\xi)$ given in (3.4)]

$$L_0(\xi) = 1, \qquad M_0(\xi) = 0,$$

$$L_1(\xi) = \frac{1}{15}\xi, \qquad M_1(\xi) = \frac{1}{60}\xi^2,$$

$$L_2(\xi) = \frac{1}{7200}\xi^5 + \frac{13}{1260}\xi^2, \qquad M_2(\xi) = \frac{1}{420}\xi^3 + \frac{1}{140}.$$

(For further results see Acta Mathematica vol. 92, 1954, p. 281). By means of these formulas the Nicholson formulas can be extended to complex argument and order

$$J_\nu(z) = \frac{1}{3}\left[\frac{2(z-\nu)}{z}\right]^{\frac{1}{2}} \left[J_{\frac{1}{3}}(y) + J_{-\frac{1}{3}}(y)\right] + O(z^{-1}),$$

$$Y_\nu(z) = \left[2\left(\frac{z-\nu}{3z}\right)\right]^{\frac{1}{2}} \left[J_{\frac{1}{3}}(y) - J_{-\frac{1}{3}}(y)\right] + O(z^{-1}),$$

$$K_{i\nu}(z) = \left[\frac{2(z-\nu)}{3z}\right]^{\frac{1}{2}} e^{-\frac{1}{2}\pi\nu} K_{\frac{1}{3}}(y) + O(z^{-1}).$$

The first two formulas valid for $|\arg z| \leq \pi - \varepsilon$, the third for $|\arg z| \leq \frac{3}{2}\pi - \varepsilon$. In all three cases

$$y = \frac{1}{3} z^{-\frac{1}{2}} [2(z - \nu)]^{\frac{3}{2}}, \quad z - \nu = O\left(z^{\frac{1}{3}}\right).$$

Similar formulas for both, variable and order real positive have been given by LANGER and WATSON.

Watson's formulas

$$J_p(x) = 3^{-\frac{1}{2}} w \left[J_{\frac{1}{3}}\left(\frac{1}{3} p w^3\right) \cos\delta - Y_{\frac{1}{3}}\left(\frac{1}{3} p w^3\right) \sin\delta \right] + O(p^{-1}),$$

$$Y_p(x) = 3^{-\frac{1}{2}} w \left[J_{\frac{1}{3}}\left(\frac{1}{3} p w^3\right) \sin\delta + Y_{\frac{1}{3}}\left(\frac{1}{3} p w^3\right) \cos\delta \right] + O(p^{-1}),$$

$$x > p, \quad w = \left(\frac{x^2}{p^2} - 1\right)^{\frac{1}{2}}, \quad \delta = p w - \frac{1}{3} p w^3 - p \arctan w + \frac{\pi}{6},$$

$$J_p(x) = 3^{-\frac{1}{2}} \pi^{-1} w e^{p\alpha} K_{\frac{1}{3}}\left(\frac{1}{3} p w^3\right) + O(p^{-1}),$$

$$Y_p(x) = -3^{-\frac{1}{2}} w e^{p\alpha} \left[I_{\frac{1}{3}}\left(\frac{1}{3} p w^3\right) + I_{-\frac{1}{3}}\left(\frac{1}{3} p w^3\right) \right] + O(p^{-1}),$$

$$x < p, \quad \alpha = p\left(w + \frac{1}{3} w^3\right) - \tanh^{-1} w, \quad w = \left(1 - \frac{x^2}{p^2}\right)^{\frac{1}{2}}.$$

Langer's formulas

$$J_p(x) = w^{-\frac{1}{2}} (w - \arctan w)^{\frac{1}{2}} \left[J_{\frac{1}{3}}(z) \cos\left(\frac{\pi}{6}\right) - Y_{\frac{1}{3}}(z) \sin\left(\frac{\pi}{6}\right) \right] + O\left(p^{-\frac{4}{3}}\right),$$

$$Y_p(x) = w^{-\frac{1}{2}} (w - \arctan w)^{\frac{1}{2}} \left[J_{\frac{1}{3}}(z) \sin\left(\frac{\pi}{6}\right) + Y_{\frac{1}{3}}(z) \cos\left(\frac{\pi}{6}\right) \right] + O\left(p^{-\frac{4}{3}}\right),$$

$$x > p, \quad w = \left(\frac{x^2}{p^2} - 1\right)^{\frac{1}{2}}, \quad z = p(w - \arctan w),$$

$$J_p(x) = \pi^{-1} w^{-\frac{1}{2}} (\tanh^{-1} w - w)^{\frac{1}{2}} K_{\frac{1}{3}}(z) + O\left(p^{-\frac{4}{3}}\right),$$

$$Y_p(x) = -w^{-\frac{1}{2}}(\tanh^{-1} w - w)^{\frac{1}{2}}\left[I_{\frac{1}{3}}(z) + I_{-\frac{1}{3}}(z)\right] + O\left(p^{-\frac{4}{3}}\right),$$

$$x < p, \quad w = \left(1 - \frac{x^2}{p^2}\right)^{\frac{1}{2}}, \quad z = p(\tanh^{-1} w - w).$$

(See also Handbook of mathematical functions. Nat. Bureau of Standards Washington D. C. 1964, p. 364. — F. W. J. OLVER, Phil. Trans. Roy. Soc. (A) 247, 1954, 307—368; 249, 1957, 65—97; 250, 1958, 479—518.)

Transitional region

It is possible to give an asymptotic expansion for $K_{ip}(x)$ valid in the so called transitional region, i.e. both x and p large (real > 0) such that $\varepsilon = 1 - \frac{p}{x} = O\left(x^{-\frac{2}{3}}\right)$. The leading term of this expansion turns out to be the formerly listed formula of the "Nicholson type". One has

$$K_{ip}(x) = -i\frac{\pi}{2} e^{-\frac{\pi}{2}p}\left(-i\frac{x}{2}\right)^{-\frac{1}{3}}\left[\sum_{l=0}^{m}(-1)^l p_l(\xi)\left(-i\frac{x}{2}\right)^{-\frac{2}{3}l}\right]$$

$$\times\,[Ai(\varrho) + i\,Bi(\varrho)] + O\left[(-ix)^{-\frac{2m+2}{3}}\right].$$

Here

$$\xi = i\varepsilon x\left(-\frac{1}{2}ix\right)^{-\frac{1}{3}},$$

$$\varrho = \sum_{k=0}^{m}(-1)^k q_k(\xi)\left(-\frac{1}{2}ix\right)^{-\frac{2}{3}k}.$$

The p_l and q_k are polynomials

$$p_0(\xi) = 1, \qquad q_0(\xi) = \xi,$$

$$p_1(\xi) = \frac{1}{15}\xi, \qquad q_1(\xi) = \frac{1}{60}\xi^2,$$

$$p_2(\xi) = \frac{13}{1260}\xi^2, \qquad q_2(\xi) = \frac{2}{1575}\xi^3 + \frac{1}{140}.$$

Or also, since $Ai(x) + i\,Bi(x) = 2e^{i\frac{\pi}{3}} \cdot Ai\left(xe^{-\frac{2}{3}\pi i}\right)$

$$K_{ip}(x) = \pi e^{-\frac{1}{2}\pi p}\left(\frac{1}{2}x\right)^{-\frac{1}{3}}\left[\sum_{l=0}^{m}(-1)^l p_l(\xi)\left(-\frac{1}{2}ix\right)^{-\frac{2}{3}l}\right]$$

$$\times\, Ai\left(\varrho e^{-\frac{2}{3}\pi i}\right) + O\left[(-ix)^{-\frac{2m+2}{3}}\right]$$

both formulas valid

$$\varepsilon = t - \frac{p}{x} = o\left(x^{-\frac{2}{3}}\right).$$

3.15 Zeros

General results. Any zero of a solution of the Bessel or the modified Bessel differential equation is a simple zero with the the origin being the only possible exception. Furthermore the real zeros of two real linearly independent solutions of Bessel's differential equation separate one another. A real solution is defined by $a\,J_\nu(x) + b\,Y_\nu(x)$ with a, b, real and $x > 0$.

3.15.1 Bessel functions of the first kind

In case the order ν is a real number the following statements hold: The zeros of $J_\nu(z)$ and $J_\nu'(z)$ are symmetrical with respect to the axes of coordinates and $J_\nu(z)$ has an infinite number of real zeros. Denote by $\gamma_{\nu,1}, \gamma_{\nu,2}, \ldots$ the positive roots of $J_\nu(x)$ arranged in ascending order of magnitude, then

$$0 < \gamma_{\nu,1} < \gamma_{\nu+1,1} < \gamma_{\nu,2} < \gamma_{\nu+1,2} < \gamma_{\nu,3} < \cdots \qquad \text{(interlacing property).}$$

If $-\frac{1}{2} < \nu < \frac{1}{2}$ and if $n\pi < x < \left(n + \frac{1}{2}\right)\pi$ then $J_\nu(x)$ is positive (negative) when n is an even (odd) integer. Let $\nu > -1$ and A, B, C, D, real numbers such that $AD - BC \neq 0$, then the positive zeros of $A\,J_\nu(x) + Bx\,J_\nu'(x)$ and $C\,J_\nu(x) + Dx\,J_\nu'(x)$ separate one another and no function of this type can have a repeated zero other than zero. Again, let $\nu > -1$ and A and B real numbers then the function $A\,J_\nu(x) + Bx\,J_\nu'(x)$ has only real zeros except that it has two purely imaginary zeros when $\nu + \frac{A}{B} < 0$.

Let $\nu > 1$ then the function $J_{-\nu}(x)$ has an infinity of real zeros and also $2[\nu]$ conjugate complex zeros, among them two purely imaginary zeros when $[\nu]$ is an odd integer.

The principal branch of the function $A\,J_\nu(z) + B\,J_{-\nu}(z)$, ($A$, B, real $B \neq 0$, $\nu > 0$) has $[\nu]$ complex zeros with a positive real part in case $[\nu]$ is even: When $[\nu]$ is odd there exist $[\nu] - 1$ or $[\nu] + 1$ complex zeros with a positive real part according as $\frac{A}{B} \gtrless 0$. The number of zeros of $z^{-\nu} J_\nu(\nu)$ between the imaginary axis and the line on which $Re\, z = m\pi + \left(\frac{1}{2}\, Re\, \nu + \frac{1}{4}\right)\pi$ is equal to m for sufficiently large m and all the zeros of $J_\nu(z)$ lie inside a strip $|\mathrm{Im}(z)| < A$, where A is bounded when ν is bounded.

Representation as an infinite product

$$\Gamma(\nu + 1)\left(\frac{z}{2}\right)^{-\nu} J_\nu(z) = \prod_{l=1}^{\infty} (1 - z^2 \gamma_{\nu,l}^{-2}), \qquad \nu \neq -1,\ 2,\ -3, \ldots$$

Here $\gamma_{\nu,l}$ are the zeros of $z^{-\nu} J_\nu(z)$ which are in the half plane $Re\, z > 0$ (those are symmetrical to the real axis) arranged according to non-decreasing real parts (in case of zeros which are purely imaginary only those with a positive imaginary part are considered). Also

$$\frac{J_{\nu+1}(z)}{J_\nu(z)} = -2z \sum_{n=1}^{\infty} (z^2 - \gamma_{\nu,l}^2)^{-1}.$$

The zeros ν_n of $J_\nu(z)$ with regard to ν for fixed positive real z are real and simple and asymptotically near to negative integers.

3.15.2 Bessel functions of the second kind

When $[\nu]$ is even, than $Y_\nu(z)$ has $[\nu]$ complex zeros in $|\arg z| \le \frac{1}{2}\pi$. When $[\nu]$ is odd then $Y_\nu(z)$ has $[\nu] - 1$ or $[\nu] + 1$ complex zeros in the same domain according as $\cos(\pi\nu) \lesseqgtr 0$. Thus $Y_{2n}(z)$ and

$$Y_{2n+1}(z) \qquad (n = 0, 1, 2, \ldots)$$

have $2n$ complex zeros in $|\arg z| \le \frac{1}{2}\pi$. $Y_n(z)$ (n integer) has complex zeros in the left half plane on all branches but the principle branch. $Y_\nu(z)$ has positive real zeros only if ν is rational but not an integer. The combination (ν real, $a, b > 0$)

$$J_\nu(ax)\, Y_\nu(bx) - J_\nu(bx)\, Y_\nu(ax)$$

is a single valued even function of x whose zeros are all real and simple.

For greater detail and numerical results see Handbook of mathematical functions. N. B. S. 1964.

3.15.3 Zeros with respect to the order

Let s be a real and positive number. Then all the zeros of $K_\nu(s)$ and $\frac{d}{ds} K_\nu(s)$ with respect to ν are purely imaginary, i.e.

$$\nu_l = \pm i\alpha_l, \text{ with } \alpha_l > 0.$$

If s is large then asymptotically

$$\alpha_l \sim s + \beta_l s^{\frac{1}{3}},$$

where β_l is the lth zero of a function $f(\beta)$ defined by

$$f(\beta) = 3 \cdot 2^{-\frac{1}{6}} Ai\left(-2^{\frac{1}{3}}\beta\right).$$

Likewise the zeros of

$$I_\nu(as)\,K_\nu(bs) - I_\nu(bs)\,K_\nu(as) \qquad (a, b, s \text{ fixed real} > 0)$$

with respect to ν are all purely imaginary and the number of zeros of both these functions is infinite.

For information about the zeros of

$H_\nu^{(1)}(z)$, $\frac{d}{dz} H_\nu^{(1)}(z)$ and $\frac{d}{dz} H_\nu^{(1)}(z) + ik\,H_\nu^{(1)}(z)$ (k=constant) with respect to the order see J. A. COCHRAN, Numerische Mathematik 7, (1965) 238—250; J. KELLER, S. I. RUBINOW, Jour. of Mathematical Physics 4, (1963) 829—832; W. MAGNUS and L. KOTIN, Numerische Mathematik 2, (1960) 228—244.

3.16 Miscellaneous

3.16.1 Kelvin's functions

In this section ν is real, x is real and non negative and n is a positive integer or zero.

$$\operatorname{ber}_\nu(x) + i\operatorname{bei}_\nu(x) = J_\nu\left(x e^{i3\frac{\pi}{4}}\right) = e^{i\nu\pi} J_\nu\left(x e^{-i\frac{\pi}{4}}\right),$$

$$\operatorname{ker}_\nu(x) + i\operatorname{kei}_\nu(x) = e^{-\frac{1}{2}i\pi\nu} K_\nu\left(x e^{i\frac{\pi}{4}}\right).$$

When $\nu = 0$ the index is usually omitted.

Functional relations

$$\operatorname{ber}_{-\nu}(x) = \cos(\nu\pi)\operatorname{ber}_\nu(x) + \sin(\nu\pi)\operatorname{bei}_\nu(x) + \frac{2}{\pi}\sin(\pi\nu)\operatorname{ker}_\nu(x),$$

$$\operatorname{bei}_{-\nu}(x) = -\sin(\pi\nu)\operatorname{ber}_\nu(x) + \cos(\pi\nu)\operatorname{bei}_\nu(x) + \frac{2}{\pi}\operatorname{kei}_\nu(x)$$

$$\operatorname{ker}_{-\nu}(x) = \cos(\pi\nu)\operatorname{ker}_\nu(x) - \sin(\nu\pi)\operatorname{kei}_\nu(x),$$

$$\operatorname{kei}_{-\nu}(x) = \sin(\nu\pi)\operatorname{ker}_\nu(x) + \cos(\nu\pi)\operatorname{kei}_\nu(x),$$

$$\operatorname{ber}_{-n}(x) = (-1)^n\operatorname{ber}_n(x), \qquad \operatorname{bei}_{-n}(x) = (-1)^n\operatorname{bei}_n(x),$$

$$\operatorname{ker}_{-n}(x) = (-1)^n\operatorname{ker}_n(x), \qquad \operatorname{kei}_{-n}(x) = (-1)^n\operatorname{kei}_n(x),$$

$$\operatorname{ber}_n(-x) = (-1)^n\operatorname{ber}_n(x), \quad \operatorname{bei}_n(-x) = (-1)^n\operatorname{bei}_n(x),$$

Series expansions

$$\operatorname{ber}_\nu(x) = \left(\frac{1}{2}x\right)^\nu \sum_{k=0}^{\infty} \frac{\cos\left[\left(\frac{3}{4}\nu + \frac{1}{2}k\right)\pi\right]}{k!\,\Gamma(\nu + k + 1)} \left(\frac{1}{2}x^2\right)^k,$$

$$\mathrm{bei}_\nu(x) = \left(\frac{1}{2}x\right)^\nu \sum_{k=0}^{\infty} \frac{\sin\left[\left(\frac{3}{4}\nu + \frac{1}{2}k\right)\pi\right]}{k!\,\Gamma(\nu+k+1)} \left(\frac{1}{2}x^2\right)^k,$$

$$\mathrm{ber}_\nu(x) + i\,\mathrm{bei}_\nu(x) = \sum_{k=0}^{\infty} 2^{-\frac{1}{2}k} e^{i\frac{\pi}{4}(3\nu+k)} \frac{x^k}{k!} J_{\nu+k}(x),$$

$$\mathrm{ber}_\nu^2(x) + \mathrm{bei}_\nu^2(x) = \left(\frac{1}{2}x\right)^{2\nu} \sum_{k=0}^{\infty} \frac{\left(\frac{1}{4}x^2\right)^{2k}}{k!\,\Gamma(\nu+k+1)\,\Gamma(\nu+2k+1)},$$

$$\ker(x) = -\log\left(\frac{1}{2}x\right)\mathrm{ber}(x) + \frac{1}{4}\pi\,\mathrm{bei}(x) + \sum_{k=0}^{\infty}(-1)^k \frac{\psi(1+2k)}{[(2k)!]^2}\left(\frac{1}{4}x^2\right)^{2k},$$

$$\mathrm{kei}(x) = -\log\left(\frac{1}{2}x\right)\mathrm{bei}(x) - \frac{1}{4}\pi\,\mathrm{ber}(x) + \sum_{k=0}^{\infty}(-1)^k \frac{\psi(2+2k)}{[(2k+1)!]^2}\left(\frac{1}{4}x^2\right)^{2k+1}.$$

Differential equations

$$x^2 w'' + x w' - (i x^2 + \nu^2)\, w = 0,$$

$$w = \mathrm{ber}_\nu(x) + i\,\mathrm{bei}_\nu(x),$$

$$w = \mathrm{ber}_{-\nu}(x) + i\,\mathrm{bei}_{-\nu}(x),$$

$$w = \ker_\nu(x) + i\,\mathrm{kei}_\nu(x),$$

$$w = \ker_{-\nu}(x) + i\,\mathrm{kei}_{-\nu}(x).$$

Recurrence relations

Let f_ν and g_ν be any of the following pair

1. $f_\nu = \mathrm{ber}_\nu(x), \quad g_\nu = \mathrm{bei}_\nu(x),$
2. $f_\nu = \mathrm{bei}_\nu(x), \quad g_\nu = -\mathrm{ber}_\nu(x),$
3. $f_\nu = \ker_\nu(x), \quad g_\nu = \mathrm{kei}_\nu(x),$
4. $f_\nu = \mathrm{kei}_\nu(x), \quad g_\nu = -\ker_\nu(x),$

then

$$f_{\nu+1} + f_{\nu-1} = -\nu\, 2^{\frac{1}{2}} x^{-1} (f_\nu - g_\nu),$$

$$f'_\nu = 2^{-\frac{3}{2}} (f_{\nu+1} + g_{\nu+1} - f_{\nu-1} - g_{\nu-1}),$$

$$f'_\nu - \nu x^{-1} f_\nu = 2^{-\frac{1}{2}} (f_{\nu+1} + g_{\nu+1}),$$

$$f'_\nu + \nu x^{-1} f_\nu = 2^{-\frac{1}{2}} (f_{\nu-1} + g_{\nu-1}).$$

3.16.2 Some inversion formulas involving Bessel functions

1. $$g(y) = \int_0^\infty f(x)\,(xy)^{\frac{1}{2}}\,J_\nu(xy)\,dx,$$

$$f(x) = \int_0^\infty g(y)\,(xy)^{\frac{1}{2}}\,J_\nu(xy)\,dy \qquad \text{HANKEL.}$$

2. $$g(y) = \int_0^\infty f(x)\,(xy)^{\frac{1}{2}}\,Y_\nu(xy)\,dx,$$

$$f(x) = \int_0^\infty g(y)\,(xy)^{\frac{1}{2}}\,\mathbf{H}_\nu(xy)\,dy \qquad \text{TITCHMARSH}$$

3. $$g(y) = \int_0^\infty f(x)\,(xy)^{\frac{1}{2}}\,K_\nu(xy)\,dx,$$

$$f(x) = \frac{1}{2\pi i}\int_{c-i\infty}^{c+i\infty} g(y)\,(xy)^{\frac{1}{2}}\,[I_\nu(xy) + I_{-\nu}(xy)]\,dy \qquad \text{MEIJER.}$$

4. $$g(y) = \int_0^\infty f(x)\,(xy)^{\frac{1}{2}}\,s_{\mu,\nu}(xy)\,dx,$$

$$f(x) = 2^{2-2\mu}\left[\Gamma\left(\frac{1}{2}+\frac{1}{2}\mu-\frac{1}{2}\nu\right)\Gamma\left(\frac{1}{2}+\frac{1}{2}\mu+\frac{1}{2}\nu\right)\right]^{-2}$$

$$\times\int_0^\infty g(y)\,(xy)^{\frac{1}{2}}\,[s_{\mu,\nu}(xy) - S_{\mu,\nu}(xy)]\,dy \qquad \text{HARDY.}$$

5. $$g(y) = \int_0^\infty f(x)\,[J_{ix}(y) + J_{-ix}(y)]\,dx,$$

$$f(x) = \frac{1}{2}x\,\mathrm{cosech}\,(\pi x)\int_0^\infty y^{-1} g(y)\,[J_{ix}(y) + J_{-ix}(y)]\,dy \qquad \text{TITCHMARSH}$$

6. $$g(y) = \int_0^\infty f(x)\,K_{ix}(y)\,dx,$$

$$f(x) = 2\pi^{-2}x\,\sinh\,(\pi x)\int_0^\infty g(y)\,K_{ix}(y)\,y^{-1}\,dy \qquad \text{LEBEDEV.}$$

7. $$g(y) = \pi^{-1}\int_{-\infty}^\infty f(x)\,e^{\frac{1}{2}\pi(x+y)}\,K_{i(x+y)}(a)\,dx,$$

$$f(x) = \pi^{-1}\int_{-\infty}^\infty g(y)\,e^{\frac{1}{2}\pi(x+y)}\,K_{i(x+y)}(a)\,dy \qquad \text{CRUM.}$$

8. $$g(y) = \int_0^\infty f(x)\,H_{ix}^{(2)}(y)\,dx,$$

$$f(x) = -\frac{1}{2}x\,\sinh\,(\pi x)\,e^{\pi x}\int_0^\infty y^{-1} f(y)\,H_{ix}^{(2)}(y)\,dy \qquad \text{LEBEDEV.}$$

See also chap. XI.

3.16.3 Some inequalities

For x positive real and $\nu > 0$, $I_\nu(x)$ is a decreasing function of ν whereas $K_\nu(x)$ is an increasing function of ν.

i. e.

$$\left.\begin{aligned} I_\nu(x) &> I_\mu(x) \\ \text{and} \quad & \\ K_\nu(x) &< K_\mu(x) \end{aligned}\right\} \quad \text{for} \quad x > 0, \quad \mu > \nu > 0.$$

Literature

ERDÉLYI, A.: Higher transcendental functions, vol. 2. New York: McGraw-Hill 1953.

GRAY, A., G. B. MATHEWS and T. M. MACROBERT: A treatise on the theory of Bessel functions. London: Macmillan 1931.

MCLACHLAN, N. W.: Bessel functions for Engineers. Oxford: Clarendon press 1955.

PETIAU, G.: La theorie des fonctions de Bessel. Paris 1955.

WATSON, G. N.: A treatise on the theory of Bessel functions. Cambridge: U. Press 1952.

WETRICH, R.: Die Zylinderfunktionen und ihre Anwendungen. Leipzig: Teubner 1937.

Chapter IV

Legendre functions

4.1 Legendre's differential equation

4.1.1 Legendre functions, notations, definitions

The Legendre functions are solutions of the Legendre differential equation

$$(1 - z^2)\frac{d^2w}{dz^2} - 2z\frac{dw}{dz} + [\nu(\nu + 1) - \mu^2(1 - z^2)^{-1}]\, w = 0.$$

This differential equation has 3 regular singular points at $z = +1$, $z = -1$, and $z = \infty$. The respective exponents at these points are

$$p_1 = \frac{1}{2}\mu, p_2 = -\frac{1}{2}\mu;\ p_1 = \frac{1}{2}\mu, p_2 = -\frac{1}{2}\mu;\ p_1 = -1 - \nu, p_2 = \nu.$$

The parameters ν, μ are in general unrestricted complex numbers. As to the variable z one distinguishes in general two cases. a) The variable z is real and between -1 and $+1$. In this case we denote a fundamental system of solutions by $P_\nu^\mu(x)$ and $Q_\nu^\mu(x)$ [or also by $P_\nu^\mu(\cos\vartheta)$ and $Q_\nu^\mu(\cos\vartheta)$

with $0 \leq \vartheta \leq \pi$]. b) The variable z is an arbitrary point in the complex z plane with the exception of points on the real axis between $+1$ and $-\infty$ (the z plane cut along the real axis from $+1$ to $-\infty$). In this case we denote a fundamental system of solutions in gothic letters as $\mathfrak{P}_\nu^\mu(z)$ and $\mathfrak{Q}_\nu^\mu(z)$. They are called Legendre function of the first and second kind respectively. The Legendre functions are special cases of the hypergeometric function $F(a, b; c; z)$ where the parameters a, b, c are such that a quadratic transform exists. Because of this property it is possible to express a solution of Legendre's differential equation in terms of the hypergeometric function with a choice of 18 different arguments in 72 different ways. The general definition is of the form

$$\mathfrak{P}_\nu^\mu(z) = A_1 F(a_1, b_1; c_1; \zeta) + A_2 F(a_2, b_2; c_2; \zeta),$$

$$e^{-i\pi\mu}\mathfrak{Q}_\nu^\mu(z) = A_3 F(a_3, b_3; c_3; \zeta) + A_4 F(a_4, b_4; c_4; \zeta),$$

where ζ can have one of the values

$$\frac{1}{2} - \frac{1}{2}z;\ \frac{1}{2} + \frac{1}{2}z;\ \frac{(z-1)}{(z+1)};$$

$$\frac{(z+1)}{(z-1)};\ \frac{2}{(1+z)};\ \frac{2}{(1-z)};\ 1-z^2;\ (1-z^2)^{-1};\ z^2;\ z^{-2};\ 1-z^{-2};\ \frac{z^2}{(z^2-1)};$$

$$\frac{1}{2}\left[-z+(z^2-1)^{\frac{1}{2}}\right](z^2-1)^{-\frac{1}{2}};\ \frac{\left[z-(z^2-1)^{\frac{1}{2}}\right]}{\left[z+(z^2-1)^{\frac{1}{2}}\right]};$$

$$\frac{2(z^2-1)^{\frac{1}{2}}}{\left[z+(z^2-1)^{\frac{1}{2}}\right]};\quad \frac{2(z^2-1)^{\frac{1}{2}}}{\left[-z+(z^2-1)^{\frac{1}{2}}\right]};$$

$$\frac{1}{2}\left[z+(z^2-1)^{\frac{1}{2}}\right](z^2-1)^{-\frac{1}{2}};\ \frac{\left[z+(z^2-1)^{\frac{1}{2}}\right]}{\left[z-(z^2-1)^{\frac{1}{2}}\right]}.$$

The following tables list only 36 of the set of 72 solutions. But if for each of the formulas in the tables the transformation formula for the hypergeometric function

$$F(a, b; c; \zeta) = (1-\zeta)^{c-a-b} F(c-a, c-b; c; \zeta)$$

is applied one obtains the complete set of solutions. From these expressions for the $\mathfrak{P}_\nu^\mu(z)$ and $e^{-i\pi\mu}\mathfrak{Q}_\nu^\mu(z)$ the corresponding formulas for the $P_\nu^\mu(x)$ and $Q_\nu^\mu(x)$ can be constructed [see 4.3. of this chapter]. In order to make the functions, listed in the following tables one valued we define

$$(z^2-1)^\alpha = (z-1)^\alpha (z+1)^\alpha$$

with

$$|\arg(z \pm 1)| < \pi \text{ and } |\arg z| < \pi.$$

Then

$$-z-1 = e^{\mp i\pi}(z+1), \quad -z+1 = e^{\mp i\pi}(z-1), \quad 1-z^2 = e^{\mp i\pi}(z^2-1),$$

where the upper or the lower sign has to be taken according as $\operatorname{Im} z \gtrless 0$.

The following tables list the Legendre functions of the first and second kind $\mathfrak{P}_\nu^\mu(z)$ and $\mathfrak{Q}_\nu^\mu(z)$ respectively, where as mentioned before the argument z is a point in the complex z-plane under exclusion of points on the real axis between $+1$ and $-\infty$.

4.1.2 Representations as hypergeometric functions

The representations most frequently used are

$$\Gamma(1-\mu)\,\mathfrak{P}_\nu^\mu(z) = (z+1)^{\frac{1}{2}\mu}(z-1)^{-\frac{1}{2}\mu} F\left(-\nu, \nu+1; 1-\mu; \frac{1}{2}-\frac{1}{2}z\right),$$

$$\Gamma(1-\mu)\,\mathfrak{P}_\nu^\mu(z) = 2^\mu (z^2-1)^{-\frac{1}{2}\mu} F\left(1-\mu+\nu, -\mu-\nu; 1-\mu; \frac{1}{2}-\frac{1}{2}z\right),$$

$$\Gamma\left(\frac{3}{2}+\nu\right)\mathfrak{Q}_\nu^\mu(z) = e^{i\pi\mu}\, 2^{-\nu-1}\pi^{\frac{1}{2}}\,\Gamma(\nu+\mu+1) z^{-\nu-\mu-1}(z^2-1)^{\frac{1}{2}\mu}$$
$$\times F\left(1+\frac{1}{2}\nu+\frac{1}{2}\mu, \frac{1}{2}+\frac{1}{2}\nu+\frac{1}{2}\mu; \frac{3}{2}+\nu; z^{-2}\right).$$

The cases when these hypergeometric functions reduce to simpler cases are listed in 4.4. The cases in which μ or (and) ν becomes a positive or negative integer are likewise listed there.

Some further relations with hypergeometric functions. It is sometimes of advantage to express one of the Legendre functions by a sum of 2 hypergeometric functions whose arguments are different. Such formulas are

$$\mathfrak{Q}_\nu^\mu(z)\, e^{-i\pi\mu} \sin[\pi(\nu+\mu)]\,\Gamma(1-\mu)$$
$$= \frac{1}{2}\pi\left[e^{\mp i\nu\pi}\left(\frac{z+1}{z-1}\right)^{\frac{1}{2}\mu} F\left(-\nu, \nu+1; 1-\mu; \frac{1}{2}-\frac{1}{2}z\right)\right.$$
$$\left.-\left(\frac{z-1}{z+1}\right)^{\frac{1}{2}\mu} F\left(-\nu, \nu+1; 1-\mu; \frac{1}{2}+\frac{1}{2}z\right)\right],$$

$$2\,\mathfrak{Q}_\nu^\mu(z)\, e^{-i\pi\mu}\,\Gamma(1+\mu) = \Gamma(1+\nu+\mu)\,\Gamma(\mu-\nu)\left(\frac{z+1}{z-1}\right)^{\frac{1}{2}\mu}$$
$$\times\left[F\left(-\nu, \nu+1; 1+\mu; \frac{1}{2}+\frac{1}{2}z\right) - e^{\mp i\pi\nu}\left(\frac{z-1}{z+1}\right)^\mu\right.$$
$$\left.\times F\left(-\nu, \nu+1; 1+\mu; \frac{1}{2}-\frac{1}{2}z\right)\right],$$

$$\pi\Gamma(1+\mu)\,\mathfrak{P}_\nu^\mu(z) = \Gamma(\nu+\mu+1)\,\Gamma(\mu-\nu)\sin(\pi\mu)\left(\frac{z+1}{z-1}\right)^{\frac{1}{2}\mu}$$

$$\times\left[F\left(-\nu,\nu+1;1+\mu;\frac{1}{2}+\frac{1}{2}z\right)-\frac{\sin(\pi\nu)}{\sin(\pi\mu)}e^{\mp i\pi\mu}\left(\frac{z-1}{z+1}\right)^{\frac{1}{2}\mu}\right.$$

$$\left.\times F\left(-\nu,\nu+1;1+\mu;\frac{1}{2}-\frac{1}{2}z\right)\right],$$

$$(2\pi)^{\frac{1}{2}}(z^2-1)^{\frac{1}{4}}\frac{\Gamma\left(\frac{3}{2}+\nu\right)}{\Gamma(1+\nu+\mu)}\mathfrak{P}_\nu^\mu(z)$$

$$=\left[z+(z^2-1)^{\frac{1}{2}}\right]^{\nu+\frac{1}{2}}F\left[\frac{1}{2}+\mu,\frac{1}{2}-\mu;\frac{3}{2}+\nu;\frac{z+(z^2-1)^{\frac{1}{2}}}{2(z^2-1)^{\frac{1}{2}}}\right]$$

$$+ie^{-i\pi\mu}\left[z-(z^2-1)^{\frac{1}{2}}\right]^{\nu+\frac{1}{2}}F\left[\frac{1}{2}+\mu,\frac{1}{2}-\mu;\frac{3}{2}+\nu;\frac{-z+(z^2-1)^{\frac{1}{2}}}{2(z^2-1)^{\frac{1}{2}}}\right],$$

$$\mathfrak{Q}_\nu^\mu(z)\sin[\pi(\nu+\mu)]\,\Gamma\left(\frac{1}{2}-\mu\right)=\left(\frac{1}{2}\pi\right)^{\frac{1}{2}}\Gamma(\mu-\nu)\,(z^2-1)^{-\frac{1}{4}}e^{i\pi\mu}$$

$$\times\left\{\cos(\pi\nu)\left[z+(z^2-1)^{\frac{1}{2}}\right]^{-\nu-\frac{1}{2}}\right.$$

$$\times F\left[\frac{1}{2}+\mu,\frac{1}{2}-\mu;\frac{1}{2}-\nu;\frac{z+(z^2-1)^{\frac{1}{2}}}{2(z^2-1)^{\frac{1}{2}}}\right]$$

$$+i\cos(\pi\mu)\,e^{-i\pi\nu}\left[z+(z^2-1)^{\frac{1}{2}}\right]^{-\nu-\frac{1}{2}}$$

$$\left.\times F\left[\frac{1}{2}+\mu,\frac{1}{2}-\mu;\frac{1}{2}-\nu;\frac{-z+(z^2-1)^{\frac{1}{2}}}{2(z^2-1)^{\frac{1}{2}}}\right]\right\}.$$

In the above formulas the upper or lower sign has to be taken according as $\operatorname{Im} z \gtrless 0$.

Expansions for $\mathfrak{P}_\nu^\mu(z)$

	A_1 / A_2	a_1 / a_2	b_1 / b_2	c_1 / c_2	ζ	Remarks
1	$(z+1)^{\frac{1}{2}\mu}(z-1)^{-\frac{1}{2}\mu}\Big/\Gamma(1-\mu)$	$-\nu$	$1+\nu$	$1-\mu$	$\frac{1}{2}-\frac{1}{2}z$	
	0	...	...	...		
2	$\Gamma(-\mu)\,(z+1)^{\frac{1}{2}\mu}\,(z-1)^{-\frac{1}{2}\mu}\Big/[\Gamma(1+\nu-\mu)\,\Gamma(-\nu-\mu)]$	$-\nu$	$1+\nu$	$1+\mu$	$\frac{1}{2}+\frac{1}{2}z$	The upper or lower sign according as Im $z \gtrless 0$
	$-\pi^{-1}\sin(\pi\nu)\,\Gamma(\mu)\,(z-1)^{\frac{1}{2}\mu}\,(z+1)^{-\frac{1}{2}\mu}\,e^{\mp i\pi\mu}$	$-\nu$	$1+\nu$	$1-\mu$		
3	$2^{-\nu}(z+1)^{\frac{1}{2}\mu+\nu}\,(z-1)^{-\frac{1}{2}\mu}\Big/\Gamma(1-\mu)$	$-\nu$	$-\nu-\mu$	$1-\mu$	$\frac{z-1}{z+1}$	
	0	...	...	...		
4	$-2^{\nu+1}\Gamma(-\mu)\,e^{\pm i\pi\nu}\,(z+1)^{\frac{1}{2}\mu}\,(z-1)^{-\frac{1}{2}\mu-\nu-1}$ $\times\,[\Gamma(1+\nu-\mu)\,\Gamma(-\nu-\mu)]^{-1}$	$1+\nu$	$1+\nu+\mu$	$1+\mu$	$\frac{z+1}{z-1}$	The upper or lower sign taken according as Im $z \gtrless 0$
	$\pi^{-1}\,2^{\nu+1}\sin(\pi\nu)\,\Gamma(\mu)\,e^{\pm i\pi(\nu-\mu)}\,(z-1)^{\frac{1}{2}\mu-\nu-1}\,(z+1)^{-\frac{1}{2}\mu}$	$1+\nu$	$1+\nu-\mu$	$1-\mu$		

Expansions for $\mathfrak{P}_\nu^\mu(z)$

	A_1 / A_2	a_1 / a_2	b_1 / b_2	c_1 / c_2	ζ	Remarks
5	$2^{\nu+1}\Gamma(-1-2\nu)(z+1)^{\frac{1}{2}\mu-\nu-1}(z-1)^{-\frac{1}{2}\mu}[\Gamma(-\nu)\Gamma(-\nu-\mu)]^{-1}$	$1+\nu$	$1+\nu-\mu$	$2+2\nu$	$\frac{2}{1+z}$	
	$2^{-\nu}\Gamma(1+2\nu)(z+1)^{\frac{1}{2}\mu+\nu}(z-1)^{-\frac{1}{2}\mu}[\Gamma(1+\nu)\Gamma(1+\nu-\mu)]^{-1}$	$-\nu$	$-\nu-\mu$	-2ν		
6	$2^{-\nu}\Gamma(1+2\nu)(z+1)^{\frac{1}{2}\mu}(z-1)^{\nu-\frac{1}{2}\mu}[\Gamma(1+\nu)\Gamma(1+\nu-\mu)]^{-1}$	$-\nu$	$-\nu+\mu$	-2ν	$\frac{2}{1-z}$	
	$2^{\nu+1}\Gamma(-1-2\nu)(z+1)^{\frac{1}{2}\mu}(z-1)^{-\frac{1}{2}\mu-\nu-1}[\Gamma(-\nu)\Gamma(-\nu-\mu)]^{-1}$	$1+\nu$	$1+\nu+\mu$	$2+2\nu$		
7	$2^\mu(z^2-1)^{-\frac{1}{2}\mu}\Big/\Gamma(1-\mu)$	$\frac{1}{2}+\frac{1}{2}\nu-\frac{1}{2}\mu$	$-\frac{1}{2}\nu-\frac{1}{2}\mu$	$1-\mu$	$1-z^2$	$Re\, z>0$
	0	$\cdots$	$\cdots$	$\cdots$		
8	$2^{-\nu-1}\pi^{-\frac{1}{2}}\Gamma\left(-\frac{1}{2}-\nu\right)(z^2-1)^{-\frac{1}{2}\nu-\frac{1}{2}}\Big/\Gamma(-\nu-\mu)$	$\frac{1}{2}+\frac{1}{2}\nu-\frac{1}{2}\mu$	$\frac{1}{2}+\frac{1}{2}\nu+\frac{1}{2}\mu$	$\nu+\frac{3}{2}$	$(1-z^2)^{-1}$	
	$2^\nu\pi^{-\frac{1}{2}}\Gamma\left(\frac{1}{2}+\nu\right)(z^2-1)^{\frac{1}{2}\nu}\Big/\Gamma(1+\nu-\mu)$	$-\frac{1}{2}\nu+\frac{1}{2}\mu$	$-\frac{1}{2}\nu-\frac{1}{2}\mu$	$\frac{1}{2}-\nu$		

9	$2^{\mu}\pi^{\frac{1}{2}}(z^2-1)^{-\frac{1}{2}\mu}\left[\Gamma\left(\frac{1}{2}-\frac{1}{2}\nu-\frac{1}{2}\mu\right)\Gamma\left(1+\frac{1}{2}\nu-\frac{1}{2}\mu\right)\right]^{-1}$	$-\frac{1}{2}\nu-\frac{1}{2}\mu$	$\frac{1}{2}+\frac{1}{2}\nu-\frac{1}{2}\mu$	$\frac{1}{2}$	z^2	
	$-\pi^{\frac{1}{2}}2^{\mu+1}z(z^2-1)^{-\frac{1}{2}\mu}\left[\Gamma\left(\frac{1}{2}+\frac{1}{2}\nu-\frac{1}{2}\mu\right)\Gamma\left(-\frac{1}{2}\nu-\frac{1}{2}\mu\right)\right]^{-1}$	$\frac{1}{2}-\frac{1}{2}\nu-\frac{1}{2}\mu$	$1+\frac{1}{2}\nu-\frac{1}{2}\mu$	$\frac{3}{2}$		
10	$2^{-\nu-1}\pi^{-\frac{1}{2}}\Gamma\left(-\frac{1}{2}-\nu\right)z^{-\nu+\mu-1}(z^2-1)^{-\frac{1}{2}\mu}\Big/\Gamma(-\nu-\mu)$	$\frac{1}{2}+\frac{1}{2}\nu-\frac{1}{2}\mu$	$1+\frac{1}{2}\nu-\frac{1}{2}\mu$	$\nu+\frac{3}{2}$	z^{-2}	
	$2^{\nu}\pi^{-\frac{1}{2}}\Gamma\left(\frac{1}{2}+\nu\right)z^{\nu+\mu}(z^2-1)^{-\frac{1}{2}\mu}\Big/\Gamma(1+\nu-\mu)$	$-\frac{1}{2}\nu-\frac{1}{2}\mu$	$\frac{1}{2}-\frac{1}{2}\nu-\frac{1}{2}\mu$	$\frac{1}{2}-\nu$		
11	$2^{\mu}(z^2-1)^{-\frac{1}{2}\mu}z^{\nu+\mu}\Big/\Gamma(1-\mu)$	$-\frac{1}{2}\nu-\frac{1}{2}\mu$	$\frac{1}{2}-\frac{1}{2}\nu-\frac{1}{2}\mu$	$1-\mu$	$1-z^{-2}$	$Re\, z>0$
	0	...	...	...		
12	$\pi^{\frac{1}{2}}2^{\mu}e^{\mp i\frac{\pi}{2}(\nu+\mu)}(z^2-1)^{\frac{1}{2}\nu}$ $\times\left[\Gamma\left(\frac{1}{2}-\frac{1}{2}\nu-\frac{1}{2}\mu\right)\Gamma\left(1+\frac{1}{2}\nu-\frac{1}{2}\mu\right)\right]^{-1}$	$-\frac{1}{2}\nu-\frac{1}{2}\mu$	$\frac{1}{2}\mu-\frac{1}{2}\nu$	$\frac{1}{2}$	$z^2(z^2-1)^{-1}$	The upper or lower sign according as Im $z \gtrless 0$
	$-\pi^{\frac{1}{2}}2^{\mu+1}e^{\mp i\frac{\pi}{2}(\mu+\nu-1)}z(z^2-1)^{\frac{1}{2}\nu-\frac{1}{2}}$ $\times\left[\Gamma\left(-\frac{1}{2}\nu-\frac{1}{2}\mu\right)\Gamma\left(\frac{1}{2}+\frac{1}{2}\nu-\frac{1}{2}\mu\right)\right]^{-1}$	$\frac{1}{2}-\frac{1}{2}\nu-\frac{1}{2}\mu$	$\frac{1}{2}-\frac{1}{2}\nu+\frac{1}{2}\mu$	$\frac{3}{2}$		

Expansions for $\mathfrak{P}_\nu^\mu(z)$

	A_1 / A_2	a_1 / a_2	b_1 / b_2	c_1 / c_2	ζ	Remarks
13	$(2\pi)^{-\frac{1}{2}}(z^2-1)^{-\frac{1}{4}}\Gamma\left(-\frac{1}{2}-\nu\right)\left[z-(z^2-1)^{\frac{1}{2}}\right]^{\nu+\frac{1}{2}}\Big/\Gamma(-\nu-\mu)$	$\frac{1}{2}+\mu$	$\frac{1}{2}-\mu$	$\frac{3}{2}+\nu$	$\frac{-z+(z^2-1)^{\frac{1}{2}}}{2(z^2-1)^{\frac{1}{2}}}$	
	$(2\pi)^{-\frac{1}{2}}(z^2-1)^{-\frac{1}{4}}\Gamma\left(\frac{1}{2}+\nu\right)\left[z-(z^2-1)^{\frac{1}{2}}\right]^{-\nu-\frac{1}{2}}\Big/\Gamma(1+\nu-\mu)$	$\frac{1}{2}+\mu$	$\frac{1}{2}-\mu$	$\frac{1}{2}-\nu$		
14	$\pi^{-\frac{1}{2}}2^\mu\Gamma\left(-\frac{1}{2}-\nu\right)(z^2-1)^{\frac{1}{2}\mu}\left[z+(z^2-1)^{\frac{1}{2}}\right]^{-\nu-\mu-1}\Big/\Gamma(-\nu-\mu)$	$\frac{1}{2}+\mu$	$1+\nu+\mu$	$\nu+\frac{3}{2}$	$\frac{z-(z^2-1)^{\frac{1}{2}}}{z+(z^2-1)^{\frac{1}{2}}}$	
	$\pi^{-\frac{1}{2}}2^\mu\Gamma\left(\frac{1}{2}+\nu\right)(z^2-1)^{\frac{1}{2}\mu}\left[z+(z^2-1)^{\frac{1}{2}}\right]^{\nu-\mu}\Big/\Gamma(1+\nu-\mu)$	$\frac{1}{2}+\mu$	$-\nu+\mu$	$\frac{1}{2}-\nu$		
15	$2^\mu(z^2-1)^{-\frac{1}{2}\mu}\left[z+(z^2-1)^{\frac{1}{2}}\right]^{\nu+\mu}\Big/\Gamma(1-\mu)$	$-\nu-\mu$	$\frac{1}{2}-\mu$	$1-2_\mu$	$\frac{2(z^2-1)^{\frac{1}{2}}}{z+(z^2-1)^{\frac{1}{2}}}$	$Re\, z > 0$
	0	…	…	…		

No.						
16	$2^{\mu}(z^2-1)^{-\frac{1}{2}\mu}\left[z-(z^2-1)^{\frac{1}{2}}\right]^{\nu+\mu}\Big/\Gamma(1-\mu)$	$-\nu-\mu$	$\frac{1}{2}-\mu$	$1-2\mu$	$\dfrac{2(z^2-1)^{\frac{1}{2}}}{-z+(z^2-1)^{\frac{1}{2}}}$	
	0	…	…	…		
17	$(2\pi)^{-\frac{1}{2}}\Gamma\left(\frac{1}{2}+\nu\right)e^{\mp i\pi\left(\mu-\frac{1}{2}\right)}(z^2-1)^{-\frac{1}{4}}\left[z+(z^2-1)^{\frac{1}{2}}\right]^{-\nu-\frac{1}{2}} \times [\Gamma(\nu-\mu+1)]^{-1}$	$\frac{1}{2}-\mu$	$\frac{1}{2}+\mu$	$\frac{1}{2}-\nu$	$\dfrac{z+(z^2-1)^{\frac{1}{2}}}{2(z^2-1)^{\frac{1}{2}}}$	The upper or lower sign according as Im $z \gtrless 0$
	$(2\pi)^{-\frac{1}{2}}\Gamma\left(-\frac{1}{2}-\nu\right)e^{\mp i\pi\left(\mu-\frac{1}{2}\right)}(z^2-1)^{-\frac{1}{4}}\left[z+(z^2-1)^{\frac{1}{2}}\right]^{\nu+\frac{1}{2}} \times [\Gamma(-\nu-\mu)]^{-1}$	$\frac{1}{2}-\mu$	$\frac{1}{2}+\mu$	$\frac{3}{2}+\nu$		
18	$\pi^{-\frac{1}{2}}2^{-\mu}\Gamma\left(\frac{1}{2}+\nu\right)(z^2-1)^{-\frac{1}{2}\mu}\left[z-(z^2-1)^{\frac{1}{2}}\right]^{\nu+\mu}\Big/\Gamma(\nu-\mu+1)$	$-\nu-\mu$	$\frac{1}{2}-\mu$	$\frac{1}{2}-\nu$	$\dfrac{z+(z^2-1)^{\frac{1}{2}}}{z-(z^2-1)^{\frac{1}{2}}}$	
	$\pi^{-\frac{1}{2}}2^{-\mu}\Gamma\left(-\frac{1}{2}-\nu\right)(z^2-1)^{-\frac{1}{2}\mu}\left[z+(z^2-1)^{\frac{1}{2}}\right]^{\nu-\mu+1}\Big/\Gamma(-\nu-\mu)$	$1+\nu-\mu$	$\frac{1}{2}-\mu$	$\frac{3}{2}+\nu$		

Expansions for $e^{-i\pi\mu}\mathfrak{Q}_\nu^\mu(z)$

	A_3 / A_4	a_3 / a_4	b_3 / b_4	c_3 / c_4	ζ	Remarks
19	$\frac{1}{2}\Gamma(1+\nu+\mu)\,\Gamma(-\mu)\,(z+1)^{-\frac{1}{2}\mu}(z-1)^{\frac{1}{2}\mu}\Big/\Gamma(1+\nu-\mu)$	$-\nu$	$1+\nu$	$1+\mu$	$\frac{1}{2}-\frac{1}{2}z$	
	$\frac{1}{2}\Gamma(\mu)\,(z+1)^{\frac{1}{2}\mu}(z-1)^{-\frac{1}{2}\mu}$	$-\nu$	$1+\nu$	$1-\mu$		
20	$-\frac{1}{2}e^{\mp i\pi\nu}\Gamma(\mu)\,(z-1)^{\frac{1}{2}\mu}(z+1)^{-\frac{1}{2}\mu}$	$-\nu$	$1+\nu$	$1-\mu$	$\frac{1}{2}+\frac{1}{2}z$	The upper or lower sign according as Im $z \gtrless 0$
	$-\frac{1}{2}e^{\mp i\pi\nu}\,\Gamma(1+\nu+\mu)\,\Gamma(-\mu)\,(z+1)^{\frac{1}{2}\mu}(z-1)^{-\frac{1}{2}\mu}\Big/\Gamma(1+\nu-\mu)$	$-\nu$	$1+\nu$	$1+\mu$		
21	$2^{-1-\nu}\Gamma(\mu)\,(z+1)^{\nu+\frac{1}{2}\mu}(z-1)^{-\frac{1}{2}\mu}$	$-\nu$	$-\nu-\mu$	$1-\mu$	$\frac{z-1}{z+1}$	
	$2^{-1-\nu}\Gamma(1+\nu+\mu)\,\Gamma(-\mu)\,(z+1)^{\nu-\frac{1}{2}\mu}(z-1)^{\frac{1}{2}\mu}\Big/\Gamma(1+\nu-\mu)$	$-\nu$	$\mu-\nu$	$1+\mu$		
22	$2^\nu\Gamma(-\mu)\,\Gamma(1+\nu+\mu)\,(z-1)^{-\nu-\frac{1}{2}\mu-1}(z+1)^{\frac{1}{2}\mu}\Big/\Gamma(1+\nu-\mu)$	$1+\nu$	$1+\nu+\mu$	$1+\mu$	$\frac{z+1}{z-1}$	
	$2^\nu\Gamma(\mu)\,(z+1)^{-\frac{1}{2}\mu}(z-1)^{\frac{1}{2}\mu-\nu-1}$	$1+\nu$	$1+\nu-\mu$	$1-\mu$		

Expansions for $e^{-i\pi\mu}\,\mathfrak{Q}_\nu^\mu(z)$

	A_3 / A_4	a_3 / a_4	b_3 / b_4	c_3 / c_4	ζ	Remarks
23	$2^\nu\Gamma(1+\nu)\,\Gamma(1+\nu+\mu)\,(z+1)^{\frac{1}{2}\mu-\nu-1}(z-1)^{-\frac{1}{2}\mu}\Big/\Gamma(2+2\nu)$	$1+\nu-\mu$	$1+\nu$	$2+2\nu$	$\frac{2}{1+z}$	
	0	...	...	...		
24	$2^\nu\Gamma(1+\nu)\,\Gamma(1+\nu+\mu)\,(z+1)^{\frac{1}{2}\mu}(z-1)^{-\frac{1}{2}\mu-\nu-1}\Big/\Gamma(2+2\nu)$	$1+\nu+\mu$	$1+\nu$	$2+2\nu$	$\frac{2}{1-z}$	
	0	...	...	...		
25	$2^{\mu-1}\Gamma(\mu)\,(z^2-1)^{-\frac{1}{2}\mu}$	$\frac{1}{2}+\frac{1}{2}\nu-\frac{1}{2}\mu$	$-\frac{1}{2}\nu-\frac{1}{2}\mu$	$1-\mu$	$1-z^2$	$Re\,z>0$
	$2^{-1-\mu}\Gamma(1+\nu+\mu)\,\Gamma(-\mu)\,(z^2-1)^{\frac{1}{2}\mu}\Big/\Gamma(1+\nu-\mu)$	$\frac{1}{2}+\frac{1}{2}\nu+\frac{1}{2}\mu$	$\frac{1}{2}\mu-\frac{1}{2}\nu$	$1+\mu$		
26	$2^{-1-\nu}\pi^{\frac{1}{2}}\Gamma(1+\nu+\mu)\,(z^2-1)^{-\frac{1}{2}-\frac{1}{2}\nu}\Big/\Gamma\left(\frac{3}{2}+\nu\right)$	$\frac{1}{2}+\frac{1}{2}\nu-\frac{1}{2}\mu$	$\frac{1}{2}+\frac{1}{2}\nu+\frac{1}{2}\mu$	$\frac{3}{2}+\nu$	$(1-z^2)^{-1}$	
	0	...	...	...		
27	$\pi^{\frac{1}{2}}\,2^{\mu-1}\Gamma\left(\frac{1}{2}+\frac{1}{2}\nu+\frac{1}{2}\mu\right)e^{\pm i\frac{\pi}{2}(\mu-\nu-1)}(z^2-1)^{-\frac{1}{2}\mu}\times\left[\Gamma\left(1+\frac{1}{2}\nu-\frac{1}{2}\mu\right)\right]^{-1}$	$-\frac{1}{2}\nu-\frac{1}{2}\mu$	$\frac{1}{2}+\frac{1}{2}\nu-\frac{1}{2}\mu$	$\frac{1}{2}$	z^2	The upper or lower sign according as Im $z \gtrless 0$
	$\pi^{\frac{1}{2}}\,2^{\mu}\Gamma\left(1+\frac{1}{2}\nu+\frac{1}{2}\mu\right)e^{\pm i\frac{\pi}{2}(\mu-\nu)}z(z^2-1)^{-\frac{1}{2}\mu}\times\left[\Gamma\left(\frac{1}{2}+\frac{1}{2}\nu-\frac{1}{2}\mu\right)\right]^{-1}$	$\frac{1}{2}-\frac{1}{2}\nu-\frac{1}{2}\mu$	$1+\frac{1}{2}\nu-\frac{1}{2}\mu$	$\frac{3}{2}$		

Expansions for $e^{-i\pi\mu}\,\mathfrak{Q}_\nu^\mu(z)$

	A_3 / A_4	a_3 / a_4	b_3 / b_4	c_3 / c_4	ζ	Remarks
28	$2^{-1-\nu}\pi^{\frac{1}{2}}\Gamma(1+\nu+\mu)\,z^{-1-\nu-\mu}(z^2-1)^{\frac{1}{2}\mu}\Big/\Gamma\left(\frac{3}{2}+\nu\right)$	$1+\frac{1}{2}\nu+\frac{1}{2}\mu$	$\frac{1}{2}+\frac{1}{2}\nu+\frac{1}{2}\mu$	$\frac{3}{2}+\nu$	z^{-2}	
	0	…	…	…		
29	$2^{\mu-1}\Gamma(\mu)\,z^{\nu+\mu}(z^2-1)^{-\frac{1}{2}\mu}$	$-\frac{1}{2}\nu-\frac{1}{2}\mu$	$\frac{1}{2}-\frac{1}{2}\nu-\frac{1}{2}\mu$	$1-\mu$	$1-z^{-2}$	
	$2^{-1-\mu}\Gamma(1+\nu+\mu)\,\Gamma(-\mu)\,z^{\nu-\mu}(z^2-1)^{\frac{1}{2}\mu}\Big/\Gamma(1+\nu-\mu)$	$\frac{1}{2}\mu-\frac{1}{2}\nu$	$\frac{1}{2}-\frac{1}{2}\nu+\frac{1}{2}\mu$	$1+\mu$		
30	$\pi^{\frac{1}{2}}\,2^{\mu-1}\,\Gamma\left(\frac{1}{2}+\frac{1}{2}\nu+\frac{1}{2}\mu\right)e^{\mp i\pi\left(\nu+\frac{1}{2}\right)}(z^2-1)^{\frac{1}{2}\nu}\Big/\Gamma\left(1+\frac{1}{2}\nu-\frac{1}{2}\mu\right)$	$-\frac{1}{2}\nu-\frac{1}{2}\mu$	$\frac{1}{2}\mu-\frac{1}{2}\nu$	$\frac{1}{2}$	$\frac{z^2}{z^2-1}$	The upper or lower sign according as Im $z \gtrless 0$
	$\pi^{\frac{1}{2}}2^{\mu}\Gamma\left(1+\frac{1}{2}\nu+\frac{1}{2}\mu\right)e^{\mp i\pi\left(\nu-\frac{1}{2}\right)}z\,(z^2-1)^{\frac{1}{2}\nu-\frac{1}{2}}\Big/\Gamma\left(\frac{1}{2}+\frac{1}{2}\nu-\frac{1}{2}\mu\right)$	$\frac{1}{2}-\frac{1}{2}\nu-\frac{1}{2}\mu$	$\frac{1}{2}-\frac{1}{2}\nu+\frac{1}{2}\mu$	$\frac{3}{2}$		
31	$\left(\frac{1}{2}\pi\right)^{\frac{1}{2}}\Gamma(1+\nu+\mu)\,(z^2-1)^{-\frac{1}{4}}\left[z-(z^2-1)^{\frac{1}{2}}\right]^{\nu+\frac{1}{2}}\Big/\Gamma\left(\frac{3}{2}+\nu\right)$	$\frac{1}{2}+\mu$	$\frac{1}{2}-\mu$	$\frac{3}{2}+\nu$	$\frac{-z+(z^2-1)^{\frac{1}{2}}}{2(z^2-1)^{\frac{1}{2}}}$	
	0	…	…	…		

No.						
32	$\pi^{\frac{1}{2}} 2^{\mu} \Gamma(1+\nu+\mu)(z^2-1)^{\frac{1}{2}\mu}\left[z+(z^2-1)^{\frac{1}{2}}\right]^{-1-\nu-\mu} \Big/ \Gamma\left(\frac{3}{2}+\nu\right)$	$\frac{1}{2}+\mu$	$1+\nu+\mu$	$\frac{3}{2}+\nu$	$\dfrac{z-(z^2-1)^{\frac{1}{2}}}{z+(z^2-1)^{\frac{1}{2}}}$	
	0	$\cdots$	$\cdots$	$\cdots$		
33	$2^{\mu-1}\Gamma(\mu)(z^2-1)^{-\frac{1}{2}\mu}\left[z+(z^2-1)^{\frac{1}{2}}\right]^{\nu+\mu}$	$-\nu-\mu$	$\frac{1}{2}-\mu$	$1-2\mu$	$\dfrac{2(z^2-1)^{\frac{1}{2}}}{z+(z^2-1)^{\frac{1}{2}}}$	$Re\, z > 0$
	$2^{-1-\mu}\Gamma(1+\nu+\mu)\Gamma(-\mu)(z^2-1)^{\frac{1}{2}\mu}\left[z+(z^2-1)^{\frac{1}{2}}\right]^{\nu-\mu} \Big/ \Gamma(1+\nu-\mu)$	$\mu-\nu$	$\frac{1}{2}+\mu$	$1-2\mu$		
34	$2^{\mu-1}\Gamma(\mu)(z^2-1)^{-\frac{1}{2}\mu}\left[z-(z^2-1)^{\frac{1}{2}}\right]^{\nu+\mu}$	$-\nu-\mu$	$\frac{1}{2}-\mu$	$1-2\mu$	$\dfrac{2(z^2-1)^{\frac{1}{2}}}{-z+(z^2-1)^{\frac{1}{2}}}$	
	$2^{-1-\mu}\Gamma(1+\nu+\mu)\Gamma(-\mu)(z^2-1)^{\frac{1}{2}\mu}\left[z-(z^2-1)^{\frac{1}{2}}\right]^{\nu-\mu} \Big/ \Gamma(1+\nu-\mu)$	$\mu-\nu$	$\frac{1}{2}+\mu$	$1+2\mu$		
35	$\left(\frac{1}{2}\pi\right)^{\frac{1}{2}}\Gamma\left(\frac{1}{2}+\nu\right)(z^2-1)^{-\frac{1}{4}}\left[z-(z^2-1)^{\frac{1}{2}}\right]^{\frac{1}{2}+\nu} \Big/ \Gamma(1+\nu-\mu)$	$\frac{1}{2}+\mu$	$\frac{1}{2}-\mu$	$\frac{1}{2}-\nu$	$\dfrac{z+(z^2-1)^{\frac{1}{2}}}{2(z^2-1)^{\frac{1}{2}}}$	
	$\pi^{-\frac{1}{2}} 2^{-1-\nu}\Gamma(1+\nu+\mu)\Gamma\left(-\frac{1}{2}-\nu\right)\cos(\pi\mu)(z^2-1)^{-\frac{1}{2}-\frac{1}{2}\nu}$	$1+\nu-\mu$	$1+\nu+\mu$	$\frac{3}{2}+\nu$		
36	$2^{\mu}\pi^{\frac{1}{2}}\Gamma\left(\frac{1}{2}+\nu\right)e^{\mp i\pi\left(\frac{1}{2}+\mu\right)}(z^2-1)^{\frac{1}{2}\mu}\left[z+(z^2-1)^{\frac{1}{2}}\right]^{\mu-\nu} \times [\Gamma(1+\nu-\mu)]^{-1}$	$\frac{1}{2}+\mu$	$\mu-\nu$	$\frac{1}{2}-\nu$	$\dfrac{z+(z^2-1)^{\frac{1}{2}}}{z-(z^2-1)^{\frac{1}{2}}}$	The upper or lower sign according as Im $z \gtrless 0$
	$-2^{-\mu}\pi^{-\frac{1}{2}}\cos(\pi\mu)\Gamma(1+\nu+\mu)\Gamma\left(-\frac{1}{2}-\nu\right)e^{\mp i\pi(\nu-\mu)} \times (z^2-1)^{-\frac{1}{2}\mu}\left[z+(z^2-1)^{\frac{1}{2}}\right]^{1+\nu-\mu}$	$\frac{1}{2}-\mu$	$1+\nu-\mu$	$\frac{3}{2}+\nu$		

4.2 Relations between Legendre functions

4.2.1 Relations between different solutions, Wronskians

Legendre's differential equation remains unchanged if μ is replaced by $-\mu$, z by $-z$, and ν by $-\nu-1$. Therefore

$$\mathfrak{P}_\nu^{\pm\mu}(\pm z),\ \mathfrak{Q}_\nu^{\pm\mu}(\pm z),\ \mathfrak{P}_{-\nu-1}^{\pm\mu}(\pm z),\ \mathfrak{Q}_{-\nu-1}^{\pm\mu}(\pm z)$$

are likewise solutions of Legendre's differential equation and it must be possibel to express each of the above expressions by a linear combination of a fundamental system of solutions. These relations are

$$\mathfrak{P}_\nu^\mu(z) = \mathfrak{P}_{-\nu-1}^\mu(z),$$

$$e^{i\pi\mu}\Gamma(\nu+\mu+1)\,\mathfrak{Q}_\nu^{-\mu}(z) = e^{-i\pi\mu}\Gamma(\nu-\mu+1)\,\mathfrak{Q}_\nu^\mu(z),$$

$$\mathfrak{Q}_\nu^\mu(z)\sin[\pi(\nu+\mu)] - \mathfrak{Q}_{-\nu-1}^\mu(z)\sin[\pi(\nu-\mu)] = \pi e^{i\pi\mu}\cos(\pi\nu)\,\mathfrak{P}_\nu^\mu(z),$$

$$\mathfrak{Q}_\nu^\mu(z)\sin(\pi\mu) = \frac{1}{2}\pi e^{i\pi\mu}\left[\mathfrak{P}_\nu^\mu(z) - \frac{\Gamma(\nu+\mu+1)}{\Gamma(\nu-\mu+1)}\mathfrak{P}_\nu^{-\mu}(z)\right],$$

$$\mathfrak{P}_\nu^{-\mu}(z) = \frac{\Gamma(\nu-\mu+1)}{\Gamma(\nu+\mu+1)}\left[\mathfrak{P}_\nu^\mu(z) - \frac{2}{\pi}e^{-i\pi\mu}\sin(\pi\mu)\,\mathfrak{Q}_\nu^\mu(z)\right],$$

$$\mathfrak{P}_\nu^\mu(z) = \frac{\Gamma(\nu+\mu+1)}{\Gamma(\nu-\mu+1)}\mathfrak{P}_\nu^{-\mu}(z) + \frac{2}{\pi}e^{-i\pi\mu}\sin(\pi\mu)\,\mathfrak{Q}_\nu^\mu(z).$$

For $\mu = m$, $(m = 1, 2, 3, \ldots)$

$$\mathfrak{P}_\nu^m(z) = \frac{\Gamma(\nu+m+1)}{\Gamma(\nu-m+1)}\mathfrak{P}_\nu^{-m}(z),$$

$$\mathfrak{P}_\nu^{-\mu}(z) = \frac{e^{-i\pi\mu}\,\Gamma(\nu-\mu+1)}{\pi\cos(\pi\nu)\,\Gamma(\nu+\mu+1)}\sin[\pi(\nu-\mu)]\,[\mathfrak{Q}_\nu^\mu(z) - \mathfrak{Q}_{-\nu-1}^\mu(z)],$$

$$\mathfrak{Q}_{-\nu-1}^\mu(z) - \mathfrak{Q}_\nu^\mu(z) = e^{i\pi\mu}\cos(\pi\nu)\,\Gamma(\nu+\mu+1)\,\Gamma(\mu-\nu)\,\mathfrak{P}_\nu^{-\mu}(z),$$

$$\mathfrak{P}_\nu^\mu(-z) = e^{\mp i\pi\nu}\,\mathfrak{P}_\nu^\mu(z) - \frac{2}{\pi}e^{-i\pi\mu}\sin[\pi(\nu+\mu)]\,\mathfrak{Q}_\nu^\mu(z),$$

$$\mathfrak{Q}_\nu^\mu(z)\,e^{-i\pi\mu}\sin[\pi(\nu+\mu)] = \frac{1}{2}\pi\,[e^{\mp i\pi\nu}\,\mathfrak{P}_\nu^\mu(z) - \mathfrak{P}_\nu^\mu(-z)],$$

$$\mathfrak{Q}_\nu^\mu(-z) = -e^{\pm i\pi\nu}\,\mathfrak{Q}_\nu^\mu(z).$$

In the last three formulas the upper or lower sign has to be taken according as $\operatorname{Im} z \gtrless 0$.

Whipple's formula

$$\mathfrak{Q}_\nu^\mu(z) = e^{i\pi\mu}\left(\frac{1}{2}\pi\right)^{\frac{1}{2}}\Gamma(\nu+\mu+1)\,(z^2-1)^{-\frac{1}{4}}\,\mathfrak{P}_{-\mu-\frac{1}{2}}^{-\nu-\frac{1}{2}}\left[z(z^2-1)^{-\frac{1}{2}}\right],$$

$$\Gamma(-\nu-\mu)\,\mathfrak{P}_\nu^\mu(z) = i\,e^{i\pi\nu}\left(\frac{1}{2}\pi\right)^{-\frac{1}{2}}(z^2-1)^{-\frac{1}{4}}\,\mathfrak{Q}_{-\mu-\frac{1}{2}}^{-\nu-\frac{1}{2}}\left[z(z^2-1)^{-\frac{1}{2}}\right].$$

Both formulas valid for $Re\, z > 0$.

The Wronskian determinant

$$W\{\mathfrak{P}_\nu^\mu(z), \mathfrak{Q}_\nu^\mu(z)\} = \mathfrak{P}_\nu^\mu(z) \frac{d}{dz} \mathfrak{Q}_\nu^\mu(z) - \mathfrak{Q}_\nu^\mu(z) \frac{d}{dz} \mathfrak{P}_\nu^\mu(z).$$

is equal to

$$W\{\mathfrak{P}_\nu^\mu(z), \mathfrak{Q}_\nu^\mu(z)\} = e^{i\pi\mu} \frac{\Gamma(1+\mu+\nu)}{\Gamma(1+\nu-\mu)} (1-z^2)^{-1}.$$

With this formula and the formulas above the Wronskians of other combinations such as for instance

$$W\{\mathfrak{P}_\nu^{-\mu}(z), \mathfrak{Q}_{-\nu-1}^\mu(z)\} \text{ etc.}$$

can be obtained. We note

$$W\{\mathfrak{P}_\nu^{-\mu}(z), \mathfrak{Q}_\nu^\mu(z)\} = e^{i\pi\mu}(1-z^2)^{-1},$$

which never vanishes.

4.2.2 Relations between contiguous Legendre functions

$$\mathfrak{P}_\nu^{\mu+2}(z) + 2(\mu+1)\, z(z^2-1)^{-\frac{1}{2}} \mathfrak{P}_\nu^{\mu+1}(z) = (\nu-\mu)(\nu+\mu+1)\, \mathfrak{P}_\nu^\mu(z),$$

$$(2\nu+1)\, z\, \mathfrak{P}_\nu^\mu(z) = (\nu-\mu+1)\, \mathfrak{P}_{\nu+1}^\mu(z) + (\nu+\mu)\, \mathfrak{P}_{\nu-1}^\mu(z).$$

$$\mathfrak{P}_{\nu-1}^\mu(z) - \mathfrak{P}_{\nu+1}^\mu(z) = -(2\nu+1)(z^2-1)^{\frac{1}{2}} \mathfrak{P}_\nu^{\mu-1}(z),$$

$$(\nu-\mu)(\nu-\mu+1)\, \mathfrak{P}_{\nu+1}^\mu(z) - (\nu+\mu)(\nu+\mu+1)\, \mathfrak{P}_{\nu-1}^\mu(z)$$
$$= (2\nu+1)(z^2-1)^{\frac{1}{2}} \mathfrak{P}_\nu^{\mu+1}(z),$$

$$\mathfrak{P}_{\nu-1}^\mu(z) - z\, \mathfrak{P}_\nu^\mu(z) = -(\nu-\mu+1)(z^2-1)^{\frac{1}{2}} \mathfrak{P}_\nu^{\mu-1}(z),$$

$$z\, \mathfrak{P}_\nu^\mu(z) - \mathfrak{P}_{\nu+1}^\mu(z) = -(\nu+\mu)(z^2-1)^{\frac{1}{2}} \mathfrak{P}_\nu^{\mu-1}(z),$$

$$(\nu-\mu)\, z\, \mathfrak{P}_\nu^\mu(z) - (\nu+\mu)\, \mathfrak{P}_{\nu-1}^\mu(z) = (z^2-1)^{\frac{1}{2}} \mathfrak{P}_\nu^{\mu+1}(z),$$

$$\frac{d\mathfrak{P}_\nu^\mu(z)}{dz} = (\nu+\mu)(\nu-\mu+1)(z^2-1)^{-\frac{1}{2}} \mathfrak{P}_\nu^{\mu-1}(z) - \frac{\mu z}{z^2-1} \mathfrak{P}_\nu^\mu(z),$$

$$(z^2-1)\frac{d\mathfrak{P}_\nu^\mu(z)}{dz} = (\nu-\mu+1)\, \mathfrak{P}_{\nu+1}^\mu(z) - (\nu+1)\, z\, \mathfrak{P}_\nu^\mu(z)$$
$$= \nu z\, \mathfrak{P}_\nu^\mu(z) - (\nu+\mu)\, \mathfrak{P}_{\nu-1}^\mu(z).$$

The same formulas are valid for the $\mathfrak{Q}_\nu^\mu(z)$.

4.3 The functions $P_\nu^\mu(x)$ and $Q_\nu^\mu(x)$. (Legendre functions on the cut)

4.3.1 Definitions and representations

If the variable in Legendre's differential equation is real and lies between -1 and $+1$ one denotes two linearly independent solutions by

$$P_\nu^\mu(x) \text{ and } Q_\nu^\mu(x).$$

They are defined by the following equivalent expressions

$$P_\nu^\mu(x) = e^{i\frac{1}{2}\pi\mu}\,\mathfrak{P}_\nu^\mu(x+i0) = e^{-i\frac{1}{2}\pi\mu}\,\mathfrak{P}_\nu^\mu(x-i0),$$

$$P_\nu^\mu(x) = \frac{1}{2}\left[e^{i\frac{1}{2}\pi\mu}\,\mathfrak{P}_\nu^\mu(x+i0) + e^{-i\frac{1}{2}\pi\mu}\,\mathfrak{P}_\nu^\mu(x-i0)\right],$$

$$P_\nu^\mu(x) = \frac{i}{\pi}e^{-i\pi\mu}\left[e^{-i\frac{1}{2}\pi\mu}\,\mathfrak{Q}_\nu^\mu(x+i0) - e^{i\frac{1}{2}\pi\mu}\,\mathfrak{Q}_\nu^\mu(x-i0)\right],$$

$$Q_\nu^\mu(x) = \frac{1}{2}e^{-i\pi\mu}\left[e^{-i\frac{1}{2}\pi\mu}\,\mathfrak{Q}_\nu^\mu(x+i0) + e^{i\frac{1}{2}\pi\mu}\,\mathfrak{Q}_\nu^\mu(x-i0)\right].$$

Here $f(x \pm i0)$ is defined as $\lim_{\varepsilon\to 0} f(x \pm i\varepsilon)$.

With these definitions, formulas for the $P_\nu^\mu(x)$ and $Q_\nu^\mu(x)$ can be given by the aid of the corresponding formulas for the $\mathfrak{P}_\nu^\mu(z)$, $\mathfrak{Q}_\nu^\mu(z)$ as listed in the tables before.

Some of these expressions are listed here. We restrict ourselves mostly to the $P_\nu^\mu(x)$ since (see the next section) the corresponding $Q_\nu^\mu(x)$ can be obtained by use of the formula

$$\frac{2}{\pi}\sin(\pi\mu)\,Q_\nu^\mu(x) = \cos(\pi\mu)\,P_\nu^\mu(x) - \frac{\Gamma(\nu+\mu+1)}{\Gamma(\nu-\mu+1)}\,P_\nu^{-\mu}(x).$$

Furthermore some of the expressions for the $P_\nu^\mu(x)$ here are only valid for $0 < x < 1$. But the use of relations between $P_\nu^\mu(x)$ and $P_\nu^\mu(-x)$ (see next section) allow to cover the whole interval $-1 < x < +1$.

$$\Gamma(1-\mu)\,P_\nu^\mu(x) = (1+x)^{\frac{1}{2}\mu}(1-x)^{-\frac{1}{2}\mu} \times F\left(-\nu, 1+\nu; 1-\mu; \frac{1}{2}-\frac{1}{2}x\right), \quad -1 < x < 1,$$

$$\Gamma(1-\mu)\,P_\nu^\mu(x) = 2^{-\nu}(1+x)^{\frac{1}{2}\mu+\nu}(1-x)^{-\frac{1}{2}\mu} \times F\left(-\nu, -\nu-\mu; 1-\mu; \frac{x-1}{x+1}\right),$$

$$\Gamma(1-\mu)\, P_\nu^\mu(x) = 2^\mu (1-x^2)^{-\frac{1}{2}\mu}$$
$$\times F\left(\frac{1}{2}+\frac{1}{2}\nu-\frac{1}{2}\mu, -\frac{1}{2}\nu-\frac{1}{2}\mu; 1-\mu; 1-x^2\right),$$

$$\Gamma(1-\mu)\, P_\nu^\mu(x) = 2^\mu (1-x^2)^{-\frac{1}{2}\mu} x^{\nu+\mu}$$
$$\times F\left(-\frac{1}{2}\nu-\frac{1}{2}\mu, \frac{1}{2}-\frac{1}{2}\nu-\frac{1}{2}\mu; 1-\mu; 1-x^{-2}\right).$$

The Last three formulas being valid for $0 < x < 1$

$$P_\nu^\mu(x) = 2^\mu \pi^{\frac{1}{2}} (1-x^2)^{-\frac{1}{2}\mu}$$
$$\times \left\{\left[\Gamma\left(\frac{1}{2}-\frac{1}{2}\nu-\frac{1}{2}\mu\right)\Gamma\left(1+\frac{1}{2}\nu-\frac{1}{2}\mu\right)\right]^{-1}\right.$$
$$\times F\left(-\frac{1}{2}\mu-\frac{1}{2}\nu, \frac{1}{2}-\frac{1}{2}\mu+\frac{1}{2}\nu; \frac{1}{2}; x^2\right)$$
$$-2x\left[\Gamma\left(\frac{1}{2}+\frac{1}{2}\nu-\frac{1}{2}\mu\right)\Gamma\left(-\frac{1}{2}\nu-\frac{1}{2}\mu\right)\right]^{-1}$$
$$\left.\times F\left(\frac{1}{2}-\frac{1}{2}\nu-\frac{1}{2}\mu, 1+\frac{1}{2}\nu-\frac{1}{2}\mu; \frac{3}{2}; x^2\right)\right\},$$

$$Q_\nu^\mu(x) = 2^\mu \pi^{\frac{3}{2}} (1-x^2)^{-\frac{1}{2}\mu}$$
$$\times \left\{\cot\left[\frac{\pi}{2}(\nu+\mu)\right]\left[\Gamma\left(\frac{1}{2}+\frac{1}{2}\nu-\frac{1}{2}\mu\right)\Gamma\left(-\frac{1}{2}\nu-\frac{1}{2}\mu\right)\right]^{-1}\right.$$
$$\times F\left(\frac{1}{2}-\frac{1}{2}\nu-\frac{1}{2}\mu, \frac{1}{2}\nu-\frac{1}{2}\mu+1; \frac{3}{2}; x^2\right)$$
$$-\frac{1}{2}\tan\left[\frac{\pi}{2}(\nu+\mu)\right]\left[\Gamma\left(1+\frac{1}{2}\nu-\frac{1}{2}\mu\right)\Gamma\left(\frac{1}{2}-\frac{1}{2}\nu-\frac{1}{2}\mu\right)\right]^{-1}$$
$$\left.\times F\left(-\frac{1}{2}\nu-\frac{1}{2}\mu, \frac{1}{2}+\frac{1}{2}\nu-\frac{1}{2}\mu; \frac{1}{2}; x^2\right)\right\}.$$

The last two expression being valid for $-1 < x < +1$

$$Q_\nu^\mu(x) = \frac{1}{2}\Gamma(1+\nu+\mu)\,\Gamma(-\mu)\,[\Gamma(1+\nu-\mu)]^{-1}(1-x)^{\frac{1}{2}\mu}(1+x)^{-\frac{1}{2}\mu}$$
$$\times F\left(-\nu, 1+\nu; 1+\mu; ; \frac{1}{2}-\frac{1}{2}x\right)$$
$$+\frac{1}{2}\Gamma(\mu)\cos(\pi\mu)\,(1+x)^{\frac{1}{2}\mu}(1-x)^{-\frac{1}{2}\mu}$$
$$\times F\left(-\nu, \nu+1; 1-\mu; \frac{1}{2}-\frac{1}{2}x\right),$$
$$0 < x < 1,$$

$$\Gamma\left(\frac{3}{2}+\nu\right) P_\nu^\mu(\cos\vartheta) = i\pi^{-\frac{1}{2}} 2^\mu \Gamma(1+\nu+\mu)(\sin\vartheta)^\mu$$
$$\times\left[e^{-i\vartheta(1+\nu+\mu)} F\left(\frac{1}{2}+\mu, 1+\nu+\mu; \frac{3}{2}+\nu; e^{-i2\vartheta}\right)\right.$$
$$\left. - e^{i\vartheta(1+\nu+\mu)} F\left(\frac{1}{2}+\mu, 1+\nu+\mu; \frac{3}{2}+\nu; e^{i2\vartheta}\right)\right],$$

$$\Gamma\left(\frac{3}{2}+\nu\right) Q_\nu^\mu(\cos\vartheta) = \pi^{\frac{1}{2}} 2^{\mu-1} \Gamma(1+\nu+\mu)(\sin\vartheta)^\mu$$
$$\times\left[e^{i\vartheta(1+\nu+\mu)} F\left(\frac{1}{2}+\mu, 1+\nu+\mu; \frac{3}{2}+\nu; e^{i2\vartheta}\right)\right.$$
$$\left. + e^{-i\vartheta(1+\nu+\mu)} F\left(\frac{1}{2}+\mu, 1+\nu+\mu; \frac{3}{2}+\nu; e^{-i2\vartheta}\right)\right],$$

$$\Gamma\left(\frac{3}{2}+\nu\right) P_\nu^\mu(\cos\vartheta) = i(2\pi\sin\vartheta)^{-\frac{1}{2}} \Gamma(1+\nu+\mu)$$
$$\times\left\{e^{-i\left[\left(\frac{1}{2}+\nu\right)\vartheta+\left(\frac{1}{2}+\mu\right)\frac{\pi}{2}\right]} F\left[\frac{1}{2}+\mu, \frac{1}{2}-\mu; \frac{3}{2}+\nu; -\frac{1}{2}\operatorname{cosec}\vartheta\, e^{-i\left(\vartheta+\frac{\pi}{2}\right)}\right]\right.$$
$$\left. - e^{i\left[\left(\frac{1}{2}+\nu\right)\vartheta+\left(\frac{1}{2}+\mu\right)\frac{\pi}{2}\right]} F\left[\frac{1}{2}+\mu, \frac{1}{2}-\mu; \frac{3}{2}+\nu; -\frac{1}{2}\operatorname{cosec}\vartheta\, e^{i\left(\vartheta+\frac{\pi}{2}\right)}\right]\right\},$$

$$\Gamma\left(\frac{3}{2}+\nu\right) Q_\nu^\mu(\cos\vartheta) = \frac{1}{2}\pi^{\frac{1}{2}} (2\sin\vartheta)^{-\frac{1}{2}} \Gamma(1+\nu+\mu)$$
$$\times\left\{e^{i\left[\left(\frac{1}{2}+\nu\right)\vartheta+\left(\frac{1}{2}+\mu\right)\frac{\pi}{2}\right]} F\left[\frac{1}{2}+\mu, \frac{1}{2}-\mu; \frac{3}{2}+\nu; -\frac{1}{2}\operatorname{cosec}\vartheta\, e^{i\left(\vartheta+\frac{\pi}{2}\right)}\right]\right.$$
$$+ e^{-i\left[\left(\frac{1}{2}+\nu\right)\vartheta+\left(\frac{1}{2}+\mu\right)\frac{\pi}{2}\right]}$$
$$\left. \times F\left[\frac{1}{2}+\mu, \frac{1}{2}-\mu; \frac{3}{2}+\nu; -\frac{1}{2}\operatorname{cosec}\vartheta\, e^{-i\left(\vartheta+\frac{\pi}{2}\right)}\right]\right\}.$$

Upon replacing the hypergeometric functions by their series representation one obtains representations of $P_\nu^\mu(x)$ and $Q_\nu^\mu(x)$ in the form of trigonometric series

$$\Gamma\left(\frac{3}{2}+\nu\right) P_\nu^\mu(\cos\vartheta) = \pi^{-\frac{1}{2}} 2^{\mu+1} (\sin\vartheta)^\mu \Gamma(\nu+\mu+1)$$
$$\times \sum_{l=0}^{\infty} \frac{\left(\frac{1}{2}+\mu\right)_l (1+\nu+\mu)}{l!\left(\frac{3}{2}+\nu\right)_l} \sin[(2l+\nu+\mu+1)\vartheta],$$

$$\Gamma\left(\frac{3}{2}+\nu\right) Q_\nu^\mu(\cos\vartheta) = \pi^{\frac{1}{2}} 2^\mu (\sin\vartheta)^\mu \Gamma(\nu+\mu+1)$$

$$\times \sum_{l=0}^{\infty} \frac{\left(\frac{1}{2}+\mu\right)_l (1+\nu+\mu)_l}{l!\left(\frac{3}{2}+\nu\right)_l} \cos\left[(2l+\nu+\mu+1)\vartheta\right].$$

Both series are convergent for $0<\vartheta<\pi$. Furthermore

$$\Gamma\left(\frac{3}{2}+\nu\right) P_\nu^\mu(\cos\vartheta) = \left(\frac{1}{2}\pi\sin\vartheta\right)^{-\frac{1}{2}} \Gamma(\nu+\mu+1)$$

$$\times \sum_{l=0}^{\infty} (-1)^l \frac{\left(\frac{1}{2}+\mu\right)_l \left(\frac{1}{2}-\mu\right)_l (2\sin\vartheta)^{-l}}{l!\left(\frac{3}{2}+\nu\right)_l}$$

$$\times \sin\left[\left(\nu+l+\frac{1}{2}\right)\vartheta + \left(\frac{1}{2}+\mu\right)\frac{\pi}{2} + \frac{1}{2}l\pi\right],$$

$$\Gamma\left(\frac{3}{2}+\nu\right) Q_\nu^\mu(\cos\vartheta) = \left(\frac{2}{\pi}\sin\vartheta\right)^{-\frac{1}{2}} \Gamma(\nu+\mu+1)$$

$$\times \sum_{l=0}^{\infty} (-1)^l \frac{\left(\frac{1}{2}+\mu\right)_l \left(\frac{1}{2}-\mu\right)_l (2\sin\vartheta)^{-l}}{l!\left(\frac{3}{2}+\nu\right)_l}$$

$$\times \cos\left[\left(\nu+l+\frac{1}{2}\right)\vartheta + \left(\frac{1}{2}+\mu\right)\frac{\pi}{2} + \frac{1}{2}l\pi\right].$$

The last two series converge when $\frac{\pi}{6}<\vartheta<5\frac{\pi}{6}$. The first two series terminate after a finite number of terms when $\frac{1}{2}+\mu=0, -1, -2, -3, \ldots$ The last two when $\frac{1}{2}\pm\mu=0, -1, -2, -3, \ldots$ Furthermore

$$\Gamma(1+\mu)\left[\tan\left(\frac{1}{2}\vartheta\right)\right]^{-\mu} P_\nu^{-\mu}(\cos\vartheta) = F\left(-\nu, 1+\nu; 1+\mu; \sin^2\frac{\vartheta}{2}\right),$$

$$0<\vartheta<\pi,$$

$$Q_\nu^\mu(\cos\vartheta) = \frac{1}{2}\Gamma(1+\nu+\mu)\,\Gamma(-\mu)\,[\Gamma(1+\nu-\mu)]^{-1}\left(\tan\frac{\vartheta}{2}\right)^\mu$$

$$\times F\left(-\nu, 1+\nu; 1+\mu; \sin^2\frac{\vartheta}{2}\right)$$

$$+\frac{1}{2}\Gamma(\mu)\cos(\pi\mu)\left(\cot\frac{\vartheta}{2}\right)^\mu F\left(-\nu, 1+\nu; 1-\mu; \sin^2\frac{\vartheta}{2}\right),$$

$$0<\vartheta<\pi.$$

4.3.2 Relations between different solutions, Wronskians

$$P_{-\nu-1}(x) = P_\nu^\mu(x),$$

$$P_\nu^{-\mu}(x) = \frac{\Gamma(\nu-\mu+1)}{\Gamma(\nu+\mu+1)} [P_\nu^\mu(x) \cos(\pi\mu) - 2\pi^{-1} \sin(\pi\mu) Q_\nu^\mu(x)].$$

For $\mu = m$, $(m = 1, 2, 3, \ldots)$

$$P_\nu^{-m}(x) = \frac{\Gamma(\nu+1-m)}{\Gamma(\nu+1+m)} (-1)^m P_\nu^m(x),$$

$$Q_\nu^{-\mu}(x) = \frac{\Gamma(\nu-\mu+1)}{\Gamma(\nu+\mu+1)} \left[Q_\nu^\mu(x) \cos(\pi\mu) + \frac{1}{2}\pi P_\nu^\mu(x) \sin(\pi\mu)\right].$$

For $\mu = m$, $(m = 1, 2, 3, \ldots)$

$$Q_\nu^{-m}(x) = \frac{\Gamma(\nu+1-m)}{\Gamma(\nu+1+m)} (-1)^m Q_\nu^m(x)$$

$$\sin[\pi(\nu-\mu)] Q_{-\nu-1}^\mu(x) = \sin[\pi(\nu+\mu)] Q_\nu^\mu(x) - \pi \cos(\pi\nu) \cos(\pi\mu) P_\nu^\mu(x),$$

$$P_\nu^\mu(-x) = P_\nu^\mu(x) \cos[\pi(\nu+\mu)] - 2\pi^{-1} Q_\nu^\mu(x) \sin[\pi(\nu+\mu)],$$

$$Q_\nu^\mu(-x) = -Q_\nu^\mu(x) \cos[\pi(\nu+\mu)] - \frac{1}{2}\pi P_\nu^\mu(x) \sin[\pi(\nu+\mu)],$$

both for, $0 < x < 1$,

$$Q_\nu^\mu(x) + Q_\nu^\mu(-x) = -\frac{1}{2}\pi \tan\left[\frac{\pi}{2}(\nu+\mu)\right][P_\nu^\mu(x) + P_\nu^\mu(-x)],$$

$$Q_\nu^\mu(x) - Q_\nu^\mu(-x) = \frac{1}{2}\pi \cot\left[\frac{\pi}{2}(\nu+\mu)\right][P_\nu^\mu(x) - P_\nu^\mu(-x)],$$

$$\pi P_\nu(-x) = -\sin(\pi\nu)[Q_\nu(x) + Q_{-\nu-1}(x)],$$

$$\pi P_\nu(x) = \tan(\pi\nu)[Q_\nu(x) - Q_{-\nu-1}(x)].$$

Wronskian determinant

The Wronskian determinant

$$W[P_\nu^\mu(x), Q_\nu^\mu(x)] = P_\nu^\mu(x) \frac{d}{dx} Q_\nu^\mu(x) - P_\nu^\mu(x) \frac{d}{dx} Q_\nu^\mu(x)$$

is equal to

$$W[P_\nu^\mu(x), Q_\nu^\mu(x)] = \frac{\Gamma(1+\nu+\mu)}{\Gamma(1+\nu-\mu)} (1-x^2)^{-1}.$$

Furthermore

$$W[P_\nu^\mu(x), P_\nu^{-\mu}(x)] = -\frac{2}{\pi} \sin(\pi\mu)(1-x^2)^{-1},$$

$$W[Q_\nu^\mu(x), Q_\nu^{-\mu}(x)] = -\frac{1}{2}\pi \sin(\pi\mu)(1-x^2)^{-1},$$

$$W[P_\nu^{-\mu}(x), Q_\nu^\mu(x)] = \cos(\pi\mu)(1-x^2)^{-1}.$$

Values for $P_\nu^\mu(x)$ and $Q_\nu^\mu(x)$ and their derivatives at $x = 0$

$$P_\nu^\mu(0) = 2^\mu \pi^{-\frac{1}{2}} \cos\left[\frac{1}{2}\pi(\nu+\mu)\right] \frac{\Gamma\left(\frac{1}{2}+\frac{1}{2}\nu+\frac{1}{2}\mu\right)}{\Gamma\left(1+\frac{1}{2}\nu-\frac{1}{2}\mu\right)},$$

$$Q_\nu^\mu(0) = -2^{\mu-1}\pi^{\frac{1}{2}} \sin\left[\frac{1}{2}\pi(\nu+\mu)\right] \frac{\Gamma\left(\frac{1}{2}+\frac{1}{2}\nu+\frac{1}{2}\mu\right)}{\Gamma\left(1+\frac{1}{2}\nu-\frac{1}{2}\mu\right)},$$

$$\left(\frac{dP_\nu^\mu(x)}{dx}\right)_{x=0} = 2^{\mu+1}\pi^{-\frac{1}{2}} \sin\left[\frac{1}{2}\pi(\nu+\mu)\right] \frac{\Gamma\left(1+\frac{1}{2}\nu+\frac{1}{2}\mu\right)}{\Gamma\left(1+\frac{1}{2}\nu-\frac{1}{2}\mu\right)},$$

$$\left(\frac{dQ_\nu^\mu(x)}{dx}\right)_{x=0} = 2^\mu \pi^{\frac{1}{2}} \cos\left[\frac{1}{2}\pi(\nu+\mu)\right] \frac{\Gamma\left(1+\frac{1}{2}\nu+\frac{1}{2}\mu\right)}{\Gamma\left(\frac{1}{2}+\frac{1}{2}\nu-\frac{1}{2}\mu\right)}.$$

4.3.3 Contiguous relations

$$P_\nu^{\mu+2}(x) + 2(\mu+1)\,x(1-x^2)^{-\frac{1}{2}} P_\nu^{\mu+1}(x) + (\nu-\mu)(\nu+\mu+1) \times P_\nu^\mu(x) = 0,$$

$$(2\nu+1)\,x P_\nu^\mu(x) = (\nu-\mu+1) P_{\nu+1}^\mu(x) + (\nu+\mu) P_{\nu-1}^\mu(x),$$

$$P_{\nu-1}^\mu(x) - P_{\nu+1}^\mu(x) = (2\nu+1)(1-x^2)^{\frac{1}{2}} P_\nu^{\mu-1}(x),$$

$$(\nu-\mu)(\nu-\mu+1) P_{\nu+1}^\mu(x) - (\nu+\mu)(\nu+\mu+1) P_{\nu-1}^\mu(x) = (2\nu+1)(1-x^2)^{\frac{1}{2}} P_\nu^{\mu+1}(x),$$

$$P_{\nu-1}^\mu(x) - x P_\nu^\mu(x) = (\nu-\mu+1)(1-x^2)^{\frac{1}{2}} P_\nu^{\mu-1}(x),$$

$$x P_\nu^\mu(x) - P_{\nu+1}^\mu(x) = (\nu+\mu)(1-x^2)^{\frac{1}{2}} P_\nu^{\mu-1}(x),$$

$$(\nu-\mu)\,x P_\nu^\mu(x) - (\nu+\mu) P_{\nu-1}^\mu(x) = (1-x^2)^{\frac{1}{2}} P_\nu^{\mu+1}(x),$$

$$(\nu-\mu+1) P_{\nu+1}^\mu(x) - (\nu+\mu+1)\,x P_\nu^\mu(x) = (1-x^2)^{\frac{1}{2}} P_\nu^{\mu+1}(x),$$

$$(1-x^2)\frac{dP_\nu^\mu(x)}{dx} = (\nu+1)\,x P_\nu^\mu(x) - (\nu-\mu+1) P_{\nu+1}^\mu(x) = -\nu x P_\nu^\mu(x) + (\nu+\mu) P_{\nu-1}^\mu(x).$$

The same formulas hold for the $Q_\nu^\mu(x)$.

4.4 Special values for the parameters

4.4.1 Value $\pm \frac{1}{2}$ for one of the parameters

$$\mathfrak{P}_\nu^{\frac{1}{2}}(z) = (2\pi)^{-\frac{1}{2}}(z^2-1)^{-\frac{1}{4}} \times \left\{\left[z+(z^2-1)^{\frac{1}{2}}\right]^{\nu+\frac{1}{2}} + [z+(z^2-1)]^{-\nu-\frac{1}{2}}\right\},$$

$$\mathfrak{Q}_\nu^{\frac{1}{2}}(z) = i\left(\frac{1}{2}\pi\right)^{\frac{1}{2}}(z^2-1)^{-\frac{1}{4}}\left[z+(z^2-1)^{\frac{1}{2}}\right]^{-\nu-\frac{1}{2}},$$

$$\mathfrak{P}_\nu^{-\frac{1}{2}}(z) = (2\pi)^{-\frac{1}{2}}\left(\frac{1}{2}+\nu\right)^{-1}(z^2-1)^{-\frac{1}{4}} \times \left\{\left[z+(z^2-1)^{\frac{1}{2}}\right]^{\nu+\frac{1}{2}} - \left[z+(z^2-1)^{\frac{1}{2}}\right]^{-\nu-\frac{1}{2}}\right\},$$

$$\mathfrak{Q}_\nu^{-\frac{1}{2}}(z) = -i\left(\frac{1}{2}\pi\right)^{\frac{1}{2}}\left(\frac{1}{2}+\nu\right)^{-1}(z^2-1)^{-\frac{1}{4}}\left[z+(z^2-1)^{\frac{1}{2}}\right]^{-\nu-\frac{1}{2}},$$

$$\mathfrak{P}_\nu^{-\nu}(z) = 2^{-\nu}\frac{(z^2-1)^{\frac{1}{2}\nu}}{\Gamma(1+\nu)},$$

$$P_\nu^{\frac{1}{2}}(x) = (2\pi)^{-\frac{1}{2}}(1-x^2)^{-\frac{1}{4}} \times \left\{\left[x+i(1-x^2)^{\frac{1}{2}}\right]^{\nu+\frac{1}{2}} + \left[x-i(1-x^2)^{\frac{1}{2}}\right]^{\nu+\frac{1}{2}}\right\},$$

$$Q_\nu^{\frac{1}{2}}(x) = -\frac{i}{2}\left(\frac{1}{2}\pi\right)^{\frac{1}{2}} 1-x^2)^{-\frac{1}{4}} \times \left\{\left[x+i(1-x^2)^{\frac{1}{2}}\right]^{-\nu-\frac{1}{2}} - \left[x-i(1-x^2)^{\frac{1}{2}}\right]^{-\nu-\frac{1}{2}}\right\},$$

$$P_\nu^{-\frac{1}{2}}(x) = -(2\pi)^{-\frac{1}{2}} i\left(\frac{1}{2}+\nu\right)^{-1}(1-x^2)^{-\frac{1}{4}} \times \left\{\left[x+i(1-x^2)^{\frac{1}{2}}\right]^{\nu+\frac{1}{2}} - \left[x-i(1-x^2)^{\frac{1}{2}}\right]^{\nu+\frac{1}{2}}\right\},$$

$$Q_\nu^{-\frac{1}{2}}(x) = \frac{1}{2}\left(\frac{1}{2}\pi\right)^{\frac{1}{2}}\left(\frac{1}{2}+\nu\right)^{-1}(1-x^2)^{-\frac{1}{4}} \times \left\{\left[x+i(1-x^2)^{\frac{1}{2}}\right]^{-\nu-\frac{1}{2}} + \left[x-i(1-x^2)^{\frac{1}{2}}\right]^{-\nu-\frac{1}{2}}\right\},$$

$$P_\nu^{-\nu}(x) = 2^{-\nu}\frac{(1-x^2)^{\frac{1}{2}\nu}}{\Gamma(1+\nu)}.$$

Reduction to elliptic integrals

$$\mathfrak{P}_{-\frac{1}{2}}(z) = 2\pi^{-1}\left(\frac{1}{2}z+\frac{1}{2}\right)^{-\frac{1}{2}} K\left[(z-1)^{\frac{1}{2}}(z+1)^{-\frac{1}{2}}\right]$$

$$= 2\pi^{-1}\left[z+(z^2-1)^{\frac{1}{2}}\right]^{-\frac{1}{2}} K\left\{(z^2-1)^{\frac{1}{4}}\left[2z-2(z^2-1)^{\frac{1}{2}}\right]^{\frac{1}{2}}\right\},$$

$$\mathfrak{Q}_{-\frac{1}{2}}(z) = \left(\frac{1}{2}z+\frac{1}{2}\right)^{-\frac{1}{2}} K\left[\left(\frac{1}{2}z+\frac{1}{2}\right)^{-\frac{1}{2}}\right]$$

$$= 2\left[z+(z^2-1)^{\frac{1}{2}}\right]^{-\frac{1}{2}} K\left[z-(z^2-1)^{\frac{1}{2}}\right],$$

$$\mathfrak{P}_{\frac{1}{2}}(z) = 2\pi^{-1}\left[z+(z^2-1)^{\frac{1}{2}}\right]^{\frac{1}{2}} E\left\{(z^2-1)^{\frac{1}{4}}\left[2z-2(z^2-1)^{\frac{1}{2}}\right]^{\frac{1}{2}}\right\},$$

$$\mathfrak{Q}_{\frac{1}{2}}(z) = z\left(\frac{1}{2}z+\frac{1}{2}\right)^{-\frac{1}{2}} K\left[\left(\frac{1}{2}z+\frac{1}{2}\right)^{\frac{1}{2}}\right] - 2(z+1)^{\frac{1}{2}}$$

$$\times E\left[\left(\frac{1}{2}z+\frac{1}{2}\right)^{-\frac{1}{2}}\right],$$

$$P_{-\frac{1}{2}}(x) = 2\pi^{-1} K\left[\left(\frac{1}{2}-\frac{1}{2}x\right)^{\frac{1}{2}}\right],\quad Q_{-\frac{1}{2}}(x) = K\left[\left(\frac{1}{2}+\frac{1}{2}x\right)^{\frac{1}{2}}\right].$$

4.4.2 Integer values of μ or (and) ν

Generally from 4.1.2 and 4.3.1

$$\Gamma(1-\mu)\,\mathfrak{P}_\nu^\mu(z) = (z+1)^{\frac{1}{2}\mu}(z-1)^{-\frac{1}{2}\mu}$$

$$\times F\left(-\nu, 1+\nu; 1-\mu; \frac{1}{2}-\frac{1}{2}z\right)$$

$$= 2^\mu(z^2-1)^{-\frac{1}{2}\mu} F\left(1-\mu+\nu, -\mu-\nu; 1-\mu; \frac{1}{2}-\frac{1}{2}z\right),$$

$$\Gamma(1-\mu)\,P_\nu^\mu(x) = (1+x)^{\frac{1}{2}\mu}(1-x)^{-\frac{1}{2}\mu}$$

$$\times F\left(-\nu, 1+\nu; 1-\mu; \frac{1}{2}-\frac{1}{2}x\right) = 2^\mu(1-x^2)^{-\frac{1}{2}\mu}$$

$$\times F\left(1-\mu+\nu, -\mu-\nu; 1-\mu; \frac{1}{2}-\frac{1}{2}x\right).$$

Since

$$P^{\mu}_{-\nu-1}(x) = P^{\mu}_{\nu}(x);\ \mathfrak{P}^{\mu}_{-\nu-1}(z) = \mathfrak{P}^{\mu}_{\nu}(z)$$

it is sufficient to consider positive integer ν only. In the following we denote by n and m arbitrary non negative integers, $n, m = 0, 1, 2, \ldots$ We consider now the cases:

1. $\nu = n,\ \mu \neq m$

Then $P^{\mu}_{\nu}(x)$ and $\mathfrak{P}^{\mu}_{\nu}(z)$ are polynomials of degree n multiplied by an elementary function.

2. $\nu = n, \mu = -m$

Then $P^{\mu}_{\nu}(x)$ and $\mathfrak{P}^{\mu}_{\nu}(z)$ are polynomials of degree $n - m$ (provided $n \geq m$) multiplied by an elementary function.

3. $\mu = m$

Then for general ν

$$\Gamma(\nu - m + 1)\, m!\, \mathfrak{P}^{m}_{\nu}(z)$$
$$= 2^{-m}\Gamma(\nu + m + 1)\, (z^2 - 1)^{\frac{1}{2}m} F\left(1 + m + \nu, m - \nu; 1 + m; \frac{1}{2} - \frac{1}{2} z\right),$$

$$\Gamma(\nu - m + 1)\, m!\, P^{m}_{\nu}(x) = (-2)^{-m}\, \Gamma(\nu + m + 1)\, (1 - x^2)^{\frac{1}{2}m}$$
$$\times F\left(1 + m + \nu, m - \nu; 1 + m; \frac{1}{2} - \frac{1}{2} x\right).$$

4. $\nu = n, \mu = m.$

Then $P^{\mu}_{\nu}(x)$ and $\mathfrak{P}^{\mu}_{\nu}(z)$ are polynomials of degree $n - m$ for $n \geq m$, multiplied by an elementary function

5. $\mu = m, \nu = n, n < m$

Then $P^{\mu}_{\nu}(x)$ and $\mathfrak{P}^{\mu}_{\nu}(z)$ vanish identically. However,

$$\Gamma(\nu - \mu + 1)\ \mathfrak{P}^{\mu}_{\nu}(z) \text{ and } \Gamma(\nu - \mu + 1)\ P^{\mu}_{\nu}(x)$$

approach finite limits as $\mu \to m$ and $\nu \to n$ and these expressions can then be taken as solutions of Legendre's differential equation. There also exist the relations:

$$\mathfrak{P}^{m}_{\nu}(z) = (z^2 - 1)^{\frac{1}{2}m} \frac{d^m}{dz^m} \mathfrak{P}_{\nu}(z),$$

$$\mathfrak{Q}^{m}_{\nu}(z) = (z^2 - 1)^{\frac{1}{2}m} \frac{d^m}{dz^m} \mathfrak{Q}_{\nu}(z),$$

$$P^{m}_{\nu}(x) = (-1)^m (1 - x^2)^{\frac{1}{2}m} \frac{d^m}{dx^m} P_{\nu}(x),$$

$$Q^{m}_{\nu}(x) = (-1)^m (1 - x^2)^{\frac{1}{2}m} \frac{d^m}{dz^m} Q_{\nu}(x),$$

$$m = 1, 2, 3, \ldots$$

Furthermore

$$\mathfrak{P}_\nu^{-m}(z) = (z^2-1)^{-\frac{1}{2}m} \int_1^z \cdots \int_1^z \mathfrak{P}_\nu(z)\,(dz)^m,$$

$$\mathfrak{Q}_\nu^{-m}(z) = (-1)^m\,(z^2-1)^{-\frac{1}{2}m} \int_z^\infty \cdots \int_z^\infty \mathfrak{Q}_\nu(z)\,(dz)^m,$$

$$P_\nu^{-m}(x) = (-1)^m\,(1-x^2)^{-\frac{1}{2}m} \int_1^x \cdots \int_1^x P_\nu(x)\,(dx)^m.$$

If μ becomes a positive integer the Legendre functions of the second kind involve logarithmic terms

$$2\pi \operatorname{cosec}(\pi\nu)\,\mathfrak{Q}_\nu^m(z) = \pi \operatorname{cosec}(\pi\nu)\,\mathfrak{P}_\nu^m(z)$$

$$\times \left\{\log\left[\frac{(z+1)}{(z-1)}\right] - 2\gamma - \psi(\nu+m+1) - \psi(\nu-m+1)\right\}$$

$$-(-1)^m\left[\frac{(z+1)}{(z-1)}\right]^{\frac{1}{2}m} \sum_{r=0}^{m-1} \Gamma(r-\nu)\,\Gamma(r+\nu+1)\,\Gamma(m-r)$$

$$\times \cos(r\pi)\,\frac{\left(\frac{1}{2}-\frac{1}{2}z\right)^r}{r!}$$

$$-\left[\frac{(z+1)}{(z-1)}\right]^{\frac{1}{2}m} \sum_{l=1}^\infty \Gamma(m+l-\nu)\,\Gamma(m+l+\nu+1)\,\sigma(l)$$

$$\times \left(\frac{1}{2}-\frac{1}{2}z\right)^{m+l}[l!\,(m+l)!]^{-1} - \frac{\Gamma(\nu+m+1)}{\Gamma(\nu-m+1)}$$

$$\times \left[\frac{(z-1)}{(z+1)}\right]^{\frac{1}{2}m} \sum_{l=0}^\infty \Gamma(l-\nu)\,\Gamma(l+\nu+1)\,\sigma(m+l)\,\frac{\left(\frac{1}{2}-\frac{1}{2}z\right)^l}{l!\,(l+m)!},$$

where

$$\sigma(l) = 1 + \frac{1}{2} + \cdots + \frac{1}{l} = \psi(l+1) - \psi(1) = \gamma + \psi(l+1)$$

$$\sigma(0) = 0.$$

If μ is a negative integer use the proceeding formula and (see 4.2.1)

$$\Gamma(\nu+m+1)\,\mathfrak{Q}_\nu^{-m}(z) = \Gamma(\nu-m+1)\,\mathfrak{Q}_\nu^m(z).$$

Especially for $m = 0$

$$\mathfrak{Q}_\nu(z) = \frac{1}{2}\,\mathfrak{P}_\nu(z)\left\{\log\left[\frac{(z+1)}{(z-1)}\right] - 2\gamma - 2\psi(\nu+1)\right\}$$

$$-\pi^{-1}\sin(\pi\nu)\sum_{l=1}^\infty \Gamma(l-\nu)\,\Gamma(l+\nu+1)\,\sigma(l)\left(\frac{1}{2}-\frac{1}{2}z\right)^l (l!)^{-2}.$$

In case ν is also a positive integer $\nu = n$ $(n = 1, 2, 3, \ldots)$

$$\mathfrak{Q}_n(z) = \frac{1}{2}\,\mathfrak{P}_n(z)\left\{\log\left[\frac{(z+1)}{(z-1)}\right] - 2\sigma(n)\right\}$$
$$+ \sum_{l=0}^{n} (-1)^l\,(n+l)!\,\sigma(l)\left(\frac{1}{2} - \frac{1}{2}z\right)^l [(l!)^2\,(n-l)!]^{-1}.$$

If one denotes

$$W_{n-1}(x) = \sum_{n=0}^{\left[\frac{n}{2}-\frac{1}{2}\right]} (2n - 4m - 1)\,[(n-m)\,(2m+1)]^{-1}\,\mathfrak{P}_{n-2m-1}(z),$$

then also

$$\mathfrak{Q}_n(z) = \frac{1}{2}\,\mathfrak{P}_n(z)\log\left[\frac{(z+1)}{(z-1)}\right] - W_{n-1}(z),$$
$$Q_n(x) = \frac{1}{2}\,P_n(x)\log\left[\frac{(1+x)}{(1-x)}\right] - W_{n-1}(x),$$
$$n = 1, 2, 3, \ldots$$

and $W_{n-1}(z)$, $(n = 1, 2, 3, \ldots)$ satisfies the differentialequation

$$(1 - z^2)\,W''_{n-1} - 2z\,W'_{n-1} + n(n+1)\,W_{n-1} = P'_n.$$

Some particular cases

$$\mathfrak{Q}_0(z) = \frac{1}{2}\log\left[\frac{(z+1)}{(z-1)}\right],$$
$$Q_0(x) = \frac{1}{2}\log\left[\frac{(1+x)}{(1-x)}\right],$$
$$\mathfrak{Q}_1(x) = \frac{1}{2}\,z\log\left[\frac{(z+1)}{(z-1)}\right] - 1,$$
$$Q_1(x) = \frac{1}{2}\,x\log\left[\frac{(1+x)}{(1-x)}\right] - 1,$$
$$\mathfrak{Q}_2(z) = \frac{1}{2}\,\mathfrak{P}_2(z)\log\left[\frac{(z+1)}{(z-1)}\right] - \frac{3}{2}z,$$
$$Q_2(x) = \frac{1}{2}\,P_2(x)\log\left[\frac{(1+x)}{(1-x)}\right] - \frac{3}{2}x,$$
$$\mathfrak{Q}_3(z) = \frac{1}{2}\,\mathfrak{P}_3(z)\log\left[\frac{(z+1)}{(z-1)}\right] - \frac{5}{2}z^2 + \frac{2}{3},$$
$$Q_3(x) = \frac{1}{2}\,P_3(x)\log\left[\frac{(1+x)}{(1-x)}\right] - \frac{5}{2}x^2 + \frac{2}{3},$$

Some generating functions

$$(1 - 2hz + h^2)^{-\frac{1}{2}} = \begin{cases} \sum_{n=0}^{\infty} h^n\,\mathfrak{P}_n(z), \\ \sum_{n=0}^{\infty} h^{-n-1}\,\mathfrak{P}_n(z) \end{cases}$$

according as

$$|h| < \text{Min}\left|z \pm (z^2-1)^{\frac{1}{2}}\right|$$

or

$$|h| > \text{Max}\left|z \pm (z^2-1)^{\frac{1}{2}}\right|$$

respectively

$$(1-2hz+h^2)^{-\frac{1}{2}} \log\left\{(z^2-1)^{-\frac{1}{2}}\left[z-h+(1-2hz+h^2)^{\frac{1}{2}}\right]\right\},$$

$$= \sum_{n=0}^{\infty} h^n \mathfrak{Q}_n(z),$$

$$Re\, z > 1, \ |t| < 1,$$

$$(1-2hx+h^2)^{-\frac{1}{2}} \log\left\{(1-x^2)^{-\frac{1}{2}}\left[x-h+(1-2hx+h^2)^{\frac{1}{2}}\right]\right\}$$

$$\times (z-t)^{-1} = \sum_{n=0}^{\infty} (2n+1)\, \mathfrak{P}_n(t)\, \mathfrak{Q}_n(z),$$

$$\left|t+(t^2-1)^{\frac{1}{2}}\right| < \left|z+(z^2-1)^{\frac{1}{2}}\right|,$$

4.4.3 Differentiation with respect to one of the parameters

$$\left[\frac{\partial}{\partial\nu}\mathfrak{P}_\nu^\mu(z)\right]_{\nu=-\frac{1}{2}} = 0, \quad \left[\frac{\partial}{\partial\nu}P_\nu^\mu(x)\right]_{\nu=-\frac{1}{2}} = 0,$$

$$e^{-i\pi\mu}\left[\frac{\partial}{\partial\nu}\mathfrak{Q}_\nu^\mu(z)\right]_{\nu=-\frac{1}{2}} = -\left(\frac{1}{2}\pi\right)^{\frac{1}{2}} \Gamma\left(\frac{1}{2}+\mu\right)(z^2-1)^{-\frac{1}{4}}$$

$$\times\, \mathfrak{Q}_{\mu-\frac{1}{2}}\left[z(z^2-1)^{-\frac{1}{2}}\right] = -\frac{1}{2}\pi\Gamma^2\left(\frac{1}{2}+\mu\right)\mathfrak{P}^{-\mu}_{-\frac{1}{2}}(z),$$

$$\left[\frac{\partial}{\partial\nu}Q_\nu^\mu(x)\right]_{\nu=-\frac{1}{2}} = -\frac{1}{2}\pi\,\Gamma^2\left(\frac{1}{2}+\mu\right)P^{-\mu}_{-\frac{1}{2}}(x).$$

Especially

$$\left[\frac{\partial}{\partial\nu}\mathfrak{Q}_\nu(x)\right]_{\nu=-\frac{1}{2}} = -\pi\left(\frac{1}{2}+\frac{1}{2}z\right)^{-\frac{1}{2}} K\left[(z-1)^{\frac{1}{2}}(z+1)^{-\frac{1}{2}}\right],$$

$$\left[\frac{\partial}{\partial\nu}Q_\nu(x)\right]_{\nu=-\frac{1}{2}} = -\pi K\left[\left(\frac{1}{2}-\frac{1}{2}x\right)^{\frac{1}{2}}\right],$$

$$\left[\frac{\partial}{\partial\nu}\mathfrak{P}_\nu^\mu(z)\right]_{\nu=0} = -\frac{1}{2}(1-z)\left[\Gamma(2-\mu)\right]^{-1}\left(\frac{z+1}{z-1}\right)^{\frac{1}{2}\mu}$$

$$\times F\left(1,1;2-\mu;\frac{1}{2}-\frac{1}{2}z\right),$$

$$\left[\frac{\partial}{\partial\nu} P_\nu^\mu(x)\right]_{\nu=0} = -\frac{1}{2}(1-x)\,[\Gamma(2-\mu)]^{-1}\left(\frac{1+x}{1-x}\right)^{\frac{1}{2}\mu} \times F\left(1,1;2-\mu;\frac{1}{2}-\frac{1}{2}x\right)$$

particularly

$$\left[\frac{\partial}{\partial\nu}\mathfrak{P}_\nu(z)\right]_{\nu=0} = \log\left(\frac{1}{2}+\frac{1}{2}z\right); \quad \left[\frac{\partial}{\partial\nu}P_\nu(x)\right]_{\nu=0} = \log\left(\frac{1}{2}+\frac{1}{2}x\right)$$

generally

$$\frac{\partial}{\partial\nu}[\mathfrak{P}_\nu^\mu(z)] = \pi\cot(\pi\nu)\,\mathfrak{P}_\nu^\mu(z) - \pi^{-1}\sin(\pi\nu)\,(z+1)^{\frac{1}{2}\mu}(z-1)^{-\frac{1}{2}\mu}$$
$$\times\sum_{n=0}^{\infty}\frac{\left(\frac{1}{2}-\frac{1}{2}z\right)^n\Gamma(n-\nu)\,\Gamma(n+1+\nu)}{n!\,\Gamma(n+1-\mu)}[\psi(n+1+\nu)-\psi(n-\nu)],$$

$$\frac{\partial}{\partial\nu}[P_\nu^\mu(z)] = \pi\cot(\pi\nu)\,P_\nu^\mu(x) - \pi^{-1}\sin(\pi\nu)\,(1+x)^{\frac{1}{2}\mu}(1-x)^{-\frac{1}{2}\mu}$$
$$\times\sum_{n=0}^{\infty}\frac{\left(\frac{1-x}{2}\right)^n\Gamma(n-\nu)\,\Gamma(n+1+\nu)}{\Gamma(n+1-\mu)}[\psi(n+1+\nu)-\psi(n-\nu)],$$

$$\frac{\partial}{\partial\mu}[\mathfrak{P}_\nu^\mu(z)]_{\mu=0} = \frac{1}{2}\{[\psi(\nu+1)+\psi(-\nu)]\,\mathfrak{P}_\nu(z)+\mathfrak{Q}_\nu(z)+\mathfrak{Q}_{-\nu-1}(z)\}$$
$$= \psi(-\nu)\,\mathfrak{P}_\nu(z)+\mathfrak{Q}_{-\nu-1}(z)$$
$$= [\psi(-\nu)-\pi\cot(\pi\nu)]\,\mathfrak{P}_\nu(z)+\mathfrak{Q}_\nu(z),$$

$$\frac{\partial}{\partial\mu}[\mathfrak{Q}_\nu^\mu(z)]_{\mu=0} = [i\pi+\psi(\nu+1)]\,\mathfrak{Q}_v(z).$$

4.5 Series involving Legendre functions

4.5.1 Addition theorems and series of the addition theorem type

$$\mathfrak{P}_\nu\left[zz'-(z^2-1)^{\frac{1}{2}}(z'^2-1)^{\frac{1}{2}}\cos\psi\right]$$
$$= \sum_{m=0}^{\infty}(-1)^m\varepsilon_m\frac{\Gamma(\nu-m+1)}{\Gamma(\nu+m+1)}\mathfrak{P}_\nu^m(z)\,\mathfrak{P}_\nu^m(z')\cos(m\psi)$$
$$= \sum_{m=0}^{\infty}(-1)^m\varepsilon_m\,\mathfrak{P}_\nu^m(z)\,\mathfrak{P}_\nu^{-m}(z')\cos(m\psi),$$

$Re\,z>0,\ Re\,z'>0,\ |\arg(z-1)|<\pi,\ |\arg(z'-1)|<\pi,$

$$\mathfrak{Q}_\nu\left[zz'-(z^2-1)^{\frac{1}{2}}(z'^2-1)^{\frac{1}{2}}\cos\psi\right]$$
$$= \sum_{m=0}^{\infty}(-1)^m\varepsilon_m\,\mathfrak{Q}_\nu^m(z)\,\mathfrak{P}_\nu^{-m}(z')\cos(m\psi),$$

z, z' real, $1 < z' < z, \nu \neq -1, -2, -3, \ldots \psi$ real,

$$P_\nu(\cos\vartheta\cos\vartheta' + \sin\vartheta\sin\vartheta'\cos\psi)$$

$$= \sum_{m=0}^{\infty} \varepsilon_m \frac{\Gamma(\nu - m + 1)}{\Gamma(\nu + m + 1)} P_\nu^m(\cos\vartheta)\, P_\nu^m(\cos\vartheta')\cos(m\psi)$$

$$= \sum_{m=0}^{\infty} (-1)^m \varepsilon_m P_\nu^m(\cos\vartheta)\, P_\nu^{-m}(\cos\vartheta')\cos(m\psi),$$

$$0 \le \vartheta < \pi,\ 0 \le \vartheta' < \pi,\ \vartheta + \vartheta' < \pi,$$

$$Q_\nu(\cos\vartheta\cos\vartheta' + \sin\vartheta\sin\vartheta'\cos\psi)$$

$$= \sum_{m=0}^{\infty} (-1)^m \varepsilon_m P_\nu^{-m}(\cos\vartheta)\, Q_\nu^m(\cos\vartheta)\cos(m\psi),$$

$$0 < \vartheta' < \frac{1}{2}\pi,\ \ 0 < \vartheta < \pi,\ \ 0,\ \ 0 < \vartheta + \vartheta' < \pi,\ \psi \text{ real},$$

$$2^{-2\mu}(z^2-1)^{\frac{\mu}{2}}\left(\sin\frac{\varphi}{2}\right)^{-\mu}\left[1 + (z^2-1)\sin^2\frac{\varphi}{2}\right]^{-\frac{1}{2}\mu}$$

$$\times\ \mathfrak{P}_\nu^\mu[z^2 - (z^2-1)\cos\varphi]$$

$$= \Gamma(\mu)\sum_{m=0}^{\infty}(-1)^m(\mu+m)\, C_m^\mu(\cos\varphi)\, \mathfrak{P}_\nu^{\mu+m}(z)\, \mathfrak{P}_\nu^{-\mu-m}(z),$$

$$2^{-2\mu}(z^2-1)^{\frac{1}{2}\mu}\left(\sin\frac{\varphi}{2}\right)^{-\mu}\left[1 + (z^2-1)\sin^2\frac{\varphi}{2}\right]^{-\frac{1}{2}\mu}$$

$$\times\ e^{-i\pi\mu}\,\mathfrak{Q}_\nu^\mu[z^2 - (z^2-1)\cos\varphi]$$

$$= e^{-i2\pi\mu}\,\Gamma(\mu)\sum_{m=0}^{\infty}(-1)^m(\mu+m)\, C_m^\mu(\cos\varphi)\, \mathfrak{P}_\nu^{-\mu-m}(z)\, \mathfrak{Q}_\nu^{\mu+m}(z),$$

$$2^{-2\mu}(z^2-1)^{\frac{1}{2}\mu}\left(\sin\frac{\varphi}{2}\right)^{-\mu}\left[1 + (z^2-1)\sin^2\frac{\varphi}{2}\right]^{-\frac{1}{2}\mu}$$

$$\times\ \mathfrak{P}_\nu^{-\mu}[z^2 - (z^2-1)\cos\varphi]$$

$$= \Gamma(\mu)\sum_{m=0}^{\infty}(\mu+m)(\mu-\nu)_m(\mu+\nu+1)_m\, C_m^\mu(\cos\varphi)\,[\mathfrak{P}_\nu^{-\mu-m}(z)]^2,$$

$$2^{-2\mu}(1-x^2)^{\frac{1}{2}\mu}\left(\sin\frac{\varphi}{2}\right)^{-\mu}\left[1 - (1-x^2)\sin^2\frac{\varphi}{2}\right]^{-\frac{1}{2}\mu}$$

$$\times\ P_\nu^\mu[x^2 + (1-x^2)\cos\varphi]$$

$$= \Gamma(\mu)\sum_{m=0}^{\infty}(-1)^m(\mu+m)\, C_m^\mu(\cos\varphi)\, P_\nu^{\mu+m}(x)\, P_\nu^{-\mu-m}(x),$$

$$2^{-2\mu}(1-x^2)^{\frac{1}{2}\mu}\left(\sin\frac{\varphi}{2}\right)^{-\mu}\left[1-(1-x^2)\sin^2\frac{\varphi}{2}\right]^{-\frac{1}{2}\mu}$$
$$\times P_\nu^{-\mu}\left[x^2+(1-x^2)\cos\varphi\right]$$
$$=\Gamma(\mu)\sum_{m=0}^{\infty}(-1)^m(\mu+m)(\mu-\nu)_m(\mu+\nu+1)_m C_m^\mu(\cos\varphi)\left[P_\nu^{-\mu-m}(x)\right]^2.$$

4.5.2 Some series of the cardinal type

$$\sum_{m=0}^{\infty}(-1)^m\frac{\varepsilon_m}{m^2-\alpha^2}\mathfrak{P}^\mu_{-\frac{1}{2}+\beta m}(\cos\vartheta)=-\pi\alpha^{-1}\operatorname{cosec}(\alpha\pi)P^\mu_{-\frac{1}{2}+\alpha\beta}(\cos\vartheta).$$
$$\operatorname{Re}\mu<\frac{5}{2},\quad 0>\beta\leq\frac{\pi}{\vartheta},$$

$$\sum_{m=0}^{\infty}(-1)^m\frac{\varepsilon_m}{m^2-\alpha^2}\mathfrak{P}^\mu_{-\frac{1}{2}+i\frac{1}{2}\beta m}(\cosh c)$$
$$=-\pi\alpha^{-1}\operatorname{cosec}(\alpha\pi)\mathfrak{P}^\mu_{-\frac{1}{2}+i\frac{1}{2}\alpha\beta}(\cosh c)$$
$$Re\,\mu<\frac{5}{2},\ 0<\beta\leq\frac{\pi}{c},$$

$$\sum_{m=0}^{\infty}(-1)^m\frac{\varepsilon_m}{m^2-\alpha^2}\mathfrak{P}_\nu^{\beta m}(x)\,\mathfrak{P}_\nu^{-\beta m}(x)=-\pi\alpha^{-1}\operatorname{cosec}(\alpha\pi)\,\mathfrak{P}_\nu^{\alpha\beta}(x)\,\mathfrak{P}_\nu^{-\alpha\beta}(x),$$
$$0<\beta\leqslant 1,$$

$$\sum_{m=0}^{\infty}(-1)^m\frac{\varepsilon_m}{m^2-\alpha^2}\mathfrak{P}_\nu^{\beta m}(z)\,\mathfrak{P}_\nu^{-\beta m}(z)=-\pi\alpha^{-1}\operatorname{cosec}(\alpha\pi)\,\mathfrak{P}_\nu^{\alpha\beta}(z)\,\mathfrak{P}_\nu^{-\alpha\beta}(z),$$
$$0<\beta\leq 1,$$

$$\sum_{m=0}^{\infty}(-1)^m\frac{\varepsilon_m\cos(\pi\beta m)}{m^2-\alpha^2}\mathfrak{Q}_{-\frac{1}{2}+\beta m}(z)\,\mathfrak{Q}_{-\frac{1}{2}-\beta m}(z)$$
$$=-\pi\alpha^{-1}\operatorname{cosec}(\pi\alpha)\cos(\pi\alpha\beta)\,\mathfrak{Q}_{-\frac{1}{2}+\alpha\beta}(z)\,\mathfrak{Q}_{-\frac{1}{2}-\alpha\beta}(z)$$
$$0<\beta\leq 1,$$

$$\sum_{m=0}^{\infty}(-1)^m\frac{\varepsilon_m}{m^2-\alpha^2}\mathfrak{P}^\mu_{-\frac{1}{2}+i\beta m}(\cosh c)\,\mathfrak{P}^{-\mu}_{-\frac{1}{2}+i\beta m}(\cosh c)$$
$$=-\pi\alpha^{-1}\operatorname{cosec}(\pi\alpha)\,\mathfrak{P}^\mu_{-\frac{1}{2}+i\beta\alpha}(\cosh c)\,\mathfrak{P}^{-\mu}_{-\frac{1}{2}+i\beta\alpha}(\cosh c),$$
$$0<\beta<\frac{\pi}{c}.$$

Here α is arbitrarily complex but $\alpha \neq 0, \pm 1, \pm 2, \ldots$

4.5.3 Some series of the Fourier type

The following occuring parameters are explained as follows:

$$\lambda(m) = e^{-i\frac{y}{d}(2\pi m + c)}, \quad m_1 = -\left[\frac{ad+c}{2\pi}\right], \quad m_2 = \left[\frac{ad-c}{2\pi}\right],$$

$$d \sum_{n=-\infty}^{\infty} e^{inc}\, \mathfrak{P}^{\mu}_{-\frac{1}{2}+id+ind}(\cosh a)$$

$$= (2\pi)^{\frac{1}{2}} \left[\Gamma\left(\frac{1}{2}-\mu\right)\right]^{-1} (\sinh a)^{\mu} \sum_{m_1}^{m_2} \lambda(m)$$

$$\times \left[\cosh a - \cosh\left(\frac{2\pi m + c}{d}\right)\right]^{-\mu-\frac{1}{2}},$$

$$Re\,\mu < \frac{1}{2}\,;\ \text{if } c = \pm a d,\ Re\,\mu < -\frac{1}{2}.$$

$$a d \sum_{n=-\infty}^{\infty} e^{inc}\, \mathfrak{P}_{\nu}^{\frac{a}{\pi}(y+nd)}(b)\, \mathfrak{P}_{\nu}^{-\frac{a}{d}(y+nd)}(b)$$

$$= \pi \sum_{m_1}^{m_2} \lambda(m)\, \mathfrak{P}_{\nu}\left\{1 + 2(b^2-1)\cos^2\left[\frac{\pi}{2ad}(2\pi m + c)\right]\right\}$$

$$b > 1,\ c \neq \pm a d,$$

$$a d \sum_{n=-\infty}^{\infty} e^{inc} \cos(an)\, \mathfrak{Q}_{-\frac{1}{2}+\frac{a}{\pi}(y+nd)}(b)\, \mathfrak{Q}_{-\frac{1}{2}-\frac{a}{\pi}(y+nd)}(b)$$

$$= \pi^2 \sum_{m_1}^{m_2} \lambda(m) \left\{b^2 - \sin^2\left[\frac{\pi}{2ad}(2\pi m + c)\right]\right\}^{-\frac{1}{2}}$$

$$\times K\left\{\cos\left[\frac{\pi}{2ad}(2\pi m + c)\right]\left[b^2 - \sin^2\left(\frac{2\pi^2 m + \pi c}{2ad}\right)\right]^{-\frac{1}{2}}\right\},$$

$$d \sum_{n=-\infty}^{\infty} e^{inc}\, \mathfrak{P}^{\mu}_{-\frac{1}{2}+\frac{1}{2}i(y+nd)}(\cosh a)\, \mathfrak{P}^{-\mu}_{-\frac{1}{2}+\frac{1}{2}i(y+nd)}(\cosh a)$$

$$= 2\,\mathrm{cosech}\, a \sum_{m_1}^{m_2} \lambda(m)\, P_{\mu-\frac{1}{2}}\left[2\,\mathrm{cosech}^2 a \sinh^2\left(\frac{2\pi m + c}{d}\right) - 1\right],$$

$$c \neq 0,\ c \neq \pm a d,$$

$$d \sum_{n=-\infty}^{\infty} e^{inc}\, P^{\mu}_{-\frac{1}{2}+y+nd}(\cos a)$$

$$= (2\pi)^{\frac{1}{2}} \left[\Gamma\left(\frac{1}{2}-\mu\right)\right]^{-1} (\sin a)^{\mu} \sum_{m_1}^{m_2} \lambda(m)$$

$$\times \left[\cos\left(\frac{2\pi m + c}{d}\right) - \cos a\right]^{-\mu-\frac{1}{2}},$$

$$0 < a < \pi, \; Re\,\mu < \frac{1}{2}\,; \text{ if } c = \pm a d, \; Re\,\mu < -\frac{1}{2},$$

$$a d \sum_{n=-\infty}^{\infty} e^{inc} P_\nu^{\frac{\pi}{a}(y+nd)}(b)\, P_\nu^{-\frac{a}{\pi}(y+nd)}(b)$$

$$= \pi \sum_{m_1}^{m_2} \lambda(m)\, P_\nu\left\{1 - 2(1-b^2)\cos^2\left[\frac{\pi}{2ad}(2\pi m + c)\right]\right\},$$

$$0 < b < 1, \; c \neq \pm a d.$$

A large number of special cases can be obtained by specializing parameters.

4.5.4 Special cases

$$\Gamma\left(\frac{1}{2} - \mu\right) \sum_{m=0}^{\infty} \varepsilon_m \cos(m v)\, P^\mu_{m-\frac{1}{2}}(\cos\vartheta)$$

$$= (2\pi)^{\frac{1}{2}} (\sin\vartheta)^\mu (\cos v - \cos\vartheta)^{-\mu-\frac{1}{2}}, \qquad 0 \leq v < \vartheta,$$

$$= 0 \qquad \vartheta < v < \pi,$$

$$Re\,\mu < \frac{1}{2},$$

$$\Gamma\left(\frac{1}{2} - \mu\right) \sum_{m=0}^{\infty} (-1)^m \varepsilon_m \cos(m v)\, P^\mu_{m-\frac{1}{2}}(\cos\vartheta)$$

$$= (2\pi)^{\frac{1}{2}} (\sin\vartheta)^\mu (\cos\vartheta - \cos v)^{-\mu-\frac{1}{2}}, \qquad \vartheta < v < \pi,$$

$$= 0 \qquad 0 < v < \vartheta,$$

$$Re\,\mu < \frac{1}{2},$$

$$\Gamma\left(\frac{1}{2} - \mu\right) \sum_{m=0}^{\infty} P^\mu_m(\cos\vartheta) \cos\left[\left(m + \frac{1}{2}\right) v\right]$$

$$= \left(\frac{1}{2}\pi\right)^{\frac{1}{2}} (\sin\vartheta)^\mu (\cos v - \cos\vartheta)^{-\mu-\frac{1}{2}}, \qquad 0 < v < \vartheta,$$

$$= 0 \qquad \vartheta < v < \pi,$$

$$Re\,\mu < \frac{1}{2},$$

$$\sum_{m=0}^{\infty} \varepsilon_m \mathfrak{Q}^\mu_{m-\frac{1}{2}}(z) \cos(m v)$$

$$= e^{i\pi\mu} \left(\frac{1}{2}\pi\right)^{\frac{1}{2}} \Gamma\left(\frac{1}{2} + \mu\right) (z^2 - 1)^{\frac{1}{2}\mu} (z - \cos v)^{-\mu-\frac{1}{2}},$$

$$Re\,\mu > -\frac{1}{2},$$

$$(x - \cos\varphi)^{-\nu-\frac{1}{2}} = 2^{\nu-\frac{1}{2}} \pi^{-\frac{1}{2}} \Gamma\left(\frac{1}{2} - \nu\right) \Gamma(\nu) \sum_{m=0}^{\infty} (-1)^m (\nu + m)$$

$$\times C_m^\nu (\cos\varphi) P_{\nu+m-\frac{1}{2}}(-x), \quad x < \cos\varphi,$$

$$= 0 \quad \text{otherwise}, \quad Re\, \nu < \frac{1}{2},$$

$$(z - \cos\varphi)^{-\nu-\frac{1}{2}} = 2^{\frac{1}{2}+\nu} \pi^{-\frac{1}{2}} \left[\Gamma\left(\frac{1}{2} + \nu\right)\right]^{-1} \Gamma(\nu)$$

$$\times \sum_{m=0}^{\infty} (\nu + m) C_m^\nu (\cos\varphi) \mathfrak{Q}_{\nu+m-\frac{1}{2}}(z),$$

$$Re\, \nu > -\frac{1}{2},$$

$$\sum_{m=0}^{\infty} \varepsilon_m \cos(mv)\, \mathfrak{Q}^\mu_{-\frac{1}{2}+mk}(z) = e^{i\pi\mu} (2\pi)^{-\frac{1}{2}} \Gamma\left(\frac{1}{2} + \mu\right) \pi k^{-1} (z^2 - 1)^{\frac{1}{2}\mu}$$

$$\times \sum_{r_1}^{r_2} \left[z - \cos\left(\frac{2\pi r + v}{m}\right)\right]^{-\mu-\frac{1}{2}}$$

$$r_1 = -\left[\frac{1}{2} k + \frac{1}{2} \frac{v}{\pi}\right], \quad r_2 = \left[\frac{1}{2} k - \frac{1}{2} \frac{v}{\pi}\right],$$

$$k = 1, 2, 3, \ldots, \quad Re\, \mu > -\frac{1}{2},$$

$$\sum_{m=0}^{\infty} \varepsilon_m \frac{\Gamma(\nu + 1)}{\Gamma(\nu + m + 1)} \mathfrak{P}_\nu^m(z) \cos[m(v - \varphi)]$$

$$= \left[z + (z^2 - 1)^{\frac{1}{2}} \cos(v - \varphi)\right]^\nu,$$

$$Re\, z > 0,$$

$$P_\nu^{-\mu}(\cos\vartheta) = \pi^{-1} \sin(\pi\nu) \sum_{m=0}^{\infty} (-1)^m [(\nu - m)^{-1} - (\nu + m + 1)^{-1}]$$

$$\times P_m^{-\mu}(\cos\vartheta),$$

$$-\pi < \vartheta < \pi, \quad \mu \geq 0,$$

$$P_\nu^{-\mu}(\cos\vartheta)\, P_\nu^{-\lambda}(\cos\vartheta') = \pi \sin(\pi\nu)$$

$$\times \sum_{m=0}^{\infty} (-1)^m [(\nu - m)^{-1} - (\nu + m + 1)^{-1}]\, P_m^{-\mu}(\cos\vartheta)\, P_m^{-\lambda}(\cos\vartheta'),$$

$$-\pi < \vartheta + \vartheta' < \pi, \quad -\pi < \vartheta - \vartheta' < \pi, \quad (\mu, \lambda) \geq 0,$$

$$P_\nu^m(x)\, P_\nu^{-m}(x') = \pi^{-1} \sin(\pi\nu)$$

$$\times \sum_{n=0}^{\infty} (-1)^n [(\nu - n)^{-1} - (\nu + n + 1)^{-1}]\, P_n^m(x)\, P_n^{-m}(x'),$$

$$0<\vartheta<\pi,\quad 0<\vartheta'<\pi,\quad \vartheta+\vartheta'<\pi,\quad x=\cos\vartheta,\quad x'=\cos\vartheta',$$

$$P_\nu(-x)=\pi^{-2}\cos(\pi\nu)\sum_{n=0}^{\infty}\left[\left(n+\nu+\frac{1}{2}\right)^{-1}-\left(n-\nu-\frac{1}{2}\right)^{-1}\right]$$
$$\times\,\varepsilon_n Q_{n-\frac{1}{2}}(x),$$
$$0<x<1.$$

4.6 Integral representations

4.6.1 General Argument

$$\mathfrak{P}_\nu^m(z)\cos(m\varphi)=\frac{\Gamma(\nu+m+1)}{2\pi\Gamma(\nu+1)}\int_0^{2\pi}\left[z+(z^2-1)^{\frac{1}{2}}\cos(t-\varphi)\right]^\nu$$
$$\times\cos(mt)\,dt,$$

$$\mathfrak{P}_\nu^m(z)\sin(m\varphi)=\frac{\Gamma(\nu+m+1)}{2\pi\Gamma(\nu+1)}\int_0^{2\pi}\left[z+(z^2-1)^{\frac{1}{2}}\cos(t-\varphi)\right]^\nu$$
$$\times\sin(mt)\,dt.$$

In both formulas $Re\, z>0,\ m=0,1,2,\ldots$

$$\Gamma\left(\frac{1}{2}-\mu\right)(z^2-1)^{\frac{1}{2}\mu}\,\pi^{\frac{1}{2}}\,2^{-\mu}\,\mathfrak{P}_\nu^\mu(z)=\int_0^{\pi}\left[z+(z^2-1)^{\frac{1}{2}}\cos t\right]^{\nu+\mu}$$
$$\times(\sin t)^{-2\mu}\,dt,$$
$$Re\,\mu<\frac{1}{2},$$

$$\Gamma\left(\frac{1}{2}-\mu\right)\left(\frac{1}{2}\pi\right)^{\frac{1}{2}}(\sinh\alpha)^{-\mu}\,\mathfrak{P}_\nu^\mu(\cosh\alpha)=\int_0^{\alpha}(\cosh\alpha-\cosh t)^{-\mu-\frac{1}{2}}$$
$$\times\cosh\left[\left(\nu+\frac{1}{2}\right)t\right]dt,$$
$$Re\,\mu<\frac{1}{2}$$

$$\mathfrak{P}_\nu^\mu(\cosh\alpha)=(2\pi)^{\frac{1}{2}}\sec(\pi\nu)\left[\Gamma(-\nu-\mu)\,\Gamma(1+\nu-\mu)\,\Gamma\left(\frac{1}{2}+\mu\right)\right]^{-1}$$
$$\times(\sinh\alpha)^{-\mu}\int_\alpha^{\infty}(\cosh t-\cosh\alpha)^{\mu-\frac{1}{2}}\sinh\left[\left(\nu+\frac{1}{2}\right)t\right]dt,$$
$$Re\,\mu<\frac{1}{2},\quad Re\,(\mu\pm\nu)>0,$$

$$\Gamma(-\nu-\mu)\,\Gamma(1+\nu-\mu)\left(\frac{1}{2}\pi\right)^{\frac{1}{2}}(z^2-1)^{\frac{1}{2}\mu}\,\mathfrak{P}_\nu^\mu(z)$$

$$=\Gamma\left(\frac{1}{2}-\mu\right)\int_0^\infty (z+\cosh t)^{\mu-\frac{1}{2}}\cosh\left[\left(\nu+\frac{1}{2}\right)t\right]dt,$$

$$Re\,(\mu-\nu)>0,\quad Re\,(\mu+\nu+1)>0,$$

$$\Gamma(-\nu-\mu)\,\Gamma(1+\nu)\,2^\nu(z^2-1)^{\frac{1}{2}\mu}\,\mathfrak{P}_\nu^\mu(z)=\int_0^\infty (z+\cosh t)^{\mu-\nu-1}$$

$$\times(\sinh t)^{2\nu+1}\,dt,$$

$$Re\,\mu>Re\,\nu>-1,$$

$$\left(\frac{1}{2}\pi\right)^{\frac{1}{2}}\Gamma\left(\frac{1}{2}-\mu\right)\mathfrak{P}_\nu^\mu(\cosh\alpha)=(\sinh\alpha)^\mu$$

$$\times\left\{\sin(\pi\mu)\int_\alpha^\infty(\cosh t-\cosh\alpha)^{-\mu-\frac{1}{2}}\cosh\left[\left(\nu+\frac{1}{2}\right)t\right]dt\right.$$

$$\left.-\sin(\pi\nu)\int_0^\infty(\cosh t+\cosh\alpha)^{-\mu-\frac{1}{2}}\cosh\left[\left(\nu+\frac{1}{2}\right)t\right]dt\right\},$$

$$\pi\Gamma(-\nu-\mu)\,\mathfrak{P}_\nu^\mu(z)=\Gamma(-\nu)\left\{\int_0^\pi\left[z-(z^2-1)^{\frac{1}{2}}\cos t\right]^\nu\cos(\mu t)\,dt\right.$$

$$\left.+\sin(\pi\mu)\int_0^\infty\left[z+(z^2-1)^{\frac{1}{2}}\cosh t\right]^\nu e^{\mu t}\,dt\right\},$$

$$Re\,(\nu+\mu)<0,\quad Re\,\nu<0,\quad Re\,z>0,$$

$$\Gamma(1-\mu)\,\Gamma(-\nu-\mu)\,\Gamma(\nu-\mu+1)\,2^{-\mu}(z^2-1)^{\frac{1}{2}\mu}\,\mathfrak{P}_\nu^\mu(z)$$

$$=\Gamma(1-2\mu)\int_0^\infty t^{-\nu-\mu-1}(1+2tz+t^2)^{\mu-\frac{1}{2}}\,dt,$$

$$Re\,(\mu+\nu)<0,\quad Re\,(\mu-\nu)<1,$$

$$\Gamma(-\mu)\,\mathfrak{P}_\nu^\mu(z)=(z^2-1)^{\frac{1}{2}\mu}\int_1^z\mathfrak{P}_\nu(t)\,(z-t)^{-\mu-1}\,dt,$$

$$Re\,\mu<0,$$

$$\mathfrak{P}_\nu\left[zz' - (z^2-1)^{\frac{1}{2}}(z'^2-1)^{\frac{1}{2}}\cos\varphi\right]$$

$$= (2\pi)^{-1}\int_0^{2\pi}\left[z + (z^2-1)^{\frac{1}{2}}\cos(t-\varphi)\right]^\nu\left[z' + (z'^2-1)^{\frac{1}{2}}\cos t\right]^{-\nu-1} dt,$$

$$Re\,(z, z') > 0,$$

$$\Gamma(\nu-\mu+1)\,\Gamma\left(\frac{1}{2}+\mu\right)\mathfrak{Q}_\nu^\mu(z) = e^{i\pi\mu}\,\pi^{\frac{1}{2}}\,2^{-\mu}\Gamma(\nu+\mu+1)\,(z^2-1)^{\frac{1}{2}\mu}$$

$$\times\int_0^\infty\left[z + (z^2-1)^{\frac{1}{2}}\cosh t\right]^{-\nu-\mu-1}(\sinh t)^{2\mu}\,dt,$$

$$Re\,(\nu \pm \mu + 1) > 0,$$

$$\Gamma\left(\frac{1}{2}-\mu\right)\mathfrak{Q}_\nu^\mu(\cosh\alpha) = \left(\frac{1}{2}\pi\right)^{\frac{1}{2}} e^{i\pi\mu}(\sinh\alpha)^\mu\int_\alpha^\infty e^{-\left(\nu+\frac{1}{2}\right)t}$$

$$\times(\cosh t - \cosh\alpha)^{-\mu-\frac{1}{2}}\,dt,$$

$$Re\,(\nu+\mu+1) > 0,\ \ Re\,\mu < \frac{1}{2},$$

$$\Gamma(1+\nu)\,\mathfrak{Q}_\nu^\mu(z)$$

$$= e^{i\pi\mu}\,2^{-\nu-1}\Gamma(\nu+\mu+1)\,(z^2-1)^{-\frac{1}{2}\mu}\int_0^\pi(z+\cos t)^{\mu-\nu-1}(\sin t)^{2\nu+1}\,dt,$$

$$Re\,\nu > -1,\ \ Re\,(\nu+\mu+1) > 0,$$

$$\Gamma(\nu-\mu+1)\,\mathfrak{Q}_\nu^\mu(z) = e^{i\pi\mu}\Gamma(\nu+1)\int_0^\infty\left[z + (z^2-1)^{\frac{1}{2}}\cosh t\right]^{-\nu-1}$$

$$\times\cosh(\mu t)\,dt,$$

$$Re\,(\nu\pm\mu) > -1,\ \ \nu \neq -1, -2, -3, \ldots,$$

$$\mathfrak{Q}_\nu^\mu(z) = e^{i\pi\mu}(2\pi)^{-\frac{1}{2}}(z^2-1)^{\frac{1}{2}\mu}\,\Gamma\left(\frac{1}{2}+\mu\right)$$

$$\times\left\{\int_0^\pi(z-\cos t)^{-\mu-\frac{1}{2}}\cos\left[\left(\nu+\frac{1}{2}\right)t\right]dt - \cos(\pi\nu)\right.$$

$$\left.\times\int_0^\infty(z+\cosh t)^{-\mu-\frac{1}{2}}e^{-\left(\nu+\frac{1}{2}\right)t}\,dt\right\},$$

$$Re\,\mu > -\frac{1}{2},\ Re\,(\nu + \mu + 1) > 0,$$

$$\Gamma(-\mu)\,\mathfrak{Q}_\nu^\mu(z) = e^{i\pi\mu}(z^2 - 1)^{\frac{1}{2}\mu} \int_z^\infty \mathfrak{Q}_\nu(t)\,(t - z)^{-\mu-1}\,dt,$$

$$Re\,\mu < 0,\ Re\,(\nu + \mu + 1) > 0,$$

$$\mathfrak{Q}_n(z) = \frac{1}{2}\int_{-1}^{1} (z - t)^{-1} P_n(t)\,dt = 2^n \int_z^\infty (t - z)^n (t^2 - 1)^{-n-1}\,dt,$$

$$|\arg(z \pm 1)| < \pi,$$

$$\Gamma\left(\frac{1}{2} - \mu\right)\mathfrak{Q}_\nu^\mu(z) = e^{i\pi\mu}\,2^\mu \pi^{\frac{1}{2}}(z^2 - 1)^{\frac{1}{2}\mu}\left[z + (z^2 - 1)^{\frac{1}{2}}\right]^{-\nu-\frac{1}{2}}$$

$$\times \int_0^\infty e^{-(\nu+\mu+1)t}\left\{(1 - e^{-t})\left[z + (z^2 - 1)^{\frac{1}{2}} - z e^{-t} + (z^2 - 1)\,e^{-t}\right]\right\}dt,$$

$$Re\,\mu < \frac{1}{2},\ Re\,(\nu + \mu + 1) > 0,$$

$$\mathfrak{Q}_\nu^\mu(z) = \frac{1}{2}e^{i\pi\mu}(z^2 - 1)^{\frac{1}{2}\mu}\int_{-1}^{1}(1 - t^2)^{-\frac{1}{2}\mu}(z - t)^{-1}\,\mathfrak{P}_\nu^\mu(t)\,dt,$$

$$\nu + \mu = 0, 1, 2, \ldots\ Re\,\mu < 1.$$

z not on the real axis between -1 and $+1$

$$\mathfrak{P}_n(z)\,\mathfrak{Q}_m(z) = \frac{1}{2}\int_{-1}^{1}(z - t)^{-1}\,\mathfrak{P}_n(t)\,\mathfrak{P}_m(t)\,dt,$$

$$n \le m,\ n, m = 0, 1, 2, \ldots$$

z not on the real axis between -1 and $+1$

$$(z^2 - 1)^{\frac{1}{2}\mu}\,\mathfrak{P}_\nu^\mu(z) = \left(\frac{1}{2}\pi\right)^{-\frac{1}{2}}[\Gamma(1 + \nu - \mu)\,\Gamma(-\nu - \mu)]^{-1}$$

$$\times \int_0^\infty t^{-\mu-\frac{1}{2}} K_{\nu+\frac{1}{2}}(t)\,e^{-zt}\,dt,$$

$$Re\, z > -1,\ Re\,(1-\mu+\nu) > 0,\ Re\,(\nu+\mu) < 0,$$

$$\mathfrak{P}_\nu\left(\frac{a^2+b^2+c^2}{2ab}\right) = -4\pi^{-2}\sin(\pi\nu)\int_0^\infty K_{\nu+\frac{1}{2}}(at)\,K_{\nu+\frac{1}{2}}(bt)\cos(ct)\,dt,$$

$$-1 < Re\,\nu < 0.$$

4.6.2 Argument real between − 1 and + 1

$$\Gamma\left(\frac{1}{2}-\mu\right)(1-x)^{\frac{1}{2}\mu}\,\pi^{\frac{1}{2}}\,2^{-\mu}P_\nu^\mu(x) = \int_0^\pi\left[x+(1-x^2)^{\frac{1}{2}}\cos t\right]^{\nu+\mu}$$
$$\times\,(\sin t)^{-2\mu}\,dt,$$

$$Re\,\mu < \frac{1}{2},$$

$$2\pi\,\Gamma(1+\nu)\,P_\nu^m(\cos\vartheta)\cos(m\varphi)$$
$$= i^m\Gamma(\nu+m+1)\int_0^{2\pi}[\cos\vartheta + i\sin\vartheta\cos(t-\varphi)]^\nu\cos(mt)\,dt,$$

$$2\pi\Gamma(1+\nu)\,P_\nu^m(\cos\vartheta)\sin(m\varphi)$$
$$= i^m\Gamma(\nu+m+1)\int_0^{2\pi}[\cos\vartheta + i\sin\vartheta\cos(t-\varphi)]^\nu\sin(mt)\,dt,$$

both formulas valid for $0 < \vartheta < \frac{1}{2}\pi$,

$$m = 0, 1, 2, \ldots$$

$$\Gamma\left(\frac{1}{2}-\mu\right)P_\nu^\mu(\cos\vartheta) = \left(\frac{1}{2}\pi\right)^{-\frac{1}{2}}(\sin\vartheta)^\mu\int_0^\vartheta(\cos t-\cos\vartheta)^{-\mu-\frac{1}{2}}$$
$$\times\cos\left[\left(\nu+\frac{1}{2}\right)t\right]dt,$$

$$0 < \vartheta < \pi,\ Re\,\mu < \frac{1}{2},$$

$$\Gamma(-\mu-\nu)\,\Gamma(1-\mu+\nu)\left(\frac{1}{2}\pi\right)^{\frac{1}{2}}(1-x^2)^{\frac{1}{2}\mu}P_\nu^\mu(x) = \Gamma\left(\frac{1}{2}-\mu\right)$$
$$\times\int_0^\infty(x+\cosh t)^{\mu-\frac{1}{2}}\cosh\left[\left(\nu+\frac{1}{2}\right)t\right]dt,$$

$$Re\,(\mu+\nu) < 0,\ Re\,(\nu-\mu+1) > 0,$$

$$\Gamma\left(\frac{1}{2}+\mu\right)\Gamma(\nu-\mu+1)\,P_\nu^\mu(\cos\vartheta)$$

$$= i\,2^{-\mu}\pi^{-\frac{1}{2}}\,\Gamma(\nu+\mu+1)\,(\sin\vartheta)^\mu$$

$$\times\left[\int_0^\infty(\cos\vartheta+i\sin\vartheta\cosh t)^{-\nu-\mu-1}(\sinh t)^{2\mu}\,dt\right.$$

$$\left.-\int_0^\infty(\cos\vartheta-i\sin\vartheta\cosh t)^{-\nu-\mu-1}(\sinh t)^{2\mu}\,dt\right],$$

$$Re\,\mu>-\frac{1}{2},\quad Re\,(\nu\pm\mu+1)>0.$$

$$\pi\Gamma(\nu-\mu+1)\,P_\nu^\mu(\cos\vartheta)=i\Gamma(\nu+1)$$

$$\times\left[e^{-i\frac{1}{2}\pi\mu}\int_0^\infty(\cos\vartheta+i\sin\vartheta\cosh t)^{-\nu-1}\cosh(\mu t)\,dt\right.$$

$$\left.-e^{i\frac{1}{2}\pi\mu}\int_0^\infty(\cos\vartheta-i\sin\vartheta\cosh t)^{-\nu-1}\cosh(\mu t)\,dt\right],$$

$$\Gamma(-\nu-\mu)\,P_\nu^\mu(\cos\vartheta)=e^{i\frac{1}{2}\pi\mu}\,\Gamma(-\nu)$$

$$\times\left[\int_0^\pi(\cos\vartheta-i\sin\vartheta\cosh t)^\nu\cos(\mu t)\,dt\right.$$

$$\left.+\sin(\pi\mu)\int_0^\infty(\cos\vartheta+i\sin\vartheta\cosh t)^\nu e^{\mu t}\,dt\right],$$

$$0<\vartheta<\frac{\pi}{2},\quad Re\,(-\nu-\mu)>0,$$

$$\Gamma\left(\frac{1}{2}+\mu\right)\Gamma(\nu-\mu+1)\,Q_\nu^\mu(\cos\vartheta)=\pi^{\frac{1}{2}}\,2^{-\mu-1}\Gamma(\nu+\mu+1)\,(\sin\vartheta)^\mu$$

$$\times\left[\int_0^\infty(\cos\vartheta+i\sin\vartheta\cosh t)^{-\nu-\mu-1}(\sinh t)^{2\mu}\,dt\right.$$

$$\left.+\int_0^\infty(\cos\vartheta-i\sin\vartheta\cosh t)^{-\nu-\mu-1}(\sinh t)^{2\mu}\,dt\right],$$

$$Re\,\mu>-\frac{1}{2},\quad Re\,(\nu\pm\mu+1)>0,$$

$$2Q_\nu^\mu(\cos\vartheta)\,\Gamma(\nu-\mu+1)$$
$$=\Gamma(\nu+1)\left[e^{-i\frac{1}{2}\pi\mu}\int_0^\infty(\cos\vartheta+i\sin\vartheta\cosh t)^{-\nu-1}\cosh(\mu t)\,dt\right.$$
$$\left.+e^{i\frac{1}{2}\pi\mu}\int_0^\infty(\cos\vartheta-i\sin\vartheta\cosh t)^{-\nu-1}\cosh(\mu t)\,dt\right],$$
$$Re\,(\nu\pm\mu)>-1,\ \nu\neq-1,-2,-3,\ldots,$$
$$\Gamma(-\mu)\,P_\nu^\mu(x)=(1-x^2)^{\frac{1}{2}\mu}\int_x^1 P_\nu(t)\,(t-x)^{-\mu-1}\,dt,$$
$$Re\,\mu<0.$$

4.6.3 Products of two Legendre functions

$$[P_\nu^\mu(x)]^2=\Gamma(1+\nu+\mu)\,[\Gamma(1+\nu-\mu)]^{-1}$$
$$\int_0^\infty\left\{J_\mu\left[\frac{1}{2}t\,(1-x^2)^{\frac{1}{2}}\right]\right\}^2 J_{2\nu+1}(t)\,dt,$$
$$Re\,(\nu+\mu)>-1,$$
$$P_\nu^\mu(x)\,P_\nu^{-\mu}(x)=\int_0^\infty J_\mu\left[\frac{1}{2}t(1-x^2)^{\frac{1}{2}}\right]J_{-\mu}\left[\frac{1}{2}t(1-x^2)^{\frac{1}{2}}\right]J_{2\nu+1}(t)\,dt,$$
$$Re\,\nu>-1,$$
$$[\mathfrak{P}_\nu^\mu(z)]^2=\left[\frac{1}{2}\pi\Gamma(1-\mu+\nu)\,\Gamma(-\mu-\nu)\right]^{-1}\int_0^\infty\left[K_{\nu+\frac{1}{2}}\left(\frac{1}{2}t\right)\right]^2$$
$$\times J_{-2\mu}\left[t(z^2-1)^{\frac{1}{2}}\right]dt,$$
$$Re\,(\nu-\mu)>-1,\ \ Re\,(\mu+\nu)<0,$$
$$[\mathfrak{P}_\nu^\mu(z)]^2=2\,[\Gamma(1+\nu-\mu)\,\Gamma(-\mu-\nu)]^{-1}\int_0^\infty\left\{J_{-\mu}\left[\frac{1}{2}t(z^2-1)^{\frac{1}{2}}\right]\right\}^2$$
$$\times K_{2\nu+1}(t)\,dt,$$
$$\mathfrak{P}_\nu^\mu(z)\,\mathfrak{P}_\nu^{-\mu}(z)=-2\pi^{-1}\sin(\pi\nu)\int_0^\infty J_\mu\left[\frac{1}{2}t(z^2-1)^{\frac{1}{2}}\right]$$
$$\times J_{-\mu}\left[\frac{1}{2}t(z^2-1)^{\frac{1}{2}}\right]K_{2\nu+1}(t)\,dt,$$
$$-1<Re\,\nu<0,$$

$$[P_\nu^\mu(x)]^2 = 2\,[\Gamma(\nu+1-\mu)\,\Gamma(-\nu-\mu)]^{-1} \int_0^\infty \left\{ I_{-\mu}\left[\frac{1}{2}t(1-x^2)^{\frac{1}{2}}\right]\right\}^2$$

$$\times K_{2\nu+1}(t)\,dt,$$

$$Re\,(\mu+\nu) < 0,\ Re\,(\nu-\mu) > -1,$$

$$P_\nu^\mu(x)\,P_\nu^{-\mu}(x) = -2\pi^{-1}\sin(\pi\nu)\int_0^\infty I_\mu\left[\frac{1}{2}t(1-x^2)^{\frac{1}{2}}\right]$$

$$\times I_{-\mu}\left[\frac{1}{2}t(1-x^2)^{\frac{1}{2}}\right] K_{2\nu+1}(t)\,dt,$$

$$-1 < Re\,\nu < 0.$$

4.7 Integrals involving Legendre functions

4.7.1 Indefinite integrals

If $w_\nu^\mu(z)$ and $w_\sigma^\varrho(z)$ denote any solution of Legendre's differential equation with the parameters ν, μ and σ, ϱ respectively then

$$\int_a^b [(\nu-\sigma)(\nu+\sigma+1) + (\varrho^2-\mu^2)(1-z^2)^{-1}]\, w_\nu^\mu(z)\, w_\sigma^\varrho(z)\,dz$$

$$= \left[(1-z^2)\left(w_\nu^\mu \frac{d}{dz} w_\sigma^\varrho - w_\sigma^\varrho \frac{d}{dz} w_\nu^\mu\right)\right]_a^b$$

$$= [z(\nu-\sigma)\, w_\nu^\mu w_\sigma^\varrho + (\sigma+\varrho)\, w_\nu^\mu w_{\sigma-1}^\varrho - (\nu+\mu)\, w_{\nu-1}^\mu w_\sigma^\varrho]_a^b.$$

For the special case $\mu = \varrho = 0$

$$\int_a^b w_\nu(z)\, w_\sigma(z)\,dz = [(\nu-\sigma)(\nu+\sigma+1)]^{-1}\left[(z^2-1)^{\frac{1}{2}}(w_\sigma w_\nu^1 - w_\nu w_\sigma^1)\right]_a^b.$$

If $v_\nu(x)$ and $v_\sigma(x)$ denote any two Legendre functions with the argument x between -1 and $+1$ then

$$\int_a^b v_\nu(x)\, v_\sigma(x)\,dx = [(\nu-\sigma)(\nu+\sigma+1)]^{-1}\left[(1-x^2)^{\frac{1}{2}}(v_\sigma v_\nu^1 - v_\nu v_\sigma^1)\right]_a^b.$$

4.7.2 Definite integrals with finite limits

$$\int_0^1 x^\sigma P_\nu(x)\,dx = \pi^{\frac{1}{2}}\, 2^{-\sigma-1}\,\Gamma(1+\sigma)\left[\Gamma\left(1+\frac{1}{2}\sigma-\frac{1}{2}\nu\right)\right.$$

$$\left.\times\,\Gamma\left(\frac{1}{2}\nu+\frac{1}{2}\sigma+\frac{3}{2}\right)\right]^{-1},$$

$$Re\,\sigma > -1,$$

$$\int_0^1 x^\sigma (1-x^2)^{-\frac{1}{2}\mu} P_\nu^\mu(x)\,dx$$

$$= \Gamma\left(\frac{1}{2}+\frac{1}{2}\sigma\right)\Gamma\left(1+\frac{1}{2}\sigma\right)2^{\mu-1}\left[\Gamma\left(1+\frac{1}{2}\sigma-\frac{1}{2}\nu-\frac{1}{2}\mu\right)\right.$$

$$\left.\times\,\Gamma\left(\frac{1}{2}\sigma+\frac{1}{2}\nu-\frac{1}{2}\mu+\frac{3}{2}\right)\right]^{-1},$$

$$Re\,\mu < 1,\ \ Re\,\sigma > -1,$$

$$\int_0^1 (1-x^2)^{-1}\,[P_\nu^\mu(x)]^2\,dx = -\frac{1}{2}\mu^{-1}\frac{\Gamma(1+\mu+\nu)}{\Gamma(1-\mu+\nu)}$$

$$Re\,\mu < 0,\ \ \nu+\mu = 1, 2, 3, \ldots,$$

$$\int_{-1}^1 P_\nu(x)\,P_\sigma(x)\,dx = 2\pi^{-2}[(\sigma-\nu)\,(\sigma+\nu+1)]^{-1}$$

$$\times\,\{2\sin(\pi\nu)\sin(\pi\sigma)\,[\psi(\nu+1)-\psi(\sigma+1)] + \pi\sin(\pi\sigma-\pi\nu)\},$$

$$\sigma+\nu+1 \neq 0.$$

Hence for $\sigma = \nu$

$$\int_{-1}^1 [P_\nu(x)]^2\,dx = \left(\frac{1}{2}+\nu\right)^{-1}[1-2\pi^{-2}\sin^2(\pi\nu)\,\psi'(\nu+1)],$$

$$\int_{-1}^1 Q_\nu(x)\,Q_\sigma(x)\,dx = [(\sigma-\nu)\,(\sigma+\nu+1)]^{-1}$$

$$\times\left\{[\psi(\nu+1)-\psi(\sigma+1)]\,[1+\cos(\sigma\pi)\cos(\nu\pi)] - \frac{1}{2}\pi\sin(\pi\nu-\pi\sigma)\right\},$$

$$\nu, \sigma \neq -1, -2, \ldots$$

$$\int_{-1}^1 [Q_\nu(x)]^2\,dx = (2\nu+1)^{-1}\left\{\frac{1}{2}\pi^2 - \psi'(\nu+1)\,[1+\cos^2(\pi\nu)]\right\},$$

$$\nu \neq -1, -2, -3, \ldots,$$

$$\int_{-1}^1 P_\nu(x)\,Q_\sigma(x)\,dx = [(\nu-\sigma)\,(\nu+\sigma+1)]^{-1}$$

$$\times\,\{[\psi(\nu+1)-\psi(\sigma+1)]\,[1-\cos(\sigma\pi-\nu\pi) - 2\pi^{-1}\sin(\pi\nu)\cos(\pi\sigma)]\}.$$

$$\sigma \neq -1, -2, -3, \ldots$$

$$\int_{-1}^1 P_\nu(x)\,Q_\nu(x)\,dx = -\pi^{-1}(2\nu+1)^{-1}\sin(2\pi\nu)\,\psi'(\nu+1),$$

$$\nu \neq -1, -2, \ldots$$

$$\int_0^1 P_\nu(x)\, P_\sigma(x)\, dx = 2\pi^{-1}[(\sigma-\nu)(\sigma+\nu+1)]^{-1}$$

$$\times \left[A \sin\left(\frac{1}{2}\sigma\pi\right)\cos\left(\frac{1}{2}\nu\pi\right) - A^{-1}\sin\left(\frac{1}{2}\nu\pi\right)\cos\left(\frac{1}{2}\sigma\pi\right)\right],$$

$$\int_0^1 Q_\nu(x)\, Q_\sigma(x)\, dx = [(\sigma-\nu)(\sigma+\nu+1)]^{-1}\cdot\{\psi(\nu+1)-\psi(\sigma+1)$$

$$-\frac{1}{2}\pi\left[(A-A^{-1})\sin\left(\frac{1}{2}\pi\sigma+\frac{1}{2}\pi\nu\right)-(A+A^{-1})\right.$$

$$\left.\left.\times\sin\left(\frac{1}{2}\sigma\pi-\frac{1}{2}\nu\pi\right)\right]\right\},$$

$$\nu,\sigma \neq -1,-2,\ldots$$

$$\int_0^1 P_\nu(x)\, Q_\sigma(x)\, dx = [(\sigma-\nu)(\sigma+\nu+1)]^{-1}$$

$$\times\left[A^{-1}\cos\left(\frac{1}{2}\pi\nu-\frac{1}{2}\pi\sigma\right)-1\right],$$

$$\sigma \neq -1,-2,\ldots$$

$$A = \Gamma\left(\frac{1}{2}+\frac{1}{2}\nu\right)\Gamma\left(1+\frac{1}{2}\sigma\right)\left[\Gamma\left(\frac{1}{2}+\frac{1}{2}\sigma\right)\Gamma\left(1+\frac{1}{2}\nu\right)\right]^{-1},$$

$$\int_{-1}^1 Q_n^m(x)\, P_l^m(x)\, dx = (-1)^m \frac{[1-(-1)^{l+n}]\,(n+m)!}{(l-n)\,(l+n+1)\,(n-m)!},$$

l, m, n positive integer

$$[\Gamma(\nu+1)\,\Gamma(\mu+1)]^2 \int_{-1}^1 P_\nu(x)\, P_\mu(x)\,(1+x)^{\nu+\mu}\, dx$$

$$= 2^{\nu+\mu+1}[\Gamma(\nu+\mu+1)]^4\,[\Gamma(2\nu+2\mu+2)]^{-1},$$

$$Re\,(\nu+\mu+1) > 0.$$

4.7.3 Infinite integrals

$$\int_1^\infty \mathfrak{P}_\nu(x)\, \mathfrak{Q}_\sigma(x)\, dx = [(\sigma-\nu)(\sigma+\nu+1)]^{-1},$$

$$Re\,\sigma > Re\,\nu > 0,$$

$$\int_1^\infty \mathfrak{Q}_\nu(x)\, \mathfrak{Q}_\sigma(x)\, dx = [(\sigma-\nu)(\sigma+\nu+1)]^{-1}\,[\psi(\sigma+1)-\psi(\nu+1)],$$

$$Re\,(\sigma+\nu) > -1,\ \sigma+\nu+1 \neq 0,\ \nu,\sigma \neq -1,-2,-3,\ldots,$$

$$\int_1^\infty [\mathfrak{Q}_\nu(x)]^2\, dx = (2\nu+1)^{-1}\,\psi'(\nu+1),$$

$$Re\,\nu > -\frac{1}{2},$$

$$\int_0^\infty t^{-\frac{1}{2}\mu}\,\mathfrak{P}_\nu^\mu\left[(1+t)^{\frac{1}{2}}\right] e^{-zt}\, dt = 2^\mu z^{\frac{1}{2}\mu-\frac{5}{4}}\, e^{\frac{1}{2}z}\, W_{\frac{1}{2}\mu+\frac{1}{4},\,\frac{1}{2}\nu+\frac{1}{4}}(z),$$

$$Re\,z > 0,$$

$$\int_0^\infty t^{-\frac{1}{2}\mu}\,(1+t)^{-\frac{1}{2}}\,\mathfrak{P}_\nu^\mu\left[(1+t)^{\frac{1}{2}}\right] e^{-zt}\, dt$$

$$= 2^\mu z^{\frac{1}{2}\mu-\frac{3}{4}}\, e^{\frac{1}{2}z}\, W_{\frac{1}{2}\mu-\frac{1}{4},\,\frac{1}{2}\nu+\frac{1}{4}}(z),$$

$$Re\,z > 0,$$

$$z^\alpha K_\nu(z) = \left(\frac{1}{2}\pi\right)^{\frac{1}{2}} \int_1^\infty (t^2-1)^{-\frac{1}{2}\alpha-\frac{1}{4}}\,\mathfrak{P}_{\nu-\frac{1}{2}}^{\alpha+\frac{1}{2}}(t)\, e^{-zt}\, dt,$$

$$Re\,\alpha < \frac{1}{2},\quad Re\,z > 0,$$

$$K_\nu(ax)\,K_\nu(bx) = \frac{1}{2}\pi(ab)^{-\frac{1}{2}}\sec(\pi\nu)\int_0^\infty \mathfrak{P}_{-\frac{1}{2}+\nu}\left(\frac{t^2+a^2+b^2}{2ab}\right)$$

$$\times\cos(xt)\, dt,$$

$$x > 0,\quad -\frac{1}{2} < Re\,\nu < \frac{1}{2},$$

$$K_\nu(az)\,K_\nu(bz) = \frac{1}{2}\pi(ab)^{\frac{1}{2}}\int_1^\infty (a^2+b^2+2abt)^{-\frac{1}{2}}\, e^{-z(a^2+b^2+2abt)^{\frac{1}{2}}}$$

$$\times\,\mathfrak{P}_{-\frac{1}{2}+\nu}(t)\, dt,$$

$$Re\,z > 0,$$

$$I_\nu(ax)\,K_\nu(bx) = \pi^{-1}(ab)^{-\frac{1}{2}}\int_0^\infty \mathfrak{Q}_{\nu-\frac{1}{2}}\left(\frac{a^2+b^2+t^2}{2ab}\right)\cos(xt)\, dt,$$

$$x > 0,\ a < b.$$

4.8 Asymptotic behavior

4.8.1 Large parameters and variable

Regarding the behavior of the Legendre functions for large (in general complex) values of their variables one has to distinguish between 3 cases.

1. One parameter is large while the other two are fixed.
2. Two parameters are large while one is fixed.
3. All three parameters are large.

Here only the first case is considered. (For the second case see R. C. Thorne, Phil. Trans. Roy. Soc. (A) 249, 1957, 585—620, where an asymptotics for $\mathfrak{P}_\nu^{-\mu}(z)$, $\mathfrak{Q}_\nu^{-\mu}(z)$, $P_\nu^{-\mu}(x)$, $Q_\nu^{-\mu}(x)$ is given when z or x respectively is fixed and ν and μ are large positive numbers with $0 < \mu < \nu$ and $\alpha = \frac{\mu}{\left(\frac{1}{2}+\nu\right)}$ is kept fixed as $\nu \to \infty$).

a) The argument z is large and ν and μ are fixed, then

$$\mathfrak{P}_\nu^\mu(z) \text{ and } \mathfrak{Q}_\nu^\mu(z)$$

are given by the convergent hypergeometric series (10) and (28) in 4.1.2.

b) The argument z is fixed and one of the parameters ν or μ is large. In this case the fact can be used that the hypergeometric series $F(a, b; c; \zeta)$ represents an asymptotic series in c valid in $-\pi < \arg c < \pi$; the series does not converge provided that ζ is a real number larger than 1. The representations of

$$\mathfrak{P}_\nu^\mu(z),\ \mathfrak{Q}_\nu^\mu(z),\ P_\nu^\mu(x),\ Q_\nu^\mu(x)$$

in 4.1.2 and 4.3 cover these cases. If, for instance μ is large and ν and z are fixed one has to choose such a representation in the form of a hypergeometric series $F(a, b; c; \zeta)$ that a and b do not depend on μ. This expansion is then valid in $|\arg c| < \pi$ for all points z such that ζ is not a real number larger than 1.

Therefore formula (1) in the tables of 4.1.2 would yield an asymptotic expansion of $\mathfrak{P}_\nu^{-\mu}(z)$ for large μ in $-\pi < \arg \mu < \pi$ for all fixed points of the z plane which are not on the real axis between -1 and $-\infty$. For the $P_\nu^\mu(\cos\vartheta)$ and the $Q_\nu^\mu(\cos\vartheta)$ one has for fixed ν and μ large (unrestricted) the convergent expansions of the $P_\nu^{-\mu}(\cos\theta)$ and the $Q_\nu^\mu(\cos\theta)$ formula on page 169. For fixed μ and large ν the 3rd and the 4th formula on page 168 represent asymptotic expansions valid in $-\pi < \arg\nu < \pi$, provided $0 < \vartheta < \pi$. The latter series are even convergent when $\frac{1}{6}\pi < \vartheta < \frac{5}{6}\pi$.

4.8.2 Behavior near the singular points

Function	Restrictions	Leading term
Behavior at $+1$		
$\mathfrak{P}_\nu^\mu(z)$	$\mu \neq 1, 2, 3, \ldots$	$2^{\frac{1}{2}\mu}(z-1)^{-\frac{1}{2}\mu} / \Gamma(1-\mu)$
$\mathfrak{P}_\nu^m(z)$	$m = 0, 1, 2, \ldots$	$2^{-\frac{1}{2}m}\Gamma(\nu+m+1)(z-1)^{\frac{1}{2}m} / [m!\,\Gamma(\nu-m+1)]$
$e^{-i\pi\mu}\,\mathfrak{Q}_\nu^\mu(z)$	$Re\,\mu > 0$	$2^{\frac{1}{2}\mu-1}\Gamma(\mu)(z-1)^{-\frac{1}{2}\mu}$
$e^{-i\pi\mu}\,\mathfrak{Q}_\nu^\mu(z)$	$Re\,\mu < 0$	$2^{-1-\frac{1}{2}\mu}\Gamma(-\mu)\Gamma(\nu+\mu+1)(z-1)^{\frac{1}{2}\mu} / \Gamma(\nu-\mu+1)$
$\mathfrak{Q}_\nu(z)$	$\nu \neq -1, -2, -3, \ldots$	$-\gamma - \psi(\nu+1) - \frac{1}{2}\log\left(\frac{1}{2}z - \frac{1}{2}\right)$
$P_\nu^\mu(x)$	$\mu \neq 1, 2, 3, \ldots$	$2^{\frac{1}{2}\mu}(1-x)^{-\frac{1}{2}\mu} / \Gamma(1-\mu)$
$P_\nu^m(x)$	$m = 0, 1, 2, \ldots$	$(-1)^m 2^{-\frac{1}{2}m}\Gamma(\nu+m+1)(1-x)^{\frac{1}{2}m} / [m!\,\Gamma(\nu-m+1)]$
$Q_\nu^\mu(x)$	$Re\,\mu > 0$	$2^{\frac{1}{2}\mu-1}\Gamma(\mu)\cos(\pi\mu)(1-x)^{-\frac{1}{2}\mu}$
$Q_\nu^\mu(x)$	$Re\,\mu < 0$	$2^{-1-\frac{1}{2}\mu}\Gamma(-\mu)\Gamma(\nu+\mu+1)(1-x)^{\frac{1}{2}\mu} / \Gamma(\nu-\mu+1)$
$Q_\nu(x)$	$\nu \neq -1, -2, -3, \ldots$	$-\frac{1}{2}\log\left(\frac{1}{2} - \frac{1}{2}x\right) - \gamma - \psi(\nu+1)$

Function	Restriction	Leading term
Behavior at -1		
$P_\nu^\mu(x)$	$Re\,\mu > 0$	$-2^{\frac{1}{2}\mu} \sin(\pi\nu)\, \pi^{-1} \Gamma(\mu)\, (1+x)^{-\frac{1}{2}\mu}$
$P_\nu^\mu(x)$	$Re\,\mu < 0$	$2^{-\frac{1}{2}\mu}\, \Gamma(-\mu)\, (1+x)^{\frac{1}{2}\mu} \big/ [\Gamma(1+\nu-\mu)\, \Gamma(-\nu-\mu)]$
$P_\nu(x)$	None	$\pi^{-1} \sin(\pi\nu) \left[\log\left(\frac{1}{2}+\frac{1}{2}x\right) + \gamma + 2\psi(\nu+1) + \pi \cot(\pi\nu)\right]$
$Q_\nu^\mu(x)$	$Re\,\mu > 0$	$-2^{\frac{1}{2}\mu-1}\, \Gamma(\mu) \cos(\pi\nu)\, (1+x)^{-\frac{1}{2}\mu}$
$Q_\nu^\mu(x)$	$Re\,\mu < 0$	$-2^{-1-\frac{1}{2}\mu} \cos[\pi(\nu+\mu)]\, \Gamma(-\mu)\, \Gamma(\nu+\mu+1)\, (1+x)^{\frac{1}{2}\mu} \big/ \Gamma(\nu-\mu+1)$
$Q_\nu(x)$	$\nu \neq -1, -2, -3, \ldots$	$\frac{1}{2} \cos(\pi\nu) \left[\log\left(\frac{1}{2}+\frac{1}{2}x\right) + \gamma + 2\psi(\nu+1) - \pi \tan(\pi\nu)\right]$
Behavior at ∞		
$\mathfrak{P}_\nu^\mu(z)$	$Re\,\nu > -\frac{1}{2}$	$2^\nu \pi^{-\frac{1}{2}}\, \Gamma\left(\frac{1}{2}+\nu\right) z^\nu \big/ \Gamma(1+\nu-\mu)$
$\mathfrak{P}_\nu^\mu(z)$	$Re\,\nu < -\frac{1}{2}$	$\pi^{-\frac{1}{2}}\, 2^{-\nu-1}\, \Gamma\left(-\frac{1}{2}-\nu\right) z^{-\nu-1} \big/ \Gamma(-\nu-\mu)$
$\mathfrak{Q}_\nu^\mu(z)$	$\nu \neq -n - \frac{3}{2}$	$e^{i\pi\mu}\, 2^{-\nu-1} \pi^{\frac{1}{2}}\, \Gamma(\nu+\mu+1)\, z^{-\nu-1} \big/ \Gamma\left(\frac{3}{2}+\nu\right)$
$\mathfrak{Q}_\nu^\mu(z)$	$\nu = -n - \frac{3}{2}$	$e^{i\pi\mu}\, 2^{-\nu-1} \pi^{\frac{1}{2}}\, \Gamma(1+\nu+\mu) \left(1+\frac{1}{2}\nu+\frac{1}{2}\mu\right)_{n+1} \left(\frac{1}{2}+\frac{1}{2}\nu+\frac{1}{2}\mu\right)_{n+1} z^{-\nu-2\mu-3}$
	$n = 0, 1, 2, \ldots$	

4.9 Associated Legendre functions and surface spherical harmonics

Legendre's differential equation for $\mu = 0$ is

$$(1 - x^2)\, w'' - 2xw' + \nu(\nu + 1)\, w = 0.$$

If one demands that a non identically vanishing solution of this equation should be continuous at $x = \pm 1$, then $\nu = n$ $(n = 0, 1, 2, \ldots)$. Such a solution is then

$$w = P_n(x).$$

These are the Legendre polynomials (see 5.4). Again, if $\mu = m$

$$(1 - x^2)\, w'' - 2xw' + [\nu(\nu + 1) - m^2 (1 - x^2)^{-1}]\, w = 0,$$
$$(m = 1, 2, 3, \ldots).$$

Then a non vanishing solution which is continuous at $x = \pm 1$ exists only if $\nu = n$ $(n = 0, 1, 2, \ldots)$.

$$w = P_n^m(x),$$

$w = P_n^{-m}(x)$ is likewise a solution and

$$P_n^{-m}(x) = 0, \quad n < m$$
$$= \frac{(-1)^m\,(n - m)!}{(n + m)!}\, P_n^m(x), \; n \geq m.$$

Also

$$P_n^m(x) = 0, \quad m > n.$$

These are the associated Legendre functions of the first kind of order n and degree m. They have the property that every function continuous in $-1 \leq x \leq 1$ which vanishes at the endpoints of this interval can be approximated with arbitrary accuracy by the associated Legendre functions $P_n^m(x)$ of arbitrary order m. Another solution of the differential-equation (associated Legendre functions of the second kind) is

$$w = Q_n^m(x).$$

They do not remain finite at $x = \pm 1$. (For these functions see also 4.4.2).

Orthogonality relations and other integral formulas

$$\int_{-1}^{1} P_n^m(x)\, P_l^m(x)\, dx = 0, \; l \neq n,$$
$$= \left(n + \frac{1}{2}\right)^{-1} (n + m)!\, [(n - m)!]^{-1} \qquad l = n,$$

$$\int_{-1}^{1} Q_n^m(x)\, P_l^m(x)\, dx$$
$$= (-1)^m\, [1 - (-1)^{l+n}\, (n + m)!]\, [(l - n)\, (l + n + 1)\, (n - m)!]^{-1},$$
$$l, m, n = 1, 2, 3, \ldots$$

$$\int_{-1}^{1} (1 - x^2)^{-1} [P_n^m(x)]^2 \, dx = m^{-1}(n + m)! \, [(n - m)!]^{-1},$$

$$\int_{-1}^{1} (1 - x^2)^{-1} P_n^m(x) \, P_n^k(x) \, dx = 0.$$

Zeros of $P_n^m(x)$.

The function $P_n^m(x)$ has $n - m$ simple zeros between -1 and $+1$ which lie symmetrical to $x = 0$.

Surface spherical harmonics

These are solutions of the 2 dimensional Laplace equation in spherical coordinates (variables φ and ϑ)

$$\operatorname{cosec} \vartheta \frac{\partial}{\partial \vartheta}\left(\sin \vartheta \frac{\partial Y}{\partial \vartheta}\right) + \operatorname{cosec}^2 \vartheta \frac{\partial^2 Y}{\partial \varphi^2} + n(n + 1) \, Y = 0.$$

A one valued solution in φ and a continuous solution in ϑ is a linear combination of the form

$$P_n^m(\cos \vartheta) \cos (m\varphi), \; P_n^m(\cos \vartheta) \sin (m\varphi).$$

The most general solution is of the form

$$Y_n(\varphi, \vartheta) = \sum_{m=0}^{n} (A_m \cos (m\varphi) + B_m \sin (m\varphi)] \, P_n^m(\cos \vartheta).$$

An expansion of this form is

$$(\cos \vartheta + i \sin \vartheta \cos \varphi)^n = P_n(\cos \vartheta) + 2 \sum_{m=1}^{n} (-i)^m \, n! \, [n + m)!]^{-1} \times P_n^m(\cos \vartheta) \cos (m\varphi),$$

(see also various types of such expansions in 4.5.1).

4.10 Gegenbauer functions, toroidal functions and conical functions

The Gegenbauer functions are solutions of Gegenbauer's differential equation

$$(z^2 - 1) \frac{d^2 w}{dz^2} + (2\nu + 1) \, z \frac{dw}{dz} - \alpha(\alpha + 2\nu) \, w = 0.$$

They can be expressed either as hypergeometric or as Legendre functions of the first kind

$$C_\alpha^\nu(z) = \Gamma(\alpha + 2\nu) \, [\Gamma(\alpha + 1) \, \Gamma(2\nu)]^{-1} \times F\left(\alpha + 2\nu, -\alpha; \nu + \frac{1}{2}; \frac{1}{2} - \frac{1}{2} z\right)$$

or

$$C_\alpha^\nu(z) = \pi^{\frac{1}{2}} 2^{\frac{1}{2}-\nu} \Gamma(\alpha+2\nu) [\Gamma(\alpha+1)\Gamma(\nu)]^{-1}$$

$$\times (z^2-1)^{\frac{1}{4}-\frac{1}{2}\nu} \mathfrak{P}^{\frac{1}{2}-\nu}_{\alpha+\nu-\frac{1}{2}}(z),$$

$$C_\alpha^\nu(0) = \pi^{\frac{1}{2}} \Gamma(\alpha+2\nu)\Gamma\left(\frac{1}{2}+\nu\right)$$

$$\times \left[\Gamma(1+\alpha)\Gamma(2\nu)\Gamma\left(\frac{1}{2}+\frac{1}{2}\alpha+\nu\right)\Gamma\left(\frac{1}{2}-\frac{1}{2}\alpha\right)\right]^{-1}.$$

Integral representations for the Gegenbauer functions $C_\alpha^\nu(z)$ can be immediately obtained from the formerly given list of integralrepresentations for the $\mathfrak{P}_\nu^\mu(z)$ as well as all previously given results for the $\mathfrak{P}_\nu^\mu(z)$ can be used to establish similar results for the $C_\alpha^\nu(z)$.

We mention especially the recurrence relations

$$(\alpha+2)\,C_{\alpha+2}^\nu(z) = 2(\nu+\alpha+1)\,z\,C_{\alpha+1}^\nu(z) - (2\nu+\alpha)\,C_\alpha^\nu(z),$$

$$\alpha\,C_\alpha^\nu(z) = 2\nu\,[z\,C_{\alpha-1}^{\nu+1}(z) - C_{\alpha-2}^{\nu+1}(z)],$$

$$(\alpha+2)\,C_\alpha^\nu(z) = 2\nu\,[C_\alpha^{\nu+1}(z) - z\,C_{\alpha-1}^{\nu+1}(z)],$$

$$\alpha\,C_\alpha^\nu(z) = (\alpha-1+2\nu)\,z\,C_{\alpha-1}^\nu(z) - 2\nu(1-z^2)\,C_{\alpha-2}^{\nu-1}(z),$$

$$\frac{d}{dz}\,C_\alpha^\nu(z) = 2\nu\,C_{\alpha-1}^{\nu+1}(z),$$

$$\sin(\pi\alpha+2\pi\nu)\,C_\alpha^\nu(z) = -\sin(\alpha\pi)\,C_{-\alpha-2\nu}^\nu(z),$$

$$\lim_{v\to 0}\Gamma(\nu)\,C_\alpha^\nu(\cos\varphi) = 2\alpha^{-1}\cos(\alpha\varphi),$$

$$\alpha \neq 0.$$

A second solution of Gegenbauer's differential equation is

$$D_\alpha^\nu(z) = 2^{-1-\alpha}\Gamma(\nu)\,\Gamma(2\nu+\alpha)\,[\Gamma(\nu+\alpha+1)]^{-1}$$

$$\times F\left(\frac{1}{2}\alpha+\nu, \frac{1}{2}+\nu+\frac{1}{2}\alpha; \nu+\alpha+1; z^2\right).$$

The most important case occurs when $\alpha = n$ ($n = 0, 1, 2, \ldots$). In this case the Gegenbauer function $C_\alpha^\nu(z)$ reduces to a polynomial of degree n. These polynomials $C_n^\nu(z)$ are denoted as Gegenbauer or ultraspherical polynomials. They are separately treated (see 5.3). The $C_n^\nu(z)$ are the only solutions of the Gegenbauer differential equation which remain finite at the points $z = \pm 1$.

Integrals

$$C_\alpha^\nu(z) = -\pi^{-1} \sin(\alpha\pi) \int_0^\infty (1 + 2tz + t^2)^{-\nu} t^{-\alpha-1},$$

$$-2 < Re\,\nu < Re\,\alpha < 0, \quad |\arg(z \pm 1)| < \pi,$$

$$(1 + 2tz + t^2)^{-\nu} = \frac{1}{2} i \int_{c-i\infty}^{c+i\infty} \operatorname{cosec}(\alpha\pi)\, t^\alpha C_\alpha^\nu(z)\, d\alpha,$$

$$-2 < Re\,\nu < c < 0.$$

Toroidal functions

The ring or toroidal functions, which arise when Laplace's differential equation is transformed into toroidal coordinates, are Legendre functions with the argument $\cosh z$. They are (see also 4.1)

$$\mathfrak{P}^\mu_{\nu-\frac{1}{2}}(\cosh z), \quad \mathfrak{Q}^\mu_{\nu-\frac{1}{2}}(\cosh z).$$

The properties of these functions can be obtained from the previous sections upon replacing ν by $\nu - \frac{1}{2}$ and z by $\cosh z$.

Conical functions

The conical functions, which arise when Laplace's differential equation is transformed into conical coordinates, are solutions of the following differential equation (see also 12.1, 12.2)

$$(1 - z^2)\, w'' - 2zw' - \left[p^2 + \frac{1}{4} + (1 - z^2)^{-1} \mu^2\right] w = 0$$

p is a parameter. Solutions of this equation are

$$\mathfrak{P}^\mu_{-\frac{1}{2}+ip}(z), \quad \mathfrak{Q}^\mu_{-\frac{1}{2}+ip}(z).$$

The conical functions for a real argument numerically less than unity are

$$P^\mu_{-\frac{1}{2}+ip}(\cos\vartheta), \quad Q^\mu_{-\frac{1}{2}+ip}(\cos\vartheta).$$

Again the principal properties can be obtained from the general results regarding $P_\nu^\mu(\cos\vartheta)$, $Q_\nu^\mu(\cos\vartheta)$ (see 4.3). The functions $P^\mu_{-\frac{1}{2}+i\lambda}(\cos\vartheta)$ and $\mathfrak{P}^\mu_{-\frac{1}{2}+i\lambda}(\cosh a)$ are even functions of λ. The leading term of the asymptotic expansions of these functions are for real $\lambda \gg 1$

$$P^\mu_{-\frac{1}{2}+i\lambda}(\cos\vartheta) \sim (2\pi \sin\vartheta)^{-\frac{1}{2}} \lambda^{\mu-1} e^{\lambda\mu},$$

$$\mathfrak{P}^\mu_{-\frac{1}{2}+i\lambda}(\cosh a) \sim \left(\frac{1}{2}\sinh a\right)^{-\frac{1}{2}} \lambda^{\mu-1} \cos\left(a\lambda + \frac{1}{2}\pi\mu - \frac{\pi}{4}\right).$$

A general expansion for the latter is

$$\mathfrak{P}^{\mu}_{-\frac{1}{2}+i\lambda}(\cosh a) = (2\pi \sinh a)^{-\frac{1}{2}}$$

$$\times \left\{ e^{-ia\lambda} \frac{\Gamma(-i\lambda)}{\Gamma\left(\frac{1}{2}-\mu-i\lambda\right)} {}_2F_1\left(\frac{1}{2}+\mu, \frac{1}{2}-\mu; 1+i\lambda; -\frac{1}{2}e^{-a}\operatorname{cosech} a\right)\right.$$

$$\left. + e^{ia\lambda} \frac{\Gamma(i\lambda)}{\Gamma\left(\frac{1}{2}-\mu+i\lambda\right)} {}_2F_1\left(\frac{1}{2}+\mu, \frac{1}{2}-\mu; 1-i\lambda; -\frac{1}{2}e^{-a}\operatorname{cosech} a\right)\right\}.$$

This function is the kernel of an integral inversion formula which is: If

$$g(y) = \int_0^\infty f(x)\, \mathfrak{P}^{\mu}_{-\frac{1}{2}+ix}(y)\, dx$$

then

$$f(x) = \pi^{-1} x \sinh(\pi x)\, \Gamma\left(\frac{1}{2}-\mu+ix\right) \Gamma\left(\frac{1}{2}-\mu-ix\right) \int_1^\infty g(y)$$

$$\times\, \mathfrak{P}^{\mu}_{-\frac{1}{2}+ix}(y)\, dy.$$

Conversely, if

$$g(y) = \int_1^\infty f(x)\, \mathfrak{P}^{\mu}_{-\frac{1}{2}+iy}(x)\, dx,$$

then

$$f(x) = \pi^{-1} \int_0^\infty y \sinh(\pi y)\, \Gamma\left(\frac{1}{2}-\mu+iy\right) \Gamma\left(\frac{1}{2}-\mu-iy\right) g(y)$$

$$\times\, \mathfrak{P}^{\mu}_{-\frac{1}{2}+iy}(x)\, dy.$$

These formulas are generalizations of the well known Mehler inversion formula (see also chap. XI).

4.11 Miscellaneous

Legendre functions expressed in terms of Bessel functions

$$\lim_{\nu\to\infty} \nu^{\mu}\, \mathfrak{P}_{\nu}^{-\mu}\left(\cosh\frac{z}{\nu}\right) = I_{\mu}(z)$$

$$\lim_{\nu\to\infty} \nu^{\mu}\, P_{\nu}^{-\mu}\left(\cos\frac{x}{\nu}\right) = J_{\mu}(z)$$

$$\lim_{\nu\to\infty} \nu^{-\mu}\, e^{-i\pi\mu}\, \mathfrak{Q}_{\nu}^{\mu}\left(\cosh\frac{z}{\nu}\right) = K_{\mu}(z)$$

$$\lim_{\nu\to\infty} \nu^{\mu}\, Q_{\nu}^{-\mu}\left(\cos\frac{x}{\nu}\right) = -\frac{1}{2}\pi Y_{\mu}(x).$$

Szego's expansion

$$P_\nu(\cos\vartheta) = (\vartheta \operatorname{cosec}\vartheta)^{\frac{1}{2}} \sum_{m=0}^{\infty} A_m(\vartheta)\left(\nu + \frac{1}{2}\right)^{-m} J_m\left[\left(\nu + \frac{1}{2}\right)\vartheta\right].$$

The $A_m(\vartheta)$ are elementary functions, regular in $0 \le Re\,\vartheta < \pi$. In particular $A_0 = 1$, $A_1 = \frac{1}{8}\left(\cot\vartheta - \frac{1}{\vartheta}\right)$.

This series is uniformly convergent in $0 \le \vartheta \le \theta_0 - \varepsilon$, where $\varepsilon > 0$ and $\vartheta_0 = 2\pi\left(\sqrt{2} - 1\right) = (0.828\ldots)\,\pi$.

Some inequalities

If ν and μ are real and $\nu \ge 1$, $\nu - \mu + 1 > 0$, $\mu > 0$ then

$$|P_\nu^{\pm\mu}(\cos\vartheta)| < \Gamma(\nu \pm \mu + 1)\left(\frac{1}{8}\pi\nu\right)^{-\frac{1}{2}} \frac{(\operatorname{cosec}\vartheta)^{\frac{1}{2}+\mu}}{\Gamma(1+\nu)},$$

$$|Q_\nu^{\pm\mu}(\cos\vartheta)| < \Gamma(\nu \pm \mu + 1)\left(2\,\frac{\pi}{\nu}\right)^{\frac{1}{2}} \frac{(\operatorname{cosec}\vartheta)^{\frac{1}{2}+\mu}}{\Gamma(1+\nu)},$$

$$|P_\nu^{\pm m}(\cos\vartheta)| < 2\Gamma(\nu \pm m + 1)\,(\pi\nu)^{-\frac{1}{2}} \frac{(\operatorname{cosec}\vartheta)^{\frac{1}{2}+m}}{\Gamma(1+\nu)},$$

$$|Q_\nu^{\pm m}(\cos\vartheta)| \le \Gamma(\nu \pm m + 1)\left(\frac{\pi}{\nu}\right)^{\frac{1}{2}} \frac{(\operatorname{cosec}\vartheta)^{\frac{1}{2}+m}}{\Gamma(1+\nu)},$$

$$m = 0, 1, 2, 3, \ldots$$

Literature

ERDÉLYI, A.: Higher transcendental functions, Vol. 1. New York: McGraw-Hill 1953.

HOBSON, E. W.: The theory of spherical and ellipsoidal harmonics. Cambridge 1931.

HEINE, E.: Theorie der Kugelfunktionen, 2. vols. Berlin: G. Riemer 1878, 1881.

LENSE, J.: Kugelfunktionen. Leipzig: Akademische Verlagsgesellschaft 1950.

MACROBERT, T. M.: Spherical harmonics. Methuen 1947.

PRASHAD, G.: A treatise on spherical harmonics and the functions of Bessel and Lamé. Benares Math. Soc. 2 vols. 1930, 1932.

ROBIN, L.: Fonctions spheriques de Legendre et fonctions sphéroidales, 3 vols. Paris: Gauthier-Villars 1957—1959.

SNOW, C.: Hypergeometric and Legendre functions with applications to integral equations and potential theory. National Bureau of Standards, Washington, D. C. 1951.

WHITTAKER, E. T., and G. N. WATSON: A course of modern analysis. Cambridge 1944.

Chapter V

Orthogonal polynomials

5.1 Orthogonal systems

5.1.1 General systems

A set of functions $\{\varphi_n(x)\}$, real or complex valued, defined over an interval (a, b) is said to be linearly independent if

$$a_1\varphi_1(x) + a_2\varphi_2(x) + \cdots a_n\varphi_n(x) \equiv 0$$

is true only when $a_1 = a_2 = \cdots = a_n = 0$. If $\alpha(x)$ is a non-decreasing function defined over (a, b), it is possible to define a numerical-valued function (φ_i, φ_j) for every pair φ_i, φ_j of functions by a stieltjes integral

$$(\varphi_i, \varphi_j) = \int_a^b \varphi_i(x)\, \bar{\varphi}_j(x)^*\, d\alpha(x)$$

provided that the integral on the r. h. s. exists. The quantity (φ_i, φ_j) is called the inner product. As a special case if $\alpha(x)$ is absolutely continuous with $\frac{d\alpha}{dx} = w(x)$, the above definition of (φ_i, φ_j) may be written as

$$(\varphi_i, \varphi_j) = \int_a^b \varphi_i(x)\, \bar{\varphi}_j(x)\, w(x)\, dx$$

provided that $\varphi_n(x)\,[w(x)]^{\frac{1}{2}}$ are square-integrable. The function $w(x)$ is called the weight function associated with the interval (a, b). Any two functions $f(x)$, $g(x)$ defined over (a, b) and with weight function $w(x)$, are said to be orthogonal if

$$(f, g) = \int_a^b f(x)\, \bar{g}(x)\, w(x)\, dx = 0.$$

A set of functions $\{\varphi_n(x)\}$, defined over an interval (a, b) and with the weight function $w(x)$, is said to be an orthogonal system if $(\varphi_i, \varphi_j) = 0$ for $i \neq j$. A family of functions $\{\varphi_n(x)\}$ is said to be orthonormal if

$$(\varphi_i, \varphi_j) = \delta_{ij}.$$

If it is assumed that the set of functions $\{\varphi_n(x)\}$ does not include null functions, then $(\varphi_i, \varphi_i) \neq 0$ and thus the set $\{\varphi_n(x)\}$ may be orthogonalized. Conversely, the functions of any finite subset of an orthogonal system is linearly independent. It is well known that any countable set of linearly independent functions $\{\varphi_n(x)\}$ can be orthonormalized with respect to the inner product. Finally, the orthonormalized set $\{\psi_n(x)\}$

---

* $\bar{\varphi}(x)$ is the complex conjugate of $\varphi(x)$.

obtained from $\{\varphi_n(x)\}$, the weight function $w(x)$ and the interval (a, b) is not necessarily unique.

Let $L^2(a, b; w)$ be the class of all functions $f(x)$ for which the integral

$$\int_a^b [f(x)]^2 w(x)\, dx$$

exists and is finite. If $\{\varphi_n(x)\}$ is any orthonormal set of functions such that $\varphi_n(x) \in L^2(a, b; w)$, then it is known that an arbitrary function $f(x) \in L^2(a, b; w)$ can be expressed in terms of a generalized Fourier series

$$f(x) \sim \sum_{n=1}^{\infty} a_n \varphi_n(x),$$

where

$$a_n = \int_a^b f(x)\, \bar{\varphi}_n(x)\, w(x)\, dx.$$

In approximating $f(x) \in L^2(a, b; w)$ by $\sum_{i=1}^{n} b_i \varphi_j(x)$ in the square-mean sense, it can be proved that the best possible choice for the coefficients b_i is that of the Fourier coefficients. In other words

$$\int_a^b \left[f(x) - \sum_{i=1}^{n} a_i \varphi_i(x)\right]^2 w(x)\, dx \leq \int_a^b [f(x) - \sum b_i \varphi_i(x)]^2 w(x)\, dx.$$

An orthonormal system $\{\varphi_n(x)\}$ is said to be complete if $f(x) \equiv 0$ whenever $(f, \varphi_n) = 0$ for all n.

An orthonormal system is said to be closed in $L^2(a, b; w)$ if

$$\sum_{n=1}^{\infty} |a_n|^2 = \int_a^b [f(x)]^2 w(x)\, dx$$

for every $f(x) \in L^2(a, b; w)$ where

$$a_n = \int_a^b f(x)\, \bar{\varphi}_n(x)\, w(x)\, dx.$$

Finally, if $\sum_{n=1}^{\infty} a_n \varphi_n(x)$ is the generalized Fourier series associated with $f(x)$, i.e.

$$a_n = \int_a^b f(x)\, \bar{\varphi}_n(x)\, w(x)\, dx,$$

then the Bessel's inequality

$$\sum_{n=1}^{\infty} |a_n|^2 \leq \int_a^b [f(x)]^2 w(x)\, dx$$

is true.

For a detailed treatment of orthogonal functions see KACZMARZ — and STEINHAUS (1951), SANSONE (1959) and TRICOMI (1955).

5.1.2 General orthogonal polynomials

This section deals with some properties of orthogonal polynomials in general.

Apart from a constant factor in each polynomial, the interval (a, b) and the weight function $w(x)$ determine the system of orthogonal polynomials uniquely. Let $\{p_n(x)\}$ be any such system with the weight function $w(x)$ over the interval (a, b), then

1. Any polynomial of degree $m < n$ is orthogonal to $p_n(x)$.

2. All zeros of $p_n(x)$ are simple and located in the open interval (a, b).

3. If α, β $(\alpha < \beta)$ be any two consecutive zeros of $p_n(x)$, i.e. $p_n(x) \neq 0$ for any value of x in the open interval (α, β), then there is exactly one zero of $p_{n+1}(x)$ in the open interval (α, β).

4. For each $m > n$, there is at least one zero of $p_m(x)$ between any two consecutive zeros of $p_n(x)$.

5. Any three consecutive polynomials are connected by a linear relation, i.e. there exist constants A_n, B_n and C_n such that

$$p_{n+1}(x) = (A_n x + B_n)\, p_n(x) - C_n p_{n-1}(x),$$

where A_n, B_n, C_n are functions of n:

6. If k_n is the coefficient of x^n in $p_n(x)$ and

$$h_n = (p_n, p_n) = \int_a^b [p_n(x)]^2\, w(x)\, dx,$$

then

$$\sum_{m=0}^{n} \frac{1}{h_m} p_m(x)\, p_m(y) = \frac{k_n [p_{n+1}(x)\, p_n(y) - p_n(x)\, p_{n+1}(y)]}{k_{n+1}\, h_n (x - y)}.$$

In the limit as $y \to x$.

$$\sum_{m=0}^{n} \frac{1}{h_m} [p_m(x)]^2 = \frac{k_n}{h_n\, k_{n+1}} [p_n(x)\, p'_{n+1}(x) - p'_n(x)\, p_{n+1}(x)].$$

7. If $\pi_n(x)$ is any polynomial of degree n with the leading term x^n, i.e. the coefficient of x^n is unity, then

$$\int_a^b [\pi_n(x)]^2\, w(x)\, dx$$

is minimum if and only if

$$\pi_n(x) = \varepsilon k_n^{-1} p_n(x),$$

where ε is a constant and $|\varepsilon| = 1$.

8. Let $k_n(x, y)$ be the polynomials defined by

$$K_n(x, y) = \sum_{m=0}^{n} \frac{1}{h_m} \overline{p_m(x)}\, p_m(y).$$

(i) For any finite value $x_0 \leq a$, the polynomials $K_n(x_0, x)$ are orthogonal over the interval (a, b) with respect to the weight function $(x - x_0)\, w(x)$.

(ii) If $\pi_n(x)$ is an arbitrary polynomial of degree n such that

$$\int_a^b [\pi_n(x)]^2\, w(x)\, dx = 1$$

and x_0 is an arbitrary complex constant, the maximum of $|\pi_n(x_0)|^2$ is given by the polynomial

$$\pi_n(x) = \varepsilon\, [K_n(x_0, x_0)]^{-\frac{1}{2}}\, K_n(x_0, x),$$

where $|\varepsilon| = 1$.

9. If the interval (a, b) is finite, the set of orthogonal polynomials $\{p_n(x)\}$, $n = 0, 1, 2, \ldots$, associated with the weight function $w(x)$ is closed in $L^2(a, b; w)$.

5.1.3 The classical polynomials

The orthogonal polynomials of JACOBI, GEGENBAUER, HERMITE, CHEBYSHEV, LAGUERRE and LEGENDRE are commonly called the "classical polynomials" and are treated individually in the sections that follow. These polynomials have a number of properties common to all of them.

(i) $\{p_n'(x)\}$ is a system of orthogonal polynomials.

(ii) $p_n(x)$ satisfies a second order linear differential equation of the form

$$A(x) \frac{d^2 y}{dx^2} + B(x) \frac{dy}{dx} + \lambda_n y = 0,$$

where $A(x)$ and $B(x)$ are independent of n, and λ_n is independent of x.

(iii) there is a generalized Rodrigues' formula

$$p_n(x) = \frac{1}{K_n \cdot w(x)} \frac{d^n}{dx^n} [w(x)\, X^n],$$

where X is a polynomial in x with coefficients independent of n and K_n is a constant (dependent on n).

(iv) If k_n is the coefficient of x^n in $p_n(x)$ and K_n is the constant in Rodrigues' formula, then

$$(-1)^n k_n K_n > 0.$$

Furthermore any one of the properties (i)—(iii) characterizes the classical polynomials in the sense that any system of orthogonal polynomials, for which one of the properties (i)—(iii) is valid, can be reduced to a system of classical polynomials.

Besides the classical polynomials represent solutions of certain Sturm-Liouville Eigenvalue problems as listed below:

Differential equation	Boundary conditions	Eigenvalues	Eigensolutions
$(1-x^2)\,w'' - 2x\,w' + \lambda w = 0$	w finite at $x = \pm 1$	$\lambda_n = n(n+1)$	$w = P_n(x)$ (Legendre)
$(1-x^2)\,w'' - (2v+1)\,x\,w' + \lambda(\lambda + 2v)\,w = 0$	w finite at $x = \pm 1$	$\lambda_n = n$	$w = C_n^{v}(x)$ (Gegenbauer)
$(1-x^2)\,w'' + [\beta - \alpha - (\alpha + \beta + 2)\,x]\,w' + \lambda(\lambda + \alpha + \beta + 1)\,w = 0$	w finite at $x = \pm 1$	$\lambda_n = n$	$w = P_n^{(\alpha,\beta)}(x)$ (Jacobi)
$(1-x^2)\,w'' - x\,w' + \lambda w = 0$	w finite at $x = \pm 1$	$\lambda_n = n^2$	$w = T_n(x)$ (Chebycheff)
$w'' - 2x\,w' + \lambda w = 0$	At $x = \pm\infty$ w behaves Like a finite power of x	$\lambda_n = 2n$	$w = He_n(x)$ (Hermite)
$x\,w'' + (\alpha + 1 - x)\,w' + \lambda w = 0$	Behavior of w; At $x = 0$, finite; at $x = +\infty$ like a finite power of x	$\lambda_n = n$	$w = L_n^{(\alpha)}(x)$ (generalised Laguerre)

The following table gives a survey about the orthogonal property of the classical polynomials. The polynomials listed in the first column are such that

$$\int_a^b w(x)\, p_n(x)\, p_m(x)\, dx = \begin{cases} 0 & m \neq n \\ 1 & m = n \end{cases}$$ (The set $p_n(x)$ form an orthonormal system)

Orthonormal polynomial $p_n(x)$	Interval of orthogonality (a, b)	Weight function $w(x)$
$\left(n + \frac{1}{2}\right)^{\frac{1}{2}} P_n(x)$ (LEGENDRE)	$(-1, +1)$	1
$2^\nu \Gamma(\nu) \left[\frac{n!\,(n+\nu)}{2\pi\, \Gamma(n+2\nu)}\right]^{\frac{1}{2}} C_n^\nu(x)$ (GEGENBAUER)	$(-1, +1)$	$(1-x^2)^{\nu - \frac{1}{2}}$
$\left[\frac{n!\, \Gamma(\alpha+\beta+n+1)\,(\alpha+\beta+2n+1)}{2^{\alpha+\beta+1}\, \Gamma(\alpha+n+1)\, \Gamma(\beta+n+1)}\right]^{\frac{1}{2}} \times P_n^{(\alpha,\beta)}(x)$ (JACOBI)	$(-1, +1)$	$(1-x)^\alpha (1+x)^\beta$
$\left(\frac{E_n}{\pi}\right)^{\frac{1}{2}} T_n(x)$ (CHEBYCHEFF)	$(-1, +1)$	$(1-x^2)^{-\frac{1}{2}}$
$(2\pi)^{-\frac{1}{4}} (n!)^{-\frac{1}{2}} He_n(x)$ (HERMITE)	$(-\infty, +\infty)$	$e^{-\frac{1}{2}x^2}$
$\left[\frac{n!}{\Gamma_n(1+\alpha+n)}\right]^{\frac{1}{2}} L_n^{(\alpha)}(x)$ (LAGUERRE)	$(0, +\infty)$	$x^\alpha e^{-x}$

5.2 Jacobi polynomials

5.2.1 Definition, notation, and special cases

The Jacobi polynomials $P_n^{(\alpha,\beta)}(x)$ are the polynomials orthogonal over the interval $(-1, +1)$ with the weight function $w(x) = (1-x)^\alpha (1+x)^\beta$ and normalized by the relation

$$P_n^{(\alpha,\beta)}(1) = \binom{n+\alpha}{n} = \frac{(\alpha+1)_n}{n!}.$$

It is assumed that $\alpha > -1$, $\beta > -1$. COURANT and HILBERT (1953) consider the orthogonal polynomials associated with the weight function $w(x) = x^{q-1}(1-x)^{p-q}$, $q > 0$, $p - q > -1$, over the interval $(0, 1)$ and denote it by $G_n(p, q, x)$. More generally, the orthogonal polynomials associated with the weight function $w(x) = (x-a)^\alpha (b-x)^\beta$ on the finite interval (a, b) can be written in the form of constant multiples of $P_n^{(\alpha,\beta)}\left(2\frac{x-a}{b-a} - 1\right)$. If $G_n(p, q, x)$ is normalized such that $k_n = 1$, the following relations connect the polynomials $P_n^{(\alpha,\beta)}(x)$ and $G_n(p, q, x)$

$$G_n(p, q, x) = \frac{n!\,\Gamma(n+p)}{\Gamma(2n+p)} P_n^{(p-q, q-1)}(2x-1),$$

$$P_n^{(\alpha,\beta)}(x = \frac{\Gamma(\alpha+\beta+2n+1)}{n!\,\Gamma(\alpha+\beta+n+1)} G_n\left(1+\alpha+\beta, 1+\beta, \frac{x+1}{2}\right).$$

Special cases and relations

$$P_n^{(\alpha,\beta)}(-x) = (-1)^n P_n^{(\beta,\alpha)}(x),$$

$$\binom{n}{l} P_n^{(-l,\beta)}(x) = \binom{n+\beta}{l}\left(\frac{x-1}{2}\right)^l P_{n-l}^{(l,\beta)}(x),$$

$$P_n^{(\alpha,\beta)}(-1) = (-1)^n \binom{n+\beta}{n},$$

$$P_0^{(\alpha,\beta)}(x) = 1,$$

$$P_1^{(\alpha,\beta)}(x) = \frac{\alpha-\beta}{2} + \frac{1}{2}(\alpha+\beta+2)\,x,$$

$$P_2^{(\alpha,\beta)}(x) = \frac{1}{4}\left\{\frac{(1+\alpha)(2+\alpha) + (1+\beta)(2+\beta) - 2(\alpha+2)(\beta+2)}{2}\right.$$
$$+ x[(1+\alpha)(2+\alpha) - (1+\beta)(2+\beta)]$$
$$\left. + x^2\left[\frac{(1+\alpha)(2+\alpha) + (1+\beta)(2+\beta) + 2(\alpha+2)(\beta+2)}{2}\right]\right\}.$$

The Gegenbauer polynomials $C_n^\lambda(x)$; the Chebyshev (Tschebyscheff) polynomials $T_n(x)$, $U_n(x)$ and the Legendre polynomials $P_n(x)$ are special cases of the Jacobi polynomials given by the following relations:

$$C_n^\lambda(x) = \frac{\Gamma\left(\lambda+\frac{1}{2}\right)\Gamma(2\lambda+n)}{\Gamma(2\lambda)\,\Gamma\left(\frac{1}{2}+\lambda+n\right)} P_n^{\left(\lambda-\frac{1}{2},\lambda-\frac{1}{2}\right)}(x),$$

$$T_n(x) = \frac{2\cdot 4\cdot 6\cdots(2n)}{1\cdot 3\cdot 5\cdots(2n-1)} P_n^{\left(-\frac{1}{2},-\frac{1}{2}\right)}(x),$$

$$= 2^{2n}\frac{(n!)^2}{(2n)!} P_n^{\left(-\frac{1}{2},-\frac{1}{2}\right)}(x),$$

$$U_n(x) = \frac{1}{2} \frac{2 \cdot 4 \cdot 6 \cdots (2n+2)}{1 \cdot 3 \cdot 5 \cdots (2n+1)} P_n^{\left(\frac{1}{2},\frac{1}{2}\right)}(x),$$

$$P_n(x) = P_n^{(0,0)}(x),$$

$$P_n^{\left(-\frac{1}{2},-\frac{1}{2}\right)}(\cos\theta) = \frac{1 \cdot 3 \cdot 5 \cdots (2n-1)}{2 \cdot 4 \cdot 6 \cdots (2n)} \cos n\theta,$$

$$P_n^{\left(-\frac{1}{2},\frac{1}{2}\right)}(\cos\theta) = \frac{1 \cdot 3 \cdot 5 \cdots (2n-1)}{2 \cdot 4 \cdot 6 \cdots (2n)} \frac{\cos\left[\left(n+\frac{1}{2}\right)\theta\right]}{\cos\left(\frac{\theta}{2}\right)},$$

$$P_n^{\left(\frac{1}{2},-\frac{1}{2}\right)}(\cos\theta) = \frac{1 \cdot 3 \cdot 5 \cdots (2n-1)}{2 \cdot 4 \cdot 6 \cdots (2n)} \frac{\sin\left[\left(n+\frac{1}{2}\right)\theta\right]}{\sin\left(\frac{\theta}{2}\right)},$$

$$P_n^{\left(\frac{1}{2},\frac{1}{2}\right)}(\cos\theta) = 2\,\frac{1 \cdot 3 \cdot 5 \cdots (2n+1)}{2 \cdot 4 \cdot 6 \cdots (2n+2)} \frac{\sin\left[(n+1)\,\theta\right]}{\sin\theta}.$$

The polynomials $P_n^{(\alpha,\alpha)}(x)$ are even or odd according as n is even or odd.

Furthermore

$$P_{2n}^{(\alpha,\alpha)}(x) = \frac{\Gamma(n+1)\,\Gamma(\alpha+2n+1)}{\Gamma(2n+1)\,\Gamma(\alpha+n+1)} P_n^{\left(\alpha,-\frac{1}{2}\right)}(2x^2-1)$$

$$= (-1)^n \frac{\Gamma(n+1)\,\Gamma(\alpha+2n+1)}{\Gamma(2n+1)\,\Gamma(\alpha+n+1)} P_n^{\left(-\frac{1}{2},\alpha\right)}(1-2x^2),$$

$$P_{2n+1}^{(\alpha,\alpha)}(x) = \frac{\Gamma(n+1)\,\Gamma(\alpha+2n+2)}{\Gamma(2n+2)\,\Gamma(\alpha+n+1)}\, x\, P_n^{\left(\alpha,\frac{1}{2}\right)}(2x^2-1)$$

$$= (-1)^n \frac{\Gamma(n+1)\,\Gamma(\alpha+2n+2)}{\Gamma(2n+2)\,\Gamma(\alpha+n+1)}\, x\, P_n^{\left(\frac{1}{2},\alpha\right)}(1-2x^2).$$

5.2.2 Elementary results

Explicit expressions for $P_n^{(\alpha,\beta)}(x)$

$$P_n^{(\alpha,\beta)}(x) = 2^{-n} \sum_{r=0}^{n} \binom{n+\alpha}{r} \binom{n+\beta}{n-r} (x+1)^r (x-1)^{n-r}$$

$$= \frac{\Gamma(\alpha+n+1)}{n!\,\Gamma(\alpha+\beta+n+1)} \sum_{r=0}^{n} \binom{n}{r} \frac{\Gamma(\alpha+\beta+n+r+1)}{\Gamma(\alpha+r+1)} \left(\frac{x-1}{2}\right)^r.$$

Representation by hypergeometric series

$$P_n^{(\alpha,\beta)}(x) = \binom{n+\alpha}{n} {}_2F_1\left(-n, \alpha+\beta+n+1; \alpha+1; \frac{1-x}{2}\right).$$

$$= \binom{n+\alpha}{n} \left(\frac{1+x}{2}\right)^n {}_2F_1\left(-n, -\beta-n; \alpha+1; \frac{x-1}{x+1}\right).$$

$$= (-1)^n \binom{n+\beta}{n} {}_2F_1\left(-n, \alpha+\beta+n+1; \beta+1; \frac{x+1}{2}\right).$$

$$= \binom{n+\beta}{n} \left(\frac{x-1}{2}\right)^n {}_2F_1\left(-n, -\alpha-n; \beta+1; \frac{x+1}{x-1}\right).$$

$$= \binom{n+\alpha}{n} \left(\frac{1+x}{2}\right)^{-\beta} {}_2F_1\left(\alpha+n+1, -\beta-n; \alpha+1; \frac{1-x}{2}\right).$$

$$= \binom{n+\beta}{n} \left(\frac{1-x}{2}\right)^{-\alpha} (-1)^n$$

$$\times {}_2F_1\left(\beta+n+1, -n-\alpha; \beta+1; \frac{x+1}{2}\right).$$

$$= \binom{\alpha+\beta+2n}{n} \left(\frac{x-1}{2}\right)^n$$

$$\times {}_2F_1\left(-n, -\alpha-n; -2n-\alpha-\beta; \frac{2}{1-x}\right).$$

$$= \binom{\alpha+\beta+2n}{n} \left(\frac{x+1}{2}\right)^n$$

$$\times {}_2F_1\left(-n, -\beta-n; -2n-\alpha-\beta; -\frac{2}{1+x}\right).$$

More results can be obtained by applying transformation formulas for ${}_2F_1(a, b; c; z)$.

An elementary integral

$$2n \int_0^x (1-t)^\alpha (1+t)^\beta P_n^{(\alpha,\beta)}(t)\, dt$$

$$= P_{n-1}^{(\alpha+1,\beta+1)}(0) - (1-x)^{\alpha+1} (1+x)^{\beta+1} P_{n-1}^{(\alpha+1,\beta+1)}(x).$$

Orthogonality relations

$$\int_{-1}^{1} P_n^{(\alpha,\beta)}(x)\, P_m^{(\alpha,\beta)}(x)\, (1-x)^\alpha (1+x)^\beta\, dx$$

$$= \begin{cases} 0, & m \neq n, \\ \dfrac{\Gamma(\alpha+n+1)\,\Gamma(\beta+n+1)}{n!\,\Gamma(\alpha+\beta+n+1)} \cdot \dfrac{2^{\alpha+\beta+1}}{\alpha+\beta+2n+1}, & m = n, \end{cases}$$

$$\int_0^1 G_n(p, q, x)\, G_m(p, q, x)\, x^{q-1} (1-x)^{p-q} dx$$

$$= \begin{cases} 0, & m \neq n, \\ n!\, \dfrac{\Gamma(p+n)\,\Gamma(q+n)\,\Gamma(p-q+n+1)}{(2n+p)\,[\Gamma(p+2n)]^2}, & m = n. \end{cases}$$

Differentiation formulas

$$\frac{d^m}{dx^m} P_n^{(\alpha,\beta)}(x) = 2^{-m}(\alpha+\beta+n+1)_m P_{n-m}^{(\alpha+m,\beta+m)}(x),$$

$$\frac{d}{dx} P_n^{(\alpha,\beta)}(x) = \frac{\alpha+\beta+n+1}{2} P_{n-1}^{(\alpha+1,\beta+1)}(x),$$

$$(1-x^2)(\alpha+\beta+2n)\frac{d}{dx} P_n^{(\alpha,\beta)}(x) = n[\alpha-\beta-(\alpha+\beta+2n)x]\,P_n^{(\alpha,\beta)}(x) + 2(\alpha+n)(\beta+n)\,P_{n-1}^{(\alpha,\beta)}(x).$$

Contiguous relations

$$\left(\frac{\alpha}{2}+\frac{\beta}{2}+n+1\right)(1-x)\,P_n^{(\alpha+1,\beta)}(x) = (\alpha+n+1)\,P_n^{(\alpha,\beta)}(x) - (n+1)\,P_{n+1}^{(\alpha,\beta)}(x),$$

$$\left(\frac{\alpha}{2}+\frac{\beta}{2}+n+1\right)(1+x)\,P_n^{(\alpha,\beta+1)}(x) = (\beta+n+1)\,P_n^{(\alpha,\beta)}(x) + (n+1)\,P_{n+1}^{(\alpha,\beta)}(x),$$

$$(1-x)\,P_n^{(\alpha+1,\beta)}(x) + (1+x)\,P_n^{(\alpha,\beta+1)}(x) = 2\,P_n^{(\alpha,\beta)}(x),$$

$$(\alpha+\beta+2n)\,P_n^{(\alpha-1,\beta)}(x) = (\alpha+\beta+n)\,P_n^{(\alpha,\beta)}(x) - (\beta+n)\,P_{n-1}^{(\alpha,\beta)}(x),$$

$$(\alpha+\beta+2n)\,P_n^{(\alpha,\beta-1)}(x) = (\alpha+\beta+n)\,P_n^{(\alpha,\beta)}(x) + (\alpha+n)\,P_{n-1}^{(\alpha,\beta)}(x),$$

$$P_n^{(\alpha,\beta-1)}(x) = P_n^{(\alpha-1,\beta)}(x) + P_{n-1}^{(\alpha,\beta)}(x).$$

Recurrence relations

$$2n(\alpha+\beta+n)(\alpha+\beta+2n-2)\,P_n^{(\alpha,\beta)}(x) = [(\alpha+\beta+2n-2)_3 + (\alpha^2-\beta^2)(\alpha+\beta+2n-1)]\,P_{n-1}^{(\alpha,\beta)}(x) - 2(\alpha+n-1)(\beta+n-1)(\alpha+\beta+2n)\,P_{n-2}^{(\alpha,\beta)}(x),$$

$$n = 2, 3, 4, \ldots$$

Rodrigues' formula

$$P_n^{(\alpha,\beta)}(x) = \frac{(-1)^n}{n!\,2^n}\,\frac{1}{(1-x)^\alpha(1+x)^\beta}\,\frac{d^n}{dx^n}[(1-x)^{\alpha+n}(1+x)^{\beta+n}].$$

Generating functions

$$\sum_{n=0}^{\infty} P_n^{(\alpha,\beta)}(x)\,z^n = \frac{2^{\alpha+\beta}}{R}(1-z+R)^{-\alpha}(1+z+R)^{-\beta},$$

$$|z| < 1,$$

where

$$R = (1 - 2\,xz + z^2)^{\frac{1}{2}}$$

and $\quad R = 1 \quad$ when $\quad z = 0$

$$\sum_{n=0}^{\infty} \frac{P_n^{(\alpha,\beta)}(x)}{(1+\alpha)_n} \frac{z^n}{(1+\beta)_n} = {}_0F_1\left(;1+\alpha;\frac{x-1}{2}z\right){}_0F_1\left(;1+\beta;\frac{x+1}{2}z\right),$$

$$\sum_{n=0}^{\infty} \frac{(1+\alpha+\beta)_n}{(1+\alpha)_n} P_n^{(\alpha,\beta)}(x)\, z^n$$

$$= (1-t)^{-\alpha-\beta-1}\, {}_2F_1\left(\frac{\alpha+\beta+1}{2}, \frac{\alpha+\beta+2}{2}; 1+\alpha; \frac{2(x-1)\,z}{(1-z)^2}\right),$$

$$\sum_{n=0}^{\infty} \frac{(\alpha+\beta+1-\gamma)_n\,(\gamma)_n\,P_n^{(\alpha,\beta)}(x)}{(\alpha+1)_n\,(\beta+1)_n} z^n$$

$$= {}_2F_1\left(\gamma, \alpha+\beta+1-\gamma; 1+\alpha; \frac{1-z-R}{2}\right)$$

$$\times\, {}_2F_1\left(\gamma, \alpha+\beta+1-\gamma; 1+\beta; \frac{1+z-R}{2}\right),$$

where $R = (1 - 2xz + z^2)^{\frac{1}{2}}$ and γ is arbitrary.

The differential equation.

The polynomials $P_n^{(\alpha,\beta)}(x)$ satisfy the differential equation

$$(1-x^2)\,y'' + [\beta - \alpha - (\alpha+\beta+2)\,x]\,y' + n(\alpha+\beta+n+1)\,y = 0.$$

The following differential equations are transformations of the differential equation of Jacobi polynomials

$$(1-x^2)\frac{d^2u}{dx^2} + [\alpha - \beta + (\alpha+\beta-2)\,x]\frac{du}{dx} + (n+1)\,(\alpha+\beta+n)\,u = 0,$$

$$u = (1-x)^{\alpha}\,(1+x)^{\beta}\,P_n^{(\alpha,\beta)}(x),$$

$$\frac{d^2u}{dx^2} + \left[\frac{1}{4}\frac{1-\alpha^2}{(1-x)^2} + \frac{1}{4}\frac{1-\beta^2}{(1+x)^2} + \frac{2n(\alpha+\beta+n+1)+(\alpha+1)\,(\beta+1)}{2(1-x^2)}\right]u = 0,$$

$$u = (1-x)^{\frac{\alpha+1}{2}}\,(1+x)^{\frac{1+\beta}{2}}\,P_n^{(\alpha,\beta)}(x),$$

$$\frac{d^2u}{dx^2} + \left[\frac{1-4\alpha^2}{16\sin^2\left(\frac{x}{2}\right)} + \frac{1-4\beta^2}{16\cos^2\left(\frac{x}{2}\right)} + \left(n + \frac{\alpha+\beta+1}{2}\right)^2\right]u = 0,$$

$$u = \left(\sin\frac{x}{2}\right)^{\alpha+\frac{1}{2}}\left(\cos\frac{x}{2}\right)^{\beta+\frac{1}{2}} P_n^{(\alpha,\beta)}(\cos x).$$

A second solution of the differential equation of Jacobi polynomials is given by the function $Q_n^{(\alpha,\beta)}(x)$ defined by

$$Q_n^{(\alpha,\beta)}(x) = \frac{\Gamma(\alpha+n+1)\,\Gamma(\beta+n+1)\,2^{\alpha+\beta+n}}{\Gamma(\alpha+\beta+2n+2)\,(x+1)^{\beta}\,(x-1)^{\alpha+n+1}} \times {}_2F_1\left(n+1, \alpha+n+1; \alpha+\beta+2n+2; \frac{2}{1-x}\right)$$

and is called the Jacobi function of the second kind

$$Q_n^{(\alpha,\beta)}(x) = -\frac{1}{2}(x-1)^{-\alpha}(1+x)^{-\beta}\,q_n^{(\alpha,\beta)}(x) + Q_0^{(\alpha,\beta)}(x)\,P_n^{(\alpha,\beta)}(x),$$

where

$$q_n^{(\alpha,\beta)}(x) = \int_{-1}^{1}(t-x)^{-1}(1-t)^{\alpha}(1+t)^{\beta}\,[P_n^{(\alpha,\beta)}(t) - P_n^{(\alpha,\beta)}(x)]\,dt,$$

and x is in the complex plane cut along the real axis from -1 to $+1$. On the cut itself

$$Q_n^{(\alpha,\beta)}(\xi) = \frac{1}{2}[Q_n^{(\alpha,\beta)}(\xi+i0) - Q_n^{(\alpha,\beta)}(\xi-i0)], \qquad -1<\xi<1.$$

Other relations connecting $P_n^{(\alpha,\beta)}(x)$ and $Q_n^{(\alpha,\beta)}(x)$ are

$$P_n^{(\alpha,\beta)}(x)\,Q_{n-1}^{(\alpha,\beta)}(x) - Q_n^{(\alpha,\beta)}(x)\,P_{n-1}^{(\alpha,\beta)}(x) = 2^{\alpha+\beta-1}(\alpha+\beta+2n)\frac{\Gamma(\alpha+n)\,\Gamma(\beta+n)}{n!\,\Gamma(\alpha+\beta+n+1)}(x-1)^{-\alpha}(x+1)^{-\beta},$$

$$P_n^{(\alpha,\beta)}(x)\frac{d}{dx}Q_n^{(\alpha,\beta)}(x) - Q_n^{(\alpha,\beta)}(x)\frac{d}{dx}P_n^{(\alpha,\beta)}(x) = -2^{\alpha+\beta}\frac{\Gamma(\alpha+n+1)\,\Gamma(\beta+n+1)}{n!\,\Gamma(\alpha+\beta+n+1)}(x-1)^{-\alpha-1}(x+1)^{-\beta-1}.$$

For more details about the function $Q_n^{(\alpha,\beta)}(x)$ see SZEGÖ (1959).

Integral representation

$$P_n^{(\alpha,\beta)}(x) = \frac{1}{2\pi i}\int^{(x+)}\left(\frac{1}{2}\frac{z^2-1}{z-x}\right)^n\left(\frac{1-z}{1-x}\right)^{\alpha}\left(\frac{1+z}{1+x}\right)^{\beta}dz,$$

where $x \neq \pm 1$ and the contour of integration is a simple closed contour, in the positive sense, around $z=x$. It is further assumed that the points $z=\pm 1$ are outside the contour, and $\left(\frac{1-z}{1-x}\right)^{\alpha}$ and $\left(\frac{1+z}{1+x}\right)^{\beta}$ are to be taken as unity when $z=x$

$$Q_n^{(\alpha,\beta)}(x) = 2^{-n-1}(x-1)^{-\alpha}(x+1)^{-\beta} \times \int_{-1}^{1}(x-t)^{-n-1}(1-t)^{\alpha+n}(1+t)^{\beta+n}\,dt,$$

where x is in the complex plane cut along the real axis from -1 to $+1$ and $n>0$, $\alpha+\beta+1 \neq 0$.

5.2.3 Asymptotic expansions

The following result give asymptotic approximations for $n \to \infty$

$$P_n^{(\alpha,\beta)}(\cos\theta) \approx \frac{\cos\left[\left(n + \frac{\alpha+\beta+1}{2}\right)\theta - \frac{\pi}{4}(1+2\alpha)\right]}{\sqrt{\pi n}\left(\sin\frac{\theta}{2}\right)^{\frac{1}{2}+\alpha}\left(\cos\frac{\theta}{2}\right)^{\frac{1}{2}+\beta}} + O\left(n^{-\frac{3}{2}}\right),$$

$$0 < \theta < \pi.$$

The result holds uniformly for

$$\varepsilon \leq \theta \leq \pi - \varepsilon, \quad \varepsilon > 0.$$

$$P_n^{(\alpha,\beta)}(x) \approx \frac{(x-1)^{-\frac{\alpha}{2}}(x+1)^{-\frac{\beta}{2}}}{(2\pi n)^{\frac{1}{2}}}\left[(x+1)^{\frac{1}{2}} + (x-1)^{\frac{1}{2}}\right]^{\alpha+\beta}$$

$$\times (x-1)^{-\frac{1}{4}}\left[x + (x^2-1)^{\frac{1}{2}}\right]^{\frac{1}{2}+n}$$

for $x > 1$ or $x < -1$.
If α, β are real and c is a fixed positive constant, then as $n \to \infty$

$$P_n^{(\alpha,\beta)}(\cos\theta) = \begin{cases} \theta^{-\frac{1}{2}-\alpha}\, O\left(n^{-\frac{1}{2}}\right), & \text{if } \frac{c}{n} \leq \theta \leq \frac{\pi}{2}, \\ O(n^{\alpha}), & \text{if } 0 \leq \theta \leq \frac{c}{n}, \end{cases}$$

$$\left\{\left(\frac{d}{dx}\right)^k P_n^{(\alpha,\beta)}(x)\right\}_{x=\cos\theta} = \begin{cases} \theta^{-\alpha-k-\frac{1}{2}}\, O\left(n^{k-\frac{1}{2}}\right), & \text{if } \frac{c}{n} \leq \theta \leq \frac{\pi}{2}, \\ O(n^{2k+\alpha}), & \text{if } 0 \leq \theta \leq \frac{c}{n}, \end{cases}$$

$$\left(\frac{d}{dx}\right)^k P_n^{(\alpha,\beta)}(x) = O(n^q),$$

$$-1 \leq x \leq +1, \quad q = \operatorname{Max}\left(\alpha + 2k, \beta + 2k, k - \frac{1}{2}\right).$$

5.2.4 Miscellaneous results

Limit relations

$$\lim_{n\to\infty} \frac{1}{n^{\alpha}} P_n^{(\alpha,\beta)}\left(\cos\frac{x}{n}\right) = \lim_{n\to\infty} \frac{1}{n^{\alpha}} P_n^{(\alpha,\beta)}\left(1 - \frac{x^2}{2n^2}\right) = \left(\frac{2}{x}\right)^{\alpha} J_{\alpha}(x),$$

which holds uniformly for $|x| \leq R$, R fixed.

Let

$$x_m^{(n)} = m\text{th zero of } P_n^{(\alpha,\beta)}(x),$$

$$x_1^{(n)} < x_2^{(n)} < \cdots < x_n^{(n)},$$

$$j_{\alpha,m} = m\text{th zero (positive) of the Bessel function } J_\alpha(x),$$
$$0 < j_{\alpha,1} < j_{\alpha,2} < \cdots,$$

then

$$\lim_{n\to\infty} n \arccos x_{n-m+1}^{(n)} = j_{\alpha,m} \text{ provided that } \alpha > -1, \beta > -1.$$

Inequalities

For $-1 \leq x \leq 1$, $\alpha \geq -\frac{1}{2}$

$$\left[\frac{1-x}{2}\right]^{\frac{\alpha+1}{2}} |P_n^{(\alpha,0)}(x)| \leq 1.$$

Let

$$x_0 = \frac{\beta - \alpha}{\alpha + \beta + 1}; \alpha > -1, \beta > -1,$$

then

$$\underset{-1\leq x\leq 1}{\operatorname{Max}} |P_n^{(\alpha,\beta)}(x)| = \begin{cases} \binom{n+q}{n} \sim n^q, & \text{if } q = \max(\alpha, \beta) \geq -\frac{1}{2}, \\ |P_n^{(\alpha,\beta)}(x')| \sim n^{-\frac{1}{2}}, & \text{if } q = \max(\alpha, \beta) < -\frac{1}{2}, \end{cases}$$

where x' is one of the two maximum points nearest x_0.

Series involving Jacobi polynomials

$$\operatorname{sgn} x = c_0 + \sum_{n=1}^{\infty} \frac{1}{n\, h_n} P_n^{(\alpha,\beta)}(x)\, P_{n-1}^{(\alpha+1,\beta+1)}(0),$$
$$-1 < x < 1,$$

where

$$\operatorname{sgn} x = \begin{cases} 1, & x > 0 \\ -1, & x < 0 \end{cases}$$

and

$$c_0 = \frac{\Gamma(\alpha+\beta+2)}{\Gamma(\alpha+1)\,\Gamma(\beta+1)} 2^{-\alpha-\beta-1} \int_0^1 [(1-t)^\alpha (1+t)^\beta - (1+t)^\alpha (1-t)^\beta]\, dt,$$

$$(1-x)^\lambda = 2^\lambda \Gamma(\alpha+\lambda+1)$$
$$\times \sum_{n=0}^{\infty} \frac{\Gamma(\alpha+\beta+2n+1)\,\Gamma(\alpha+\beta+n+1)}{\Gamma(\alpha+n+1)\,\Gamma(\alpha+\beta+\lambda+n+2)} (-\lambda)_n\, P_n^{(\alpha,\beta)}(x),$$
$$-\lambda < \operatorname{mini}\left(\alpha+1, \frac{2\alpha+3}{4}\right), -1 < x < 1,$$

$$e^{ixy} = (2iy)^{-1-\frac{\alpha+\beta}{2}} \sum_{n=0}^{\infty} \frac{\Gamma(\alpha+\beta+n+1)}{\Gamma(\alpha+\beta+2n+1)} M_{\varkappa,\mu}(2iy)\, P_n^{(\alpha,\beta)}(x),$$
$$-1 < x < 1,$$

where

$$\varkappa = \frac{\alpha-\beta}{2}, \quad \mu = n + \frac{\alpha+\beta+1}{2}.$$

5.3 Gegenbauer or ultraspherical polynomials

5.3.1 Definition, notation, and special cases

The Gegenbauer polynomials $C_n^\lambda(x)$ are a special case of the Jacobi polynomials $P_n^{(\alpha,\beta)}(x)$ when $\alpha = \beta = \lambda - \frac{1}{2}$ and are normalized by the relation

$$C_n^\lambda(1) = \frac{(2\lambda)_n}{n!}, \quad 2\lambda \neq 0, -1, -2, \dots$$

The polynomials $C_n^\lambda(x)$ are orthogonal over the interval $(-1, 1)$ with the weight function $w(x) = (1 - x^2)^{\lambda - \frac{1}{2}}$, $\lambda > -\frac{1}{2}$. In terms of $P_n^{(\alpha,\beta)}(x)$,

$$C_n^\lambda(x) = \frac{(2\lambda)_n}{\left(\lambda + \frac{1}{2}\right)_n} P_n^{\left(\lambda - \frac{1}{2}, \lambda - \frac{1}{2}\right)}(x), \quad \lambda > -\frac{1}{2}.$$

Special cases, and relations

$$C_0^\lambda(x) = 1, \qquad \lambda \neq 0,$$

$$C_1^\lambda(x) = 2\lambda x, \qquad \lambda \neq 0,$$

$$C_2^\lambda(x) = 2\lambda(\lambda + 1)\, x^2 - \lambda, \qquad \lambda \neq 0,$$

$$C_n^\lambda(0) = \begin{cases} 0, & n \text{ odd} \\ (-1)^m \dfrac{\Gamma(\lambda + m)}{\Gamma(\lambda)\, \Gamma(m + 1)}, & n \text{ even and } n = 2m, \end{cases}$$

$$C_n^\lambda(1) = \binom{n + 2\lambda - 1}{n} = \frac{(2\lambda)_n}{n!},$$

$$C_n^\lambda(-x) = (-1)^n\, C_n^\lambda(x),$$

$$C_n^0(x) = \lim_{\lambda \to 0} \frac{1}{\lambda}\, C_n^\lambda(x) = \frac{2}{n}\, T_n(x),$$

$$C_n^0(-x) = (-1)^n\, C_n^0(x),$$

$$C_n^0(0) = \begin{cases} 0, & n \text{ odd} \\ \dfrac{(-1)^m}{m}, & n \text{ even and } n = 2m, \end{cases}$$

$$C_n^0(1) = \frac{2}{n}, \quad n \neq 0;\ C_0^0(1) = 1,$$

$$C_0^0(x) = 1,$$

$$C_1^0(x) = 2x,$$

$$C_2^0(x) = 2x^2 - 1,$$

$$C_n^0(x) = \frac{1}{n}\left\{\left(x + \sqrt{x^2 - 1}\right)^n + \left(x - \sqrt{x^2 - 1}\right)^n\right\}, \qquad -1 < x < 1,$$

$$C_n^0(\cos\theta) = \frac{2}{n} \cos n\theta; \quad C_n^1(\cos\theta) = \frac{\sin\,[(n + 1)\,\theta]}{\sin\theta}.$$

Connection with other polynomials

$$P_n^{\left(\lambda-\frac{1}{2},\lambda-\frac{1}{2}\right)}(x) = \frac{\Gamma(2\lambda)\,\Gamma\left(\lambda+n+\frac{1}{2}\right)}{\Gamma\left(\frac{1}{2}+\lambda\right)\Gamma(2\lambda+n)} C_n^\lambda(x), \quad \lambda > -\frac{1}{2},$$

$$P_n^{\left(\lambda-\frac{1}{2},\frac{1}{2}\right)}(x) = \frac{\left(\frac{1}{2}\right)_{n+1}}{(\lambda)_{n+1}\sqrt{\frac{x+1}{2}}} C_{2n+1}^\lambda\left(\sqrt{\frac{x+1}{2}}\right), \quad \lambda > -\frac{1}{2},$$

$$P_n^{\left(\lambda-\frac{1}{2},-\frac{1}{2}\right)}(x) = \frac{\left(\frac{1}{2}\right)_n}{(\lambda)_n} C_{2n}^\lambda\left(\sqrt{\frac{x+1}{2}}\right),$$

$$C_{2n}^\lambda(x) = \frac{2^{2n}\,n!\,\Gamma(\lambda+n)}{(2n)!\,\Gamma(\lambda)} P_n^{\left(\lambda-\frac{1}{2},-\frac{1}{2}\right)}(2x^2-1),$$

$$C_{2n+1}^\lambda(x) = \frac{2^{2n+1}\,n!\,\Gamma(\lambda+n+1)}{(2n+1)!\,\Gamma(\lambda)} x\cdot P_n^{\left(\lambda-\frac{1}{2},\frac{1}{2}\right)}(2x^2-1),$$

$$C_n^\lambda(x) = \frac{\Gamma\left(\lambda+\frac{1}{2}\right)(2\lambda)_n}{n!}\left(\frac{x^2-1}{4}\right)^{\frac{1}{4}-\frac{\lambda}{2}} P_{n+\lambda-\frac{1}{2}}^{\left(\frac{1}{2}-\lambda\right)}(x),$$

where $P_n^\lambda(x)$ are Legendre functions [see chap. IV].

$$C_n^{\frac{1}{2}}(x) = P_n(x),$$

where $P_n(x)$ are Legendre polynomials

$$C_n^1(x) = U_n(x),$$

$U_n(x)$ are Chebyshev (Tchebichef) polynomials of the second kind.

$$C_n^0(x) = \frac{2}{n} T_n(x),$$

where $T_n(x)$ are Chebyshev polynomial of the first kind.

5.3.2 Elementary results

Explicit expressions for $C_n^\lambda(x)$

$$C_n^\lambda(x) = \frac{1}{\Gamma(\lambda)} \sum_{m=0}^{\left[\frac{n}{2}\right]} \frac{(-1)^m\,\Gamma(\lambda+n-m)}{m!\,(n-2m)!}(2x)^{n-2m}, \quad \lambda > 0,$$

$$C_n^0(x) = \sum_{m=0}^{\left[\frac{n}{2}\right]} (-1)^m \frac{\Gamma(n-m)}{\Gamma(m+1)\,\Gamma(n-2m+1)}(2x)^{n-2m}, \quad n \neq 0.$$

Representation by hypergeometric functions

$$C_n^\lambda(x) = \frac{\Gamma(n+2\lambda)}{n!\,\Gamma(2\lambda)}\,{}_2F_1\left(-n, n+2\lambda; \lambda+\frac{1}{2}; \frac{1-x}{2}\right)$$

$$= (-1)^n \frac{\Gamma(n+2\lambda)}{\Gamma(n+1)}\,{}_2F_1\left(-n, n+2\lambda; \lambda+\frac{1}{2}; \frac{1+x}{2}\right)$$

$$= \frac{2^n(\lambda)_n}{n!}(x-1)^n\,{}_2F_1\left(-n, -n-\lambda+\frac{1}{2}; -2n-2\lambda+1; \frac{2}{1-x}\right)$$

$$= \frac{(2\lambda)_n}{n!}\left(\frac{1+x}{2}\right)^n {}_2F_1\left(-n, -n-\lambda+\frac{1}{2}; \lambda+\frac{1}{2}; \frac{x-1}{x+1}\right)$$

$$= 2^n\frac{(\lambda)_n}{n!}x^n\,{}_2F_1\left(-\frac{n}{2}, \frac{-n+1}{2}; -n-\lambda+1; x^{-2}\right)$$

$$= \frac{(\lambda)_n}{n!}\left(x+\sqrt{x^2-1}\right)^n {}_2F_1\left(-n, \lambda; -n-\lambda+1; \left(x-\sqrt{x^2-1}\right)^2\right),$$

$$C_{2n}^\lambda(x) = (-1)^n\frac{(\lambda)_n}{n!}\,{}_2F_1\left(-n, n+\lambda; \frac{1}{2}; x^2\right)$$

$$= \frac{(2\lambda)_{2n}}{(2n)!}\,{}_2F_1\left(-n, n+\lambda; \lambda+\frac{1}{2}; 1-x^2\right),$$

$$C_{2n+1}^\lambda(x) = (-1)^n\frac{(\lambda)_{n+1}}{n!}\,2x\,{}_2F_1\left(-n, n+\lambda+1; \frac{3}{2}; x^2\right)$$

$$= x\frac{(2\lambda)_{2n+1}}{(2n+1)!}\,{}_2F_1\left(-n, n+\lambda+1; \lambda+\frac{1}{2}; 1-x^2\right).$$

Also see chap. II.

$$C_n^\lambda(\cos\theta) = \sum_{m=0}^{n}\frac{\Gamma(\lambda+m)\,\Gamma(\lambda+n-m)}{m!\,(n-m)!\,[\Gamma(\lambda)]^2}\cos[(n-2m)\theta],$$

$$\lambda \neq 0,$$

$$C_n^\lambda(\cos\theta) = \frac{2}{\Gamma(\lambda)}\sum_{m=0}^{\infty}\frac{(\lambda)_m}{m!}\,\frac{\Gamma(n+2\lambda+m)}{\Gamma(n+\lambda+m+1)}\cos[(n+2\lambda+2m)\theta-\lambda\pi],$$

$$0<\lambda<1,\; 0<\theta<\pi,$$

$$C_n^\lambda(\cos\theta) = \frac{2}{\pi}\sin(\pi\lambda)\frac{\Gamma(n+2\lambda)}{\Gamma(\lambda)}$$

$$\times\sum_{m=0}^{\infty}\frac{\Gamma(\lambda+m)\,\Gamma(m-\lambda+1)}{m!\,\Gamma(n+\lambda+m+1)}\cdot\frac{\cos\left[(n+\lambda+m)\theta-\frac{\pi}{2}(\lambda+m)\right]}{(2\sin\theta)^{\lambda+m}},$$

$$0<\lambda<1,\; \frac{\pi}{6}<\theta<\frac{5\pi}{6}.$$

Integrals involving $C_n^\lambda(x)$

$$n\left(1+\frac{n}{2\lambda}\right)\int_0^x (1-t^2)^{\lambda-\frac{1}{2}}C_n^\lambda(t)\,dt$$

$$= C_{n-1}^{\lambda+1}(0) - (1-x^2)^{\lambda+\frac{1}{2}}C_{n-1}^{\lambda+1}(x),$$

$$\int_0^\pi e^{iz\cos\theta} C_n^\lambda(\cos\theta)(\sin\theta)^{2\lambda}\,d\theta = \frac{\Gamma\left(\lambda+\frac{1}{2}\right)\Gamma(2\lambda+n)}{n!\,\Gamma(2\lambda)}\,\pi^{\frac{1}{2}}\, i^n \left(\frac{2}{z}\right)^\lambda J_{\lambda+n}(z),$$

$$2(n+\lambda)\int C_n^\lambda(x)\,dx = C_{n+1}^\lambda(x) - C_{n-1}^\lambda(x),$$

$$\int C_n^\lambda(x)\,dx = \frac{1}{2(\lambda-1)}\,C_{n+1}^{\lambda-1}(x).$$

Orthogonality relation

$$\int_{-1}^{+1} C_n^\lambda(x)\,C_m^\lambda(x)\,(1-x^2)^{\lambda-\frac{1}{2}}\,dx$$

$$= \begin{cases} 0, & n \neq m, \\ \dfrac{\pi\, 2^{1-2\lambda}\,\Gamma(n+2\lambda)}{n!\,(\lambda+n)\,[\Gamma(\lambda)]^2}, & n = m, \end{cases} \qquad \lambda \neq 0,$$

$$= \begin{cases} 0, & n \neq m, \\ \dfrac{\pi^{\frac{1}{2}}\,(2\lambda)_n\,\Gamma\left(\lambda+\frac{1}{2}\right)}{n!\,(\lambda+n)\,\Gamma(\lambda)}, & n = m, \end{cases} \qquad \lambda \neq 0.$$

Differentiation formulas

$$\frac{d}{dx} C_n^\lambda(x) = 2\lambda\, C_{n-1}^{\lambda+1}(x),$$

$$\frac{d^m}{dx^m} C_n^\lambda(x) = 2^m (\lambda)_m\, C_{n-m}^{\lambda+m}(x),$$

$$\frac{d}{dx} C_{n-1}^\lambda(x) = x\frac{d}{dx} C_n^\lambda(x) - n\, C_n^\lambda(x),$$

$$\frac{d}{dx} C_{n+1}^\lambda(x) = x\frac{d}{dx} C_n^\lambda(x) + (2\lambda+n)\, C_n^\lambda(x),$$

$$\begin{aligned}(1-x^2)\frac{d}{dx} C_n^\lambda(x) &= (n+2\lambda-1)\, C_{n-1}^\lambda(x) - n x\, C_n^\lambda(x) \\ &= (n+2\lambda)\, x\, C_n^\lambda(x) - (n+1)\, C_{n+1}^\lambda(x) \\ &= 2\lambda(1-x^2)\, C_{n-1}^{\lambda+1}(x),\end{aligned}$$

$$\frac{d}{dx}\left[C_{n+1}^\lambda(x) - C_{n-1}^\lambda(x)\right] = 2(n+\lambda)\, C_n^\lambda(x).$$

Rodrigues' formula

$$C_n^\lambda(x) = \frac{(-1)^n (2\lambda)_n}{2^n\, n!\left(\lambda+\frac{1}{2}\right)_n}(1-x^2)^{\frac{1}{2}-\lambda}\frac{d^n}{dx^n}\left[(1-x^2)^{n+\lambda-\frac{1}{2}}\right]$$

$$= \frac{(-1)^n\,\Gamma\left(\lambda+\frac{1}{2}\right)\Gamma(n+2\lambda)}{2^n\, n!\,\Gamma(2\lambda)\,\Gamma\left(n+\lambda+\frac{1}{2}\right)}(1-x^2)^{\frac{1}{2}-\lambda}\frac{d^n}{dx^n}\left[(1-x^2)^{n+\lambda-\frac{1}{2}}\right],$$

$$C_n^\lambda\left[x(x^2-1)^{-\frac{1}{2}}\right] = \frac{(-1)^n (x^2-1)^{\lambda+\frac{1}{2}n}}{n!} \frac{d^n}{dx^n}[(x^2-1)^{-\lambda}].$$

Let D_x^α be the fractional derivative operator defined by

$$D_x^\alpha(x^\beta) = \left\{ \frac{\Gamma(\beta+1)}{\Gamma(\beta-\alpha+1)} x^{\beta-\alpha}, \quad \beta-\alpha+1 \neq 0, -1, -2, \ldots, \right.$$

where $\beta + 1 > 0$.

Then

$$C_n^\lambda(\sqrt{x}) = \frac{\pi^{\frac{1}{2}}(-2)^{n+1} x^{\frac{n+3}{2}}}{(n+1)!\,\Gamma(\lambda)} D_x^{n+\lambda+\frac{1}{2}}\left[x^{\frac{n}{2}+\lambda-1}\left(1-x^{\frac{1}{2}}\right)^{n+1}\right],$$

where $\lambda > -1$; $\lambda \neq 0$.

Recurrence relations

$$(n+1)\,C_{n+1}^\lambda(x) = 2(n+\lambda)\,x\,C_n^\lambda(x) - (n+2\lambda-1)\,C_{n-1}^\lambda(x),$$

$$n = 1, 2, 3, \ldots,$$

$$(n+\lambda)\,C_{n+1}^{\lambda-1}(x) = (\lambda-1)\,[C_{n+1}^\lambda(x) - C_{n-1}^\lambda(x)],$$

$$2\lambda(1-x^2)\,C_{n-1}^{\lambda+1}(x) = (n+2\lambda-1)\,C_{n-1}^\lambda(x) - n\,x\,C_n^\lambda(x),$$

$$= (n+2\lambda)\,x\,C_n^\lambda(x) - (n+1)\,C_{n+1}^\lambda(x),$$

$$n\,C_n^\lambda(x) = 2\lambda x\,C_{n-1}^{\lambda+1}(x) - 2\lambda\,C_{n-2}^{\lambda+1}(x),$$

$$C_n^{\lambda+1}(x) = x\,C_{n-1}^{\lambda+1}(x) + (n+2\lambda)\,C_{n-1}^{\lambda+1}(x)$$

$$= (n+2\lambda+x)\,C_{n-1}^{\lambda+1}(x).$$

Generating functions

$$\frac{2^{\lambda-\frac{1}{2}}}{R}(1-xz+R)^{\frac{1}{2}-\lambda} = \sum_{n=0}^\infty \frac{\left(\lambda+\frac{1}{2}\right)_n}{(2\lambda)_n} C_n^\lambda(x)\,z^n,$$

where

$$R = (1-2xz+z^2)^{\frac{1}{2}},$$

$$|z| < 1, \ \lambda \neq 0,$$

$$(1-2xz+z^2)^{-\lambda} = \sum_{n=0}^\infty C_n^\lambda(x)\,z^n,$$

$$|z| < 1, \ \lambda \neq 0,$$

$$\Gamma\left(\lambda+\frac{1}{2}\right) e^{z\cos\theta}\left(\frac{1}{2}z\sin\theta\right)^{\frac{1}{2}-\lambda} J_{\lambda-\frac{1}{2}}(z\sin\theta)$$

$$= \sum_{n=0}^\infty \frac{C_n^\lambda(\cos\theta)}{(2\lambda)_n} z^n,$$

$$\lambda \neq 0,$$

$$(1-xz)^{-\nu}\,{}_2F_1\left(\frac{1}{2}\nu;\frac{\nu+1}{2};\lambda+\frac{1}{2};\frac{z^2(x^2-1)}{(1-xz)^2}\right)=\sum_{n=0}^{\infty}\frac{(\nu)_n\,C_n^\lambda(x)}{(2\lambda)_n}z^n,$$

$$\lambda\neq 0,$$

$${}_0F_1\left(;\lambda+\frac{1}{2};\frac{z(x-1)}{2}\right){}_0F_1\left(;\lambda+\frac{1}{2};\frac{z(x+1)}{2}\right)=\sum_{n=0}^{\infty}\frac{C_n^\lambda(x)}{(2\lambda)_n\left(\lambda+\frac{1}{2}\right)_n}z^n,$$

$$\lambda\neq 0,$$

$$(1-z)^{-2\lambda}\,{}_1F_0\left(\lambda;\,;\frac{2z(x-1)}{(1-z)^2}\right)=\sum_{n=0}^{\infty}C_n^\lambda(x)\,z^n,$$

$$-\log(1-2xz+z^2)=\sum_{n=0}^{\infty}C_n^0(x)\,z^n,$$

$$|z|<1.$$

Addition theorem

$$C_n^\lambda(\cos\theta\cos\varphi+\sin\theta\sin\varphi\cos\psi)$$

$$=\frac{\Gamma(2\lambda-1)}{[\Gamma(\lambda)]^2}\sum_{m=0}^{n}\frac{2^{2m}\Gamma(n-m+1)\,[\Gamma(\lambda+m)]^2}{\Gamma(n+2\lambda+m)}(2\lambda+2m-1)$$

$$\times(\sin\theta)^m(\sin\varphi)^m\,C_{n-m}^{\lambda+m}(\cos\theta)\,C_{n-m}^{\lambda+m}(\cos\varphi)\,C_m^{\lambda-\frac{1}{2}}(\cos\psi).$$

The differential equation

$$(1-x^2)\,y''-(2\lambda+1)\,xy'+n(n+2\lambda)\,y=0,$$

$$y(x)=C_n^\lambda(x).$$

Other transformations of the differential equation are:

$$(1-x^2)\,y''+(2\lambda-3)\,xy'+(n+1)(n+2\lambda-1)\,y=0,$$

$$y(x)=(1-x^2)^{\lambda-\frac{1}{2}}\,C_n^\lambda(x),$$

$$(1-x^2)\,y''+\left[(\lambda+n)^2+\frac{2+4\lambda-4\lambda^2+x^2}{4(1-x^2)}\right]y=0,$$

$$y(x)=(1-x^2)^{\frac{2\lambda+1}{4}}\,C_n^\lambda(x),$$

$$y''+\left[(\lambda+n)^2+\frac{\lambda(1-\lambda)}{\sin^2 x}\right]y=0,$$

$$y(x)=(\sin x)^\lambda\,C_n^\lambda(\cos x).$$

Integral representations

$$C_n^\lambda(x)=\frac{1}{2\pi i}\int^{(0+)}t^{-n-1}(1-2xt+t^2)^{-\lambda}\,dt,$$

where the contour is taken around the point $t = 0$ in the positive sense and excludes both the zeros of the function $(1 - 2xt + t^2)$. Furthermore it is assumed that $\lambda > 0$.

$$C_n^\lambda(x) = \frac{2^{1-2\lambda}\,\Gamma(n+2\lambda)}{n!\,[\Gamma(\lambda)]^2} \int_0^\pi \left[x + \sqrt{x^2-1}\cos t\right]^n (\sin t)^{2\lambda-1}\,dt,$$

$$\lambda > 0,$$

$$C_n^\lambda(\cos\theta) = \frac{2^{1-\lambda}\,\Gamma(n+2\lambda)}{n!\,[\Gamma(\lambda)]^2} (\sin\theta)^{1-2\lambda} \int_0^\theta \frac{\cos[(n+\lambda)\,t]}{(\cos t - \cos\theta)^{1-\lambda}}\,dt,$$

$$\lambda > 0,$$

$$(x^2-1)^{\frac{2\lambda-1}{4}}\, C_n^\lambda(x)$$

$$= \frac{2^{\lambda-\frac{1}{2}}}{\pi} \frac{(2\lambda)_n}{\left(\lambda+\frac{1}{2}\right)_n} \int_0^\pi \left[x + \sqrt{x^2-1}\cos\varphi\right]^{n+\lambda-\frac{1}{2}} \cos\left[\left(\lambda - \frac{1}{2}\right)\varphi\right] d\varphi,$$

$$x > 1, \quad (x^2-1)^{\frac{1}{2}} > 0,$$

$$\lambda - \frac{1}{2} = \text{non-negative integer}.$$

5.3.3 Asymptotic expansions

Let

$$x = \frac{1}{2}(z + z^{-1}), \quad |z| > 1,$$

and

$$\lambda > 0 \quad \text{or} \quad \lambda < 0, \ \lambda \neq -1, -2, -3, \ldots$$

Then

$$C_n^\lambda(x) = \binom{n+\lambda-1}{n} z^n \sum_{m=0}^{N} \binom{m+\lambda-1}{m} \frac{(1-\lambda)_m}{(n+\lambda-m)_m}$$
$$\times z^{-2m}(1-z^{-2})^{-\lambda-m} + O(n^{\lambda-N-2}|z|^n)$$

and

$$C_n^\lambda(\cos\theta) = 2\binom{n+\lambda-1}{n} \sum_{m=0}^{N} \binom{m+\lambda-1}{m} \frac{(1-\lambda)_m}{(n+\lambda-m)_m}$$
$$\times \frac{\cos\left[(n+\lambda-m)\,\theta - \frac{\pi}{2}(\lambda+m)\right]}{(2\sin\theta)^{\lambda+m}} + O(n^{\lambda-N-2}),$$

$$0 < \theta < \pi.$$

The result is valid uniformly for

$$\varepsilon \le \theta \le \pi - \varepsilon, \ \varepsilon > 0.$$

Let

$$0 < \lambda < 1;\ 0 < \theta < \pi,$$

and

$$M = \max(|\sec\theta|, 2\sin\theta).$$

Then

$$C_n^\lambda(\cos\theta) = \frac{2}{\pi}\sin(\pi\lambda)\frac{\Gamma(n+2\lambda)}{\Gamma(\lambda)}\sum_{m=0}^{N-1}\frac{\Gamma(\lambda+m)\,\Gamma(m-\lambda+1)}{m!\,\Gamma(n+\lambda+m+1)}$$

$$\times\frac{\cos\left[(n+\lambda+m)\,\theta-\frac{\pi}{2}(\lambda+m)\right]}{(2\sin\theta)^{\lambda+m}}+R_N(\theta),$$

where

$$|R_N(\theta)| < \frac{2}{\pi}\sin(\lambda\pi)\frac{\Gamma(n+2\lambda)\,\Gamma(N+\lambda)\,\Gamma(N-\lambda+1)}{N!\,\Gamma(\lambda)\,\Gamma(n+N+\lambda+1)}\frac{M}{(2\sin\theta)^{N+\lambda}}.$$

5.3.4 Miscellaneous results

Limit relations

$$\lim_{\lambda\to 0}\Gamma(\lambda)\,C_n^\lambda(\cos\theta) = \frac{2}{n}\cos(n\theta), \qquad n = 1, 2, 3, \ldots,$$

$$\lim_{\lambda\to\infty}\lambda^{-\frac{n}{2}}\,C_n^{\frac{\lambda}{2}}\left(\frac{x}{\sqrt{\lambda}}\right) = \frac{1}{n!}H_n(x),$$

$$\lim_{\lambda\to 0}\frac{1}{\lambda}C_n^\lambda(x) = \frac{2}{n}T_n(x),$$

$$\lim_{x\to\infty}x^{-n}C_n^\lambda(x) = 2^n\binom{n+\lambda-1}{n}.$$

Inequalities

$$(\sin\theta)^\lambda\,|C_n^\lambda(\cos\theta)|$$

$$\leq\begin{cases}\binom{\lambda+m-1}{m}, & n = 2m,\ n \text{ even}\\[2ex] [\lambda(1-\lambda)+(n+\lambda)^2]^{-\frac{1}{2}}(n+1)\binom{\lambda+m}{m+1}, & n = 2m=+1,\ n\,\text{odd},\end{cases}$$

$$0 < \lambda < 1,\ 0 \leq \theta \leq \pi.$$

Also

$$(\sin\theta)^\lambda\,|C_n^\lambda(\cos\theta)| < 2^{1-\lambda}\frac{n^{\lambda-1}}{\Gamma(\lambda)},$$

$$0 < \lambda < 1,\ 0 \leq \theta \leq \pi,$$

$$\operatorname*{Max}_{-1\leq x\leq 1}|C_n^\lambda(x)| = \begin{cases}\binom{n+2\lambda-1}{n}, & \text{if}\quad \lambda > 0\\[1ex] |C_n^\lambda(x')|, & \text{if}\ \lambda < 0,\ \lambda \text{ not an integer}.\end{cases}$$

The point x' is one of the two maximum points nearest zero if n is odd; $x' = 0$ if n is even. Let

$$a > 0; \; \lambda \neq 0, -1, -2, -3, \ldots$$

If λ is real, then

$$C_n^\lambda(\cos\theta) = \begin{cases} \theta^{-\lambda} O(n^{\lambda-1}), & \frac{a}{n} \leq \theta \leq \frac{\pi}{2}, \\ O(n^{2\lambda-1}), & 0 \leq \theta \leq \frac{a}{n}, \end{cases}$$

$$[C_n^\lambda(x)]^2 - C_{n+1}^\lambda(x)\, C_{n-1}^\lambda(x) \geq 0, \quad < 0,$$

according as

$$|x| \geq 1 \text{ or } |x| < 1.$$

The result is valid for $\lambda > 0$. The following inequalities are valid for

$$0 < \lambda \leq 1, \; -1 < x < 1,$$

$$|C_{n+2}^\lambda(x) - C_n^\lambda(x)| < \frac{2}{\Gamma(2\lambda)\, n^{2-2\lambda}} + \frac{4}{\Gamma(\lambda)\, (n+2)^{1-\lambda}},$$

$$|C_{n+1}^\lambda(x) + C_n^\lambda(x)| < \frac{1}{\Gamma(2\lambda)} [n^{2\lambda-2} + 2(n-1)^{2\lambda-2}] + \frac{2}{\Gamma(\lambda)} \left[(n+1)^{\lambda-1} + \frac{2^{-\lambda}(1+x)}{n^{1-\lambda}(1-x^2)^{\frac{1}{(2\lambda)}}} \right],$$

$$|C_{n+1}^\lambda(x) - x C_n^\lambda(x)| < \frac{1}{\Gamma(2\lambda)} [n^{2\lambda-2} + 2(n-1)^{2\lambda-2}] + \frac{2}{\Gamma(\lambda)} (n+1)^{\lambda-1},$$

$$\left| \int_{-1}^{x} C_n^\lambda(t)\, dt \right| < \frac{2}{n+\lambda} \left[\frac{(n-1)^{2\lambda-2}}{\Gamma(2\lambda)} + \frac{(n+1)^{\lambda-1}}{\Gamma(\lambda)} \right],$$

$$\left| \frac{d}{dx} C_n^\lambda(x) \right| = 2\lambda\, |C_{n-1}^{\lambda-1}(x)|$$

$$< \frac{n+2\lambda}{1-x^2} \left[\frac{1}{\Gamma(2\lambda)} \{n^{2\lambda-2} + 2(n-1)^{2\lambda-2}\} + \frac{2}{\Gamma(\lambda)} \left\{ (n+1)^{\lambda-1} + \frac{(2\lambda-1)\, 2^{-\lambda}(n+1)^{\lambda-1}}{(1-x^2)^{1+\frac{1}{2\lambda}}} \right\} \right].$$

Series involving Gegenbauer polynomials

$$R^{-2\lambda} \exp\left[-\frac{yz(x-z)}{R^2} \right] {}_0F_1\left(; \lambda + \frac{1}{2} ; \frac{y^2 z^2 (x^2-1)}{4R^4} \right)$$

$$= \sum_{n=0}^{\infty} \frac{n!}{(2\lambda)_n} L_n^{2\lambda-1}(y)\, C_n^\lambda(x)\, z^n,$$

where $R^2 = 1 - 2xz + z^2$

$$x^n = 2^{-n} n! \sum_{m=0}^{\left[\frac{n}{2}\right]} \frac{(n + \lambda - 2m)}{m!\,(\lambda)_{n+1-m}} C_{n-2m}^{\lambda}(x),$$

$$e^{xy} = \left(\frac{y}{2}\right)^{-\lambda} \Gamma(\lambda) \sum_{n=0}^{\infty} (\lambda + n)\, I_{\lambda+n}(y)\, C_n^{\lambda}(x),$$

$$\lambda > 0,$$

$$e^{ixy} = \Gamma(\lambda) \left(\frac{y}{2}\right)^{-\lambda} \sum_{n=0}^{\infty} i^n (n + \lambda)\, J_{n+\lambda}(y)\, C_n^{\lambda}(x),$$

$$-1 < x < 1,\ \lambda > 0,$$

$$(1 - x)^p = 2^{2\lambda+p} \pi^{-\frac{1}{2}} \Gamma(\lambda)\, \Gamma\left(\lambda + p + \frac{1}{2}\right)$$

$$\times \sum_{n=0}^{\infty} \frac{(n + \lambda)\,(-p)_n}{\Gamma(n + 2\lambda + p + 1)} C_n^{\lambda}(x),$$

$$-1 < x < 1;\ \lambda > -1 - 2p \ \text{ if } \ \lambda \geq 0 \ \text{ and}$$

$$\lambda > -p - \frac{1}{2} \ \text{ if } \ -\frac{1}{2} < \lambda \leq 0,$$

$$(y \sin\theta \sin\varphi)^{\frac{1}{2}-\lambda} J_{\lambda-\frac{1}{2}}(y \sin\theta \sin\varphi)\, e^{iy\cos\theta\cos\varphi}$$

$$= 2^{\frac{1}{2}} y^{-\lambda} \Gamma(\lambda) \sum_{n=0}^{\infty} i^n \frac{n!\,(n + \lambda)}{(2\lambda)_n \Gamma(2n + \lambda)} J_{n+\lambda}(y)\, C_n^{\lambda}(\cos\theta)\, C_n^{\lambda}(\cos\varphi),$$

$$\lambda > 0;\ 0 < \theta < \pi,\ 0 < \varphi < \pi,$$

$$(1 + 2xy + y^2)^{\frac{n}{2}} \frac{C_n^{\lambda}\left(\dfrac{x + y}{\sqrt{1 + 2xy + y^2}}\right)}{C_n^{\lambda}(1)} = \sum_{m=0}^{n} \binom{n}{m} \frac{C_m^{\lambda}(x)}{C_m^{\lambda}(1)} y^{n-m}.$$

see also 3.9; 3.13.3; 3.13.5

5.4 Legendre polynomials

5.4.1 Definition, notation, and special cases

Legendre polynomials $P_n(x)$ are the polynomials orthogonal over the interval $(-1, +1)$ with the weight function $w(x) = 1$ and normalized by the relation

$$P_n(1) = 1.$$

These polynomials can also be considered as a special of the Jacobi polynomials $P_n^{(\alpha,\beta)}(x)$ when $\alpha = \beta = 0$, i.e.

$$P_n(x) = P_n^{(0,0)}(x) = C_n^{\frac{1}{2}}(x).$$

Special cases and relations

$$P_n(\pm 1) = (\pm 1)^n,$$

$$P_{2n}(0) = (-1)^n \frac{\left(\frac{1}{2}\right)_n}{n!} = (-1)^n \frac{(2n)!}{2^{2n}(n!)^2},$$

$$P_{2n+1}(0) = 0,$$

$$P'_{2n}(0) = 0,$$

$$P'_{2n+1}(0) = (-1)^n \frac{\left(\frac{3}{2}\right)_n}{n!} = (-1)^n \, 2 \frac{\left(\frac{1}{2}\right)_{n+1}}{n!} = (-1)^n (2n+1) \frac{\left(\frac{1}{2}\right)_n}{n!},$$

where

$$\frac{d}{dx} P_n(x) \equiv P'_n(x).$$

Let $x = \cos\theta$. Then

$$P_0(x) = 1,$$

$$P_1(x) = x = \cos\theta,$$

$$P_2(x) = \frac{1}{2}(3x^2 - 1) = \frac{1}{4}(1 + 3\cos 2\theta),$$

$$P_3(x) = \frac{1}{2}(5x^3 - 3x) = \frac{1}{8}(3\cos\theta + 5\cos 3\theta),$$

$$P_4(x) = \frac{1}{8}(35x^4 - 30x^2 + 3) = \frac{1}{64}(9 + 20\cos 2\theta + 35\cos 4\theta),$$

$$P_n(-x) = (-1)^n P_n(x).$$

5.4.2 Elementary results

Explicit expressions for $P_n(x)$

$$P_n(x) = \sum_{m=0}^{\left[\frac{n}{2}\right]} (-1)^m \frac{(2n-2m)!\, x^{n-2m}}{2^n m!\,(n-m)!\,(n-2m)!}$$

$$= 2^{-n} \sum_{m=0}^{\left[\frac{n}{2}\right]} (-1)^m \binom{n}{m} \binom{2n-2m}{n} x^{n-2m}$$

$$= \frac{(2n)!}{2^n (n!)^2} \left\{ x^n - \frac{n(n-1)}{2(2n-1)} x^{n-2} + \frac{n(n-1)(n-2)(n-3)}{2 \cdot 4 \cdot (2n-1)(2n-3)} x^{n-4} + \cdots \right\},$$

$$P_n(\cos\theta) = \frac{(2n)!}{2^{2n}(n!)^2}\left\{\cos n\theta + \frac{1}{1}\,\frac{n}{2n-1}\cos(n-2)\,\theta \right.$$
$$+ \frac{1\cdot 3}{1\cdot 2}\,\frac{n(n-1)}{(2n-1)(2n-3)}\cos(n-4)\,\theta$$
$$\left. + \frac{1\cdot 3\cdot 5}{1\cdot 2\cdot 3}\,\frac{n(n-1)(n-2)}{(2n-1)(2n-3)(2n-5)}\cos(n-6)\,\theta + \cdots\right\},$$

$$P_n(\cos\theta) = \frac{2}{\pi}\,\frac{n!}{\left(\frac{1}{2}\right)_{n+1}}\sum_{m=0}^{\infty} a_m \sin\left[(n+1+2m)\,\theta\right], \qquad 0 < \theta < \pi,$$

where

$$a_0 = 1,$$
$$a_m = \frac{1\cdot 3\cdot 5\cdots(2m-1)}{2\cdot 4\cdot 6\cdots 2m}\;\frac{(n+1)(n+2)\cdots(n+m)}{\left(n+\frac{3}{2}\right)\left(n+\frac{5}{2}\right)\cdots\left(n+m+\frac{1}{2}\right)}.$$

The result is valid for $n = 1, 2, 3, \ldots$

$$P_0(\cos\theta) = \frac{4}{\pi}\sum_{m=0}^{\infty} a_m \sin\left[(n+1+2m)\,\theta\right], \qquad 0 < \theta < \pi,$$

where a_m is defined above.

Another trigonometric representation is given by:

$$P_n(\cos\theta) = \frac{2}{\pi}\,\frac{n!}{\left(\frac{1}{2}\right)_{n+1}}\sum_{m=0}^{\infty}\frac{\left[\left(\frac{1}{2}\right)_m\right]^2}{m!\left(n+\frac{3}{2}\right)_m}$$
$$\times\frac{\cos\left[\left(n+m+\frac{1}{2}\right)\theta - \frac{\pi}{2}\left(m+\frac{1}{2}\right)\right]}{(2\sin\theta)^{m+\frac{1}{2}}},$$
$$\frac{\pi}{6} < \theta < \frac{5\pi}{6},$$

$$P_n(\cos\theta) = 2^{-2n}\sum_{m=0}^{n}\binom{2m}{m}\binom{2n-2m}{n-m}\cos\left[(n-2m)\,\theta\right],$$
$$0 < \theta < \pi.$$

Representation by hypergeometric series

$$P_n(x) = {}_2F_1\left(-n, n+1; 1; \frac{1-x}{2}\right)$$
$$= \frac{(2n)!}{2^n(n!)^2}\,x^n\,{}_2F_1\left(-\frac{n}{2}, \frac{1-n}{2}; \frac{1}{2}-n; x^{-2}\right)$$
$$= \binom{2n}{n}\left(\frac{x-1}{2}\right)^n {}_2F_1\left(-n, -n; -2n; \frac{2}{1-x}\right),$$

$$P_{2n}(x) = (-1)^n\binom{2n}{n}2^{-2n}\,{}_2F_1\left(-n, n+\frac{1}{2}; \frac{1}{2}; x^2\right),$$

$$P_{2n+1}(x) = (-1)^n \frac{(2n+1)!}{2^{2n}(n!)^2} x\, {}_2F_1\left(-n, n+\frac{3}{2}; \frac{3}{2}; x^2\right),$$

$$P_n(\cos\theta) = \binom{2n}{n} 2^{-2n} e^{\mp in\theta} {}_2F_1\left(\frac{1}{2}, -n; \frac{1}{2}-n; e^{\pm i2\theta}\right)$$

$$= {}_2F_1\left(n+1, -n; 1; \sin^2\left(\frac{\theta}{2}\right)\right)$$

$$= (-1)^n {}_2F_1\left(n+1, -n, 1; \cos^2\left(\frac{\theta}{2}\right)\right).$$

Integrals involving Legendre polynomials

$$\int_0^{2\pi} P_{2n}(\cos\theta)\, d\theta = 2\pi \left[\frac{\left(\frac{1}{2}\right)_n}{n!}\right]^2$$

$$= 2\pi \left[\binom{2n}{n} 2^{-2n}\right]^2,$$

$$\int_0^{2\pi} P_{2n+1}(\cos\theta)\cos\theta\, d\theta = \pi\, 2^{-4n-2}\binom{2n}{n}\binom{2n+2}{n+1},$$

$$\int_{-1}^{1} \frac{P_n(x)}{\sqrt{1-x^2}}\, dx = \frac{2^{\frac{3}{2}}}{2n+1},$$

$$\int_{-1}^{1} P_n(x) \arcsin x\, dx = \begin{cases} 0, & n \text{ even} \\ \pi\left[\frac{1\cdot 3\cdots(n-2)}{2\cdot 4\cdots(n+1)}\right]^2, & n \text{ odd}, \end{cases}$$

$$\int_0^1 P_m(x)\, P_n(x)\, dx = \begin{cases} \frac{1}{2n+1}, & m = n \\ 0, & m-n \text{ even} \\ \frac{(-1)^{\alpha+\beta} 2^{-m-n+1} m!\, n!}{(n-m)(n+m+1)(\alpha!)^2(\beta!)^2}, & n = 2\alpha+1,\ m = 2\beta, \end{cases}$$

$$\int_0^1 x^{n+2\lambda} P_n(x)\, dx = \frac{(n+2\lambda)!}{2^{n+2\lambda}\lambda!\left(\frac{3}{2}\right)_{n+\lambda}},$$

$$\int_0^1 x^n P_{n-2\lambda}(x)\, dx = \frac{1}{2}\int_{-1}^{+1} x^n P_{n-2\lambda}(x)\, dx$$

$$= \frac{n!}{2^n \lambda! \left(\frac{3}{2}\right)_{n-\lambda}},$$

where λ is a positive integer, $0 < 2\lambda < n$.

$$\int_0^{+1} x^2 P_{n-1}(x)\, P_{n+1}(x)\, dx = \frac{n(n+1)}{(2n-1)(2n+1)(2n+3)},$$

$$\int_{-1}^{+1} (\cosh 2x - t)^{-\frac{1}{2}} P_n(t)\, dt = 2^{\frac{3}{2}} (2n+1)^{-1} e^{-(2n+1)x},$$

$$\int_{-1}^{+1} x^k P_n(x)\,dx = 0, \quad \text{for} \quad k = 0, 1, \ldots, n-1,$$

$$\int_0^1 z^\lambda P_n(z)\,dz = \begin{cases} \dfrac{\lambda(\lambda-2)\cdots(\lambda-n+2)}{(\lambda+1)(\lambda+3)\cdots(\lambda+n+1)}, & n \text{ even} \\ \dfrac{(\lambda-1)(\lambda-3)\cdots(\lambda-n+2)}{(\lambda+2)(\lambda+4)\cdots(\lambda+n+1)}, & n \text{ odd}, \end{cases}$$

$$Re\,\lambda > -1,$$

$$\int_0^\pi P_n(\cos\theta)\sin m\theta\,d\theta = \begin{cases} \dfrac{\left(\frac{m-n+1}{2}\right)_n}{\left(\frac{m-n}{2}\right)_{n+1}}, & \text{if } m > n \text{ and } m+n \text{ is odd}, \\ 0 & \text{otherwise}, \end{cases}$$

$$\int_0^\pi P_n(1 - 2\sin^2\alpha\sin^2\theta)\sin\alpha\,d\alpha = 2\,\frac{\sin(2n+1)\,\theta}{(2n+1)\sin\theta},$$

$$\int_{-1}^{+1} P_{2n}(x)\,(1+kx^2)^{-n-\frac{3}{2}}\,dx = \frac{2}{2n+1}\,\frac{(-k)^n}{(1+k)^{n+\frac{1}{2}}}, \qquad |k| < 1,$$

$$\int_{-1}^{+1} (1-z^2)\,\{P_n'(z)\}^2\,dz = \frac{2n(n+1)}{2n+1},$$

$$\int_0^1 x^\lambda P_{2n+1}(x)\,dx = \frac{(-1)^n\left(\frac{1}{2}-\frac{\lambda}{2}\right)_n}{2\left(1+\frac{\lambda}{2}\right)_{n+1}}, \quad \lambda > -2,$$

$$\int_{-1}^{x} (x-t)^{-\frac{1}{2}} P_n(t)\,dt = \frac{2}{2n+1}(1+x)^{-\frac{1}{2}}\,[T_n(x) + T_{n+1}(x)],$$

$$-1 < x < 1,$$

$$\int_x^1 (t-x)^{-\frac{1}{2}} P_n(t)\,dt = \frac{2(1-x)^{-\frac{1}{2}}}{2n+1}\,[T_n(x) - T_{n+1}(x)],$$

$$\int_{-1}^{+1} P_n(x)\,e^{-i\lambda x}\,dx = i^{-n}\left(\frac{2\pi}{\lambda}\right)^{\frac{1}{2}} J_{n+\frac{1}{2}}(\lambda),$$

$$\int_{-\infty}^{\infty} \lambda^{-\frac{1}{2}} J_{n+\frac{1}{2}}(\lambda)\,e^{i\lambda x}\,d\lambda = \begin{cases} (2\pi)^{\frac{1}{2}}\,i^n P_n(x), & -1 < x < 1 \\ 0, & \text{otherwise}, \end{cases}$$

$$(2n+1)\int P_n(x)\,dx = P_{n+1}(x) - P_{n-1}(x).$$

Orthogonality relation

$$\int_{-1}^{+1} P_n(x)\,P_m(x)\,dx = \begin{cases} 0, & n \neq m \\ \dfrac{2}{2n+1}, & n = m. \end{cases}$$

Differentiation formula

$$(1 - x^2)\frac{d}{dx} P_n(x) = n\,[P_{n-1}(x) - x\,P_n(x)],$$

$$\frac{d^m}{dx^m} P_n(x) = 2^m \left(\frac{1}{2}\right)_m C_{n-m}^{m+\frac{1}{2}}(x), \quad n \geq m.$$

Recurrence relations

$$(n+1)\,P_n(x) = P'_{n+1}(x) - x\,P'_n(x),$$

$$n\,P_n(x) = x\,P'_n(x) - P'_{n-1}(x),$$

$$(n+1)\,P_{n+1}(x) = (2n+1)\,x\,P_n(x) - n\,P_{n-1}(x),$$

$$(2n+1)\,P_n(x) = P'_{n+1}(x) - P'_{n-1}(x),$$

$$(x^2-1)\,P'_n(x) = n\,x\,P_n(x) - n\,P_{n-1}(x) = \frac{n(n+1)}{2n+1}\big(P_{n+1}(x) - P_{n-1}(x)\big).$$

Rodrigues' formula

$$P_n(x) = \frac{1}{2^n n!}\frac{d^n}{dx^n}[(x^2-1)^n].$$

In terms of the fractional derivatives (see 5.2.2)

$$P_n\left(x^{\frac{1}{2}}\right) = \frac{(-1)^{n+1}\,x^{\frac{n}{2}+1}}{\Gamma(n+2)}\,D_x^{n+1}\left[x^{\frac{n-1}{2}}\left(1 - x^{\frac{1}{2}}\right)^{n+1}\right],$$

$$x \neq 0; \quad n = 0, 1, 2, \ldots$$

Generating functions

$$(1 - 2xz + z^2)^{-\frac{1}{2}} = \sum_{n=0}^{\infty} P_n(x)\,z^n, \quad -1 < x < 1,\ |z| < 1,$$

$$e^{z\cos\theta} J_0(z\sin\theta) = \sum_{n=0}^{\infty} \frac{1}{n!} P_n(\cos\theta)\,z^n,$$

$$F\left(\sin\frac{\theta}{2}, \varphi\right) = \sum_{n=0}^{\infty} \frac{2(-1)^n}{2n+1} P_n(\cos\theta)\,x^{2n+1},$$

where $x = \tan\left(\frac{\varphi}{2}\right)$, $0 < \varphi < \frac{\pi}{2}$, $0 < \theta < \pi$ and $F(k, \varphi)$ is Legendre's incomplete integral of the first kind

$${}_0F_1\left(;1;\frac{z}{2}(x-1)\right){}_0F_1\left(;1;\frac{z}{2}(x+1)\right) = \sum_{n=0}^{\infty} \frac{P_n(x)}{(n!)^2} z^n, \quad -1 < x < 1,$$

$$\sum_{n=0}^{\infty} \frac{(a)_n P_n(x)}{n!} z^n = (1 - zx)^{-a}\,{}_2F_1\left(\frac{a}{2}, \frac{a+1}{2}; 1; \frac{z^2(x^2-1)}{(1-zx)^2}\right) = R^{-a}\,{}_2F_1\left(a, 1-a; 1; \frac{1}{2} - \frac{1-zx}{2R}\right),$$

where

$$R = \sqrt{1 - 2xz + z^2},$$

$$\sum_{n=0}^{\infty} \frac{P_n(x)}{n!} z^n = e^{zx} {}_0F_1\left(; 1; \frac{z^2}{4}(x^2 - 1)\right),$$

$$\sum_{n=0}^{\infty} \frac{(a)_n (1-a)_n P_n(x)}{(n!)^2} z^n$$

$$= {}_2F_1\left(a, 1-a; 1; \frac{1-z-R}{2}\right) {}_2F_1\left(a, 1-a; 1; \frac{1+z-R}{2}\right).$$

$$\sum_{n=0}^{\infty} t^n P_n(\cos\theta) P_n(\cos\varphi)$$

$$= \frac{4\frac{K}{\pi}}{\sqrt{1-2t\cos(\theta+\varphi)+t^2}\sqrt{1-2t\cos(\theta-\varphi)+t^2}}$$

where

$$K = K(k)$$

$$k = \frac{\sqrt{1-2t\cos(\theta+\varphi)+t^2} - \sqrt{1-2t\cos(\theta-\varphi)+t^2}}{\sqrt{1-2t\cos(\theta+\varphi)+t^2} + \sqrt{1-2t\cos(\theta-\varphi)+t^2}}$$

For more generating functions see RAINVILLE (1963).

The differential equation. The differential equation for the spherical harmonics [chap. IV]

$$(1-x^2)\frac{d^2u}{dx^2} - 2x\frac{du}{dx} + \left[\nu(\nu+1) - \frac{\mu^2}{1-x^2}\right]u = 0,$$

becomes, for $\mu = 0$, the "Legendre differential equation"

$$(1-x^2)\frac{d^2u}{dx^2} - 2x\frac{du}{dx} + \nu(\nu+1)u = 0.$$

The Legendre differential equation has a polynomial as a solution if and only if ν is an integer. For $\nu = n$, $n = 0, 1, 2, \ldots$ one solution of the differential equation is the polynomial $P_n(x)$ of degree n in x.

A second solution is given by $Q_n(x)$, Legendre function of the second kind

$$Q_n(x) = Q_n^{(0,0)}(x).$$

The function $Q_n(x)$ is not a polynomial and is defined in the complex x-plane cut along the real axis from -1 to $+1$. In terms the hypergeometric series the function $Q_n(x)$ is given by

$$\frac{2^{-n}(2n+1)!}{(n!)^2} Q_n(x) = (x-1)^{-n-1} {}_2F_1\left(n+1, n+1; 2n+2; \frac{2}{1-x}\right)$$

$$= (x+1)^{-n-1} {}_2F_1\left(n+1, n+1; 2n+2; \frac{2}{1+x}\right)$$

$$= x^{-n-1} {}_2F_1\left(\frac{n+1}{2}, 1+\frac{n}{2}; \frac{3}{2}+n; x^{-2}\right).$$

Some properties of $Q_n(x)$

$$Q_n(-x) = (-1)^{n+1} Q_n(x),$$

$$Q_0(x) = \frac{1}{2} \log\left(\frac{x+1}{x-1}\right),$$

$$Q_1(x) = \frac{x}{2} \log\left(\frac{x+1}{x-1}\right) - 1,$$

$$(1-x^2) \frac{d}{dx} Q_n(x) = n\,[Q_{n-1}(x) - x Q_n(x)]$$

$$= (n+1)\,[x Q_n(x) - Q_{n+1}(x)],$$

$$n Q_n(x) = x \frac{d}{dx} Q_n(x) - \frac{d}{dx} Q_{n-1}(x),$$

$$(n+1)\, Q_n(x) = \frac{d}{dx} Q_{n+1}(x) - x \frac{d}{dx} Q_n(x),$$

$$(2n+1)\, Q_n(x) = \frac{d}{dx} [Q_{n+1}(x) - Q_{n-1}(x)],$$

$$(2n+1) \int Q_n(x)\, dx = Q_{n+1}(x) - Q_{n-1}(x).$$

The function $Q_n(x)$ is related to the Legendre polynomials by

$$Q_n(x) = \frac{1}{2} \int_{-1}^{1} (x-t)^{-1} P_n(t)\, dt,$$

$$Q_n(x) = Q_0(x)\, P_n(x) - \sum_{m=0}^{\left[\frac{n+1}{2}\right]} \frac{2n-4m+3}{(2m-1)(n-m+1)} P_{n-2m+1}(x).$$

Integral representations for $Q_n(x)$

$$Q_n(x) = 2^{-n-1} \int_{-1}^{1} (1-t^2)^n (x-t)^{-n-1}\, dt$$

$$= \int_0^\infty \left[x + \sqrt{x^2-1} \cosh t\right]^{-n-1} dt,$$

$$Q_n(\cosh \xi) = \int_\xi^\infty [2(\cosh t - \cosh \xi)]^{-\frac{1}{2}} e^{-t\left(n+\frac{1}{2}\right)} dt,$$

$$\operatorname{Im} \xi = \operatorname{Im} t, \quad Re\, t \geq Re\, \xi.$$

On the branch cut, i.e. the segment of the real axis from -1 to $+1$,

$$Q_n(\xi + i0) - Q_n(\xi - i0) = -i\pi P_n(\xi),$$

where

$$-1 < \xi < 1.$$

If

$$Q_n(\xi) \equiv \frac{1}{2} [Q_n(\xi + i0) + Q_n(\xi - i0)],$$

then

$$Q_n(\xi) = \frac{1}{2} \int_{-1}^{+1} (\xi - t)^{-1} P_n(t)\, dt,$$

where the integral on the r. h. s. is evaluated as a Cauchy principal value.

Integral representations

$$P_n(\cos\theta) = \frac{1}{\pi} \int_0^{\pi} (\cos\theta + i \sin\theta \cos t)^n\, dt$$

$$= \frac{\sqrt{2}}{\pi} \int_0^{\theta} (\cos t - \cos\theta)^{-\frac{1}{2}} \cos\left[\left(n + \frac{1}{2}\right) t\right] dt,$$

$$0 < \theta < \pi,$$

$$= \frac{\sqrt{2}}{\pi} \int_{\theta}^{\pi} (\cos\theta - \cos t)^{-\frac{1}{2}} \sin\left[\left(n + \frac{1}{2}\right) t\right] dt,$$

$$0 < \theta < \pi,$$

$$= \frac{1}{\pi} \int_0^{\pi} (\cos\theta + i \sin\theta \cos t)^{-n-1}\, dt,$$

$$P_n(x) = \frac{1}{2\pi i} \int^{(0+)} z^{-n-1} (1 - 2xz + z^2)^{-\frac{1}{2}}\, dz,$$

where the contour is taken around $z = 0$ in the positive direction and excludes both zeros of $(1 - 2xz + z^2)$

$$P_n(x) = \frac{1}{\pi} \int_0^{\pi} \left[x + \sqrt{x^2 - 1} \cos t\right]^n dt,$$

$$-1 < x < 1,$$

$$2^n P_n(x) = \frac{1}{2\pi i} \int^{(x+)} (t^2 - 1)^n (t - x)^{-n-1}\, dt.$$

5.4.3 Asymptotic expansions

If x is an arbitrary real or complex number which does not belong to the closed interval $[-1, +1]$, then as $n \to \infty$

$$P_n(x) \approx (2\pi n)^{-\frac{1}{2}} (x^2 - 1)^{-\frac{1}{4}} \left[x + \sqrt{x^2 - 1}\right]^{n+\frac{1}{2}},$$

where $(x^2 - 1)^{-\frac{1}{4}}$, $(x^2 - 1)^{\frac{1}{2}}$ and $\left(x + \sqrt{x^2 - 1}\right)^{n+\frac{1}{2}}$ are real and positive if x is real and greater than 1. The result is valid uniformly in the exterior of an arbitrary closed curve which encloses the interval $[-1, 1]$ in the sense that the ratio tends to unity uniformly.

Let

$$2x = z + \frac{1}{z},$$

where x is in the complex x-plane with a branch cut along the closed interval $[-1, +1]$ and $|z| > 1$. Then

$$P_n(x) = \frac{\left(\frac{1}{2}\right)_n}{n!} z^n \sum_{m=0}^{N-1} \frac{\left[\left(\frac{1}{2}\right)_m\right]^2}{m! \left(n - m + \frac{1}{2}\right)_m} z^{-2m} \left(1 - \frac{1}{z^2}\right)^{-m-\frac{1}{2}} + O\left(n^{-N-\frac{1}{2}} |z|^n\right)$$

and holds uniformly in the same sense as the preceeding case.

$$P_n(\cos\theta) = 2(2\pi n \sin\theta)^{-\frac{1}{2}} \cos\left[\left(n + \frac{1}{2}\right)\theta - \frac{\pi}{4}\right] + O\left(n^{-\frac{3}{2}}\right),$$

$$0 < \theta < \pi.$$

The result is valid uniformly in the interval $\varepsilon \leq \theta \leq \pi - \varepsilon$

$$P_n(\cos\theta) = \left(\frac{\theta}{\sin\theta}\right)^{\frac{1}{2}} J_0\left[\left(n + \frac{1}{2}\right)\theta\right] + O\left(n^{-\frac{3}{2}}\right),$$

holds uniformly for $0 \leq \theta \leq \pi - \varepsilon$. The error terms, in a more precise form, are:

$$O\left(\theta^{\frac{1}{2}} n^{-\frac{3}{2}}\right) \qquad \frac{c}{n} \leq \theta \leq \pi - \varepsilon,$$

$$\theta^2 O(1), \qquad 0 < \theta \leq \frac{c}{n},$$

where c is an arbitrary positive constant.

$$P_n(\cos\theta) = 2\frac{\left(\frac{1}{2}\right)_n}{n!} \sum_{m=0}^{N-1} \frac{\left[\left(\frac{1}{2}\right)_m\right]^2}{m! \left(n - m + \frac{1}{2}\right)_m} \times \frac{\cos\left[\left(n - m + \frac{1}{2}\right)\theta - \frac{\pi}{4}(2m + 1)\right]}{(2\sin\theta)^{m+\frac{1}{2}}} + O\left(n^{-N-\frac{1}{2}}\right),$$

is valid for $0 < \theta < \pi$ and holds uniformly if $\varepsilon \leq \theta \leq \pi - \varepsilon$.

5.4.4 Miscellaneous results

Inequalities

For x real and greater than 1

$$P_0(x) < P_1(x) < P_2(x) \cdots < P_n(x) \cdots.$$

For x real and $x > -1$

$$P_0(x) + P_1(x) + \cdots + P_n(x) > 0,$$

$$|P_n(x)| \leq 1, \quad -1 \leq x \leq +1,$$

$$|P_n(\cos\theta)| \leq 1, \quad 0 < \theta < \pi,$$

$$[P_n(\cos\theta)]^2 > \frac{\sin(2n+1)\,\theta}{(2n+1)\sin\theta}, \quad 0 < \theta < \pi,$$

$$\sum_{n=0}^{\infty} e^{in\varphi} P_n(\cos\theta) \text{ converges for } 0 < \theta < \pi,$$

$$(\sin\theta)^{\frac{1}{2}} |P_n(\cos\theta)| < \left(\frac{2}{\pi}\right)^{\frac{1}{2}} n^{-\frac{1}{2}}, \quad 0 \leq \theta \leq \pi,$$

where the estimate cannot be improved in the sense that if $\left(\frac{2}{\pi}\right)^{\frac{1}{2}}$ is replaced by a smaller constant the inequality no longer holds.

$P_n(x)$ has exactly n zeroes all of which lie in the interval $(-1, +1)$. For $n \geq 2$, let $\mu_1, \mu_2, \ldots, \mu_{\left[\frac{n}{2}\right]}$ denote the successive relative maxima of $|P_n(x)|$ as x decreases from 1 to 0. Then

$$1 > \mu_1 > \mu_2 > \cdots > \mu_{\left[\frac{n}{2}\right]}.$$

The successive relative maxima of $(\sin\theta)^{\frac{1}{2}} |P_n(\cos\theta)|$, when θ increases from 0 to $\frac{\pi}{2}$, form an increasing sequence provided that $n \geq 2$. Let x be a parameter, $-1 < x < 1$. The zeros of the polynomial in z

$$\sum_{m=0}^{n} \binom{n}{m} P_m(x)\, z^m$$

are all real.

Series involving Legendre polynomials

$$|x|^\lambda = \sum_{m=0}^{\infty} (-1)^m \frac{\left(2m+\frac{1}{2}\right)\left(-\frac{\lambda}{2}\right)_m}{\left(\frac{1}{2}\lambda+\frac{1}{2}\right)_{m+1}} P_{2m}(x),$$

$$-1 < x < 1, \quad \lambda > -1,$$

$$|x|^{\lambda}\operatorname{sgn} x = \sum_{m=0}^{\infty}(-1)^m\left(2m+\frac{3}{2}\right)\frac{\left(\frac{1}{2}-\frac{\lambda}{2}\right)_m}{\left(1+\frac{\lambda}{2}\right)_{m+1}}P_{2m+1}(x),$$

$$-1<x<1,\quad \lambda>-1,$$

$$(1-x)^{\lambda} = 2^{\lambda}\sum_{m=0}^{\infty}\frac{2m+1}{m+\lambda+1}\frac{(-\lambda)_m}{(1+\lambda)_m}P_m(x),$$

$$-1<x<1,\quad \lambda>-\frac{3}{4},$$

$$\frac{1}{\sqrt{2}}\frac{e^{-i\frac{\varphi}{2}}}{(\cos\varphi-\cos\theta)^{\frac{1}{2}}} = \sum_{n=0}^{\infty}e^{in\varphi}P_n(\cos\theta),$$

$$0\le\theta<\pi,\quad 0\le\varphi<\pi,$$

$$\log\left[1+\operatorname{cosec}\left(\frac{\theta}{2}\right)\right] = \sum_{n=0}^{\infty}\frac{1}{n+1}P_n(\cos\theta),$$

$$\sum_{0}^{\infty}\frac{\alpha^n P_n(x)}{n+1} = \frac{1}{\alpha}\log\left[\frac{\alpha-x+\sqrt{1-2\alpha x+\alpha^2}}{1-x}\right],\quad 1\alpha 1<1$$

$$\sum_{1}^{\infty}\frac{P_n(x)}{n(n+1)} = 1-2\log\left(1+\sqrt{\frac{1-x}{2}}\right)$$

$$\sum_{n=0}^{\infty}\cos\left(n+\frac{1}{2}\right)\beta\, P_n(\cos\theta) = \begin{cases}\dfrac{1}{\sqrt{2(\cos\beta-\cos\theta)}}, & 0\le\beta<\theta<\pi,\\ 0, & 0<\theta<\beta<\pi,\end{cases}$$

$$(2n+1)\,x^{2n} = P_0(x)+5\frac{2n}{2n+3}P_2(x)$$
$$+9\frac{2n(2n-2)}{(2n+3)(2n+5)}P_4(x)+\cdots,$$

$$(2n+3)\,x^{2n+1} = 3P_1(x)+7\frac{2n}{2n+5}P_3(x)$$
$$+11\frac{2n(2n-2)}{(2n+5)(2n+7)}P_5(x)+\cdots,$$

$$\frac{2}{\pi}(1-x^2)^{-\frac{1}{2}} = P_0(x)+5\left(\frac{1}{2}\right)^2P_2(x)+9\left(\frac{1\cdot 3}{2\cdot 4}\right)^2P_4(x)+\cdots,$$

$$x \text{ real},\quad |x|<1,$$

$$\sum_{n=0}^{\infty}\frac{(-1)^n}{(2n+1)!}\zeta^{2n+1}P_{2n+1}(z) = \sin(2\zeta)\,J_0\left(\zeta\sqrt{z^2-1}\right)$$

$$\sum_{n=0}^{\infty}\frac{(-1)^n}{(2n)!}\zeta^{2n}P_{2n}(z) = \cos(2\zeta)\,J_0\left(\zeta\sqrt{z^2-1}\right)$$

$$\sum_{n=0}^{\infty}\frac{\zeta^n}{n!}P_n(z) = e^{z\zeta}I_0\left(\zeta\sqrt{z^2-1}\right)$$

$$\sum_{m=0}^{n} \binom{n}{m} P_m(x)\, y^{n-m} = (1 + 2xy + y^2)^{\frac{n}{2}} P_n\left(\frac{x+y}{\sqrt{1+2xy+y^2}}\right).$$

Addition theorem

$$P_n(\cos\theta\cos\varphi + \sin\theta\sin\varphi\cos\delta) = P_n(\cos\theta)\, P_n(\cos\varphi)$$

$$+ 2\sum_{m=1}^{n} \frac{(n-m)!}{(n+m)!} P_n^m(\cos\theta)\, P_n^m(\cos\varphi)\cos n\delta.$$

Bilinear expansion

$$\sum_{n=1}^{\infty} \frac{(2n+1)}{n(n+1)} P_n(x)\, P_n(y) = 2\log 2 - 1 - \log[(1-x)(1+y)],$$

$$-1 < x \leq y < 1.$$

see also 4.5.4

5.5 Generalized Laguerre polynomials

5.5.1 Definition, notation, and special cases

The generalized Laguerre polynomials $L_n^{(\alpha)}(x)$, for $\alpha > -1$, are the orthogonal polynomials associated with the interval $(0, \infty)$, the weight function $w(x) = x^\alpha e^{-x}$, and standardized by the condition that the coefficient of x^n in $L_n^{(\alpha)}(x)$ equal $\frac{(-1)^n}{n!}$. The special case $\alpha = 0$ is denoted by $L_n(x)$, i.e. $L_n^{(0)}(x) \equiv L_n(x)$.

Special cases and relations

$$L_0^{(\alpha)}(x) = 1,$$

$$L_1^{(\alpha)}(x) = \alpha + 1 - x,$$

$$L_2^{(\alpha)}(x) = \frac{1}{2}[(\alpha+1)(\alpha+2) - 2(\alpha+2)x + x^2],$$

$$L_0(x) = 1,$$

$$L_1(x) = 1 - x,$$

$$L_2(x) = 1 - 2x + \frac{x^2}{2},$$

$$L_3(x) = 1 - 3x + \frac{3}{2}x^2 - \frac{x^3}{6},$$

$$L_4(x) = 1 - 4x + 3x^2 - \frac{2}{3}x^3 + \frac{x^4}{24},$$

$$L_5(x) = 1 - 5x + 5x^2 - \frac{5}{3}x^3 + \frac{5}{24}x^4 - \frac{x^5}{120},$$

$$L_n(x) = 1 - \binom{n}{1} x + \binom{n}{2} \frac{x^2}{2!} - \binom{n}{3} \frac{x^3}{3!} + \cdots + (-1)^n \frac{x^n}{n!},$$

$$L_n^{(\alpha)}(0) = \binom{n+\alpha}{n} = \frac{(\alpha+1)_n}{n!},$$

$$L_n^{(m)}(x) = (-1)^m \frac{d^m}{dx^m} [L_{m+n}(x)].$$

Hermite polynomials $H_n(x)$ are a special case of $L_n^{(\alpha)}(x)$ given by

$$L_n^{\left(\frac{1}{2}\right)}(x) = \frac{(-1)^n x^{-\frac{1}{2}}}{n!\, 2^{2n+1}} H_{2n+1}\left(\sqrt{x}\right),$$

$$L_n^{\left(-\frac{1}{2}\right)}(x) = \frac{(-1)^n}{n!\, 2^{2n}} H_{2n}\left(\sqrt{x}\right).$$

Another form of these relations is

$$H_{2n}(x) = (-1)^n\, 2^{2n} n!\, L_n^{\left(-\frac{1}{2}\right)}(x^2),$$

$$H_{2n+1}(x) = (-1)^n\, 2^{2n+1} n!\, x L_n^{\left(\frac{1}{2}\right)}(x^2).$$

5.5.2 Elementary results

Explicit expressions

$$L_n^{(\alpha)}(x) = \sum_{m=0}^{n} (-1)^m \binom{n+\alpha}{n-m} \frac{x^m}{m!}$$

$$= \binom{n+\alpha}{n} {}_1F_1(-n; \alpha+1; x).$$

These expressions can be used to extend the definition of $L_n^{(\alpha)}(x)$ to arbitrary complex values of α.

If $\alpha = -k$, $1 \leq k \leq n$

$$L_n^{(-k)}(x) = (-x)^k \frac{(n-k)!}{n!} L_{n-k}^{(k)}(x).$$

For $n \geq 1$, $L_n^{(\alpha)}(0) = 0$ if and only if $\alpha = -k$, k an integer and $1 \leq k \leq n$. Furthermore this zero is of order k exactly.

Some integrals associated with $L_n^{(\alpha)}(x)$

$$\int_0^x L_m(t)\, L_n(x-t)\, dt = \int_0^x L_{m+n}(t)\, dt$$

$$= L_{m+n}(x) - L_{m+n+1}(x),$$

$$\int_0^x L_n(t)\, dt = \frac{x}{n+1} \left\{ L_n(x) - \frac{d}{dx} L_n(x) \right\},$$

$$\int_x^\infty e^{-t} L_n^{(\alpha)}(t)\, dt = e^{-x} [L_n^{(\alpha)}(x) - L_{n-1}^{(\alpha)}(x)],$$

$$\int_0^x (x-t)^{\beta-1} t^\alpha L_n^{(\alpha)}(t)\, dt = \frac{\Gamma(\alpha+n+1)\,\Gamma(\beta)}{\Gamma(\alpha+\beta+n+1)} x^{\alpha+\beta} L_n^{(\alpha+\beta)}(x),$$

$$Re\,\alpha > -1,\; Re\,\beta > 0.$$

Orthogonality relation

$$\int_0^\infty e^{-x} x^\alpha L_m^{(\alpha)}(x)\, L_n^{(\alpha)}(x)\, dx = \begin{cases} 0, & m \neq n, \\ \Gamma(1+\alpha)\binom{n+\alpha}{n}, & m = n. \end{cases}$$

Differentiation formulas

$$\frac{d}{dx} L_n^{(\alpha)}(x) = -L_{n-1}^{(\alpha+1)}(x)$$

$$= \frac{n L_n^{(\alpha)}(x) - (n+\alpha) L_{n-1}^{(\alpha)}(x)}{x},$$

$$\frac{d}{dx}\left[L_n^{(\alpha)}(x) - L_{n+1}^{(\alpha)}(x)\right] = L_n^{(\alpha)}(x),$$

$$\frac{d}{dx}\left[e^{-x} x^\alpha L_n^{(\alpha)}(x)\right] = (n+1)\, e^{-x} x^{\alpha-1} L_{n+1}^{(\alpha-1)}(x),$$

$$\frac{d^m}{dx^m}\left[x^\alpha e^{-x} L_n^{(\alpha)}(x)\right] = \frac{(m+n)!}{n!} e^{-x} x^{\alpha-m} L_{n+m}^{(\alpha-m)}(x),$$

$$\frac{d}{dx}\left[x^\alpha L_n^{(\alpha)}(x)\right] = (n+\alpha)\, x^{\alpha-1} L_n^{(\alpha-1)}(x),$$

$$\frac{d^m}{dx^m}\left[x^\alpha L_n^{(\alpha)}(x)\right] = (n+\alpha-m+1)_m\, x^{\alpha-m} L_n^{(\alpha-m)}(x).$$

Rodrigues' formula

$$L_n^{(\alpha)}(x) = \frac{(-1)^n}{n!} x^{n+\alpha+1} e^t \frac{d^n}{dt^n}\left[t^{\alpha+1} e^{-t}\right],$$

where

$$xt = 1,$$

$$L_n^{(\alpha)}(x) = x^{-\alpha} \frac{e^x}{n!} \frac{d^n}{dx^n}\left[e^{-x} x^{n+\alpha}\right].$$

Recurrence relations

$$n L_n^{(\alpha)}(x) = (2n+\alpha-1-x)\, L_{n-1}^{(\alpha)}(x) - (n+\alpha-1)\, L_{n-2}^{(\alpha)}(x),$$

$$n = 2, 3, 4, \ldots,$$

$$x L_n^{(\alpha+1)}(x) = (n+\alpha+1)\, L_n^{(\alpha)}(x) - (n+1)\, L_{n+1}^{(\alpha)}(x)$$

$$= (n+\alpha)\, L_{n-1}^{(\alpha)}(x) - (n-x)\, L_n^{(\alpha)}(x),$$

$$L_n^{(\alpha)}(x) = L_n^{(\alpha+1)}(x) - L_{n-1}^{(\alpha+1)}(x),$$

$$(n+\alpha)\, L_n^{(\alpha-1)}(x) = (n+1)\, L_{n+1}^{(\alpha)}(x) - (n+1-x)\, L_n^{(\alpha)}(x).$$

Generating functions

$$\sum_{n=0}^{\infty} L_n^{(\alpha)}(x)\, z^n = (1-z)^{-\alpha-1} \exp\left[\frac{xz}{z-1}\right],$$
$$|z| < 1,$$

$$\sum_{n=0}^{\infty} \frac{L_n^{(\alpha)}(x)}{\Gamma(\alpha+n+1)} z^n = e^z (zx)^{-\frac{\alpha}{2}} J_\alpha\left(2\sqrt{zx}\right),$$

where $J_\alpha(z)$ is the Bessel function,

$$\sum_{n=0}^{\infty} \frac{L_n^{(\alpha)}(x)}{\Gamma(\alpha+n+1)} z^n = \frac{1}{\Gamma(\alpha+1)} e^z\, {}_0F_1(;1+\alpha;-zt),$$

$$= e^z (zt)^{-\frac{\alpha}{2}} J_\alpha\left(2\sqrt{zt}\right)$$

$$\sum_{n=0}^{\infty} \frac{(c)_n L_n^{(\alpha)}(x)}{(1+\alpha)_n} z^n = (1-z)^{-c}\, {}_1F_1\left(c; 1+\alpha; \frac{-zx}{1-z}\right),$$

$$\sum_{n=0}^{\infty} L_n^{(\alpha-n)}(x)\, z^n = e^{-zx}(1+z)^\alpha,$$

$$\sum_{n=0}^{\infty} \frac{n!}{(1+\alpha)_n} L_n^{(\alpha)}(x)\, L_n^{(\alpha)}(y)\, z^n$$

$$= (1-z)^{-\alpha-1} \exp\left[-\frac{z(x+y)}{1-z}\right] {}_0F_1\left(;1+\alpha; \frac{xy}{z(1-z)^2}\right),$$

$$= \frac{\Gamma(\alpha+1)}{1-z} \left(\frac{z}{xy}\right)^{\frac{\alpha}{2}} \exp\left[-\frac{z(x+y)}{1-z}\right] I_\alpha\left(\frac{2}{1-z}\sqrt{\frac{xy}{z}}\right)$$

$$\sum_{n=0}^{\infty} \frac{L_n(x)\, L_n(y)}{n+1} = e^{x+y} \int_{\max(x,y)}^{\infty} \frac{1}{t} e^{-t}\, dt,$$
$$x > 0, \; y > 0.$$

$$K(x,y,t) = \frac{t^{-\frac{\alpha}{2}}}{1-t} \exp\left[-\frac{1}{2}(x+y)\frac{1+t}{1-t}\right] I_\alpha\left(2\frac{\sqrt{xyt}}{1-t}\right)$$

$$= \sum_{n=0}^{\infty} \frac{t^n\, n!\, e^{-\frac{1}{2}(x+y)}}{\Gamma(n+\alpha+1)} \cdot (xy)^{\frac{\alpha}{2}} L_n^{(\alpha)}(x)\, L_n^{(\alpha)}(y).$$

For more generating functions see RAINVILLE (1963).

The differential equation. The polynomials $L_n^{(\alpha)}(x)$ satisfy the differential equation

$$x\frac{d^2y}{dx^2} + (\alpha+1-x)\frac{dy}{dx} + ny = 0.$$

For $\alpha > -1$, it is known that a necessary and sufficent condition for the differential equation

$$xy'' + (\alpha + 1 - x)\, y' + \lambda y = 0 \tag{1}$$

to have a polynomial solution is that $\lambda = n$. Furthermore $L_n^{(\alpha)}(x)$ is the only polynomial solution. Finally, for $\alpha > -1$, the polynomials $L_n^{(\alpha)}(x)$ are the only solutions of (1) which are analytic near $x = 0$.

Other differential equations connected with $L_n^{(\alpha)}(x)$ are:

$$xy'' + (x + 1)\, y' + \left(n + \frac{\alpha}{2} + 1 - \frac{\alpha^2}{4x}\right) y = 0,$$

$$y(x) = e^{-x} x^{\frac{\alpha}{2}}\, L_n^{(\alpha)}(x),$$

$$z'' + \left(\frac{2n + \alpha + 1}{2x} - \frac{1}{4} + \frac{1 - \alpha^2}{4x^2}\right) z = 0,$$

$$z(x) = e^{-\frac{x}{2}} x^{\frac{(\alpha+1)}{2}}\, L_n^{(\alpha)}(x),$$

$$u'' + \left(4n + 2\alpha + 2 - x^2 + \frac{1 - 4\alpha^2}{4x^2}\right) u = 0,$$

$$u(x) = e^{-\frac{x^2}{2}} x^{\alpha + \frac{1}{2}}\, L_n^{(\alpha)}(x^2),$$

$$xv'' + (x + 1)\, v' + \left(n + \frac{\alpha}{2} + 1 - \frac{\alpha^2}{4x}\right) v = 0,$$

$$v(x) = e^{-x} x^{-\frac{\alpha}{2}}\, L_n^{(\alpha)}(x).$$

For non-integral values of α, a linearly independent solution of (1) is given by

$$y(x) = x^{-\alpha}\, {}_1F_1(-n - \alpha; 1 - \alpha; x).$$

The same is true if α is a negative integer less than $-n$.

Integral representations

$$L_n^{(\alpha)}(x) = \frac{x^{-\frac{\alpha}{2}} e^x}{n!} \int_0^\infty e^{-z} z^{n + \frac{\alpha}{2}} J_\alpha\left(2\sqrt{zx}\right) dz,$$

$$n = 0, 1, 2, \ldots; \quad n + \alpha > -1,$$

$$L_n^{(\alpha)}(x) = \frac{(-1)^n\, \Gamma(n + \alpha + 1)}{\Gamma\left(\frac{1}{2}\right) \Gamma\left(\alpha + \frac{1}{2}\right) (2n)!} \int_{-1}^{+1} (1 - z^2)^{\alpha - \frac{1}{2}} H_{2n}\left(\sqrt{x}\, z\right) dz,$$

$$\alpha > -\frac{1}{2},$$

$H_n(x)$ are Hermite polynomials,

$$L_n^{(\alpha)}(x) = \frac{(-1)^n x^{-\frac{\alpha}{2}} e^{\frac{x}{2}}}{2} \int_0^\infty e^{-\frac{t}{2}} t^{\frac{\alpha}{2}} L_n^{(\alpha)}(t) J_\alpha\left(\sqrt{xt}\right) dt$$

$$= \frac{\Gamma(n+\alpha+1)}{\Gamma(\alpha-\beta)} x^{-\alpha} \int_0^x \frac{(x-t)^{\alpha-\beta-1}}{\Gamma(n+\beta+1)} t^\beta L_n^{(\beta)}(t)\, dt,$$

$$\alpha > \beta > -1,$$

$$L_n^{(\alpha)}(x) = \frac{e^{\frac{x}{2}}}{2\pi i} \int_\infty^{(1-)} \exp\left[-\frac{x}{2}\frac{1+z}{1-z}\right] (1-z)^{-\alpha-1} z^{-n-1}\, dz$$

$$= \frac{e^{\frac{x}{2}}}{2\pi i} \int_{-\infty}^{(0+)} \exp\left[-\frac{x}{2}\frac{1+e^{-t}}{1-e^{-t}}\right] (1-e^{-t})^{-\alpha-1} e^{nt}\, dt$$

$$= \frac{e^x x^{-\frac{\alpha}{2}} e^{i\pi\left(\alpha-\frac{1}{2}\right)}}{n!\, 2\sin(\pi\alpha)} \int_\infty^{(0+)} e^{-z} z^{n+\frac{\alpha}{2}} J_\alpha\left(2\sqrt{zx}\right) dz$$

$$= \frac{(-1)^n 2^{-\alpha} e^{\frac{x}{2}}}{2\pi i} \int^{(1+)} e^{-z\frac{x}{2}} (1-z^2)^{\frac{\alpha-1}{2}} \left(\frac{1+z}{1-z}\right)^{n+\frac{\alpha+1}{2}} dz$$

$$= \frac{e^x x^{-\alpha}}{2\pi i} \int_C e^{-z} z^{n+\alpha} (z-x)^{-n-1}\, dz,$$

where the contour C encloses the point $z = x$ but not $z = 0$

$$L_n^{(\alpha)}(x) = \frac{x^{-\alpha}}{2\pi i} \int^{(0+)} z^{-n-1} (z+x)^{n+\alpha} e^{-z}\, dz,$$

where the contour excludes $z = -x$.

Integral transforms of $L_n^{(\alpha)}(x)$

$$\int_0^\infty e^{-zt} t^\alpha L_n^{(\alpha)}(t)\, dt = \frac{\Gamma(n+\alpha+1)}{\Gamma(n+1)} z^{-n-\alpha-1} (z-1)^n,$$

$$Re\, z > 0, \quad Re\, \alpha > -1,$$

$$e^{-\frac{x}{2}} x^{\frac{\alpha}{2}} L_n^{(\alpha)}(x) = \frac{(-1)^n}{2} \int_0^\infty J_\alpha\left(\sqrt{xy}\right) e^{-\frac{y}{2}} y^{\frac{\alpha}{2}} L_n^{(\alpha)}(y)\, dy,$$

$$\int_0^\infty e^{-zx} x^{n+\frac{\alpha}{2}} J_\alpha\left[2\sqrt{bx}\right] dx = \frac{n!\, b^{\frac{\alpha}{2}} z^{-n-\alpha-1}}{e^{\frac{b}{z}}} L_n^{(\alpha)}\left(\frac{b}{z}\right),$$

$$\int_0^\infty e^{-zx} x^\lambda L_n^{(\alpha)}(x)\, dx = \frac{\Gamma(\lambda+1)\,\Gamma(\alpha+n+1)}{n!\,\Gamma(\alpha+1)} z^{-\lambda-1}$$
$$\times\, {}_2F_1(-n, \lambda+1; \alpha+1; z^{-1}),$$
$$Re\,\lambda > -1, \quad Re\, z > 0.$$

5.5.3 Asymptotic expansions

For large n, $x > 0$

$$L_n^{(\alpha)}(x) = \pi^{-\frac{1}{2}} e^{\frac{x}{2}} x^{-\frac{1}{2}\alpha-\frac{1}{4}} n^{\frac{\alpha}{2}-\frac{1}{4}} \cos\left[2\sqrt{nx} - \frac{\pi}{2}\left(\alpha+\frac{1}{2}\right)\right]$$
$$+\, O\left(n^{\frac{\alpha}{2}-\frac{3}{4}}\right).$$

The result holds uniformly in $a \leq x \leq b$, $a > 0$.

$$L_n^{(\alpha)}(x) = e^{\frac{x}{2}} x^{-\frac{\alpha}{2}} A^{-\frac{\alpha}{2}} \frac{\Gamma(n+\alpha+1)}{n!} J_\alpha\left(2\sqrt{Ax}\right) + O\left(n^{\frac{\alpha}{2}-\frac{3}{4}}\right),$$

where

$$A = n + \frac{\alpha+1}{2}; \quad w \geq x > 0; \quad \alpha > -1, \ \alpha \neq 0.$$

Finally

$$n^{-\frac{1}{2}} \log\left|L_n^{(\alpha)}(x)\right| \approx 2\, Re\left\{(-x)^{\frac{1}{2}}\right\}$$

as $n \to \infty$ and x is in the complex plane cut along the negative real axis. The value of $(-x)^{\frac{1}{2}}$ is taken real and positive for $x < 0$.

The results given below are due to Erdelyi (Golden Jubilee Commeration Volume of the Indian Math. Soc. (1960) 235—250). To give asymptotic approximation, the real axis is divided into five distinct regions:

(1) x near 0, (2) $0 < x < \nu$ (3) x near ν (4) $x > \nu$ (5) $x < 0$

where

$$\nu = 4n + 2\alpha + 2.$$

Case 1:

a)

$$n x \to 0$$

$$L_n^{(\alpha)}(x) = \frac{\Gamma(n+\alpha+1)}{n!\,\Gamma(\alpha+1)}\left[1 + O(n x)\right]$$

b)

$$\nu^{-\frac{1}{3}} x \to 0$$

$$L_n^{(\alpha)}(x) = \frac{\Gamma(n+\alpha+1)}{n!}\left(\frac{\nu x}{4}\right)^{-\frac{\alpha}{2}}$$

$$\times \exp\left(\frac{x}{2}\right)\left\{J_\alpha\left(\sqrt{\nu x}\right) + \tilde{J}_\alpha\left(\sqrt{\nu x}\right)\left[O\left(\frac{x^{\frac{3}{2}}}{\nu^{\frac{1}{2}}}\right) + O\left(\frac{x^{\frac{1}{2}}}{\nu^{\frac{3}{2}}}\right)\right]\right\}$$

where

$$\tilde{J}_\alpha(z) = \begin{cases} J_\alpha(z), & z \text{ imaginary or } 0 \leq z \leq \delta \\ \left(|J_\alpha(z)|^2 + |Y_\alpha(z)|^2\right)^{\frac{1}{2}}, & z > \delta \end{cases}$$

and δ is such that $J_\alpha(z) \neq 0$ when $0 < |z| \leq \delta$, $\alpha \geq 0$.

Case 2:

$$0 < \varrho < \frac{\pi}{2}, \quad N\varrho^3 \to \infty, \quad N\left(\frac{\pi}{2} - \varrho\right) \to \infty, \quad x = 4N\cos^2\varrho, \quad \nu = 4N$$

$$L_n^{(\alpha)}(4N\cos^2\varrho) = \frac{(-1)^n \exp(4N\cos^2\varrho)}{(2\cos\varrho)^\alpha (\pi n \sin 2\varrho)^{\frac{1}{2}}}\left\{\cos\left[N(\sin 2\varrho - 2\varrho) + \frac{\pi}{4}\right]\right.$$

$$\left. + O\left(\frac{1}{N\varrho^3}\right) + O\left(\frac{1}{N\left(\frac{\pi}{2} - \varrho\right)}\right)\right\}.$$

For the case $0 < \delta \leq \varrho \leq \frac{\pi}{2} - \delta$, the error term becomes $O\left(\frac{1}{n}\right)$.

Case 3:

(a)
$$\nu \to \infty, \quad x - \nu = 0\left(\nu^{\frac{1}{3}}\right),$$

$$L_n^{(\alpha)}(x) = \frac{(-1)^n e^{\frac{x}{2}} n^{-\frac{1}{3}}}{2^{\alpha+\frac{1}{3}} 3^{\frac{2}{3}} \Gamma\left(\frac{2}{3}\right)}\left[1 + O\left(\frac{1}{\nu}\right) + O\left(\frac{x - \nu}{\nu^{\frac{1}{3}}}\right)\right]$$

(b)
$$\nu \to \infty, \quad x - \nu = 0\left(\nu^{\frac{3}{5}}\right)$$

$$L_n^{(\alpha)}(x) = \frac{(-1)^n e^{\frac{x}{2}}}{2^{\alpha+\frac{1}{3}} n^{\frac{1}{3}}}\left\{Ai\left(\frac{x-\nu}{(4\nu)^{\frac{1}{3}}}\right) + \tilde{Ai}\left(\frac{x-\nu}{(4\nu)^{\frac{1}{3}}}\right)\left[O\left(\frac{1}{\nu}\right) + O\left(\frac{x-\nu}{\nu}\right)\right.\right.$$

$$\left.\left. + O\left(\frac{(x-\nu)^{\frac{5}{2}}}{\nu^{\frac{3}{2}}}\right)\right]\right\}$$

where

$$\tilde{Ai}(z) = \begin{cases} Ai(z), & z \geq 0 \\ [|Ai(z)|^2 + |Bi(z)|^2]^{\frac{1}{2}}, & z < 0 \end{cases}$$

and $Ai(z)$, $Bi(z)$ are the Airy functions defined in 4.3

Case 4:

$$\sigma > 0, \quad \nu\sigma^3 \to \infty, \quad x = \nu \cosh^2 \sigma$$

$$L_n^{(\alpha)}(\nu \cosh^2 \sigma) = \frac{(-1)^n \exp\left[\frac{\nu}{4}(1 + 2\sigma + e^{-2\sigma})\right]}{(2\cosh\sigma)^{\alpha+1}\left(\frac{\pi\nu}{2}\tanh\sigma\right)^{\frac{1}{2}}}\left[1 + O\left(\frac{1+\sigma^3}{\nu\sigma^3}\right)\right]$$

Case 5:

$$\sigma > 0, \quad \nu\sigma \to \infty, \quad x = -\nu \sinh^2 \sigma$$

$$L_n^{(\alpha)}(-\nu \sinh^2 \sigma) = \frac{\exp\left[\frac{\nu}{4}(1 + 2\sigma - e^{-2\sigma})\right]}{(2\sinh\sigma)^{\alpha+1}\left(\frac{\pi\nu}{2}\coth\sigma\right)^{\frac{1}{2}}}\left[1 + O\left(\frac{1+\sigma}{\nu\sigma}\right)\right].$$

5.5.4 Miscellaneous results

Limit relations

$$L_n^{(\alpha)}(x) = \lim_{\beta\to\infty} P_n^{(\alpha,\beta)}\left(1 - \frac{2x}{\beta}\right),$$

$$\lim_{\alpha\to\infty} \alpha^{-\frac{n}{2}} L_n^{(\alpha)}\left(\sqrt{\alpha}\, x + \alpha\right) = \frac{(-1)^n 2^{-\frac{n}{2}}}{n!} H_n\left(\frac{x}{\sqrt{2}}\right),$$

$$\lim_{\alpha\to\infty} \alpha^{-n} L_n^{(\alpha)}(x) = \frac{(1-x)^n}{n!}.$$

Zeros of $L_n^{(\alpha)}(x)$

Let

$j_{\alpha,m}$ = mth positive zero of $J_\alpha(x)$,

$x_{n,m}$ = mth zero of $L_n^{(\alpha)}(x)$ in the increasing order.

Then, for $\alpha > -1$

$$\lim_{n\to\infty} n x_{n,m} = \left(\frac{1}{2} j_{\alpha,m}\right)^2$$

and

$$\frac{(j_{\alpha,m})^2}{4\left(n+\frac{\alpha}{2}+\frac{1}{2}\right)} < x_{n,m} < \left[m+\frac{\alpha+1}{2}\right]$$

$$\times\frac{2m+\alpha+1+\left[\frac{1}{4}-\alpha^2+\sqrt{2m+\alpha+1}\right]^{\frac{1}{2}}}{n+\frac{\alpha}{2}+\frac{1}{2}},$$

$$m=1,2,3,\ldots,n;\ n=1,2,3,\ldots$$

For more results on the zeros of $L_n^{(\alpha)}(x)$ see SZEGÖ (1959).

Inequalities

Let α be an arbitrary real number, c and w fixed positive constants then, as $n\to\infty$

$$L_n^{(\alpha)}(x)=\begin{cases} x^{-\frac{\alpha}{2}-\frac{1}{4}}\,O\left(n^{\frac{\alpha}{2}-\frac{1}{4}}\right), & \frac{c}{n}\le x\le w,\\ O(n^{\alpha}), & 0\le x\le\frac{c}{n}.\end{cases}$$

The following results give information with respect to the relative maxima of the functions associated with $L_n^{(\alpha)}(x)$.

1. For α real, the sequence formed by the relative maxima of $|L_n^{(\alpha)}(x)|$ and by the value of this function at $x=0$, is increasing or decreasing according as

$$x>\alpha+\frac{1}{2}\quad\text{or}\quad x<\alpha+\frac{1}{2}.$$

2. For α real, the successive relative maxima of

$$\exp\left(-\frac{x}{2}\right)x^{\frac{\alpha+1}{2}}L_n^{(\alpha)}(x)$$

form an increasing sequence for $x>x_0$, where

$$x_0=\begin{cases}0, & \alpha^2\le 1,\\ \frac{\alpha^2-1}{2n+\alpha+1}, & \alpha^2>1.\end{cases}$$

3. For α real, the successive relative maxima of

$$\exp\left(-\frac{x}{2}\right)x^{\frac{2\alpha+1}{4}}\left|L_n^{(\alpha)}(x)\right|$$

form an increasing sequence if
$x > x_1$ where

$$x_1 = \begin{cases} 0, & 4\alpha^2 \leq 1, \\ \left(\alpha - \frac{1}{4}\right)^{\frac{1}{2}}, & 4\alpha^2 > 1. \end{cases}$$

Finite sums associated with $L_n^{(\alpha)}(x)$

$$\sum_{m=0}^{n} L_m^{(\alpha)}(x) = L_n^{(\alpha+1)}(x),$$

$$\sum_{m=0}^{n} \frac{(\alpha-\beta)_m}{m!} L_{n-m}^{(\beta)}(x) = L_n^{(\alpha)}(x),$$

$$\sum_{m=0}^{n} \binom{n+\alpha}{n-m} (-1)^m L_m^{(\alpha)}(x) = \frac{x^n}{n!},$$

$$\sum_{m=0}^{n} \binom{n}{m} \frac{L_m^{(\alpha)}(x)}{L_m^{(\alpha)}(0)} y^{n-m} = (y+1)^n \frac{L_n^{(\alpha)}\left(\frac{x}{y+1}\right)}{L_n^{(\alpha)}(0)},$$

$$\sum_{m=0}^{n} \frac{(xy)^m}{m!\,\Gamma(n+\alpha+1)} L_{n-m}^{(\alpha+2m)}(x+y) = \frac{\Gamma(n+1)}{\Gamma(n+\alpha+1)} L_n^{(\alpha)}(x)\, L_n^{(\alpha)}(y),$$

$$\sum_{m=0}^{n} \binom{n+\alpha}{m} z^{n-m} (1-z)^m L_{n-m}^{(\alpha)}(x) = L_n^{(\alpha)}(zx),$$

$$\sum_{m=0}^{n} L_n^{(\alpha)}(x)\, L_{n-m}^{(\beta)}(y) = L_n^{(\alpha+\beta+1)}(x+y),$$

$$\sum_{m=0}^{n} \frac{m!}{\Gamma(\alpha+m+1)} L_m^{(\alpha)}(x)\, L_m^{(\alpha)}(y)$$

$$= \frac{(n+1)!}{\Gamma(n+\alpha+1)} \frac{L_n^{(\alpha)}(x)\, L_{n+1}^{(\alpha)}(y) - L_{n+1}^{(\alpha)}(x)\, L_n^{(\alpha)}(y)}{x-y}.$$

5.6 Hermite polynomials

5.6.1 Definition, notation, and special cases

The Hermite polynomials $H_n(x)$ are defined as the orthogonal polynomials associated with the weight function $w(x) = e^{-x^2}$, over the interval $(-\infty, \infty)$ and standardized by the relation

$$\int_{-\infty}^{\infty} e^{-x^2} H_m(x)\, H_n(x)\, dx = \begin{cases} 0, & m \neq n, \\ \pi^{\frac{1}{2}}\, 2^n n!, & n = m. \end{cases}$$

The orthogonal polynomials $He_n(x)$, also called Hermite polynomials are associated with the weight function $w(x) = e^{-\frac{x^2}{2}}$, over the interval $(-\infty, \infty)$ and are standardized by the relation

$$\int_{-\infty}^{\infty} e^{-\frac{x^2}{2}} He_n(x)\, He_m(x)\, dx = \begin{cases} 0, & m \neq n, \\ n!\sqrt{2\pi}, & m = n. \end{cases}$$

These polynomials are related to each other by the following results

$$He_n(x) = 2^{-\frac{n}{2}} H_n\left(\frac{x}{\sqrt{2}}\right),$$

$$H_n(x) = 2^{\frac{n}{2}} H_n\left(x\sqrt{2}\right).$$

Special cases and relations

$$H_0(x) = 1, \qquad He_0(x) = 1,$$

$$H_1(x) = 2x, \qquad He_1(x) = x,$$

$$H_2(x) = 4x^2 - 2, \qquad He_2(x) = x^2 - 1,$$

$$H_3(x) = 8x^3 - 4x, \qquad He_3(x) = x^3 - 3x,$$

$$H_4(x) = 16x^4 - 48x^2 + 12, \qquad He_4(x) = x^4 - 6x^2 + 3,$$

$$H_5(x) = 32x^5 - 160x^3 + 120x, \qquad He_5(x) = x^5 - 10x^3 + 15x,$$

$$H_n(-x) = (-1)^n H_n(x), \qquad He_n(-x) = (-1)^n He_n(x),$$

$$H_{2n+1}(0) = 0, \qquad He_{2n+1}(0) = 0,$$

$$H_{2n}(0) = (-1)^n \frac{(2n)!}{n!}, \qquad He_{2n}(0) = \frac{(-1)^n (2n)!}{n!\, 2^n}.$$

5.6.2 Elementary results

Explicit expressions

$$H_n(x) = n! \sum_{m=0}^{\left[\frac{n}{2}\right]} \frac{(-1)^m (2x)^{n-2m}}{m!\,(n-2m)!},$$

$$He_n(x) = n! \sum_{m=0}^{\left[\frac{n}{2}\right]} \frac{(-1)^m x^{n-2m}}{m!\, 2^m (n-2m)!}.$$

Relation with other polynomials

$$H_{2n}(x) = (-1)^n 2^{2n} n!\, L_n^{\left(-\frac{1}{2}\right)}(x^2),$$

$$H_{2n+1}(x) = (-1)^n 2^{2n+1} n!\, x L_n^{\left(\frac{1}{2}\right)}(x^2),$$

$$H_n\left(\frac{x}{\sqrt{2}}\right) = (-1)^n \, 2^{\frac{n}{2}} \, n! \lim_{\alpha\to\infty} \left[\alpha^{-\frac{n}{2}} L_n^{(\alpha)}\left(\sqrt{\alpha}\, x + \alpha\right)\right],$$

$$L_n(x^2 + y^2) = \frac{(-1)^n}{n!} \sum_{m=0}^{n} \binom{n}{m} H_{2m}(x)\, H_{2n-2m}(y),$$

$$H_n(x) = n! \lim_{\lambda\to\infty} \lambda^{-\frac{n}{2}} C_n^{\lambda}\left(\frac{x}{\sqrt{\lambda}}\right).$$

Other representations

$$H_{2n}(x) = \frac{(-1)^n\,(2n)!}{n!}\, {}_1F_1\left(-n; \frac{1}{2}; x^2\right),$$

$$H_{2n+1}(x) = \frac{(-1)^n\,(2n+1)!}{n!}\, 2x\, {}_1F_1\left(-n; \frac{3}{2}; x^2\right),$$

$$H_n(x) = 2^{\frac{n}{2}}\, e^{\frac{x^2}{2}}\, D_n\left(x\sqrt{2}\right),$$

$$He_{2n}(x) = \frac{(-1)^n\,(2n)!}{2^n\, n!}\, {}_1F_1\left(-n; \frac{1}{2}; \frac{x^2}{2}\right),$$

$$He_{2n+1}(x) = \frac{(-1)^n\,(2n+1)!}{2^n\, n!}\, x\, {}_1F_1\left(-n; \frac{3}{2}; \frac{x^2}{2}\right).$$

Some definite integrals

$$\int_0^x e^{-t^2} H_n(t)\, dt = H_{n-1}(0) - e^{-x^2} H_{n-1}(x),$$

$$\int_0^x H_n(t)\, dt = \frac{H_{n+1}(x) - H_{n+1}(0)}{2(n+1)},$$

$$\int_{-\infty}^{\infty} e^{-t^2} H_{2n}(xt)\, dt = \pi^{\frac{1}{2}} \frac{(2n)!}{n!} (x^2 - 1)^n,$$

$$\int_{-\infty}^{\infty} t e^{-t^2} H_{2n+1}(xt)\, dt = \pi^{\frac{1}{2}} \frac{(2n+1)!}{n!} x (x^2 - 1)^n,$$

$$\int_{-\infty}^{\infty} t^n e^{-t^2} H_n(xt)\, dt = \pi^{\frac{1}{2}}\, n!\, P_n(x),$$

$$\int_0^{\infty} e^{-t^2} [H_n(t)]^2 \cos\left(xt\sqrt{2}\right) dt = \pi^{\frac{1}{2}}\, 2^{n-1} n!\, L_n(x^2)\, e^{-x^2},$$

$$\int_{-1}^{+1} (1 - t^2)^{\alpha - \frac{1}{2}} H_{2n}\left(\sqrt{x}\, t\right) dt = \frac{(-1)^n\,(2n)!}{\Gamma(n + \alpha + 1)} \pi^{\frac{1}{2}}\, \Gamma\left(\alpha + \frac{1}{2}\right) L_n^{(\alpha)}(x).$$

Orthogonality relations

$$\int_{-\infty}^{\infty} e^{-x^2} H_m(x)\, H_n(x)\, dt = \begin{cases} 0, & m \neq n, \\ \pi^{\frac{1}{2}}\, 2^n n!, & m = n, \end{cases}$$

$$\int_{-\infty}^{\infty} e^{-\frac{x^2}{2}} He_m(x)\, He_n(x)\, dx = \begin{cases} 0, & m \neq n, \\ n!\,(2\pi)^{\frac{1}{2}}, & m = n. \end{cases}$$

Differentiation formulas

$$\frac{d}{dx} H_n(x) = 2n H_{n-1}(x),$$

$$\frac{d}{dx} He_n(x) = n He_{n-1}(x).$$

Recurrence relations

$$H_{n+1}(x) - 2x H_n(x) + 2n H_{n-1}(x) = 0,$$

$$n = 1, 2, 3, \ldots,$$

$$He_{n+1}(x) - x He_n(x) + n He_{n-1}(x) = 0,$$

$$H_n(x) = 2x H_{n-1}(x) - H'_{n-1}(x),$$

$$n H_n(x) = -n H'_{n-1}(x) + x H'_n(x),$$

$$He_{n+1}(x) = x He_n(x) - He'_n(x).$$

Rodrigues' formulas

$$H_n(x) = (-1)^n e^{x^2} \frac{d^n}{dx^n} (e^{-x^2}),$$

$$He_n(x) = (-1)^n e^{-\frac{x^2}{2}} \frac{d^n}{dx^n} \left(e^{-\frac{x^2}{2}}\right).$$

Generating functions

$$\exp(2xz - z^2) = \sum_{n=0}^{\infty} \frac{H_n(x)}{n!} z^n,$$

$$e^{z^2} \cos(2xz) = \sum_{n=0}^{\infty} \frac{(-1)^n}{(2n)!} H_{2n}(x)\, z^{2n},$$

$$e^{z^2} \sin(2xz) = \sum_{n=0}^{\infty} \frac{(-1)^n}{(2n+1)!} H_{2n+1}(x)\, z^{2n+1},$$

$$(1 - 4z^2)^{-\frac{1}{2}} \exp\left[y^2 - \frac{(y - 2zx)^2}{1 - 4z^2}\right] = \sum_{n=0}^{\infty} \frac{H_n(x)\, H_n(y)}{n!} z^n,$$

$$(1 - z^2)^{-\frac{1}{2}} \exp\left[y^2 - \frac{(y - zx)^2}{1 - z^2}\right] = \sum_{n=0}^{\infty} 2^{-n} \frac{H_n(x)\, H_n(y)}{n!} z^n,$$

$$\frac{1 + 2xz + 4z^2}{\sqrt{1 + 4z^2}} \exp\left(\frac{4x^2 z^2}{1 + 4z^2}\right) = \sum_{n=0}^{\infty} \frac{H_n(x)}{\left(\left[\frac{n}{2}\right]\right)!} z^n,$$

$$\exp\left(zx - \frac{z^2}{2}\right) = \sum_{n=0}^{\infty} He_n(x) \frac{z^n}{n!}.$$

$$e^{-1} \cosh\left(x\sqrt{2}\right) = \sum_{n=0}^{\infty} He_{2n}(x) \frac{2^n}{(2n)!},$$

$$e^{-1} \sinh\left(x\sqrt{2}\right) = \sum_{n=0}^{\infty} He_{2n+1}(x) \frac{2^{n+\frac{1}{2}}}{(2n+1)!},$$

$$e \cos\left(x\sqrt{2}\right) = \sum_{n=0}^{\infty} (-1)^n He_{2n}(x) \frac{2^n}{(2n)!},$$

$$e \sin\left(x\sqrt{2}\right) = \sum_{n=0}^{\infty} (-1)^n He_{2n+1}(x) \frac{2^{n+\frac{1}{2}}}{(2n+1)!},$$

$$(1 - z^2)^{-\frac{1}{2}} \exp\left[\frac{x^2 - y^2}{2} - \frac{(x - yz)^2}{1 - z^2}\right]$$

$$= \sum_{n=0}^{\infty} e^{-\frac{x^2+y^2}{2}} He_n\left(x\sqrt{2}\right) He_n\left(y\sqrt{2}\right) \frac{z^n}{n!},$$

$$|z| < 1.$$

The differential equation. The polynomials $H_n(x)$ satisfy the differential equation

(1) $$y'' - 2xy' + 2ny = 0.$$

The polynomials $He_n(x)$ satisfy the differential equation

$$y'' - xy' + ny = 0.$$

Other related differential equation are:

$$y'' + (2n + 1 - x^2)\, y = 0,$$

$$y(x) = e^{-\frac{x^2}{2}} H_n(x),$$

$$u'' + 2xu' + 2(n + 1)\, u = 0,$$

$$u(x) = e^{-x^2} H_n(x).$$

Integral representations

$$H_n(x) = \frac{\Gamma(n+1)}{2\pi i} \int^{(0+)} e^{-t^2} t^{-n-1} \exp(2xt)\, dt$$

$$= \frac{2^{n+1} e^{x^2}}{\Gamma\left(\frac{1}{2}\right)} \int_0^\infty e^{-t^2} t^n \cos\left(2xt - \frac{n\pi}{2}\right) dt,$$

$$He_n(x) = \frac{1}{\sqrt{2\pi}} \int_{-\infty}^{\infty} (x+it)^n e^{-\frac{t^2}{2}}\, dt.$$

Fourier transform of $H_n(x)$

$$\frac{1}{\sqrt{2\pi}} \int_{-\infty}^{\infty} e^{-\frac{x^2}{2}} He_n\left(x\sqrt{2}\right) e^{ixy}\, dx = i^n e^{-\frac{y^2}{2}} He_n\left(y\sqrt{2}\right).$$

Addition theorems

$$\sum_{m=0}^{n} \binom{n}{m} H_m\left(x\sqrt{2}\right) H_{n-m}\left(y\sqrt{2}\right) = 2^{\frac{n}{2}} H_n(x+y),$$

$$\sum_{m=0}^{\infty} \binom{n}{m} H_m(x)\, H_{n-m}(y) = 2^{\frac{n}{2}} H_n\left(\frac{x+y}{\sqrt{2}}\right),$$

$$\sum_{m_1+m_2+\cdots+m_r=n} \frac{\alpha_1^{m_1}}{m!} \frac{\alpha_2^{m_2}}{m_2!} \cdots \frac{\alpha_r^{m_r}}{m_r!} H_{m_1}(x_1)\, H_{m_2}(x_2) \cdots H_{m_r}(x_r)$$

$$= \frac{(\alpha_1^2 + \alpha_2^2 + \cdots + \alpha_r^2)^{\frac{n}{2}}}{n!} H_n\left[\frac{\alpha_1 x_1 + \alpha_2 x_2 + \cdots + \alpha_r x_r}{(\alpha_1^2 + \alpha_2^2 + \cdots + \alpha_r^2)^{\frac{1}{2}}}\right].$$

where the summation is taken over all non negative integers $m_1, m_2, \ldots$ $\ldots, m_r$ for which

$$\sum_{i=1}^{r} m_i = n.$$

5.6.3 Asymptotic expansions

For x real and $n \to \infty$

$$\frac{\Gamma\left(\frac{n}{2}+1\right)}{\Gamma(n+1)} e^{-\frac{x^2}{2}} H_n(x) = \cos\left[(2n+1)^{\frac{1}{2}} x - \frac{n\pi}{2}\right]$$

$$+ \frac{x^3}{6}(2n+1)^{-\frac{1}{2}} \sin\left[(2n+1)^{\frac{1}{2}} x - \frac{n\pi}{2}\right]$$

$$+ O(n^{-1}).$$

The estimate is valid for any real value of x and holds uniformly in every finite real interval

$$He_{2n}\left(x\sqrt{2}\right) = (-1)^n\, 2^n (2n-1)!\, e^{\frac{x^2}{2}} \left[\cos\left(\sqrt{4n+1}\,x\right) + O\left(n^{-\frac{1}{4}}\right)\right],$$

$$He_{2n+1}\left(x\sqrt{2}\right) = (-1)^n\, 2^{n+\frac{1}{2}} (2n-1)!\,(2n+1)\, e^{\frac{x^2}{2}} \times \left[\sin\left(\sqrt{4n+3}\,x\right) + O\left(n^{-\frac{1}{4}}\right)\right].$$

If x is not real and $n \to \infty$, then

$$(2n)^{-\frac{1}{2}} \log\left\{\frac{\Gamma\left(\frac{n}{2}+1\right)}{\Gamma(n+1)}\,|H_n(x)|\right\} \approx |\operatorname{Im} x|.$$

5.6.4 Miscellaneous results

Limit relations

$$\lim_{n\to\infty}\left[\frac{(-1)^n\, n^{\frac{1}{2}}}{2^{2n}\, n!}\, H_{2n}\left(\frac{x}{2\sqrt{n}}\right)\right] = \frac{1}{\sqrt{\pi}}\cos x,$$

$$\lim_{n\to\infty}\left[\frac{(-1)^n}{2^{2n}\, n!}\, H_{2n+1}\left(\frac{x}{2\sqrt{n}}\right)\right] = \frac{2}{\sqrt{\pi}}\sin x,$$

$$\lim_{n\to\infty}\left[\left(\frac{x}{n}\right)^n H_n\left(\frac{n}{2x}\right)\right] = e^{-x^2}.$$

Sums involving $H_n(x)$

$$\sum_{m=0}^{n}\binom{n}{m} H_m(x)\,(2y)^{n-m} = H_n(x+y),$$

$$\sum_{m=0}^{\left[\frac{n}{2}\right]} \frac{1}{m!}\,\frac{H_{n-2m}(x)}{(n-2m)!} = \frac{(2x)^n}{n!},$$

$$\sum_{m=0}^{n}\frac{2^{-m}}{m!}\,[H_m(x)]^2 = \frac{2^{-n-1}}{n!}\left[(H_{n+1}(x))^2 - H_n(x)\,H_{n+2}(x)\right],$$

$$\sum_{r=0}^{\min(m,n)} (-2)^r\, r!\binom{m}{r}\binom{n}{r} H_{m-r}(x)\, H_{n-r}(x) = H_{m+n}(x),$$

$$\sum_{r=0}^{\min(m,n)} 2^r\, r!\binom{m}{r}\binom{n}{r} H_{m+n-2r}(x) = H_m(x)\, H_n(x),$$

$$\sum_{m=0}^{n}\frac{2^{-m}}{m!} H_m(x)\, H_m(y) = \frac{2^{-n-1}}{n!}\left\{\frac{H_{n+1}(x)\,H_n(y) - H_n(x)\,H_{n+1}(y)}{x-y}\right\},$$

$$\sum_{m=0}^{n}\binom{2n}{2m} H_{2m}\left(x\sqrt{2}\right) H_{2n-2m}\left(y\sqrt{2}\right) = 2^{n-1}\{H_{2n}(x+y) + H_{2n}(x-y)\},$$

$$\sum_{m=0}^{\infty}\frac{(-12)^{-m}}{(2m+1)}\,\frac{H_{2m+1}(x)}{m!} = \sqrt{6}\int_0^x e^{-\frac{t^2}{2}}\,dt.$$

5.7 Chebyshev (Tchebichef) polynomials

5.7.1 Definition, notation, and special cases

Chebyshev polynomials $T_n(x)$ and $U_n(x)$ are the suitably normalize orthogonal polynomials associated with the interval $(-1, +1)$ and the weight functions $(1-x^2)^{-\frac{1}{2}}$ and $(1-x^2)^{\frac{1}{2}}$ respectively. They are standardized by the relations $T_n(1) = 1$ and $U_n(1) = n + 1$. The polynomials $T_n(x)$ and $U_n(x)$ are called Chebyshev polynomials of the first and second kind respectively.

Closely related are the polynomiated $T_n^*(x)$ and $U_n^*(x)$.

$T_n^*(x)$ and $U_n^*(x)$ are associated with the interval $(0, 1)$ and the weight functions $(1-x^2)^{-\frac{1}{2}}$ and $(1-x^2)^{\frac{1}{2}}$ respectively. The standardizing relations are:

$$T_n^*(1) = 1, \quad U_n^*(1) = n + 1.$$

$T_n^*(x)$ and $U_n^*(x)$ are sometimes called the shifted Chebyshev polynomials of the first and second kind respectively.

Special cases, and relations

$$\begin{aligned}
T_0(x) &= 1, & U_0(x) &= 1,\\
T_1(x) &= x, & U_1(x) &= 2x,\\
T_2(x) &= 2x^2 - 1, & U_2(x) &= 4x^2 - 1,\\
T_3(x) &= 4x^3 - 3x, & U_3(x) &= 8x^3 - 4x,\\
T_4(x) &= 8x^4 - 8x^2 + 1, & U_4(x) &= 16x^4 - 12x^2 + 1,\\
T_5(x) &= 16x^5 - 20x^3 + 5x, & U_5(x) &= 32x^5 - 32x^3 + 6x,\\
T_n(1) &= 1, & U_n(1) &= n + 1,\\
T_n(-1) &= (-1)^n, & U_n(-1) &= (-1)^n (n + 1),\\
T_{2n}(0) &= (-1)^n, & U_{2n}(0) &= (-1)^n,\\
T_{2n+1}(0) &= 0, & U_{2n+1}(0) &= 0.
\end{aligned}$$

The polynomials $T_n(x)$, $U_n(x)$, $T_n^*(x)$ and $U_n^*(x)$ are connected with each other by the following relations:

$$\begin{aligned}
T_n(x) &= T_n^*\left(\frac{x+1}{2}\right)\\
&= U_n(x) - x\,U_{n-1}(x)\\
&= x\,U_{n-1}(x) - U_{n-2}(x)\\
&= \frac{1}{2}\left[U_n(x) - U_{n-2}(x)\right],
\end{aligned}$$

$$U_n(x) = U_n^*\left(\frac{x+1}{2}\right)$$
$$= \frac{1}{1-x^2}\left[x\,T_{n+1}(x) - T_{n+2}(x)\right],$$
$$T_n^*(x) = T_n(2x-1),$$
$$U_n^*(x) = U_n(2x-1).$$

For relations connecting these polynomials with $P_n^{(\alpha,\beta)}(x)$ and $C_n^\lambda(x)$ see 5.2.1 and 5.3.1.

5.7.2 Elementary results

Explicit expressions for $T_n(x)$, $U_n(x)$

$$T_n(\cos\theta) = \cos n\theta,$$

$$T_n(x) = \frac{n}{2}\sum_{m=0}^{\left[\frac{n}{2}\right]} (-1)^m \frac{\Gamma(n-m)}{m!\,(n-2m)!}(2x)^{n-2m}$$
$$= \frac{1}{2}\left[\left(x + i\sqrt{1-x^2}\right)^n + \left(x - i\sqrt{1-x^2}\right)^n\right]$$
$$= \sum_{m=0}^{\left[\frac{n}{2}\right]} \binom{n}{2m} x^{n-2m}(x^2-1)^m,$$

$$U(\cos\theta) = \frac{\sin\left[(n+1)\,\theta\right]}{\sin\theta}$$
$$= \sum_{m=0}^{\left[\frac{n}{2}\right]} (-1)^m \binom{n-m}{m} (2x)^{n-2m}$$
$$= \frac{1}{2i}\,\frac{\left[(x+i\sqrt{1-x^2})^n - (x-i\sqrt{1-x^2})^n\right]}{\sqrt{1-x^2}}$$
$$= \sum_{m=0}^{\left[\frac{n}{2}\right]} \binom{n+1}{2m+1} x^{n-2m}(x^2-1)^m.$$

Representation by hypergeometric series

$$T_n(x) = {}_2F_1\left(-n, n; \frac{1}{2}; \frac{1-x}{2}\right),$$
$$U_n(x) = (n+1)\,{}_2F_1\left(-n, n+2; \frac{3}{2}; \frac{1-x}{2}\right).$$

Further representations may be obtained from these by applying transformation formulas for ${}_2F_1(a, b; c; z)$ [see chap. II]. Also see (5.2.2) and (5.3.2).

Orthogonality relations

$$\int_{-1}^{+1} T_m(x)\, T_n(x)\, (1-x^2)^{-\frac{1}{2}}\, dx = \begin{cases} 0, & m \neq n, \\ \frac{\pi}{2}, & m = n \neq 0, \\ \pi, & m = n = 0, \end{cases}$$

$$\int_{-1}^{+1} U_n(x)\, U_m(x)\, (1-x^2)^{\frac{1}{2}}\, dx = \begin{cases} 0, & m \neq n, \\ \frac{\pi}{8}, & m = n. \end{cases}$$

Differentiation formulas

$$\frac{d}{dx} T_n(x) = n\, U_{n-1}(x),$$

$$\frac{d^m}{dx^m} T_n(x) = 2^{m-1} \Gamma(m)\, n\, C_{n-m}^m(x),$$

$$(1-x^2) \frac{d}{dx} T_n(x) = n\,[T_{n-1}(x) - x\, T_n(x)] = n\,[x\, T_n(x) - T_{n+1}(x)],$$

$$\frac{d}{dx} U_n(x) = 2 C_{n-1}^2(x),$$

$$\frac{d^m}{dx^m} U_n(x) = 2^m m!\, C_{n-m}^{m+1}(x),$$

$$(1-x^2) \frac{d}{dx} U_n(x) = (n+1)\, U_{n-1}(x) - n x\, U_n(x) = (n+2)\, x\, U_n(x) - (n+1)\, U_{n+1}(x).$$

Recurrence relations

$$T_{n+1}(x) = 2x\, T_n(x) - T_{n-1}(x),$$

$$U_{n+1}(x) = 2x\, U_n(x) - U_{n-1}(x),$$

$$T_{n+1}^*(x) = 2(2x-1)\, T_n^*(x) - T_{n-1}^*(x),$$

$$U_{n+1}^*(x) = 2(2x-1)\, U_n^*(x) - U_{n-1}^*(x).$$

Rodrigues' formulas

$$T_n(x) = \frac{(-1)^n \pi^{\frac{1}{2}} (1-x^2)^{\frac{1}{2}}}{2^{n+1} \Gamma\left(n+\frac{1}{2}\right)} \frac{d^n}{dx^n}\left[(1-x^2)^{n-\frac{1}{2}}\right],$$

$$U_n(x) = \frac{(-1)^n \pi^{\frac{1}{2}} (n+1)}{2^{n+1} \Gamma\left(n+\frac{3}{2}\right)} (1-x^2)^{-\frac{1}{2}} \frac{d^n}{dx^n}\left[(1-x^2)^{n+\frac{1}{2}}\right].$$

In terms of the fractional derivatives operator D_x^α defined by

$$D_x^\alpha(x^\beta) = \frac{\Gamma(\beta+1)}{\Gamma(\beta-\alpha+1)}\, x^{\beta-\alpha},$$
$$\beta+1>0,\ \beta-\alpha+1 \neq 0, -1, -2, \ldots,$$

$$T_n\left(x^{\frac{1}{2}}\right) = (-1)^{n+1}\, n \left(\frac{\pi}{2}\right)^{\frac{1}{2}} \frac{x^{\frac{n}{2}+\frac{3}{2}}}{\Gamma(n+2)}\, D_x^{n+\frac{1}{2}}\left[x^{\frac{n}{2}-1}\left(1-x^{\frac{1}{2}}\right)^{n+1}\right],$$
$$n = 1, 2, \ldots$$

Generating functions

$$\frac{1-z^2}{1-2xz+z^2} = 1 + 2\sum_{n=1}^{\infty} T_n(x)\, z^n,$$
$$-1<x<1,\ |z|<1.$$

Let

$$R = \sqrt{1-2xz+z^2}.$$

Then

$$1-\frac{1}{2}\log R^2 = 1 + \sum_{n=1}^{\infty} \frac{T_n(x)}{n}\, z^n,$$
$$-1<x<1,\ |z|<1,$$

$$\frac{1-xz}{R^2} = \sum_{n=0}^{\infty} T_n(x)\, z^n,$$
$$-1<x<1,\ |z|<1,$$

$$\frac{1}{R}(1-xz+R)^{\frac{1}{2}} = \sqrt{2}\sum_{n=0}^{\infty} 2^{-2n}\binom{2n}{n} T_n(x)\, z^n,$$
$$-1<x<1,\ |z|<1,$$

$$e^{zx}\cosh\left(z\sqrt{x^2-1}\right) = \sum_{n=0}^{\infty}\frac{T_n(x)}{n!}\, z^n,$$
$$-1<x<1,$$

$$\frac{1}{R^2} = \sum_{n=0}^{\infty} U_n(x)\, z^n,$$
$$-1<x<1,\ |z|<1,$$

$$\frac{1}{R}(1-xz+R)^{\frac{1}{2}} = \sqrt{2}\sum_{n=0}^{\infty} 2^{-2n-2}\binom{2n+2}{n+1} U_n(x)\, z^n,$$
$$-1<x<1,\ |z|<1,$$

$$\frac{e^{zx}}{\sqrt{x^2-1}}\sinh\left(z\sqrt{x^2-1}\right) = \sum_{n=0}^{\infty}\frac{U_n(x)}{(n+1)!}\, z^n.$$

The differential equation. The polynomials $T_n(x)$ satisfy the differential equation

$$(1-x^2)\frac{d^2y}{dx^2} - x\frac{dy}{dx} + n^2 y = 0$$

and the polynomials $U_n(x)$ satisfy the differential equation

$$(1-x^2)\frac{d^2y}{dx^2} - 3x\frac{dy}{dx} + n(n+2)\,y = 0.$$

Other related differential equations are

$$\frac{d^2y}{dx^2} + n^2 y = 0,$$

$$y = T_n(\cos x),$$

$$(1-x^2)\frac{d^2y}{dx^2} - 3x\frac{dy}{dx} + (n^2-1)\,y = 0,$$

$$y = \frac{1}{\sqrt{1-x^2}}\,T_n(x) = U_{n-1}(x).$$

Integral representations

$$2\,T_n(x) = \frac{1}{2\pi i}\int^{(0+)} \frac{z^{-n-1}(1-z^2)}{1-2xz+z^2}\,dz,$$

the contour of integration is so chosen that both zeros of $1-2xz+z^2$ lie outside it

$$U_n(x) = \frac{1}{2\pi i}\int^{(0+)} \frac{z^{-n-1}}{1-2xz+z^2}\,dz.$$

Again the contour must be such that both zeros of $1-2xz+z^2$ lie outside it.

5.7.3 Miscellaneous results

Some identities

$$2\,T_m(x)\,T_n(x) = T_{m+n}(x) + T_{m-n}(x),$$

$$m \geq n,$$

$$2\,[T_n(x)]^2 = 1 + T_{2n}(x),$$

$$2\,T_m(x)\,U_n(x) = U_{m+n}(x) - U_{m-n-2}(x),$$

$$m > n+1,$$

$$2\,T_n(x)\,U_m(x) = U_{m+n}(x) + U_{m-n}(x),$$

$$m \geq n,$$

$$2\,T_n(x)\,U_{n-1}(x) = U_{2n-1}(x),$$

$$2(x^2-1)\,U_{m-1}(x)\,U_{n-1}(x) = T_{m+n}(x) - T_{m-n}(x),$$

$$m \geq n,$$

$$2(1-x^2)\,[U_{n-1}(x)]^2 = 1 - T_{2n}(x),$$

$$\sum_{m=0}^{n} T_{2m}(x) = \frac{1}{2}\,[1 + U_{2n}(x)],$$

$$\sum_{m=0}^{n-1} T_{2m+1}(x) = \frac{1}{2}\,U_{2n-1}(x),$$

$$2(1-x^2)\sum_{m=0}^{n} U_{2m}(x) = 1 - T_{2n+2}(x),$$

$$2(1-x^2)\sum_{m=0}^{n-1} U_{2m+1}(x) = x - T_{2n+1}(x),$$

$$2\sum_{m=0}^{n} U_m(x)\,U_m(y) = \frac{U_{n+1}(x)\,U_n(y) - U_n(x)\,U_{n+1}(y)}{x-y},$$

$$2\sum_{m=0}^{n} T_{m+1}(x)\,U_m(y) = \frac{T_{n+2}(x)\,U_n(y) - T_{n+1}(x)\,U_{n+1}(y) + 1}{x-y},$$

$$\sum_{m=0}^{n} \varepsilon_m T_m(x)\,T_m(y) = \frac{T_{n+1}(x)\,T_n(y) - T_n(x)\,T_{n+1}(y)}{x-y},$$

$$\sum_{m=0}^{n} \varepsilon_m T_m(x)\,U_m(y) = \frac{\frac{1}{2}\,U_1(y) + T_{n+1}(x)\,U_n(y) - T_n(x)\,U_{n+1}(y)}{x-y}.$$

Some integrals associated with $T_n(x)$, $U_n(x)$

$$\left(n+\frac{1}{2}\right)(1+x)^{\frac{1}{2}} \int_{-1}^{x} \frac{P_n(t)}{\sqrt{x-t}}\,dt = T_n(x) + T_{n+1}(x),$$

$$\left(n+\frac{1}{2}\right)(1-x)^{\frac{1}{2}} \int_{x}^{1} (t-x)^{-\frac{1}{2}}\,P_n(t)\,dt = T_n(x) - T_{n+1}(x),$$

$$-1 < x < 1,$$

$$U_{n-1}(x) = \frac{1}{\pi}\int_{-1}^{+1} (t-x)^{-1}\,(1-t^2)^{-\frac{1}{2}}\,T_n(t)\,dt,$$

$$T_n(x) = -\frac{1}{\pi}\int_{-1}^{+1} (t-x)^{-1}\,(1-t^2)^{\frac{1}{2}}\,U_{n-1}(t)\,dt,$$

where the last two integrals are considered as Cauchy principal value.

Literature

COURANT, R., and D. HILBERT: Methods of mathematical physics (Vol. 1). New York: Interscience Publishers 1953.
ERDÉLYI, A., W. MAGNUS, F. OBERHETTINGER and F. G. TRICOMI: [1] Higher transcendental functions. Inc. New York: McGraw-Hill 1953.
— [2] Vol. 2.
KACZMARZ, S., and H. STEINHAUS: Theorie der Orthogonalreihen. New York: Chelsea publishing 1951.
RAINVILLE, E. D.: Special functions. New York: Macmillan 1963.
SANSONE, G.: Orthogonal functions. New York: Interscience publishers 1959.
SZEGÖ, G.: Orthogonal polynomials. American Math. Soc., Providence, R. I., 1959.
TRICOMI, F. G.: Vorlesungen über Orthogonalreihen. Berlin/Göttingen/Heidelberg: Springer 1955.

Chapter VI

VI. Kummer's function

6.1 Definitions and some elementary results

6.1.1 Definitions

Kummer's function[1] ${}_1F_1(a; c; z)$ is defined by the function, and all its analytic continuations, represented by the infinite series $\sum_{n=0}^{\infty} \frac{(a)_n}{(c)_n} \frac{z^n}{n!}$. That is,

$$
\begin{aligned}
{}_1F_1(a; c; z) &= \sum_{n=0}^{\infty} \frac{(a)_n}{(c)_n} \frac{z^n}{n!}, \quad c \neq 0, -1, -2, \ldots \\
&= \frac{\Gamma(c)}{\Gamma(a)} \sum_{n=0}^{\infty} \frac{\Gamma(a+n)}{\Gamma(c+n)} \frac{z^n}{n!}.
\end{aligned}
$$

Other symbols used instead of ${}_1F_1(a; c; z)$ are $\Phi(a, c; z)$ and $M(a, c, z)$.

Another approach to Kummer's function is to obtain it from the hypergeometric function ${}_2F_1(a, b; c; z)$ by replacing z by $\frac{z}{b}$ and then taking the limit as $b \to \infty$.

The series $\sum_{n=0}^{\infty} \frac{(a)_n}{(c)_n} \frac{z^n}{n!}$ converges for all z provided that $c \neq -m$, $m = 0, 1, 2, \ldots$ The following cases result in considering the convergence of the series.

a) $c \neq -m$, $m = 0, 1, 2, \ldots$,
The series converges for all z.

[1] Also called confluent hypergeometric function.

b) $c \neq -m$, $a = -n$, m, n non-negative integers. The series reduces to a polynomial of degree n in z.

c) $a \neq -n$, $c = -m$ or $a = -n$, $c = -m$ and $m < n$ where m, n are non-negative integers. The series does not converge.

d) Although the function ${}_1F_1(a; c; z)$ is undefined if $c = -m$, $m = 0, -1, -2, \ldots$, the limit of $\frac{1}{\Gamma(c)}\,{}_1F_1(a; c; z)$ as $c \to -m$ exists and is given by

$$\lim_{c \to -m} \frac{1}{\Gamma(c)}\,{}_1F_1(a; c; z) = \frac{z^{m+1}(a)_{m+1}}{(m+1)!}\,{}_1F_1(a+m+1; m+2; z).$$

Another function closely related to ${}_1F_1(a; c; z)$ is $U(a, c, z)$ defined by

$$U(a, c, z) = \frac{\pi}{\sin(\pi c)}\left[\frac{{}_1F_1(a; c; z)}{\Gamma(c)\,\Gamma(1+a-c)} - z^{1-c}\frac{{}_1F_1(a+1-c; 2-c; z)}{\Gamma(a)\,\Gamma(2-c)}\right],$$

$$-\pi < \arg z \leq \pi.$$

$U(a, c, z)$ is a multiple-valued function with its principal branch given by

$$-\pi < \arg z \leq \pi.$$

The following relations give the analytic continuation of $U(a, c, z)$

$$U(a, c, z e^{i\pi}) = \frac{\pi}{\sin(\pi c)}\, e^{-z}$$

$$\times \left[\frac{{}_1F_1(c-a; c; z)}{\Gamma(c)\,\Gamma(1+a-c)} - \frac{e^{i\pi(1-c)}\, z^{1-c}}{\Gamma(a)\,\Gamma(2-c)}\,{}_1F_1(1-a; 2-c; z)\right]$$

$$U(a, c, z e^{-i\pi}) = \frac{\pi}{\sin(\pi c)}\, e^{-z}$$

$$\times \left[\frac{{}_1F_1(c-a; c; z)}{\Gamma(c)\,\Gamma(1+a-c)} - (z e^{-i\pi})^{1-c}\frac{{}_1F_1(1-a; 2-c; z)}{\Gamma(a)\,\Gamma(2-c)}\right],$$

$$c \neq 0, -1, -2, \ldots,$$

$$U(a. c, z e^{i2\pi n}) = [1 - e^{-i2\pi c n}]\frac{\Gamma(1-c)}{\Gamma(1+a-c)}\,{}_1F_1(a; c; z)$$

$$+ e^{-i2\pi c n}\, U(a. c, z).$$

Just as the function ${}_1F_1(a; c; z)$, the function $U(a, c, z)$ can also be obtained as a limiting case of the hypergeometric function ${}_2F_1(a, b; c; z)$ and is given by the relation

$$\lim_{b \to \infty} z^{-a}\,{}_2F_1\left(a, c; b; 1 - \frac{b}{z}\right) = U(a, a-c+1, z),$$

$${}_1F_1(a; c; z) = \frac{\Gamma(c)}{\Gamma(c-a)}\, e^{i\pi\varepsilon a}\, U(a, c, z)$$

$$+ \frac{\Gamma(c)}{\Gamma(a)}\, e^{z} e^{i\pi\varepsilon(a-c)}\, U(c-a, c, -z),$$

$$U(a, c, z) = \frac{\Gamma(1-c)}{\Gamma(1+a-c)}\, {}_1F_1(a; c; z) + \frac{\Gamma(c-1)}{\Gamma(a)}\, z^{1-c}\, {}_1F_1(1+a-c; 2-c; z)$$

$$\varepsilon = \begin{cases} 1 & \operatorname{Im} z > 0, \\ -1 & \operatorname{Im} z < 0. \end{cases}$$

6.1.2 Some elementary results

Differentiation formulas

$$\frac{d}{dz}\, {}_1F_1(a; c; z) = \frac{a}{c}\, {}_1F_1(a+1; c+1; z),$$

$$\frac{d^n}{dz^n}\, {}_1F_1(a; c; z) = \frac{(a)_n}{(c)_n}\, {}_1F_1(a+n; c+n; z),$$

$$\frac{d}{dz}\, [z^a\, {}_1F_1(a; c; z)] = a z^{a-1}\, {}_1F_1(a+1; c; z),$$

$$\frac{d^n}{dz^n}\, [z^{a+n-1}\, {}_1F_1(a; c; z)] = (a)_n\, z^{a-1}\, {}_1F_1(a+n; c; z),$$

$$\frac{d}{dz}\, [z^{c-1}\, {}_1F_1(a; c; z)] = (c-1)\, z^{c-2}\, {}_1F_1(a; c-1; z),$$

$$c \neq 1, 0, -1, -2, \ldots,$$

$$\frac{d^n}{dz^n}\, [z^{c-1}\, {}_1F_1(a; c; z)] = (-1)^n\, (1-c)_n\, z^{c-1-n}\, {}_1F_1(a; c-n; z),$$

$$c \neq n, n-1, \ldots, 0, -1, -2, \ldots.$$

$$\frac{d^n}{dz^n}\, [e^{-z}\, {}_1F_1(a; c; z)] = \frac{(-1)^n\, (c-a)_n}{(c)_n}\, {}_1F_1(c-a+n; c+n; -z) = \frac{(-1)^n (c-a)_n}{(c)_n}\, e^{-z}\, {}_1F_1(a; c+n; z),$$

$$\frac{d^n}{dz^n}\, [e^{-z} z^{c+n-a-1}\, {}_1F_1(a; c; z)] = (c-a)_n\, z^{c-a-1}\, {}_1F_1(c-a+n; c; -z) = (c-a)_n\, e^{-z} z^{c-a-1}\, {}_1F_1(a-n; c; z),$$

$$\frac{d^n}{dz^n}\, [z^{c-1} e^{-z}\, {}_1F_1(a; c; z)] = (-1)^n\, (1-c)_n\, z^{c-n-1}\, {}_1F_1(c-a; c-n; -z) = (-1)^n\, (1-c)_n\, e^{-z} z^{c-n-1}\, {}_1F_1(a-n; c-n; z).$$

If ${}_1F_1(a \pm 1; c; z)$ and ${}_1F_1(a; c \pm 1; z)$ are called the functions contiguous to ${}_1F_1(a; c; z)$, then there is a relation between any one of the four contiguous functions, the function ${}_1F_1(a; c; z)$ and the derivative ${}_1F_1'(a; c; z)$

$$\frac{d}{dz}\,{}_1F_1(a; c; z) = {}_1F_1'(a; c; z)\,,$$

$$(c-a)\,{}_1F_1(a-1; c; z) = (c-a-z)\,{}_1F_1(a; c; z) + z\,{}_1F_1'(a; c; z),$$

$$(c-1)\,{}_1F_1(a; c-1; z) = (c-1)\,{}_1F_1(a; c; z) + z\,{}_1F_1'(a; c; z),$$

$$a\,{}_1F_1(a+1; c; z) = a\,{}_1F_1(a; c; z) + z\,{}_1F_1'(a; c; z),$$

$$(c-a)\,{}_1F_1(a; c+1; z) = c\,{}_1F_1(a; c; z) - c\,{}_1F_1'(a; c; z).$$

Furthermore

$$(c-1)\,{}_1F_1(a-1; c-1; z) = (c-1-z)\,{}_1F_1(a; c; z) + z\,{}_1F_1'(a; c; z).$$

Combining these five results with the differentiation formulas, it is possible to obtain five more results.

For the function $U(a, c, z)$ the corresponding relations are given by:

$$\frac{d}{dz}\,U(a, c, z) = -a\,U(a+1, c+1, z),$$

$$\frac{d^n}{dz^n}\,U(a, c, z) = (-1)^n\,(a)_n\,U(a+n, c+n, z),$$

$$\frac{d^n}{dz^n}\,[z^{c-1}\,U(a, c, z)] = (-1)^n\,(a-c+1)_n\,z^{c-n-1}\,U(a, c-n, z),$$

$$\frac{d^n}{dz^n}\,[z^{a+n-1}\,U(a, c, z)] = (a)_n\,(a-c+1)_n\,z^{a-1}\,U(a+n, c, z),$$

$$\frac{d^n}{dz^n}\,[e^{-z}\,U(a, c, z)] = (-1)^n\,e^{-z}\,U(a, c+n, z),$$

$$\frac{d^n}{dz^n}\,[e^{-z}z^{c-a+n-1}\,U(a, c, z)] = (-1)^n\,e^{-z}z^{c-a-1}\,U(a-n, c, z),$$

$$U(a-1, c, z) + (c-2a-z)\,U(a, c, z) + a(a-c+1)\,U(a+1, c, z) = 0,$$

$$(a+1-c)\,U(a, c-1, z) - U(a-1, c, z) + (a-1+z)\,U(a, c, z) = 0,$$

$$U(a, c, z) - U(a, c-1, z) - a\,U(a+1, c, z) = 0,$$

$$U(a-1, c, z) + (c-a)\,U(a, c, z) - z\,U(a, c+1, z) = 0,$$

$$(c-a-1)\,U(a, c-1, z) + (1-c-z)\,U(a, c, z) + z\,U(a, c+1, z) = 0,$$

$$(a+z)\,U(a, c, z) + a(c-a-1)\,U(a+1, c, z) - z\,U(a, c+1, z) = 0.$$

Some indefinite integrals

$$\int {}_1F_1(a;c;z)\,dz = \frac{c-1}{a-1}\,{}_1F_1(a-1;c-1;z), \qquad a \neq 1,$$

$$\int z^{c-1}\,{}_1F_1(a;c;z)\,dz = \frac{z^c}{c}\,{}_1F_1(a;c+1;z), \qquad c \neq 0,-1,-2,\ldots,$$

$$\int z^{a-2}\,{}_1F_1(a;c;z)\,dz = \frac{z^{a-1}}{a-1}\,{}_1F_1(a-1;c;z), \qquad a \neq 1,$$

$$\int e^{-z}\,{}_1F_1(a;c;z)\,dz = \frac{(c-1)\,e^{-z}}{1+a-c}\,{}_1F_1(a;c-1;z), \qquad c-a \neq 1,$$

$$\int e^{-z} z^{c-1}\,{}_1F_1(a;c;z)\,dz = \frac{z^c}{c}\,e^{-z}\,{}_1F_1(a+1;c+1;z), \quad c \neq 0,-1,-2,\ldots,$$

$$\int z^{c-a-2} e^{-z}\,{}_1F_1(a;c;z)\,dz = \frac{z^{c-a-1}}{c-a-1}\,e^{-z}\,{}_1F_1(a+1;c;z),$$

$$c-a \neq 1,$$

$$\int U(a,c,z)\,dz = \frac{1}{1-a}\,U(a-1,c-1,z), \quad a \neq 1,$$

$$\int z^{c-1}\,U(a,c,z)\,dz = \frac{z^c}{c-a}\,U(a,c+1,z), \quad c-a \neq 0,$$

$$\int z^{a-2}\,U(a,c,z)\,dz = \frac{z^{a-1}}{(a-1)(a-c)}\,U(a-1,c,z),$$

$$a-1 \neq 0, \quad c-a \neq 0$$

$$\int e^{-z}\,U(a,c,z)\,dz = -e^{-z}\,U(a,c-1,z),$$

$$\int e^{-z} z^{c-1}\,U(a,c,z)\,dz = -e^{-z} z^c\,U(a+1,c+1,z),$$

$$\int e^{-z} z^{c-a-2}\,U(a,c,z)\,dz = -e^{-z} z^{c-a-1}\,U(a+1,c,z).$$

Some definite integrals

$$\int_0^\infty t^{c'-1}\,{}_1F_1(a;c;-t)\,dt = \frac{\Gamma(c)\,\Gamma(c')\,\Gamma(a-c')}{\Gamma(a)\,\Gamma(c-c')}$$

$$\mathrm{Re}\,a > \mathrm{Re}\,c' > 0,$$

$$\int_0^\infty t^{c'-1}\,U(a,c,t)\,dt = \frac{\Gamma(c')\,\Gamma(a-c')\,\Gamma(c'-c+1)}{\Gamma(a)\,\Gamma(a-c+1)}$$

$$\mathrm{Re}\,a > \mathrm{Re}\,c' > 0, \ \mathrm{Re}\,c' > -1 + \mathrm{Re}\,c,$$

$$\int_0^\infty {}_1F_1(a;c;-t^2)\cos(2zt)\,dt$$

$$= \sqrt{\frac{\pi}{2}}\,\frac{\Gamma(c)}{\Gamma(a)}\,z^{2a-1} e^{-z^2}\,U\left(c-\frac{1}{2}, a+\frac{1}{2}, z^2\right),$$

$$\int_0^\infty e^{-t} t^{c+n-1} (t+z)^{-1} {}_1F_1(a;c;t)\,dt$$
$$= (-1)^n \Gamma(c)\,\Gamma(1-a)\, z^{n-1+c}\, U(c-a, c, z),$$
$$|\arg z| < \pi, \quad 1 - Re\, a > n > -Re\, c,$$
$$n = 0, 1, 2, \ldots,$$

$$\frac{1}{2\pi i} \int_{-i\infty}^{i\infty} \Gamma(-t)\,\Gamma(c-t)\, U(t, c, z)\, U(c-t, c, z')\,dt$$
$$= \Gamma(c)\, U(c, 2c, z+z'),$$
$$\Gamma(a) \int_0^\infty t^{a-1} (1+t)^{c'-a'-1}\, U(a, c, zt)\,dt$$
$$= \Gamma(a') \int_0^\infty t^{a-1} (1+t)^{c-a-1}\, U(a', c', zt)\,dt,$$
$$Re\, a > 0, \quad Re\, a' > 0, \quad Re\, a > Re\, c' - 1,$$
$$Re\, a' > Re\, c - 1.$$

Transformation formulas

$${}_1F_1(a;c;z) = e^z\, {}_1F_1(c-a;c;-z),$$
$$U(a,c,z) = z^{1-c}\, U(a+1-c, 2-c, z),$$
$${}_1F_1(1+a-c;2-c;z) = e^z\, {}_1F_1(1-a;2-c;-z),$$
$$U(c-a, c, -z) = e^{\pm i\pi(1-c)} z^{1-c}\, U(1-a, 2-c, -z),$$

where the sign is positive or negative according as $\operatorname{Im} z > 0$ or $\operatorname{Im} z < 0$.

6.2 Recurrence relations

$$(c-a)\, {}_1F_1(a-1;c;z) + (2a-c+z)\, {}_1F_1(a;c;z)$$
$$- a\, {}_1F_1(a+1;c;z) = 0,$$
$$c(c-1)\, {}_1F_1(a;c-1;z) - c(c-1+z)\, {}_1F_1(a;c;z)$$
$$+ (c-a)\, {}_1F_1(a;c+1;z) = 0,$$
$$(c-1)\, {}_1F_1(a;c-1;z) + (a+1-c)\, {}_1F_1(a;c;z)$$
$$- a\, {}_1F_1(a+1;c;z) = 0,$$
$$c\, {}_1F_1(a;c;z) - c\, {}_1F_1(a-1;c;z) - z\, {}_1F_1(a;c+1;z) = 0,$$
$$c(a+z)\, {}_1F_1(a;c;z) - (c-a)\, z\, {}_1F_1(a;c+1;z)$$
$$- ac\, {}_1F_1(a+1;c;z) = 0,$$

$$(c - a)\,{}_1F_1(a - 1; c; z) - (c - 1)\,{}_1F_1(a; c - 1; z) + (a - 1 + z)\,{}_1F_1(a; c; z) = 0,$$

$$c(c - a)\,{}_1F_1(a - 1; c; z) - c(c - a - z)\,{}_1F_1(a; c; z) - az\,{}_1F_1(a + 1; c + 1; z) = 0,$$

$$c\,{}_1F_1(a; c; z) - (c - a)\,{}_1F_1(a; c + 1; z) - a\,{}_1F_1(a + 1; c + 1; z) = 0,$$

$$c(c - 1)\,{}_1F_1(a; c - 1; z) - c(c - 1)\,{}_1F_1(a; c; z) - az\,{}_1F_1(a + 1; c + 1; z) = 0,$$

$$c(c - 1)\,{}_1F_1(a - 1; c - 1; z) + c(1 - c + z)\,{}_1F_1(a; c; z) - az\,{}_1F_1(a + 1; c + 1; z) = 0,$$

$$U(a - 1, c, z) + (c - 2a - z)\,U(a, c, z) + a(1 + a - c)\,U(a + 1, c, z) = 0,$$

$$U(a - 1, c, z) + (c - a)\,U(a, c, z) - zU(a, c + 1, z) = 0,$$

$$(c - a - 1)\,U(a, c - 1, z) + (1 - c - z)\,U(a, c, z) + zU(a, c + 1, z) = 0,$$

$$(-1)\,U(a, c - 1, z) + U(a, c, z) - aU(a + 1, c, z) = 0,$$

$$(1 + a - c)\,U(a, c - 1, z) - U(a - 1, c, z) + (a - 1 + z)\,U(a, c, z) = 0,$$

$$(a + z)\,U(a, c, z) - zU(a, c + 1, z) + a(c - a - 1)\,U(a + 1, c, z) = 0.$$

6.3 The differential equation

Kummer's function ${}_1F_1(a; c; z)$ satisfies the differential equation

$$z\frac{d^2w}{dz^2} + (c - z)\frac{dw}{dz} - az = 0.$$

The differential equation has a regular singularity at $z = 0$ and an irregular singularity at $z = \infty$. The irregular behavior of the singularity at $z = \infty$ is brought out clearly if Kummer's function is obtained by a limiting process from the hypergeometric function ${}_2F_1(a, b; c; z)$. This approach shows that the irregular singularity, at $z = \infty$, of Kummer's differential equation is formed by the confluence of the two regular singularities of the hypergeometric (function) differential equation at

$$z = b \text{ and } z = \infty.$$

Two linearly independent solutions of Kummer's differential equation, in the neighborhood of $z = 0$, are given by

$$w_1^{(0)}(z) = {}_1F_1(a; c; z) = \sum_{n=0}^{\infty} \frac{(a)_n}{(c)_n} \frac{z^n}{n!},$$

$$w_2^{(0)}(z) = z^{1-c} {}_1F_1(a - c + 1; 2 - c; z),$$

provided that $c \neq 0, -1, -2, \ldots$

In the neighborhood of $z = \infty$, the differential equation also has two linearly independent solutions $w_1^{(\infty)}$ and $w_2^{(\infty)}$ which are related to $w_1^{(0)}(z)$ and $w_2^{(0)}(z)$ by means of the following equations:

$$w_1^{(0)}(z) = e^{-i\pi a} \frac{\Gamma(c)}{\Gamma(c-a)} w_1^{(\infty)}(z) + \frac{\Gamma(c)}{\Gamma(a)} w_2^{(\infty)}(z),$$

$$w_2^{(0)}(z) = e^{-i\pi(a-c+1)} \frac{\Gamma(2-c)}{\Gamma(1-a)} w_1^{(\infty)}(z) + \frac{\Gamma(2-c)}{\Gamma(a-c+1)} w_2^{(\infty)}(z).$$

In the neighborhood of $z = \infty$, the solutions $w_1^{(\infty)}(z)$ and $w_2^{(\infty)}(z)$ have a simple asymptotic expansion given by

$$w_1^{(\infty)}(z) \approx z^{-a} \sum_{n=0}^{\infty} \frac{(a)_n (a-c+1)_n}{n!} (-z)^{-n},$$

$$w_2^{(\infty)}(z) \approx z^{a-c} e^z \sum_{n=0}^{\infty} \frac{(c-a)_n (1-a)_n}{n!} z^{-n},$$

where

$$|z| \to \infty, \ -\frac{3\pi}{2} < \arg z < \frac{\pi}{2}.$$

6.3.1 The logarithmic case

When c is an integer, the method of FRÖBENIUS gives only one linearly independent solution of Kummer's differential. In order to give a general solution of the differential it is necessary to find another linearly independent solution.

Case 1: If $c = 1$, the two linearly independent solutions are

$$w_1(z) = {}_1F_1(a; 1; z),$$

$$w_2(z) = w_1(z) \log z + \frac{az}{(1!)^2}\left(\frac{1}{a} - \frac{2}{1}\right) + \cdots$$

$$+ \cdots + \frac{(a)_r}{(r!)^2} z^r \left(\frac{1}{a} + \frac{1}{a+1} + \cdots + \frac{1}{a+r-1} - 2\sum_{l=1}^{r} \frac{1}{l}\right) + \cdots.$$

Case 2: If $c = n + 1$, the two linearly independent solutions are

$$w_1(z) = {}_1F_1(a; n + 1; z),\ n = 1, 2, 3, \ldots,$$

$$w_2(z) = \frac{(n-1)!}{\Gamma(a)} \sum_{r=0}^{n-1} \frac{(a-n)_r}{(1-n)_r} \frac{z^{r-n}}{r!}$$

$$+ \frac{(-1)^{n+1}}{\Gamma(a-n)\,\Gamma(n+1)} \Bigg[{}_1F_1(a; n+1; z) \log z$$

$$+ \sum_{r=0}^{\infty} \frac{(a)_r}{(n+1)_r} \frac{z^r}{r!} \{\psi(a+r) - \psi(1+r) - \psi(n+1+r)\}\Bigg].$$

Case 3: If $c = -n + 1$, $n = 0, 1, 2, \ldots$

$$w_1(z) = z^n\, {}_1F_1(a + n; n + 1; z),$$

$$w_2\,(z) = U(a, 1 - n, z)$$

$$= z^n U(a + n, n + 1, z).$$

6.3.2 Solutions and their Wronskians

If $c \neq -n$, $n = 0, 1, 2, 3, \ldots$ the complete solution of Kummer's differential equation is given by

$$a_1\, {}_1F_1(a; c; z) + a_2 U(a, c, z),$$

where a_1, a_1 are arbitrary constants.

Let

$$w_1 = {}_1F_1(a; c; z),$$

$$w_2 = e^z\, {}_1F_1(c - a; c; -z),$$

$$w_3 = z^{1-c}\, {}_1F_1(a + 1 - c; 2 - c; z),$$

$$w_4 = z^{1-c} e^z\, {}_1F_1(1 - a; 2 - c; -z),$$

$$w_5 = U(a, c, z),$$

$$w_6 = e^z U(c - a, c, -z),$$

$$w_7 = z^{1-c} U(a + 1 - c, 2 - c, z),$$

$$w_8 = z^{1-c} e^z U(1 - a, 2 - c, -z).$$

If

$$W(w_i, w_j) = w_i \frac{dw_j}{dz} - w_j \frac{dw_i}{dz}$$

then

$$W(w_1, w_2) = W(w_3, w_4) = W(w_5, w_6)$$

$$= W(w_7, w_8) = 0,$$

$$\begin{aligned}
W(w_1, w_3) &= W(w_2, w_4) = W(w_2, w_3) \\
&= W(w_1, w_4) = (1 - c)\, z^{-c} e^z, \\
W(w_1, w_5) &= -\frac{\Gamma(c)}{\Gamma(a)}\, z^{-c} e^z = W(w_1, w_6), \\
W(w_1, w_7) &= W(w_1, w_8) = \frac{\Gamma(c)}{\Gamma(c-a)}\, z^{-c} e^z e^{i\pi c\, \mathrm{sgn}(\mathrm{Im} z)}, \\
W(w_2, w_5) &= W(w_2, w_6) = -\frac{\Gamma(2-c)}{\Gamma(1+a-c)}\, z^{-c} e^z, \\
W(w_2, w_7) &= W(w_2, w_8) = -\frac{\Gamma(2-c)}{\Gamma(1-a)}\, z^{-c} e^z, \\
W(w_5, w_7) &= z^{-c} e^z \exp\left[i\pi(c-a)\, \mathrm{sgn}\,(\mathrm{Im}\, z)\right].
\end{aligned}$$

6.4 Addition and multiplication theorems

6.4.1 Addition theorems for ${}_1F_1(a; c; z)$

$$\begin{aligned}
{}_1F_1(a; c; z + z') &= \sum_{n=0}^{\infty} \frac{(a)_n}{(c)_n} \frac{z^n}{n!}\, {}_1F_1(a+n; c+n; z') \\
&= \sum_{n=0}^{\infty} \frac{(a)_n}{(c)_n} \frac{(z')^n}{n!}\, {}_1F_1(a+n; c+n; z) \\
&= \left(\frac{z}{z+z'}\right)^{c-1} \sum_{n=0}^{\infty} \frac{(-1)^n (1-c)_n}{n!\, z^n} (z')^n\, {}_1F_1(a; c-n; z) \\
&= \left(\frac{z'}{z+z'}\right)^{c-1} \sum_{n=0}^{\infty} \frac{(-1)^n (1-c)_n}{n!\, (z')^n} z^n\, {}_1F_1(a; c-n; z') \\
&= \left(\frac{z'}{z+z'}\right)^{a} \sum_{n=0}^{\infty} \frac{(a)_n}{n!} \left(\frac{z}{z+z'}\right)^{n} {}_1F_1(a+n; c; z') \\
&= \left(\frac{z}{z+z'}\right)^{a} \sum_{n=0}^{\infty} \frac{(a)_n}{n!} \left(\frac{z'}{z+z'}\right)^{n} {}_1F_1(a+n; c; z) \\
&= e^z \sum_{n=0}^{\infty} \frac{(-1)^n (c-a)_n}{(c)_n\, n!} z^n\, {}_1F_1(a; c+n; z') \\
&= e^{z'} \sum_{n=0}^{\infty} \frac{(-1)^n (c-a)_n}{(c)_n\, n!} (z')^n\, {}_1F_1(a; c+n; z) \\
&= e^z \left(\frac{z'}{z+z'}\right)^{c-a} \sum_{n=0}^{\infty} \frac{(c-a)_n}{n!} z^n (z+z')^n\, {}_1F_1(a-n; c; z') \\
&= e^{z'} \left(\frac{z}{z+z'}\right)^{c-a} \sum_{n=0}^{\infty} \frac{(c-a)_n}{n!} (z')^n (z+z')^n\, {}_1F_1(a-n; c; z) \\
&= e^z \left(\frac{z'}{z+z'}\right)^{c-1} \sum_{n=0}^{\infty} \frac{(-1)^n (1-c)_n}{n!\, (z')^n} z^n\, {}_1F_1(a-n; c-n; z') \\
&= e^{z'} \left(\frac{z}{z+z'}\right)^{c-1} \sum_{n=0}^{\infty} \frac{(-1)^n (1-c)_n}{n!\, z^n} (z')^n\, {}_1F_1(a-n; c-n; z).
\end{aligned}$$

Finally

$$\begin{aligned}{}_1F_1(a;c;z+z') &= \sum_{n=0}^{\infty} \frac{(-1)^n (a)_n}{n!\,(\lambda+n)_n} z^n\, {}_2F_1\left(-n, \lambda+n; c; 1+\frac{z'}{z}\right) \\ &\quad \times {}_1F_1(a+n; \lambda+2n+1; z) \\ &= \sum_{n=0}^{\infty} \frac{(-1)^n (a)_n}{n!\,(\lambda+n)_n} (z')^n\, {}_2F_1\left(-n, \lambda+n; c; 1+\frac{z}{z'}\right) \\ &\quad \times {}_1F_1(a+n; \lambda+2n+1; z'),\end{aligned}$$

where λ is an arbitrary parameter different from an odd negative integer. Further results may be obtained by applying transformation formulas on right hand side to Kummer's functions inside the summation sign.

6.4.2 Addition theorems for $U(a, c, z)$

$$\begin{aligned}U(a,c,z+z') &= e^z\left(\frac{z'}{z+z'}\right)^{c-1} \sum_{n=0}^{\infty} \frac{(-1)^n z^n}{n!\,(z')^n} U(a-n, c-n, z') \\ &= e^{z'}\left(\frac{z}{z+z'}\right)^{c-1} \sum_{n=0}^{\infty} \frac{(-1)^n (z')^n}{n!\, z^n} U(a-n, c-n, z) \\ &= e^z \sum_{n=0}^{\infty} \frac{(-1)^n}{n!} z^n U(a, c+n, z') \\ &= e^{z'} \sum_{n=0}^{\infty} \frac{(-1)^n}{n!} (z')^n U(a, c+n, z) \\ &= e^z\left(\frac{z'}{z+z'}\right)^{c-a} \sum_{n=0}^{\infty} \frac{(-1)^n}{n!} \left(\frac{z}{z+z'}\right)^n U(a-n, c, z') \\ &= e^{z'}\left(\frac{z}{z+z'}\right)^{c-a} \sum_{n=0}^{\infty} \frac{(-1)^n}{n!} \left(\frac{z'}{z+z'}\right)^n U(a-n, c, z) \\ &= \sum_{n=0}^{\infty} \frac{(-1)^n (a)_n}{n!} z^n U(a+n, c+n, z') \\ &= \sum_{n=0}^{\infty} \frac{(-1)^n (a)_n}{n!} (z')^n U(a+n, c+n, z) \\ &= \left(\frac{z'}{z+z'}\right)^{c-1} \sum_{n=0}^{\infty} \frac{(-1)^n (1+a-c)_n}{n!} \left(\frac{z}{z'}\right)^n U(a, c-n, z') \\ &= \left(\frac{z}{z+z'}\right)^{c-1} \sum_{n=0}^{\infty} \frac{(-1)^n (1+a-c)_n}{n!} \left(\frac{z'}{z}\right)^n U(a, c-n, z) \\ &= \left(\frac{z'}{z+z'}\right)^{a} \sum_{n=0}^{\infty} \frac{(a)_n (1+a-c)_n}{n!} \left(\frac{z}{z+z'}\right)^n U(a+n, c, z') \\ &= \left(\frac{z}{z+z'}\right)^{a} \sum_{n=0}^{\infty} \frac{(a)_n (1+a-c)_n}{n!} \left(\frac{z'}{z+z'}\right)^n U(a+n, c, z).\end{aligned}$$

6.4.3 Multiplication theorems for ${}_1F_1(a; c; z)$

$$
\begin{aligned}
{}_1F_1(a; c; zz') &= \sum_{n=0}^{\infty} \frac{(a)_n\,(z')^n}{n!\,(c)_n}\,(z-1)^n\,{}_1F_1(a+n; c+n; z') \\
&= \sum_{n=0}^{\infty} \frac{(a)_n}{(c)_n}\frac{(z)^n}{n!}\,(z'-1)^n\,{}_1F_1(a+n; c+n; z) \\
&= z^{-a} \sum_{n=0}^{\infty} \frac{(a)_n}{n!}\left(\frac{z-1}{z}\right)^n {}_1F_1(a+n; c; z') \\
&= (z')^{-a} \sum_{n=0}^{\infty} \frac{(a)_n}{n!}\left(1-\frac{1}{z'}\right)^n {}_1F_1(a+n; c; z)
\end{aligned}
$$

$$
\begin{aligned}
&{}_1F_1(a; c; zz') \\
&= z^{1-c} \sum_{n=0}^{\infty} \frac{(1-c)_n}{n!}\,(1-z)^n\,{}_1F_1(a; c-n; z') \\
&= (z')^{1-c} \sum_{n=0}^{\infty} \frac{(1-c)_n}{n!}\,(1-z')^n\,{}_1F_1(a; c-n; z) \\
&= e^{z'(z-1)} \sum_{n=0}^{\infty} \frac{(c-a)_n}{n!\,(c)_n}\,(z')^n\,(1-z)^n\,{}_1F_1(a; c+n; z') \\
&= e^{z(z'-1)} \sum_{n=0}^{\infty} \frac{(c-a)_n}{n!\,(c)_n}\,z^n\,(1-z')^n\,{}_1F_1(a; c+n; z) \\
&= e^{z'(z-1)}\,z^{1-c} \sum_{n=0}^{\infty} \frac{(1-c)_n}{n!}\,(1-z)^n\,{}_1F_1(a-n; c-n; z') \\
&= e^{z(z'-1)}\,(z')^{1-c} \sum_{n=0}^{\infty} \frac{(1-c)_n}{n!}\,(1-z')^n\,{}_1F_1(a-n; c-n; z) \\
&= e^{z'(z-1)}\,z^{a-c} \sum_{n=0}^{\infty} \frac{(c-a)_n}{n!}\left(1-\frac{1}{z}\right)^n {}_1F_1(a-n; c; z') \\
&= e^{z(z'-1)}\,(z')^{a-c} \sum_{n=0}^{\infty} \frac{(c-a)_n}{n!}\left(1-\frac{1}{z'}\right)^n {}_1F_1(a-n; c; z).
\end{aligned}
$$

Finally

$$
\begin{aligned}
{}_1F_1(a; c; zz') = \sum_{n=0}^{\infty} \frac{(-1)^n\,(a)_n}{n!\,(\lambda+n)_n}\,z^n\,{}_2F_1(-n, \lambda+n; c; z') \\
\times\,{}_1F_1(a+n; \lambda+2n+1; z)
\end{aligned}
$$

$$
= \sum_{n=0}^{\infty} \frac{(-1)^n\,(a)_n}{n!\,(\lambda+n)_n}\,(z')^n\,{}_2F_1(-n, \lambda+n; c; z)\,{}_1F_1(a+n; \lambda+2n+1; z'),
$$

where λ is an arbitrary parameter different from an odd negative integer. A statement similar to the one at the end of 6.4.1. can be made here.

6.4.4 Multiplication theorems for $U(a, c, z)$

$$U(a, c, zz') = e^{z'(z-1)} z^{1-c} \sum_{n=0}^{\infty} \frac{(1-z)^n}{n!} U(a-n, c-n, z')$$

$$= e^{z(z'-1)} (z')^{1-c} \sum_{n=0}^{\infty} \frac{(1-z')^n}{n!} U(a-n, c-n, z)$$

$$= e^{z'(z-1)} \sum_{n=0}^{\infty} \frac{(z')^n}{n!} (1-z)^n U(a, c+n, z')$$

$$= e^{z(z'-1)} \sum_{n=0}^{\infty} \frac{(z)^n}{n!} (1-z')^n U(a, c+n, z)$$

$$= e^{z'(z-1)} z^{a-c} \sum_{n=0}^{\infty} \frac{1}{n!} \left(\frac{1}{z} - 1\right)^n U(a-n, c, z')$$

$$= e^{z(z'-1)} (z')^{a-c} \sum_{n=0}^{\infty} \frac{1}{n!} \left(\frac{1}{z'} - 1\right)^n U(a-n, c, z)$$

$$U(a, c, zz') = \sum_{n=0}^{\infty} \frac{(a)_n}{n!} (z')^n (1-z)^n U(a+n, c+n, z')$$

$$= \sum_{n=0}^{\infty} \frac{(a)_n}{n!} z^n (1-z')^n U(a+n, c+n, z)$$

$$= z^{1-c} \sum_{n=0}^{\infty} \frac{(1+a-c)_n}{n!} (1-z)^n U(a, c-n, z')$$

$$= (z')^{1-c} \sum_{n=0}^{\infty} \frac{(1+a-c)_n}{n!} (1-z')^n U(a, c-n, z)$$

$$= z^{-a} \sum_{n=0}^{\infty} \frac{(a)_n (1+a-c)_n}{n!} \left(1 - \frac{1}{z}\right)^n U(a+n, c, z')$$

$$= (z')^{-a} \sum_{n=0}^{\infty} \frac{(a)_n (1+a-c)_n}{n!} \left(1 - \frac{1}{z'}\right)^n U(a+n, c, z).$$

6.5 Integral representations

6.5.1 Integral representations for ${}_1F_1(a; c; z)$

$${}_1F_1(a; c; z)$$

$$= \frac{\Gamma(c)}{\Gamma(a)\,\Gamma(c-a)} \int_0^1 e^{zt} t^{a-1} (1-t)^{c-a-1}\, dt,$$

$$Re\, c > Re\, a > 0,$$

$$= \frac{\Gamma(c)\, z^{1-c}}{\Gamma(a)\,\Gamma(c-a)} \int_0^z e^t t^{a-1} (z-t)^{c-a-1}\, dt$$

$$= \frac{\Gamma(c)\, e^z}{\Gamma(a)\,\Gamma(c-a)} \int_0^1 e^{-zt} t^{c-a-1} (1-t)^{a-1}\, dt$$

$$= \frac{2^{1-c} e^{\frac{z}{2}}\, \Gamma(c)}{\Gamma(a)\,\Gamma(c-a)} \int_{-1}^1 e^{-\frac{z}{2}t} (1+t)^{c-2} \left(\frac{1+t}{1-t}\right)^{1-a} dt$$

$$= \frac{2^{1-c} e^{\frac{z}{2}}\, \Gamma(c)}{\Gamma(a)\,\Gamma(c-a)} \int_{-1}^1 e^{\frac{z}{2}t} (1+t)^{c-2} \left(\frac{1-t}{1+t}\right)^{c-a-1} dt$$

$$= \frac{2^{1-c} e^{\frac{z}{2}}\, \Gamma(c)}{\Gamma(a)\,\Gamma(c-a)} \int_{-1}^1 e^{\frac{zt}{2}} (1-t)^{c-2} \left(\frac{1+t}{1-t}\right)^{a-1} dt$$

$$= \frac{2^{1-c}\, \Gamma(c)\, e^{\frac{z}{2}}}{\Gamma(a)\,\Gamma(c-a)} \int_0^\pi \exp\left[-\frac{z}{2}\cos t\right] (\sin t)^{c-1} \cot^{c-2a}\left(\frac{t}{2}\right) dt$$

$$= \frac{2^{1-c} e^{\frac{z}{2}}\, \Gamma(c)}{\Gamma(a)\,\Gamma(c-a)} \int_0^\pi \exp\left[\frac{z}{2}\cos t\right] (\sin t)^{c-1} \tan^{c-2a}\left(\frac{t}{2}\right) dt,$$

$$= \frac{\Gamma(c)}{\Gamma(a)\,\Gamma(c-a)}\, e^{-\alpha z} \int_\alpha^{\alpha+1} e^{zt} (t-\alpha)^{a-1} (1+\alpha-t)^{c-a-1}\, dt,$$

$$Re\, c > Re\, a > 0,$$

$${}_1F_1(a;c;z) = \frac{\Gamma(c) \exp\left(-\frac{\alpha z}{\beta-\alpha}\right)}{\Gamma(a)\,\Gamma(c-a)} (\beta-\alpha)^{1-c} \int_\alpha^\beta (t-\alpha)^{a-1} (-t+\beta)^{c-a-1}$$

$$\times \exp\left[\frac{zt}{\beta-\alpha}\right] dt$$

$$= \frac{1}{\Gamma(a)} \int_0^\infty e^{-t} t^{a-1}\, {}_0F_1(;c;zt)\, dt$$

$$Re\, a > 0$$

$$= \frac{\Gamma(c)}{\Gamma(a)} \int_0^\infty t^{a-1} (zt)^{\frac{1-c}{2}} e^{-t}\, I_{c-1}(2\sqrt{zt})\, dt,$$

$$Re\, a > 0,$$

$$ {}_1F_1\left(a; 1-a+\alpha; -\frac{z^2}{2}\right) = \sqrt{\pi}\, \frac{(1-a+\alpha)}{\Gamma(a)}\, 2^{\frac{1}{2}+\alpha-2a}\, z^{2a-\alpha-\frac{1}{2}} $$

$$ \times \int_0^\infty e^{-\frac{t^2}{4}}\, t^{2a-\alpha-\frac{1}{2}}\, I_{a-\frac{1}{2}}\left(\frac{t^2}{4}\right) J_\alpha(zt) \sqrt{zt}\, dt $$

$$ {}_1F_1\left(2a-\mu; 1+a; -\frac{z^2}{2}\right) = \frac{\Gamma(2a-\mu)}{\sqrt{\pi}\,\Gamma(1+a)}\, z^{-\mu-\frac{1}{2}}\, 2^{\mu-a+\frac{1}{2}} $$

$$ \times \int_0^\infty t^{2a-\mu-\frac{1}{2}} \exp\left(-\frac{t^2}{4}\right) K_{a-\mu-\frac{1}{2}}\left(\frac{t^2}{4}\right) J_\mu(zt) \sqrt{zt}\, dt $$

$$ Re\, \mu > -1,\ Re\,(4a-3\mu) > \frac{1}{2}, $$

$$ {}_1F_1(a; c; z) $$

$$ = \frac{\sqrt{\pi}\,\Gamma(c)}{\Gamma(a)\,\Gamma(c-a)}\, z^{\frac{1}{2}-a} \int_0^1 t^{a-\frac{1}{2}} (1-t)^{c-2a-1} \exp\left(\frac{z}{2}t\right) I_{a-\frac{1}{2}}\left(\frac{z}{2}t\right) dt, $$

$$ Re\, a > 0,\ Re\, c > 2\, Re\, a, $$

$$ {}_1F_1(a; c; z) = \frac{\Gamma(c)}{\Gamma(c-a)}\, e^z z^{\frac{1-c}{2}} \int_0^\infty e^{-t} t^{\frac{1}{2}c-a-\frac{1}{2}} J_{c-1}(2\sqrt{zt})\, dt, $$

$$ Re\, c > Re\, a > 0,\ Re\, z > 0. $$

Contour integrals

$$ {}_1F_1(a; c; z) = \frac{\Gamma(c)}{2\pi i\, \Gamma(a)} \int_{\lambda-i\infty}^{\lambda+i\infty} \frac{\Gamma(-t)\,\Gamma(a+t)}{\Gamma(c+t)} (-z)^t\, dt, $$

$$ Re\, a > -\lambda > 0,\ |\arg(-z)| < \frac{\pi}{2},\ c \neq 0, -1, -2, \ldots $$

The contour must be such that it separates the poles of $\Gamma(-t)$ from the poles of $\Gamma(a+t)$

$$ {}_1F_1(a; c; z) = \frac{\Gamma(1+a-c)\,\Gamma(c)}{\Gamma(a)\, 2\pi i} \int_0^{(1+)} e^{zt} t^{a-1} (t-1)^{c-a-1}\, dt, $$

$$ Re\, a > 0, $$

$$ = -\frac{\Gamma(c)\,\Gamma(1-a)}{2\pi i\, \Gamma(c-a)} \int_1^{(0+)} e^{zt} (-t)^{a-1} (1-t)^{c-a-1}\, dt, $$

$$ Re\,(c-a) > 0 $$

$$ {}_1F_1(a; c; z) $$

$$ = -\frac{\Gamma(c)\,\Gamma(1-a)\,\Gamma(1+a-c)}{(2\pi)^2 \exp(i\pi c)} \int^{(1+,0+,1-,0-)} e^{zt} t^{a-1} (1-t)^{c-a-1}\, dt, $$

$$ Re\, a > 0,\ c-a \neq \text{positive integer}. $$

$$ {}_1F_1(a; c; z) = \frac{\Gamma(b)\, z^{1-b}}{2\pi i} \int_{\gamma-i\infty}^{\gamma+i\infty} e^{zt} t^{-b}\, {}_2F_1(a, b; c; t^{-1})\, dt, $$

where the principal value of the integral is taken and $Re\, b > 0$, $\gamma > 1$. Taking $b = n + 1$

$$ {}_1F_1(a; c; z) = \frac{n!\, z^{-n}}{2\pi i} \int_{\gamma - i\infty}^{\gamma + i\infty} e^{zt} t^{-n-1} {}_2F_1\left(a, n+1;\; c; \frac{1}{t}\right) dt, $$

$$ n = 0, 1, 2, 3, \ldots, \gamma > 1, $$

$$ {}_1F_1(a; c; z) = \frac{\Gamma(c)\, z^{1-c}}{2\pi i} \int_{\gamma - i\infty}^{\gamma + i\infty} e^{zt} t^{-c} \left(1 - \frac{1}{t}\right)^{-a} dt, $$

$$ Re\, c > 0,\ \gamma > 1, $$

$$ {}_1F_1(a; c; kz) = \frac{e^{\alpha z}}{2\pi i} \int_{\gamma - i\infty}^{\gamma + i\infty} z^{-t} \alpha^{-t} \Gamma(t)\, {}_2F_1\left(a, t; c; \frac{k}{\alpha}\right) dt, $$

$$ Re\, \gamma > 0,\ Re\, \alpha > 0,\ Re\, \alpha > Re\, k. $$

6.5.2 Integral representations for $U(a, c, z)$

$$ U(a, c, z) = \frac{1}{\Gamma(a)} \int_0^\infty e^{-zt} t^{a-1} (1+t)^{c-a-1}\, dt, $$

$$ Re\, a > 0,\ Re\, z > 0, $$

$$ = \frac{z^{1-c}}{\Gamma(a)} \int_0^\infty e^{-t} t^{a-1} (z+t)^{c-a-1} dt, $$

$$ Re\, a > 0,\ Re\, z > 0, $$

$$ = \frac{e^z}{\Gamma(a)} \int_1^\infty e^{-zt} (t-1)^{a-1} t^{c-a-1}\, dt $$

$$ = \frac{2^{1-c} e^z}{\Gamma(a)} \int_1^\infty e^{-z\frac{t}{2}} (t-1)^{a-1} (1+t)^{c-a-1}\, dt $$

$$ = \frac{2^{1-c} e^{\frac{z}{2}}}{\Gamma(a)} \int_0^\infty \exp\left(-\frac{z}{2} \cosh t\right) (\sinh t)^{c-1} \left(\coth \frac{t}{2}\right)^{c-2a} dt $$

$$ = \frac{e^z}{\Gamma(a)} \int_\alpha^{\alpha+1} \exp\left[-\frac{z}{\alpha + 1 - t}\right] (t-\alpha)^{a-1} (\alpha + 1 - t)^{-c}\, dt $$

$$ = \frac{(\beta - \alpha)^{c-a}}{\Gamma(a)} e^z \int_\alpha^\beta \exp\left[-\frac{(\beta - \alpha)\, z}{\beta - t}\right] (t-\alpha)^{a-1} (\beta - t)^{-c}\, dt $$

$$ = \frac{\alpha^{1-c}}{\Gamma(a)} e^z \int_\alpha^\infty e^{-z\frac{u}{\alpha}} u^{c-a-1} (u-\alpha)^{a-1}\, du, $$

α real and positive.

In terms of a contour integral

$$U(a, c, z) = \frac{1}{\Gamma(a)} \int_0^{\infty e^{i\delta}} e^{-zt} t^{a-1} (1+t)^{c-a-1} \, dt,$$

$$Re\, a > 0, \quad -\frac{\pi}{2} - \delta < \arg z < \frac{\pi}{2} - \delta.$$

Also the functions t^{a-1} and $(1+t)^{c-a-1}$ assume principal values.

6.6 Integral transforms associated with ${}_1F_1(a; c; z)$, $U(a, c, z)$

6.6.1 Laplace transforms

$$\int_0^\infty t^{\lambda-1} {}_AF_B[(\alpha); (\beta); kt] \, e^{-zt} \, dt = \Gamma(\lambda) \, z^{-\lambda} \, {}_{A+1}F_B\left[(\alpha), \lambda; (\beta); \frac{k}{z}\right]$$

where

$$(\alpha) \equiv \alpha_1, \alpha_2, \ldots, \alpha_A,$$

$$(\beta) \equiv \beta_1, \beta_2, \ldots, \beta_B,$$

$$A < B, \quad Re\, \lambda > 0, \quad Re\, z > 0$$

or

$$A = B, \quad Re\, \lambda > 0, \quad Re\, z > Re\, k > 0.$$

In particular for $A = B = 1$

$$\int_0^\infty t^{\lambda-1} e^{-zt} \, {}_1F_1(a; c; kt) \, dt = \Gamma(\lambda) \, z^{-\lambda} \, {}_2F_1\left(a, \lambda; c; \frac{k}{z}\right)$$

$$= \Gamma(\lambda) \, (z-k)^{-\lambda} \, {}_2F_1\left(c-a, \lambda; c; \frac{k}{k-z}\right),$$

$$\int_0^\infty t^{\lambda-1} e^{-zt} \, {}_1F_1(a; \lambda; kt) \, dt = \Gamma(\lambda) \, z^{-\lambda} \, {}_2F_1\left(a, \lambda; \lambda; \frac{k}{z}\right)$$

$$= \Gamma(\lambda) \, z^{-\lambda} \left(1 - \frac{k}{z}\right)^{-a},$$

$$Re\, z > Re\, k, \quad |z| > |k|, \quad Re\, \lambda > 0,$$

$$\int_0^\infty e^{-\alpha t} t^{z-1} \, {}_1F_1(a; c; kt) \, dt = \alpha^{-z} \Gamma(z) \, {}_2F_1\left(a, z; c; \frac{k}{\alpha}\right),$$

$$Re\, \alpha > 0, \quad Re\, \alpha > Re\, k, \quad Re\, z > 0,$$

$$\int_0^\infty t^{-a} e^{-zt} \, {}_1F_1\left(a; c; \frac{z}{2} t\right) dt = z^{a-1} \frac{\Gamma(1-a) \, \Gamma\left(\frac{c}{2}\right) \Gamma\left(\frac{c+1}{2}\right)}{\Gamma\left(\frac{a+c}{2}\right) \Gamma\left(\frac{1+c-a}{2}\right)}$$

$$Re\, z > 0, \quad Re\,(1-a) > 0,$$

$$\int_0^\infty t^{\lambda-1}e^{-zt}\,{}_1F_1\left(a;\frac{a+\lambda+1}{2};\frac{z}{2}t\right)dt = z^{-\lambda}\frac{\Gamma(\lambda)\,\Gamma\left(\frac{1}{2}\right)\Gamma\left(\frac{1+a+\lambda}{2}\right)}{\Gamma\left(\frac{1+a}{2}\right)\Gamma\left(\frac{1+\lambda}{2}\right)}$$

$$Re\,\lambda > 0,\ Re\,z > 0,$$

$$\int_0^\infty e^{-zt}t^{\lambda-1}\,{}_1F_1(a;\lambda;kt)\,dt\int_0^\infty e^{-zt}t^{\lambda'-1}\,{}_1F_1(a';\lambda';kt)\,dt$$

$$= \frac{\Gamma(\lambda)\,\Gamma(\lambda')}{\Gamma(\lambda+\lambda')}\int_0^\infty t^{\lambda+\lambda'-1}e^{-zt}\,{}_1F_1(a+a';\lambda+\lambda';kt)\,dt$$

$$= \Gamma(\lambda)\,\Gamma(\lambda')\,z^{-\lambda-\lambda'}\left(1-\frac{k}{z}\right)^{-a-a'}$$

$$\int_0^\infty e^{-zt}t^{a-1}\,{}_0F_1(;c;kt)\,dt = \Gamma(a)\,z^{-a}\,{}_1F_1\left(a;c;\frac{k}{z}\right)$$

$$Re\,a > 0,\ Re\,z > 0,$$

$$\int_0^\infty e^{-zt}t^{a-1}\,{}_0F_1(;a;kt)\,dt = \Gamma(a)\,z^{-a}e^{\frac{k}{z}}.$$

Also

$${}_0F_1(;\alpha;-z) = \Gamma(\alpha)\,z^{\frac{1-\alpha}{2}}\,J_{\alpha-1}(2\sqrt{z}).$$

Thus

$${}_1F_1\left(a;a;-\frac{u}{z}\right) = e^{-\frac{u}{z}}$$

$$= u^{\frac{1}{2}(1-a)}\,z^a\int_0^\infty e^{-zt}t^{\frac{1}{2}(a-1)}\,J_{a-1}\left(2\sqrt{ut}\right)dt,$$

$${}_1F_1\left(a;c;-\frac{u}{z}\right) = \frac{\Gamma(c)}{\Gamma(a)}\,u^{\frac{1}{2}(1-c)}\,z^a\int_0^\infty e^{-zt}J_{c-1}\left(2\sqrt{ut}\right)t^{a-\frac{1}{2}-\frac{c}{2}}\,dt,$$

$${}_1F_1\left(a;c;\frac{u}{z}\right) = \frac{\Gamma(c)}{\Gamma(a)}\,z^a u^{\frac{1-c}{2}}\int_0^\infty e^{-zt}I_{c-1}\left(2\sqrt{ut}\right)t^{a-\frac{1}{2}-\frac{c}{2}}\,dt$$

$$= \frac{\Gamma(c)}{\Gamma(c-a)}\,e^{\frac{u}{z}}\,u^{\frac{1-c}{2}}\,z^{c-a}\int_0^\infty e^{-zt}J_{c-1}\left(2\sqrt{ut}\right)t^{\frac{c}{2}-\frac{1}{2}-a}\,dt,$$

$$\int_0^\infty e^{-zt}\left[\int_0^t u^{\lambda-1}\,{}_1F_1(a;c;ku)\,(t-u)^{\mu-1}\,{}_1F_1(a';c';k'(t-u))\,du\right]dt$$

$$= \Gamma(\lambda)\,\Gamma(\mu)\,z^{-\lambda-\mu}\,{}_2F_1\left(a,\lambda;c;\frac{k}{z}\right){}_2F_1\left(a',\mu;c';\frac{k'}{z}\right)$$

$$Re\,c > 0,\ Re\,c' > 0,\ Re\,z > 0,\ Re\,z > Re\,k,$$

$$Re\,z > Re\,k',\ |z| > |k|,\ |z| > |k'|.$$

$$\int_0^\infty e^{-zt} t^{c-1} {}_1F_1(a;c;kt)\, {}_1F_1(a';c';zt)\, dt$$
$$= z^{-c} \frac{\Gamma(c)\,\Gamma(c')\,\Gamma(c'-a'-c)}{\Gamma(c'-a')\,\Gamma(c'-c)} {}_2F_1\left(a, 1+c-c'; 1+a'+c-c'; \frac{k}{z}\right),$$
$$Re\, c > 0,\ Re\, z > 0,\ |z| > |k|$$

or
$$z = k \text{ and } Re\,(1 + a' - a) > 0.$$

$$\int_0^\infty e^{-zt} t^{c-1} {}_1F_1(a;c;zt)\, {}_1F_1(a';c';zt)\, dt$$
$$= z^{-c} \frac{\Gamma(c)\,\Gamma(c')\,\Gamma(c'-c-a')\,\Gamma(1+a'-c'+c)\,\Gamma(a'-a)}{\Gamma(c'-a')\,\Gamma(c'-c)\,\Gamma(1+a'-a-c'+c)\,\Gamma(a')},$$
$$Re\, c > 0,\ Re\, z > 0,\ Re\,(1+a'-a) > 0,$$

$$\int_0^\infty t^{z_1-1}(1+t)^{-z_2-a'}\, {}_1F_1(a;c;-zt)\, dt$$
$$= \frac{\sin(\pi z_2)\,\Gamma(z_1)\,\Gamma(1-a'-z_2)}{\sin[\pi(z_2-z_1)]\,\Gamma(1+z_1-z_2-a')} {}_2F_2(a, z_1; c, 1+z_1-z_2-a'; z),$$
$$Re\, z_1 > 0, \quad Re\, z_2 > 0,$$

$$\int_0^\infty t^{\alpha-1}\, {}_1F_1(a;c;kt)\, {}_1F_1(a';c';zt)\, e^{-zt}\, dt$$
$$= z^{-\alpha} \frac{\Gamma(\alpha)\,\Gamma(c')\,\Gamma(c'-a'-\alpha)}{\Gamma(c'-a')\,\Gamma'(c'-\alpha)} {}_3F_2\left(a, \alpha, 1+\alpha-c'; c, 1+\alpha+a'-c'; \frac{k}{z}\right),$$
$$Re\, \alpha > 0,\ Re\, z > 0,\ |z| > k$$

or
$$z = k,\ Re\,(1 + a' + c - a - \alpha) > 0.$$

$${}_1F_1(a;c;z) = \frac{\pi^{\frac{1}{2}}\,\Gamma(c)}{\Gamma(a)\,\Gamma(c-2a)} z^{\frac{1}{2}-a} \int_0^1 e^{\frac{z}{2}t}\, t^{a-\frac{1}{2}} (1-t)^{c-2a-1}$$
$$\times I_{a-\frac{1}{2}}\left(\frac{1}{2}zt\right) dt,$$
$$Re\, c > 2\, Re\, a > 0,$$

$${}_1F_1(a;c;z) = \frac{1}{\Gamma(a)} \int_0^\infty e^{-t} t^{a-1}\, {}_0F_1(;c;zt)\, dt$$
$$= \frac{\Gamma(c)}{\Gamma(a)} \int_0^\infty e^{-t} I_{c-1}\left(2\sqrt{zt}\right) t^{a-1} (zt)^{\frac{1-c}{2}}\, dt,$$

$${}_1F_1\left(a; \frac{1}{2}; z\right) = \frac{e^z}{\Gamma\left(\frac{1}{2}-a\right)} \int_0^\infty e^{-t} t^{-a-\frac{1}{2}} \cos\left(2\sqrt{zt}\right) dt,$$
$$Re\, a < \frac{1}{2},$$

$$\int_0^\infty t^{z-1}\, {}_1F_1(a;c;-t)\, dt = \frac{\Gamma(c)\,\Gamma(z)\,\Gamma(a-z)}{\Gamma(a)\,\Gamma(c-z)}$$

(Mellin transform),

$$\int_0^t u^{c-1}(t-u)^{c'-1}\, {}_1F_1(a; c; ku)\, {}_1F_1[a'; c'; k(t-u)]\, du$$

$$= \frac{\Gamma(c)\,\Gamma(c')}{\Gamma(c+c')}\, t^{c+c'-1}\, {}_1F_1(a+a'; c+c'; kt),$$

$$Re\, c > 0, \quad Re\, c' > 0.$$

$$\int_0^t u^{b-1}(t-u)^{c-b-1}\, {}_1F_1(a; b; ku)\, du$$

$$= \frac{\Gamma(b)\,\Gamma(c-b)}{\Gamma(c)}\, t^{c-1}\, {}_1F_1(a; c; kt),$$

$$\frac{1}{2\pi i}\int_{\gamma-i\infty}^{\gamma+i\infty} e^{zt} z^{-\lambda}\, {}_1F_1\left(a; c; \frac{k}{z}\right) dz = \frac{t^{\lambda-1}}{\Gamma(\lambda)}\, {}_1F_2(a; c, \lambda; kt),$$

$$\frac{1}{2\pi i}\int_{\gamma-i\infty}^{\gamma+i\infty} e^{zt} z^{-a}\, {}_1F_1\left(a; c; \frac{k}{z}\right) dz = \frac{t^{a-1}}{\Gamma(a)}\, {}_0F_1(; c; kt)$$

$$= \frac{\Gamma(c)}{\Gamma(a)}\, \frac{t^{a-\frac{c}{2}-\frac{1}{2}}}{k^{\frac{1}{2}c-\frac{1}{2}}}\, I_{c-1}\left(2\sqrt{kt}\right)$$

$$= \frac{\Gamma(c)}{\Gamma(a)}\, k^{\frac{1}{2}-\frac{1}{2}c}\, t^{a-\frac{1}{2}-\frac{c}{2}}\, I_{c-1}\left(2\sqrt{kt}\right),$$

$$Re\, z > Re\, k, \quad Re\, a > 0.$$

$$\frac{1}{2\pi i}\int_{\gamma-i\infty}^{\gamma+i\infty} e^{zt} z^{-\lambda}\, {}_2F_1\left(a, \lambda; c; \frac{k}{z}\right) dz = \frac{t^{\lambda-1}}{\Gamma(\lambda)}\, {}_1F_1(a; c; kt),$$

$$Re\, z > Re\, k, \quad Re\, \lambda > 0,$$

$$\frac{1}{2\pi i}\int_{\gamma-i\infty}^{\gamma+i\infty} e^{zt} z^{-\lambda}\, {}_1F_0\left(; a; \frac{k}{z}\right) dz = \frac{1}{2\pi i}\int_{\gamma-i\infty}^{\gamma+i\infty} e^{zt} z^{-\lambda}\left(1-\frac{k}{z}\right)^{-a} dz$$

$$= \frac{t^{\lambda-1}}{\Gamma(\lambda)}\, {}_1F_1(a; \lambda; kt),$$

$$Re\, z > Re\, k, \; Re\, \lambda > 0,$$

$$\frac{1}{2\pi i}\int_{\gamma-i\infty}^{\gamma+i\infty} (\alpha z)^{-t}\, \Gamma(t)\, {}_2F_1\left(a, t; c; \frac{k}{\alpha}\right) dt = e^{-\alpha z}\, {}_1F_1(a; c; kz),$$

$$\frac{1}{2\pi i}\int_{\gamma-i\infty}^{\gamma+i\infty} z^{-t}\, \Gamma(t)\, {}_1F_1(t; c; y)\, dt = \Gamma(c)\, e^{-z}(yz)^{\frac{c-1}{2}}\, I_{c-1}\left(2\sqrt{yz}\right),$$

$$Re\, y > 0,$$

$$\int_0^\infty t^{\alpha-1}(1+t)^{-1}\,{}_1F_1(a;c;zt)\,{}_1F_1(a';c';z't)\,dt$$
$$=\frac{\pi}{\sin(\pi\alpha)}\,{}_1F_1(a;c;-z)\,{}_1F_1(a';c';-z'),$$
$$Re\,\alpha>0.$$

6.6.2 Hankel transforms

$$\int_0^\infty t^{2c-\alpha-\frac{3}{2}}\,{}_1F_1(a;c;-kt^2)\,(zt)^{\frac{1}{2}}\,J_\alpha(zt)\,dt$$
$$=\frac{\Gamma(c)\,k^{-a}\,2^{2c-2a-\alpha-1}}{\Gamma(a-c+\alpha+1)}\,x^{2a-2c+\alpha+\frac{1}{2}}\,{}_1F_1\left(a;1+a-c+\alpha;-\frac{z^2}{4k}\right),$$
$$Re\,k>0,\quad Re\,z>0,\ Re\left(a+\frac{1}{2}\alpha+\frac{3}{4}\right)>Re\,c>0,$$

$$\int_0^\infty t^{a-1}\,{}_1F_1\left(a;\frac{a+\nu+1}{2};-\frac{t^2}{2}\right)\sqrt{zt}\,J_\nu(zt)\,dt$$
$$=z^{a-\frac{1}{2}}\,{}_1F_1\left(a;\frac{1+\nu+a}{2};-\frac{z^2}{2}\right),$$
$$Re\,a>-\frac{1}{2},\ Re\,(a+\nu)>-\frac{1}{2},$$

$$\int_0^\infty t^{\nu+\frac{1}{2}-a}\,{}_1F_1\left(\frac{a}{2};1+\nu-\frac{a}{2};-\frac{t^2}{2}\right)J_\nu(zt)\sqrt{zt}\,dt$$
$$=\sqrt{\pi}\,2^{-a+\nu+\frac{1}{2}}\,\frac{\Gamma\left(1+\nu-\frac{a}{2}\right)}{\Gamma\left(\frac{a}{2}\right)}\,z^{a-\nu-\frac{1}{2}}\,e^{-\frac{z^2}{4}}\,I_{a-\frac{1}{2}}\left(\frac{z^2}{4}\right),$$
$$Re\left(\frac{1}{2}a+\frac{1}{2}\nu+\frac{3}{4}\right)>Re\left(1+\nu-\frac{1}{2}a\right)>0,$$

$$\int_0^\infty t^{\alpha+\frac{1}{2}}\,{}_1F_1\left(a-\alpha;1+\frac{a}{2};-\frac{1}{2}t^2\right)J_\alpha(zt)\sqrt{zt}\,dt$$
$$=\frac{\Gamma(a-\alpha)\,2^{\alpha-\frac{a}{2}+\frac{1}{2}}}{\sqrt{\pi}\,\Gamma\left(1+\frac{a}{2}\right)}\,z^{a-\nu-\frac{1}{2}}\,e^{-\frac{z^2}{4}}\,K_{\frac{a}{2}-\alpha-\frac{1}{2}}\left(\frac{z^2}{4}\right),$$
$$Re\,\alpha>-1,\quad Re\,(2a-3\alpha)>\frac{1}{2},$$

$$\int_0^\infty t^{\frac{c+c'}{2}-1}\,{}_1F_1(a';c';-t)\,U(a,c,t)\,J_{c+c'-2}\left(2\sqrt{zt}\right)dt$$
$$=\frac{\Gamma(c')}{\Gamma(a+a')}\,z^{\frac{c+c'}{2}-1}\,{}_1F_1(a';a+a';-z)$$
$$\times\,U(c'-a',c+c'-a-a',z),$$
$$Re\,c'>0,\ 2\,Re(a+a')+\frac{1}{2}>Re\,(c+c')>1.$$

6.7 Special cases and its relation to other function

6.7.1 Relation to the Bessel's functions

For fixed values of c and z, the Bessel functions can be obtained from ${}_1F_1(a; c; z)$ and $U(a, c, z)$ as limiting cases.

$$\lim_{a\to\infty} \frac{1}{\Gamma(c)}\, {}_1F_1\left(a; c; \frac{z}{a}\right) = z^{\frac{1}{2}-\frac{1}{2}c}\, I_{c-1}\left(2\sqrt{z}\right),$$

$$\lim_{a\to\infty} \frac{1}{\Gamma(c)}\, {}_1F_1\left(a; c; -\frac{z}{a}\right) = z^{\frac{1}{2}-\frac{1}{2}c}\, J_{c-1}\left(2\sqrt{z}\right),$$

$$\lim_{a\to\infty} \Gamma(a+1-c)\, U\left(a, c, \frac{z}{a}\right) = 2z^{\frac{1}{2}-\frac{1}{2}c}\, K_{c-1}\left(2\sqrt{z}\right),$$

$$\lim_{a\to\infty} \Gamma(a+1-c)\, U\left(a, c, -\frac{z}{a}\right) = \begin{cases} -i\pi e^{i\pi c} z^{\frac{1}{2}-\frac{1}{2}c}\, H^{(1)}_{c-1}\left(2\sqrt{z}\right), & \operatorname{Im} z > 0, \\ i\pi e^{-i\pi c} z^{\frac{1}{2}-\frac{1}{2}c}\, H^{(2)}_{c-1}\left(2\sqrt{z}\right), & \operatorname{Im} z < 0. \end{cases}$$

Bessel functions as special cases.

$${}_1F_1\left(\nu+\frac{1}{2}; 2\nu+1; 2iz\right) = \Gamma(1+\nu)\, e^{iz}\left(\frac{z}{2}\right)^{-\nu} J_\nu(z),$$

$${}_1F_1\left(-\nu+\frac{1}{2}; -2\nu+1; 2iz\right) = \Gamma(1-\nu)\, e^{iz}\left(\frac{z}{2}\right)^{\nu} \times \left[\cos(\pi\nu)\, J_\nu(z) - \sin(\pi\nu)\, Y_\nu(z)\right].$$

$${}_1F_1\left(\nu+\frac{1}{2}; 2\nu+1; 2z\right) = \Gamma(1+\nu)\, e^{z}\left(\frac{z}{2}\right)^{-\nu} I_\nu(z),$$

$${}_1F_1(n+1; 2n+2; 2iz) = \Gamma\left(n+\frac{3}{2}\right) e^{iz}\left(\frac{z}{2}\right)^{-n-\frac{1}{2}} J_{n+\frac{1}{2}}(z),$$

$${}_1F_1(-n, -2n; 2iz) = \Gamma\left(\frac{1}{2}-n\right) e^{iz}\left(\frac{z}{2}\right)^{n+\frac{1}{2}} J_{-n-\frac{1}{2}}(z),$$

$${}_1F_1(n+1; 2n+2; 2z) = \frac{1}{2}\Gamma\left(n+\frac{3}{2}\right) e^{z}\left(\frac{z}{2}\right)^{-n-\frac{1}{2}} I_{n+\frac{1}{2}}(z),$$

$${}_1F_1\left(n+\frac{1}{2}; 2n+1; -2\sqrt{iz}\right) = \Gamma(n+1)\, e^{-2z\pi}\left(i\frac{\pi}{2}z\right)^{-n} (\operatorname{ber}_n z + i\operatorname{bei}_n z),$$

$$U\left(\nu+\frac{1}{2}, 2\nu+1, 2z\right) = \pi^{-\frac{1}{2}}\, e^{z}(2z)^{-\nu} K_\nu(z),$$

$$U\left(\nu+\frac{1}{2},\,2\nu+1,\,2iz\right)=\frac{i\pi^{\frac{1}{2}}}{2}\,e^{-i\pi(\nu-z)}(2\,z)^{-\nu}\,H_\nu^{(2)}(z),$$

$$U\left(\nu+\frac{1}{2},\,2\nu+1\,,-2iz\right)=\frac{i\pi^{\frac{1}{2}}}{2}\,e^{i\pi(\nu-z)}(2z)^{-\nu}\,H_\nu^{(1)}(z),$$

$$U(n+1,\,2n+2,\,2z)=\pi^{-\frac{1}{2}}\,e^{z}(2z)^{-n-\frac{1}{2}}\,K_{n+\frac{1}{2}}(z),$$

$$U\left(n+\frac{1}{2},\,2n+1,\sqrt{iz}\right)=\pi^{-\frac{1}{2}}\,e^{in\frac{\pi}{2}}\,e^{\sqrt{iz}}\left(2\sqrt{iz}\right)^{-n}$$
$$\times\left[\operatorname{ker}_n(z)+i\operatorname{kei}_n(z)\right].$$

Expansion of ${}_1F_1(a;c;z)$ in series of Bessel functions

$${}_1F_1(a;c;z)=\Gamma\left(c-a-\frac{1}{2}\right)e^{\frac{1}{2}z}\left(\frac{z}{4}\right)^{a-c+\frac{1}{2}}$$
$$\times\sum_{n=0}^{\infty}(-1)^n\frac{(c-2a)_n\,(2c-2a-1)_n}{(c)_n\,n!}\,I_{c-a+n-\frac{1}{2}}\left(\frac{z}{2}\right),$$
$$c\neq 0,-1,-2,\ldots,$$

$${}_1F_1(a;c;z)=\Gamma(c)\,e^{\frac{z}{2}}\left(\frac{1}{2}cz-az\right)^{\frac{1}{2}-\frac{1}{2}c}$$
$$\times\sum_{n=0}^{\infty}A_n\left(\frac{1}{2}z\right)^{\frac{n}{2}}(c-2a)^{-\frac{n}{2}}\,J_{c+n-1}\left(\sqrt{2cz-4az}\right),$$

where

$$A_0=1,\,A_1=0,\,A_2=\frac{c}{2}$$

and

$$(n+1)\,A_{n+1}=(n+c-1)\,A_{n-1}+(2a-c)\,A_{n-2},$$

$${}_1F_1(a;c;z)=\sum_{n=0}^{\infty}A_n(a,c)\,I_n(z),$$

where

$$A_0(a,c)=1,\;A_1(a,c)=\frac{2a}{c},$$

$$A_{n+1}(a,c)=\frac{2a}{c}A_n(a+1,c+1)-A_{n-1}(a,c),$$

$${}_1F_1(a;c;z)=\Gamma(c)\,e^{hz}\sum_{n=0}^{\infty}A_nz^n(-az)^{\frac{(1-c-n)}{2}}$$
$$\times J_{c+n-1}\left(2\sqrt{-az}\right)$$

where

$$A_0 = 1,\ A_1 = -ch,\ A_2 = -\frac{1}{2}(2h-1)\,a + \frac{c}{2}(c+1)\,h_2,$$

$$(n+1)\,A_{n+1} = [(1-2h)\,n - ch]\,A_n + [(1-2h)\,a - h(h-1)\,(c+n-1)]\,A_{n-1} - h(h-1)\,a\,A_{n-2}$$

and h is real.

6.7.2 Other special cases

$${}_1F_1(a;a;z) = e^z,$$

$${}_1F_1(1;2;-2iz) = \frac{e^{-iz}}{z}\sin z,$$

$${}_1F_1(1;2;2z) = \frac{e^z}{z}\sinh z,$$

$${}_1F_1(a;a+1;-z) = a z^{-a}\gamma(a,z) = \int_0^z t^{a-1}e^{-t}\,dt$$

(incomplete gamma function),

$$U(1-a,1-a,z) = e^z\Gamma(a,z) = \int_z^\infty t^{a-1}e^{-t}\,dt$$

(incomplete gamma function),

$${}_1F_1\left(\frac{1}{2};\frac{3}{2};-z^2\right) = \frac{\sqrt{\pi}}{2z}\,\mathrm{Erf}\,(z),$$

$${}_1F_1\left(1;\frac{3}{2};z^2\right) = \frac{\sqrt{\pi}}{2z}\,e^{z^2}\,\mathrm{Erf}(z)$$ (error function or integral),

$$U\left(\frac{1}{2},\frac{1}{2},z^2\right) = \sqrt{\pi}\,e^{z^2}\,\mathrm{Erfc}\,(z)$$

(complementary error function),

$$U(1,1,z) = -e^z\,Ei(-z) = e^z\int_z^\infty t^{-1}e^{-t}\,dt,$$

$$U(1,1,-z) = -e^{-z}\,Ei(z)$$

(exponential integrals),

$$U(1,1,-\log z) = -\frac{1}{z}\,li(z) = -\frac{1}{z}\int_0^z \frac{dt}{\log t}$$

(logarithmic integral),

$$U(1, 1, iz) = e^{iz}\left[-i\frac{\pi}{2} - Ci(z) + i\,Si(z)\right],$$

$$U(1, 1, -iz) = e^{-iz}\left[i\frac{\pi}{2} - Ci(z) - i\,Si(z)\right]$$

(sine and cosine integrals),

$$C_2(x) = \frac{1}{\sqrt{2\pi}}\int_0^x \frac{\cos t}{\sqrt{t}}\,dt$$

$$= \sqrt{\frac{x}{2\pi}}\left[{}_1F_1\left(\frac{1}{2}; \frac{3}{2}; -xe^{i\frac{\pi}{2}}\right) + {}_1F_1\left(\frac{1}{2}; \frac{3}{2}; -xe^{-i\frac{\pi}{2}}\right)\right],$$

$$S_2(x) = \frac{1}{\sqrt{2\pi}}\int_0^x \frac{\sin t}{\sqrt{t}}\,dt = \sqrt{\frac{x}{2\pi}}\,e^{i\frac{\pi}{2}}$$

$$\times\left[{}_1F_1\left(\frac{1}{2}; \frac{3}{2}; -ix\right) - {}_1F_1\left(\frac{1}{2}; \frac{3}{2}; ix\right)\right].$$

The Whittaker functions $M_{\varkappa,\mu}(z)$ and $W_{\varkappa,\mu}(z)$ [see chap. VII] are given in terms of Kummer's function by the following relations

$$M_{\varkappa,\mu}(z) = e^{-\frac{z}{2}} z^{\mu+\frac{1}{2}}\,{}_1F_1\left(\frac{1}{2} + \mu - \varkappa; 1 + 2\mu; z\right)$$

$$= \Gamma(1 + 2\mu)\, e^{-\frac{z}{2}} z^{\mu+\frac{1}{2}}$$

$$\times\left[\frac{e^{i\frac{\pi}{2}\varepsilon(1+2\mu-2\varkappa)}}{\Gamma\left(\frac{1}{2} + \mu + \varkappa\right)} U\left(\frac{1}{2} + \mu - \varkappa, 1 + 2\mu, z\right)\right.$$

$$\left. + \frac{e^z e^{-i\frac{\pi}{2}\varepsilon(1+2\mu+2\varkappa)}}{\Gamma\left(\frac{1}{2} + \mu - \varkappa\right)} U\left(\frac{1}{2} + \mu + \varkappa, 1 + 2\mu, -z\right)\right],$$

$$\varepsilon = \begin{cases} 1 & \operatorname{Im} z > 0, \\ -1 & \operatorname{Im} z < 0, \end{cases}$$

$$W_{\varkappa,\mu}(z) = \exp\left(-\frac{z}{2}\right) z^{\frac{1}{2}+\mu}\, U\left(\frac{1}{2} + \mu - \varkappa, 1 + 2\mu, z\right)$$

$$-\pi < \arg z \leq \pi,\ \mu \neq 0, -\frac{1}{2}, -\frac{3}{2}, \ldots$$

$$= \frac{\Gamma(-2\mu)}{\Gamma\left(\frac{1}{2} - \mu - \varkappa\right)} M_{\varkappa,\mu}(z) + \frac{\Gamma(2\mu)}{\Gamma\left(\frac{1}{2} + \mu - \varkappa\right)} M_{\varkappa,-\mu}(z).$$

The parabolic cylinders functions $D_\nu(z)$ [see chap. VIII] are given by

$$D_\nu(z) = 2^{\frac{\nu}{2}} e^{-\frac{z^2}{4}} U\left(-\frac{\nu}{2}, \frac{1}{2}, \frac{z^2}{2}\right)$$

$$= 2^{\frac{1}{2}\nu - \frac{1}{2}} z e^{-\frac{z^2}{4}} U\left(\frac{1}{2} - \frac{1}{2}\nu, \frac{3}{2}, \frac{z^2}{2}\right).$$

The Laguerre and Hermite polynomials [see chap. V] are given by

$$L_n^{(\alpha)}(z) = \frac{(\alpha+1)_n}{n!} {}_1F_1(-n; \alpha+1; z) = \frac{(-1)^n}{n!} U(-n, \alpha+1, z),$$

$$H_n(z) = 2^n U\left(-\frac{n}{2}, \frac{1}{2}, z^2\right),$$

$$H_{2n}(z) = \frac{(-1)^n}{n!} (2n)!\, {}_1F_1\left(-n; \frac{1}{2}; z^2\right),$$

$$H_{2n+1}(z) = \frac{(-1)^n}{n!} (2n+1)!\, 2z\, {}_1F_1\left(-n; \frac{3}{2}; z^2\right).$$

Other special cases are Laguerre functions [see chap. VIII]

$$L_\nu^{(\alpha)}(z) = \frac{(\alpha+1)_n}{\Gamma(\nu+1)} {}_1F_1(-\nu; \alpha+1; z).$$

Poisson-Charlier polynomials $p_n(z)$ [also written as $\varrho_n(\nu, z)$]

$$p_n(z) = a^{-\frac{n}{2}} (n!)^{\frac{1}{2}} L_n^{(z-n)}(a)$$

$$= a^{-\frac{n}{2}} (n!)^{-\frac{1}{2}} (z-n+1)_n\, {}_1F_1(-n; z-n+1; a).$$

Toronto functions $T(m, n, x)$

$$T(m, n, x) = x^{2n-m+1} \frac{\Gamma\left(\frac{1}{2}m + \frac{1}{2}\right)}{\Gamma(n+1)} e^{-x^2}\, {}_1F_1\left(\frac{m+1}{2}; n+1; x^2\right).$$

Bateman's function $k_\nu(x)$

$$k_\nu(x) = \frac{e^{-x}}{\Gamma\left(1+\frac{\nu}{2}\right)} U\left(-\frac{1}{2}\nu, 0, 2x\right),$$

$$x > 0.$$

MacRobert's E-function

$$E(\alpha, \beta::z) = \Gamma(\alpha)\,\Gamma(\beta)\, z^\alpha U(\alpha, \alpha-\beta+1, z).$$

Meixner's function $F_1(\alpha, \beta, z)$

$$F_1(\alpha, \beta, z) = e^{i\pi\alpha} \frac{\Gamma(\beta)}{\Gamma(\beta-\alpha)} U(\alpha, \beta, z),$$

$${}_2F_0(\alpha, \beta; -z^{-1}) = z^\alpha U(\alpha, \alpha-\beta+1, z).$$

6.8 Asymptotic expansions

6.8.1 Small argument

The behavior of the functions ${}_1F_1(a; c; z)$ and $U(a, c, z)$, as $z \to 0$, is given by the following results.

$${}_1F_1(a; c; z) = \sum_{n=0}^{N} \frac{(a)_n}{(c)_n} \frac{z^n}{n!} + O(|z|^{N+1}),$$

$$c \neq 0, -1, -2, \ldots,$$

$$U(a, c, z) = \frac{\Gamma(1-c)}{\Gamma(a+1-c)} + O(|z|)$$

for $\qquad Re\, c \leq 0, c \neq 0$

$$= \frac{1}{\Gamma(a+1)} + O(|z \log z|)$$

for $\qquad c = 0$

$$= \frac{\Gamma(1-c)}{\Gamma(a+1-c)} + O(|z|^{1-Re c})$$

for $\qquad 0 < Re\, c < 1$

$$= -\frac{1}{\Gamma(a)} [\log z + \psi(a) - 2\gamma] + O(|z \log z|)$$

for $\qquad c = 1.$

$$\gamma = \text{Euler's constant}$$

$$U(a, c, z) = \frac{\Gamma(1-c)}{\Gamma(a+1-c)} + \frac{\Gamma(c-1)}{\Gamma(a)} z^{1-c} + O(|z|),$$

$$Re\, c = 1, c \neq 1,$$

$$= \frac{\Gamma(c-1)}{\Gamma(a)} z^{1-c} + O(1),$$

$$1 < Re\, c < 2,$$

$$= \frac{\Gamma(c-1)}{\Gamma(a)} z^{1-c} + O(|\log z|),$$

$$c = 2$$

$$= \frac{\Gamma(c-1)}{\Gamma(a)} z^{1-c} + O(|z|^{Re c - 2}),$$

$$Re\, c \geq 2, \ c \neq 2.$$

6.8.2 Large argument

For fixed values of a, c and as $|z| \to \infty$

$$
\begin{aligned}
{}_1F_1(a;c;z) = {} & \frac{\Gamma(c)}{\Gamma(c-a)} e^{\pm i\pi a} z^{-a} \sum_{r=0}^{N} \frac{(a)_r (a-c+1)_r}{r!} (-z)^{-r} \\
& + O(|z|^{-N-a-1}) \\
& + \frac{\Gamma(c)}{\Gamma(a)} e^z z^{a-c} \sum_{r=0}^{M} \frac{(1-a)_r\,(c-a)_r}{r!} z^{-r} \\
& + O(|e^z z^{-N+a-c-1}|),
\end{aligned}
$$

where the sign is taken +ve or −ve according as $\operatorname{Im} z > 0$ or $\operatorname{Im} z < 0$.

$$-\pi < \arg z < \pi.$$

If $Re\, z \to \infty$

$${}_1F_1(a;c;z) = \frac{\Gamma(c)}{\Gamma(a)} e^z z^{a-c} [1 + O(|z|^{-1})]$$

and

$${}_1F_1(a;c;z) = \frac{\Gamma(c)}{\Gamma(c-a)} (-z)^{-a} [1 + O(|z|^{-1})]$$

for

$$Re\, z \to -\infty.$$

$$
\begin{aligned}
U(a,c,z) = {} & \sum_{r=0}^{N} (-1)^r \frac{(a)_r\,(a+1-c)_r}{r!} z^{-r-a} \\
& + O(|z|^{-N-a-1}) \\
& -\frac{3}{2}\pi < \arg z < \frac{3}{2}\pi.
\end{aligned}
$$

6.8.3 Large parameters

Case 1: $c \to \infty$ but a, z are bounded

$$
\begin{aligned}
{}_1F_1(a;c;z) &= 1 + O(|c|^{-1}) \\
&= \sum_{r=0}^{N} \frac{(a)_r}{(c)_r} \frac{z^r}{r!} + O(|c|^{-N-1}),
\end{aligned}
$$

$$
\begin{aligned}
U(a,c,z) = {} & (-c)^{-a} [1 + O(|c|^{-1})] \\
& + (2\pi)^{\frac{1}{2}} z^{1-c} \exp\left[z - c + \left(c - \frac{3}{2}\right)\log c\right] [1 + O(|c|^{-1})], \\
& |\arg c| \le \pi - \eta,\ |\arg(-c)| \le \pi - \eta,\ \eta > 0.
\end{aligned}
$$

Case 2: $c \to \infty$, $a \to \infty$, but $c - a$, z bounded

$${}_1F_1(a;c;z) = e^z \sum_{n=0}^{N} \frac{(c-a)_n}{n!\,(c)_n} (-z)^n + O(|c|^{-N-1}).$$

The expansion giving the leading term and the first two terms are:

$$ {}_1F_1(a;c;z) = e^z + O(|c|^{-1}), $$

$$ {}_1F_1(a;c;z) = e^z\left(1 - \frac{c-a}{c}z\right) + O(|c|^{-2}). $$

The corresponding result for $U(a,c,z)$ can be obtained from the transformation formula

$$ U(a,c,z) = z^{1-c}\,U(a-c+1,\,2-c,\,z) $$

and the result for $U(a,c,z)$ given in 6.8.2. The leading term estimate is given by

$$ \begin{aligned} U(a,c,z) &= \sqrt{2\pi}\exp(z+c-2)\,(2-c)^{\frac{1}{2}-c}\,[1+O(|c|^{-1})] \\ &\quad + z^{1-c}\exp[-i\pi(1+a-c)]\,(2-c)^{c-a-1}\,[1+O(|c|^{-1})] \end{aligned} $$

provided that

$$ |\arg(\pm c)| < \pi. $$

Case 3: $a\to\infty$ while c, z are fixed

$$ \begin{aligned} {}_1F_1(a;c;z) &= \Gamma(c)\exp\left(\frac{z}{2}\right)\left(\frac{cz}{2}-az\right)^{\frac{1}{2}-\frac{1}{2}c} \\ &\quad\times J_{c-1}\left[\sqrt{2cz-4az}\right]\left[1+O\left(\left|\frac{c}{2}-a\right|^{-\lambda}\right)\right], \end{aligned} $$

where

$$ \lambda = \min\left(1-\mu,\frac{1-3\mu}{2}\right), \qquad 0\le\mu<\frac{1}{3} $$

and μ is given by

$$ |z| = \left|\frac{1}{2}c-a\right|^{\mu}, $$

$$ \begin{aligned} U(a,c,z) &= \Gamma\left(\frac{c+1}{2}-a\right)\exp\left(\frac{z}{2}\right)z^{\frac{(1-c)}{2}} \\ &\quad\times\left[\cos(\pi a)\,J_{c-1}\left(\sqrt{2cz-4az}\right)-\sin(\pi a)\,Y_{c-1}\left(\sqrt{2cz-4az}\right)\right] \\ &\quad\times\left[1+O\left(\left|\frac{c}{2}-a\right|^{-\lambda}\right)\right], \end{aligned} $$

where λ is defined above.

$$ \begin{aligned} {}_1F_1(a;c;z) &= \Gamma(c)\exp\left(\frac{z}{2}\right)\left(\frac{zc}{2}-az\right)^{\frac{1}{4}-\frac{c}{2}} \\ &\quad\times\cos\left(\sqrt{2cz-4az}-\frac{\pi c}{2}+\frac{\pi}{4}\right)\left[1+O\left(\left|\frac{c}{2}-a\right|^{-\frac{1}{2}}\right)\right], \end{aligned} $$

as $$a \to -\infty,\ x \text{ real},\ c \text{ bounded}.$$

$$U(a, c, z) = \Gamma\left(\frac{c}{2} + \frac{1}{4} - a\right) \pi^{-\frac{1}{2}} \exp\left(\frac{z}{2}\right) z^{\frac{1}{4} - \frac{c}{2}}$$

$$\times \cos\left(\sqrt{2cz - 4az} - \frac{c\pi}{2} + a\pi + \frac{\pi}{4}\right)\left[1 + O\left(\left|\frac{c}{2} - a\right|^{-\frac{1}{2}}\right)\right],$$

as $$a \to -\infty, \quad x \text{ real}, \quad c \text{ bounded}.$$

The followings are first order approximations:

as $$a \to \infty; \quad c, z \text{ fixed}$$

$${}_1F_1\left(a; c; \frac{z}{a}\right) \sim \Gamma(c)\, z^{\frac{1}{2} - \frac{1}{2}c} I_{c-1}\left(2\sqrt{z}\right)$$

$${}_1F_1\left(a; c; -\frac{z}{a}\right) \sim \Gamma(c)\, z^{\frac{1}{2} - \frac{1}{2}c} J_{c-1}\left(2\sqrt{z}\right),$$

$${}_1F_1\left(c - a, c, \frac{z}{a}\right) \sim \Gamma(c)\, z^{\frac{1}{2} - \frac{1}{2}c} J_{c-1}\left(2\sqrt{z}\right)$$

$${}_1F_1\left(c - a; c; -\frac{z}{a}\right) \sim \Gamma(c)\, z^{\frac{1}{2} - \frac{1}{2}c} I_{c-1}\left(2\sqrt{z}\right).$$

6.8.4 Argument and parameters large

Case 1: a bounded, $z = kc$ and $c \to \infty$, $0 < |k| < 1$

$${}_1F_1(a; c; z) = (1 - k)^{-a}\left[1 - \frac{a(a+1)}{2c}\left(\frac{k}{1-k}\right)^2 + O(|c|^{-2})\right].$$

The result for $z = kc$, $c \to \infty$, $c - a$ bounded, $0 < k < 1$, may be obtained from the transformation formula.

Case 2: $z > 2c - a > 1$, $\frac{c}{2} - a \to \infty$, $z \to \infty$, (z real)

$${}_1F_1(a, c, z) = \Gamma(c) \sin(\pi a)$$

$$\times \exp\left[(c - 2a)\left(\frac{1}{2}\sinh 2\alpha + \cosh^2\alpha - \alpha\right)\right]$$

$$\times [(c - 2a)\cosh\alpha]^{1-c}\left[\pi\left(\frac{c}{2} - a\right)\sinh 2\alpha\right]^{-\frac{1}{2}}$$

$$\times \left[1 + O\left(\left|\frac{c}{2} - a\right|^{-1}\right)\right],$$

$$U(a, c, z) = [(c - 2a)\cosh\alpha]^{1-c}\left[\frac{\pi}{2}(c - 2a)\sinh\alpha\right]^{\frac{1}{2}}$$

$$\times \exp\left[(c - 2a)(\sinh\alpha\cosh\alpha + \cosh^2\alpha - \alpha)\right]$$

$$\times \left[1 + O\left(\left|\frac{c}{2} - a\right|^{-1}\right)\right],$$

where

$$\cosh\alpha = \sqrt{\frac{z}{2(c-2a)}}.$$

Case 3: $z = 2(c-2a)\left[1 + \frac{t}{(c-2a)^{\frac{2}{3}}}\right]$, $c - 2a \to \infty, z \to \infty$,

$z \sim 2(c-2a)$, z real

$$_1F_1(a;c;z) = \Gamma(c)\exp\left(\frac{z}{2}\right)(c-2a)^{\frac{2}{3}-c}$$
$$\times\left[\cos(\pi a)\,Ai(t) + \sin(\pi a)\,Bi(t) + O\left(|c-2a|^{-\frac{2}{3}}\right)\right],$$

where $Ai(t)$, $Bi(t)$ are the Airy functions given by

$$Ai(-t) = \frac{1}{3}\sqrt{t}\left[J_{\frac{1}{3}}(\xi) + J_{-\frac{1}{3}}(\xi)\right]$$
$$= \frac{1}{\pi}\sqrt{\frac{t}{3}}\,K_{\frac{1}{3}}(\xi),$$
$$\xi = \frac{2}{3}t^{\frac{3}{2}},$$
$$Bi(-t) = \frac{i}{2}\sqrt{\frac{t}{3}}\left[e^{i\frac{\pi}{6}}H^{(1)}_{\frac{1}{3}}(\xi) - e^{-i\frac{\pi}{6}}H^{(2)}_{\frac{1}{3}}(\xi)\right],$$

$$U(a,c,z) = \frac{\Gamma\left(\frac{1}{3}\right)}{\pi^{\frac{1}{2}}}6^{-\frac{1}{6}}\exp\left[\frac{z}{2} + a - \frac{c}{2}\right]$$
$$\times\left[1 - t\Gamma\left(\frac{5}{6}\right)\pi^{-\frac{1}{2}}(cz - 2az)^{\frac{1}{3}}3^{\frac{1}{3}} + O\left(|c-2a|^{-\frac{2}{3}}\right)\right].$$

Case 4: $z \to \infty$, $c - 2a \to \infty$, $2(c-2a) > z > 0$

$\cos\theta = \frac{z}{2(c-2a)}$, z real,

$$_1F_1(a;c;z) = \Gamma(c)\exp\left[(c-2a)\cos^2\theta\right]$$
$$\times\left[(c-2a)\cos\theta\right]^{1-c}\left[\pi(c-2a)\sin\theta\cos\theta\right]^{-\frac{1}{2}}$$
$$\times\left[\sin(\pi a) + \sin\left\{(c-2a)(\theta - \sin\theta\cos\theta) + \frac{\pi}{4}\right\}\right.$$
$$\left. + O(|c-2a|^{-1})\right],$$

$$U(a,c,z) = \exp\left[(c-2a)\cos^2\theta\right]\left[(c-2a)\cos\theta\right]^{1-c}$$
$$\times\left[(c-2a)\sin\theta\cos\theta\right]^{-\frac{1}{2}}$$
$$\times\left[\sin\left\{\frac{\pi}{4} + (c-2a)(\theta - \sin\theta\cos\theta)\right\} + O(|c-2a|^{-1})\right].$$

6.9 Products of Kummer's functions

$$_1F_1(a; c; z)\, _1F_1(a; c; -z)$$

$$= \frac{[\Gamma(c)]^2 z^{1-c}}{\Gamma(a)\,\Gamma(c-a)} \int_{-\infty}^{\infty} \exp\left[(c-2a)\,t\right] \operatorname{sech} t\, I_{c-1}(z \operatorname{sech} t)\, dt\,,$$

$$Re\, a > 0, \quad Re\,(c-a) > 0,$$

$$_1F_1(a; c; -z)\, U(a, c, z)$$

$$= \frac{\Gamma(c)}{\Gamma(a)} z^{1-c} \int_0^{\infty} \left(\tanh \frac{t}{2}\right)^{2a-c} J_{c-1}(z \sinh t)\, dt,$$

$$Re\, a > 0, \quad Re\, z > 0,$$

$$U(a, c, z)\, U(c-a, c, z)$$

$$= \frac{2}{\pi} e^z z^{1-c} \int_0^{\frac{\pi}{2}} K_{c-1}(z \sec t) \cos\left[(c-2a)\,t\right] \sec t\, dt,$$

$$Re\, z > 0,$$

$$U(a, c, z)\, U(a', c, z') = \frac{1}{\Gamma(a+a'-c+1)}$$

$$\times \int_0^{\infty} e^{-t} t^{a+a'-c} (z+t)^{-a} (z'+t)^{-a'}$$

$$\times\, _2F_1\left[a, a'; a+a'-c+1; \frac{t(z+z'+t)}{(z+t)(z'+t)}\right] dt,$$

$$z, z' \neq 0, \quad Re\,(a+a'-c+1) > 0,$$

$$\int_0^{\infty} e^{-zt} t^{c-1}\, _1F_1(a; c; t)\, _1F_1(a'; c; \lambda t)\, dt$$

$$= \Gamma(c)\, (z-1)^{-a} (z-\lambda)^{-a'} z^{a+a'-c}\, _2F_1\left[a, a'; c; \frac{\lambda}{(z-1)(z-\lambda)}\right],$$

$$Re\, c > 0, \quad Re\, z > Re\, \lambda + 1,$$

$$\int_0^{\infty} e^{-t} t^{c-1}\, _1F_1(a; c; t)\, U(a', c', \lambda t)\, dt$$

$$= \frac{\Gamma(a'-a)\,\Gamma(c)\,\Gamma(c-c'+1)}{\Gamma(a')\,\Gamma(c-a+a'-c'+1)} \lambda^{-c}$$

$$\times\, _2F_1\left[c-a, c-c'+1; c-a+a'-c'+1; 1-\frac{1}{\lambda}\right],$$

$$Re\, c > 0,$$

$$\int_0^\infty e^{-t} t^{c+c'-2} {}_1F_1(a; c; t)\, U(a', c', \lambda t)\, dt$$

$$= \frac{\Gamma(a'-a-c'+1)}{\Gamma(a'-c'+1)} \frac{\Gamma(c)\,\Gamma(c+c'-1)}{\Gamma(a'-a+c)} \lambda^{1-c-c'}$$

$$\times {}_2F_1\left[c-a, c+c'-1; a'-a+c; 1-\frac{1}{\lambda}\right],$$

$$Re\, c > 0, \quad Re\,(c+c') > 1.$$

For more results see ERDÉLYI [Tables of integral transforms Vol. I, II]. Some infinite series which involve products of Kummer's functions are:

$$\sum_{n=0}^{\infty} \frac{(c-a)_n\,(c'-a')_n}{(c)_n\,(c')_n\,n!} {}_1F_1(a; n-c; z_1)\, {}_1F_1(a'; n+c'; z_2)\, z^n$$

$$= e^z \sum_{n=0}^{\infty} \frac{(a)_n\,(a')_n}{(c)_n\,(c')_n} {}_1F_1(a+n; n+c; z_1-z)$$

$$\times {}_1F_1(a'+n; n+c'; z_2-z)$$

$$\sum_{n=0}^{\infty} \frac{(a)_n\,(a')_n}{(c)_n\,(c')_n\,n!} {}_1F_1(a+a'-c; c+2n; z)\, z^{2n}$$

$$= {}_1F_1(a; c; z)\, {}_1F_1(a'; c; z),$$

$$\sum_{n=0}^{\infty} \frac{(c-a)_n\,(a')_n}{(c)_n\,(c')_n\,n!} {}_1F_1(a; n+c; z_1)\, {}_1F_1(a'+n; n+c'; z_2-z)\, z^n$$

$$= \sum_{n=0}^{\infty} \frac{(a)_n\,(c'-a')_n}{(c)_n\,(c')_n\,n!} {}_1F_1(a+n; n+c; z_1-z)\, {}_1F_1(a'; n+c'; z_2)\, z^n,$$

$$\sum_{n=0}^{\infty} \frac{\Gamma(\lambda+n)}{\Gamma(n+1)} {}_1F_1(a-n; c; z_1)\, {}_1F_1(a-n; c; z_2)$$

$$= \left[\frac{\Gamma(c)}{\Gamma(c-\lambda)}\right]^2 (z_1 z_2)^{1-c} \int_0^{\min(z_1,z_2)} {}_1F_1(a; c-\lambda; z_1-t)$$

$$\times {}_1F_1(a; c-\lambda; z_2-t)\, e^t t^{\lambda-1} [(z_1-t)\,(z_2-t)]^{c-\lambda-1}\, dt,$$

$$\sum_{n=0}^{\infty} \frac{(\lambda)_n\,n!}{(c)_n\,(c')_n} L_n^{(c-1)}(z_1)\, L_n^{(c'-1)}(z_2)\, z^n$$

$$= \frac{1}{(1-z)^\lambda} \sum_{n=0}^{\infty} \frac{(\lambda)_n}{(c)_n\,(c')_n\,n!} {}_1F_1\left(\lambda+n; c+n; \frac{z\,z_1}{z-1}\right)$$

$$\times {}_1F_1\left(\lambda+n; c'+n; \frac{z\,z_2}{z-1}\right) \left(\frac{z\,z_1\,z_2}{(1-z)^2}\right)^n.$$

Literature

BUCHHOLZ, H.: Die konfluente hypergeometrische Funktion. Berlin/Göttingen/Heidelberg: Springer 1953.

ERDÉLYI, A.: Higher transcendental functions, Vol. 1. New York: McGraw-Hill Book Co. 1953.

ERDÉLYI, A., and C. A. SWANSON: Asymptotic forms of Confluent hypergeometric functions, memior 25. Amer. Math. Soc., Providence, R. I., 1957.

SLATER, L. J.: Confluent hypergeometric functions. Cambridge: Cambridge Univ. Press 1960.

TRICOMI, F. G.: Funzioni ipergeometriche Confluenti. Rome: Edizioni Cremonese 1954.

WHITTAKER, E. T., and G. N. WATSON: A course of modern analysis. Cambridge: Cambridge Univ. Press 1952.

Chapter VII

Whittaker functions

7.1 Whittaker's differential equation

7.1.1 Whittaker functions, notations, definitions

Kummer's differential equation, discussed in chap. VI, can be so normalized that the differential equation in the new dependent variable does not involve the first derivative term. This can be done by the substitution $w = e^{\frac{z}{2}} z^{-\frac{c}{2}} u$ in the differential equation

$$z \frac{d^2 w}{dz^2} + (c - z) \frac{dw}{dz} - a w = 0$$

to yield

$$\frac{d^2 u}{dz^2} + \left[-\frac{1}{4} + \frac{\frac{c}{2} - a}{z} + \frac{\frac{c}{2}\left(1 - \frac{c}{2}\right)}{z^2} \right] u = 0.$$

Putting

$$\frac{c}{2} - a = \varkappa,$$

$$\frac{c}{2}\left(1 - \frac{c}{2}\right) = \frac{1}{4} - \mu^2$$

or

$$a = \mu + \frac{1}{2} - \varkappa,$$

$$c = 1 + 2\mu,$$

the equation becomes

$$\frac{d^2u}{dz^2} + \left(-\frac{1}{4} + \frac{\varkappa}{z} + \frac{\frac{1}{4} - \mu^2}{z^2}\right) u = 0.$$

The parameters $\varkappa, \mu$ are given in terms of a, c by the relations

$$\varkappa = \frac{c}{2} - a,$$

$$\mu = \frac{1}{2}(c-1).$$

In order to obtain a solution of this equation in the neighborhood of the point at infinity the substitution $z' = \frac{1}{z}$ gives

$$\frac{d^2u}{dz'^2} + \frac{2}{z'}\frac{du}{dz'} + \left\{-\frac{1}{4z'^4} + \frac{\varkappa}{z'^3} + \frac{\frac{1}{4} - \mu^2}{z'^2}\right\} u = 0.$$

The functions

$$M_{\varkappa,\mu}(z) = e^{-\frac{z}{2}} z^{\mu+\frac{1}{2}} {}_1F_1\left(\mu + \frac{1}{2} - \varkappa; 1 + 2\mu; z\right),$$

$$M_{\varkappa,-\mu}(z) = e^{-\frac{z}{2}} z^{-\mu+\frac{1}{2}} {}_1F_1\left(-\mu + \frac{1}{2} - \varkappa; 1 - 2\mu; z\right).$$

Constitute a system of linearly independent solutions of the differential equation

$$\frac{d^2u}{dz^2} + \left(-\frac{1}{4} + \frac{\varkappa}{z} + \frac{\frac{1}{4} - \mu^2}{z^2}\right) u = 0.$$

Called "Whittaker's differential equation", provided that $\mu \neq -\frac{1}{2}, -\frac{3}{2}, -\frac{5}{2}, \ldots$ In order to obtain a solution valid for all μ several functions have been introduced by various authors.

$$N_{\varkappa,\mu}(z) = \frac{z^{\mu-\frac{1}{2}}}{\Gamma(1+2\mu)} M_{\varkappa,\mu}(z) = \frac{e^{-\frac{z}{2}} z^{2\mu}}{\Gamma(1+2\mu)} {}_1F_1\left(\mu + \frac{1}{2} - \varkappa; 1 + 2\mu; z\right)$$

[Erdélyi],

$$W_{\varkappa,\mu}(z) = \frac{\Gamma(2\mu)}{\Gamma\left(\mu + \frac{1}{2} - \varkappa\right)} M_{\varkappa,-\mu}(z) + \frac{\Gamma(-2\mu)}{\Gamma\left(-\mu + \frac{1}{2} - \varkappa\right)} M_{\varkappa,\mu}(z)$$

[Whittaker].

Buchholz introduced the function

$$\mathcal{M}_{\varkappa,\mu}(z) = \frac{1}{\Gamma(1+2\mu)} M_{\varkappa,\mu}(z).$$

As μ approaches any one of the numbers $-\frac{1}{2}, -\frac{3}{2}, -\frac{5}{2}, \ldots$, the functions $N_{\varkappa,\mu}(z)$, $W_{\varkappa,\mu}(z)$ and $\mathcal{M}_{\varkappa,\mu}(z)$ remain well defined. Clearly the functions $N_{\varkappa,\mu}(z)$, $W_{\varkappa,\mu}(z)$ and $\mathcal{M}_{\varkappa,\mu}(z)$ are solutions of Whittaker's differential equation. Furthermore the differential equation is unchanged if μ is changed to $-\mu$ or $\varkappa, z$ are changed to $-\varkappa, -z$ so that $W_{\varkappa,-\mu}(z)$, $W_{-\varkappa,\mu}(-z)$ and $W_{-\varkappa,-\mu}(-z)$ are also solutions.

Only $M_{\varkappa,\mu}(z)$ and $W_{\varkappa,\mu}(z)$ are treated in this chapter and will be referred to as Whittaker's first and second function respectively. From the expressions for $M_{\varkappa,\mu}(z)$ and $W_{\varkappa,\mu}(z)$ in terms of Kummer's function it is clear that these functions are multiple valued in the complex z plane. The origin is a branch point and the point at infinity an essential singularity for these functions. Unless otherwise specified, these functions will be considered to be for z in the principal branch $|\arg z| < \pi$. The values of $M_{\varkappa,\mu}(z)$ and $W_{\varkappa,\mu}(z)$ when z lies outside the principal branch, can be expressed in terms of the values in the principal branch by means of the following relations:

$$M_{\varkappa,\mu}(z e^{\pm i\pi}) = e^{\pm i\frac{\pi}{2}(1+2\mu)} M_{-\varkappa,\mu}(z),$$

$$M_{-\varkappa,\mu}(z e^{\pm i\pi}) = e^{\pm i\frac{\pi}{2}(1+2\mu)} M_{\varkappa,\mu}(z),$$

$$M_{\varkappa,\mu}(z e^{i2n\pi}) = e^{i\pi n(1+2\mu)} M_{\varkappa,\mu}(z),$$

$$M_{\varkappa,\mu}(z e^{i\pi(2n\pm1)}) = e^{i\frac{\pi}{2}(2n\pm1)(1+2\mu)} M_{-\varkappa,\mu}(z),$$

$$\begin{aligned}
W_{\varkappa,\mu}(z e^{\pm i\pi}) = & \frac{\Gamma(2\mu)}{\Gamma\left(\mu+\frac{1}{2}-\varkappa\right)} e^{\pm i\frac{\pi}{2}(1-2\mu)} M_{-\varkappa,-\mu}(z) \\
& + \frac{\Gamma(-2\mu)}{\Gamma\left(-\mu+\frac{1}{2}-\varkappa\right)} e^{\pm i\frac{\pi}{2}(1+2\mu)} M_{-\varkappa,+\mu}(z) \\
= & \frac{\Gamma\left(\mu+\frac{1}{2}+\varkappa\right)}{\Gamma\left(\mu+\frac{1}{2}-\varkappa\right)} e^{\pm i\frac{\pi}{2}(1-2\mu)} W_{-\varkappa,\mu}(z) \\
& + \frac{\Gamma(-2\mu)}{\Gamma\left(\frac{1}{2}-\mu-\varkappa\right)} e^{\pm i\frac{\pi}{2}(1+2\mu)} \left[1 - \frac{\cos\pi(\mu-\varkappa)}{\cos\pi(\mu+\varkappa)} e^{\mp i2\pi\mu}\right] \\
& \times M_{-\varkappa,\mu}(z).
\end{aligned}$$

More generally.

$$W_{\varkappa,\mu}(z e^{i2\pi n}) = \frac{\pi}{\sin(2\pi\mu)}\left\{\frac{e^{in\pi(1-2\mu)}}{\Gamma\left(\frac{1}{2}+\mu-\varkappa\right)\Gamma(1+2\mu)} M_{k,-\mu}(z)\right.$$

$$\left.-\frac{e^{in\pi(1+2\mu)}}{\Gamma\left(\frac{1}{2}-\mu-\varkappa\right)\Gamma(1+2\mu)} M_{\varkappa,\mu}(z)\right\}$$

$$= (-1)^n e^{-i2\pi n\mu} W_{\varkappa,\mu}(z)$$

$$+\frac{2\pi i(-1)^{n+1}}{\Gamma\left(\frac{1}{2}-\mu-\varkappa\right)} \frac{\sin(2\pi n\mu)}{\sin(2\pi\mu)} \frac{M_{\varkappa,\mu}(z)}{\Gamma(1+2\mu)},$$

$$W_{-\varkappa,\mu}(z e^{\pm i\pi}) = \frac{\pi}{\sin(2\pi\mu)}\left\{\frac{M_{-\varkappa,-\mu}(z e^{\pm i\pi})}{\Gamma(1-2\mu)\,\Gamma\left(\frac{1}{2}+\mu+\varkappa\right)}\right.$$

$$\left.-\frac{M_{-\varkappa,\mu}(z e^{\pm i\pi})}{\Gamma(1+2\mu)\,\Gamma\left(\frac{1}{2}-\mu+\varkappa\right)}\right\}$$

$$= \frac{\pi}{\sin(2\pi\mu)}\left\{\frac{e^{\pm i\frac{\pi}{2}(1-2\mu)} M_{\varkappa,-\mu}(z)}{\Gamma(1-2\mu)\,\Gamma\left(\frac{1}{2}+\mu+\varkappa\right)}\right.$$

$$\left.-\frac{e^{\pm i\frac{\pi}{2}(1+2\mu)}}{\Gamma(1+2\mu)\,\Gamma\left(\frac{1}{2}-\mu+\varkappa\right)} M_{\varkappa,\mu}(z)\right\},$$

$$W_{-\varkappa,\mu}[z e^{i\pi(2n\pm1)}] = (-1)^n e^{-i2\pi n\mu} W_{-\varkappa,\mu}(z e^{\pm i\pi})$$

$$+\frac{2\pi i(-1)^{n+1}}{\Gamma(1+2\mu)\,\Gamma\left(\frac{1}{2}-\mu+\varkappa\right)} \frac{\sin(2\pi n\mu)}{\sin(2\pi\mu)} M_{-\varkappa,\mu}(z e^{\pm i\pi})$$

$$= \frac{\pi}{\sin(2\pi\mu)}\left[\frac{e^{i\frac{\pi}{2}(2n\pm1)(1-2\mu)}}{\Gamma(1-2\mu)\,\Gamma\left(\frac{1}{2}+\mu+\varkappa\right)} M_{\varkappa,-\mu}(z)\right.$$

$$\left.-\frac{e^{i\frac{\pi}{2}(2n\pm1)(1+2\mu)}}{\Gamma(1+2\mu)\,\Gamma\left(\frac{1}{2}-\mu+\varkappa\right)} M_{\varkappa,\mu}(z)\right].$$

It is possible to express $M_{\varkappa,\mu}(z)$ and $W_{\varkappa,\mu}(z)$ in terms of Whittaker functions with argument outside the principal branch by means of the following relations:

$$M_{\varkappa,\mu}(z) = \frac{\Gamma\left(\frac{1}{2} - \mu + \varkappa\right)\Gamma(1 + 2\mu)}{2\pi i}\left\{e^{i\frac{\pi}{2}(1+2\mu)} W_{-\varkappa,\mu}(z e^{i\pi})\right.$$
$$\left. - e^{-i\frac{\pi}{2}(1+2\mu)} W_{-\varkappa,\mu}(z e^{-i\pi})\right\},$$
$$W_{\varkappa,\mu}(z) = \frac{\Gamma\left(\frac{1}{2} + \mu + \varkappa\right)\Gamma\left(\frac{1}{2} - \mu + \varkappa\right)}{2\pi i}\{e^{+i\pi\varkappa} W_{-\varkappa,\mu}(z e^{i\pi})$$
$$- e^{-i\pi\varkappa} W_{-\varkappa,\mu}(z e^{-i\pi})\}.$$

Transformation formulas.

$$M_{\varkappa,\mu}(z) = e^{\pm i\frac{\pi}{2}(2\mu+1)} M_{-\varkappa,\mu}(-z),$$

where the sign is positive or negative according as $\operatorname{Im} z > 0$ or $\operatorname{Im} z < 0$

$$W_{\varkappa,\mu}(z) = W_{\varkappa,-\mu}(z),$$
$$W_{-\varkappa,\mu}(-z) = W_{-\varkappa,-\mu}(-z).$$

7.1.2 Solutions and the Wronskians

Let

$$u_1 = M_{\varkappa,\mu}(z),$$
$$u_2 = M_{\varkappa,-\mu}(z),$$
$$u_3 = M_{-\varkappa,\mu}(-z) = e^{-i\frac{\pi}{2}\varepsilon(1+2\mu)} u_1,$$
$$u_4 = M_{-\varkappa,-\mu}(-z) = e^{-i\frac{\pi}{2}\varepsilon(1-2\mu)} u_2,$$
$$u_5 = W_{\varkappa,\mu}(z),$$
$$u_6 = W_{\varkappa,-\mu}(z) = u_5,$$
$$u_7 = W_{-\varkappa,\mu}(-z),$$
$$u_8 = W_{-\varkappa,-\mu}(-z) = u_7,$$
$$W(u_i, u_j) = u_i \frac{du_j}{dz} - u_j \frac{du_i}{dz}.$$

Then

$$W(u_1, u_3) = W(u_2, u_4) = W(u_5, u_6) = W(u_7, u_8) = 0,$$
$$W(u_1, u_2) = -2\mu,$$

$$W(u_1, u_4) = -2\mu e^{-i\frac{\pi}{2}\varepsilon(1-2\mu)},$$

$$W(u_1, u_5) = -\frac{\Gamma(1+2\mu)}{\Gamma\left(\frac{1}{2}+\mu-\varkappa\right)} = W(u_1, u_6),$$

$$W(u_1, u_7) = W(u_1, u_8) = \frac{\Gamma(1+2\mu)}{\Gamma\left(\frac{1}{2}+\mu+\varkappa\right)} e^{i\frac{\pi}{2}\varepsilon(1-2\mu)},$$

$$W(u_2, u_3) = 2\mu \exp\left[-i\frac{\pi}{2}\varepsilon(1+2\mu)\right],$$

$$W(u_2, u_5) = -\frac{\Gamma(1-2\mu)}{\Gamma\left(\frac{1}{2}-\mu-\varkappa\right)} = W(u_2, u_6),$$

$$W(u_2, u_7) = W(u_2, u_8) = -\frac{\Gamma(1-2\mu)}{\Gamma\left(\frac{1}{2}-\mu+\varkappa\right)} e^{i\frac{\pi}{2}\varepsilon(1+2\mu)},$$

$$W(u_3, u_4) = -2\mu e^{i\pi\varepsilon},$$

$$W(u_3, u_5) = W(u_3, u_6) = -\frac{\Gamma(1+2\mu)}{\Gamma\left(\frac{1}{2}+\mu-\varkappa\right)} e^{-i\frac{\pi}{2}\varepsilon(1+2\mu)},$$

$$W(u_3, u_7) = W(u_3, u_8) = \frac{\Gamma(1+2\mu)}{\Gamma\left(\frac{1}{2}+\mu-\varkappa\right)} e^{-i\pi\varepsilon\mu},$$

$$W(u_4, u_5) = W(u_4, u_6) = -\frac{\Gamma(1-2\mu)}{\Gamma\left(\frac{1}{2}-\mu-\varkappa\right)} e^{-i\frac{\pi}{2}\varepsilon(1-2\mu)},$$

$$W(u_4, u_7) = W(u_4, u_8) = -\frac{\Gamma(1-2\mu)}{\Gamma\left(\frac{1}{2}-\mu+\varkappa\right)} e^{i\pi\varepsilon\mu},$$

$$W(u_5, u_7) = W(u_5, u_8) = W(u_6, u_7) = W(u_6, u_8) = e^{-i\pi\varepsilon\varkappa},$$

where $\varepsilon = 1$, for $\operatorname{Im} z > 0$; $\varepsilon = -1$, $\operatorname{Im} z < 0$. From the above table of values of the Wronskian determinants it is clear that the only pair of solutions which always give a complete solution of Whittaker's differential equation are u_5 or u_6 and u_7 or u_8.

(i) If 2μ and $2(\varkappa \pm \mu)$ are not zero or integers, any two of u_1, u_2, u_5, u_7 may be chosen to give a complete system of solutions.

(ii) If $2(\mu - \varkappa)$ is a negative integer the pairs (u_1, u_5), (u_3, u_5) cannot be taken to give two linearly independent solutions. In this case one possible pair of linearly independent solutions is (u_2, u_7).

(iii) If $2(\mu - \varkappa)$ is a positive integer, the pairs (u_2, u_7), (u_4, u_7) are not linearly independent, however the pair (u_1, u_5) is valid.

7.2 Some elementary results

7.2.1 Differentiation formulas

$$\frac{d}{dz}\left(e^{\frac{z}{2}} z^{-\mu-\frac{1}{2}} M_{\varkappa,\mu}(z)\right) = \frac{\frac{1}{2}+\mu-\varkappa}{1+2\mu} e^{\frac{z}{2}} z^{-\mu-1} M_{\varkappa-\frac{1}{2},\mu+\frac{1}{2}}(z),$$

$$\frac{d^n}{dz^n}\left[e^{\frac{z}{2}} z^{-\mu-\frac{1}{2}} M_{\varkappa,\mu}(z)\right] = \frac{\left(\frac{1}{2}+\mu-\varkappa\right)_n}{(1+2\mu)_n} e^{\frac{z}{2}} z^{-\mu-\frac{1}{2}-\frac{n}{2}} M_{\varkappa-\frac{n}{2},\mu+\frac{n}{2}}(z),$$

$$\frac{d}{dz}\left[e^{\frac{z}{2}} z^{-\varkappa} M_{\varkappa,\mu}(z)\right] = \left(\frac{1}{2}+\mu-\varkappa\right) z^{-\varkappa-1} e^{\frac{z}{2}} M_{\varkappa-1,\mu}(z),$$

$$\frac{d^n}{dz^n}\left[e^{\frac{z}{2}} z^{n-\varkappa-1} M_{\varkappa,\mu}(z)\right] = \left(\frac{1}{2}+\mu-\varkappa\right)_n e^{\frac{z}{2}} z^{-\varkappa-1} M_{\varkappa-n,\mu}(z),$$

$$\frac{d^n}{dz^n}\left[e^{\frac{z}{2}} z^{\mu-\frac{1}{2}} M_{\varkappa,\mu}(z)\right] = (-1)^n (-2\mu)_n e^{\frac{z}{2}} z^{\mu-\frac{n}{2}-\frac{1}{2}} M_{\varkappa-\frac{n}{2},\mu-\frac{n}{2}}(z),$$

$$\frac{d}{dz}\left[e^{\frac{z}{2}} z^{\mu-\frac{1}{2}} M_{\varkappa,\mu}(z)\right] = 2\mu e^{\frac{z}{2}} z^{\mu-1} M_{\varkappa-\frac{1}{2},\mu-\frac{1}{2}}(z),$$

$$\frac{d^n}{dz^n}\left[e^{-\frac{z}{2}} z^{\mu-\frac{1}{2}} M_{\varkappa,\mu}(z)\right] = (-1)^n (-2\mu)_n e^{-\frac{z}{2}} z^{\mu-\frac{n}{2}-\frac{1}{2}} M_{\varkappa+\frac{n}{2},\mu-\frac{n}{2}}(z),$$

$$\frac{d^n}{dz^n}\left[e^{-\frac{z}{2}} z^{-\mu-\frac{1}{2}} M_{\varkappa,\mu}(z)\right] = \frac{(-1)^n\left(\frac{1}{2}+\mu+\varkappa\right)_n}{(1+2\mu)_n} e^{-\frac{z}{2}} z^{-\mu-\frac{n}{2}-\frac{1}{2}}$$
$$\times M_{\varkappa+\frac{n}{2},\mu+\frac{n}{2}}(z),$$

$$\frac{d^n}{dz^n}\left[e^{-\frac{z}{2}} z^{\varkappa+n-1} M_{\varkappa,\mu}(z)\right] = \left(\frac{1}{2}+\mu+\varkappa\right)_n e^{-\frac{z}{2}} z^{\varkappa-1} M_{\varkappa+n,\mu}(z),$$

$$\frac{d^n}{dz^n}\left[e^{\frac{z}{2}} z^{\mu-\frac{1}{2}} W_{\varkappa,\mu}(z)\right] = (-1)^n \left(\frac{1}{2}-\mu-\varkappa\right)_n e^{\frac{z}{2}} z^{\mu-\frac{n}{2}-\frac{1}{2}}$$
$$\times W_{\varkappa-\frac{n}{2},\mu-\frac{n}{2}}(z),$$

$$\frac{d^n}{dz^n}\left[e^{\frac{z}{2}} z^{-\mu-\frac{1}{2}} W_{\varkappa,\mu}(z)\right] = (-1)^n \left(\frac{1}{2}+\mu-\varkappa\right)_n e^{\frac{z}{2}} z^{-\mu-\frac{n}{2}-\frac{1}{2}}$$
$$\times W_{\varkappa-\frac{n}{2},\mu+\frac{n}{2}}(z),$$

$$\frac{d^n}{dz^n}\left[e^{\frac{z}{2}} z^{n-\varkappa-1} W_{\varkappa,\mu}(z)\right] = \left(\frac{1}{2}+\mu-\varkappa\right)_n \left(\frac{1}{2}-\mu-\varkappa\right)_n z^{-\varkappa-1} e^{\frac{z}{2}}$$
$$\times W_{\varkappa-n,\mu}(z),$$

$$\frac{d^n}{dz^n}\left[e^{-\frac{z}{2}} z^{\mu-\frac{1}{2}} W_{\varkappa,\mu}(z)\right] = (-1)^n e^{-\frac{z}{2}} z^{\mu-\frac{n}{2}-\frac{1}{2}} W_{\varkappa+\frac{n}{2},\mu-\frac{n}{2}}(z),$$

$$\frac{d^n}{dz^n}\left[e^{-\frac{z}{2}} z^{-\mu-\frac{1}{2}} W_{\varkappa,\mu}(z)\right] = (-1)^n e^{-\frac{z}{2}} z^{-\mu-\frac{n}{2}-\frac{1}{2}} W_{\varkappa+\frac{n}{2},\mu+\frac{n}{2}}(z),$$

$$\frac{d^n}{dz^n}\left[e^{-\frac{z}{2}} z^{\varkappa+n-1} W_{\varkappa,\mu}(z)\right] = (-1)^n e^{-\frac{z}{2}} z^{\varkappa-1} W_{\varkappa+n,\mu}(z).$$

The following results are analogous to the contiguous relations for Kummer's functions [see 6.1.2] and can be obtained from the differentiation formulas above by taking $n = 1$.

$$2\mu M_{\varkappa-\frac{1}{2},\mu-\frac{1}{2}}(z) = \sqrt{z}\, M'_{\varkappa,\mu}(z) + \frac{2\mu - 1 + z}{2\sqrt{z}} M_{\varkappa,\mu}(z),$$

$$\left(\frac{1}{2} + \mu - \varkappa\right) M_{\varkappa-\frac{1}{2},\mu+\frac{1}{2}}(z) = (1 + 2\mu)\sqrt{z}\, M'_{\varkappa,\mu}(z) + \frac{(1+2\mu)\left(-\mu - \frac{1}{2} + \frac{z}{2}\right)}{\sqrt{z}} M_{\varkappa,\mu}(z),$$

$$\left(\frac{1}{2} + \mu - \varkappa\right) M_{\varkappa-1,\mu}(z) = z M'_{\varkappa,\mu}(z) + \left(\frac{z}{2} - \varkappa\right) M_{\varkappa,\mu}(z),$$

$$2\mu M_{\varkappa+\frac{1}{2},\mu-\frac{1}{2}}(z) = \sqrt{z}\, M'_{\varkappa,\mu}(z) + \frac{2\mu - 1 - z}{2\sqrt{z}} M_{\varkappa,\mu}(z),$$

$$\left(\frac{1}{2} + \mu + \varkappa\right) M_{\varkappa+1,\mu}(z) = z M'_{\varkappa,\mu}(z) + \left(\varkappa - \frac{z}{2}\right) M_{\varkappa,\mu}(z),$$

$$\left(\frac{1}{2} + \mu + \varkappa\right) M_{\varkappa+\frac{1}{2},\mu+\frac{1}{2}}(z) = -(1 + 2\mu)\sqrt{z}\, M'_{\varkappa,\mu}(z) + \frac{(1+2\mu)\left(\frac{1}{2} + \mu + \frac{z}{2}\right)}{\sqrt{z}} M_{\varkappa,\mu}(z),$$

$$(2\varkappa + 2\mu - 1)\, W_{\varkappa-\frac{1}{2},\mu-\frac{1}{2}}(z) = 2\sqrt{z}\, W'_{\varkappa,\mu}(z) + \frac{2\mu + z - 1}{\sqrt{z}} W_{\varkappa,\mu}(z),$$

$$2 W_{\varkappa+\frac{1}{2},\mu-\frac{1}{2}}(z) = \frac{1 - 2\mu + z}{\sqrt{z}} W_{\varkappa,\mu}(z) - \sqrt{z}\, W'_{\varkappa,\mu}(z),$$

$$(2\varkappa - 2\mu - 1)\, W_{\varkappa-\frac{1}{2},\mu+\frac{1}{2}}(z) = \sqrt{z}\, W'_{\varkappa,\mu}(z) - \frac{1 + 2\mu - z}{\sqrt{z}} W_{\varkappa,\mu}(z),$$

$$\left(\frac{1}{2} + \mu - \varkappa\right)\left(\frac{1}{2} - \mu - \varkappa\right) W_{\varkappa-1,\mu}(z) = z W'_{\varkappa,\mu}(z) + \left(\frac{z}{2} - \varkappa\right) W_{\varkappa,\mu}(z),$$

$$2 W_{\varkappa+\frac{1}{2},\mu+\frac{1}{2}}(z) = \frac{\frac{1}{2} + \mu + \frac{z}{2}}{\sqrt{z}} W_{\varkappa,\mu}(z) - \sqrt{z}\, W'_{\varkappa,\mu}(z),$$

$$W_{\varkappa+1,\mu}(z) = \left(\frac{z}{2} - \varkappa\right) W_{\varkappa,\mu}(z) - z W'_{\varkappa,\mu}(z),$$

where

$$M'_{\varkappa,\mu}(z) = \frac{d}{dz} M_{\varkappa,\mu}(z),$$
$$W'_{\varkappa,\mu}(z) = \frac{d}{dz} W_{\varkappa,\mu}(z).$$

7.2.2 Recurrence relations

$$4\mu(1+2\mu)\sqrt{z}\, M_{\varkappa-\frac{1}{2},\mu-\frac{1}{2}}(z) - 4\mu(1+2\mu)\, M_{\varkappa,\mu}(z) - (1+\mu-2\varkappa)\sqrt{z}\, M_{\varkappa-\frac{1}{2},\mu+\frac{1}{2}}(z) = 0,$$

$$2\mu M_{\varkappa-\frac{1}{2},\mu-\frac{1}{2}}(z) - 2\mu M_{\varkappa+\frac{1}{2},\mu-\frac{1}{2}}(z) - \sqrt{z}\, M_{\varkappa,\mu}(z) = 0,$$

$$4\mu(1+2\mu)\sqrt{z}\, M_{\varkappa-\frac{1}{2},\mu-\frac{1}{2}}(z) - 2(1+2\mu)(2\mu-z)\, M_{\varkappa,\mu}(z) - (1+2\mu-2\varkappa)\sqrt{z}\, M_{\varkappa-\frac{1}{2},\mu+\frac{1}{2}}(z) = 0,$$

$$4\mu(1+2\mu)\sqrt{z}\, M_{\varkappa-\frac{1}{2},\mu-\frac{1}{2}}(z) + 2(1+2\mu)(2\mu+z)\, M_{\varkappa,\mu}(z) - (1+2\mu+2\varkappa)\sqrt{z}\, M_{\varkappa+\frac{1}{2},\mu+\frac{1}{2}}(z) = 0,$$

$$4\mu(1+2\mu)\sqrt{z}\, M_{\varkappa+\frac{1}{2},\mu-\frac{1}{2}}(z) - 4\mu(1+2\mu)\, M_{\varkappa,\mu}(z) + (1+2\mu+2\varkappa)\sqrt{z}\, M_{\varkappa+\frac{1}{2},\mu+\frac{1}{2}}(z) = 0,$$

$$(1+2\mu-2\varkappa)\, M_{\varkappa-\frac{1}{2},\mu+\frac{1}{2}}(z) - 2(1+2\mu)\sqrt{z}\, M_{\varkappa,\mu}(z) + (1+2\mu+2\varkappa)\, M_{\varkappa+\frac{1}{2},\mu+\frac{1}{2}}(z) = 0,$$

$$(1+2\mu-2\varkappa)\, M_{\varkappa-1,\mu}(z) + 2(2\varkappa-z)\, M_{\varkappa,\mu}(z) - (1+2\mu+2\varkappa)\, M_{\varkappa+1,\mu}(z) = 0,$$

$$(\mu-\varkappa)\, W_{\varkappa-\frac{1}{2},\mu}(z) + \sqrt{z}\, W_{\varkappa,\mu-\frac{1}{2}}(z) - W_{\varkappa+\frac{1}{2},\mu}(z) = 0,$$

$$\left(\frac{1}{2}+\mu-\varkappa\right)\sqrt{z}\, W_{\varkappa-\frac{1}{2},\mu+\frac{1}{2}}(z) - (2\mu-z)\, W_{\varkappa,\mu}(z) - \sqrt{z}\, W_{\varkappa+\frac{1}{2},\mu-\frac{1}{2}}(z) = 0,$$

$$\left(\frac{1}{2}-\mu-\varkappa\right)\sqrt{z}\, W_{\varkappa-\frac{1}{2},\mu-\frac{1}{2}}(z) + \left(\varkappa-\mu-\frac{1}{2}\right)\sqrt{z}\, W_{\varkappa-\frac{1}{2},\mu+\frac{1}{2}}(z) + 2\mu W_{\varkappa,\mu}(z) = 0,$$

$$\left(\mu - \frac{1}{2} + \varkappa\right)\sqrt{z}\, W_{\varkappa-\frac{1}{2},\mu-\frac{1}{2}}(z) - (2\mu + z)\, W_{\varkappa,\mu}(z) + \sqrt{z}\, W_{\varkappa+\frac{1}{2},\mu+\frac{1}{2}}(z) = 0,$$

$$\sqrt{z}\, W_{\varkappa+\frac{1}{2},\mu-\frac{1}{2}}(z) - 2\mu W_{\varkappa,\mu}(z) + \sqrt{z}\, W_{\varkappa+\frac{1}{2},\mu+\frac{1}{2}}(z) = 0,$$

$$(\mu + \varkappa)\, W_{\varkappa-\frac{1}{2},\mu}(z) - \sqrt{z}\, W_{\varkappa,\mu+\frac{1}{2}}(z) + W_{\varkappa+\frac{1}{2},\mu}(z) = 0,$$

$$\left(\frac{1}{2} + \mu - \varkappa\right)\left(\frac{1}{2} - \mu - \varkappa\right) W_{\varkappa-1,\mu}(z) + (2\varkappa - z)\, W_{\varkappa,\mu}(z) + W_{\varkappa+1,\mu}(z) = 0.$$

7.2.3 Relations between Whittaker's functions and Kummer's functions

$${}_1F_1(a; c; z) = e^{\frac{z}{2}} z^{-\frac{c}{2}} M_{\frac{c}{2}-a,\frac{c}{2}-\frac{1}{2}}(z)$$

$$= e^{\frac{z}{2}} z^{-\frac{c}{2}} \left[\frac{\Gamma(c)}{\Gamma(a)} e^{-i\frac{\pi}{2}\varepsilon(c-2a)} W_{a-\frac{c}{2},\frac{c}{2}-\frac{1}{2}}(-z) + \frac{\Gamma(c)}{\Gamma(c-a)} e^{i\pi\varepsilon a} W_{\frac{c}{2}-a,\frac{c}{2}-\frac{1}{2}}(z)\right],$$

$\varepsilon = \pm 1$ according as $Im\ z \gtrless 0$

$$U(a, c, z) = e^{\frac{z}{2}} z^{-\frac{c}{2}} W_{\frac{c}{2}-a,\frac{c}{2}-\frac{1}{2}}(z)$$

$$= e^{\frac{z}{2}} z^{-\frac{c}{2}} \left[\frac{\Gamma(1-c)}{\Gamma(1+a-c)} M_{\frac{c}{2}-a,\frac{c}{2}-\frac{1}{2}}(z) + \frac{\Gamma(c-1)}{\Gamma(a)} M_{\frac{c}{2}-a,\frac{1}{2}-\frac{c}{2}}(z)\right].$$

For other relations see 6.7.2 and 7.1.1.

7.2.4 Special cases

$$M_{\alpha,0}(z) = e^{-\frac{z}{2}} z^{\frac{1}{2}} L_{\alpha-\frac{1}{2}}(z),$$

$$M_{0,\frac{1}{2}}(z) = 2\sinh\left(\frac{z}{2}\right),$$

$$M_{0,\frac{1}{2}}(-iz) = -2i\sin\left(\frac{z}{2}\right),$$

$$M_{0,\mu}(z) = z^{\frac{1}{2}+\mu}\, {}_0F_1\left(; \frac{1}{2} + \mu; \frac{z^2}{4}\right),$$

$$M_{0,\mu}(z) = \Gamma(1+\mu)\, 2^{2\mu} e^{-i\frac{\pi}{2}\mu} z^{\frac{1}{2}} J_\mu\left(\frac{1}{2} z e^{i\frac{\pi}{2}}\right)$$
$$= \Gamma(1+\mu)\, 2^{2\mu} z^{\frac{1}{2}} I_\mu\left(\frac{1}{2} z\right),$$
$$M_{0,\mu}(iz) = \Gamma(1+\mu)\, 2^{2\mu} e^{-i\frac{\mu}{4}(2\mu-1)} z^{\frac{1}{2}} J_\mu\left(-\frac{1}{2} z\right),$$
$$M_{0,\mu}(-iz) = \Gamma(1+\mu)\, 2^{2\mu} e^{-i\frac{\pi}{4}(2\mu+1)} z^{\frac{1}{2}} J_\mu\left(\frac{1}{2} z\right),$$
$$M_{\mu+\frac{1}{2},\mu}(z) = e^{-\frac{z}{2}} z^{\mu+\frac{1}{2}},$$
$$M_{-\mu-\frac{1}{2},\mu}(z) = e^{\frac{z}{2}} z^{\mu+\frac{1}{2}},$$
$$M_{\mu+\frac{1}{2}+n,\mu}(z) = \frac{e^{\frac{z}{2}} z^{\frac{1}{2}-\mu}}{(1+2\mu)_n} \frac{d^n}{dz^n}\left(e^{-z} z^{2\mu+n}\right),$$
$$n = 0, 1, 2, \ldots,$$
$$2\mu \neq -1, -2, -3, \ldots$$

It may be observed at this point that the functions $M_{\pm\frac{n}{2},\mu\pm\frac{n}{2}}(z)$ may be expressed in terms of $I_\mu\left(\frac{z}{2}\right)$ by making use of the differentiation formulas given above. For the function $W_{\varkappa,\mu}(z)$

$$W_{0,\frac{1}{2}}(z) = e^{-\frac{z}{2}}, \qquad W_{0,\frac{1}{2}}(-z) = e^{\frac{z}{2}},$$
$$W_{0,\frac{1}{2}}(\pm iz) = e^{\mp i\frac{z}{2}},$$
$$W_{0,\mu}(z) = z^{\frac{1}{2}} \pi^{-\frac{1}{2}} K_\mu\left(\frac{z}{2}\right)$$
$$= \frac{i}{2} \pi^{\frac{1}{2}} z^{\frac{1}{2}} \exp\left(-i\frac{\pi}{2}\mu\right) H_\mu^{(1)}\left(\frac{1}{2} z e^{i\frac{\pi}{2}}\right),$$
$$W_{0,\mu}(iz) = \frac{\pi^{\frac{1}{2}}}{2} z^{\frac{1}{2}} e^{-i\frac{\pi}{4}(1+2\mu)} H_\mu^{(2)}\left(\frac{z}{2}\right),$$
$$W_{0,\mu}(-iz) = \frac{\pi^{\frac{1}{2}}}{2} z^{\frac{1}{2}} e^{i\frac{\pi}{4}(1+2\mu)} H_\mu^{(1)}\left(\frac{z}{2}\right),$$
$$W_{\mu+\frac{1}{2},\mu}(z) = W_{\mu+\frac{1}{2},-\mu}(z) = \exp\left(-\frac{z}{2}\right) z^{\mu+\frac{1}{2}}.$$

The Bessel functions as limiting cases of Whittaker functions are given by the following relations

$$\lim_{\varkappa\to\infty}\left(\frac{z}{\varkappa}\right)^{-\mu-\frac{1}{2}} M_{\varkappa,\mu}\left(\frac{z}{\varkappa}\right)=\Gamma(1+2\mu)\, z^{-\mu} I_{2\mu}(2\sqrt{z}),$$

$$\lim_{\varkappa\to\infty}\left(-\frac{z}{\varkappa}\right)^{-\mu-\frac{1}{2}} M_{\varkappa,\mu}\left(-\frac{z}{\varkappa}\right)=\Gamma(1+2\mu)\, z^{-\mu} J_{2\mu}(2\sqrt{z}),$$

$$\lim_{\varkappa\to\infty}\Gamma\left(\frac{1}{2}-\mu-\varkappa\right)\left(\frac{z}{\varkappa}\right)^{-\mu-\frac{1}{2}} W_{\varkappa,\mu}\left(\frac{z}{\varkappa}\right)=2z^{-\mu}K_{2\mu}(2\sqrt{z}),$$

$$\lim_{\varkappa\to\infty}\Gamma\left(\frac{1}{2}-\mu-\varkappa\right)\left(-\frac{z}{\varkappa}\right)^{-\mu-\frac{1}{2}} W_{\varkappa,\mu}\left(-\frac{z}{\varkappa}\right)$$
$$=\begin{cases}-i\pi\exp\left[i\pi(1+2\mu)\right] z^{-\mu}H^{(1)}_{2\mu}(2\sqrt{z}), & \operatorname{Im} z>0,\\ i\pi\exp\left[-i\pi(1+2\mu)\right] z^{-\mu}H^{(2)}_{2\mu}(2\sqrt{z}), & \operatorname{Im} z<0.\end{cases}$$

Finally, the following results give expansions of $M_{\varkappa,\mu}(z)$ in terms of the Bessel functions.

$$M_{\varkappa,\mu}(z)=\Gamma(\mu+\varkappa)\, 2^{2\mu+2\varkappa} z^{\frac{1}{2}-\varkappa}$$
$$\times\sum_{r=0}^{\infty}\frac{(-1)^r\,(2\varkappa)_r\,(2\mu+2\varkappa)_r}{(1+2\mu)_r\, r!} I_{\mu+\varkappa+r}\left(\frac{z}{2}\right),$$

$$M_{\varkappa,\mu}(-iz)=\Gamma(\mu+\varkappa)\, 2^{2\mu+2\varkappa} z^{\frac{1}{2}-\varkappa} e^{-i\frac{\pi}{4}(1+2\mu)}$$
$$\times\sum_{r=0}^{\infty}\frac{i^r\,(2\varkappa)_r\,(2\mu+2\varkappa)_r}{(1+2\mu)_r\, r!} J_{\mu+\varkappa+r}\left(\frac{z}{2}\right).$$

Both the expansions are valid for all values of arg z.

7.3 Addition and multiplication theorems

7.3.1 Addition theorems for $M_{\varkappa,\mu}(z)$

$$M_{\varkappa,\mu}(z+z')=e^{-\frac{z}{2}}\left(\frac{z'}{z+z'}\right)^{\mu-\frac{1}{2}}\sum_{n=0}^{\infty}\frac{(-1)^n\,(-2\mu)_n}{n!\,(z')^{\frac{n}{2}}} z^n M_{\varkappa-\frac{n}{2},\mu-\frac{n}{2}}(z')$$

$$=e^{-\frac{z'}{2}}\left(\frac{z}{z'+z}\right)^{\mu-\frac{1}{2}}\sum_{n=0}^{\infty}\frac{(-1)^n\,(-2\mu)_n}{n!\,(z)^{\frac{n}{2}}} (z')^n M_{\varkappa-\frac{n}{2},\mu-\frac{n}{2}}(z)$$

$$=e^{-\frac{z}{2}}\left(\frac{z'}{z+z'}\right)^{-\frac{1}{2}-\mu}\sum_{n=0}^{\infty}\frac{\left(\frac{1}{2}+\mu-\varkappa\right)_n}{n!\,(1+2\mu)_n}(z')^{-\frac{n}{2}} z^n$$
$$\times M_{\varkappa-\frac{n}{2},\mu+\frac{n}{2}}(z')$$

$$= e^{-\frac{z'}{2}} \left(\frac{z}{z+z'}\right)^{-\frac{1}{2}-\mu} \sum_{n=0}^{\infty} \frac{\left(\frac{1}{2}+\mu-\varkappa\right)_n}{n!\,(1+2\mu)_n} (z)^{-\frac{n}{2}} (z')^n$$
$$\times M_{\varkappa-\frac{n}{2},\mu+\frac{n}{2}}(z)$$

$$= e^{-\frac{z}{2}} \left(\frac{z'}{z+z'}\right)^{-\varkappa} \sum_{n=0}^{\infty} \frac{\left(\frac{1}{2}+\mu-\varkappa\right)_n}{n!} z^n M_{\varkappa-n,\mu}(z')$$

$$= e^{-\frac{z'}{2}} \left(\frac{z}{z+z'}\right)^{-\varkappa} \sum_{n=0}^{\infty} \frac{\left(\frac{1}{2}+\mu-\varkappa\right)_n}{n!} (z')^n M_{\varkappa-n,\mu}(z)$$

$$= e^{\frac{z}{2}} \left(\frac{z'}{z+z'}\right)^{\mu-\frac{1}{2}} \sum_{n=0}^{\infty} \frac{(-1)^n\,(-2\mu)_n}{n!\,(z')^{\frac{n}{2}}} z^n M_{\varkappa+\frac{n}{2},\mu-\frac{n}{2}}(z')$$

$$= e^{\frac{z'}{2}} \left(\frac{z}{z+z'}\right)^{\mu-\frac{1}{2}} \sum_{n=0}^{\infty} \frac{(-1)^n\,(-2\mu)_n}{n!\,(z)^{\frac{n}{2}}} (z')^n M_{\varkappa+\frac{n}{2},\mu-\frac{n}{2}}(z)$$

$$= e^{\frac{z}{2}} \left(\frac{z'}{z+z'}\right)^{-\mu-\frac{1}{2}} \sum_{n=0}^{\infty} \frac{(-1)^n\left(\frac{1}{2}+\mu+\varkappa\right)_n}{n!\,(1+2\mu)_n} (z')^{-\frac{n}{2}} z^n$$
$$\times M_{\varkappa+\frac{n}{2},\mu+\frac{n}{2}}(z')$$

$$= e^{\frac{z'}{2}} \left(\frac{z}{z+z'}\right)^{-\mu-\frac{1}{2}} \sum_{n=0}^{\infty} \frac{(-1)^n\left(\frac{1}{2}+\mu+\varkappa\right)_n}{n!\,(1+2\mu)_n} \frac{(z')^n}{z^{\frac{n}{2}}}$$
$$\times M_{\varkappa+\frac{n}{2},\mu+\frac{n}{2}}(z)$$

$$= e^{\frac{z}{2}} \left(\frac{z'}{z+z'}\right)^{\varkappa} \sum_{n=0}^{\infty} \frac{\left(\frac{1}{2}+\mu+\varkappa\right)_n}{n!} z^n M_{\varkappa+n,\mu}(z')$$

$$= e^{\frac{z'}{2}} \left(\frac{z}{z+z'}\right)^{\varkappa} \sum_{n=0}^{\infty} \frac{\left(\frac{1}{2}+\mu+\varkappa\right)_n}{n!} (z')^n M_{\varkappa+n,\mu}(z).$$

7.3.2 Addition theorems for $W_{\varkappa,\mu}(z)$

$$W_{\varkappa,\mu}(z+z') = e^{\frac{z}{2}} \left(\frac{z'}{z+z'}\right)^{\varkappa} \sum_{n=0}^{\infty} \frac{(-1)^n}{n!} \left(\frac{z}{z+z'}\right)^n W_{\varkappa+n,\mu}(z')$$

$$= e^{\frac{z}{2}} \left(\frac{z'}{z+z'}\right)^{\mu-\frac{1}{2}} \sum_{n=0}^{\infty} \frac{(-1)^n}{n!} \frac{z^n}{(z')^{\frac{n}{2}}} W_{\varkappa+\frac{n}{2},\mu-\frac{n}{2}}(z')$$

$$= e^{\frac{z}{2}} \left(\frac{z'}{z+z'}\right)^{-\mu-\frac{1}{2}} \sum_{n=0}^{\infty} \frac{(-1)^n}{n!} \frac{z^n}{(z')^{\frac{n}{2}}} W_{\varkappa+\frac{n}{2},\mu+\frac{n}{2}}(z')$$

$$= e^{\frac{z'}{2}} \left(\frac{z}{z+z'}\right)^{-\mu-\frac{1}{2}} \sum_{n=0}^{\infty} \frac{(-1)^n}{n!} \frac{(z')^n}{z^{\frac{n}{2}}} W_{\varkappa+\frac{n}{2},\mu+\frac{n}{2}}(z)$$

$$= e^{\frac{z'}{2}} \left(\frac{z}{z+z'}\right)^{\mu-\frac{1}{2}} \sum_{n=0}^{\infty} \frac{(-1)^n}{n!} \frac{(z')^n}{z^{\frac{n}{2}}} W_{\varkappa+\frac{n}{2},\mu-\frac{n}{2}}(z)$$

$$= e^{\frac{z'}{2}} \left(\frac{z}{z+z'}\right)^{\varkappa} \sum_{n=0}^{\infty} \frac{(-1)^n}{n!} \left(\frac{z'}{z+z'}\right)^n W_{\varkappa+n,\mu}(z)$$

$$= e^{-\frac{z}{2}} \left(\frac{z'}{z+z'}\right)^{-\varkappa} \sum_{n=0}^{\infty} \frac{\left(\frac{1}{2}+\mu-\varkappa\right)_n \left(\frac{1}{2}-\mu-\varkappa\right)_n}{n!} \left(\frac{z}{z+z'}\right)^n \times W_{\varkappa-n,\mu}(z')$$

$$= e^{-\frac{z'}{2}} \left(\frac{z}{z+z'}\right)^{-\varkappa} \sum_{n=0}^{\infty} \frac{\left(\frac{1}{2}+\mu-\varkappa\right)_n \left(\frac{1}{2}-\mu-\varkappa\right)_n}{n!} \left(\frac{z'}{z+z'}\right)^n \times W_{\varkappa-n,\mu}(z)$$

$$= e^{-\frac{z}{2}} \left(\frac{z'}{z+z'}\right)^{-\frac{1}{2}-\mu} \sum_{n=0}^{\infty} \frac{(-1)^n \left(\frac{1}{2}+\mu-\varkappa\right)_n}{n!\,(z')^{\frac{n}{2}}} z^n \times W_{\varkappa-\frac{n}{2},\mu+\frac{n}{2}}(z')$$

$$= e^{-\frac{z'}{2}} \left(\frac{z}{z+z'}\right)^{-\frac{1}{2}-\mu} \sum_{n=0}^{\infty} \frac{(-1)^n \left(\frac{1}{2}+\mu-\varkappa\right)_n}{n!} \frac{(z')^n}{z^{\frac{n}{2}}} \times W_{\varkappa-\frac{n}{2},\mu+\frac{n}{2}}(z)$$

$$= e^{-\frac{z}{2}} \left(\frac{z'}{z+z'}\right)^{\mu-\frac{1}{2}} \sum_{n=0}^{\infty} \frac{(-1)^n \left(\frac{1}{2}-\mu-\varkappa\right)_n}{n!\,(z')^{\frac{n}{2}}} z^n \times W_{\varkappa-\frac{n}{2},\mu-\frac{n}{2}}(z')$$

$$= e^{-\frac{z'}{2}} \left(\frac{z}{z+z'}\right)^{\mu-\frac{1}{2}} \sum_{n=0}^{\infty} \frac{(-1)^n \left(\frac{1}{2}-\mu-\varkappa\right)_n}{n!\,z^{\frac{n}{2}}} (z')^n \times W_{\varkappa-\frac{n}{2},\mu-\frac{n}{2}}(z).$$

7.3.3 Multiplication theorems for $M_{\varkappa,\mu}(z)$

$$M_{\varkappa,\mu}(zz') = e^{\frac{z'(z-1)}{2}} z^{-\varkappa} \sum_{n=0}^{\infty} \frac{\left(\frac{1}{2}+\mu+\varkappa\right)_n}{n!\, z^n} (z-1)^n M_{\varkappa+n,\mu}(z')$$

$$= e^{\frac{z(z'-1)}{2}} (z')^{-\varkappa} \sum_{n=0}^{\infty} \frac{\left(\frac{1}{2}+\mu+\varkappa\right)_n}{n!} \left(\frac{z'-1}{z'}\right)^n M_{\varkappa+n,\mu}(z)$$

$$= e^{\frac{z'(z-1)}{2}} z^{\frac{1}{2}+\mu} \sum_{n=0}^{\infty} \frac{(-1)^n \left(\frac{1}{2}+\mu+\varkappa\right)_n}{n!\,(1+2\mu)_n} (z')^{\frac{n}{2}} (z-1)^n$$
$$\times M_{\varkappa+\frac{n}{2},\mu+\frac{n}{2}}(z')$$

$$= e^{\frac{z(z'-1)}{2}} (z')^{\frac{1}{2}+\mu} \sum_{n=0}^{\infty} \frac{(-1)^n \left(\frac{1}{2}+\mu+\varkappa\right)_n}{n!\,(1+2\mu)_n} (z)^{\frac{n}{2}} (z'-1)^n$$
$$\times M_{\varkappa+\frac{n}{2},\mu+\frac{n}{2}}(z)$$

$$= e^{\frac{z'}{2}(z-1)} z^{\frac{1}{2}-\mu} \sum_{n=0}^{\infty} \frac{(-1)^n (-2\mu)_n}{n!} (z')^{\frac{n}{2}} (z-1)^n$$
$$\times M_{\varkappa+\frac{n}{2},\mu-\frac{n}{2}}(z')$$

$$= e^{\frac{z}{2}(z'-1)} (z')^{\frac{1}{2}-\mu} \sum_{n=0}^{\infty} \frac{(-1)^n (-2\mu)_n}{n!} z^{\frac{n}{2}} (z'-1)^n$$
$$\times M_{\varkappa+\frac{n}{2},\mu-\frac{n}{2}}(z)$$

$$= e^{\frac{z'(1-z)}{2}} z^{\varkappa} \sum_{n=0}^{\infty} \frac{\left(\frac{1}{2}+\mu-\varkappa\right)_n}{n!} (z')^n (z-1)^n M_{\varkappa-n,\mu}(z')$$

$$= e^{\frac{z(1-z')}{2}} (z')^{\varkappa} \sum_{n=0}^{\infty} \frac{\left(\frac{1}{2}+\mu-\varkappa\right)_n}{n!} z^n (z'-1)^n M_{\varkappa-n,\mu}(z).$$

$$= e^{\frac{z'(1-z)}{2}} z^{-\mu+\frac{1}{2}} \sum_{n=0}^{\infty} \frac{(-1)^n (-2\mu)_n}{n!} (z')^{\frac{n}{2}} (z-1)^n$$
$$\times M_{\varkappa-\frac{n}{2},\mu-\frac{n}{2}}(z')$$

$$= e^{\frac{z(1-z')}{2}} (z')^{-\mu+\frac{1}{2}} \sum_{n=0}^{\infty} \frac{(-1)^n (-2\mu)_n}{n!} z^{\frac{n}{2}} (z'-1)^n$$
$$\times M_{\varkappa-\frac{n}{2},\mu-\frac{n}{2}}(z)$$

$$= e^{\frac{z'(1-z)}{2}} z^{\mu+\frac{1}{2}} \sum_{n=0}^{\infty} \frac{\left(\frac{1}{2}+\mu-\varkappa\right)_n}{n!\,(1+2\mu)_n} (z')^{\frac{n}{2}} (z-1)^n$$
$$\times M_{\varkappa-\frac{n}{2},\mu+\frac{n}{2}}(z')$$

$$= e^{\frac{z(1-z')}{2}} (z')^{\mu+\frac{1}{2}} \sum_{n=0}^{\infty} \frac{\left(\frac{1}{2}+\mu-\varkappa\right)_n}{n!\,(1+2\mu)_n} z^{\frac{n}{2}} (z'-1)^n$$
$$\times M_{\varkappa-\frac{n}{2},\mu+\frac{n}{2}}(z).$$

7.3.4 Multiplication theorems for $W_{\varkappa,\mu}(z)$

$$W_{\varkappa,\mu}(zz') = e^{\frac{z'(z-1)}{2}} z^{\varkappa} \sum_{n=0}^{\infty} \frac{(-1)^n}{n!} z^{-n} (z-1)^n W_{\varkappa+n,\mu}(z')$$

$$= e^{\frac{z(z'-1)}{2}} (z')^{\varkappa} \sum_{n=0}^{\infty} \frac{(-1)^n}{n!} (z')^{-n} (z'-1)^n W_{\varkappa+n,\mu}(z)$$

$$= e^{\frac{z'(z-1)}{2}} z^{\frac{1}{2}-\mu} \sum_{n=0}^{\infty} \frac{(z')^{\frac{n}{2}}}{n!} (1-z)^n W_{\varkappa+\frac{n}{2},\mu-\frac{n}{2}}(z')$$

$$= e^{\frac{z(z'-1)}{2}} (z')^{\frac{1}{2}-\mu} \sum_{n=0}^{\infty} \frac{z^{\frac{n}{2}}}{n!} (1-z')^n W_{\varkappa+\frac{n}{2},\mu-\frac{n}{2}}(z)$$

$$= e^{\frac{z'(z-1)}{2}} (z)^{\mu+\frac{1}{2}} \sum_{n=0}^{\infty} \frac{(z')^{\frac{n}{2}}}{n!} (1-z)^n W_{\varkappa+\frac{n}{2},\mu+\frac{n}{2}}(z')$$

$$= e^{\frac{z(z'-1)}{2}} (z')^{\mu+\frac{1}{2}} \sum_{n=0}^{\infty} \frac{z^{\frac{n}{2}}}{n!} (1-z')^n W_{\varkappa+\frac{n}{2},\mu+\frac{n}{2}}(z)$$

$$= e^{\frac{z'(1-z)}{2}} z^{\varkappa} \sum_{n=0}^{\infty} \frac{\left(\frac{1}{2}+\mu-\varkappa\right)_n \left(\frac{1}{2}-\mu-\varkappa\right)_n}{n!\,(z')^{-n}} (z-1)^n$$
$$\times W_{\varkappa-n,\mu}(z')$$

$$= e^{\frac{z(1-z')}{2}} (z')^{\varkappa} \sum_{n=0}^{\infty} \frac{\left(\frac{1}{2}+\mu-\varkappa\right)_n \left(\frac{1}{2}-\mu-\varkappa\right)_n}{n!\,z^{-n}} (z'-1)^n$$
$$\times W_{\varkappa-n,\mu}(z)$$

$$= e^{\frac{z'(1-z)}{2}} z^{\mu+\frac{1}{2}} \sum_{n=0}^{\infty} \frac{\left(\frac{1}{2}+\mu-\varkappa\right)_n}{n!} (z')^{\frac{n}{2}} (1-z)^n$$
$$\times W_{\varkappa-\frac{n}{2},\mu+\frac{n}{2}}(z')$$

$$= e^{\frac{z(1-z')}{2}} (z')^{\mu+\frac{1}{2}} \sum_{n=0}^{\infty} \frac{\left(\frac{1}{2}+\mu-\varkappa\right)_n}{n!} z^{\frac{n}{2}} (1-z')^n$$

$$\times W_{\varkappa-\frac{n}{2}, \mu+\frac{n}{2}}(z)$$

$$= e^{\frac{z'(1-z)}{2}} z^{\frac{1}{2}-\mu} \sum_{n=0}^{\infty} \frac{\left(\frac{1}{2}-\mu-\varkappa\right)_n}{n!} (z')^{\frac{n}{2}} (1-z)^n$$

$$\times W_{\varkappa-\frac{n}{2}, \mu-\frac{n}{2}}(z')$$

$$= e^{\frac{z(1-z')}{2}} (z')^{\frac{1}{2}-\mu} \sum_{n=0}^{\infty} \frac{\left(\frac{1}{2}-\mu-\varkappa\right)_n}{n!} z^{\frac{n}{2}} (1-z')^n$$

$$\times W_{\varkappa-\frac{n}{2}, \mu-\frac{n}{2}}(z).$$

7.4 Integral representations

7.4.1 Integral representations for $M_{\varkappa,\mu}(z)$

The following integral representations for $M_{\varkappa,\mu}(z)$ can be obtained from the corresponding integrals for Kummer's function and the definition of $M_{\varkappa,\mu}(z)$. If

$$Re\left(\frac{1}{2}+\mu \pm \varkappa\right) > 0, \qquad |\arg z| < \pi,$$

$$\frac{\Gamma\left(\frac{1}{2}+\mu+\varkappa\right)\Gamma\left(\frac{1}{2}+\mu-\varkappa\right)}{\Gamma(1+2\mu)} M_{\varkappa,\mu}(z)$$

$$= e^{-\frac{z}{2}} z^{\mu+\frac{1}{2}} \int_0^1 e^{zt} t^{\mu-\varkappa-\frac{1}{2}} (1-t)^{\mu+\varkappa-\frac{1}{2}} dt$$

$$= e^{\frac{z}{2}} z^{\mu+\frac{1}{2}} \int_0^1 e^{-zt} t^{\mu+\varkappa-\frac{1}{2}} (1-t)^{\mu-\varkappa-\frac{1}{2}} dt$$

$$= 2^{-2\mu} z^{\mu+\frac{1}{2}} \int_{-1}^1 e^{\frac{1}{2}zt} (1+t)^{\mu-\varkappa-\frac{1}{2}} (1-t)^{\mu-\varkappa-\frac{1}{2}} dt$$

$$= 2^{-2\mu} z^{\mu-\frac{1}{2}} \int_{-1}^1 e^{-\frac{1}{2}zt} (1+t)^{\mu+\varkappa-\frac{1}{2}} (1-t)^{\mu+\varkappa-\frac{1}{2}} dt,$$

$$M_{\varkappa,\mu}(z) = \frac{\Gamma(1+2\mu)\, z^{\frac{1}{2}-\mu}}{\Gamma(2\varkappa)\,\Gamma(1+2\mu-2\varkappa)} \int_0^z e^{-\frac{1}{2}(z-t)} t^{\mu-\varkappa-\frac{1}{2}} (z-t)^{2\varkappa-1}$$

$$\times M_{0,\mu-\varkappa}(t)\, dt$$

$$= \frac{\Gamma(1+2\mu)\,\Gamma(1+\mu-\varkappa)}{\Gamma(2\varkappa)\,\Gamma(1+2\mu-2\varkappa)} \exp\left[-\frac{z}{2} - \frac{i\pi}{2}(\mu-\varkappa)\right] \cdot 2^{2(\mu-\varkappa)}$$

$$\times z^{\frac{1}{2}+\varkappa} \int_0^1 e^{\frac{zt}{2}} (1-t)^{2\varkappa-1}\, t^{\mu-\varkappa} J_{\mu-\varkappa}\left(\frac{izt}{2}\right) dt.$$

$Re(\mu-\varkappa) > -\frac{1}{2}$, $\quad Re\,\varkappa > 0$; $\quad z$ real and positive.

$$= 2^{-2\mu} z^{\mu+\frac{1}{2}} \int_0^\pi \exp\left[-\frac{z}{2}\cos t\right] \sin^{2\mu} t \cot^{2\varkappa}\left(\frac{t}{2}\right) dt$$

$$= 2^{-2\mu} z^{\mu+\frac{1}{2}} \int_0^\pi \exp\left[\frac{z}{2}\cos t\right] \sin^{2\mu} t \tan^{2\varkappa}\left(\frac{t}{2}\right) dt$$

$$= e^{-z\left(\alpha+\frac{1}{2}\right)} z^{\mu+\frac{1}{2}} \int_\alpha^{\alpha+1} e^{zt} (t-\alpha)^{\mu-\varkappa-\frac{1}{2}} (\alpha+1-t)^{\mu+\varkappa-\frac{1}{2}} dt$$

$$= \frac{\exp\left[-z\left(\frac{\alpha}{\beta-\alpha}+\frac{1}{2}\right)\right]}{(\beta-\alpha)^{2\mu}} z^{\mu+\frac{1}{2}} \int_\alpha^\beta \exp\left[\frac{zt}{\beta-\alpha}\right] (t-\alpha)^{\mu-\varkappa-\frac{1}{2}}$$

$$\times (\beta-t)^{\mu+\varkappa-\frac{1}{2}} dt$$

$$= e^{-\frac{z}{2}} z^{\frac{1}{2}-\mu} \int_0^z e^t t^{\mu-\varkappa-\frac{1}{2}} (z-t)^{\mu+\varkappa-\frac{1}{2}} dt,$$

$$M_{\varkappa,\mu}(z) = \frac{\exp\left(-\frac{z}{2}\right) z^{\frac{1}{2}}}{\Gamma\left(\frac{1}{2}+\mu-\varkappa\right)} \Gamma(1+2\mu) \int_0^\infty e^{-t} t^{-\varkappa-\frac{1}{2}} I_{2\mu}\left(2\sqrt{zt}\right) dt$$

$$= \frac{\Gamma(1+2\mu)}{\Gamma\left(\frac{1}{2}+\mu+\varkappa\right)} e^{+\frac{z}{2}} z^{\frac{1}{2}} \int_0^\infty e^{-t} t^{\varkappa-\frac{1}{2}} J_{2\mu}\left(2\sqrt{zt}\right) dt,$$

$$Re\left(-\frac{1}{2}-\mu+\varkappa\right) > 0.$$

The function $M_{\varkappa,\mu}(z)$ can be given in terms of a contour integral as:

$$\frac{\Gamma\left(\frac{1}{2}+\mu-\varkappa\right)}{\Gamma(1+2\mu)} M_{\varkappa,\mu}(z) = \frac{1}{2\pi i} \exp\left(-\frac{z}{2}\right) z^{\mu+\frac{1}{2}}$$

$$\times \int_{\gamma-i\infty}^{\gamma+i\infty} \frac{\Gamma(-t)\,\Gamma\left(\frac{1}{2}+\mu-\varkappa+t\right)}{\Gamma(1+2\mu+t)} (-z)^t\, dt,$$

where

$$|\arg z| < \frac{\pi}{2}, \quad 2\mu \neq 0, -1, -2, \ldots$$

7.4.2 Integral representations for $W_{\varkappa,\mu}(z)$

$$\Gamma\left(\frac{1}{2}+\mu-\varkappa\right) W_{\varkappa,\mu}(z)$$

$$= \exp\left(-\frac{z}{2}\right) z^{\mu+\frac{1}{2}} \int_0^\infty e^{-zt} t^{\mu-\varkappa-\frac{1}{2}} (1+t)^{\mu+\varkappa-\frac{1}{2}}\, dt$$

$$|\arg z| < \frac{\pi}{2}, \quad Re(\mu-\varkappa) > -\frac{1}{2}$$

$$= \exp\left(-\frac{z}{2}\right) z^{\frac{1}{2}-\mu} \int_0^\infty e^{-t} t^{\mu-\varkappa-\frac{1}{2}} (z+t)^{\mu+\varkappa-\frac{1}{2}}\, dt$$

$$|\arg z| < \pi, \quad Re(\mu-\varkappa) > -\frac{1}{2}$$

$$= \exp\left(\frac{z}{2}\right) z^{\mu+\frac{1}{2}} \int_1^\infty e^{-zt} t^{\mu+\varkappa-\frac{1}{2}} (t-1)^{\mu-\varkappa-\frac{1}{2}}\, dt$$

$$|\arg z| < \frac{\pi}{2}, \quad Re(\mu-\varkappa) > -\frac{1}{2}$$

$$= \exp\left(\frac{z}{2}\right) z^{\mu+\frac{1}{2}} \int_\alpha^{\alpha+1} \exp\left\{\frac{z}{\alpha+1-t}\right\} (t-\alpha)^{\mu-\varkappa-\frac{1}{2}} (\alpha+1-t)^{-2\mu-1}\, dt$$

$$|\arg z| < \pi, \quad Re\left(\frac{1}{2}+\mu-\varkappa\right) > 0,$$

$$= \exp\left(\frac{z}{2}\right) z^{\mu+\frac{1}{2}} (\beta-\alpha)^{\mu+\varkappa+\frac{1}{2}}$$

$$\times \int_\alpha^\beta \exp\left\{\frac{z(\beta-\alpha)}{\beta-t}\right\} (t-\alpha)^{\mu-\varkappa-\frac{1}{2}} (\beta-t)^{-2\mu-1}\, dt$$

$$|\arg z| < \pi, \quad Re\left(\frac{1}{2}+\mu-\varkappa\right) > 0$$

$$= 2^{-2\mu} z^{\mu+\frac{1}{2}} \int_1^\infty \exp\left(-\frac{zt}{2}\right) (1+t)^{\mu+\varkappa-\frac{1}{2}} (t-1)^{\mu-\varkappa-\frac{1}{2}}\, dt$$

$$|\arg z| < \frac{\pi}{2}, \quad Re(\mu-\varkappa) > -\frac{1}{2}$$

$$= 2^{-2\mu} z^{\mu+\frac{1}{2}} \int_0^\infty \exp\left(-\frac{z}{2}\cosh t\right) (\sinh t)^{2\mu} \left(\coth\frac{t}{2}\right)^{2\varkappa} dt.$$

$$|\arg z| < \frac{\pi}{2}, \quad Re(\mu-\varkappa) > -\frac{1}{2}$$

$$W_{\varkappa,\mu}(z) = \frac{2 z^{\frac{1}{2}} \exp\left(-\frac{z}{2}\right)}{\Gamma\left(\frac{1}{2}+\mu-\varkappa\right)\Gamma\left(\frac{1}{2}-\mu-\varkappa\right)} \int_0^\infty e^{-t} t^{-\varkappa-\frac{1}{2}} K_{2\mu}\left(2\sqrt{zt}\right) dt,$$

$$Re\left(\frac{1}{2}\pm\mu-\varkappa\right) > 0.$$

In terms of a Barnes type integral, the function $W_{\varkappa,\mu}(z)$ is given by:

$$\Gamma\left(\frac{1}{2}+\mu-\varkappa\right)\Gamma\left(\frac{1}{2}-\mu-\varkappa\right)W_{\varkappa,\mu}(z)$$

$$=\frac{1}{2\pi i}\exp\left(-\frac{z}{2}\right)z^{\varkappa}\int_{\gamma-i\infty}^{\gamma+i\infty}\Gamma(-t)\,\Gamma\left(\frac{1}{2}+\mu-\varkappa+t\right)\Gamma\left(\frac{1}{2}-\mu-\varkappa+t\right)\times(z)^{-t}\,dt,$$

where

$$\frac{1}{2}+\varkappa\pm\mu\neq 0,1,2,\ldots,\qquad |\arg z|<\frac{3\pi}{2},$$

and the path of integration is to be chosen in such a way as to separate the poles of $\Gamma(-t)$,

$$\Gamma\left(\frac{1}{2}+\mu-\varkappa+t\right)\text{ and }\Gamma\left(\frac{1}{2}-\mu-\varkappa+t\right).$$

7.5 Integral transforms

7.5.1 Laplace transforms associated with $M_{\varkappa,\mu}(z)$

$$\int_0^\infty e^{-zt}t^{\alpha-1}M_{\varkappa,\mu}(\beta t)\,dt=\beta^{\mu+\frac{1}{2}}\left(z+\frac{\beta}{2}\right)^{-\mu-\frac{1}{2}-\alpha}\Gamma\left(\mu+\frac{1}{2}+\alpha\right)$$
$$\times\,{}_2F_1\left[\frac{1}{2}+\mu-\varkappa,\ \frac{1}{2}+\mu+\alpha;\ 1+2\mu;\ \frac{2\beta}{\beta+2z}\right],$$
$$Re\left(\frac{1}{2}+\mu+\alpha\right)>0,\qquad 2\,Re\,z>Re\,\beta,$$

$$\int_0^\infty\exp\left(-\beta\frac{t}{2}\right)t^{\alpha-1}M_{\varkappa,\mu}(\beta t)\,dt=\frac{\Gamma(1+2\mu)\,\Gamma(\varkappa-\alpha)\,\Gamma\left(\frac{1}{2}+\mu+\alpha\right)}{\Gamma\left(\frac{1}{2}+\mu+\varkappa\right)\Gamma\left(\frac{1}{2}+\mu-\alpha\right)\cdot\beta^{\alpha}}$$
$$2\,Re\,z>|Re\,\beta|,\qquad Re\,(1+2\mu)>0,$$

$$\int_0^\infty e^{-zt}t^{\mu-\frac{1}{2}}M_{\varkappa,\mu}(\lambda t)\,dt=\Gamma(1+2\mu)\,\lambda^{\frac{1}{2}+\mu}\left(z-\frac{\lambda}{2}\right)^{\varkappa-\mu-\frac{1}{2}}$$
$$\times\left(z+\frac{\lambda}{2}\right)^{-\mu-\frac{1}{2}-\varkappa},$$
$$Re\,(1+2\mu)>0,\qquad 2\,Re\,z>|Re\,\lambda|,$$

$$\int_0^\infty\exp\,(-\lambda t)\,t^{-\varkappa-\frac{1}{2}}I_{2\mu}\left(2z^{\frac{1}{2}}t^{\frac{1}{2}}\right)dt$$
$$=\frac{\Gamma\left(\frac{1}{2}+\mu-\varkappa\right)}{\Gamma(1+2\mu)}z^{-\frac{1}{2}}\lambda^{\varkappa}\exp\left(\frac{1}{2}\frac{z}{\lambda}\right)M_{\varkappa,\mu}\left(\frac{z}{\lambda}\right),$$

$$Re\left(\frac{1}{2}+\mu-\varkappa\right)<0,$$

$$\int_0^\infty e^{-t}t^{\varkappa-\frac{1}{2}}J_{2\mu}\left(2\sqrt{zt}\right)dt=\frac{\Gamma\left(\frac{1}{2}+\mu+\varkappa\right)}{\Gamma(1+2\mu)}z^{-\frac{1}{2}}\exp\left(-\frac{z}{2}\right)M_{\varkappa,\mu}(z),$$

$$\frac{1}{2\pi i}\int_{\gamma-i\infty}^{\gamma+i\infty}\exp\left[zt+\frac{\lambda}{2t}\right]t^{\frac{1}{2}+\mu-\nu}M_{\varkappa,\mu}\left(\frac{\lambda}{t}\right)dt$$

$$=\frac{1}{\Gamma(\nu)}\lambda^{\frac{1}{2}+\mu}z^{\nu-1}\,{}_1F_2\left(\frac{1}{2}+\mu-\varkappa;1+2\mu,\nu;\lambda z\right),$$

$$\frac{1}{2\pi i}\int_{\gamma-i\infty}^{\gamma+i\infty}\exp\left[zt+\frac{\lambda}{2t}\right]t^{\varkappa}M_{\varkappa,\mu}\left(\frac{\lambda}{t}\right)dt$$

$$=\frac{\Gamma(1+2\mu)}{\Gamma\left(\frac{1}{2}+\mu-\varkappa\right)}z^{-\varkappa-\frac{1}{2}}I_{2\mu}\left(2\sqrt{\lambda z}\right)$$

$$=\frac{z^{\mu-\frac{1}{2}-\varkappa}}{\Gamma\left(\frac{1}{2}+\mu-\varkappa\right)}\,{}_0F_1(;1+2\mu;\lambda z),$$

$$\frac{1}{2\pi i}\int_{\gamma-i\infty}^{\gamma+i\infty}e^{zt}t^{-2\mu-1}\left(1-\frac{\lambda}{t}\right)^{-\mu-\frac{1}{2}+\varkappa}dt$$

$$=\frac{\lambda^{-\mu-\frac{1}{2}}}{\Gamma(1+2\mu)}z^{\mu-\frac{1}{2}}\exp\left(\frac{\lambda z}{2}\right)M_{\varkappa,\mu}(\lambda z),$$

$$Re\,(1+2\mu)>0,\qquad Re\,(z-\varkappa)>0,$$

$$\frac{1}{2\pi i}\int_{\gamma-i\infty}^{\gamma+i\infty}e^{zt}t^{-\nu}\,{}_2F_1\left[\frac{1}{2}+\mu-\varkappa,\nu;1+2\mu;\frac{\lambda}{t}\right]dt$$

$$=\frac{1}{\Gamma(\nu)}\lambda^{-\mu-\frac{1}{2}}z^{-\mu+\nu-\frac{3}{2}}\exp\left(\frac{\lambda z}{2}\right)M_{\varkappa,\mu}(\lambda z).$$

The last two integrals may also be considered as contour integral representations of $M_{\varkappa,\mu}(z)$.

7.5.2 Laplace transforms associated with $W_{\varkappa,\mu}(z)$

$$\int_0^\infty e^{-zt}t^{\nu-1}W_{\varkappa,\mu}(t)\,dt=\frac{\Gamma\left(\frac{1}{2}+\mu+\nu\right)\Gamma\left(\frac{1}{2}-\mu+\nu\right)}{\Gamma(1+\nu-\varkappa)}$$

$$\times\,{}_2F_1\left[\frac{1}{2}-\mu+\nu,\frac{1}{2}+\mu+\nu;1-\varkappa+\nu;\frac{1}{2}-z\right]$$

$$=\frac{\Gamma\left(\frac{1}{2}+\mu+\nu\right)\Gamma\left(\frac{1}{2}-\mu+\nu\right)}{\Gamma(1-\varkappa+\nu)}\left(z+\frac{1}{2}\right)^{-\frac{1}{2}-\mu-\nu}$$

$$\times\,{}_2F_1\left[\frac{1}{2}+\mu-\varkappa,\frac{1}{2}+\mu+\nu;1-\varkappa+\nu;\frac{2z-1}{2z+1}\right],$$

$$Re\,(2z+1)>0,\quad Re\left(\frac{1}{2}+\nu\pm\mu\right)>0,$$

$$\int_0^\infty \exp\left(-\frac{3}{2}t\right)t^{2\mu-1+\varkappa}W_{\varkappa,\mu}(t)\,dt=\frac{\Gamma\left(\frac{1}{2}+\mu+\varkappa\right)\Gamma\left(\frac{5}{4}+\frac{3}{2}\mu+\frac{1}{2}\varkappa\right)}{\left(\frac{1}{2}+3\mu+\varkappa\right)\Gamma\left(\frac{3}{4}+\frac{1}{2}\mu-\frac{1}{2}\varkappa\right)}$$

$$\int_0^\infty \exp\left(-\frac{t}{2}\right)t^{\nu-1}W_{\varkappa,\mu}(t)\,dt=\frac{\Gamma\left(\frac{1}{2}+\mu+\nu\right)\Gamma\left(\frac{1}{2}-\mu+\nu\right)}{\Gamma(1-\varkappa+\nu)},$$

$$\int_0^\infty \exp\left(\frac{t}{2}\right)t^{\nu-1}W_{\varkappa,\mu}(t)\,dt=\frac{\Gamma\left(\frac{1}{2}+\mu-\nu\right)\Gamma\left(\frac{1}{2}-\mu-\nu\right)\Gamma(-\varkappa-\nu)}{\Gamma\left(\frac{1}{2}+\mu-\varkappa\right)\Gamma\left(\frac{1}{2}-\mu-\varkappa\right)},$$

$$\int_0^\infty \exp\,(-zt)\,t^{c-1}\,{}_2F_1\left[\frac{1}{2}+\mu-\varkappa,\frac{1}{2}-\mu-\varkappa;c;-t\right]dt$$

$$=\Gamma(c)\exp\left(\frac{z}{2}\right)z^{-\varkappa-c}\,W_{\varkappa,\mu}(z),$$

$$\int_0^\infty \exp\,(-zt)\,t^{-\varkappa-\frac{1}{2}}K_{2\mu}\left(2\sqrt{\lambda t}\right)dt=\frac{\Gamma\left(\frac{1}{2}+\mu-\varkappa\right)\Gamma\left(\frac{1}{2}-\mu-\varkappa\right)}{2\sqrt{\lambda}}$$

$$\times z^{\varkappa}\exp\left(\frac{\lambda}{2z}\right)W_{\varkappa,\mu}\left(\frac{\lambda}{z}\right),$$

$$\frac{1}{2\pi i}\int_{\gamma-i\infty}^{\gamma+i\infty} t^{\varkappa}\exp\left[zt+\frac{\lambda}{2t}\right]W_{\varkappa,\mu}\left(\frac{\lambda}{t}\right)dt$$

$$=\frac{2\sqrt{\lambda}\,z^{-\varkappa-\frac{1}{2}}}{\Gamma\left(\frac{1}{2}-\mu-\varkappa\right)\Gamma\left(\frac{1}{2}+\mu-\varkappa\right)}K_{2\mu}\left(2\sqrt{\lambda z}\right),$$

$$\frac{1}{2\pi i}\int_{\gamma-i\infty}^{\gamma+i\infty}\exp\left[\left(z+\frac{1}{2}\right)t\right]t^{c-\mu-\frac{1}{2}+\nu}\,W_{\varkappa,\mu}(t)\,dt$$

$$=\frac{z^{c-1}}{\Gamma(c)}\,{}_2F_1\left[\frac{1}{2}+\mu-\varkappa,\frac{1}{2}-\mu-\varkappa;c;-z\right],$$

$$\frac{1}{2\pi i}\int_{\gamma-i\infty}^{\gamma+i\infty}\exp\left[z\left(t+\frac{1}{2}\right)\right]t^{-\mu-\frac{1}{2}}\,W_{\varkappa,\mu}(t)\,dt=\frac{z^{\mu-\frac{1}{2}-\varkappa}(1+z)^{\mu-\frac{1}{2}+\varkappa}}{\Gamma\left(\frac{1}{2}+\mu-\varkappa\right)},$$

$$Re\left(\frac{1}{2}+\mu-\varkappa\right)>0,$$

$$\frac{\Gamma(1-\varkappa+\nu)\,z^{\nu-1}}{\Gamma\left(\frac{1}{2}+\mu+\nu\right)\Gamma\left(\frac{1}{2}-\mu+\nu\right)}W_{\varkappa,\mu}(z)$$

$$= \frac{1}{2\pi i} \int_{\gamma-i\infty}^{\gamma+i\infty} e^{zt}\, {}_2F_1\left(\frac{1}{2} - \mu + \nu, \frac{1}{2} + \mu + \nu; 1 - \varkappa + \nu; \frac{1}{2} - t\right) dt$$

$$= \frac{1}{2\pi i} \int_{\gamma-i\infty}^{\gamma+i\infty} \exp(zt) \left(t + \frac{1}{2}\right)^{-\mu - \frac{1}{2} - \nu}$$

$$\times\, {}_2F_1\left[\frac{1}{2} + \mu - \varkappa, \frac{1}{2} + \mu + \nu; 1 - \varkappa + \nu; \frac{2t-1}{2t+1}\right] dt,$$

$$Re\left(\frac{1}{2} \pm \mu + \nu\right) > 0, \quad Re\, t > -\frac{1}{2}.$$

For more results on integral transforms of Whittaker functions see ERDÉLYI, Tables of integral transforms, Vol. I, and II.

7.6 Asymptotic expansions

7.6.1 Large argument

For fixed values of $\varkappa, \mu$ and as $|z| \to \infty$

$$M_{\varkappa,\mu}(z) \approx \frac{\Gamma(1+2\mu)}{\Gamma\left(\frac{1}{2} + \mu - \varkappa\right)} \exp\left(\frac{z}{2}\right) z^{-\varkappa}\, {}_2F_0\left(\mu + \frac{1}{2} + \varkappa, \frac{1}{2} - \mu + \varkappa;; \frac{1}{z}\right)$$

$$+ \frac{\Gamma(1+2\mu)}{\Gamma\left(\frac{1}{2} + \mu - \varkappa\right)} \exp\left(-\frac{z}{2}\right) z^{\varkappa} e^{i \pm \frac{\pi}{2}(2\varkappa - 2\mu - 1)}$$

$$\times\, {}_2F_0\left(\frac{1}{2} + \mu - \varkappa, \frac{1}{2} - \mu - \varkappa;; -\frac{1}{z}\right),$$

where the sign is taken positive or negative according as

$$-\frac{3\pi}{2} < \arg z < \frac{\pi}{2}$$

or

$$-\frac{\pi}{2} < \arg z < \frac{3\pi}{2},$$

$$W_{\varkappa,\mu}(z) \approx \exp\left(-\frac{z}{2}\right) z^{\varkappa}\, {}_2F_0\left(\frac{1}{2} + \mu - \varkappa, \frac{1}{2} - \mu - \varkappa;; -\frac{1}{z}\right),$$

$$-\frac{3\pi}{2} < \arg z < \frac{3\pi}{2},$$

$$W_{\varkappa,\mu}(z) \approx \exp\left[\frac{z}{2} \pm i\pi\varkappa\right] z^{-\varkappa}\, {}_2F_0\left(\frac{1}{2} + \mu + \varkappa, \frac{1}{2} - \mu + \varkappa;; \frac{1}{z}\right),$$

where the sign is taken positive or negative according as

$$-\frac{\pi}{2} < \arg z < \frac{5\pi}{2}$$

or

$$-\frac{5\pi}{2} < \arg z < \frac{\pi}{2}.$$

7.6.2 Large parameters

Case 1: $\varkappa$, z bounded, $|\mu| \to \infty$

$$M_{\varkappa,\mu}(z) = z^{\mu+\frac{1}{2}}[1 + O(|\mu|^{-1})],$$
$$|\arg \mu| < \frac{\pi}{2},$$

$$M_{\varkappa\pm\mu,\alpha+\mu}(z) \approx \exp\left(\mp \frac{z}{2}\right) z^{\mu+\frac{1}{2}+\alpha}\left[1 + O\left(\frac{1}{|2\mu + 1 + 2\alpha|}\right)\right],$$
$$|\arg \mu| < \frac{\pi}{2},$$

$$M_{\varkappa,\mu}(z) = \frac{\left[\Gamma\left(\frac{1}{2} + \mu\right)\right]^2 z^{\mu+\frac{1}{2}}}{\Gamma\left(\frac{1}{2} + \mu - \varkappa\right)\Gamma\left(\frac{1}{2} + \mu + \varkappa\right)}\left\{\sum_{n=0}^{N} a_n \frac{n!}{(1+\mu)_n} + O\left(\frac{1}{|\mu|^{N+1}}\right)\right\}$$

where a_n are functions of $\varkappa$, z

$$a_0 = 1, \quad a_1 = \varkappa^2\left(1 - \frac{z}{4\varkappa}\right)^2$$
$$a_2 = \frac{\varkappa^2}{2}\left(1 - \frac{z}{4\varkappa}\right)^2 + \frac{\varkappa^4}{4}\left(1 - \frac{z}{4\varkappa}\right)^4,$$
$$a_n = \frac{1}{2\pi i}\,\frac{(2n)!\,2^{-2n}}{(n!)^2}\int^{(0+)} \exp\left(-\frac{zt}{2}\right)\left(\frac{1-t}{1+t}\right)^{-\varkappa} t^{-2n-1}\,dt.$$

The corresponding results for $W_{\varkappa,\mu}(z)$ can be obtained from its definition in terms of $M_{\varkappa,\mu}(z)$ and the above results.

Case 2: $\varkappa$ large, $|\arg(\varkappa z)| < 2\pi$

$$M_{\varkappa,\mu}(z) = \Gamma(1 + 2\mu)\,\pi^{-\frac{1}{2}} z^{\frac{1}{4}} \varkappa^{-\mu-\frac{1}{4}}$$
$$\times \cos\left[2\varkappa^{\frac{1}{2}} z^{\frac{1}{2}} - \frac{\pi}{4}(1 + 4\mu)\right] + O\left(|\varkappa|^{-\mu-\frac{3}{4}}\right),$$

$$M_{-\varkappa,\mu}(z) = \Gamma(1 + 2\mu)\,\pi^{-\frac{1}{2}} z^{\frac{1}{4}} \varkappa^{-\mu-\frac{1}{4}} \cdot \exp\left[\mp i \frac{\pi}{4}(1 + 4\mu)\right]$$
$$\times \cos\left[\pm 2i\varkappa^{\frac{1}{2}} z^{\frac{1}{2}} - \frac{\pi}{4}(1 + 4\mu)\right] + O\left(|\varkappa|^{-\mu-\frac{3}{4}}\right),$$

where upper or lower sign must be taken throughout according as

$$-3\pi < \arg(\varkappa z) < \pi$$

or

$$-\pi < \arg(\varkappa z) < 3\pi.$$

The leading term for the function $W_{\varkappa,\mu}$ are given by the following results:

$$W_{\varkappa,\mu}(z) \approx \sqrt{2}\, z^{\frac{1}{4}} \exp(-\varkappa)\, \varkappa^{\varkappa - \frac{1}{4}} \cos\left[2\varkappa^{\frac{1}{2}} z^{\frac{1}{2}} + \frac{\pi}{4}(1 - 4\varkappa)\right],$$

$$|\arg \varkappa| < \pi, \quad |\arg(\varkappa z)| < 2\pi,$$

$$W_{\varkappa,\mu}(z) \approx \frac{z^{\frac{1}{4}}}{\sqrt{2}} \varkappa^{\varkappa - \frac{1}{4}} \exp(-\varkappa) \exp\left[\pm i\left(\frac{\pi}{4}[4\varkappa - 1] - 2\varkappa^{\frac{1}{2}} z^{\frac{1}{2}}\right)\right],$$

the sign being taken positive or negative according as

$$\operatorname{Im} \varkappa \lesseqgtr 0,$$

$$W_{-\varkappa,\mu}(z) \approx \frac{z^{\frac{1}{4}}}{\sqrt{2}} \varkappa^{-\varkappa - \frac{1}{4}} \exp\left[\varkappa - 2\varkappa^{\frac{1}{2}} z^{\frac{1}{2}}\right],$$

where either

$$-\pi < \arg(\varkappa z) < 3\pi, \ \operatorname{Im} \varkappa > 0$$

or

$$-3\pi < \arg(\varkappa z) < \pi, \ \operatorname{Im} \varkappa < 0.$$

For more detailed results see Buchholz [1953], Slater [1960], Kazarinoff [1], [2].

7.6.3 Argument, parameters large

Asymptotic expansions, as well as the asymptotic approximations, for the Whittaker functions, when z and $\varkappa$ are both large, were obtained by Buchholz (Z. angew. Math. Mech. **30**, 133—148). These are listed in Buchholz (1953) and Slater (1960). Kazarinoff (1957) obtained some further results; however, most of these results are too complicated to be listed here. The results given below are due to Jorna [1964]. Let u_n be defined by the following integro-differential equation

$$u_{n+1} + \frac{(t^2 - 1)^2}{8\varkappa} \frac{du_n}{dt} = -\frac{1}{32\varkappa} \int^{t} \left(5t^2 - 2 - \frac{16\mu^2 - 1}{t^2}\right) u_n \, dt,$$

where

$$t = \frac{z^{\frac{1}{2}}}{(z - 4\varkappa)^{\frac{1}{2}}}.$$

Then

$$W_{\varkappa,\mu}(z) = \frac{\left\{\Gamma\left(\mu + \frac{1}{2} + \varkappa\right)\Gamma\left(\frac{1}{2} - \mu + \varkappa\right)\right\}^{\frac{1}{2}}}{(2\pi)^{\frac{1}{2}}} \left(1 - \frac{4\varkappa}{z}\right)^{-\frac{1}{4}}$$

$$\times \exp\left\{-\frac{z}{2}\left(1 - \frac{4\varkappa}{z}\right)^{\frac{1}{2}} + 2\varkappa \cosh^{-1}\left(\frac{z}{4\varkappa}\right)^{\frac{1}{2}}\right\} \sum_{n=0}^{\infty} u_n,$$

$$W_{-\varkappa,\mu}(z e^{\pm i\pi}) = \frac{(2\pi)^{\frac{1}{2}} \exp(\mp i\pi\varkappa)\left(1-\frac{4\varkappa}{z}\right)^{-\frac{1}{4}}}{\left\{\Gamma\left(\frac{1}{2}+\mu+\varkappa\right)\Gamma\left(\frac{1}{2}-\mu+\varkappa\right)\right\}^{\frac{1}{2}}}$$

$$\times \exp\left\{\frac{z}{2}\left(1-\frac{4\varkappa}{z}\right)^{\frac{1}{2}} - 2\varkappa \cosh^{-1}\left(\frac{z}{4\varkappa}\right)^{\frac{1}{2}}\right\} \sum_{n=0}^{\infty} (-1)^n u_n,$$

where the sign is taken +ve or −ve according as

$$0 \leq \arg z \leq \frac{\pi}{2}, \quad 0 \leq \arg \varkappa \leq \frac{\pi}{2}$$

or

$$-\frac{\pi}{2} \leq \arg z \leq 0, \quad -\frac{\pi}{2} \leq \arg \varkappa \leq 0.$$

The results are valid for $1 < \frac{z}{4\varkappa} < \infty$, and may be applied to the cases $0 < \frac{z}{4\varkappa} < 1$ and $-\infty < \frac{z}{4\varkappa} < 0$ after some appropriate changes. For details see Jorna. The results do not apply to the case $\frac{z}{4\varkappa} \sim 1$.

Case 1: $\varkappa^2 - \mu^2 \to \infty$; $z, \varkappa, \mu$ all reall.

$$W_{\varkappa,\mu}(z) = \frac{\left[\Gamma\left(\frac{1}{2}+\mu+\varkappa\right)\Gamma\left(\frac{1}{2}-\mu+\varkappa\right)\right]^{\frac{1}{2}}}{(2\pi)^{\frac{1}{2}}}\left(1-\frac{4\varkappa}{z}+\frac{4\mu^2}{z^2}\right)^{-\frac{1}{4}}$$

$$\times \exp\left\{-\frac{z}{2}\left(1-\frac{4\varkappa}{z}+\frac{4\mu^2}{z^2}\right)^{\frac{1}{2}} + \varkappa \cosh^{-1}\left(\frac{z-2\varkappa}{2\varkappa}\right)\right.$$

$$\left. - |\mu| \cosh^{-1}\left(\frac{\varkappa z - 2\mu^2}{\varkappa z}\right)\right\} \sum_{n=0}^{\infty} u_n,$$

$$W_{-\varkappa,\mu}(z e^{\pm i\pi}) = \frac{(2\pi)^{\frac{1}{2}} \exp(\mp i\pi\varkappa)}{\left[\Gamma\left(\frac{1}{2}+\mu+\varkappa\right)\Gamma\left(\frac{1}{2}-\mu+\varkappa\right)\right]^{\frac{1}{2}}}\left(1-\frac{4\varkappa}{z}+\frac{4\mu^2}{z^2}\right)^{-\frac{1}{4}}$$

$$\times \exp\left\{\frac{z}{2}\left(1-\frac{4\varkappa}{z}+\frac{4\mu^2}{z^2}\right)^{\frac{1}{2}} - \varkappa \cosh^{-1}\left(\frac{z-2\varkappa}{2\varkappa}\right)\right.$$

$$\left. + |\mu| \cosh^{-1}\left(\frac{\varkappa z - 2\mu^2}{\varkappa z}\right)\right\} \sum_{n=0}^{\infty} (-1)^n u_n.$$

The results are applicable for $1 < \frac{z-2\varkappa}{2\varkappa} < \infty$.

Case 2: $z = 4\varkappa$, $\varkappa \to \infty$, μ small.

$$W_{\varkappa,\mu}(z) = W_{\varkappa,\mu}(4\varkappa) = \frac{\left[\Gamma\left(\frac{1}{2}+\mu+\varkappa\right)\Gamma\left(\frac{1}{2}-\mu+\varkappa\right)\right]^{\frac{1}{2}}}{\pi}$$

$$\times\left(\frac{\varkappa}{12}\right)^{\frac{1}{6}}\left\{1+\frac{\Gamma\left(\frac{5}{6}\right)}{840\,\pi^{\frac{1}{2}}}(9-35\xi)\left(\frac{3}{4\varkappa}\right)^{\frac{4}{3}}-\cdots\right\},$$

where

$$\xi = 16\mu^2 - 1,$$

$$W_{-\varkappa,\mu}(4\varkappa e^{\pm i\pi}) = \frac{\Gamma\left(\frac{1}{3}\right)\exp\left[\mp i\pi\left(\varkappa-\frac{1}{6}\right)\right]}{\left[\Gamma\left(\frac{1}{2}+\mu+\varkappa\right)\Gamma\left(\frac{1}{2}-\mu+\varkappa\right)\right]^{\frac{1}{2}}}$$

$$\times\left(\frac{16\varkappa}{3}\right)^{\frac{1}{6}}\left\{1-\frac{\exp\left(\pm i\frac{2}{3}\pi\right)\Gamma\left(\frac{5}{6}\right)(9-35\xi)}{840\,\pi^{\frac{1}{2}}}\left(\frac{3}{4\varkappa}\right)^{\frac{4}{3}}\cdots\right\}.$$

Case 3: $|z| \gg |\varkappa|$.

$$W_{\varkappa,\mu}(z) \sim \frac{(2\pi)^{\frac{1}{2}}e^{-\frac{z}{2}}z^{\varkappa}e^{\varkappa}\varkappa^{-\varkappa}}{\left[\Gamma\left(\frac{1}{2}+\mu-\varkappa\right)\Gamma\left(\frac{1}{2}-\mu-\varkappa\right)\right]^{\frac{1}{2}}}\left(1-\frac{3\xi-1}{96\varkappa}+\cdots\right)$$

if $\varkappa < 0$,

$$W_{\varkappa,\mu}(z) \sim e^{-\frac{z}{2}}z^{\varkappa}\frac{\left[\Gamma\left(\frac{1}{2}+\mu+\varkappa\right)\Gamma\left(\frac{1}{2}-\mu+\varkappa\right)\right]^{\frac{1}{2}}}{(2\pi)^{\frac{1}{2}}e^{\varkappa}\varkappa^{-\varkappa}}\left(1-\frac{3\xi-1}{96\varkappa}\cdots\right)$$

if $\varkappa > 0$.

7.7 Products of Whittaker functions

$$\int_0^{\infty}\exp\left[\frac{1}{2}(z+z')\,t\right]t^{\lambda-\mu-\mu'-2}(1+t)^{-1}M_{\varkappa,\mu}(zt)\,M_{\varkappa',\mu'}(z't)\,dt$$

$$= \frac{\pi}{\sin(\pi\lambda)}e^{-\frac{1}{2}(z+z')}M_{-\varkappa,\mu}(z)\,M_{-\varkappa',\mu'}(z'),$$

$$Re\,\lambda > 0,$$

$$\int_0^\infty \exp\left[\frac{1}{2}(z-z')\,t\right] t^{\lambda-\mu-\mu'-2} M_{\varkappa,\mu}(zt)\, M_{\varkappa',\mu'}(z't)\, dt$$

$$= z^{\mu+\frac{1}{2}} (z')^{\mu'+\frac{1}{2}-\lambda} \frac{\Gamma(\lambda)\,\Gamma(1+2\mu')\,\Gamma\left(\frac{1}{2}+\mu'+\varkappa'-\lambda\right)}{\Gamma\left(\frac{1}{2}+\mu'+\varkappa'\right)\Gamma(1+2\mu'-\lambda)}$$

$$\times {}_3F_2\left[\frac{1}{2}+\mu-\varkappa, \lambda-2\mu, \lambda; 1+2\mu, \frac{1}{2}-\mu'-\varkappa'+\lambda; \frac{z}{z'}\right],$$

$$Re\,\lambda > 0, \quad Re\,z' > 0, \quad |z| > |z'|$$

or

$$z = z' \text{ and } Re\,(\mu+\varkappa+\mu'-\varkappa'-\lambda) > -2.$$

In these two integral, the parameters may be specialized to give particular cases.

$$W_{\varkappa,\mu}(z)\, W_{-\varkappa,\mu}(z) = -z \int_0^\infty \left(\coth\frac{t}{2}\right)^{-2\varkappa}$$

$$\times \{J_{2\mu}(z\sinh t)\sin[\pi(\mu-\varkappa)] + Y_{2\mu}(z\sinh t)\cos[\pi(\mu-\varkappa)]\}\, dt,$$

$$z \text{ real}; \quad \frac{1}{2} + Re\,\varkappa > |Re\,\mu|,$$

$$W_{\varkappa,\mu}(iz)\, W_{\varkappa,\mu}(-iz) = \frac{2z}{\Gamma\left(\frac{1}{2}+\mu-\varkappa\right)\Gamma\left(\frac{1}{2}-\mu-\varkappa\right)} \int_0^\infty \left(\coth\frac{t}{2}\right)^{2\varkappa}$$

$$\times K_{2\mu}(z\sinh t)\, dt,$$

$$|\arg z| \leq \frac{\pi}{2}; \quad |Re\,\mu| < \frac{1}{2} - Re\,\varkappa,$$

$$W_{\varkappa,\mu}(z)\, W_{\lambda,\mu}(z') = \frac{(z\,z')^{\mu+\frac{1}{2}} \exp\left[-\frac{1}{2}(z+z')\right]}{\Gamma(1-\varkappa-\lambda)}$$

$$\times \int_0^\infty e^{-t} t^{-\varkappa-\lambda} (z+t)^{-\frac{1}{2}-\mu+\varkappa} (z'+t)^{-\frac{1}{2}-\mu+\lambda}$$

$$\times {}_2F_1\left[\frac{1}{2}+\mu-\varkappa, \frac{1}{2}+\mu-\lambda; 1-\varkappa-\lambda; \Theta\right] dt,$$

where

$$\Theta = \frac{t(z+z'+t)}{(z+t)(z'+t)}; \; z \neq 0, \; z' \neq 0,$$

$$Re\,(\varkappa+\lambda) < 1, \quad |\arg z| < \pi, \quad |\arg z'| < \pi,$$

$$W_{\varkappa,\mu}(z)\, W_{\lambda,\mu}(z) = \frac{\Gamma(-2\mu)\,\Gamma(1-\varkappa-\lambda)}{\Gamma\left(\frac{1}{2}-\mu-\varkappa\right)\Gamma\left(\frac{1}{2}-\mu-\lambda\right)} z^{\mu+\frac{1}{2}} \exp\left(-\frac{z}{2}\right)$$

$$\times \sum_{n=0}^{\infty} \frac{\left(\frac{1}{2}+\mu-\varkappa\right)_n \left(\frac{1}{2}+\mu-\lambda\right)_n}{n!\,(1+2\mu)_n} z^n$$

$$\times W_{\varkappa+\lambda-\mu-n-\frac{1}{2},\,\mu+n}(z)$$

$$+ \frac{\Gamma(2\mu)\,\Gamma(1-\varkappa-\lambda)}{\Gamma\left(\frac{1}{2}+\mu-\varkappa\right)\Gamma\left(\frac{1}{2}+\mu-\lambda\right)} z^{-\mu+\frac{1}{2}} \exp\left(-\frac{z}{2}\right)$$

$$\times \sum_{n=0}^{\infty} \frac{\left(\frac{1}{2}-\mu-\varkappa\right)_n \left(\frac{1}{2}-\mu-\lambda\right)_n}{n!\,(1-2\mu)_n} z^n$$

$$\times W_{\varkappa+\lambda+\mu-n-\frac{1}{2},\,n-\mu}(z),$$

$$2\mu \neq 0, \pm 1, \pm 2, \ldots, \qquad \varkappa + \lambda \neq 1, 2, 3, \ldots,$$

$$Re\,(\varkappa+\lambda) > 0, \quad |\arg z| < \pi.$$

Literature

BUCHHOLZ, H.: Die konfluente hypergeometrische Funktion. Berlin/Göttingen/Heidelberg: Springer 1953.

ERDÉLYI, A.: Higher transcendental functions, Vol. 1. New York: McGraw-Hill 1953.

JORNA, S.: Proc. Roy. Soc., series A, Vol. **281**, 111—129.

KAZARINOFF, N. D.: [1] Trans. Amer. Math. Soc. **78**, 305—328.

— [2] J. Math. Mech. **6**, 341—360.

SLATER, L. J.: Confluent hypergeometric functions. Cambridge: Univ. Press 1960.

TRICOMI, F. G.: Funzioni ipergeometriche Confluenti. Rome: Edizioni Cremonese 1954.

Chapter VIII

Parabolic cylinder functions and parabolic functions

8.1 Parabolic cylinder functions

8.1.1 The differential equation

The parabolic cylinder functions may, in general, be considered as solutions of the differential equation

$$\frac{d^2y}{dx^2} + (a x^2 + b x + c)\, y = 0, \tag{1}$$

which, by a simple change of variable, reduces to the form

$$\frac{d^2y}{dz^2} + \left(\nu + \frac{1}{2} - \frac{1}{4} z^2\right) = 0. \tag{2}$$

In the present work the solutions of this transformed equation are called the parabolic cylinder functions.

The symbol $D_\nu(z)$ is used to denote the solutions of (2) given by

$$D_\nu(z) = 2^{\frac{1}{2}\left(\nu+\frac{1}{2}\right)} z^{-\frac{1}{2}} W_{\frac{\nu}{2}+\frac{1}{4}, \pm\frac{1}{4}}\left(\frac{z^2}{2}\right)$$

$$= 2^{\frac{1}{2}(\nu-1)} e^{-\frac{z^2}{4}} z\, U\left(\frac{1-\nu}{2}, \frac{3}{2}, \frac{z^2}{2}\right)$$

$$= 2^{\frac{\nu}{2}} e^{-\frac{z^2}{4}} \left\{ \frac{\Gamma\left(\frac{1}{2}\right)}{\Gamma\left(\frac{1-\nu}{2}\right)}\, {}_1F_1\left(-\frac{\nu}{2}; \frac{1}{2}; \frac{z^2}{2}\right) + \frac{z}{\sqrt{2}} \frac{\Gamma\left(-\frac{1}{2}\right)}{\Gamma\left(-\frac{\nu}{2}\right)}\, {}_1F_1\left(\frac{1-\nu}{2}; \frac{3}{2}; \frac{z^2}{2}\right) \right\},$$

and are called the "Parabolic cylinder functions". The values of $D_\nu(0)$ and $D'_\nu(0)$ are given by

$$D_\nu(0) = \frac{\Gamma\left(\frac{1}{2}\right) 2^{\frac{\nu}{2}}}{\Gamma\left(\frac{1-\nu}{2}\right)}, \qquad D'_\nu(0) = \frac{\Gamma\left(-\frac{1}{2}\right) 2^{\frac{\nu-1}{2}}}{\Gamma\left(-\frac{\nu}{2}\right)}.$$

Since a solution of (2) is completely determined by its value and the value of its first derivative at $z = 0$, the following statement gives an equivalent definition of $D_\nu(z)$:

The functions $D_\nu(z)$ are those solutions of (2) for which

$$D_\nu(0) = \frac{\Gamma\left(\frac{1}{2}\right) 2^{\frac{\nu}{2}}}{\Gamma\left(\frac{1-\nu}{2}\right)}, \qquad D'_\nu(0) = \frac{\Gamma\left(-\frac{1}{2}\right) 2^{\frac{\nu-1}{2}}}{\Gamma\left(-\frac{\nu}{2}\right)}.$$

The differential equation (2) has the solutions

$$D_\nu(z),\ D_\nu(-z),\ D_{-\nu-1}(iz),\ D_{-\nu-1}(-iz),$$

which are connected by the linear relations:

$$D_\nu(z) = \frac{\Gamma(\nu+1)}{\sqrt{2\pi}} \left\{ e^{i\frac{\pi}{2}\nu} D_{-\nu-1}(iz) + e^{-i\frac{\pi}{2}\nu} D_{-\nu-1}(-iz) \right\}$$

$$= e^{-i\pi\nu} D_\nu(-z) + \frac{\sqrt{2\pi}}{\Gamma(-\nu)} e^{-i\frac{\pi}{2}(\nu+1)} D_{-\nu-1}(iz)$$

$$= e^{i\pi\nu} D_\nu(-z) + \frac{\sqrt{2\pi}}{\Gamma(-\nu)} e^{i\frac{\pi}{2}(\nu+1)} D_{-\nu-1}(-iz).$$

Three more relations may be obtained from these by replacing z by $-z$ and thus get all six possible relations connecting any three of the four solutions given above.

The parabolic cylinder functions are entire functions of z and are related to the Hermite polynomials (see chap. V) by

$$H_n(z) = 2^{\frac{n}{2}} e^{\frac{z^2}{2}} D_n\left(z\sqrt{2}\right).$$

Furthermore, the values of $D_\nu(z)$ are real if ν and z are real.

The following differential equations, together with their solutions, are closely related to (2):

$$\frac{d^2u}{dz^2} + z\frac{du}{dz} + (\nu+1)\,u = 0, \tag{3}$$

$$u(z) = e^{-\frac{z^2}{4}} D_\nu(z),$$

$$\frac{d^2w}{dz^2} + \left(\frac{1}{4}z^2 - \lambda\right) w = 0, \tag{4}$$

$$w(z) = D_{i\lambda-\frac{1}{2}}\left(\pm z e^{i\frac{\pi}{4}}\right).$$

It may however be necessary to find a solution of (4) which is real on the real axis. In this case one may choose any one of the following solutions.

$$w_{\frac{1}{2}}(z) = Re\, D_{i\lambda-\frac{1}{2}}\left(\pm z e^{i\frac{\pi}{4}}\right),$$

$$w_{\frac{3}{4}}(z) = \mathrm{Im}\, D_{i\lambda-\frac{1}{2}}\left(\pm z e^{i\frac{\pi}{4}}\right),$$

$$w_5(z) = 2^{-i\frac{\lambda}{2}} \left\{ D_{i\lambda-\frac{1}{2}}\left(z e^{i\frac{\pi}{4}}\right) + D_{i\lambda-\frac{1}{2}}\left(-z e^{i\frac{\pi}{4}}\right) \right\}.$$

For more discussion see ERDÉLYI (1953).

8.1.2 Special values of the parameter

$$D_0(z) = \frac{1}{\sqrt{2\pi}} z K_{\frac{1}{2}}\left(\frac{1}{4} z^2\right) = e^{-\frac{z^2}{4}} H_0\left(\frac{z}{\sqrt{2}}\right) = e^{-\frac{z^2}{4}},$$

$$D_n(z) = 2^{-\frac{n}{2}} e^{-\frac{z^2}{4}} H_n\left(\frac{z}{\sqrt{2}}\right),$$

$$D_{\frac{1}{2}}(z) = \frac{1}{\sqrt{\pi}} \left(\frac{z}{2}\right)^{\frac{3}{2}} \left\{K_{\frac{1}{4}}\left(\frac{1}{4} z^2\right) + K_{\frac{3}{4}}\left(\frac{1}{4} z^2\right)\right\},$$

$$D_{\frac{3}{2}}(z) = \frac{1}{\sqrt{\pi}} \left(\frac{z}{2}\right)^{\frac{5}{2}} \left[2K_{\frac{1}{4}}\left(\frac{1}{4} z^2\right) + 3K_{\frac{3}{4}}\left(\frac{1}{4} z^2\right) - K_{\frac{5}{4}}\left(\frac{1}{4} z^2\right)\right],$$

$$D_{\frac{5}{2}}(z) = \frac{1}{\pi} \left(\frac{z}{2}\right)^{\frac{7}{2}} \left\{5K_{\frac{1}{4}}\left(\frac{1}{4} z^2\right) + 9K_{\frac{3}{4}}\left(\frac{1}{4} z^2\right) - 5K_{\frac{5}{4}}\left(\frac{1}{4} z^2\right) - K_{\frac{7}{4}}\left(\frac{1}{4} z^2\right)\right\},$$

$$D_{-\frac{1}{2}}(z) = \sqrt{\frac{z}{2\pi}} K_{\frac{1}{4}}\left(\frac{1}{4} z^2\right),$$

$$D_{-\frac{3}{2}}(z) = \frac{1}{\sqrt{2\pi}} z^{\frac{3}{2}} \left[K_{\frac{3}{4}}\left(\frac{1}{4} z^2\right) - K_{\frac{1}{4}}\left(\frac{1}{4} z^2\right)\right],$$

$$D_{-\frac{5}{2}}(z) = \frac{1}{\sqrt{2\pi}} \frac{z^{\frac{5}{2}}}{3} \left\{K_{\frac{5}{4}}\left(\frac{1}{4} z^2\right) - 3K_{\frac{3}{4}}\left(\frac{1}{4} z^2\right) + 2K_{\frac{1}{4}}\left(\frac{1}{4} z^2\right)\right\},$$

$$D_{-n-1}(z) = \sqrt{\frac{\pi}{2}} \frac{(-1)^n}{n!} e^{-\frac{z^2}{4}} \frac{d^n}{dz^n}\left[e^{\frac{z^2}{2}} \operatorname{Erfc}\left(\frac{z}{\sqrt{2}}\right)\right],$$

$$n = 0, 1, 2, \ldots,$$

where

$$\operatorname{Erfc}(z) = 1 - \operatorname{Erf}(z) = 1 - \frac{2}{\sqrt{\pi}} \int_0^x e^{-t^2} dt.$$

8.1.3 Recurrence relations, differentiation formulas and Wronskians

$$\frac{d^m}{dz^m}\left[e^{\frac{z^2}{4}} D_\nu(z)\right] = (-1)^m (-\nu)_m e^{\frac{z^2}{4}} D_{\nu-m}(z),$$

$$m = 1, 2, 3, 4, \ldots,$$

$$\frac{d}{dz}\left[e^{\frac{z^2}{4}} D_\nu(z)\right] = +\nu e^{\frac{z^2}{4}} D_{\nu-1}(z),$$

$$\frac{d^m}{dz^m}\left[e^{-\frac{z^2}{4}} D_\nu(z)\right] = (-1)^m e^{-\frac{z^2}{4}} D_{\nu+m}(z),$$

$$m = 1, 2, 3, 4, \ldots,$$

$$\frac{d}{dz}\left[e^{-\frac{z^2}{4}} D_\nu(z)\right] = -e^{\frac{z^2}{4}} D_{\nu+1}(z).$$

Recurrence relations

$$D_{\nu+1}(z) - z D_\nu(z) + \nu D_{\nu-1}(z) = 0,$$

$$D'_\nu(z) + \frac{1}{2} z D_\nu(z) - \nu D_{\nu-1}(z) = 0,$$

$$D'_\nu(z) - \frac{z}{2} D_\nu(z) + D_{\nu+1}(z) = 0,$$

$$2D'_\nu(z) - \nu D_{\nu-1}(z) + D_{\nu+1}(z) = 0.$$

Wronskians

Let

$$W[u, v] = u\frac{dv}{dz} - v\frac{du}{dz}.$$

Then

$$W[D_\nu(z), D_\nu(-z)] = \frac{\sqrt{\pi}}{\Gamma(-\nu)},$$

$$W[D_\nu(z), D_{-\nu-1}(iz)] = e^{-i\frac{\pi}{2}(\nu+1)},$$

$$W[D_\nu(z), D_{-\nu-1}(-iz)] = e^{i\frac{\pi}{2}(\nu+1)},$$

$$W[D_{-\nu-1}(iz), D_{-\nu-1}(-iz)] = \frac{\sqrt{2\pi}}{\Gamma(1+\nu)}.$$

It is clear from the values of the Wronskian determinants that $D_\nu(z)$ and $D_{-\nu-1}(iz)$ are linearly independent for all values of ν and hence form a fundamental system of solutions of (2). The following Taylor series expansions follow immediately from the differentiation formulas above:

$$D_\nu(x+y) = e^{-\frac{1}{2}\left(xy+\frac{1}{2}y^2\right)} \sum_{n=0}^{\infty} \binom{\nu}{n} y^n D_{\nu-n}(x)$$

$$= e^{-\frac{1}{2}\left(xy+\frac{1}{2}x^2\right)} \sum_{n=0}^{\infty} \binom{\nu}{n} x^n D_{+\nu-n}(y)$$

$$= e^{\frac{1}{2}\left(xy+\frac{1}{2}y^2\right)} \sum_{n=0}^{\infty} \frac{(-y)^n}{n!} D_{\nu+n}(x)$$

$$= e^{\frac{1}{2}\left(xy+\frac{1}{2}x^2\right)} \sum_{n=0}^{\infty} \frac{(-x)^n}{n!} D_{\nu+n}(y).$$

8.1.4 Integral representations

$$D_\nu(z) = \frac{e^{-\frac{z^2}{4}}}{\Gamma(-\nu)} \int_0^\infty t^{-\nu-1} e^{-\left(\frac{t^2}{2}+zt\right)} dt,$$

$$Re\,\nu < 0,$$

$$= \frac{e^{-\frac{z^2}{4}}}{2^{-\frac{\nu}{2}} \Gamma\left(-\frac{\nu}{2}\right)} \int_0^\infty e^{-\frac{z^2}{2}t} t^{-\frac{\nu}{2}-1} (1+t)^{\frac{1}{2}(\nu-1)} dt,$$

$$Re\,\nu < 0, \quad |\arg z| \leq \frac{\pi}{4},$$

$$D_\nu\left(z\sqrt{2}\right) = \frac{e^{\frac{z^2}{2}}}{2^{\frac{\nu}{2}} \Gamma(-\nu)} \int_z^\infty e^{-t^2} (t-z)^{-\nu-1} dt,$$

$$Re\,\nu < 0,$$

$$D_\nu(z) = \frac{2^{\frac{1}{2}(\nu-1)}}{\Gamma\left(\frac{1-\nu}{2}\right)} z e^{-\frac{z^2}{4}} \int_0^\infty e^{-\frac{1}{2}z^2 t} t^{-\frac{1+\nu}{2}} (1+t)^{\frac{\nu}{2}} dt,$$

$$Re\,\nu < 1, \quad |\arg z| \leq \frac{\pi}{4},$$

$$D_\nu\left(a^{\frac{1}{2}} z^{-\frac{1}{2}}\right) = \frac{z^{-\frac{\nu}{2}} 2^{-\frac{\nu}{2}-1}}{\Gamma(-\nu)} e^{-\frac{1}{4}\frac{a}{z}} \int_0^\infty e^{-zt} t^{-\frac{\nu}{2}-1} e^{-(2at)^{\frac{1}{2}}} dt,$$

$$Re\,\nu < 0,$$

$$D_\nu(z) = 2^{\frac{1}{2}} \pi^{-\frac{1}{2}} e^{\frac{z^2}{4}} \int_0^\infty e^{-\frac{t^2}{2}} \cos\left(\frac{\pi\nu}{2} - zt\right) t^\nu dt,$$

$$Re\,\nu > -1,$$

$$= -\frac{\Gamma(\nu+1)}{2\pi i} e^{-\frac{z^2}{4}} \int_\infty^{(0+)} e^{-zt-\frac{t^2}{2}} (-t)^{-\nu-1} dt,$$

$$\arg(-t) = -\pi,$$

$$= \frac{2^{\frac{\nu}{2}+\frac{1}{2}}}{2\pi i} e^{-\frac{z^2}{4}} z^{-1} \int_{-i\infty}^{i\infty} \frac{\Gamma(t)\,\Gamma\left(\frac{1}{2}-\frac{\nu}{2}-t\right)\Gamma\left(-\frac{\nu}{2}-t\right)}{\Gamma\left(\frac{1}{2}-\frac{\nu}{2}\right)\Gamma\left(-\frac{\nu}{2}\right)} \left(\frac{z^2}{2}\right)^t dt,$$

$$\nu \neq 0, 1, 2, 3, \ldots, \quad |\arg z| < \frac{3\pi}{4}.$$

The path of integration must be such that it separates the poles of $\Gamma(t)$ from those of

$$\Gamma\left(\frac{1}{2}-\frac{\nu}{2}-t\right) \text{ and } \Gamma\left(-\frac{\nu}{2}-t\right).$$

$$D_\nu(z)\, D_{-\nu-1}(z) = -\frac{1}{\sqrt{\pi}} \int_0^\infty \left(\coth\frac{t}{2}\right)^{\nu+\frac{1}{2}} \frac{\sin\left[\frac{1}{2}(z^2 \sinh t + \pi\nu)\right]}{\sqrt{\sinh t}}\, dt,$$

z real, $Re\,\nu < 0$,

$$D_\nu(z)\, D_{-\nu-1}(z) = 2\int_0^\infty e^{-zt} J_{\nu+\frac{1}{2}}(t^2) \cos\left(zt - \frac{\pi}{2}\nu\right) dt,$$

$Re\, z > 0,\ Re\,\nu > -1$,

$$D_{-\nu-1}\left(z e^{i\frac{\pi}{4}}\right) D_{-\nu-1}\left(z e^{-i\frac{\pi}{4}}\right)$$

$$= \frac{\sqrt{\pi}}{\Gamma(\nu+1)} \int_0^\infty e^{-zt} J_{\nu+\frac{1}{2}}\left(\frac{1}{2}t^2\right) dt,$$

$Re\, z > 0,\ Re\,\nu > -1$,

$$= \frac{\pi^{-\frac{1}{2}}\, 2^{\frac{3}{2}}}{\Gamma(\nu+1)} \int_0^\infty e^{-zt} K_{\nu+\frac{1}{2}}(t^2) \cos\left[zt + \frac{\pi}{2}(\nu+1)\right] dt,$$

$-1 < Re\,\nu < 0$,

$$= -\frac{1}{\pi}\int_0^\infty (\cosh t)^{-\nu-1} (\sinh t)^{-\nu-2} \exp\left[-\frac{z^2}{2}\sinh t\right] dt,$$

$Re\,\nu < -1,\ |\arg z| < \frac{\pi}{4}$,

$$D_\nu\left(z e^{i\frac{\pi}{4}}\right) D_\nu\left(z e^{-i\frac{\pi}{4}}\right) = \frac{1}{\Gamma(-\nu)} \int_0^\infty (\coth t)^\nu \frac{e^{-\frac{z^2}{2}\sinh 2t}}{\sinh t}\, dt,$$

$Re\,\nu < 0,\ |\arg z| < \frac{\pi}{4}$,

$$D_{-\nu-1}[(1+i)\,x] = e^{-i\frac{x^2}{2}} \frac{2^{\frac{1}{2}(1-\nu)}}{\Gamma\left(\frac{1+\nu}{2}\right)} \int_0^\infty \frac{e^{-ix^2t^2}}{(1+t^2)^{\frac{\nu}{2}+1}}\, t^\nu\, dt,$$

$Re\,\nu > -1,\ Re\, ix^2 \geq 0$,

$$D_\nu[(1+i)\,x] = \frac{2^{\frac{(\nu+1)}{2}}}{\Gamma\left(-\frac{\nu}{2}\right)} \int_1^\infty e^{-i\frac{x^2}{2}t} \frac{(1+t)^{\frac{\nu-1}{2}}}{(t-1)^{\frac{\nu}{2}+1}}\, dt,$$

$Re\,\nu < 0,\ Re\, ix^2 \geq 0$,

$$\frac{\Gamma(-\nu)\, e^{i\frac{\pi}{4}\nu}}{2^{1+\frac{\nu}{2}}} \exp\left(i\,\frac{x^2}{2}\right)\{D_\nu[(1+i)\,x] + D_\nu[(-1-i)\,x]\}$$

$$= \int_0^\infty e^{-i\frac{t^2}{4}}\, t^{-\nu-1} \cos(xt)\, dt,$$

$$x \text{ real}; \quad -2 < Re\,\nu < 0.$$

8.1.5 Integrals involving parabolic cylinder functions

$$\int_0^\infty [D_{2\nu}(t)]^2\, dt = \frac{\sqrt{2\pi}}{4}\, \frac{\psi\left(\frac{1}{2}-\nu\right) - \psi(-\nu)}{\Gamma(-2\nu)},$$

where

$$\psi(z) = \frac{\Gamma'(z)}{\Gamma(z)},$$

$$\int_0^\infty e^{-\frac{3}{4}t^2}\, t^\nu\, D_{\nu+1}(t)\, dt = 2^{-\frac{(\nu+1)}{2}}\, \Gamma(\nu+1) \sin\left[\frac{\pi}{4}(1-\nu)\right],$$

$$Re\,\nu > -1,$$

$$\int_0^\infty e^{-\frac{t^2}{4}}\, t^{2\mu-1}\, D_{2\nu}(t)\, dt = \frac{\pi^{\frac{1}{2}}\, 2^{\nu-\mu}\, \Gamma(2\mu)}{\Gamma\left(\frac{1}{2}+\mu-\nu\right)},$$

$$Re\,\mu > 0,$$

$$\int_0^\infty e^{-\frac{t^2}{4}}\, \frac{t^{-\nu-1}}{x^2+t^2}\, D_{-\nu-1}(t)\, dt = \left(\frac{\pi}{2}\right)^{\frac{1}{2}} \Gamma(-\nu)\, e^{\frac{x^2}{4}}\, x^{-\nu-2}\, D_\nu(x),$$

$$Re\,\nu < 0,$$

$$\int_0^\infty e^{\frac{t^2}{4}}\, D_{-2\nu}(t)\, t^{+\nu-\frac{1}{2}}\, (xt)^{\frac{1}{2}}\, J_{+\nu}(xt)\, dt = e^{\frac{x^2}{4}}\, x^{\nu-\frac{1}{2}}\, D_{-2\nu}(x),$$

$$Re\,\nu > -\frac{1}{2},$$

$$\int_0^\infty e^{-xt}\, t^{\beta-1}\, D_{-2\nu}\left(2\sqrt{at}\right) dt$$

$$= \frac{\pi^{\frac{1}{2}}\, \Gamma(2\beta)}{\Gamma\left(\frac{1}{2}+\beta+\nu\right)}\, 2^{1-\nu-2\beta}\, (x+a)^{-\beta}\, {}_2F_1\left(\nu, \beta; \frac{1}{2}+\nu+\beta; \frac{x-a}{x+a}\right),$$

$$Re\,\beta > 0,\; Re\left(\frac{x}{a}\right) > 0,$$

$$\int_0^\infty e^{-\frac{t^2}{4}} t^{\nu+1} D_{2n+2}(t)\, J_\nu(xt)\, dt$$

$$= \pi^{-\frac{1}{2}} 2^{\nu+1} \Gamma\left(\nu + \frac{3}{2}\right) x^\nu e^{-\frac{x^2}{4}} {}_1F_1\left(-\nu - 1; \frac{1}{2}; \frac{x^2}{2}\right),$$

$$Re\, \nu > -\frac{3}{2},$$

$$\int_0^\infty e^{\frac{t^2}{4}} t^{2c-1} {}_1F_1\left(a; c; -\frac{1}{2} x t^2\right) D_{-2\nu}(t)\, dt$$

$$= \frac{\pi^{\frac{1}{2}}}{2^{\nu+c}} \frac{\Gamma(2c)\, \Gamma(\nu - c + a)}{\Gamma(\nu)\, \Gamma\left(\frac{1}{2} + a + \nu\right)} {}_2F_1\left[a, c + \frac{1}{2}; \frac{1}{2} + a + \nu; 1 - x\right],$$

$$|1 - x| < 1, \quad Re\, c > 0, \quad Re\,(\nu - c + a) > 0,$$

$$\int_0^\infty e^{\frac{t^2}{4}} t^{2c-2} {}_1F_1\left(a; c; -\frac{x}{2} t^2\right) D_{-2\nu}(t)\, dt$$

$$= \frac{\sqrt{2\pi}}{2^{c+\nu}} \frac{\Gamma(2c - 1)\, \Gamma\left(\frac{1}{2} + a - c + \nu\right)}{\Gamma(a + \nu)\, \Gamma\left(\frac{1}{2} + \nu\right)} {}_2F_1\left(a, c - \frac{1}{2}; a + \nu; 1 - x\right),$$

$$|1 - x| < 1, \quad Re\, c > \frac{1}{2}, \quad Re\,(\nu - c + a) > -\frac{1}{2},$$

$$\int_{-\infty}^\infty e^{-\frac{1}{2\lambda}(x-t)^2} e^{\frac{t^2}{4}} D_\nu(t)\, dt = (1 - \lambda)^{\frac{\nu}{2}} e^{\frac{x^2}{4(1-\lambda)}} D_\nu\left[\frac{x}{\sqrt{1 - \lambda}}\right],$$

$$0 < Re\, \lambda < 1.$$

Finally

$$\int_0^\infty [D_n(t)]^2\, dt = (2\pi)^{\frac{1}{2}} n!,$$

$$n = 0, 1, 2, \ldots$$

8.1.6 Asymptotic expansions

The following asymptotic expansions for $D_\nu(z)$are valid for large values of $|z|$ and $|z| \gg |\nu|$

$$D_\nu(z) \approx e^{-\frac{z^2}{4}} z^\nu \left\{1 - \frac{\nu(\nu - 1)}{2 z^2} + \frac{\nu(\nu - 1)(\nu - 2)(\nu - 3)}{2 \cdot 4 \cdot z^4} \mp \cdots\right\}$$

$$= e^{-\frac{z^2}{4}} z^\nu \left\{\sum_{n=0}^N \frac{\left(-\frac{\nu}{2}\right)_n \left(\frac{1 - \nu}{2}\right)_n}{n!} \left(-\frac{z^2}{2}\right)^{-n} + O(|z^2|^{-N-1})\right\},$$

$$|\arg z| < \frac{3\pi}{4},$$

$$D_\nu(z) \approx e^{-\frac{z^2}{4}} z^\nu \left\{1 - \frac{\nu(\nu-1)}{2z^2} + \frac{\nu(\nu-1)(\nu-2)(\nu-3)}{2\cdot 4z^4} \mp \cdots\right\}$$

$$-\frac{\sqrt{2\pi}}{\Gamma(-\nu)} e^{-i\pi\nu} e^{\frac{z^2}{4}} z^{-\nu-1} \left\{1 + \frac{(\nu+1)(\nu+2)}{2z^2} + \frac{(\nu+1)(\nu+2)(\nu+3)(\nu+4)}{2\cdot 4z^4} + \cdots\right\}$$

$$= z^\nu e^{-\frac{z^2}{4}} \left\{\sum_{n=0}^{N} \frac{\left(-\frac{\nu}{2}\right)_n \left(\frac{1-\nu}{2}\right)_n}{n!} \left(-\frac{z^2}{2}\right)^{-n} + O(|z^2|^{-N-1})\right\}$$

$$-\frac{\sqrt{2\pi}}{\Gamma(-\nu)} e^{-i\pi\nu} e^{\frac{z^2}{4}} z^{-\nu-1} \left\{\sum_{n=0}^{N} \frac{\left(\frac{\nu}{2}\right)_n \left(\frac{1+\nu}{2}\right)_n}{n!} \left(\frac{z^2}{2}\right)^{-n} + O(|z^2|^{-N-1})\right\},$$

$$\frac{\pi}{4} < \arg z < \frac{5\pi}{4},$$

$$D_\nu(z) \approx e^{-\frac{z^2}{4}} z^\nu \left\{1 - \frac{\nu(\nu-1)}{2z^2} + \frac{\nu(\nu-1)(\nu-2)(\nu-3)}{2\cdot 4z^2} \mp \cdots\right\}$$

$$-\frac{\sqrt{2\pi}}{\Gamma(-\nu)} e^{-i\pi\nu} e^{\frac{z^2}{4}} z^{-\nu-1} \left\{1 + \frac{(\nu+1)(\nu+2)}{2z^2} + \frac{(\nu+1)(\nu+2)(\nu+3)(\nu+4)}{2\cdot 4z^2} + \cdots\right\}$$

$$= e^{-\frac{z^2}{4}} z^\nu \left\{\sum_{n=0}^{N} \frac{\left(-\frac{\nu}{2}\right)_n \left(\frac{1-\nu}{2}\right)_n}{n!} \left(-\frac{z^2}{2}\right)^{-n} + O(|z^2|^{-N-1})\right\}$$

$$-\frac{\sqrt{2\pi}}{\Gamma(-\nu)} e^{-i\pi\nu} e^{\frac{z^2}{4}} z^{-\nu-1} \left\{\sum_{n=0}^{\infty} \frac{\left(\frac{\nu}{2}\right)_n \left(\frac{1+\nu}{2}\right)_n}{n!} \left(\frac{z^2}{2}\right)^{-n} + O(|z^2|^{-N-1})\right\},$$

$$-\frac{5\pi}{4} < \arg z < -\frac{\pi}{4}.$$

For $|\nu| \to \infty$, $|z|$ bounded

$$D_\nu(z) \approx \frac{1}{\sqrt{2}} \exp\left[\frac{\nu}{2}\log(-\nu) - \frac{\nu}{2} - z\sqrt{-\nu}\right]\left\{1 + O\left(|\nu|^{-\frac{1}{2}}\right)\right\},$$

$$|\arg(-\nu)| \leq \frac{\pi}{2}.$$

For the behavior of $D_\nu(z)$ when $|z|$ and $|\nu|$ are both large see SCHWID (1935) and Handbook of mathematical functions, National Bureau of Standards (1964).

8.1.7 Miscellaneous

Zeros of $D_\nu(z)$.

If ν is real, $D_\nu(z)$ has $[\nu + 1]$ real zeros where $[\nu + 1]$ is the largest positive integer less than $(\nu + 1)$. In particular $D_n(z)$ has exactly n zeros, all being real. Furthermore $D_n(z)$ has no other zero.

The functions $D_\nu(z)$ satisfy the following addition theorem:

$$D_\nu(ax + by) = \exp\left[\frac{1}{4}(bx - ay)^2\right]\left(\frac{a}{\sqrt{a^2 + b^2}}\right)^\nu$$

$$\times \sum_{n=0}^{\infty} \binom{\nu}{n} D_{\nu-n}\left(\sqrt{a^2 + b^2}\, x\right) D_n\left(\sqrt{a^2 + b^2}\, y\right)\left(\frac{b}{a}\right)^n,$$

$$Re\, \nu \geq 0;\ a, b, x, y \text{ real and positive};\ a > b.$$

For a discussion on the representation of arbitrary functions in terms of the series and the integrals involving $D_\nu(z)$, see ERDÉLYI (1953).

8.2 Parabolic functions

The parabolic functions are the Whittaker functions of a special type and are encountered as solutions of the wave equation by the method of separation of variables in parabolic coordinates. To a large extent the properties of these functions can be derived from these of the Whittaker functions by specializing the parameters and the variable. The wave equation in parabolic coordinates (see chap. XII) becomes

$$\frac{1}{\xi^2 + \eta^2}\left\{\frac{\partial^2 u}{\partial \xi^2} + \frac{1}{\xi}\frac{\partial u}{\partial \xi} + \frac{\partial^2 u}{\partial \eta^2} + \frac{1}{\eta}\frac{\partial u}{\partial \eta}\right\} + \frac{1}{\xi^2 \eta^2}\frac{\partial^2 u}{\partial \varphi^2} + k^2 u = 0.$$

If it is assumed that the wave equation in parabolic coordinates has a particular solution of the form

$$u(\xi, \eta, \varphi) = u_1(\xi)\, u_2(\eta)\, u_3(\varphi),$$

then

$$\frac{d^2 u_1}{d\xi^2} + \frac{1}{\xi}\frac{du_1}{d\xi} + (k^2\xi^2 - 4\mu_0^2\xi^{-2} + \lambda_0)\, u_1 = 0,$$

$$\frac{d^2 u_2}{d\eta^2} + \frac{1}{\eta}\frac{du_2}{d\eta} + (k^2\eta^2 - 4\mu_0^2\eta^{-2} - \lambda_0)\, u_2 = 0,$$

$$\frac{d^2 u_3}{d\varphi^2} + 4\mu_0^2 u_3 = 0, \quad u_3(\varphi) = e^{\pm i 2\mu_0 \varphi}.$$

For solutions which are single-valued and continuous on the surfaces

$$\xi = \xi_0 \text{ (const) and } \eta = \eta_0 \text{ (const)},\ 2\mu_0$$

must be an integer.

The differential equation for u_1 is equivalent to Kummer's differential equation (chap. VI) if k, λ_0, μ_0 are arbitrary complex parameters. However, if k, λ_0 are real and $2\mu_0$ is an integer, the differential equation for u_1 may be written as

$$\frac{d^2 u_1}{d\xi^2} + \frac{1}{\xi}\frac{du_1}{d\xi} + (\xi^2 - 4\mu^2\xi^{-2} - 2\lambda)\, u_1 = 0,$$

where ξ, λ are real and $2\mu = 0, 1, 2, 3, \ldots$ The solutions of this differential equation are

$$\xi^{-1} M_{\pm i\lambda,\mu}\left(\pm i \frac{\xi^2}{2}\right), \ \xi^{-1} W_{\pm i\lambda,\mu}\left(\pm i \frac{\xi^2}{2}\right),$$

and are connected to each other by the relations:

$$M_{-i\lambda,\mu}(-i\xi) = e^{-i\frac{\pi}{2}(2\mu+1)} M_{i\lambda,\mu}(i\xi),$$

$$M_{i\lambda,\mu}(i\xi) = \frac{\Gamma(2\mu+1)\, e^{\pi\lambda}}{\Gamma\left(\frac{1}{2}+\mu-i\lambda\right)} W_{-i\lambda,\mu}(-i\xi)$$

$$+ \frac{\Gamma(2\mu+1)\, e^{\pi\lambda + i\frac{\pi}{2}(2\mu+1)}}{\Gamma\left(\frac{1}{2}+\mu+i\lambda\right)} W_{i\lambda,\mu}(i\xi),$$

where ξ is real, positive and $\arg(\pm i\xi) = \pm\frac{\pi}{2}$.

For the solution of certain boundary-value problems, it is advantageous to use the functions

$$\left(\frac{\pi}{2\xi}\right)^{\frac{1}{2}} M_{i\lambda_n,\mu}(i\xi), \quad n = 1, 2, 3, \ldots,$$

which are orthogonal over the interval $(0, \xi_0)$. The sequence of real numbers λ_n, $n = 1, 2, 3, \ldots$ is determined such that

$$M_{i\lambda_n,\mu}(i\xi_0) = 0 \text{ and } \lambda_n < \lambda_{n+1}.$$

For more details see BUCHHOLZ (1943). There are a large number of integrals and series involving parabolic function but most of the results are too complicated to be given here. The following result is the expression for the spherical wave in terms of parabolic functions

$$\frac{2e^{i\frac{(\xi^2+\eta^2)}{2}}}{\xi^2+\eta^2} = \frac{-i}{\xi\eta}\int_{-i\infty}^{i\infty} W_{z,0}(-i\xi^2)\, W_{-z,0}(-i\eta^2)\frac{dz}{\cos(\pi z)}$$

$$= \frac{2}{\xi\eta}\sum_{n=0}^{\infty}(-1)^n\, W_{-n-\frac{1}{2},0}(-i\xi^2)\, W_{-n-\frac{1}{2},0}(-i\eta^2).$$

For more results see BUCHHOLZ (1943, 1948, 1949, 1953) and ERDÉLYI (1953).

Asymptotic expansions

For $\xi \to \infty$ and λ, μ fixed

$$W_{i\lambda,\mu}(i\xi) = \xi^{i\lambda} \exp\left[-i\frac{\xi}{2} - \pi\frac{\lambda}{2}\right][1 + O(\xi^{-1})],$$

$$W_{-i\lambda,\mu}(-i\xi) = \xi^{-i\lambda} \exp\left[i\frac{\xi}{2} - \pi\frac{\lambda}{2}\right][1 + O(\xi^{-1})].$$

For $\lambda \to \infty$ and μ, ξ fixed

$$W_{\pm i\lambda,\mu}(\pm i\xi) = \sqrt{2}\,(\lambda\xi)^{-\frac{1}{4}} \lambda^{\pm i\lambda} \times \exp\left[-\frac{\pi\lambda}{2} \mp i\frac{\pi}{4} \mp i\lambda\right] \cosh\left[\lambda - 2(\lambda\xi)^{\frac{1}{2}} \pm i\frac{\pi}{4}\right] \times \left[1 + O\left(\lambda^{-\frac{1}{2}}\right)\right],$$

$$W_{\pm i\lambda,\mu}(\mp i\xi) = \sqrt{2}\,(\lambda\xi)^{-\frac{1}{4}} \lambda^{\pm i\lambda} \times \exp\left[-\frac{\pi\lambda}{2} \mp i\lambda\right] \cos\left[\pm i\lambda - 2(\lambda\xi)^{\frac{1}{2}} - \frac{\pi}{4}\right] \times \left[1 + O\left(\lambda^{-\frac{1}{2}}\right)\right].$$

For $|\xi| \to \infty$, $|\lambda| \to \infty$ and $\frac{\xi}{\lambda}$, *a* fixed negative number

$$M_{-i\lambda,\mu}[4ik^2\lambda] = \Gamma(2\mu + 1) \exp\left[i\frac{\pi}{2}\left(\mu + \frac{1}{2}\right)\right] \times \lambda^{-\mu}\left(\frac{2}{\pi}\right)^{\frac{1}{2}} k^{\frac{1}{2}} (1 + k^2)^{-\frac{1}{4}} \times \sin\left[\lambda\left(2k\sqrt{1 + k^2} + 2\sinh^{-1} k\right) - \pi\left(\mu - \frac{1}{4}\right)\right] \times \left[1 + O\left(\lambda^{-\frac{1}{2}}\right)\right],$$

where $\lambda, k, \mu + \frac{1}{2}$ are real and positive.

Literature

BUCHHOLZ, H.: Z. angew. Math. Mech. **23** (1943) 47—58, 100—118.
— Z. phys. **24** (1948) 196—218.
— Ann. phys. **2** (1948) 185—210.
— Math. Z. **52** (1949) 355—383.
— Die konfluente hypergeometrische Funktion. Berlin/Göttingen/Heidelberg: Springer 1953.
ERDÉLYI, A.: Higher transcendental functions, Vol. 2. New York: McGraw-Hill 1953.
OLVER, F. W. J.: Jour. Res. NBS **63 B**, 2, 1959, 131—169.
SCHWID, N.: Trans. Amer. math. Soc. **37** (1935) 339—362.

Appendix to Chapter VIII

Laguerre functions

The functions

$$w(z) = L_\nu^{(\alpha)}(z) = \frac{\Gamma(\alpha+\nu+1)}{\Gamma(1+\alpha)\,\Gamma(1+\nu)} z^{-\frac{1}{2}(\alpha+1)} \exp\left(\frac{z}{2}\right) M_{\nu+\frac{(\alpha+1)}{2},\frac{\alpha}{2}}(z)$$

$$= \frac{\Gamma(\alpha+\nu+1)}{\Gamma(1+\alpha)\,\Gamma(1+\nu)} {}_1F_1(-\nu; 1+\alpha; z)$$

are called "generalized Laguerre functions" and satisfy the Laguerre differential equation

$$z\frac{d^2w}{dz^2} + (1+\alpha-z)\frac{dw}{dz} + \nu w = 0,$$

$$L_\nu^{(0)}(z) \equiv L_\nu(z).$$

The generalized Laguerre functions, in their totality, are not a special case but comprise all the confluent hypergeometric functions.

For $\nu = n$, $n = 0, 1, 2, 3, \ldots$, the functions $L_n(z)$ and $L_n^{(\alpha)}(z)$ are the Laguerre and generalized Laguerre polynomials respectively. For the functions $L_\nu(z)$ there exists the following integral representation

$$L_\nu(z) = \frac{e^z}{2\Gamma(\nu+1)} \int_0^\infty \left(\frac{t^2}{4}\right)^\nu \exp\left(-\frac{t^2}{4}\right) J_0\left(t\sqrt{z}\right) t\,dt,$$

$$Re\,\nu > -1.$$

For large values of n one may use the following series expansion to calculate the value of $L_n^{(\alpha)}(x)$

$$L_n^{(\alpha)}(x) = \frac{\Gamma(\alpha+n+1)}{n!\,n^\alpha} \exp(hx) \sum_{m=0}^\infty A_m(h) \left(\frac{x}{n}\right)^{\frac{m-\alpha}{2}} J_{\alpha+m}\left(2\sqrt{nx}\right),$$

where

$$\sum_{m=0}^\infty A_m(h)\, z^m = e^{nz} \frac{[1+(h-1)z]^n}{(1+zh)^{\alpha+n+1}}$$

$$(x, h \text{ real and positive}).$$

More explicitly

$$A_m(h) = h^{m-n} \sum_{r=0}^n \binom{n}{r} (h-1)^{n-r} L_m^{-(\alpha+m+r+1)}\left(-\frac{n}{h}\right),$$

$$A_m(1) = L_m^{-(\alpha+m+n+1)}(-n).$$

Chapter IX

The incomplete gamma function and special cases

9.1 The incomplete gamma functions

9.1.1 Definition and notation

The incomplete gamma functions $\gamma(a, x)$ and $\Gamma(a, x)$ are defined by

$$\gamma(a, x) = \int_0^x t^{a-1} e^{-t}\, dt, \quad Re\, a > 0,$$

$$\Gamma(a, x) = \int_x^\infty t^{a-1} e^{-t}\, dt = \Gamma(a) - \gamma(a, x).$$

Closely related to $\gamma(a, x)$ and $\Gamma(a, x)$ are the functions[1] $\gamma^*(a, x)$ and $E_n(x)$ given by

$$\gamma^*(a, x) = \frac{x^{-a}}{\Gamma(a)} \gamma(a, x) = \frac{x^{-a}}{\Gamma(a)} \int_0^x t^{a-1} e^{-t}\, dt.$$

$$E_n(x) = x^{n-1} \Gamma(1-n, x)$$

$$= \int_1^\infty t^{-n} e^{-xt}\, dt.$$

The function $P(a, x)$, treated in N. B. S. handbook is defined by

$$P(a, x) = \frac{1}{\Gamma(a)} \int_0^x t^{a-1} e^{-t}\, dt$$

$$= \frac{1}{\Gamma(a)} \gamma(a, x)$$

$$= x^a \gamma^*(a, x).$$

The functions $\gamma(a, x)$ and $\Gamma(a, x)$ are related to Kummer's functions ${}_1F_1(a; c; z)$ and $U(a, c, z)$ by the following:

$$\gamma(a, x) = a^{-1} x^a e^{-x}\, {}_1F_1(1; a+1; x)$$

$$= a^{-1} x^a\, {}_1F_1(a; 1+a; -x),$$

$$\Gamma(a, x) = x^a e^{-x} U(1, a+1, x)$$

$$= e^{-x} U(1-a, 1-a, +x).$$

---

[1] At no other place in the text does the symbol $E_n(x)$ mean the function defined here. This symbol is reserved for another function discussed in chap. 1. The function $E_n(x)$ as defined here is discussed in the next section (9.2) and is denoted there by $\mathcal{E}_n(x)$.

The functions $\gamma^*(a, x)$ and $E_n(x)$ are given in terms of the Kummer functions by

$$\gamma^*(a, x) = \frac{e^{-x}}{\Gamma(a+1)}\, {}_1F_1(1; a+1; x)$$
$$= \frac{1}{\Gamma(a+1)}\, {}_1F_1(a; a+1; -x),$$
$$E_n(x) = e^{-x} U(1, 2-n, x)$$
$$= x^{n-1} e^{-x} U(n, n, x).$$

For $Re\, a \leq 0$, it is clear that the definition of $\gamma(a, x)$ cannot be used. In this case, and as a matter of fact in general, it is advantageous to define $\gamma(a, x)$ in terms of the uniquely defined functions ${}_1F_1(1; a+1; x)$ and x^a. The functions $\gamma(a, x)$ is undefined for $a = 0, -1, -2, \ldots$, whereas the function $\gamma^*(a, x)$ is an entire function of both x and a. Except when a is an integer, the function $\Gamma(a, x)$ is *a* multivalued function of x with a branch point at the origin; however, it is an entire function of a.

9.1.2 Elementary properties

Other representations of $\gamma(a, x)$ and $\Gamma(a, x)$ are

$$\gamma(a, x) = e^{-x} \sum_{n=0}^{\infty} \frac{x^{a+n}}{(a)_{n+1}}$$
$$= \sum_{n=0}^{\infty} \frac{(-1)^n}{n!} \frac{x^{a+n}}{a+n}$$
$$= \Gamma(a) - e^{-\frac{x}{2}} x^{\frac{1}{2}(a-1)} W_{\frac{1}{2}a-\frac{1}{2}, \frac{a}{2}}(x)$$
$$= \Gamma(a)\, e^{-x} x^{\frac{a}{2}} \sum_{n=0}^{\infty} e_n(-1)\, x^{\frac{n}{2}} I_{n+a}\left(2\sqrt{x}\right),$$

where

$$e_n(x) = \sum_{m=0}^{n} \frac{x^m}{m!},$$
$$\Gamma(a, x) = e^{-\frac{x}{2}} x^{\frac{a-1}{2}} W_{\frac{a-1}{2}, \frac{a}{2}}(x)$$
$$= e^{-x} x^a \sum_{n=0}^{\infty} \frac{L_n^{(a)}(x)}{n+1}$$
$$= \Gamma(a) - \sum_{n=0}^{\infty} \frac{(-1)^n}{n!} \frac{x^{a+n}}{a+n},$$
$$a \neq 0, -1, -2, \ldots$$

Special values

$$\gamma(1, x) = 1 - e^{-x},$$

$$\gamma(n+1, x) = n!\left[1 - e^{-x}\sum_{m=0}^{n}\frac{x^m}{m!}\right] = n!\,[1 - e^{-x}e_n(x)],$$

$$\gamma\left(\frac{1}{2}, x^2\right) = \sqrt{\pi}\,\mathrm{Erf}(x),$$

$$\Gamma(0, x) = -Ei(-x),$$

$$\Gamma(1, x) = e^{-x}, \quad \Gamma\left(\frac{1}{2}, x^2\right) = \sqrt{\pi}\,\mathrm{Erf}(x),$$

$$\Gamma(n+1, x) = n!\,e^{-x}\sum_{m=0}^{n}\frac{x^m}{m!} = n!\,e^{-x}e_n(x),$$

$$\Gamma(a, ix) = \exp\left(i\frac{\pi}{2}a\right)[C(x, a) - iS(x, a)],$$

$$\gamma^*(-n, x) = x^n,$$

$$\gamma^*\left(\frac{1}{2}, -x^2\right) = \frac{2}{\sqrt{\pi}}\,x^{-1}\int_0^x e^{t^2}\,dt.$$

Differentiation formulas

$$\frac{\partial}{\partial x}\gamma(a, x) = x^{a-1}e^{-x},$$

$$\frac{\partial^n}{\partial x^n}[x^{-a}\gamma(a, x)] = (-1)^n x^{-n-a}\gamma(a+n, x),$$

$$\frac{\partial^n}{\partial x^n}[e^x\gamma(a, x)] = (-1)^n(1-a)_n e^x\gamma(a-n, x),$$

$$\frac{\partial^n}{\partial x^n}[e^x x^{n-a}\gamma(a, x)] = \frac{n!}{a}\,{}_1F_1(n+1; a+1; x),$$

$$n = 0, 1, 2, \ldots,$$

$$\frac{\partial}{\partial x}\Gamma(a, x) = -x^{a-1}e^{-x},$$

$$\frac{\partial^n}{\partial x^n}[x^{-a}\Gamma(a, x)] = (-1)^n x^{-a-n}\Gamma(a+n, x),$$

$$\frac{\partial^n}{\partial x^n}[e^x\Gamma(a, x)] = (-1)^n(1-a)_n e^x\Gamma(a-n, x),$$

$$\frac{\partial^n}{\partial x^n}[e^x x^{n-a}\Gamma(a, x)] = n!\,(1-a)_n\,U(n+1, a+1, x),$$

$$\frac{\partial}{\partial a}\gamma^*(a, x)\Big|_{a=0} = -E_1(x) - \log x = -\int_x^\infty \frac{e^{-t}}{t}\,dt - \log x,$$

$$\frac{\partial^n}{\partial x^n}[x^a e^x\gamma^*(a, x)] = x^{a-n}e^x\gamma^*(a-n, x).$$

Recurrence relations

$$\gamma(a+1, x) = a\gamma(a, x) - x^a e^{-x},$$

$$\Gamma(a+1, x) = a\Gamma(a, x) + x^a e^x,$$

$$x\gamma^*(a+1, x) = \gamma^*(a, x) - \frac{e^{-x}}{\Gamma(a+1)}.$$

9.1.3 Integral representations

$$\gamma(a, x) = x^a \operatorname{cosec}(\pi a) \int_0^\pi e^{x\cos t} \cos(at + x \sin t)\, dt,$$

$$x \neq 0, \quad a \text{ not an integer}$$

$$= x^{\frac{a}{2}} \int_0^\infty e^{-t} t^{\frac{a}{2}-1} J_a\left[2\sqrt{xt}\right] dt,$$

$$Re\, a > 0,$$

$$= \frac{-x^a}{2i \sin(\pi a)} \int_1^{(0+)} e^{-xt} (-t)^{a-1}\, dt,$$

$$-\pi \leq \arg(-t) \leq \pi, \; x \neq 0, \; a \text{ not an integer},$$

$$\Gamma(a, x) = \frac{e^{-x} x^a}{\Gamma(1-a)} \int_0^\infty \frac{t^{-a}}{x+t} e^{-t}\, dt,$$

$$Re\, a < 1; \; x \text{ not a negative real},$$

$$\Gamma(a, x) = \frac{2x^{\frac{a}{2}}}{\Gamma(1-a)} e^{-x} \int_0^\infty e^{-t} t^{-\frac{a}{2}} K_a\left[2\sqrt{xt}\right] dt,$$

$$Re\, a < 1.$$

9.1.4 Some integrals and series connected with the incomplete gamma functions

$$\int_0^\infty t^{a-1} \Gamma(c, t)\, dt = \frac{\Gamma(a+c)}{a},$$

$$Re\, a > 0, \; Re\,(a+c) > 0,$$

$$\int_0^\infty e^{-at} \Gamma(b, ct)\, dt = \frac{\Gamma(b)}{a}\left[1 + \left(\frac{c}{a+c}\right)^b\right],$$

$$Re\, b > -1, \; Re\,(a+c) > 0,$$

$$\int_0^1 e^{-xt} t^{a-\beta-1} \gamma(\beta, x - xt)\, dt = \frac{\Gamma(\beta)}{\Gamma(a)} \Gamma(a-\beta)\, x^{\beta-a} \gamma(a, x),$$

$$Re\, a > Re\, \beta > -1, \; a\beta \neq 0,$$

$$\int_0^\infty e^{-zt} t^{\alpha-1} \gamma(a, t)\, dt = \frac{\Gamma(a+\alpha)}{a(1+z)^{a+\alpha}} \, {}_2F_1\left(1, a+\alpha; a+1; \frac{1}{z+1}\right),$$
$$Re\, z > 0,\ Re\,(a+\alpha) > 0,$$

$$\int_0^\infty e^{-zt} t^{\beta-1} \Gamma(\alpha, t)\, dt = \frac{\Gamma(\alpha+\beta)}{\beta(1+z)^{\alpha+\beta}} \, {}_2F_1\left(1, \alpha+\beta; 1+\beta; \frac{z}{1+z}\right),$$
$$Re\, z > -\frac{1}{2},\quad Re\,\beta > 0,\quad Re\,(\alpha+\beta) > 0,$$

$$\gamma(a, x+y) - \gamma(a, x) = \Gamma(a, x) - \Gamma(a, x+y)$$
$$= e^{-x} x^{a-1} \sum_{n=0}^{\infty} \frac{(-1)^n (1-a)_n}{x^n} \left[1 - e^{-y} e_n(y)\right],$$
$$|y| < |x|,$$

where

$$e_n(y) = \sum_{m=0}^{n} \frac{y^m}{m!},$$

$$\gamma^*(a, x) = e^{-x} \sum_{n=0}^{\infty} \frac{x^n}{\Gamma(a+n+1)}$$
$$= \frac{1}{\Gamma(a)} \sum_{n=0}^{\infty} \frac{(-x)^n}{(a+n)\, n!},$$
$$|x| < \infty.$$

For non-negative real values of x and y, the following result holds

$$\sum_{n=0}^{\infty} \frac{n!\, L_n^{(a)}(x)\, L_n^{(a)}(y)}{(n+1)\,\Gamma(n+a+1)} = \begin{cases} \dfrac{e^{x+y}}{(xy)^a} \dfrac{\gamma(a, x)\Gamma(a, x)}{\Gamma(a)}, & x \geq y > 0,\ a \text{ arbitrary}, \\ \dfrac{e^x}{x^a} \dfrac{\Gamma(a, x)}{\Gamma(a+1)}, & x > 0,\ y = 0,\ Re\, a < \dfrac{1}{2}, \\ \dfrac{-1}{a\,\Gamma(a+1)}, & x = y = 0,\ Re\, a < 0. \end{cases}$$

9.1.5 Asymptotic expansions

For $x \to \infty$, $|\arg x| < \frac{3\pi}{2}$

$$\Gamma(a, x) = e^{-x} U(1-a, 1-a, x)$$
$$= e^{-x} x^{a-1} \left[\sum_{n=0}^{N-1} (1-a)_n (-x)^{-n} + O\left(|x|^{-N}\right)\right].$$

The corresponding result for $\gamma(a, x)$ can obtained from the relation

$$\gamma(a, x) = \Gamma(a) - \Gamma(a, x)$$

and the above result.

For $a \to \infty$, $|\arg a| < \frac{\pi}{2}$

$$\Gamma(a+1, a) \sim e^{-a} a^a \left(\sqrt{\frac{\pi}{2}}\, a + \frac{2}{3} + \frac{\sqrt{2\pi}}{24}\, a^{-\frac{1}{2}} + \cdots \right).$$

9.2 Special cases

9.2.1 The exponential and logarithmic integrals

The exponential integrals $E_1(x), Ei(x), E^*(x)$ and the logarithmic integral $li(x)$ are defined by

$$E_1(x) = -Ei(-x) = \int_x^\infty t^{-1} e^{-t}\, dt, \qquad |\arg x| < \pi,$$

where it is assumed that the path of integration excludes the origin and does not cross the negative real axis. The functions one single valued in the x-plane with a branch cut along the negative real axis. Analytic continuation gives multi-valued functions.

$$E^*(x) = -\,⨍_{-x}^{\infty} t^{-1} e^{-t}\, dt = ⨍_{-\infty}^{x} t^{-1} e^{-t}\, dt, \quad x > 0,$$

where the symbol $⨍$ is used to mean the Cauchy principal value of the integral, e.g.

$$E^*(x) = -\lim_{\varepsilon \to 0} \left[\int_{-x}^{-\varepsilon} t^{-1} e^{-t}\, dt + \int_{\varepsilon}^{\infty} t^{-1} e^{-t}\, dt \right],$$

$$li(x) = ⨍_0^x \frac{dt}{\log t} = E^*(\log x).$$

The function $li(x)$ has branch points at $x = 0, 1, \infty$.

Other functions considered here are

$$\mathcal{E}_n(x) = \int_1^\infty t^{-n} e^{-xt}\, dt,$$

$$Re\, x > 0, \quad n = 0, 1, 2, \ldots,$$

$$\alpha_n(x) = \int_1^\infty t^n e^{-xt}\, dt,$$

$$Re\, x > 0, \; n = 0, 1, 2, \ldots,$$

$$\beta_n(x) = \int_{-1}^{+1} t^n e^{-xt}\, dt,$$

$$n = 0, 1, 2, \ldots$$

For $n > 0$, the function $\mathcal{E}_n(x)$ is multi-valued with branch points at $x = 0$ and $x = \infty$.

The function $\mathcal{E}_n(x)$ can be considered as a generalization of $E_1(x)$ since

$$\mathcal{E}_1(x) = E_1(x).$$

The function $E_1(x)$ and $E^*(x)$ are related to each other by the following:

$$-E_1(-x \pm i0) = E^*(x) \pm i\pi,$$

$$E^*(x) = -\frac{1}{2}\left[E_1(-x + i0) + E_1(-x - i0)\right].$$

Other representations are

$$E_1(x) = -\gamma - \log x - \sum_{n=1}^{\infty} \frac{(-x)^n}{n\, n!}, \qquad |\arg x| < \pi$$

$$= \Gamma(0, x) = x^{-\frac{1}{2}} e^{-\frac{x}{2}} W_{-\frac{1}{2},0}(x)$$

$$= e^{-x} U(1, 1, x)$$

$$= e^{-x} \sum_{n=0}^{\infty} \frac{L_n(x)}{n+1}, \quad x > 0,$$

$$E^*(x) = \gamma + \log x + \sum_{n=1}^{\infty} \frac{x^n}{n\, n!}, \quad x > 0,$$

$$\mathcal{E}_n(x) = \frac{(-x)^{n-1}}{(n-1)!}\left[\psi(n) - \log x\right] - \sum_{m=0}^{\infty}{}' \frac{(-x)^m}{(m-n+1)\, m!},$$

where the prime indicates the omission of the term $m = n - 1$ and $\psi(x)$ is the function discussed in chap. I.

$$\alpha_n(x) = n!\, e^{-x} x^{-n-1} e_n(x) = n!\, e^{-x} x^{-n-1} \sum_{m=0}^{n} \frac{x^m}{m!},$$

$$\beta_n(x) = n!\, x^{-n-1} \left\{ e^x \sum_{m=0}^{n} \frac{(-x)^m}{m!} - e^{-x} \sum_{m=0}^{n} \frac{x^m}{m!} \right\},$$

$$li(x) = -(-\log x)^{-\frac{1}{2}} x^{\frac{1}{2}} W_{-\frac{1}{2},0}(-\log x),$$

$$\alpha_0(x) = \sqrt{\frac{2}{\pi x}}\, K_{\frac{1}{2}}(x),$$

$$\alpha_1(x) = \sqrt{\frac{2}{\pi x}}\, K_{\frac{3}{2}}(x),$$

$$\beta_0(x) = \sqrt{\frac{2\pi}{x}}\, I_{\frac{1}{2}}(x),$$

$$\beta_1(x) = -\sqrt{\frac{2\pi}{x}}\, I_{\frac{3}{2}}(x),$$

$$\mathcal{E}_n(x) = x^{n-1}\Gamma(1-n, x),$$

$$\alpha_n(x) = x^{-n-1}\Gamma(n+1, x),$$

$$\beta_n(x) = x^{-n-1}\{\Gamma(n+1, -x) - \Gamma(n+1, x)\}.$$

Special values

$$\mathcal{E}_n(0) = \frac{1}{n-1}, \quad n > 1,$$

$$\mathcal{E}_0(x) = x^{-1}e^{-x} = \alpha_0(x),$$

$$\beta_0(x) = \frac{2}{x}\sinh x.$$

Differentiation formulas

$$\frac{d}{dx}\mathcal{E}_n(x) = -\mathcal{E}_{n-1}(x), \quad n = 1, 2, 3, \ldots,$$

$$\frac{d^n}{dx^n}E_1(x) = -\frac{d^n}{dx^n}[Ei(-x)] = (-1)^n e^{-x}U(1, n+1, x),$$

$$\frac{d^n}{dx^n}[e^x E_1(x)] = -\frac{d^n}{dx^n}[e^x\, Ei(-x)]$$

$$= -e^x Ei(-x) - \sum_{m=0}^{n-1}\frac{(-1)^m m!}{x^{m+1}},$$

$$n = 1, 2, 3, \ldots$$

Recurrence relations

$$n\,\mathcal{E}_{n+1}(x) = e^{-x} - x\,\mathcal{E}_n(x), \quad n = 1, 2, 3, \ldots,$$

$$x\alpha_{n+1}(x) = e^{-x} + (n+1)\,\alpha_n(x), \quad n = 0, 1, 2, 3, \ldots,$$

$$x\beta_{n+1}(x) = (n+1)\,\beta_n(x) - (-1)^n e^x - e^{-x}, \quad n = 0, 1, 2, 3, \ldots$$

Some definite and indefinite integrals

$$\int_0^\infty \frac{e^{-at} - e^{-bt}}{t}\, dt = \log\frac{b}{a},$$

$$\int_0^\infty [E_1(x)]^2\, dx = 2\log 2,$$

$$\int_0^\infty \frac{e^{-at}}{t+b}\, dt = e^{ab}E_1(ab), \quad Re\, a > 0,$$

$$= -\exp(ab)\, Ei(-ab), \quad b \text{ not negative real},$$

$$\int_0^\infty \frac{t - ia}{t^2 + a^2} e^{ibt}\, dt = e^{ab} E_1(ab), \quad a > 0, b > 0,$$

$$= - \exp(ab)\, Ei(-ab)$$

$$\int_0^\infty \frac{t + ia}{t^2 + a^2} e^{ibt}\, dt = e^{-ab} E_1(-ab - i0),$$

$$= e^{-ab}[+i\pi - E^*(ab)], \quad a > 0, b > 0,$$

$$\int_0^\infty \frac{e^{iat}}{b + t}\, dt = e^{-iab} E_1(-iab), \quad a > 0, b > 0,$$

$$= - \exp(-iab)\, Ei(iab)$$

$$\int_1^\infty e^{-xt} \log t\, dt = \frac{1}{x} E_1(x)$$

$$= - \frac{1}{x} Ei(-x),$$

$$- \int_0^x \frac{1 - e^{+t}}{t}\, dt = E^*(x) - \log x - \gamma, \quad x > 0,$$

$$\int_0^x \frac{1 - e^{-t}}{t}\, dt = -Ei(-x) + \log x + \gamma,$$

$$\int_a^\infty (b + t)^{-1} e^{-ct}\, dt = - e^{bc}\, Ei[-c(a + b)], \quad Re\, c > 0,$$

$$\int_0^\infty \frac{\sin(xt)}{t^2 + y^2}\, dt = \frac{1}{2y}[-e^{xy} Ei(-xy) + e^{-xy} E^*(xy)],$$

$$x > 0, \ y > 0,$$

$$\int_0^\infty \frac{t \cos(xt)}{t^2 + y^2}\, dt = \frac{1}{2}[-e^{xy} Ei(-xy) - e^{-xy} E^*(xy)],$$

$$x > 0, y > 0,$$

$$\int_0^1 \frac{e^{xt}}{t} \sin(yt)\, dt = \pi - \arctan\frac{y}{x} - \operatorname{Im} Ei(x - iy),$$

$$x > 0, \ y > 0,$$

$$\int_0^1 \frac{e^{-xt}}{t} \sin(yt)\, dt = \arctan\frac{y}{x} - \operatorname{Im} Ei(-x - iy),$$

$$x > 0, \ y \text{ real},$$

$$\int_0^1 \frac{e^{-xt}}{t}(1 - \cos yt)\, dt = \log\left(1 + \frac{y^2}{x^2}\right)^{\frac{1}{2}} + Ei(-x) - Re\, Ei(-x - iy),$$

$$x > 0, \ y \text{ real},$$

$$\int_0^1 (1 - \cos xt) \frac{e^{yt}}{t} dt = \log\left(1 + \frac{x^2}{y^2}\right)^{\frac{1}{2}} + E^*(y) - Re\, Ei(y - ix),$$

$$y > 0, \quad x \text{ real},$$

$$\int_x^\infty t^{a-1} Ei(-t)\, dt = -\frac{1}{a} [\Gamma(a, x) + x^a\, Ei(-x)],$$

$$Re\, x > 0, \quad a \neq 0,$$

$$\int_0^\infty e^{-xt} t^{a-1} Ei(-t)\, dt = -\frac{\Gamma(a)}{a(1+x)^a}\, {}_2F_1\left(1, a; 1 + a; \frac{x}{1+x}\right),$$

$$Re\, a > 0, \quad Re\, x > -\frac{1}{2},$$

$$\int_0^\infty e^{-xt}\, \mathcal{E}_n(t)\, dt = \frac{(-1)^{n-1}}{x^n} \left\{\log(1 + x) + \sum_{m=1}^{n-1} \frac{(-x)^m}{m}\right\},$$

$$x > -1,$$

$$\int \frac{x e^x}{x^2 + y^2} dx = Re\, [e^{iy}\, Ei(x - iy)],$$

$$y > 0,$$

$$\int \frac{e^x}{x^2 + y^2} dx = \frac{1}{y} \operatorname{Im} [e^{iy}\, Ei(x - iy)], \quad y > 0,$$

$$\int \frac{e^{ix}}{x^2 + y^2} dx = \frac{i}{2y} [e^y Ei(-y + ix) - e^{-y} Ei(y + ix)],$$

$$\int \frac{x e^{ix}}{x^2 + y^2} dx = \frac{1}{2} [e^{-y}\, Ei(y + ix) + e^y\, Ei(-y + ix)].$$

In each one of the four indefinite integrals a constant of integration may be added on the right hand side.

Addition theorem

$$-Ei(-x - y) = -Ei(-x) + e^{-x} \sum_{n=0}^\infty \frac{n!\,[1 - e^{-y}\, e_n(y)]}{(-x)^{n+1}},$$

$$|y| < |x|.$$

Asymptotic expansions

For $|x| \to \infty, \ -\frac{3\pi}{2} < \arg x < \frac{3\pi}{2}$

$$E_1(x) = -Ei(-x) = x^{-1} e^{-x} \left\{\sum_{n=0}^N \frac{n!}{(-x)^n} + O(|x|^{-N-1})\right\},$$

$$\mathcal{E}_n(x) = x^{-1} e^{-x} \left\{1 - \frac{n}{x} + O(|x|^{-2})\right\},$$

where $|x| \gg n$.

For $x > 0$, $x \to \infty$

$$E^*(x) = x^{-1} e^{-x} \left[\sum_{n=0}^{N} \frac{n!}{x^n} + O(x^{-N-1}) \right].$$

9.2.2 Sine and cosine integrals

The functions Si, $si(x)$ and $Ci(x)$ are defined by

$$Si(x) = \int_0^x \frac{\sin t}{t} dt,$$

$$si(x) = \int_\infty^x \frac{\sin t}{t} dt = -\frac{\pi}{2} + Si(x),$$

$$Ci(x) = -\int_x^\infty \frac{\cos t}{t} dt$$

$$= \gamma + \log x - \int_0^x \frac{1 - \cos t}{t} dt,$$

$$|\arg x| < \pi.$$

The functions $Si(x)$ and $si(x)$ are entire functions of x; $Ci(x)$ is a multi-valued function of x with a logarithmic singularity at the origin. The corresponding functions when the trigonometric sine and cosine are replaced by $\sinh t$ and $\cosh t$ are given by

$$Shi(x) = \int_0^x \frac{\sinh t}{t} dt,$$

$$Chi(x) = -\int_0^x \frac{1 - \cosh t}{t} dt + \gamma + \log x,$$

$$|\arg x| < \pi.$$

The function $Shi(x)$ is an entire function of x, but the function $Chi(x)$ has a branch point at $x = 0$.

In the complex x-plane, the following relations give the symmetry properties

$$Si(-x) = -Si(x),$$

$$Si(\bar{x}) = \overline{Si(x)},$$

$$Ci(-x) = -i\pi + Ci(x), \quad |\arg x| < \pi,$$

$$Ci(\bar{x}) = \overline{Ci(x)}.$$

In terms of the exponential integral, the functions $Si(x)$ and $Ci(x)$ are given by

$$Si(x) = \frac{1}{2i}\{Ei(ix) - Ei(-ix)\} + \frac{\pi}{2},$$
$$|\arg x| < \frac{\pi}{2},$$
$$Ci(x) = +\frac{1}{2}[Ei(ix) + Ei(-ix)],$$
$$|\arg x| < \frac{\pi}{2}.$$

Other representations for these functions are

$$Si(x) = \sum_{n=0}^{\infty} \frac{(-1)^n x^{2n+1}}{(2n+1)(2n+1)!}$$
$$= \pi \sum_{n=0}^{\infty} \left[J_{n+\frac{1}{2}}\left(\frac{x}{2}\right)\right]^2,$$
$$Ci(x) = \gamma + \log x + \sum_{n=1}^{\infty} \frac{(-1)^n x^{2n}}{2n\,\Gamma(2n+1)},$$
$$Shi(x) = \sum_{n=0}^{\infty} \frac{x^{2n+1}}{(2n+1)\,\Gamma(2n+2)},$$
$$Chi(x) = \gamma + \log x + \sum_{n=1}^{\infty} \frac{x^{2n}}{2n\,\Gamma(2n+1)}.$$

Integral representations

$$si(x) = -\frac{\pi}{2} + Si(x)$$
$$= -\int_0^{\frac{\pi}{2}} e^{-x\cos t} \cos(x \sin t)\, dt,$$
$$Ci(x) = Ei(-x) + \int_0^{\frac{\pi}{2}} e^{-x\cos t} \sin(x \sin t)\, dt.$$

Some integrals connected with these functions are

$$\int_0^{\infty} [si(x)]^2\, dx = \int_0^{\infty} [Ci(x)]^2\, dx = \frac{\pi}{4},$$
$$\int_0^{\infty} Ci(x)\, si(x)\, dx = \log\left(\frac{1}{2}\right),$$
$$\int_0^{\infty} \sin x\, si(x)\, dx = \int_0^{\infty} \cos x\, Ci(x)\, dx = -\frac{\pi}{4},$$

$$\int_0^\infty e^{-xt}\, si(t)\, dt = -x^{-1} \arctan x, \quad Re\, x > 0,$$

$$\int_0^\infty e^{-xt}\, Ci(t)\, dt = \frac{1}{2x} \log (1 + x^2)^{-1}, \quad Re\, x > 0,$$

$$\int_0^\infty t^{-1} e^{-xt} \log (1 + t^2)\, dt = [Ci(x)]^2 + [si(x)]^2, \quad Re\, x > 0,$$

$$\int_0^1 \frac{(1 - e^{-xt})}{t} \cos (yt)\, dt = \log \left(1 + \frac{x^2}{y^2}\right)^{\frac{1}{2}} + Ci(y) - Re\, Ei(-x - iy),$$

$$y > 0, \; x \text{ real}.$$

Asymptotic expansions

For $x \to \infty$

$$Ci(x) + i\, si(x) \sim e^{ix} \sum_{n=0}^\infty \Gamma(n+1)\, (ix)^{-n-1}.$$

For $|x| \to \infty$, $|\arg x| < \pi$

$$\sin x\, Ci(x) - \cos x\, si(x) \sim x^{-1} \sum_{n=0}^\infty (-1)^n \frac{(2n)!}{x^{2n}},$$

$$-\cos x\, Ci(x) - \sin x\, si(x) \sim x^{-2} \sum_{n=0}^\infty (-1)^n \frac{(2n+1)!}{x^{2n}}.$$

9.2.3 The error functions

The functions Erf(x) and Erfc(x) are defined by[1]

$$\mathrm{Erf}(x) = \frac{2}{\sqrt{\pi}} \int_0^x e^{-t^2}\, dt,$$

$$\mathrm{Erfc}(x) = \frac{2}{\sqrt{\pi}} \int_x^\infty e^{-t^2}\, dt$$

$$= 1 - \mathrm{Erf}(x),$$

where the contour of integration, in the definition of Erfc(x), is subject to the restriction $|\arg t| < \frac{\pi}{4}$ as $t \to \infty$. The error functions are entire functions of x. In the complex x-plane

$$\mathrm{Erf}(-x) = -\mathrm{Erf}(x),$$

$$\mathrm{Erf}(\bar{x}) = \overline{\mathrm{Erf}(x)}.$$

[1] Also written as erf(x) and erf c(x).

Some other representations of the error functions are

$$\operatorname{Erf}(x) = \frac{2}{\sqrt{\pi}} \sum_{n=0}^{\infty} \frac{(-1)^n x^{2n+1}}{(2n+1)\,\Gamma(n+1)}$$

$$= \frac{2}{\sqrt{\pi}} e^{-x^2} \sum_{n=0}^{\infty} \frac{x^{2n+1}}{\left(\frac{3}{2}\right)_n}$$

$$= \sqrt{2} \sum_{n=0}^{\infty} (-1)^n \left[I_{2n+\frac{1}{2}}(x^2) - I_{2n+\frac{3}{2}}(x^2) \right]$$

$$= e^{-x^2} x^{\frac{1}{2}} \sum_{n=0}^{\infty} e_n(-1)\, x^n I_{n+\frac{1}{2}}(2x)$$

$$= \frac{2x}{\sqrt{\pi}}\, {}_1F_1\left(\frac{1}{2}; \frac{3}{2}; -x\right)$$

$$= \frac{2x}{\sqrt{\pi}} e^{-x^2}\, {}_1F_1\left(1; \frac{3}{2}; x\right)$$

$$= \frac{1}{\sqrt{\pi}} \gamma\left(\frac{1}{2}, x^2\right),$$

$$\operatorname{Erfc}(x) = \frac{1}{\sqrt{\pi}} \Gamma\left(\frac{1}{2}, x^2\right)$$

$$= \frac{1}{\sqrt{\pi}} e^{-x^2} U\left(\frac{1}{2}, \frac{1}{2}, x^2\right).$$

Differentiation formula

$$\frac{d^{n+1}}{dx^{n+1}} \operatorname{Erf}(x) = (-1)^n \frac{2}{\sqrt{\pi}} e^{-x^2} H_n(x),$$

where $H_n(x)$ are the Hermite polynomials (see chap. V).

Some integrals associated with the error functions

$$\int_0^1 e^{-x^2t^2} \frac{dt}{1+t^2} = \frac{\pi}{2} e^{x^2} \left[1 - \{\operatorname{Erf}(x)\}^2\right], \qquad Re\, x > 0,$$

$$\int_0^\infty e^{-x^2t^2 - yt}\, dt = \frac{\sqrt{\pi}}{2x} \exp\left(\frac{y^2}{4x^2}\right) \operatorname{Erfc}\left(\frac{y}{2x}\right), \qquad Re\, x > 0,$$

$$\int_0^x \operatorname{Erf}(t)\, dt = x \operatorname{Erf}(x) - \frac{1}{\sqrt{\pi}} (1 - e^{-x^2}),$$

$$\int_0^\infty \operatorname{Erf}(xt)\, e^{-yt}\, dt = \frac{1}{y} \exp\left(\frac{y^2}{4x^2}\right) \operatorname{Erf}\left(\frac{y}{2x}\right), \qquad Re\, y > 0,\ |\arg x| < \frac{\pi}{4},$$

$$\int_0^\infty \frac{e^{-xt}}{\sqrt{y+t}}\,dt = \pi^{\frac{1}{2}} x^{-\frac{1}{2}} e^{xy} \operatorname{Erfc}\left(\sqrt{xy}\right),$$

$$Re\, x > 0,\ |\arg y| < \pi,$$

$$\int_0^\infty \frac{e^{-xt^2}}{t^2+y^2}\,dt = \frac{\pi}{2y} e^{xy^2} \operatorname{Erfc}\left(\sqrt{x}\, y\right),$$

$$x > 0,\ y > 0,$$

$$\int_0^\infty \frac{e^{-xt}}{t+y} t^{-\frac{1}{2}}\,dt = \frac{\pi}{\sqrt{y}} e^{xy} \operatorname{Erfc}(xy),$$

$$Re\, x > 0,\ y \neq 0,\ |\arg y| < \pi,$$

$$\int_0^\infty e^{-xt} \operatorname{Erf}\left(\sqrt{yt}\right) dt = x^{-1} y^{\frac{1}{2}} (x+y)^{-\frac{1}{2}},$$

$$Re\,(x+y) > 0,$$

$$\int_0^\infty e^{-xt} \operatorname{Erf}\left(y^{\frac{1}{2}} t^{-\frac{1}{2}}\right) dt = \frac{1}{x} e^{-2\sqrt{xy}},$$

$$Re\, x > 0,\ Re\, y > 0,$$

$$\int_0^\infty e^{(x-a)t} \operatorname{Erfc}\left(\sqrt{xt} + \sqrt{\frac{y}{t}}\right) dt$$

$$= a^{-\frac{1}{2}} \left(\sqrt{x} + \sqrt{a}\right) \exp\left[-2\left(\sqrt{xy} + \sqrt{ay}\right)\right],$$

$$Re\, a > 0,\ Re\, y > 0,$$

$$\int_0^\infty \sin(xt) \operatorname{Erfc}(yt)\,dt = \frac{1}{x}\left[1 - \exp\left(-\frac{x^2}{4y^2}\right)\right],$$

$$x > 0,\ Re\, y > 0,$$

$$\int \operatorname{Erf}(t)\,dt = t \operatorname{Erf}(t) + \pi^{-\frac{1}{2}} e^{-t^2},$$

$$\int e^{xt} \operatorname{Erf}(yt)\,dt = \frac{1}{x}\left[e^{xt} \operatorname{Erf}(yt) - e^{\frac{x^2}{4y^2}} \operatorname{Erf}\left(yt - \frac{x}{2y}\right)\right],$$

$$x \neq 0.$$

Repeated integrals of the error function

For $n = 1, 2, 3, \ldots$, define the functions $\{1\}^n \operatorname{Erfc}(x)$ by the relations

$$\{1\}^0 \operatorname{Erfc}(x) = \operatorname{Erfc}(x),$$

$$\{1\} \operatorname{Erfc}(x) = \int_x^\infty \operatorname{Erfc}(t)\,dt,$$

$$\{1\}^n \operatorname{Erfc}(x) = \int_x^\infty \left[\{1\}^{n-1} \operatorname{Erfc}(t)\right] dt.$$

The function $\{1\}^n \operatorname{Erfc}(x)$ can be written as a single integral as

$$\{1\}^n \operatorname{Erfc}(x) = \frac{2}{\sqrt{\pi}} \int_x^\infty \frac{(t-x)^n}{n!} e^{-t^2} \, dt.$$

Other representations of $\{1\}^n \operatorname{Erfc}(x)$ are

$$\{1\}^n \operatorname{Erfc}(x) = \sum_{m=0}^\infty \frac{(-x)^m}{2^{n-m} \, m! \, \Gamma\left(1 + \frac{n-m}{2}\right)},$$

$$\{1\}^n \operatorname{Erfc}(x) = \frac{e^{-\frac{x^2}{2}}}{\sqrt{\pi}} \, 2^{-\frac{n}{2}+\frac{1}{2}} D_{-n-1}\left(x\sqrt{2}\right),$$

where $D_\nu(z)$ are the parabolic cylinder functions (see chap. VIII)

$$\{1\}^n \operatorname{Erfc}(x) = e^{-x^2} \left\{ \frac{2^{-n}}{\Gamma\left(\frac{n}{2}+1\right)} \, {}_1F_1\left(\frac{n+1}{2}; \frac{1}{2}; x^2\right) - \frac{x \, 2^{-n+1}}{\Gamma\left(\frac{n+1}{2}\right)} \, {}_1F_1\left(\frac{n}{2}+1; \frac{3}{2}; x^2\right) \right\}.$$

The Hermite polynomials $H_n(x)$ can be expressed in terms of these functions by the relation

$$H_n(iz) = + i^n \, 2^{n-1} \, n! \left\{ (-1)^n \{1\}^n \operatorname{Erfc}(z) + \{1\}^n \operatorname{Erfc}(-z) \right\}.$$

Derivatives and recurrence relations

$$\frac{d}{dx}\left[\{1\}^n \operatorname{Erfc}(x)\right] = - \{1\}^{n-1} \operatorname{Erfc}(x),$$

$$n = 0, 1, 2, 3, \ldots,$$

$$\frac{d^n}{dx^n}\left[e^{x^2} \operatorname{Erfc}(x)\right] = (-1)^n \, 2^n n! \, e^{x^2} \{1\}^n \operatorname{Erfc}(x),$$

$$\{1\}^n \operatorname{Erfc}(x) = -\frac{x}{n} \{1\}^{n-1} \operatorname{Erfc}(x) + \frac{1}{2n} \{1\}^{n-2} \operatorname{Erfc}(x),$$

$$2n(n-1) \{1\}^n \operatorname{Erfc}(x) = (2n - 3 + 2x^2) \{1\}^{n-2} \operatorname{Erfc}(x) - \frac{1}{2} \{1\}^{n-4} \operatorname{Erfc}(x),$$

$$n = 4, 5, 6, \ldots$$

Asymptotic approximation

For $|x| \to \infty$, $|\arg x| < \frac{3\pi}{4}$

$$n! \, \{1\}^n \operatorname{Erfc}(x) \sim \frac{2}{\sqrt{\pi}} \frac{e^{-x^2}}{(2x)^{n+1}} \sum_{m=0}^\infty \frac{(-1)^m (n+2m)!}{m! \, (2x)^{2m}}.$$

9.2.4 Fresnel integrals

The Fresnel integrals $C(x)$ and $S(x)$ are defined by

$$C(x) = \int_0^x \cos\left(\frac{\pi}{2} t^2\right) dt = \frac{1}{\sqrt{2\pi}} \int_0^z t^{-\frac{1}{2}} \cos t \, dt,$$

$$S(x) = \int_0^x \sin\left(\frac{\pi}{2} t^2\right) dt = \frac{1}{\sqrt{2\pi}} \int_0^z t^{-\frac{1}{2}} \sin t \, dt,$$

where

$$z = \frac{\pi}{2} x^2.$$

Closely related are the functions given by

$$C_1(x) = \sqrt{\frac{2}{\pi}} \int_0^x \cos t^2 \, dt,$$

$$C_2(x) = \frac{1}{\sqrt{2\pi}} \int_0^x t^{-\frac{1}{2}} \cos t \, dt,$$

$$S_1(x) = \sqrt{\frac{2}{\pi}} \int_0^x \sin t^2 \, dt,$$

$$S_2(x) = \frac{1}{\sqrt{2\pi}} \int_0^x t^{-\frac{1}{2}} \sin t \, dt.$$

The six functions defined above are related to each other by the following

$$C(x) = C_1\left(\sqrt{\frac{\pi}{2}}\, x\right) = C_2\left(\frac{\pi}{2} x^2\right),$$

$$S(x) = S_1\left(\sqrt{\frac{\pi}{2}}\, x\right) = S_2\left(\frac{\pi}{2} x^2\right),$$

$$C(-x) = -C(x),$$

$$C(ix) = i\,C(x),$$

$$C(\bar{x}) = \overline{C(x)},$$

$$S(-x) = -S(x),$$

$$S(ix) = -i\,S(x),$$

$$S(\bar{x}) = \overline{S(x)}.$$

Other representations of these functions are

$$C(x) = \sum_{n=0}^{\infty} \frac{(-1)^n \left(\frac{\pi}{2}\right)^{2n}}{(4n+1)\,\Gamma(2n+1)} x^{4n+1}$$

$$= \sum_{n=0}^{\infty} J_{2n+\frac{1}{2}}\left(\frac{\pi}{2} x^2\right),$$

$$C(x) = \cos\left(\frac{\pi}{2} x^2\right) \sum_{n=0}^{\infty} \frac{(-1)^n \left(\frac{\pi}{2}\right)^{2n}}{\left(\frac{3}{2}\right)_{2n}} x^{4n+1}$$

$$+ \sin\left(\frac{\pi}{2} x^2\right) \sum_{n=0}^{\infty} \frac{(-1)^n \left(\frac{\pi}{2}\right)^{2n+1}}{\left(\frac{3}{2}\right)_{2n+1}} x^{4n+3},$$

$$S(x) = \sum_{n=0}^{\infty} \frac{(-1)^n \left(\frac{\pi}{2}\right)^{2n+1}}{(4n+3)\,\Gamma(2n+2)} x^{4n+3}$$

$$= -\cos\left(\frac{\pi}{2} x^2\right) \sum_{n=0}^{\infty} \frac{(-1)^n \left(\frac{\pi}{2}\right)^{2n+1}}{\left(\frac{3}{2}\right)_{2n+1}} x^{4n+3}$$

$$+ \sin\left(\frac{\pi}{2} x^2\right) \sum_{n=0}^{\infty} \frac{(-1)^n \left(\frac{\pi}{2}\right)^{2n}}{\left(\frac{3}{2}\right)_{2n}} x^{4n+1}$$

$$= \sum_{n=1}^{\infty} J_{2n-\frac{1}{2}}\left(\frac{\pi}{2} x^2\right),$$

$$C(x) + iS(x) = \frac{1+i}{2} \operatorname{Erf}\left[\frac{(1-i)}{2} x \sqrt{\pi}\right]$$

$$= x\, {}_1F_1\left(\frac{1}{2}; \frac{3}{2}; i\frac{\pi}{2} x^2\right)$$

$$= x e^{i\frac{\pi}{2}x^2}\, {}_1F_1\left(1; \frac{3}{2}; -i\frac{\pi}{2} x^2\right),$$

$$C_2(x) = \sum_{n=0}^{\infty} J_{2n+\frac{1}{2}}(x)$$

$$= \frac{1}{2} \int_0^x J_{-\frac{1}{2}}(t)\, dt,$$

$$S_2(x) = \sum_{n=1}^{\infty} J_{2n-\frac{1}{2}}(x)$$

$$= \frac{1}{2} \int_0^x J_{\frac{1}{2}}(t)\, dt.$$

Some integrals connected with Fresnel integrals

$$\int_0^x C(t)\, dt = x\, C(x) - \frac{1}{\pi} \sin\left(\frac{\pi}{2} x^2\right),$$

$$\int_0^x S(t)\, dt = x\, S(x) - \frac{1}{\pi}\left[1 - \cos\left(\frac{\pi}{2} x^2\right)\right],$$

$$\pi \int_0^{\infty} \left\{[1 - C(t)]^2 + \left[\frac{1}{2} - S(t)\right]^2\right\} dt = 1,$$

$$\int_0^{\infty} e^{-xt}\, C(t)\, dt = \frac{1}{x}\left\{\left[\frac{1}{2} - S\left(\frac{x}{\pi}\right)\right]\cos\left(\frac{x^2}{2\pi}\right) - \left[\frac{1}{2} - C\left(\frac{x}{\pi}\right)\right]\sin\left(\frac{x^2}{2\pi}\right)\right\},$$

$$Re\, x > 0,$$

$$\int_0^{\infty} e^{-xt} C(yt)\, dt = \frac{1}{x}\left\{\left[\frac{1}{2} - S\left(\frac{x}{\pi y}\right)\right]\cos\left(\frac{x^2}{2\pi y^2}\right) - \left[\frac{1}{2} - C\left(\frac{x}{\pi y}\right)\right]\sin\left(\frac{x^2}{2\pi y^2}\right)\right\},$$

$$Re\, x > 0,\ y > 0,$$

$$\int_0^{\infty} e^{-xt}\, S(t)\, dt = \frac{1}{x}\left\{\left[\frac{1}{2} - C\left(\frac{x}{\pi}\right)\right]\cos\left(\frac{x^2}{2\pi}\right) + \left[\frac{1}{2} - S\left(\frac{x}{\pi}\right)\right]\sin\left(\frac{x^2}{2\pi}\right)\right\},$$

$$Re\, x > 0,$$

$$\int_0^{\infty} e^{-xt}\, S(yt)\, dt = \frac{1}{x}\left\{\left[\frac{1}{2} - C\left(\frac{x}{\pi y}\right)\right]\cos\left(\frac{x^2}{2\pi y^2}\right) + \left[\frac{1}{2} - S\left(\frac{x}{\pi y}\right)\right]\sin\left(\frac{x^2}{2\pi y^2}\right)\right\},$$

$$Re\, x > 0,\ y > 0,$$

$$\int_0^{\infty} e^{-xt}\, C\left(\sqrt{\frac{2bt}{\pi}}\right) dt = \frac{b\,(x^2 + b^2)^{-\frac{1}{2}}}{2x\,(\sqrt{x^2 + b^2} - x)^{\frac{1}{2}}},$$

$$Re\, x > 0,\ b > 0,$$

$$\int_0^{\infty} e^{-xt}\, S\left(\sqrt{\frac{2yt}{\pi}}\right) dt = \frac{y\,(x^2 + y^2)^{-\frac{1}{2}}}{2x\,(\sqrt{x^2 + y^2} + x)^{\frac{1}{2}}},$$

$$Re\, x > 0,\ y > 0,$$

$$\int \cos(ax^2+bx+c)\,dx = \sqrt{\frac{\pi}{2a}}\left\{C\left[\frac{2ax+b}{\sqrt{2\pi a}}\right]\cos\left(\frac{b^2-4ac}{4a}\right) + S\left(\frac{2ax+b}{\sqrt{2\pi a}}\right)\sin\left(\frac{b^2-4ac}{4a}\right)\right\},$$

$$\int \sin(ax^2+bx+c)\,dx = \sqrt{\frac{\pi}{2a}}\left[S\left(\frac{2ax+b}{\sqrt{2\pi a}}\right)\cos\left(\frac{b^2-4ac}{4a}\right) - C\left(\frac{2ax+b}{\sqrt{2\pi a}}\right)\sin\left(\frac{b^2-4ac}{4a}\right)\right].$$

Asymptotic expansions

For large values of x

$$C(x) = \frac{1}{2} + \frac{1}{\pi x}\sin\left(\frac{\pi}{2}x^2\right) + O\left(\frac{1}{x^2}\right),$$

$$S(x) = \frac{1}{2} - \frac{1}{\pi x}\cos\left(\frac{\pi}{2}x^2\right) + O\left(\frac{1}{x^2}\right).$$

9.2.5 Incomplete beta function

The function $B(x, y, \alpha)$ defined by

$$B(x, y, \alpha) = \int_0^\alpha t^{x-1}(1-t)^{y-1}\,dt, \quad 0 < \alpha < 1$$

is called the incomplete beta function.

$$B(x, y, 1) = B(x, y).$$

In the normalized form

$$I(x, y, \alpha) = \frac{B(x, y, \alpha)}{B(x, y)}.$$

Thus

$$I(x, y, 1) = 1,$$

$$I(x, y, \alpha) = 1 - I(y, x, 1-\alpha).$$

Recurrence relations

$$I(x, y, \alpha) = \alpha I(x-1, y, \alpha) + (1-\alpha)\,I(x, y-1, \alpha),$$

$$(x+y)\,I(x, y, \alpha) = x I(x+1, y, \alpha) + y I(x, y+1, \alpha),$$

$$(x+y-xy)\,I(x, y, \alpha) = x(1-\alpha)\,I(x+1, y-1, \alpha) + y I(x, y+1, \alpha).$$

In terms of the hypergeometric function

$$B(x, y, \alpha) = x^{-1}\alpha^x\,{}_2F_1(x, 1-y; x+1; \alpha).$$

The function $I(x, y, \alpha)$ has the binomial expansion

$$I(x, n+1-x, \alpha) = \sum_{r=x}^{n}\binom{n}{r}\alpha^r(1-\alpha)^{n-r},$$

x, *a* positive integer.

Literature

ARTIN, E.: Einführung in die Theorie der Gammafunktion. Leipzig 1931.
ERDÉLYI, A.: Higher Transcendental functions, Vol. II. New York: McGraw-Hill 1953.
— Tables of integral transforms, Vols. 1, 2. New York: McGraw-Hill 1954.
NIELSEN, N.: Theorie des Integrallogarithmus. Leipzig 1906.
OBERHETTINGER, F.: Tabellen zur Fourier-Transformation. Berlin/Göttingen/Heidelberg: Springer 1957.
TRICOMI, F. G.: Funzioni ipergeometriche confluenti. Rome: Edizioni Cremonese 1954.
WHITTAKER, E. T., and G. N. WATSON: A course of modern analysis. Cambridge: Cambridge Univ. Press 1952.

Chapter X

Elliptic integrals, theta functions and elliptic functions

General remarks. Any integral of the type $\int R\left(z, Z^{\frac{1}{2}}\right)$, where $R(x, y)$ is a rational function of x and y and Z is a polynomial of the third or fourth degree in z with real coefficients and no repeated factors is called an elliptic integral. One can show using suitable substitutions that all elliptic integrals can be expressed in terms of three standardintegrals which are called Legendre's Normal elliptic integrals of the first, second and third kind and which are respectively given by

$$\int_0^x [(1-x^2)(1-k^2x^2)]^{-\frac{1}{2}}\,dx,$$

$$\int_0^x \left[\frac{(1-k^2x^2)}{(1-x^2)}\right]^{\frac{1}{2}} dx,$$

$$\int_0^x (x^2-v^2)^{-1}[(1-x^2)(1-k^2x^2)]^{-\frac{1}{2}}\,dx.$$

Doubly periodic functions. A uniform function $f(z)$ which has two primitive periods p_1 and p_2 is said to be doubly periodic. For all values of z

$$f(z+np_1+mp_2)=f(z), \quad (m, n = 0, 1, -1, 2, -2, \ldots).$$

The primitive periods p_1 and p_2 are in general arbitrarily complex but these two primitive periods can not have the same argument. It is therefore sufficient to investigate the behavior of such doubleperiodic functions for points of the complex z plane which are in one period parallelogram with the corners

$$z_0, z_0+p_1, z_0+p_2, \text{ and } z_0+p_1+p_2.$$

Elliptic functions. A doubly periodic function with no singularities in the period parallelogram but isolated poles is called an elliptic function. The Jacobian and the Weierstrass elliptic functions which will be considered here can be expressed as inverse functions of an elliptic integral as quoted above.

10.1 Elliptic integrals

The elliptic integral of the first kind in Legendre's normal form is given by

$$F(k,\varphi)=\int_0^{\varphi}(1-k^2\sin^2 t)^{-\frac{1}{2}}\,dt=\int_0^{\sin\varphi}[(1-t^2)(1-k^2t^2)]^{-\frac{1}{2}}\,dt.$$

The corresponding elliptic normal integral of the second kind is defined by

$$E(k,\varphi)=\int_0^{\varphi}(1-k^2\sin^2 t)^{\frac{1}{2}}\,dt=\int_0^{\sin\varphi}\left[\frac{(1-k^2t^2)}{(1-t^2)}\right]^{\frac{1}{2}}dt.$$

The property k is called the modulus of the elliptic integral. For $\varphi=\frac{\pi}{2}$ the above defined integrals are called the complete elliptic normal integrals:

$$F\left(k,\frac{\pi}{2}\right)=K(k)=\int_0^{\frac{\pi}{2}}(1-k^2\sin^2 t)^{-\frac{1}{2}}\,dt=\int_0^{1}[(1-t^2)(1-k^2t^2)]^{-\frac{1}{2}}\,dt,$$

$$E\left(k,\frac{\pi}{2}\right)=E(k)=\int_0^{\frac{\pi}{2}}(1-k^2\sin^2 t)^{\frac{1}{2}}\,dt=\int_0^{1}\left[\frac{(1-k^2t^2)}{(1-t^2)}\right]^{\frac{1}{2}}dt.$$

Or, expressed as hypergeometric functions

$$K(k)=\frac{\pi}{2}F\left(\frac{1}{2},\frac{1}{2};1;k^2\right),$$

$$E(k)=\frac{\pi}{2}F\left(-\frac{1}{2},\frac{1}{2};1;k^2\right).$$

The definition by hypergeometric functions is then also valid when the modulus k is a complex number (see chap. II). The modulus complementary to k is denoted by k' and defined by

$$k'=(1-k^2)^{\frac{1}{2}}.$$

The complete elliptic integrals with the modulus k' are denoted by

$$K(k')=K\left[(1-k^2)^{\frac{1}{2}}\right]=K'(k),\quad E(k')=E\left[(1-k^2)^{\frac{1}{2}}\right]=E'(k).$$

Between these integrals there exists Legendre's relation

$$E K' + E' K = \frac{\pi}{2}.$$

Series expansions. The following series expansions converge for $k < 1$

$$K(k) = \frac{\pi}{2}\left[1 + 2\frac{k^2}{8} + 9\left(\frac{k^2}{8}\right)^2 + \cdots\right] = \frac{\pi}{2}\sum_{n=0}^{\infty}\frac{\left(\frac{1}{2}\right)_n\left(\frac{1}{2}\right)_n}{(n!)^2}k^{2n},$$

$$E(k) = \frac{\pi}{2}\left[1 - 2\frac{k^2}{8} - 3\left(\frac{k^2}{8}\right)^2 \cdots\right] = \frac{\pi}{2}\sum_{n=0}^{\infty}\frac{\left(-\frac{1}{2}\right)_n\left(\frac{1}{2}\right)_n}{(n!)^2}k^{2n},$$

$$K(k) = \sum_{n=0}^{\infty}\frac{\left(\frac{1}{2}\right)_n\left(\frac{1}{2}\right)_n}{(n!)^2} \times \left[\psi(n+1) - \psi\left(n+\frac{1}{2}\right) - \frac{1}{2}\log(1-k^2)\right](1-k^2)^n,$$

$$E(k) = 1 + \frac{1}{4}\sum_{n=0}^{\infty}\frac{\left(\frac{1}{2}\right)_n\left(\frac{3}{2}\right)_n}{n!\,(n+1)!}$$
$$\times\left[\psi(n+2) + \psi(n+1) - \psi\left(n+\frac{3}{2}\right) - \psi\left(n+\frac{1}{2}\right) - \log(1-k^2)\right]$$
$$\times(1-k^2)^{n+1}.$$

Special values for the modulus

$$k = 2^{-\frac{1}{2}}, \quad K = K' = \frac{1}{4}\pi^{-\frac{1}{2}}\left[\Gamma\left(\frac{1}{4}\right)\right]^2,$$

$$k = 2^{\frac{1}{2}} - 1,\; K' = 2^{\frac{1}{2}}K,\; k = \sin\left(\frac{\pi}{18}\right),\; K' = 3^{\frac{1}{2}}K,$$

$$k = \frac{\left(2^{\frac{1}{2}} - 1\right)}{\left(2^{\frac{1}{2}} + 1\right)},\; K' = 2K.$$

Transformation formulas. The transformation formulas for the hypergeometric function admit the following transformation formulas for the elliptic integrals. Let the modulus k be replaced by a new quantity $k_1 = k_1(k)$. Then $K(k_1)$ and $E(k_1)$ can be expressed in terms of $K(k)$ and $E(k)$ for a suitable choice of the transformation $k_1(k)$. The following table lists these cases.

Similarly, when k and φ in $F(k,\varphi)$ and $E(k,\varphi)$ are suitably replaced by $k_1 = k_1(k)$ and $\varphi_1 = \varphi_1(\varphi)$, then $F(k_1,\varphi_1)$ and $E(k_1,\varphi_1)$ can be expressed in terms of $F(k,\varphi)$ and $E(k,\varphi)$. These transformations are listed below. The abbreviation $\Delta = (1 - k^2\sin^2\varphi)^{\frac{1}{2}}$ is used in this table.

k_1	$K(k_1)$	$K'(k_1)$	$E(k_1)$	$E'(k_1)$
$\frac{1}{k}$	$k(K+iK')$	kK'	$\frac{E+iE'-k'^2K-ik^2K'}{k}$	$\frac{E'}{k}$
k'	K'	K	E'	E
$\frac{1}{k'}$	$k'(K'+iK)$	$k'K$	$\frac{E'+iE-k^2K'-ik'^2K}{k'}$	$\frac{E}{k'}$
$\frac{ik}{k'}$	$k'K$	$k'(K'-iK)$	$\frac{E}{k'}$	$\frac{E'+iE-k^2K'-ik'^2K}{k'}$
$\frac{k'}{ik}$	kK'	$k(K+iK')$	$\frac{E'}{k}$	$\frac{E-iE'-k'^2K+ik^2K'}{k}$
$\frac{1-k'}{1+k'}$	$\left(\frac{1}{2}+\frac{1}{2}k'\right)K$	$(1+k')K'$	$\frac{E+k'K}{1+k'}$	$\frac{2E'-k^2K'}{1+k'}$
$\frac{2k^{\frac{1}{2}}}{1+k}$	$(1+k)K$	$\left(\frac{1}{2}+\frac{1}{2}k\right)K'$	$\frac{2E-k'^2K}{1+k}$	$\frac{E'+kK'}{1+k}$

k_1	$\sin\varphi_1$	$\cos\varphi_1$	$F(k_1, \varphi_1)$	$E(k_1, \varphi_1)$
$\frac{1}{k}$	$k\sin\varphi$	Δ	$kF(k,\varphi)$	$\frac{[E(k,\varphi) - k'F(k,\varphi)]}{k}$
k'	$-i\tan\varphi$	$\sec\varphi$	$-iF(k,\varphi)$	$i[E(k,\varphi) - F(k,\varphi) - \Delta\tan\varphi]$
$\frac{1}{k'}$	$-ik'\tan\varphi$	$\Delta\sec\varphi$	$-ik'F(k,\varphi)$	$\frac{i[E(k,\varphi) - k'^2F(k,\varphi) - \Delta\tan\varphi]}{k'}$
$\frac{ik}{k'}$	$k'\Delta^{-1}\sin\varphi$	$\Delta^{-1}\cos\varphi$	$k'F(k,\varphi)$	$\frac{\left[E(k,\varphi) - \frac{1}{2}k^2\Delta^{-1}\sin 2\varphi\right]}{k'}$
$-\frac{ik'}{k}$	$-ik\Delta^{-1}\sin\varphi$	Δ^{-1}	$-ikF(k,\varphi)$	$\frac{i\left[E(k,\varphi) - F(k,\varphi) - \frac{1}{2}k^2\sin 2\varphi\right]}{k}$
$\frac{1-k'}{1+k'}$	$\left(\frac{1}{2} + \frac{1}{2}k'\right)\Delta^{-1}\sin 2\varphi$	$\Delta^{-1}(\cos^2\varphi - k'\sin^2\varphi)$	$(1+k')F(k,\varphi)$	$2(1+k')^{-1}\Big[E(k,\varphi) + k'F(k,\varphi)$ $-\left(\frac{1}{2} - \frac{1}{2}k\right)\Delta^{-1}\sin 2\varphi\Big]$
$\frac{2k^{\frac{1}{2}}}{1+k}$	$(1+k\sin^2\varphi)^{-1}$ $\times(1+k)\sin\varphi$	$(1+k\sin^2\varphi)^{-1}$ $\times\Delta\cos\varphi$	$(1+k)F(k,\varphi)$	$(1+k)^{-1}[2E(k,\varphi) - k'^2F(k,\varphi)$ $+ (1+k\sin^2\varphi)^{-1}k\Delta\sin 2\varphi]$

Reduction of some elliptic integrals to Legendre's normal form

$A\,F(k, \varphi)$	A	k	$\cos \varphi$
$\int_x^\infty (t^3 - 1)^{-\frac{1}{2}}\, dt$	$3^{-\frac{1}{4}}$	$\sin\left(\frac{\pi}{12}\right)$	$\frac{x - 1 - 3^{\frac{1}{2}}}{x - 1 + 3^{\frac{1}{2}}}$
$\int_1^x (t^3 - 1)^{-\frac{1}{2}}\, dt$	$3^{-\frac{1}{4}}$	$\sin\left(\frac{\pi}{12}\right)$	$\frac{3^{\frac{1}{2}} + 1 - x}{3^{\frac{1}{2}} - 1 + x}$
$\int_x^1 (1 - t^3)^{-\frac{1}{2}}\, dt$	$3^{-\frac{1}{4}}$	$\sin\left(5\frac{\pi}{12}\right)$	$\frac{x - 1 + 3^{\frac{1}{2}}}{1 - x + 3^{\frac{1}{2}}}$
$\int_{-\infty}^x (1 - t^3)^{-\frac{1}{2}}\, dt$	$3^{-\frac{1}{4}}$	$\sin\left(5\frac{\pi}{12}\right)$	$\frac{1 - x - 3^{\frac{1}{2}}}{1 - x + 3^{\frac{1}{2}}}$
$\int_x^1 (1 + t^4)^{-\frac{1}{2}}\, dt$	$\frac{1}{2}$	$2^{-\frac{1}{2}}$	$\frac{x\, 2^{\frac{1}{2}}}{(1 + x^4)^{\frac{1}{2}}}$
$\int_0^x (1 + t^4)^{-\frac{1}{2}}\, dt$	$\frac{1}{2}$	$2^{-\frac{1}{2}}$	$\frac{1 - x^2}{1 + x^2}$
$\int_1^x (t^4 - 1)^{-\frac{1}{2}}\, dt$	$\frac{1}{2}$	$2^{-\frac{1}{2}}$	$\frac{1}{x}$

$\int_x^1 (1-t^4)^{-\frac{1}{2}}$	$\frac{1}{2}$	$2^{-\frac{1}{2}}$	x
$\int_x^\infty (1+t^4)^{-\frac{1}{2}}\,dt$	$\frac{1}{2}$	$2^{-\frac{1}{2}}$	$\frac{x^2-1}{x^2+1}$
$\int_0^x [(a^2-t^2)(b^2-t^2)]^{-\frac{1}{2}}\,dt$	$\frac{1}{a}$	$\frac{b}{a}$	$\left[1-\left(\frac{x}{b}\right)^2\right]^{\frac{1}{2}}$
$\int_x^b [(a^2-t^2)(b^2-t^2)]^{-\frac{1}{2}}\,dt$	$\frac{1}{a}$	$\frac{b}{a}$	$\left[\frac{\left(\frac{a}{b}\right)^2-1}{\left(\frac{a}{x}\right)^2-1}\right]^{\frac{1}{2}}$
$\int_b^x [(a^2-t^2)(t^2-b^2)]^{-\frac{1}{2}}\,dt$	$\frac{1}{a}$	$\left[1-\left(\frac{b}{a}\right)^2\right]^{\frac{1}{2}}$	$\left[\frac{\left(\frac{a}{x}\right)^2-1}{\left(\frac{a}{b}\right)^2-1}\right]^{\frac{1}{2}}$
$\int_x^a [(a^2-t^2)(t^2-b^2)]^{-\frac{1}{2}}\,dt$	$\frac{1}{a}$	$\left[1-\left(\frac{b}{a}\right)^2\right]^{\frac{1}{2}}$	$\left[\frac{\left(\frac{x}{b}\right)^2-1}{\left(\frac{a}{b}\right)^2-1}\right]^{\frac{1}{2}}$
$\int_a^x [(t^2-a^2)(t^2-b^2)]^{-\frac{1}{2}}\,dt$	$\frac{1}{a}$	$\frac{b}{a}$	$\left[\frac{\left(\frac{a}{b}\right)^2-1}{\left(\frac{x}{b}\right)^2-1}\right]^{\frac{1}{2}}$
$\int_x^\infty [(t^2-a^2)(t^2-b^2)]^{-\frac{1}{2}}\,dt$	$\frac{1}{a}$	$\frac{b}{a}$	$\left[1-\left(\frac{a}{x}\right)^2\right]^{\frac{1}{2}}$

$A\,F(k,\varphi)$	A	k	$\cos\varphi$
$\int_0^x [(a^2+t^2)(b^2+t^2)]^{-\frac{1}{2}}\,dt$	$\frac{1}{a}$	$\left[1-\left(\frac{b}{a}\right)^2\right]^{\frac{1}{2}}$	$\left[1+\left(\frac{x}{b}\right)^2\right]^{-\frac{1}{2}}$
$\int_x^\infty [(a^2+t^2)(b^2+t^2)]^{-\frac{1}{2}}\,dt$	$\frac{1}{a}$	$\left[1-\left(\frac{b}{a}\right)^2\right]^{\frac{1}{2}}$ $b<a$	$\left[1+\left(\frac{a}{x}\right)^2\right]^{-\frac{1}{2}}$
$\int_0^x [(a^2-t^2)(b^2+t^2)]^{-\frac{1}{2}}\,dt$	$(a^2+b^2)^{-\frac{1}{2}}$	$a(a^2+b^2)^{-\frac{1}{2}}$	$\left[\frac{1-\left(\frac{x}{a}\right)^2}{1+\left(\frac{x}{b}\right)^2}\right]^{\frac{1}{2}}$
$\int_x^a [(a^2-t^2)(t^2+b^2)]^{-\frac{1}{2}}\,dt$	$(a^2+b^2)^{-\frac{1}{2}}$	$a(a^2+b^2)^{-\frac{1}{2}}$	$\frac{x}{a}$
$\int_b^x [(a^2+t^2)(t^2-b^2)]^{-\frac{1}{2}}\,dt$	$(a^2+b^2)^{-\frac{1}{2}}$	$a(a^2+b^2)^{-\frac{1}{2}}$	$\frac{b}{x}$
$\int_x^\infty [(a^2+t^2)(t^2-b^2)]^{-\frac{1}{2}}\,dt$	$(a^2+b^2)^{-\frac{1}{2}}$	$a(a^2+b^2)^{-\frac{1}{2}}$	$\left[\frac{1-\left(\frac{b}{x}\right)^2}{1+\left(\frac{a}{x}\right)^2}\right]^{\frac{1}{2}}$

Similar reductions to the elliptic integral of the second kind $E(k, \varphi)$ in Legendre's normal form are listed below

$A\,E(k, \varphi)$	A	k	$\cos\varphi$
$\int_0^x (a^2 - t^2)^{\frac{1}{2}} (b^2 - t^2)^{-\frac{1}{2}}\, dt$	a	$\frac{b}{a}$	$\left[1 - \left(\frac{x}{b}\right)^2\right]^{\frac{1}{2}}$
$\int_x^a (b^2 + t^2)^{\frac{1}{2}} (a^2 - t^2)^{-\frac{1}{2}}\, dt$	$(a^2 + b^2)^{\frac{1}{2}}$	$\left[1 + \left(\frac{a}{b}\right)^2\right]^{-\frac{1}{2}}$	$\frac{x}{a}$
$\int_b^x t^{-2} (t^2 + a^2)^{\frac{1}{2}} (t^2 - b^2)^{-\frac{1}{2}}\, dt$	$b^{-2}(a^2 + b^2)^{\frac{1}{2}}$	$\left[1 + \left(\frac{b}{a}\right)^2\right]^{-\frac{1}{2}}$	$\frac{b}{x}$
$\int_b^x t^2 (t^2 + a^2)^{-\frac{3}{2}} (t^2 - b^2)^{-\frac{1}{2}}\, dt$	$(a^2 + b^2)^{-\frac{1}{2}}$	$\left[1 + \left(\frac{b}{a}\right)^2\right]^{-\frac{1}{2}}$	$\left[\frac{1 - \left(\frac{b}{x}\right)^2}{1 + \left(\frac{a}{x}\right)^2}\right]^{\frac{1}{2}}$
$\int_0^x (a^2 + t^2)^{\frac{1}{2}} (b^2 + t^2)^{-\frac{3}{2}}\, dt$	$a\,b^{-2}$	$\left[1 - \left(\frac{b}{a}\right)^2\right]^{\frac{1}{2}}$	$\left[1 + \left(\frac{x}{b}\right)^2\right]^{-\frac{1}{2}}$
$\int_b^x t^{-2} [(t^2 - b^2)(a^2 - t^2)]^{-\frac{1}{2}}\, dt$	$(a\,b^2)^{-1}$	$\left[1 - \left(\frac{b}{a}\right)^2\right]^{\frac{1}{2}}$	$\left[\frac{\left(\frac{a}{x}\right)^2 - 1}{\left(\frac{a}{b}\right)^2 - 1}\right]^{\frac{1}{2}}$

The following integrals admit likewise a reduction to Legendre's normal form

$$\int_0^\varphi (1 - k^2 \sin^2 t)^{-\frac{1}{2}} \sin^2 t \, dt = k^{-2}(F - E),$$

$$\int_0^\varphi (1 - k^2 \sin^2 t)^{-\frac{1}{2}} \cos^2 t \, dt = k^{-2}(E - k'^2 F),$$

$$\int_0^\varphi (1 - k^2 \sin^2 t)^{-\frac{3}{2}} dt = k'^{-2} \left[E - k^2 \sin\varphi \cos\varphi \, (1 - k^2 \sin^2 \varphi)^{-\frac{1}{2}}\right],$$

$$\int_0^\varphi \tan^2 t \, (1 - k^2 \sin^2 t)^{-\frac{1}{2}} dt = k'^{-2} \, [\tan\varphi \, (1 - k^2 \sin^2 \varphi) - E],$$

$$\int_0^\varphi \sin^2 t \, (1 - k^2 \sin^2 t)^{-\frac{3}{2}} dt = (k k')^{-2} \, (E - k'^2 F) - k'^{-2} \sin\varphi \times \cos\varphi \, (1 - k^2 \sin^2 \varphi)^{-\frac{1}{2}},$$

$$\int_0^\varphi \cos^2 t \, (1 - k^2 \sin^2 t)^{-\frac{3}{2}} dt = k^{-2}(F - E) + \sin\varphi \cos\varphi \, (1 - k^2 \sin^2 \varphi)^{-\frac{1}{2}},$$

$$\int_{\varphi_1}^{\varphi_2} (\sin t)^{-2} \, (1 - k^2 \sin^2 t)^{-\frac{1}{2}} dt = \left[F - E - (1 - k^2 \sin^2 t)^{\frac{1}{2}} \cot t\right]_{\varphi_1}^{\varphi_2},$$

$$\int_0^\varphi (\cos t)^{-2} (1 - k^2 \sin^2 t)^{-\frac{1}{2}} dt = k'^{-2} \times \left[(1 - k^2 \sin^2 \varphi)^{\frac{1}{2}} \tan\varphi - E + k'^2 F\right],$$

$$\int_0^\varphi (1 - k^2 \sin^2 t)^{\frac{1}{2}} \tan^2 t \, dt = \tan\varphi \, (1 - k^2 \sin^2 t)^{\frac{1}{2}} + F - 2E.$$

Reduction of some further elliptic integrals to Legendre's normal form

Let

$$X(t) = (t - a)\,(t - b)\,(t - c), \quad a, b, c \text{ real}, \; a > b > c.$$

With the abbreviations

$$C = 2(a - c)^{\frac{1}{2}}, \quad k = \left[\frac{(b - c)}{(a - c)}\right]^{\frac{1}{2}}, \quad k' = \left[\frac{(a - b)}{(a - c)}\right]^{\frac{1}{2}}$$

one has

$$\int_x^\infty X^{-\frac{1}{2}} dt = C \int_0^y [(1 - t^2)\,(1 - k^2 t^2)]^{-\frac{1}{2}} dt, \quad y = \left[\frac{(a - c)}{(x - c)}\right]^{\frac{1}{2}},$$

$$\int_{-\infty}^x (-X)^{-\frac{1}{2}} dt = C \int_0^y [(1 - t^2)\,(1 - k'^2 t^2)]^{-\frac{1}{2}} dt, \quad y = \left[\frac{(a - c)}{(a - x)}\right]^{\frac{1}{2}},$$

$$\int_a^x X^{-\frac{1}{2}}\,dt = C\int_y^1 [(1-t^2)\,(k'^2+k^2t^2)]^{-\frac{1}{2}}\,dt,\quad y=\left[\frac{(a-b)}{(x-c)}\right]^{\frac{1}{2}},$$

$$\int_x^a (-X)^{-\frac{1}{2}}\,dt = C\int_0^y [(1-t^2)\,(1-k'^2t^2)]^{-\frac{1}{2}}\,dt,\quad y=\left[\frac{(a-x)}{(a-b)}\right]^{\frac{1}{2}},$$

$$\int_x^b X^{-\frac{1}{2}}\,dt = C\int_y^1 [(1-t^2)\,(t^2-k'^2)]^{-\frac{1}{2}}\,dt,\quad y=\left[\frac{(a-b)}{(a-x)}\right]^{\frac{1}{2}},$$

$$\int_b^x (-X)^{-\frac{1}{2}}\,dt = C\int_y^1 [(1-t^2)\,(t^2-k^2)]^{-\frac{1}{2}}\,dt,\quad y=\left[\frac{(b-c)}{(x-c)}\right]^{\frac{1}{2}},$$

$$\int_c^x X^{-\frac{1}{2}}\,dt = C\int_0^y [(1-t^2)\,(1-k^2t^2)]^{-\frac{1}{2}}\,dt,\quad y=\left[\frac{(x-c)}{(b-c)}\right]^{\frac{1}{2}},$$

$$\int_x^c (-X)^{-\frac{1}{2}}\,dt = \int_y^1 [(1-t^2)\,(k^2+k'^2t^2)]^{-\frac{1}{2}}\,dt,\quad y=\left[\frac{(b-c)}{(b-x)}\right]^{\frac{1}{2}}.$$

Now let $Y(t)=(t-a)\,(t-b)\,(t-c)\,(t-d)$; a, b, c, d real, $a>b>c>d$. Then if $a<t<\infty$,

$$\int Y^{-\frac{1}{2}}\,dt = 2\,[(a-c)\,(b-d)]^{-\frac{1}{2}}\int [(1-x^2)\,(1-k^2x^2)]^{-\frac{1}{2}}\,dx.$$

with

$$k^2=\frac{(a-d)\;(b-c)}{[(a-c)\;(b-d)]}.$$

The limits in the integral over the variable x are determined by the limits of the t integration by means of the equation

$$x=\left[\frac{(b-d)\;(t-a)}{(a-d)\;(t-b)}\right]^{\frac{1}{2}}.$$

The same formula is valid when

$$c<t<b \quad\text{and}\quad x=\left[\frac{(b-d)\;(t-c)}{(b-c)\;(t-d)}\right]^{\frac{1}{2}}.$$

For $d<t<c$ one has

$$\int (-Y)^{-\frac{1}{2}}\,dt = 2\,[(a-c)\,(b-d)]^{-\frac{1}{2}}\int [(1-x^2)\,(1-k^2x^2)]^{-\frac{1}{2}}\,dx,$$

where

$$k^2=\frac{(a-b)\;(c-d)}{[(a-c)\;b-d)]}.$$

The limits in the x integration are determined by

$$x=\left[\frac{(a-c)\;(t-d)}{(c-d)\;(a-t)}\right]^{\frac{1}{2}}.$$

For $b < t < a$ the preceding formula is valid with

$$x = \left[\frac{(a - c)(t - b)}{(a - b)(t - c)}\right]^{\frac{1}{2}}.$$

Now let

$$Y(t) = (t - \alpha)(t - \beta)[(t - \mu)^2 + \nu^2]; \; \alpha, \beta, \mu, \nu \text{ real}, \; \alpha > \beta; \; \nu > 0.$$

Put

$$a^2 = \nu^2 + (\mu - \alpha)^2, \; b^2 = \nu^2 + (\mu - \beta)^2.$$

Then for

$$a < t < \infty \text{ and } -\infty < t < \beta,$$

$$\int Y^{-\frac{1}{2}} dt = (ab)^{-\frac{1}{2}} \int [(1 - x^2)(1 - k^2 x^2)]^{-\frac{1}{2}} dx,$$

where

$$k^2 = \frac{[(a + b)^2 - (\alpha - \beta)^2]}{(4ab)}.$$

The connection between the t limits and the x limits is given by

$$t = \frac{\left[(\alpha b - a\beta) + (\alpha b + a\beta)(1 - x^2)^{\frac{1}{2}}\right]}{\left[b - a + (b + a)(1 - x^2)^{\frac{1}{2}}\right]}.$$

For $\beta < t < \alpha$

$$\int (-Y)^{-\frac{1}{2}} dt = (ab)^{-\frac{1}{2}} \int [(1 - x^2)(1 - k^2 x^2)]^{-\frac{1}{2}} dx,$$

with

$$k^2 = \frac{[(\alpha - \beta)^2 - (a - b)^2]}{(4ab)},$$

$$t = \frac{\left[(\alpha b + a\beta) - (\alpha b - a\beta)(1 - x^2)^{\frac{1}{2}}\right]}{\left[b + a - (b - a)(1 - x^2)^{\frac{1}{2}}\right]}.$$

With $k^2 = \frac{(a - b)}{(a - c)}$ and $\sin^2 \varphi = \frac{(a - c)}{(a + x)}$ the following integrals reduce to (provided $a > b > c$)

$$\int_x^\infty [(a + t)(b + t)(c + t)]^{-\frac{1}{2}} dt = 2(a - c)^{-\frac{1}{2}} F(k, \varphi),$$

$$\int_x^\infty (a + t)^{-1} [(a + t)(b + t)(c + t)]^{-\frac{1}{2}} dt = 2k^{-2}(a - c)^{-\frac{3}{2}}$$
$$\times [F(k, \varphi) - E(k, \varphi)],$$

$$\int_x^\infty (b+t)^{-1} [(a+t)(b+t)(c+t)]^{-\frac{1}{2}} dt = 2(k k')^{-2} (a-c)^{-\frac{3}{2}}$$

$$\times \left[E(k,\varphi) - k'^2 F(k,\varphi) - k^2 \sin\varphi \cos\varphi \,(1-k^2\sin^2\varphi)^{-\frac{1}{2}}\right],$$

$$\int_x^\infty (c+t)^{-1} [(a+t)(b+t)(c+t)]^{-\frac{1}{2}} dt = 2k'^{-2} (a-c)^{\frac{3}{2}}$$

$$\times \left[(1-k^2\sin^2\varphi)^{\frac{3}{2}} - E(k,\varphi)\right].$$

Reduction of some triple integrals to complete elliptic integrals

$$\int_0^\pi \int_0^\pi \int_0^\pi (1 - \cos u \cos v \cos w)^{-1}\, du\, dv\, dw = 4\pi K^2\left(\sin\frac{\pi}{4}\right),$$

$$\int_0^\pi \int_0^\pi \int_0^\pi (3 - \cos v \cos w - \cos u \cos w - \cos u \cos v)^{-1}\, du\, dv\, dw$$

$$= 3^{\frac{1}{2}} \pi K^2\left(\sin\frac{\pi}{12}\right),$$

$$\int_0^\pi \int_0^\pi \int_0^\pi (3 - \cos u - \cos v - \cos w)^{-1}\, du\, dv\, dw$$

$$= 4\left[18 + 12\cdot 2^{\frac{1}{2}} - 10\cdot 3^{\frac{1}{2}} - 7\cdot 6^{\frac{1}{2}}\right] K^2\left[\left(2 - 3^{\frac{1}{2}}\right)\left(3^{\frac{1}{2}} - 2^{\frac{1}{2}}\right)\right].$$

Elliptic normal integral of the third kind

The elliptic normal integral of the third kind is denoted by

$$\Pi(\varphi, n, k) = \int_0^\varphi (1 + n\sin^2 t)^{-1} (1 - k^2\sin^2 t)^{-\frac{1}{2}} dt$$

$$= \int_0^{\sin\varphi} (1 + n x^2)\, [(1 - x^2)(1 - k^2 x^2)]^{-\frac{1}{2}} dx.$$

All indenifite integrals of the form

$$R\left[[t, (a_0 t^4 + a_1 t^3 + a_2 t^2 + a_3 t + a_4)^{\frac{1}{2}}\right],$$

where R is a rational function of the variables indicated above (a_0, a_1, a_2, a_3, a_4 are constants) can be expressed linearly in terms of elementary functions and elliptic normal integrals of the first, second and third kind.

Relations by Legendre

$$\sin\varphi\cos\varphi\,(1-k'^2\sin^2\varphi)^{\frac{1}{2}}\int_0^{\frac{1}{2}\pi}(1-\cos^2\varphi\cos^2 t)^{-1}\cos^2 t\,(1-k^2\sin^2 t)^{-\frac{1}{2}}\,dt$$
$$=\frac{1}{2}\pi-[KE(k',\varphi)+EF(k',\varphi)-KF(k',\varphi)],$$

$$\frac{1}{2}k'^2\sin(2\varphi)\,(1-k'^2\sin^2\varphi)^{\frac{1}{2}}\int_0^{\frac{1}{2}\pi}\sin^2 t$$
$$\times[1-(1-k'^2\sin^2\varphi)\sin^2 t]^{-1}(1-k^2\sin^2 t)^{-\frac{1}{2}}\,dt$$
$$=\frac{1}{2}\pi-[KE(k',\varphi)+EF(k',\varphi)-KF(k',\varphi)],$$

$$\frac{1}{2}k^2\sin(2\varphi)\,(1-k^2\sin^2\varphi)^{\frac{1}{2}}\int_0^{\frac{1}{2}\pi}\sin^2 t$$
$$\times(1-k^2\sin^2\varphi\sin^2 t)^{-1}(1-k^2\sin^2 t)^{-\frac{1}{2}}\,dt$$
$$=KE(k,\varphi)-EF(k,\varphi).$$

Differentiation with respect to the modulus

$$\frac{\partial F}{\partial k}=k'^{-2}\left[\frac{E-k'^2F}{k}-\sin\varphi\cos\varphi\,(1-k^2\sin^2\varphi)^{-\frac{1}{2}}\right],$$
$$\frac{\partial E}{\partial k}=\frac{(E-F)}{k}.$$

Elliptic integrals as special cases of Legendre functions

$$K(k)=\frac{1}{2}\pi(1-k^2)^{-\frac{1}{2}}\mathfrak{P}_{-\frac{1}{2}}\left[\frac{(1+k^2)}{(1-k^2)}\right]$$
$$=\frac{1}{2}\pi P_{-\frac{1}{2}}(1-2k^2)$$
$$=Q_{-\frac{1}{2}}(2k^2-1)$$
$$=k^{-1}\mathfrak{Q}_{-\frac{1}{2}}(2k^{-2}-1),$$

$$E(k)=-(k^{-2}-1)^{\frac{1}{2}}\mathfrak{Q}^1_{-\frac{1}{2}}(2k^{-2}-1)$$
$$=\frac{1}{4}\pi\left[P_{\frac{1}{2}}(1-2k^2)+P_{-\frac{1}{2}}(1-2k^2)\right]$$
$$=\frac{1}{2}\left[Q_{-\frac{1}{2}}(2k^2-1)-Q_{\frac{1}{2}}(2k^2-1)\right].$$

For further connection see also 4.4.

10.2 The theta functions

Let z be an arbitrary complex number and τ a parameter with a positive imaginary part. It is customary to write $\tau = i\pi t$ in case τ purely imaginary (t real and positive). If we write

$$q = e^{i\pi\tau},$$

then the theta functions are defined as

$$\vartheta_0(z,\tau) = \sum_{n=0}^{\infty} (-1)^n \varepsilon_n q^{n^2} \cos(2\pi n z)$$

$$= (-i\tau)^{-\frac{1}{2}} \sum_{n=-\infty}^{\infty} e^{\frac{-i\pi\left(z+n-\frac{1}{2}\right)^2}{\tau}},$$

$$\vartheta_1(z,\tau) = 2\sum_{n=0}^{\infty} q^{\left(n+\frac{1}{2}\right)^2} (-1)^n \sin[(2n+1)\pi z]$$

$$= (-i\tau)^{-\frac{1}{2}} \sum_{n=-\infty}^{\infty} (-1)^n e^{\frac{-i\pi\left(z+n-\frac{1}{2}\right)^2}{\tau}},$$

$$\vartheta_2(z,\tau) = 2\sum_{n=0}^{\infty} q^{\left(n+\frac{1}{2}\right)^2} \cos[(2n+1)\pi z]$$

$$= (-i\tau)^{-\frac{1}{2}} \sum_{n=-\infty}^{\infty} (-1)^n e^{\frac{-i\pi(z+n)^2}{\tau}},$$

$$\vartheta_3(z,\tau) = \sum_{n=0}^{\infty} \varepsilon_n q^{n^2} \cos(2\pi n z)$$

$$= (-i\tau)^{-\frac{1}{2}} \sum_{n=-\infty}^{\infty} e^{\frac{-i\pi(z+n)^2}{\tau}}.$$

One often writes

$$\vartheta_0(z,\tau) \equiv \vartheta_4(z,\tau).$$

When the parameter τ is held constant we write instead of $\vartheta(z,\tau)$ briefly $\vartheta(z)$. Between the theta functions there exist the relations

$$\left[k = \left(\frac{\vartheta_2}{\vartheta_3}\right)^2, \quad k' = \left(\frac{\vartheta_0}{\vartheta_3}\right)^2, \quad k^2 + k'^2 = 1\right],$$

$$\vartheta_0^2(z) = k\vartheta_1^2(z) + k'\vartheta_3^2(z),$$

$$\vartheta_2^2(z) = -k'\vartheta_1^2(z) + k\vartheta_3^2(z),$$

$$\vartheta_1^2(z) = k\vartheta_0^2(z) - k'\vartheta_2^2(z),$$

$$\vartheta_0^4(z) + \vartheta_2^4(z) = \vartheta_1^4(z) + \vartheta_3^4(z).$$

For $\tau \to 0$ i.e. in the limit $q \to 1 - 0$ one has

$$\lim_{\tau\to 0} \tau^{\frac{1}{2}} \vartheta_2(0,\tau) = \lim_{\tau\to 0} \tau^{\frac{1}{2}} \vartheta_3(0,\tau) = e^{i\frac{\pi}{4}}.$$

Increasing of the argument	$\vartheta_0(z,\tau)$	$\vartheta_1(z,\tau)$	$\vartheta_2(z,\tau)$	$\vartheta_3(z,\tau)$	Exponential factor
$m + n\tau$	$(-1)^n \vartheta_0$	$(-1)^{m+n} \vartheta_1$	$(-1)^m \vartheta_2$	ϑ_3	$e^{-n\pi i(2z+n\tau)}$
$m - \frac{1}{2} + n\tau$	ϑ_3	$(-1)^{m+1} \vartheta_2$	$(-1)^{m+n} \vartheta_1$	$(-1)^n \vartheta_0$	
$m + \left(n + \frac{1}{2}\right)\tau$	$(-1)^n i\vartheta_1$	$(-1)^{m+n} i\vartheta_0$	$(-1)^m \vartheta_3$	ϑ_2	$e^{-\left(n+\frac{1}{2}\right)\pi i\left[2z+\left(n+\frac{1}{2}\right)\tau\right]}$
$m - \frac{1}{2} + \left(n + \frac{1}{2}\right)\tau$	ϑ_2	ϑ_3	$(-1)^{m+n} i\vartheta_0$	$(-1)^n i\vartheta_1$	

The theta functions are periodic functions of z with the periods 1 and 2. Moreover there exist relations which admit to express the theta functions of argument $z + \frac{1}{2}n + \frac{1}{2}m$ with arbitrary integers m and n again by a theta function multiplied by an exponential function. The preceeding table shows these reduction formulas.

In this table, for instance ϑ_0 means $\vartheta_0(z,\tau)$ etc. The location of the zeroes follow from the next table.

Function	$\vartheta_0(z,\tau)$	$\vartheta_1(z,\tau)$	$\vartheta_2(z,\tau)$	$\vartheta_3(z,\tau)$
zero at $z =$	$n + m\tau + \frac{1}{2}\tau$	$n + m\tau$	$n + \frac{1}{2} + m\tau$	$n + \frac{1}{2} + \left(m + \frac{1}{2}\right)\tau$

Let a, b, c, d integers such that $ad - bc = 1$. Then the theta functions with the argument $z' = z(c\tau + d)^{-1}$ and with the parameter

$$\tau' = (a\tau + b)(c\tau + d)^{-1}$$

can be expressed as theta functions of argument z and parameter τ. This can be accomplished by the formulas below.

$$\vartheta_1(z,\tau+1) = e^{i\frac{\pi}{4}} \vartheta_1(z,\tau); \quad \vartheta_1\left(\frac{z}{\tau}, -\frac{1}{\tau}\right) = -i\left(\frac{\tau}{i}\right)^{\frac{1}{2}} e^{\frac{i\pi z^2}{\tau}} \vartheta_1(z,\tau),$$

$$\vartheta_1(z,\tau+1) = e^{i\frac{\pi}{4}} \vartheta_2(z,\tau); \quad \vartheta_2\left(\frac{z}{\tau}, -\frac{1}{\tau}\right) = \left(\frac{\tau}{i}\right)^{\frac{1}{2}} e^{\frac{i\pi z^2}{\tau}} \vartheta_0(z,\tau),$$

$$\vartheta_3(z, \tau + 1) = \vartheta_0(z, \tau); \; \vartheta_3\left(\frac{z}{\tau}, -\frac{1}{\tau}\right) = \left(\frac{\tau}{i}\right)^{\frac{1}{2}} e^{\frac{i\pi z^2}{\tau}} \vartheta_3(z, \tau),$$

$$\vartheta_0(z, \tau + 1) = \vartheta_3(z, \tau); \; \vartheta_0\left(\frac{z}{\tau}, -\frac{1}{\tau}\right) = \left(\frac{\tau}{i}\right)^{\frac{1}{2}} e^{\frac{i\pi z^2}{\tau}} \vartheta_2(z, \tau).$$

Representations of the theta functions by infinite products

$$\vartheta_0(z) = Q_0 \prod_{n=1}^{\infty} [1 - 2q^{2n-1}\cos(2\pi z) + q^{4n-2}],$$

$$\vartheta_1(z) = 2Q_0\, q^{\frac{1}{4}} \sin(\pi z) \prod_{n=1}^{\infty} [1 - 2q^{2n}\cos(2\pi z) + q^{4n}],$$

$$\vartheta_2(z) = 2Q_0\, q^{\frac{1}{4}} \cos(\pi z) \prod_{n=1}^{\infty} [1 + 2q^{2n}\cos(2\pi z) + q^{4n}],$$

$$\vartheta_3(z) = Q_0 \prod_{n=1}^{\infty} [1 + 2q^{2n-1}\cos(2\pi z) + q^{4n-2}].$$

Or also

$$\vartheta_0(z) = Q_0 \prod_{n=1}^{\infty} 4q^{2n-1} \sin\left[\pi z + \left(n - \frac{1}{2}\right)\pi\tau\right] \sin\left[\pi z - \left(n - \frac{1}{2}\right)\pi\tau\right],$$

$$\vartheta_1(z) = 2Q_0\, q^{\frac{1}{4}} \sin(\pi z) \prod_{n=1}^{\infty} 4q^{2n} \sin(\pi z + \mu\pi\tau) \sin(\pi z - n\pi\tau),$$

$$\vartheta_2(z) = 2Q_0\, q^{\frac{1}{4}} \cos(\pi z) \prod_{n=1}^{\infty} 4q^{2n} \cos(\pi z + n\pi\tau) \cos(\pi z - n\pi\tau),$$

$$\vartheta_3(z) = Q_0 \prod_{n=1}^{\infty} 4q^{2n-1} \cos\left[\pi z + \left(n - \frac{1}{2}\right)\pi\tau\right] \cos\left[\pi z - \left(n - \frac{1}{2}\right)\pi\tau\right],$$

$$Q_0 = \prod_{n=1}^{\infty} (1 - q^{2n}).$$

From this one can finally deduce

$$\vartheta_0(z) = Q_0 Q_3^2 \prod_{n=1}^{\infty} \left[1 - \sin^2(\pi z)\operatorname{cosec}^2\left(n - \frac{1}{2}\right)\pi\tau\right],$$

$$\vartheta_1(z) = 2Q_0^3\, q^{\frac{1}{4}} \sin(\pi z) \prod_{n=1}^{\infty} [1 - \sin^2(\pi z)\operatorname{cosec}^2(n\pi\tau)],$$

$$\vartheta_2(z) = 2Q_0 Q_1^2\, q^{\frac{1}{4}} \cos(\pi z) \prod_{n=1}^{\infty} [1 - \sin^2(\pi z)\sec^2(n\pi\tau)],$$

$$\vartheta_3(z) = Q_0 Q_2^2 \prod_{n=1}^{\infty} \left[1 - \sin^2(\pi z)\sec^2\left(n - \frac{1}{2}\right)\pi\tau\right].$$

Furthermore

$$\vartheta_0(z) = Q_0 Q_3 \prod_{n=1}^{\infty} [1 - \cos(2\pi z) \sec(2n-1)\pi\tau],$$

$$\vartheta_1(z) = 2 Q_0 Q_4 q^{\frac{1}{4}} \sin(\pi z) \prod_{n=1}^{\infty} [1 - \cos(2\pi z) \sec(2\pi n\tau)],$$

$$\vartheta_2(z) = 2 Q_0 Q_4 q^{\frac{1}{4}} \cos(\pi z) \prod_{n=1}^{\infty} [1 + \cos(2\pi z) \sec(2\pi n\tau)],$$

$$\vartheta_3(z) = Q_0 Q_5 \prod_{n=1}^{\infty} [1 + \cos(2\pi z) \sec(2n-1)\pi\tau].$$

Here

$$Q_0 = \prod_{n=1}^{\infty} (1 - q^{2n}), \qquad Q_1 = \prod_{n=1}^{\infty} (1 + q^{2n}),$$

$$Q_2 = \prod_{n=1}^{\infty} (1 + q^{2n-1}), \qquad Q_3 = \prod_{n=1}^{\infty} (1 - q^{2n-1}),$$

$$Q_4 = \prod_{n=1}^{\infty} (1 + q^{4n}), \qquad Q_5 = \prod_{n=1}^{\infty} (1 + q^{4n-2}).$$

Values for the argument $z = 0$

$$\vartheta_0(0) = \left(\frac{2Kk'}{\pi}\right)^{\frac{1}{2}}, \quad \vartheta_1(0) = 0, \quad \vartheta_1'(0) = 2K\left(\frac{2Kkk'}{\pi}\right)^{\frac{1}{2}},$$

$$\vartheta_2(0) = \left(\frac{2Kk}{\pi}\right)^{\frac{1}{2}}, \quad \vartheta_3(0) = \left(\frac{2K}{\pi}\right)^{\frac{1}{2}},$$

$$\vartheta_0'(m + n\tau) = (-1)^{n+1}\, 2\pi i n q^{-n^2} \vartheta_0(0),$$

$$\vartheta_1'(m + n\tau) = (-1)^{m+n}\, q^{-n^2} \vartheta_1'(0),$$

$$\vartheta_2'(m + n\tau) = (-1)^{m+1} 2\pi i n q^{-n^2} \vartheta_2(0),$$

$$\vartheta_3'(m + n\tau) = -\, 2\pi i n q^{-n^2} \vartheta_3(0).$$

If we denote

$$\vartheta(z) = \vartheta(z, \tau); \quad \vartheta = \vartheta(0, \tau),$$

then

$$\vartheta_1' = \pi \vartheta_2 \vartheta_3 \vartheta_0,$$

$$\frac{\vartheta_1'''}{\vartheta_1} = \frac{\vartheta_2''}{\vartheta_2} + \frac{\vartheta_3''}{\vartheta_3} + \frac{\vartheta_0''}{\vartheta_0},$$

$$\frac{d^2 \log \vartheta_0(z)}{dz^2} = \frac{\vartheta_0''}{\vartheta_0} - \left(\frac{\vartheta_1'}{\vartheta_0}\right)^2 \left(\frac{\vartheta_1(z)}{\vartheta_0(z)}\right)^2,$$

$$\frac{d^2 \log \vartheta_1(z)}{dz^2} = \frac{\vartheta_0''}{\vartheta_0} - \left(\frac{\vartheta_1'}{\vartheta_0}\right)^2 \left(\frac{\vartheta_0(z)}{\vartheta_1(z)}\right)^2,$$

$$\frac{d^2 \log \vartheta_2(z)}{dz^2} = \frac{\vartheta_0''}{\vartheta_0} - \left(\frac{\vartheta_1'}{\vartheta_0}\right)^2 \left(\frac{\vartheta_3(z)}{\vartheta_2(z)}\right)^2,$$

$$\frac{d^2 \log \vartheta_3(z)}{dz^2} = \frac{\vartheta_0''}{\vartheta_0} - \left(\frac{\vartheta_1'}{\vartheta_0}\right)^2 \left(\frac{\vartheta_2(z)}{\vartheta_3(z)}\right)^2.$$

The theta functions are solutions of the following partial differential-equation

$$\frac{\partial^2 \vartheta(z)}{\partial z^2} = 4\pi i \frac{\partial \vartheta(z)}{\partial \tau}.$$

The logarithm of the theta functions admits the following Fourier series expansions

$$\log \vartheta_0(z) = \log \gamma - 2 \sum_{n=1}^{\infty} \frac{q^n}{1-q^{2n}} \frac{\cos(2n\pi z)}{n},$$

$$\log \vartheta_1(z) = \frac{1}{4} \log \gamma + \log(2 \sin \pi z) - 2 \sum_{n=1}^{\infty} \frac{q^{2n}}{1-q^{2n}} \frac{\cos(2n\pi z)}{n},$$

$$\log \vartheta_2(z) = \frac{1}{4} \log \gamma + \log(2 \cos \pi z) - 2 \sum_{n=1}^{\infty} \frac{(-1)^n q^{2n}}{1-q^{2n}} \frac{\cos(2n\pi z)}{n},$$

$$\log \vartheta_3(z) = \log \gamma - 2 \sum_{n=1}^{\infty} \frac{(-1)^n q^n}{1-q^{2n}} \frac{\cos(2n\pi z)}{n}.$$

For the coefficients of the r.h.s. Fourier expansions see also *p.* 371

Addition theorems of the theta functions

We denote by

$$u' = \frac{1}{2}(u+v+w+z), \quad w' = \frac{1}{2}(u-v+w-z),$$

$$v' = \frac{1}{2}(u+v-w-z), \quad z' = \frac{1}{2}(u-v-w+z),$$

then

$$u'^2 + v'^2 + w'^2 + z'^2 = u^2 + v^2 + w^2 + z^2$$

and the following relations exist

$$\vartheta_3(u)\,\vartheta_3(w)\,\vartheta_3(v)\,\vartheta_3(z) + \vartheta_2(u)\,\vartheta_2(w)\,\vartheta_2(v)\,\vartheta_2(z)$$
$$= \vartheta_3(u')\,\vartheta_3(w')\,\vartheta_3(v')\,\vartheta_3(z') + \vartheta_2(u')\,\vartheta_2(w')\,\vartheta_2(v')\,\vartheta_2(z'),$$

$$\vartheta_3(u)\,\vartheta_3(v)\,\vartheta_3(w)\,\vartheta_3(z) - \vartheta_2(u)\,\vartheta_2(v)\,\vartheta_2(w)\,\vartheta_2(z)$$
$$= \vartheta_0(u')\,\vartheta_0(v')\,\vartheta_0(w')\,\vartheta_0(z') + \vartheta_1(u')\,\vartheta_1(v')\,\vartheta_1(w')\,\vartheta_1(z').$$

If we use the abbreviations

$(\lambda, \mu, \nu, \varrho)$ for $\vartheta_\lambda(u), \vartheta_\mu(v), \vartheta_\nu(w), \vartheta_\varrho(z)$,

$(\lambda, \mu, \nu, \varrho)'$ for $\vartheta_\lambda(u'), \vartheta_\mu(v'), \vartheta_\nu(w'), \vartheta_\varrho(z')$,

$[\lambda, \mu, \nu, \varrho]$ for $\vartheta_\lambda, \vartheta_\mu, \vartheta_\nu(u+v), \vartheta_\varrho(u-v)$,

$\{\lambda, \mu, \nu, \varrho\}$ for $\vartheta_\lambda(u), \vartheta_\mu(u), \vartheta_\nu(v), \vartheta_\varrho(v)$, where $\lambda, \mu, \nu, \varrho$ denotes any of the numbers 0, 1, 2, 3 [$\vartheta = \vartheta(0)$] and u', v', w', z' are defined above, the following relations hold

$$(3\,3\,3\,3) + (2\,2\,2\,2) = (3\,3\,3\,3)' + (2\,2\,2\,2)',$$
$$(3\,3\,3\,3) - (2\,2\,2\,2) = (0\,0\,0\,0)' + (1\,1\,1\,1)',$$
$$(0\,0\,0\,0) + (1\,1\,1\,1) = (3\,3\,3\,3)' - (2\,2\,2\,2)',$$
$$(0\,0\,0\,0) - (1\,1\,1\,1) = (0\,0\,0\,0)' - (1\,1\,1\,1)',$$
$$(0\,0\,3\,3) + (1\,1\,2\,2) = (0\,0\,3\,3)' + (1\,1\,2\,2)',$$
$$(0\,0\,3\,3) - (1\,1\,2\,2) = (3\,3\,0\,0)' + (2\,2\,1\,1)',$$
$$(0\,0\,2\,2) + (1\,1\,3\,3) = (0\,0\,2\,2)' + (1\,1\,3\,3)',$$
$$(0\,0\,2\,2) - (1\,1\,3\,3) = (2\,2\,0\,0)' + (3\,3\,1\,1)',$$
$$(3\,3\,2\,2) + (0\,0\,1\,1) = (3\,3\,2\,2)' + (0\,0\,1\,1)',$$
$$(3\,3\,2\,2) - (0\,0\,1\,1) = (2\,2\,3\,3)' + (1\,1\,0\,0)',$$
$$(3\,2\,0\,1) + (2\,3\,1\,0) = (1\,0\,2\,3)' - (0\,1\,3\,2)',$$
$$(3\,2\,0\,1) - (2\,3\,1\,0) = (3\,2\,0\,1)' - (2\,3\,1\,0)';$$

$$[3\,3\,3\,3] = \{3\,3\,3\,3\} + \{1\,1\,1\,1\} = \{0\,0\,0\,0\} + \{2\,2\,2\,2\},$$
$$[3\,3\,0\,0] = \{0\,0\,3\,3\} + \{2\,2\,1\,1\} = \{3\,3\,0\,0\} + \{1\,1\,2\,2\},$$
$$[3\,3\,2\,2] = \{2\,2\,3\,3\} - \{0\,0\,1\,1\} = \{3\,3\,2\,2\} - \{1\,1\,0\,0\},$$
$$[3\,3\,1\,1] = \{1\,1\,3\,3\} - \{3\,3\,1\,1\} = \{0\,0\,2\,2\} - \{2\,2\,0\,0\},$$
$$[0\,0\,3\,3] = \{0\,0\,3\,3\} - \{1\,1\,2\,2\} = \{3\,3\,0\,0\} - \{2\,2\,1\,1\},$$
$$[0\,0\,0\,0] = \{3\,3\,3\,3\} - \{2\,2\,2\,2\} = \{0\,0\,0\,0\} - \{1\,1\,1\,1\},$$
$$[0\,0\,2\,2] = \{0\,0\,2\,2\} - \{1\,1\,3\,3\} = \{2\,2\,0\,0\} - \{3\,3\,1\,1\},$$
$$[0\,0\,1\,1] = \{3\,3\,2\,2\} - \{2\,2\,3\,3\} = \{1\,1\,0\,0\} - \{0\,0\,1\,1\},$$
$$[2\,2\,3\,3] = \{3\,3\,2\,2\} + \{0\,0\,1\,1\} = \{2\,2\,3\,3\} + \{1\,1\,0\,0\},$$
$$[2\,2\,0\,0] = \{0\,0\,2\,2\} + \{3\,3\,1\,1\} = \{1\,1\,3\,3\} + \{2\,2\,0\,0\},$$
$$[2\,2\,2\,2] = \{2\,2\,2\,2\} - \{1\,1\,1\,1\} = \{3\,3\,3\,3\} - \{0\,0\,0\,0\},$$
$$[2\,2\,1\,1] = \{1\,1\,2\,2\} - \{2\,2\,1\,1\} = \{0\,0\,3\,3\} - \{3\,3\,0\,0\},$$
$$[0\,2\,0\,2] = \{0\,2\,0\,2\} + \{1\,3\,1\,3\}, \quad [0\,2\,2\,0] = \{0\,2\,0\,2\} - \{1\,3\,1\,3\},$$
$$[3\,2\,3\,2] = \{3\,2\,3\,2\} + \{0\,1\,0\,1\}, \quad [3\,2\,2\,3] = \{3\,2\,3\,2\} - \{0\,1\,0\,1\},$$
$$[0\,3\,0\,3] = \{0\,3\,0\,3\} + \{1\,2\,1\,2\}, \quad [0\,3\,3\,0] = \{0\,3\,0\,3\} - \{1\,2\,1\,2\},$$
$$[0\,2\,1\,3] = \{1\,3\,0\,2\} + \{0\,2\,1\,3\}, \quad [0\,2\,3\,1] = \{1\,3\,0\,2\} - \{0\,2\,1\,3\},$$
$$[3\,2\,1\,0] = \{0\,1\,3\,2\} + \{3\,2\,0\,1\}, \quad [3\,2\,0\,1] = \{0\,1\,3\,2\} - \{3\,2\,0\,1\},$$
$$[0\,3\,1\,2] = \{1\,2\,0\,3\} + \{0\,3\,1\,2\}, \quad [0\,3\,2\,1] = \{1\,2\,0\,3\} - \{0\,3\,1\,2\},$$

From these tables one obtains among others for $u = v$

$$\vartheta_0^2\vartheta_3\vartheta_3(2u) = \vartheta_0^2(u)\,\vartheta_3^2(u) - \vartheta_1^2(u)\,\vartheta_2^2(u),$$
$$\vartheta_0^3\vartheta_0(2u) = \vartheta_3^4(u) - \vartheta_2^4(u),$$
$$\vartheta_0^2\vartheta_2\vartheta_2(2u) = \vartheta_0^2(u)\,\vartheta_2^2(u) - \vartheta_1^2(u)\,\vartheta_3^2(u),$$
$$\vartheta_0\vartheta_2\vartheta_3\vartheta_1(2u) = 2\vartheta_0(u)\,\vartheta_1(u)\,\vartheta_2(u)\,\vartheta_3(u).$$

10.3 Definition of the Jacobi elliptic functions by the theta functions

Two properties depending on K and K' are defined by

$$K(\tau) \equiv K = \frac{1}{2}\pi\vartheta_3^2, \quad K'(\tau) \equiv K' = -i\tau K.$$

If instead of τ new parameters k and k' are introduced given by

$$k = \left(\frac{\vartheta_2}{\vartheta_3}\right)^2, \quad k' = \left(\frac{\vartheta_0}{\vartheta_3}\right)^2 \quad [\text{where } k^2 + k'^2 = 1],$$

then $K(\tau)$ and $K'(\tau)$ can be regarded as functions of k or (and) k' [$K(k)$ and $K(k')$]. These are the properties defined in 10.1 Then the Jacobian elliptic functions are defined by

$$\operatorname{sn}(z, k) = k^{-\frac{1}{2}} \frac{\vartheta_1\left(\frac{\frac{1}{2}z}{K}\right)}{\vartheta_0\left(\frac{\frac{1}{2}z}{K}\right)},$$

$$\operatorname{cn}(z, k) = \left(\frac{k'}{k}\right)^{\frac{1}{2}} \frac{\vartheta_2\left(\frac{\frac{1}{2}z}{K}\right)}{\vartheta_0\left(\frac{\frac{1}{2}z}{K}\right)},$$

$$\operatorname{dn}(z, k) = k'^{\frac{1}{2}} \frac{\vartheta_3\left(\frac{\frac{1}{2}z}{K}\right)}{\vartheta_0\left(\frac{\frac{1}{2}z}{K}\right)}.$$

The parameter q which occurs in the theta functions can be expressed by the new parameters as

$$q = e^{i\pi\tau} = e^{-\frac{\pi K'}{K}},$$

with

$$K = \int_0^{\frac{\pi}{2}} (1 - k^2 \sin^2 t)^{-\frac{1}{2}}\, dt, \quad K' = K\left[(1 - k^2)^{-\frac{1}{2}}\right].$$

The connection between q and k can also be given by a series valid for $|k| \leq 1$

$$q^{\frac{1}{4}} = \left(\frac{k}{4}\right)^{\frac{1}{2}} \left[1 + 2\left(\frac{k}{4}\right)^2 + 15\left(\frac{k}{4}\right)^4 + 150\left(\frac{k}{4}\right)^6 + 1707\left(\frac{k}{4}\right)^8 + \cdots\right].$$

If one puts

$$L = \frac{1 - (1 - k^2)^{\frac{1}{4}}}{1 + (1 - k^2)^{\frac{1}{4}}}$$

one obtains the rapidly convergent series

$$q = \frac{1}{2} L + \frac{2}{2^5} L^5 + \frac{15}{2^9} L^9 + \frac{150}{2^{13}} L^{13} + \frac{1707}{2^{17}} L^{17} + \cdots.$$

The functions $\operatorname{sn} z$, $\operatorname{cn} z$, $\operatorname{dn} z$ are elliptic functions with the periods $4K$ and $2iK'$ for sn, $4K$ and $2K + i\,2K'$ for cn, $2K$ and $i\,4K'$ for dn.

It is customary to use separate symbols for quotients and reciprocals of these functions

$$\operatorname{ns} z = \frac{1}{\operatorname{sn} z}, \qquad \operatorname{nc} z = \frac{1}{\operatorname{cn} z}, \qquad \operatorname{ds} z = \frac{\operatorname{dn} z}{\operatorname{sn} z},$$

$$\operatorname{sc} z = \frac{\operatorname{sn} z}{\operatorname{cn} z}, \qquad \operatorname{sd} z = \frac{\operatorname{sn} z}{\operatorname{dn} z}, \qquad \operatorname{dc} z = \frac{\operatorname{dn} z}{\operatorname{cn} z},$$

$$\operatorname{cs} z = \frac{\operatorname{cn} z}{\operatorname{sn} z}, \qquad \operatorname{cd} z = \frac{\operatorname{cn} z}{\operatorname{dn} z}, \qquad \operatorname{nd} z = \frac{1}{\operatorname{dn} z}.$$

Power series expansions

$$\operatorname{sn} z = z - (1 + k^2)\frac{z^3}{3!} + (1 + 14k^2 + k^4)\frac{z^5}{5!} - \cdots,$$

$$\operatorname{cn} z = 1 - \frac{z^2}{2!} + (1 + 4k^2)\frac{z^4}{4!} - (1 + 44k^2 + 16k^4)\frac{z^6}{6!} + \cdots,$$

$$\operatorname{dn} z = 1 - k^2\frac{z^2}{2!} + k^2(4 + k^2)\frac{z^4}{4!} - k^2(16 + 44k^2 + k^4)\frac{z^6}{6!} + \cdots.$$

Addition theorems and related formulas

$$\operatorname{sn}(u + v) = \frac{\operatorname{sn} u \operatorname{cn} v \operatorname{dn} v + \operatorname{sn} v \operatorname{cn} u \operatorname{dn} u}{1 - k^2 \operatorname{sn}^2 u \operatorname{sn}^2 v},$$

$$\operatorname{cn}(u + v) = \frac{\operatorname{cn} u \operatorname{cn} v - \operatorname{sn} u \operatorname{dn} u \operatorname{sn} v \operatorname{dn} v}{1 - k^2 \operatorname{sn}^2 u \operatorname{sn}^2 v},$$

$$\operatorname{dn}(u + v) = \frac{\operatorname{dn} u \operatorname{dn} v - k^2 \operatorname{sn} u \operatorname{cn} u \operatorname{sn} v \operatorname{cn} v}{1 - k^2 \operatorname{sn}^2 u \operatorname{sn}^2 v},$$

$$\operatorname{sn}^2 u = (1 - \operatorname{cn} 2u)(1 + \operatorname{dn} 2u)^{-1},$$

$$\operatorname{cn}^2 u = (\operatorname{cn} 2u + \operatorname{dn} 2u)(1 + \operatorname{dn} 2u)^{-1},$$

$$\operatorname{dn}^2 u = (\operatorname{dn} 2u + k^2 \operatorname{cn} 2u + k'^2)(1 + \operatorname{dn} 2u)^{-1},$$

$$(1 - \operatorname{dn} 2u)(1 + \operatorname{dn} 2u)^{-1} = k^2 \operatorname{sn}^2 u \operatorname{cn}^2 u \operatorname{nd}^2 u,$$

$$(1 - \operatorname{cn} 2u)(1 + \operatorname{cn} 2u)^{-1} = \operatorname{sn}^2 u \operatorname{dn}^2 u \operatorname{nc}^2 u.$$

Differential quotients and differential equations

$$(\mathrm{sn}\,u)' = \mathrm{cn}\,u\,\mathrm{dn}\,u, \quad (\mathrm{cn}\,u)' = -\,\mathrm{sn}\,u\,\mathrm{dn}\,u,$$
$$(\mathrm{dn}\,u)' = -\,k^2\,\mathrm{sn}\,u\,\mathrm{cn}\,u,$$
$$[(\mathrm{sn}\,u)']^2 = (1-\mathrm{sn}^2\,u)\,(1-k^2\,\mathrm{sn}^2\,u),$$
$$[(\mathrm{cn}\,u)']^2 = (1-\mathrm{cn}^2\,u)\,(k'^2+k^2\,\mathrm{cn}^2\,u),$$
$$[(\mathrm{dn}\,u)']^2 = -\,(1-\mathrm{dn}^2\,u)\,(k'^2-\mathrm{dn}^2\,u).$$

Fourier series expansions for the elliptic functions and some of their combinations

$$2Kk\,\mathrm{sn}\,(2Kz) = 4\pi\sum_{n=0}^{\infty}\frac{q^{n+\frac{1}{2}}}{1-q^{2n+1}}\sin\,[(2n+1)\,\pi z],$$

$$2Kk\,\mathrm{cn}\,(2Kz) = 4\pi\sum_{n=0}^{\infty}\frac{q^{n+\frac{1}{2}}}{1+q^{2n+1}}\cos\,[(2n+1)\,\pi z],$$

$$2K\,\mathrm{dn}\,(2Kz) = \pi + 4\pi\sum_{n=1}^{\infty}\frac{q^n}{1+q^{2n}}\cos\,(2n\pi z).$$

These expansions are valid for $|\mathrm{Im}\,(z)| < \mathrm{Im}\,(\tau)$

$$2K\,\mathrm{ns}\,(2Kz) = \pi\,\mathrm{cosec}\,(\pi z) + 4\pi\sum_{n=0}^{\infty}\frac{q^{2n+1}}{1-q^{2n+1}}\sin\,[(2n+1)\,\pi z],$$

$$2Kk'\,\mathrm{nc}\,(2Kz) = \pi\sec\,(\pi z) - 4\pi\sum_{n=0}^{\infty}\frac{(-1)^n\,q^{2n+1}}{1-q^{2n+1}}\cos\,[(2n+1)\,\pi z],$$

$$2Kk'\,\mathrm{nd}\,(2Kz) = \pi + 4\pi\sum_{n=1}^{\infty}\frac{(-1)^n\,q^n}{1+q^{2n}}\cos\,(2n\pi z),$$

$$2K\,\mathrm{cs}\,(2Kz) = \pi\cot\,(\pi z) - 4\pi\sum_{n=0}^{\infty}\frac{q^{2n}}{1+q^{2n}}\sin\,(2\pi n z),$$

$$2K\,\mathrm{ds}\,(2Kz) = \pi\,\mathrm{cosec}\,(\pi z) - 4\pi\sum_{n=0}^{\infty}\frac{q^{2n+1}}{1+q^{2n+1}}\sin\,[(2n+1)\,\pi z],$$

$$2K\,\mathrm{dc}\,(2Kz) = \pi\sec\,(\pi z) + 4\pi\sum_{n=0}^{\infty}\frac{(-1)^n\,q^{2n+1}}{1-q^{2n+1}}\cos\,[(2n+1)\,\pi z],$$

$$2k'K\,\mathrm{sc}\,(2Kz) = \tan\,(\pi z) + 4\sum_{n=1}^{\infty}\frac{(-1)^n\,q^{2n}}{1+q^{2n}}\sin\,(2\pi n z),$$

$$2kk'K\,\mathrm{sd}\,(2Kz) = \sum_{n=0}^{\infty}\frac{(-1)^n\,q^{n+\frac{1}{2}}}{1+q^{2n+1}}\sin\,[(2n+1)\,\pi z],$$

$$kK\,\mathrm{cd}\,(2Kz) = 2\pi\sum_{n=0}^{\infty}\frac{(-1)^n\,q^{n+\frac{1}{2}}}{1-q^{2n+1}}\cos\,[(2n+1)\,\pi z].$$

These series are valid for $\mathrm{Im}\,(z) < \mathrm{Im}\,(\tau)$.

If one considers that $q = \exp\left(-\frac{\pi K'}{K}\right)$ one can also write the coefficients occuring in the series above in the form

$$\frac{q^{n+\frac{1}{2}}}{1 - q^{2n+1}} = \frac{1}{2} \operatorname{cosech}\left[\frac{\left(n + \frac{1}{2}\right)\pi K'}{K}\right],$$

$$\frac{q^{n+\frac{1}{2}}}{1 + q^{2n+1}} = \frac{1}{2} \operatorname{sech}\left[\frac{\left(n + \frac{1}{2}\right)\pi K'}{K}\right],$$

$$\frac{q^n}{1 + q^{2n}} = \frac{1}{2} \operatorname{sech}\left(\frac{n\pi K'}{K}\right),$$

$$\frac{q^n}{1 - q^{2n}} = \frac{1}{2} \operatorname{cosech}\left(\frac{n\pi K'}{K}\right).$$

Further trigonometric expansions are

$$2kK \operatorname{sn}(2Kz) = \pi \sum_{n=-\infty}^{\infty} \operatorname{cosec}\left[\pi z - \frac{\left(n + \frac{1}{2}\right) i\pi K'}{K}\right],$$

$$2kK \operatorname{cn}(2Kz) = i\pi \sum_{n=-\infty}^{\infty} (-1)^n \operatorname{cosec}\left[\pi z - \frac{\left(n + \frac{1}{2}\right) i\pi K'}{K}\right],$$

$$2K \operatorname{dn}(2Kz) = i\pi \sum_{n=-\infty}^{\infty} (-1)^n \left\{\tan\left[\pi z - \frac{\left(n + \frac{1}{2}\right) i\pi K'}{K}\right]\right\}^{-1}.$$

Increasing of the arguments of the Jacobian elliptic functions. If the argument of one of the Jacobian elliptic functions is increased by an even multiple of K or iK' the resulting function can again be expressed in terms of Jacobian elliptic functions. This is demonstrated in the tables below. Here m and n is one of the numbers $0, \pm 1, \pm 2, \ldots$

Increasing of the argument	sn z	cn z	dn z
$2mK + 2niK'$	$(-1)^m \operatorname{sn} z$	$(-1)^{m+n} \operatorname{cn} z$	$(-1)^n \operatorname{dn} z$
$(2m-1)K + 2niK'$	$(-1)^{m+1} \operatorname{cd} z$	$(-1)^{m+n} k' \operatorname{sd} z$	$(-1)^n \operatorname{nd} z$
$2mK + (2n+1)iK'$	$(-1)^m k^{-1} \operatorname{ns} z$	$(-1)^{m+n+1} ik^{-1} \operatorname{ds} z$	$i(-1)^{n+1} \operatorname{cs} z$
$(2m-1)K + (2n+1)iK'$	$(-1)^{m+1} k^{-1} \operatorname{dc} z$	$(-1)^{m+n} ik'k^{-1} \operatorname{nc} z$	$(-1)^n ik' \operatorname{sc} z$

Values of the Jacobian elliptic functions at $z = \frac{n}{2}K + i\frac{m}{2}K'$

		0	$\frac{1}{2}K$	K	$\frac{3}{2}K$	$2K$
0	sn	0	$(1+k')^{-\frac{1}{2}}$	1	$(1+k')^{-\frac{1}{2}}$	0
	cn	1	$(1+k'^{-1})^{-\frac{1}{2}}$	0	$-(1+k'^{-1})^{-\frac{1}{2}}$	-1
	dn	1	$k'^{\frac{1}{2}}$	k'	$k'^{\frac{1}{2}}$	1
$\frac{1}{2}iK'$	sn	$ik^{-\frac{1}{2}}$	$(2k)^{-\frac{1}{2}}\left[(1+k)^{\frac{1}{2}}+i(1-k)^{\frac{1}{2}}\right]$	$k^{-\frac{1}{2}}$	$(2k)^{-\frac{1}{2}}\left[(1+k)^{\frac{1}{2}}-i(1-k)^{\frac{1}{2}}\right]$	$-ik^{-\frac{1}{2}}$
	cn	$(1+k^{-1})^{\frac{1}{2}}$	$\left[\frac{k'}{(2k)}\right]^{\frac{1}{2}}(1-i)$	$-i(k^{-1}-1)^{\frac{1}{2}}$	$-\left[\frac{k'}{(2k)}\right]^{\frac{1}{2}}(1+i)$	$-(1+k^{-1})^{\frac{1}{2}}$
	dn	$(1+k)^{\frac{1}{2}}$	$\left(\frac{k'}{2}\right)^{\frac{1}{2}}\left[(1+k')^{\frac{1}{2}}-(1-k')^{\frac{1}{2}}\right]$	$(1-k)^{\frac{1}{2}}$	$\left(\frac{k'}{2}\right)\left[(1+k')^{\frac{1}{2}}+i(1-k')^{\frac{1}{2}}\right]$	$(1+k)^{\frac{1}{2}}$
iK'	sn	Res. $=k^{-1}$	$(1-k')^{-\frac{1}{2}}$	k^{-1}	$(1-k')^{-\frac{1}{2}}$	Res. $=-k^{-1}$
	cn	Res. $=-ik^{-1}$	$-i(k'^{-1}-1)^{-\frac{1}{2}}$	$-\frac{ik'}{k}$	$-i(k'^{-1}-1)^{-\frac{1}{2}}$	Res. $=ik^{-1}$
	dn	Res. $=-i$	$-ik'^{\frac{1}{2}}$	0	$ik'^{\frac{1}{2}}$	Res. $=-i$

		0	$\frac{1}{2}K$	K	$\frac{3}{2}K$	$2K$
$\frac{3}{2}iK'$	sn	$-ik^{-\frac{1}{2}}$	$(2k)^{-\frac{1}{2}}\left[(1+k)^{\frac{1}{2}}-i(1-k)^{\frac{1}{2}}\right]$	$k^{-\frac{1}{2}}$	$(2k)^{-\frac{1}{2}}\left[(1+k)^{\frac{1}{2}}+i(1-k)^{\frac{1}{2}}\right]$	$ik^{-\frac{1}{2}}$
	cn	$-(1+k^{-1})^{\frac{1}{2}}$	$-\left[\frac{k'}{(2k)}\right]^{\frac{1}{2}}(1+i)$	$-i(k^{-1}-1)^{-\frac{1}{2}}$	$\left[\frac{k'}{(2k)}\right]^{\frac{1}{2}}(1-i)$	$(1+k^{-1})^{\frac{1}{2}}$
	dn	$-(1+k)^{\frac{1}{2}}$	$-\left(\frac{1}{2}k'\right)^{\frac{1}{2}} \times \left[(1+k')^{\frac{1}{2}}+i(1-k')^{\frac{1}{2}}\right]$	$-(1-k')^{\frac{1}{2}}$	$-\left(\frac{1}{2}k'\right)^{\frac{1}{2}} \times \left[(1+k')^{\frac{1}{2}}-i(1-k')^{\frac{1}{2}}\right]$	$-(1+k)^{\frac{1}{2}}$
$2iK'$	sn	0	$(1+k')^{-\frac{1}{2}}$	1	$(1+k')^{-\frac{1}{2}}$	0
	cn	-1	$-(1+k'^{-1})^{-\frac{1}{2}}$	0	$(1+k'^{-1})^{-\frac{1}{2}}$	1
	dn	-1	$-k'^{\frac{1}{2}}$	$-k'$	$-k'^{\frac{1}{2}}$	-1

The remark Res. $= k^{-1}$ for instance in the third column means that the function $\operatorname{sn} z$ has a pole of the first order with residue k^{-1} at the point $z = iK'$ etc.

The location of the zeros and of the poles can be obtained from the table below

Function	Zeros	Poles
$\operatorname{sn} z$	$2nK + 2miK'$	$2nK + (2m+1)iK'$
$\operatorname{cn} z$	$(2n+1)K + 2miK'$	$2nK + (2m+1)iK'$
$\operatorname{dn} z$	$(2n+1)K + (2m+1)iK'$	$2nK + (2m+1)iK'$

Imaginary argument

$$\operatorname{sn}(iz, k) = i \operatorname{sn}(z, k') \operatorname{nc}(z, k'),$$

$$\operatorname{cn}(iz, k) = \operatorname{nc}(z, k'),$$

$$\operatorname{dn}(iz, k) = \operatorname{dn}(z, k') \operatorname{nc}(z, k').$$

Integrals involving Jacobian elliptic functions

$$\int \operatorname{sn} z \, dz = k^{-1} \log(\operatorname{dn} z + k \operatorname{cn} z),$$

$$\int \operatorname{cn} z \, dz = i k^{-1} \log(\operatorname{dn} z - i k \operatorname{sn} z),$$

$$\int \operatorname{dn} z \, dz = i \log(\operatorname{cn} z - i \operatorname{sn} z),$$

$$\int \operatorname{ns} z \, dz = \log[\operatorname{ns} z (\operatorname{dn} z - \operatorname{cn} z)],$$

$$\int \operatorname{nc} z \, dz = k'^{-1} \log[\operatorname{nc} z (\operatorname{dn} z + k' \operatorname{sn} z)],$$

$$\int \operatorname{nd} z \, dz = -i k'^{-1} \log[\operatorname{nd} z (\operatorname{cn} z + i k' \operatorname{sn} z)],$$

$$\int \operatorname{sn} z \operatorname{nc} z = k'^{-1} \log[\operatorname{nc} z (\operatorname{dn} z + k')],$$

$$\int \operatorname{sn} z \operatorname{nd} z = i (k k')^{-1} \log[\operatorname{nd} z (i k' - k \operatorname{cn} z)],$$

$$\int \operatorname{cn} z \operatorname{nd} z = -k^{-1} \log[\operatorname{nd} z (1 - k \operatorname{sn} z)],$$

$$\int \operatorname{cn} z \operatorname{ns} z \, dz = \log[\operatorname{ns} z (1 - \operatorname{dn} z)],$$

$$\int \operatorname{dn} z \operatorname{ns} z \, dz = \log[\operatorname{ns} z (1 - \operatorname{cn} z)],$$

$$\int \mathrm{dn}\, z\, \mathrm{nc}\, z\, \mathrm{dz} = \log [\mathrm{nc}\, z\, (1 + \mathrm{sn}\, z)],$$

$$\int \mathrm{sn}\, z\, \mathrm{cn}\, z\, \mathrm{dz} = -k^{-2}\, \mathrm{dn}\, z,$$

$$\int \mathrm{sn}\, z\, \mathrm{dn}\, z\, \mathrm{dz} = -\mathrm{cn}\, z,$$

$$\int \mathrm{cn}\, z\, \mathrm{dn}\, z\, \mathrm{dz} = \mathrm{sn}\, z,$$

$$\int \mathrm{ns}\, z\, \mathrm{nc}\, z\, \mathrm{dz} = \int \mathrm{cn}\, z\, \mathrm{ns}\, z\, \mathrm{dz} + \int \mathrm{sn}\, z\, \mathrm{nc}\, z\, \mathrm{dz},$$

$$\int \mathrm{dn}^2\, z\, \mathrm{dz} = \mathrm{zn}\, z + \frac{E}{K}\, z,$$

$$\int \mathrm{ns}^2\, z\, \mathrm{dz} = \left(1 - \frac{E}{K}\right) z - \frac{d}{dz} \ln \vartheta_1 \left(\frac{z}{2K}\right),$$

$$\int \mathrm{nc}^2\, z\, \mathrm{dz} = k'^{-2} \left[\left(k'^2 - \frac{E}{K}\right) z - \frac{d}{dz} \log \vartheta_2 \left(\frac{z}{2K}\right)\right],$$

$$\int \mathrm{nd}^2\, z\, \mathrm{dz} = k'^{-2} \left[\frac{E}{K}\, z + \frac{d}{dz} \log \vartheta_3 \left(\frac{z}{2K}\right)\right],$$

$$\int \mathrm{cn}\, z\, \mathrm{dn}\, z\, \mathrm{ns}\, z\, \mathrm{dz} = \log \mathrm{sn}\, z,$$

$$\int \mathrm{sn}\, z\, \mathrm{dn}\, z\, \mathrm{nc}\, z\, \mathrm{dz} = -\log \mathrm{cn}\, z,$$

$$\int \mathrm{sn}\, z\, \mathrm{cn}\, z\, \mathrm{nd}\, z = -k^{-2} \log \mathrm{dn}\, z,$$

$$\int \mathrm{sn}\, z\, \mathrm{nc}\, z\, \mathrm{nd}\, z\, \mathrm{dz} = k'^{-2} \log (\mathrm{dn}\, z\, \mathrm{nc}\, z),$$

$$\int \mathrm{cn}\, z\, \mathrm{ns}\, z\, \mathrm{nd}\, z = \log (\mathrm{sn}\, z\, \mathrm{nd}\, z),$$

$$\int \mathrm{dn}\, z\, \mathrm{ns}\, z\, \mathrm{nc}\, z\, \mathrm{dz} = \log (\mathrm{sn}\, z\, \mathrm{nc}\, z),$$

$$\int \mathrm{sn}\, z\, \mathrm{nc}^2\, z\, \mathrm{dz} = k'^{-2}\, \mathrm{dn}\, z\, \mathrm{nc}\, z,$$

$$\int \mathrm{cn}\, z\, \mathrm{ns}^2\, z\, \mathrm{dz} = -\mathrm{dn}\, z\, \mathrm{ns}\, z,$$

$$\int \mathrm{dn}\, z\, \mathrm{ns}^2\, z\, \mathrm{dz} = -\mathrm{cn}\, z\, \mathrm{ns}\, z,$$

$$\int \mathrm{sn}\, z\, \mathrm{nd}^2\, z\, \mathrm{dz} = -k'^{-2}\, \mathrm{cn}\, z\, \mathrm{nd}\, z,$$

$$\int \mathrm{cn}\, z\, \mathrm{nd}^2\, z\, \mathrm{dz} = \mathrm{sn}\, z\, \mathrm{nd}\, z,$$

$$\int \mathrm{dn}\, z\, \mathrm{nc}^2\, z\, \mathrm{dz} = \mathrm{sn}\, z\, \mathrm{nc}\, z.$$

The table below shows the transformation of the Jacobian elliptic function of the argument z and the modulus k when these properties are changed into different values z_1 and k_1.

z_1	k_1	$\mathrm{sn}\,(z_1, k_1)$	$\mathrm{cn}\,(z_1, k_1)$	$\mathrm{dn}\,(z_1, k_1)$
$k\,z$	$\frac{1}{k}$	$k\,\mathrm{sn}$	dn	cn
$i\,z$	k'	$i\,\mathrm{sc}$	nc	dc
$k'\,z$	$i\,\frac{k}{k'}$	$k'\,\mathrm{sd}$	cd	nd
$i k\,z$	$i\,\frac{k'}{k}$	$i k\,\mathrm{sd}$	nd	cd
$i k'\,z$	$\frac{1}{k'}$	$i k'\,\mathrm{sc}$	dc	nc
$(1+k)\,z$	$\frac{2\sqrt{k}}{1+k}$	$\frac{(1+k)\,\mathrm{sn}}{1+k\,\mathrm{sn}^2}$	$\frac{\mathrm{cn}\,\mathrm{dn}}{1+k\,\mathrm{sn}^2}$	$\frac{1-k\,\mathrm{sn}^2}{1+k\,\mathrm{sn}^2}$
$(1+k')\,z$	$\frac{1-k'}{1+k'}$	$(1+k')\,\mathrm{sd}\,\mathrm{cn}$	$\frac{1-(1+k')\,\mathrm{sn}^2}{\mathrm{dn}}$	$1-(1-k')\,\frac{\mathrm{sn}^2}{\mathrm{dn}}$
$\frac{(1+k')^2\,z}{2}$	$\left(\frac{1-\sqrt{k'}}{1+\sqrt{k'}}\right)^2$	$\frac{k^2\,\mathrm{sn}\,\mathrm{cn}}{\sqrt{k_1}\,(1+\mathrm{dn})\,(k'+\mathrm{dn})}$	$\frac{\mathrm{dn}-\sqrt{k'}}{1-\sqrt{k'}}\sqrt{\frac{2(1+k')}{(1+\mathrm{dn})\,(\mathrm{dn}+k')}}$	$\frac{\sqrt{1+k_1}\,(\mathrm{dn}+\sqrt{k'})}{\sqrt{1+\mathrm{dn}}\,\sqrt{k'+\mathrm{dn}}}$

Definition of the Jacobian elliptic functions by the inverse function of the elliptic integral of the first kind. Consider in the elliptic integral of the first kind in Legendre's normal form

$$u(k,\varphi) \equiv F(k,\varphi) = \int_0^{\varphi} (1 - k^2 \sin^2 t)^{-\frac{1}{2}} dt$$

φ as function of u, then φ is called the amplitude of u

$$\varphi = \operatorname{am}(u,k).$$

φ is a periodic function of u with the period $i\,4K'$. The Jacobian elliptic functions can then be expressed in the form

$$\operatorname{sn}(u,k) = \sin\varphi = \sin\operatorname{am}(u,k),$$

$$\operatorname{cn}(u,k) = \cos\varphi = \cos\operatorname{am}(u,k),$$

$$\operatorname{dn}(u,k) = [(1 - k^2\sin^2\varphi)]^{\frac{1}{2}} = [1 - k^2\operatorname{sn}^2(u,k)]^{\frac{1}{2}}.$$

If one introduces in the elliptic integral of the first kind a new variable $x = \sin\varphi$ then

$$u(k,x) = \int_0^x [(1-t^2)(1-k^2t^2)]^{-\frac{1}{2}},$$

then

$$x = \operatorname{sn}(u,k).$$

For the function am (u, k) the following Fourier expansion is valid

$$\operatorname{am}(z,k) = \frac{\frac{1}{2}\pi z}{K} + \sum_{n=1}^{\infty} \frac{2q^n}{n(1+q^{2n})} \sin\left(\frac{n\pi z}{K}\right),$$

$$|\operatorname{Im}(z)| < \frac{1}{2}\pi \operatorname{Im}(\tau).$$

10.4 The Jacobian zeta function

The Jacobi zeta function is defined by

$$\operatorname{zn}(z,k) = \int_0^z \operatorname{dn}^2(t,k)\,\mathrm{dt}\,dt - \frac{E(k)}{K(k)} z$$

or

$$\operatorname{zn}(z,k) = E[k, \operatorname{am}(z,k)] - \frac{E(k)}{K(k)} z.$$

Connection with the theta function

$$\operatorname{zn}(z, k) = \frac{d}{dz} \log \vartheta_0\left(\frac{\frac{1}{2} z}{K}\right), \quad \operatorname{zn}(z + 2K) = \operatorname{zn} z.$$

Additions theorem

$$\operatorname{zn}(u + v) = \operatorname{zn} u + \operatorname{zn} v - k \operatorname{sn} u \operatorname{sn} v \operatorname{sn}(u + v).$$

Imaginary argument

$$\operatorname{zn}(iz, k) = i\left[\operatorname{sc}(z, k') \operatorname{dc}(z, k') - \operatorname{zn}(z, k') - \frac{\frac{1}{2}\pi z}{(K K')}\right].$$

Fourier expansion

$$\operatorname{zn}(z, k) = 2\pi K^{-1} \sum_{n=1}^{\infty} \frac{q^n}{1 - q^{2n}} \sin\left(\frac{n\pi z}{K}\right),$$

$$= \pi K^{-1} \sum_{n=1}^{\infty} \operatorname{cosech}\left(\frac{n\pi K'}{K}\right) \sin\left(\frac{n\pi z}{K}\right),$$

$$\left|\operatorname{Im}\left(\frac{\frac{1}{2}\pi z}{K}\right)\right| < \frac{1}{2} \operatorname{Im}(\tau).$$

Furthermore

$$\operatorname{zn}(nK, k) = 0 \quad \text{for} \quad n = 0, 1, 2, \ldots$$

10.5 The elliptic functions of Weierstrass

The Weierstrass $\wp$ function is an elliptic function with the periods 2ω and $2\omega'$. The quotient $\tau = \frac{\omega'}{\omega}$ of these two periods is not real and its imaginary part is positive. The $\wp$ function can be defined by

$$\wp(z) = z^{-2} + \sum_{n,m}{}' \left[(z - 2n\omega - 2m\omega')^{-2} - (2n\omega + 2m\omega')^{-2}\right].$$

The double sum on the r. h. s. has to be taken over all positive and negative integers n and m with the exception of $n = m = 0$. The Laurent expansion around the point $z = 0$ is:

$$\wp(z) = z^{-2} + \frac{g_2}{20} z^2 + \frac{g_3}{28} z^4 + \frac{g_2^2 z^6}{1200} + \frac{3 g_2 g_3}{6160} z^8 + \cdots.$$

Here two new parameters g_2 and g_3 are used instead of the periods 2ω and $2\omega'$. These properties are given by

$$g_2 = \frac{15}{4} \sum_{n,m}{}' (n\omega + m\omega')^{-4}, \quad g_3 = \frac{35}{16} \sum_{n,m}{}' (n\omega + m\omega')^{-6}.$$

If one wishes to emphasize the dependency on the parameters 2ω, $2\omega'$ or g_2, g_3 one writes respectively $p(z; 2\omega, 2\omega')$ or $p(z; g_2, g_3)$. The function $p(z)$ is a solution of the following differential equation of the first order

$$\begin{aligned} p'^2(z) &= \left(\frac{dp}{dz}\right)^2 = 4\{[p(z)-e_1]\,[p(z)-e_2]\,[p(z)-e_3]\} \\ &= 4p^3(z) - g_2\,p(z) - g_3, \end{aligned}$$

where the properties e_1, e_2, e_3 are given by

$$e_1 = p(\omega), \quad e_2 = p(\omega+\omega'), \quad e_3 = p(\omega'),$$

$$0 = p'(\omega), \quad 0 = p'(\omega+\omega'), \quad 0 = p'(\omega'),$$

$$p''(\omega) = 2(e_1-e_2)(e_1-e_3),$$

$$p''(\omega+\omega') = -2(e_1-e_2)(e_2-e_3),$$

$$p''(\omega') = 2(e_1-e_3)(e_2-e_3).$$

The relation between the parameters g and e is given by

$$e_1+e_2+e_3 = 0, \quad e_1e_2e_3 = \frac{1}{4}g_3, \quad e_1e_2+e_1e_3+e_2e_3 = -\frac{1}{4}g_2.$$

From the differential equation for $p(z)$ it follows that $y = p(z)$ is the inverse function of

$$z = \int_y^\infty (4t^3 - g_2t - g_3)^{-\frac{1}{2}}\,dt = \frac{1}{2}\int_y^\infty [(t-e_1)(t-e_2)(t-e_3)]^{-\frac{1}{2}}\,dt.$$

Addition theorem

$$p(z_1+z_2) = -p(z_1) - p(z_2) + \frac{1}{4}\left[\frac{p'(z_1)-p'(z_2)}{p(z_1)-p(z_2)}\right]^2.$$

Hence

$$p(z+\omega) = e_1 + \frac{(e_1-e_2)(e_1-e_3)}{p(z)-e_1},$$

$$p(z+\omega+\omega') = e_2 + \frac{(e_2-e_1)(e_2-e_3)}{p(z)-e_2},$$

$$p(z+\omega') = e_3 + \frac{(e_3-e_1)(e_3-e_2)}{p(z)-e_3}.$$

Double argument

$$p(2z) = \frac{\left[p^2(z) + \frac{1}{4}g_2\right]^2 + 2g_3p(z)}{4p^3(z) - g_2p(z) - g_3}.$$

Connection with the theta functions

$$\wp(z) = -\frac{\eta_1}{\omega} - \frac{d^2}{dz^2}\left[\log\left(\frac{\frac{1}{2}z}{\omega}\right)\right],$$

where

$$\eta_1 = -\frac{1}{12}\,\omega^{-1}\frac{\vartheta_1'''}{\vartheta_1'}, \qquad q = e^{\frac{i\pi\omega'}{\omega}}, \qquad \tau = \frac{\omega'}{\omega}.$$

Furthermore

$$[\wp(z)] - e_k]^{\frac{1}{2}} = \frac{1}{2}\,\omega^{-1}\frac{\vartheta_1'}{\vartheta_{k+1}}\,\frac{\vartheta_{k+1}\left(\frac{\frac{1}{2}z}{\omega}\right)}{\vartheta_1\left(\frac{\frac{1}{2}z}{\omega}\right)}, \quad k = 1, 2, 3; \;\; \vartheta_4 \equiv \vartheta_0$$

$$\wp'(z) = -\frac{1}{4}\,\omega^{-3}\,\frac{\vartheta_0\left(\frac{\frac{1}{2}z}{\omega}\right)\vartheta_2\left(\frac{\frac{1}{2}z}{\omega}\right)\vartheta_3\left(\frac{\frac{1}{2}z}{\omega}\right)\vartheta_1'^3}{\vartheta_1^3\left(\frac{\frac{1}{2}z}{\omega}\right)\vartheta_0\,\vartheta_2\,\vartheta_3}.$$

Connection with the Jacobian elliptic functions

$$\begin{aligned}\wp\left[z(e_1 - e_3)^{-\frac{1}{2}}\right] &= e_1 + (e_1 - e_3)\,\mathrm{cn}^2 z\,\mathrm{ns}^2 z\\ &= e_2 + (e_1 - e_3)\,\mathrm{dn}^2 z\,\mathrm{ns}^2 z\\ &= e_3 + (e_1 - e_3)\,\mathrm{ns}^2 z.\end{aligned}$$

The modulus of the Jacobian elliptic functions and the corresponding elliptic normal integral of the first kind are then

$$k = \left[\frac{(e_2 - e_3)}{(e_1 - e_3)}\right]^{\frac{1}{2}}; \; K(k) = \omega(e_1 - e_3)^{\frac{1}{2}}.$$

Expansions in trigonometric form

$$\wp(z) = -\eta_1\omega^{-1} + \left(\frac{\frac{1}{2}\pi}{\omega}\right)^2 \sum_{n=-\infty}^{\infty} \mathrm{cosec}^2\left(\frac{\frac{1}{2}\pi z}{\omega} + \frac{n\pi\omega'}{\omega}\right),$$

$$\wp(z) = -\eta_1\omega^{-1} + \left(\frac{\frac{1}{2}\pi}{\omega}\right)^2 \mathrm{cosec}^2\left(\frac{\frac{1}{2}\pi z}{\omega}\right) - 2\left(\frac{\pi}{\omega}\right)^2 \sum_{n=1}^{\infty} \frac{n q^{2n}}{1 - q^{2n}}\cos\left(\frac{n\pi z}{\omega}\right),$$

$$q = e^{\frac{i\pi\omega'}{\omega}}, \qquad \eta_1 = -\frac{1}{12}\,\omega^{-1}\frac{\vartheta_1'''}{\vartheta_1'}.$$

The Weierstrass zeta function

$$\zeta(z) = -\int \wp(z)\,dz, \; \zeta'(z) = -\wp(z).$$

It is sometimes customary to write $\zeta(z;\omega,\omega')$ or $\zeta(z;g_2,g_3)$ according as one wishes to emphasize the dependency on the parameters ω,ω' or g_2, g_3.

Expansion in partial fractions

$$\zeta(z) = z^{-1} + \sum_{n,m}{}' [(z-2n-2m)^{-1} + z(2n+2m)^{-2} + (2n+2m)^{-1}]$$

or, if one summation is carried out

$$\zeta(z) = z\frac{\eta_1}{\omega} + \frac{1}{2}\pi\omega^{-1}\cot\left(\frac{\frac{1}{2}\pi z}{\omega}\right) + \frac{1}{2}\pi\omega^{-1}$$

$$\times \sum_{1}^{\infty}\left[\cot\left(\frac{\frac{1}{2}\pi z}{\omega} + \frac{m\pi\omega'}{\omega}\right) + \cot\left(\frac{\frac{1}{2}\pi z}{\omega} - \frac{m\pi\omega'}{\omega}\right)\right].$$

Laurent expansion

$$\zeta(z) = z^{-1} - \frac{g_2}{60}z^3 - \frac{g_3}{140}z^5 - \cdots$$

If one increases the argument z by $2n\omega$ or $2nm\omega'$ one has

$$\zeta(z + 2n\omega + 2m\omega') = \zeta(z) + 2n\eta_1 + 2m\eta_2$$

$$(n, m = 0, \pm 1, \pm 2, \ldots),$$

$$\eta_1 = \zeta(\omega) = -\frac{1}{12}\omega^{-1}\frac{\vartheta_1'''}{\vartheta_1'},$$

$$\eta_2 = \zeta(\omega'), \quad \eta_3 = \zeta(\omega+\omega') = \eta_1 + \eta_2,$$

where (Legendre's relation)

$$\omega'\eta_1 + \omega\eta_2 = i\frac{1}{2}\pi.$$

Additions theorem

$$\zeta(z_1+z_2) = \zeta(z_1) + \zeta(z_2) + \frac{1}{2}\frac{\zeta''(z_1) - \zeta''(z_2)}{\zeta'(z_1) - \zeta'(z_2)}.$$

Connection with the theta function

$$\zeta(z) = \frac{\eta_1 z}{\omega} + \frac{d}{dz}\log\vartheta_1\left(\frac{\frac{1}{2}z}{\omega}\right).$$

Fourier expansion

$$\zeta(z) = \frac{\eta_1 z}{\omega} + \frac{1}{2}\pi\omega^{-1}\cot\left(\frac{\frac{1}{2}\pi z}{\omega}\right) + 2\pi\omega^{-1}\sum_{n=1}^{\infty}\frac{q^{2n}}{1-q^{2n}}\sin\left(\frac{n\pi z}{\omega}\right).$$

The Weierstrass sigma function

$$\sigma(z) = e^{\int \zeta(z)\,dz}, \quad \zeta(z) = \frac{\sigma'(z)}{\sigma(z)}.$$

Representation as an infinite product

$$\sigma(z) = z \prod_{n,m}{}' \left(1 - \frac{z}{v}\right) e^{\frac{z}{v} + \frac{1}{2}\left(\frac{z}{v}\right)^2},$$

$$v = 2n\omega + 2m\omega';\, n, m = 0, \pm 1, \pm 2, \ldots,$$

n, m not both zero.

Taylor series expansion around $z = 0$

$$\sigma(z) = z - \frac{g_2}{2^4 \cdot 3 \cdot 5} z^5 - \frac{g_3}{2^3 \cdot 3 \cdot 5 \cdot 7} z^7 - \frac{g_2^2}{2^9 \cdot 3^2 \cdot 5 \cdot 7} z^9 - \cdots.$$

If the argument z is increased by multiples of the periods of the function one obtains

$$\sigma(z + 2n\omega + 2m\omega') = \pm\, e^{2(n\eta_1 + m\eta_2)(z + n\omega + m\omega')}\, \sigma(z).$$

The positive sign has to be chosen when both, m and n are even. Connection with the theta function

$$\sigma(z) = 2\omega\, (\vartheta_1')^{-1}\, e^{\frac{\frac{1}{2}\eta_1 z^2}{\omega}}\, \vartheta_1\!\left(\frac{\frac{1}{2} z}{\omega}\right), \qquad \tau = \frac{\omega'}{\omega}.$$

Definition of $\sigma_\alpha(z)$

$$\sigma_\alpha(z) = (\vartheta_\alpha)^{-1}\, e^{\frac{\frac{1}{2}\eta_1 z^2}{\omega}}\, \vartheta_\alpha\!\left(\frac{\frac{1}{2} z}{\omega}\right), \quad \alpha = 0, 2, 3.$$

Substitution of half periods

$$\sigma(z + \omega) = e^{\eta_1 z}\, \sigma(\omega)\, \sigma_2(z),$$

$$\sigma(z + \omega') = e^{\eta_2 z}\, \sigma(\omega')\, \sigma_0(z),$$

$$\sigma(z + \omega + \omega') = e^{\eta_3 z} \sigma(\omega + \omega')\, \sigma_3(z).$$

Connection with the function $p(z)$

$$\frac{\sigma_2(z)}{\sigma(z)} = [p(z) - e_1]^{\frac{1}{2}}; \quad \frac{\sigma_3(z)}{\sigma(z)} = [p(z) - e_2]^{\frac{1}{2}},$$

$$\frac{\sigma_0(z)}{\sigma(z)} = [p(z) - e_3]^{\frac{1}{2}},$$

$$p(z_1) - p(z_2) = -\frac{\sigma(z_1 - z_2)\, \sigma(z_1 + z_2)}{\sigma^2(z_1)\, \sigma^2(z_2)}.$$

Connection with the Jacobian elliptic functions

$$\wp'(2z) = -\frac{2\sigma_0(z)\,\sigma_2(z)\,\sigma_3(z)}{\sigma^3(z)} = -\frac{\sigma(2z)}{\sigma^4(z)}.$$

Connection with the Jacobian elliptic functions

$$\frac{\sigma(z)}{\sigma_0(z)} = \omega K^{-1} \operatorname{sn}\left(\frac{Kz}{\omega}\right), \quad \frac{\sigma_2(z)}{\sigma_0(z)} = \operatorname{cn}\left(\frac{Kz}{\omega}\right),$$

$$\frac{\sigma_3(z)}{\sigma_0(z)} = \operatorname{dn}\left(\frac{Kz}{\omega}\right).$$

10.6 Connections between the parameters and special cases

The functions $\vartheta_\alpha(z)$, sn z, cn z, dn z, am z, zn z, $\wp(z)$, $\zeta(z)$, $\sigma(z)$ depend in addition to their variable z also on one or in the last three cases on 2 parameters. Since all functions can be expressed in terms of the theta functions or, the latter 3 also in terms of the Jacobian elliptic functions, the formulas that connect the parameters of these functions are listed below.

Function	Parameter
$\vartheta_\alpha(z)$	τ or $q = \exp(i\pi\tau)$
am z sn z cn z dn z zn z	$k = \left(\frac{\vartheta_2}{\vartheta_3}\right)^2$
$\wp(z)$ $\zeta(z)$ $\sigma(z)$	ω, ω' or g_2, g_3 or e_1, e_2, e_3; $(e_1 + e_2 + e_3 = 1)$

1. Given ω, ω'. Then

$$g_2 = \frac{15}{4}\sum_{n,m}{}'(n\omega + m\omega')^{-4}, \quad g_3 = \frac{35}{16}\sum_{n,m}{}'(n\omega + m\omega')^{-6},$$

$$e_1 = \wp(\omega), \quad e_3 = \wp(\omega'), \quad e_2 = \wp(\omega + \omega'),$$

or, with

$$q = \exp\left(\frac{i\pi\omega'}{\omega}\right)$$

also

$$e_1 = -\frac{\eta_1}{\omega} + \left(\frac{\frac{1}{2}\pi}{\omega}\right)^2 \left[1 + 2\sum_{n=1}^{\infty} \sec^2 (n\pi\tau)\right],$$

$$e_2 = -\frac{\eta_1}{\omega} + \frac{1}{2}\left(\frac{\pi}{\omega}\right)^2 \sum_{n=1}^{\infty} \sec^2\left[\left(n - \frac{1}{2}\right)\pi\tau\right],$$

$$e_3 = -\frac{\eta_1}{\omega} + \frac{1}{2}\left(\frac{\pi}{\omega}\right)^2 \sum_{n=1}^{\infty} \operatorname{cosec}^2\left[\left(n - \frac{1}{2}\right)\pi\tau\right],$$

$$\eta_1 = \frac{\frac{1}{2}\pi^2}{\omega}\left[\frac{1}{6} + \sum_{n=1}^{\infty}\operatorname{cosec}^2 (n\pi\tau)\right].$$

The properties above can also be obtained using the theta functions

$$\tau = \frac{\omega'}{\omega}, \quad q = \exp\left(\frac{i\pi\omega'}{\omega}\right), \quad \eta_1 = -\frac{1}{12}\omega^{-1}\frac{\vartheta_1'''}{\vartheta_1'},$$

$$g_2 = \frac{2}{3}\left(\frac{\frac{1}{2}\pi}{\omega}\right)^4 (\vartheta_0^8 + \vartheta_2^8 + \vartheta_3^8),$$

$$g_3 = \frac{4}{27}\left(\frac{\frac{1}{2}\pi}{\omega}\right)^6 (\vartheta_2^4 + \vartheta_3^4)(\vartheta_0^4 - \vartheta_2^4),$$

$$e_1 = \frac{\pi^2}{12}\omega^{-2}(\vartheta_0^4 + \vartheta_3^4); \quad e_2 = \frac{\pi^2}{12}\omega^{-2}(\vartheta_2^4 - \vartheta_0^4); \quad e_3 = -\frac{\pi^2}{12}\omega^{-2}(\vartheta_2^4 + \vartheta_3^4).$$

The connection of the parameters above with those of the Jacobian elliptic functions is

$$k = \left(\frac{\vartheta_2}{\vartheta_3}\right)^2 = \left[\frac{(e_2 - e_3)}{(e_1 - e_3)}\right]^{\frac{1}{2}},$$

$$k' = \left(\frac{\vartheta_0}{\vartheta_3}\right)^2 = \left[\frac{(e_1 - e_2)}{(e_1 - e_3)}\right]^{\frac{1}{2}},$$

$$K = \omega(e_1 - e_3)^{\frac{1}{2}}, \quad K' = -i\omega'(e_1 - e_3)^{\frac{1}{2}}$$

or

$$K = \frac{1}{2}\pi\vartheta_3^2, \quad K' = -i\tau K.$$

2. Given e_1, e_2, e_3. Then

$$\omega = \frac{1}{2}\int_{e_3}^{e_2} [(t-e_1)(t-e_2)(t-e_3)]^{-\frac{1}{2}}\, dt,$$

$$\omega' = \frac{1}{2}\int_{e_2}^{e_1} [(t-e_1)(t-e_2)(t-e_3)]^{-\frac{1}{2}}\, dt,$$

$$g_2 = -4(e_2 e_3 + e_1 e_3 + e_1 e_2); \; g_3 = 4 e_1 e_2 e_3$$

or also

$$\omega = (e_1 - e_3)^{-\frac{1}{2}} K, \; \omega' = (e_1 - e_3)^{-\frac{1}{2}} i K',$$

$$k = \left[\frac{(e_2 - e_3)}{(e_1 - e_3)}\right]^{\frac{1}{2}}.$$

Then again

$$\tau = \frac{\omega'}{\omega} = \frac{i K'}{K}.$$

Special cases

1. $k = 0,\; k' = 1$ $\qquad \omega = \pi (6 e_1)^{-\frac{1}{2}}, \; \omega' = i\infty,$

$\operatorname{sn} z = \sin z$

$\operatorname{cn} z = \cos z$

$\operatorname{dn} z = 1$

$$\wp(z) = -\frac{1}{3}\left(\frac{\frac{1}{2}\pi}{\omega}\right)^2 + \left(\frac{\frac{1}{2}\pi}{\omega}\right)^2 \operatorname{cosec}^2\left(\frac{\frac{1}{2}\pi z}{\omega}\right),$$

$K = \frac{1}{2}\pi$ $\qquad \zeta(z) = \frac{1}{3} z \left(\frac{\frac{1}{2}\pi}{\omega}\right)^2 + \frac{1}{2}\pi\omega^{-1}\cot\left(\frac{\frac{1}{2}\pi z}{\omega}\right),$

$K' = \infty$ $\qquad \sigma(z) = 2\pi^{-1}\omega \sin\left(\frac{\frac{1}{2}\pi z}{\omega}\right) \exp\left[\frac{1}{6}\left(\frac{\frac{1}{2}\pi z}{\omega}\right)^2\right],$

$$\lim_{k\to 0} k^{-2} \exp\left(-\frac{\pi K'}{K}\right) = \frac{1}{16}$$

$$q = 0,\; \vartheta_0(z) = \vartheta_3(z) = 1,$$

$$\vartheta_1(z) = \vartheta_2(z) = 0,$$

$$e_2 = e_3 = -\frac{1}{2} e_1,\; g_2 = 3 e_1^2,\; g_3 = e_1^3.$$

2. $k = 1,\; k' = 0$ $\qquad \omega = \infty, \; \omega' = \frac{1}{2}\pi i (3 e_1)^{-\frac{1}{2}},$

$$\operatorname{sn} z = \tanh z \qquad \operatorname{cn} z = \operatorname{dn} z = \operatorname{sech} z \qquad \wp(z) = +\, e_1 + 3e_1 \operatorname{cosech}\left[z(3\,e_1)^{\frac{1}{2}}\right],$$

$$K = \infty \qquad \zeta(z) = -e_1 z + (3e_1)^{\frac{1}{2}} \coth\left[z(3\,e_1)^{\frac{1}{2}}\right],$$

$$K' = \frac{1}{2}\pi \qquad \sigma(z) = (3e_1)^{-\frac{1}{2}} \sinh\left[z(3\,e_1)^{\frac{1}{2}}\right] e^{-\frac{1}{2}e_1 z^2},$$

$$\lim_{k\to 1} k'^{-2} \exp\left(-\frac{\pi K}{K'}\right) = \frac{1}{16} \qquad q = 1,\ \ e_1 = e_2 = -\frac{1}{2}e_3,$$

$$g_2 = 3e_3^2,\ \ g_3 = e_3^3.$$

Literature

Bellman, R.: A brief introduction to theta functions. New York: Holt, Rinehart and Winston 1961.

Byrd, P., and M. Friedman: Handbook of elliptic integrals for engineers and physicists. Berlin/Göttingen/Heidelberg: Springer 1954.

Erdélyi, A.: Higher transcendental functions, Vol. 2. New York: Mc-Graw-Hill 1953.

Hancock, H.: Theory of elliptic functions. New York: Dover 1958.

— Elliptic integrals. New York: Wiley 1917.

Jahnke, E., F. Emde and F. Loesch: Tables of higher functions. New York: McGraw-Hill 1960.

Milne-Thomson, L. M.: Jacobian elliptic function tables. New York: Dover 1950.

MacRobert, T. M.: Functions of a complex variable. London: MacMillan 1958.

Neville, E. H.: Jacobian elliptic functions. Oxford 1951.

Oberhettinger, F., and W. Magnus: Anwendung der elliptischen Funktionen in Physik und Technik. Berlin/Göttingen/Heidelberg: Springer 1949.

Schuler, M., und H. Gebelein: Five place tables of elliptic functions. Berlin/Göttingen/Heidelberg: Springer 1955.

Tricomi, F.: Elliptische Funktionen. Berlin/Göttingen/Heidelberg: Springer 1948.

Whittaker, E. T., and G. N. Watson: A course of modern analysis. Cambridge 1944.

Chapter XI

XI. Integral transforms

A function $g(y)$ of a variable y (which may be complex) is called the integral transform of a function $f(x)$ with respect to a kernel $K(x, y)$ when

$$g(y) = \int_a^b K(x, y)\, f(x)\, dx.$$

The kernel functions $K(x, y)$ and the limits a, b of the interval considered here are the following most frequently encountered in applications.

a	b	$K(x, y)$	Type of transform	
0	∞	$\left(\frac{2}{\pi}\right)^{\frac{1}{2}} \cos(xy)$	Fourier cosine	1
0	∞	$\left(\frac{2}{\pi}\right)^{\frac{1}{2}} \sin(xy)$	Fourier sine	2
$-\infty$	∞	$(2\pi)^{-\frac{1}{2}} e^{ixy}$	exponential Fourier	3
0	∞	e^{-xy}	One sided Laplace	4
$-\infty$	∞	e^{-xy}	Two sided Laplace	5
0	∞	x^{y-1}	Mellin	6
0	∞	$(xy)^{\frac{1}{2}} J_\nu(xy)$	Hankel	7
0	∞	$K_{ix}(y)$	Lebedev	8
1	∞	$\mathfrak{P}_{-\frac{1}{2}+ix}(y)$	Mehler	9
1	∞	$\mathfrak{P}^{\mu}_{-\frac{1}{2}+ix}(y)$	generalised Mehler	10
$-\infty$	∞	$e^{-h(x-y)^2}$	Gauss (or Weierstrass)	11

These transform types listed before admit (under certain conditions) explicit inversion formulas:

1. Fourier cosine transform

$$g(y) = \left(\frac{2}{\pi}\right)^{\frac{1}{2}} \int_0^\infty f(x) \cos(xy)\, dx,$$

$$f(x) = \left(\frac{2}{\pi}\right)^{\frac{1}{2}} \int_0^\infty g(y) \cos(xy)\, dy.$$

2. Fourier sine transform

$$g(y) = \left(\frac{2}{\pi}\right)^{\frac{1}{2}} \int_0^\infty f(x) \sin(xy)\, dx,$$

$$f(x) = \left(\frac{2}{\pi}\right)^{\frac{1}{2}} \int_0^\infty g(y) \sin(xy)\, dy.$$

3. Exponential Fourier transform

$$g(y) = \left(\frac{1}{2\pi}\right)^{\frac{1}{2}} \int_{-\infty}^\infty f(x)\, e^{ixy}\, dx,$$

$$f(x) = \left(\frac{1}{2\pi}\right)^{\frac{1}{2}} \int_{-\infty}^\infty g(y)\, e^{-ixy}\, dy.$$

4. One sided Laplace transformation

$$g(y) = \int_0^\infty f(x)\, e^{-xy}\, dx,$$

$$f(x) = \frac{1}{2\pi i} \int_{c-i\infty}^{c+i\infty} g(y)\, e^{xy}\, dy.$$

5. Two sided Laplace transform

$$g(y) = \int_{-\infty}^\infty f(x)\, e^{-xy}\, dx,$$

$$f(x) = \frac{1}{2\pi i} \int_{c-i\infty}^{c+i\infty} g(y)\, e^{xy}\, dy.$$

6. Mellin transform

$$g(y) = \int_0^\infty f(x)\, x^{y-1}\, dx,$$

$$f(x) = \frac{1}{2\pi i} \int_{c-i\infty}^{c+i\infty} g(y)\, x^{y}\, dy.$$

7. Hankel transform

$$g(y) = \int_0^\infty f(x)\, (xy)^{\frac{1}{2}}\, J_\nu(xy)\, dx,$$

$$f(x) = \int_0^\infty g(y)\, (xy)^{\frac{1}{2}}\, J_\nu(xy)\, dy.$$

For $\nu = \frac{1}{2}$ the Hankel transform reduces to the Fourier sine transform and for $\nu = -\frac{1}{2}$ it reduces to the Fourier cosine transform.

8. Lebedev transform

$$g(y) = \int_0^\infty f(x)\, K_{ix}(y)\, dx,$$

$$f(x) = 2\pi^{-2} x \sinh(\pi x) \int_0^\infty y^{-1} K_{ix}(y)\, g(y)\, dy.$$

9. Mehler transform

$$g(y) = \int_0^\infty f(x)\, \mathfrak{P}_{-\frac{1}{2}+ix}(y)\, dx,$$

$$f(x) = x \tanh(\pi x) \int_1^\infty \mathfrak{P}_{-\frac{1}{2}+ix}(y)\, g(y)\, dy.$$

10. Generalised Mehler transform

$$g(y) = \int_0^\infty f(x)\, \mathfrak{P}^\mu_{-\frac{1}{2}+ix}(y)\, dx,$$

$$f(x) = \pi^{-1} x \sinh(\pi x)\, \Gamma\left(\frac{1}{2} - \mu + ix\right) \Gamma\left(\frac{1}{2} - \mu - ix\right) \times \int_1^\infty g(y)\, \mathfrak{P}^\mu_{-\frac{1}{2}+ix}(y)\, dy.$$

11. Gauss (or Weierstrass) transform

$$f(s) = \int_{-\infty}^\infty e^{-h(x-s)^2} F(x)\, dx.$$

$F(x)$ can be gained as follows.

Substitute $2hx = -t$ and put

$$e^{-\frac{t^2}{4h}} F\left(-\frac{t}{2h}\right) = G(t).$$

Then

$$f(s)\, e^{hs^2} = \frac{1}{2h} \int_{-\infty}^\infty e^{-st} G(t)\, dt,$$

which is a two sided Laplace transform and $G(t)$ and hence $F(x)$ can be obtained.

The theory of these transforms is well established. The most elementary facts of the function $g(y)$ as defined by the integrals under 4., 5., 6. (provided the integrals exist) are

With respect to 4.

a) There exists a $\bar{y}$ such that the integral for $g(y)$ diverges to the left of $\bar{y}$ and converges to the right of it. Then $\bar{y}$ is called the abscissa of convergence.

b) The function $g(y)$ represents an analytic function of the complex variable y. The domain of analyticity is the half plane $Re\, y > \bar{y}$.

With respect to 5.

The integral in 5. can be written as superposition of two one sided Laplace transforms

$$\int_{-\infty}^{\infty} f(x)\, e^{-xy}\, dx = \int_{0}^{\infty} f(x)\, e^{-xy}\, dx + \int_{0}^{\infty} f(-x)\, e^{xy}\, dx,$$

each of which has an abscissa of convergence $\bar{y}_1, \bar{y}_2$. Hence $g(-y)$ as defined by 5. is analytic in the strip

$$\bar{y}_1 < Re\, y < \bar{y}_2.$$

With respect to 6.

The substitution $x = e^{-t}$ transforms $g(y)$ into

$$g(y) = \int_{-\infty}^{\infty} F(e^{-t})\, e^{-ty}\, dt,$$

which is of the type 5.

Therefore $g(y)$ represents an analytic function in the strip

$$\bar{Y}_1 < Re\, Y < \bar{Y}_2$$

in the complex y plane.

For the case of absolute convergence of the integrals $g(y)$ given by 4., 5., 6., the properties $\bar{y}{:}\bar{y}_1,\ \bar{y}_2{:}\bar{Y}_1,\ \bar{Y}_2$ have to be replaced by new parameters $y^*{:}y_1^*,\ y_2^*{:}Y_1^*,\ Y_2^*$, say. The integration in the inversion formulas has to be taken along a line parallel to the imaginary y-axis in the distance c from the origin, where c is arbitrary as long as the straight line is in the domain of absolute convergence.

In the following tables examples of the transform type mentioned before are given. It should be observed that an integral of the form

$$\int_{0}^{\infty} f(t)\, e^{-at} \cos(bt) \sin(ct)\, J_\nu(dt)\, K_{it}(x)\, \mathfrak{P}^{\mu}_{-\frac{1}{2}+it}(x)\, dt$$

can be regarded as Laplace transform or as Fourier cosine transform or as Fourier sine transform or as Hankel, Lebedev or Mehler transform.

Examples for the Fourier cosine transform

$f(x)$	$g(y) = \left(\frac{2}{\pi}\right)^{\frac{1}{2}} \int_0^\infty f(x) \cos(xy)\, dx$
$0 \quad 0 < x < a$ $x^{-1} \quad x > a$	$-\left(\frac{2}{\pi}\right)^{\frac{1}{2}} Ci(ay)$
$x^{-\frac{1}{2}} \quad 0 < x < a$ $0 \quad x > a$	$2y^{-\frac{1}{2}} C(ay)$
$0 \quad 0 < x < a$ $x^{-\frac{1}{2}} \quad x > a$	$2y^{-\frac{1}{2}} \left[\frac{1}{2} - C(ay)\right]$
$0 \quad 0 < x < b$ $(a+x)^{-n} \quad x > b$ $n = 1, 2, 3, \ldots$ if $n = 1$, the sum at the r. h. s. has to be omitted	$\left(\frac{2}{\pi}\right)^{\frac{1}{2}} \sum_{m=1}^{n} \frac{(m-1)!}{(n-1)!} (a+b)^{-m} (-y)^{n-m-1} \sin\left[\frac{1}{2}\pi(n-m) - by\right]$ $-\frac{(-y)^{n-1}}{(n-1)!}\left[\sin\left(ay + \frac{1}{2} n\pi\right) Ci(ay+by)\right.$ $\left. - \cos\left(ay + \frac{1}{2} n\pi\right) si(ay+by)\right]$
$0 \quad 0 < x < b$ $(a+x)^{-\frac{1}{2}} \quad x > b$	$y^{-\frac{1}{2}} [\sin(ay) + \cos(ay) - 2\cos(ay)\, C(ay+by) - 2\sin(ay)\, S(ay+by)]$

$f(x)$	$F_c(y)$
$(a^2 + x^2)^{-1}$	$\left(\frac{1}{2}\pi\right)^{\frac{1}{2}} a^{-1} e^{-ay}$
$(a^2 + x^2)^{-\nu-\frac{1}{2}}$ $Re\,\nu > -\frac{1}{2}$	$2^{\frac{1}{2}-\nu}\left(\frac{y}{a}\right)^{\nu}\left[\Gamma\left(\frac{1}{2}+\nu\right)\right]^{-1} K_\nu(ay)$
$(a^2 - x^2)^{\nu-\frac{1}{2}}, \quad 0 < x < a$ $0 \quad x > a$ $Re\,\nu > -\frac{1}{2}$	$2^{\nu-\frac{1}{2}}\Gamma\left(\frac{1}{2}+\nu\right)\left(\frac{a}{y}\right)^{\nu} J_\nu(ay)$
$0 \quad 0 < x < a$ $(x^2 - a^2)^{-\nu-\frac{1}{2}}, \quad x > a$ $-\frac{1}{2} < Re\,\nu < \frac{1}{2}$	$-2^{-\nu-\frac{1}{2}}\Gamma\left(\frac{1}{2}-\nu\right)\left(\frac{y}{a}\right)^{\nu} Y_\nu(ay)$
$0, \quad 0 < x < 2a$ $(x^2 - 2ax)^{-\nu-\frac{1}{2}}, \quad x > 2a$ $-\frac{1}{2} < Re\,\nu < \frac{1}{2}$	$-2^{-\nu-\frac{1}{2}}\Gamma\left(\frac{1}{2}-\nu\right)\left(\frac{y}{a}\right)^{\nu}[J_\nu(ay)\sin(ay) + Y_\nu(ay)\cos(ay)]$

$f(x)$	$g(y) = \left(\frac{2}{\pi}\right)^{\frac{1}{2}} \int_0^\infty f(x) \cos(xy)\, dx$
$(x^2 + 2ax)^{-\nu-\frac{1}{2}}$ $-\frac{1}{2} < Re\,\nu < \frac{1}{2}$	$-2^{-\nu-\frac{1}{2}} \Gamma\left(\frac{1}{2} - \nu\right) \left(\frac{y}{a}\right)^\nu [Y_\nu(ay) \cos(ay) - J_\nu(ay) \sin(ay)]$
$(2ax - x^2)^{\nu-\frac{1}{2}}, \quad 0 < x < 2a$ $0 \quad x > 2a$ $Re\,\nu > -\frac{1}{2}$	$2^{\nu+\frac{1}{2}} \Gamma\left(\frac{1}{2} + \nu\right) \cos(ay)\, J_\nu(ay)$
e^{-ax}	$\left(\frac{2}{\pi}\right)^{\frac{1}{2}} a (a^2 + y^2)^{-1}$
$x^{\nu-1} e^{-ax}$ $Re\,\nu > 0$	$\left(\frac{2}{\pi}\right)^{\frac{1}{2}} \Gamma(\nu)\, (a^2 + y^2)^{-\frac{1}{2}\nu} \cos\left[\nu \arctan\left(\frac{y}{a}\right)\right]$
e^{-ax^2}	$(2a)^{-\frac{1}{2}} e^{-\frac{y^2}{4a}}$
$x^{2n} e^{-a^2x^2}$ $n = 0, 1, 2, \ldots$	$(-1)^n\, 2^{-n-\frac{1}{2}} a^{-2n-1} e^{-\frac{y^2}{4a^2}} He_{2n}\left(2^{-\frac{1}{2}} \frac{y}{a}\right)$

$f(x)$	$F_c(y)$
$x^{\nu} e^{-ax^2}$ $Re\,\nu > -1$	$(2\pi a)^{-\frac{1}{2}} a^{-\frac{1}{2}\nu}\,\Gamma\left(\frac{1}{2}+\frac{1}{2}\nu\right)\,{}_1F_1\left(\frac{1}{2}+\frac{1}{2}\nu;\,\frac{1}{2};\,-\frac{y^2}{4a}\right)$
$x^{\nu-1} e^{-\frac{a}{x}}$ $Re\,\nu > -1$	$\left(\frac{2}{\pi}\right)^{\frac{1}{2}}\left(\frac{y}{a}\right)^{\frac{1}{2}\nu}\left\{e^{i\pi\frac{\nu}{4}} K_\nu\left[2(iay)^{\frac{1}{2}}\right] + e^{-i\pi\frac{\nu}{4}} K_\nu\left[2(-iay)^{\frac{1}{2}}\right]\right\}$
$x^{\nu-1} e^{-ax^{\frac{1}{2}}}$ $Re\,\nu > 0$	$\left(\frac{2}{\pi}\right)^{\frac{1}{2}}\Gamma(2\nu)\,(2y)^{-\nu}\left\{e^{-i\left(\frac{1}{2}\pi\nu+\frac{a^2}{8y}\right)} D_{-2\nu}\left[\frac{1}{2}ay^{-\frac{1}{2}}(1-i)\right] + e^{i\left(\frac{1}{2}\pi\nu+\frac{a^2}{8y}\right)} D_{-2\nu}\left[\frac{1}{2}ay^{-\frac{1}{2}}(1+i)\right]\right\}$
$(a^2+x^2)^{-\frac{1}{2}} e^{-b(a^2+x^2)^{\frac{1}{2}}}$	$\left(\frac{2}{\pi}\right)^{\frac{1}{2}} K_0\left[a(b^2+y^2)^{\frac{1}{2}}\right]$
$\log x,\quad 0 < x < 1$ $0 \qquad x > 1$	$-\left(\frac{2}{\pi}\right)^{\frac{1}{2}} y^{-1}\,Si(y)$
$x^{\nu-1}\sin(ax)$ $-1\,Re\,\nu < 1$	$\frac{1}{2}\left(\frac{1}{2}\pi\right)^{\frac{1}{2}}\left[\cos\left(\frac{1}{2}\pi\nu\right)\Gamma(1-\nu)\right]^{-1}\left[(y+a)^{-\nu} - \operatorname{sgn}(y-a)\,\lvert y-a\rvert^{-\nu}\right]$
$\left(\frac{\sin ax}{x}\right)^{2m}$ $m = 1, 2, 3, \ldots$	$(-1)^m 2^{-2m+\frac{1}{2}} m\pi^{-\frac{1}{2}}\left\{(m!)^{-2} y^{2m-1} + \sum_{n=1}^{m}\frac{(-1)^n\left[(2an+y)^{2m-1} + (\lvert 2an-y\rvert)^{2m-1}\right]}{(m+n)!\,(m-n)!}\right\}\quad y \le 2am$ $0 \qquad y \ge 2am$

$f(x)$	$g(y) = \left(\frac{2}{\pi}\right)^{\frac{1}{2}} \int_0^\infty f(x) \cos(xy)\, dx$
$\left(\frac{\sin ax}{x}\right)^{2m+1}$ $m = 0, 1, 2, \ldots$	$(-1)^m \pi^{\frac{1}{2}} 2^{-2m-\frac{3}{2}} (2m+1) F(a)$ $F(a) = \sum_{n=0}^{m} (-1)^n \frac{[(2n+1)a + y]^{2m} + [(2n+1)a - y]^{2m}}{(m+1+n)!\,(m-n)!}$ $0 \le y \le a$ $F(a) = \sum_{n=0}^{k-1} (-1)^n \frac{[y + (2n+1)a]^{2m} - [(y - (2n+1)a]^{2m}}{(m+1+n)!\,(m-n)!}$ $+ \sum_{n=k}^{m} (-1)^n \frac{[(2n+1)a + y]^{2m} + [(2n+1)a - y]^{2m}}{(m+1+n)!\,(m-n)!}$ $(2k-1)a \le y \le (2k+1)a$ $F(a) = 0 \qquad y \ge (2m+1)a$ $k = 1, 2, 3, \ldots$
$e^{-ax^2} \sin(bx^2)$	$2^{-\frac{1}{2}} (a^2+b^2)^{-\frac{1}{4}} e^{-\frac{1}{4}ay^2/(a^2+b^2)} \sin\left[\frac{1}{2}\arctan\left(\frac{b}{a}\right) - \frac{1}{4} by^2 \big/ (a^2+b^2)\right]$
$e^{-ax^2} \cos(bx^2)$	$2^{-\frac{1}{2}} (a^2+b^2)^{-\frac{1}{4}} e^{-\frac{1}{4}ay^2/(a^2+b^2)} \cos\left[\frac{1}{2}\arctan\left(\frac{b}{a}\right) - \frac{1}{4} by^2 \big/ (a^2+b^2)\right]$
$\cos(a^3x^3)$	$\left(\frac{2\pi}{y}\right)^{-\frac{1}{2}} (3a)^{-\frac{3}{2}} \left[3^{\frac{1}{2}} K_{\frac{1}{3}}(z) + \pi J_{\frac{1}{3}}(z) + \pi J_{-\frac{1}{3}}(z)\right] \qquad z = 2\left(\frac{y}{3a}\right)^{\frac{3}{2}}$

$\sin(a^3x^3)$	$\frac{1}{9}\left(\frac{1}{2}\pi y\right)^{\frac{1}{2}} a^{-\frac{3}{2}} \Big\{ I_{-\frac{1}{3}}(z) + I_{\frac{1}{3}}(z) + J_{-\frac{1}{3}}(z) - J_{\frac{1}{3}}(z) - 2\Big[\boldsymbol{J}_{\frac{1}{3}}(z) - \boldsymbol{J}_{-\frac{1}{3}}(z)$ $- i\boldsymbol{J}_{\frac{1}{3}}(iz) + i\boldsymbol{J}_{-\frac{1}{3}}(iz)\Big]\Big\}, \qquad z = 2\left(\frac{y}{3a}\right)^{\frac{3}{2}}$
$x^{\nu-1}\sin\left(\frac{a}{x}\right)$ $-1 < Re\,\nu < 2$	$\frac{1}{2}\left(\frac{1}{2}\pi\right)^{\frac{1}{2}} a^{\frac{1}{2}\nu} \sec\left(\frac{1}{2}\pi\nu\right) y^{-\frac{1}{2}\nu}$ $\times\left[J_\nu(2\sqrt{ay}) + J_{-\nu}(2\sqrt{ay}) + I_\nu(2\sqrt{ay}) - I_{-\nu}(2\sqrt{ay})\right]$ $= \left(\frac{1}{2}\pi\right)^{\frac{1}{2}}\left(\frac{y}{a}\right)^{-\frac{1}{2}\nu}\Big\{\cos\left(\frac{1}{2}\pi\nu\right) J_\nu(2\sqrt{ay})$ $- \sin\left(\frac{1}{2}\pi\nu\right)\left[Y_\nu(2\sqrt{ay}) + \frac{2}{\pi}K_\nu(2\sqrt{ay})\right]\Big\}$
$x^{\nu-1}\cos\left(\frac{a}{x}\right)$ $-1 < Re\,\nu < 1$	$\frac{1}{2}\left(\frac{1}{2}\pi\right)^{\frac{1}{2}} \operatorname{cosec}\left(\frac{1}{2}\pi\nu\right)\left(\frac{y}{a}\right)^{-\frac{1}{2}\nu}$ $\times\left[J_{-\nu}(2\sqrt{ay}) - J_\nu(2\sqrt{ay}) + I_{-\nu}(2\sqrt{ay}) - I_\nu(2\sqrt{ay})\right]$ $= \left(\frac{1}{2}\pi\right)^{\frac{1}{2}}\left(\frac{y}{a}\right)^{-\frac{1}{2}\nu}\Big\{\cos\left(\frac{1}{2}\pi\nu\right)\left[\frac{2}{\pi}K_\nu(2\sqrt{ay}) - Y_\nu(2\sqrt{ay})\right]$ $- \sin\left(\frac{1}{2}\pi\nu\right) J_\nu(2\sqrt{ay})\Big\}$

$f(x)$	$g(y) = \left(\frac{2}{\pi}\right)^{\frac{1}{2}} \int_0^\infty f(x) \cos(xy)\, dx$
$(a^2 + x^2)^{-\frac{1}{2}} \sin\left[b(a^2 + x^2)^{\frac{1}{2}}\right]$	$\left(\frac{1}{2}\pi\right)^{\frac{1}{2}} J_0\left[a(b^2 - y^2)^{\frac{1}{2}}\right]$ $\quad 0 < y < b$ $0 \quad y > b$
$(a^2 + x^2)^{-\frac{1}{2}} \cos\left[b(a^2 + x^2)^{\frac{1}{2}}\right]$	$-\left(\frac{1}{2}\pi\right)^{\frac{1}{2}} Y_0\left[a(b^2 - y^2)^{\frac{1}{2}}\right]$ $\quad 0 < y < b$ $\left(\frac{2}{\pi}\right)^{\frac{1}{2}} K_0\left[a(y^2 - b^2)^{\frac{1}{2}}\right]$ $\quad y > b$
$\sin\left[b(a^2 - x^2)^{\frac{1}{2}}\right]$ $\quad 0 < x < a$ $0 \quad x > a$	$\left(\frac{1}{2}\pi\right)^{\frac{1}{2}} ab(b^2 + y^2)^{-\frac{1}{2}} J_1\left[a(b^2 + y^2)^{\frac{1}{2}}\right]$
$(a^2 - x^2)^{-\frac{1}{2}} \cos\left[b(a^2 - x^2)^{\frac{1}{2}}\right]$ $\quad 0 < x < a$ $0 \quad x > a$	$\left(\frac{1}{2}\pi\right)^{\frac{1}{2}} J_0\left[a(b^2 + y^2)^{\frac{1}{2}}\right]$
$0 \quad 0 < x < a$ $(x^2 - a^2)^{-\frac{1}{2}} \cos\left[b(x^2 - a^2)^{\frac{1}{2}}\right]$ $\quad x > a$	$\left(\frac{2}{\pi}\right)^{\frac{1}{2}} K_0\left[a(b^2 - y^2)^{\frac{1}{2}}\right]$ $\quad 0 < y < b$ $-\left(\frac{1}{2}\pi\right)^{\frac{1}{2}} Y_0\left[a(y^2 - b^2)^{\frac{1}{2}}\right]$ $\quad y > b$

$\operatorname{sech}(ax)$	$\left(\frac{1}{2}\pi\right)^{\frac{1}{2}} a^{-1} \operatorname{sech}\left(\frac{\frac{1}{2}\pi y}{a}\right)$
$[\operatorname{sech}(ax)]^{\nu}$ $Re\, \nu > 0$	$2^{\nu-2} a^{-1} \left(\frac{2}{\pi}\right)^{\frac{1}{2}} [\Gamma(\nu)]^{-1} \Gamma\left(\frac{1}{2}\nu + i\frac{\frac{1}{2}y}{a}\right) \Gamma\left(\frac{1}{2}\nu - i\frac{y}{2a}\right)$
$(\cosh x + z)^{-\mu}$ $Re\, \mu > 0$ z not on the real axis between -1 and $-\infty$	$[\Gamma(\mu)]^{-1} (z^2 - 1)^{\frac{1}{4} - \frac{1}{2}\mu} \Gamma(\mu + iy)\, \Gamma(\mu - iy)\, \mathfrak{P}_{-\frac{1}{2}+iy}^{\frac{1}{2}-\mu}(z)$
$(\cosh a - \cosh x)^{-\mu}$ $\quad 0 < x < a$ 0 $\quad x > a$ $Re\, \mu < 1$	$\Gamma(1-\mu)\, (\sinh a)^{\frac{1}{2}-\mu}\, \mathfrak{P}_{-\frac{1}{2}+iy}^{\mu-\frac{1}{2}}(\cosh a)$
$\cosh(ax) \operatorname{sech}(bx)$ $0 < a < b$	$(2\pi)^{\frac{1}{2}} b^{-1} \cos\left(\frac{\frac{1}{2}\pi a}{b}\right) \cosh\left(\frac{\frac{1}{2}\pi y}{b}\right) \left[\cos\left(\frac{\pi a}{b}\right) + \cosh\left(\frac{\pi y}{b}\right)\right]^{-1}$
$\sinh(ax) \operatorname{cosech}(bx)$ $0 < a < b$	$\left(\frac{1}{2}\pi\right)^{\frac{1}{2}} b^{-1} \sin\left(\frac{\pi a}{b}\right) \left[\cos\left(\frac{\pi a}{b}\right) + \cosh\left(\frac{\pi y}{b}\right)\right]^{-1}$
$\sinh(ax) \operatorname{sech}(bx)$	$\frac{1}{2}(2\pi)^{-\frac{1}{2}} b^{-1} \Big\{ 2\pi \sin\left(\frac{\pi a}{b}\right) \left[\cos\left(\frac{\pi a}{b}\right) + \cosh\left(\frac{\pi y}{b}\right)\right]^{-1}$ $+ \psi\left(\frac{3b - a + iy}{4b}\right) + \psi\left(\frac{3b - a - iy}{4b}\right) - \psi\left(\frac{3b + a - iy}{4b}\right) - \psi\left(\frac{3b + a + iy}{4b}\right) \Big\}$

$f(x)$	$g(y) = \left(\frac{2}{\pi}\right)^{\frac{1}{2}} \int_0^\infty f(x) \cos(xy)\, dx$
$e^{-a\sinh x}$	$\left(\frac{2}{\pi}\right)^{\frac{1}{2}} S_{0,iy}(a)$
$e^{-a\cosh x} \cos(b \sinh x)$	$\left(\frac{2}{\pi}\right)^{\frac{1}{2}} \cosh\left[y \arctan\left(\frac{b}{a}\right)\right] K_{iy}\left[(a^2+b^2)^{\frac{1}{2}}\right]$
$e^{-bx^2} He_{2n}(ax)$ $n = 0, 1, 2, \ldots$	$(2b)^{-\frac{1}{2}} \left(1 - \frac{a^2}{2b}\right)^n e^{-\frac{1}{4}y^2/b} He_{2n}\left[\frac{1}{2} a y \left(\frac{1}{2} a^2 b - b^2\right)^{-\frac{1}{2}}\right]$
$0, \quad 0 < x < 1$ $(x^2-1)^{-\frac{1}{2}\mu} \mathfrak{P}_\nu^\mu(x)$ $x > 1$ $Re\,\mu > Re\,\nu > -1 - Re\,\mu$	$-y^{\mu-\frac{1}{2}} \left[\cos\left(\frac{1}{2}\pi\mu - \frac{1}{2}\pi\nu\right) J_{\nu+\frac{1}{2}}(y) + \sin\left(\frac{1}{2}\pi\mu - \frac{1}{2}\pi\nu\right) Y_{\nu+\frac{1}{2}}(y)\right]$
$(1-x^2)^{-\frac{1}{2}\mu} P_\nu^\mu(x)$ $0 < x < 1$ $0 \quad x > 1$ $Re\,\mu < 1$	$2^{\mu-\frac{1}{2}} \left[\Gamma\left(\frac{3-\mu+\nu}{2}\right) \Gamma\left(\frac{2-\mu-\nu}{2}\right)\right]^{-1}$ $\times (\mu+\nu)(\mu-\nu-1)\, y^{\mu-\frac{1}{2}} s_{-\mu-\frac{1}{2},\nu+\frac{1}{2}}(y)$

$\Gamma(\mu + ix)\,\Gamma(\mu - ix)\,\mathfrak{P}^{\frac{1}{2}-\mu}_{-\frac{1}{2}+ix}(z)$ $z > 1, \quad Re\,\mu > 0$	$\Gamma(\mu)\,(z^2 - 1)^{\frac{1}{2}\mu - \frac{1}{4}}\,(z + \cosh y)^{-\mu}$
$J_{2n}(ax)$ $n = 0\,1, 2, \ldots$	$\left(\frac{2}{\pi}\right)^{\frac{1}{2}} (-1)^n\,(a^2 - y^2)^{-\frac{1}{2}}\,T_{2n}\left(\frac{y}{a}\right) \quad 0 < y < a$ $0 \quad y > a$
$J_\nu(ax)$ $Re\,\nu > -1$	$\left(\frac{2}{\pi}\right)^{\frac{1}{2}} (a^2 - y^2)^{-\frac{1}{2}} \cos\left[\nu \arcsin\left(\frac{y}{a}\right)\right] \quad 0 < y < a$ $-\left(\frac{2}{\pi}\right)^{\frac{1}{2}} a^\nu \sin\left(\frac{1}{2}\pi\nu\right)(y^2 - a^2)^{-\frac{1}{2}}\left[y + (y^2 - a^2)^{\frac{1}{2}}\right]^{-\nu} \quad y > a$
$x^\nu J_\nu(ax)$ $-\frac{1}{2} < Re\,\nu < \frac{1}{2}$	$2^{\frac{1}{2}}\pi^{-1}\Gamma\left(\frac{1}{2} + \nu\right)(2a)^\nu\,(a^2 - y^2)^{-\nu-\frac{1}{2}}, \quad 0 < y < a$ $-2^{\frac{1}{2}}\pi^{-1}\Gamma\left(\frac{1}{2} + \nu\right)(2a)^\nu \sin(\pi\nu)\,(y^2 - a^2)^{-\nu-\frac{1}{2}} \quad y > a$
$x^{-\nu} J_{\nu+2n}(ax)$ $n = 0, 1, 2, \ldots$ $Re\,\nu > -\frac{1}{2}$	$(-1)^n\,\pi^{-\frac{1}{2}}\,2^{\nu-\frac{1}{2}}\,a^{-\nu}(2n)!\,\Gamma(\nu)\,[\Gamma(2\nu + 2n)]^{-1}$ $\times\,(a^2 - y^2)^{\nu-\frac{1}{2}}\,C^\nu_{2n}\left(\frac{y}{a}\right) \quad 0 < y < a$ $0 \quad y > a$

$f(x)$	$g(y) = \left(\frac{2}{\pi}\right)^{\frac{1}{2}} \int_0^\infty f(x) \cos(xy)\, dx$
$J_\nu(ax)\, J_\nu(bx)$ $Re\, \nu > -\frac{1}{2}$	$\left(\frac{2}{\pi}\right)^{\frac{1}{2}} \pi^{-1} (ab)^{-\frac{1}{2}} \mathfrak{Q}_{\nu-\frac{1}{2}}\left(\frac{a^2+b^2-y^2}{2ab}\right), \quad 0 < y < a-b$ $\left(\frac{2}{\pi}\right)^{\frac{1}{2}} \pi^{-1} (ab)^{-\frac{1}{2}} Q_{\nu-\frac{1}{2}}\left(\frac{a^2+b^2-y^2}{2ab}\right), \quad a-b < y < a+b$ $-\left(\frac{2}{\pi}\right)^{\frac{1}{2}} \pi^{-1} \sin(\pi\nu)\, (ab)^{-\frac{1}{2}} \mathfrak{Q}_{\nu-\frac{1}{2}}\left(\frac{y^2-a^2-b^2}{2ab}\right) \quad y > a+b$
$x^\nu Y_\nu(ax)$ $-\frac{1}{2} < Re\, \nu < \frac{1}{2}$	$0 \quad 0 < y < a$ $-2^{\frac{1}{2}+\nu} a^\nu \left[\Gamma\left(\frac{1}{2}-\nu\right)\right]^{-1} (y^2-a^2)^{-\nu-\frac{1}{2}}, \quad y > a$
$x^{-\nu} Y_\nu(ax)$ $-\frac{1}{2} < Re\, \nu < \frac{1}{2}$	$-2^{\frac{1}{2}-\nu} a^{-\nu} \Gamma\left(\frac{1}{2}-\nu\right) \sin(\pi\nu)\, (a^2-y^2)^{\nu-\frac{1}{2}}, \quad 0 < y < a$ $-2^{\frac{1}{2}-\nu} a^{-\nu} \pi^{-1} \Gamma\left(\frac{1}{2}-\nu\right) (y^2-a^2)^{\nu-\frac{1}{2}}, \quad y > a$
$J_\nu\left(ax^{\frac{1}{2}}\right) J_\nu\left(bx^{\frac{1}{2}}\right)$ $Re\, \nu > -1$	$\left(\frac{2}{\pi}\right)^{\frac{1}{2}} y^{-1} J_\nu\left(\frac{\frac{1}{2}ab}{y}\right) \sin\left[\frac{a^2+b^2}{4y} - \frac{1}{2}\pi\nu\right]$

$(a^2+x^2)^{-\frac{1}{2}} J_\nu\left[b(a^2+x^2)^{\frac{1}{2}}\right]$	$\left(\frac{2}{\pi}\right)^{\frac{1}{2}} \cos\left(\frac{1}{2}\pi\nu\right) I_{\frac{1}{2}\nu}\left\{\frac{1}{2}a\left[y-(y^2-b^2)^{\frac{1}{2}}\right]\right\}$ $\times K_{\frac{1}{2}\nu}\left\{\frac{1}{2}a\left[y+(y^2-b^2)^{\frac{1}{2}}\right]\right\}$ $\quad y > b$
$(a^2+x^2)^{\frac{1}{2}\nu} J_\nu\left[b(a^2+x^2)^{\frac{1}{2}}\right]$ $Re\,\nu < \frac{1}{2}$	$-a^{\frac{1}{2}}(ab)^\nu (b^2-y^2)^{-\frac{1}{2}\nu-\frac{1}{4}} Y_{\nu+\frac{1}{2}}\left[a(b^2-y^2)^{\frac{1}{2}}\right],$ $\quad 0 < y < b$ $-a^{\frac{1}{2}}\, 2\pi^{-1} \sin(\pi\nu)\,(ab)^\nu (y^2-b^2)^{-\frac{1}{2}\nu-\frac{1}{4}} K_{\nu+\frac{1}{2}}\left[a(y^2-b^2)^{\frac{1}{2}}\right],$ $\quad y > b$
$(a^2+x^2)^{-\frac{1}{2}\nu} J_\nu\left[b(a^2+x^2)^{\frac{1}{2}}\right]$ $Re\,\nu > -\frac{1}{2}$	$a^{\frac{1}{2}}(ab)^{-\nu} (b^2-y^2)^{\frac{1}{2}\nu-\frac{1}{4}} J_{\nu-\frac{1}{2}}\left[a(b^2-y^2)^{\frac{1}{2}}\right],$ $\quad 0 < y < b$ 0 $\quad y > b$
$(a^2-x^2)^{\frac{1}{2}\nu} J_\nu\left[b(a^2-x^2)^{\frac{1}{2}}\right]$ $\quad 0 < x < a$ 0 $\quad x > a$ $Re\,\nu > -1$	$a^{\frac{1}{2}}(ab)^\nu (b^2+y^2)^{-\frac{1}{2}\nu-\frac{1}{4}} J_{\nu+\frac{1}{2}}\left[a(b^2+y^2)^{\frac{1}{2}}\right]$
$(a^2-x^2)^{-\frac{1}{2}} J_\nu\left[b(a^2-x^2)^{\frac{1}{2}}\right]$ $\quad 0 < x < a$ 0 $\quad x > a$ $Re\,\nu > -1$	$\left(\frac{1}{2}\pi\right)^{\frac{1}{2}} J_{\frac{1}{2}\nu}\left\{\frac{1}{2}a\left[(b^2+y^2)^{\frac{1}{2}}+y\right]\right\} J_{\frac{1}{2}\nu}\left\{\frac{1}{2}a\left[(b^2+y^2)^{\frac{1}{2}}-y\right]\right\}$

$f(x)$	$g(y) = \left(\frac{2}{\pi}\,\pi\right)^{\frac{1}{2}} \int\limits_0^\infty f(x) \cos(xy)\, dx$
$0 \qquad 0 < x < a$ $(x^2 - a^2)^{-\frac{1}{2}} J_\nu \left[b (x^2 - a^2)^{\frac{1}{2}}\right] \quad x > a$ $Re\,\nu > -1$	$-\left(\frac{1}{2}\pi\right)^{\frac{1}{2}} J_{\frac{1}{2}\nu} \left\{\frac{1}{2} a \left[y - (y^2 - b^2)^{\frac{1}{2}}\right]\right\}$ $\times\, Y_{-\frac{1}{2}\nu} \left\{\frac{1}{2} a \left[y + (y^2 - b^2)^{\frac{1}{2}}\right]\right\}$ $y > b$
$0 \qquad 0 < x < a$ $(x^2 - a^2)^{\frac{1}{2}\nu} J_\nu \left[b (x^2 - a^2)^{\frac{1}{2}}\right] \quad x > a$ $-1 < Re\,\nu < \frac{1}{2}$	$2 a^{\frac{1}{2}} \pi^{-1} (ab)^\nu (b^2 - y^2)^{-\frac{1}{2}\nu - \frac{1}{4}} K_{\nu + \frac{1}{2}} \left[a (b^2 - y^2)^{\frac{1}{2}}\right], \quad 0 < y < b$ $-a^{\frac{1}{2}} (ab)^\nu (y^2 - b^2)^{-\frac{1}{2}\nu - \frac{1}{4}} Y_{-\nu - \frac{1}{2}} \left[a (y^2 - b^2)^{\frac{1}{2}}\right], \quad y > b$
$K_\nu(ax)$ $-1 < Re\,\nu < 1$	$\frac{1}{4} (2\pi)^{\frac{1}{2}} \sec\left(\frac{1}{2}\pi\nu\right) (a^2 + y^2)^{-\frac{1}{2}}$ $\times \left\{a^{-\nu} \left[y + (a^2 + y^2)^{\frac{1}{2}}\right]^\nu + a^\nu \left[y + (a^2 + y^2)^{\frac{1}{2}}\right]^{-\nu}\right\}$
$x^\nu K_\nu(ax)$ $Re\,\nu > -\frac{1}{2}$	$2^{\nu - \frac{1}{2}} a^\nu \Gamma\left(\frac{1}{2} + \nu\right) (a^2 + y^2)^{-\nu - \frac{1}{2}}$

$f(x)$	$g(y)$
$I_\nu(ax)\, K_\nu(bx)$ $b > a, \quad Re\, \nu > -\frac{1}{2}$	$(2\pi ab)^{-\frac{1}{2}} \mathfrak{Q}_{\nu-\frac{1}{2}}\left(\frac{a^2+b^2+y^2}{2ab}\right)$
$K_\nu(ax)\, K_\nu(bx)$ $-\frac{1}{2} < Re\, \nu < \frac{1}{2}$	$\left(\frac{1}{2}\pi\right)^{\frac{3}{2}} (ab)^{-\frac{1}{2}} \sec(\pi\nu)\, \mathfrak{P}_{\nu-\frac{1}{2}}\left(\frac{a^2+b^2+y^2}{2ab}\right)$
$(a^2+x^2)^{-\frac{1}{2}} K_\nu\left[b(a^2+x^2)^{\frac{1}{2}}\right]$	$(2\pi)^{-\frac{1}{2}} K_{\frac{1}{2}\nu}\left\{\frac{1}{2}a\left[(y^2+b^2)^{\frac{1}{2}}-y\right]\right\} K_{\frac{1}{2}\nu}\left\{\frac{1}{2}a\left[(y^2+b^2)^{\frac{1}{2}}+y\right]\right\}$
$(a^2+x^2)^{\mp\frac{1}{2}\nu} K_\nu\left[b(a^2+x^2)^{\frac{1}{2}}\right]$	$a^{\frac{1}{2}} (ab)^{\mp\nu} (b^2+y^2)^{\pm\frac{1}{2}\nu-\frac{1}{4}} K_{\pm\nu-\frac{1}{2}}\left[a(b^2+y^2)^{\frac{1}{2}}\right]$
$K_{ix}(a)\, K_{ix}(b)$	$\left(\frac{1}{2}\pi\right)^{\frac{1}{2}} K_0\left[(a^2+b^2+2ab\cosh y)^{\frac{1}{2}}\right]$

Examples for the Fourier sine transform

$f(x)$	$g_s(y) = \left(\frac{2}{\pi}\right)^{\frac{1}{2}} \int\limits_0^\infty f(x) \sin(xy)\, dx$
$1 \quad 0 < x < a$ $0 \quad x > a$	$\left(\frac{2}{\pi}\right)^{\frac{1}{2}} y^{-1}[1 - \cos(ay)]$
x^{-1}	$\left(\frac{1}{2}\pi\right)^{\frac{1}{2}}$
$x^{-1}, \quad 0 < x < a$ $0 \quad x > a$	$\left(\frac{2}{\pi}\right)^{\frac{1}{2}} Si(ay)$
$0 \quad 0 < x < a$ $x^{-1} \quad x > a$	$-\left(\frac{2}{\pi}\right)^{\frac{1}{2}} si(ay)$
$x^{-\frac{1}{2}} \quad 0 < x < a$ $0 \quad x > a$	$2y^{-\frac{1}{2}} S(ay)$
$0 \quad 0 < x < b$ $(a+x)^{-n} \quad x > b$ $n = 1, 2, 3, \ldots$	$\left(\frac{2}{\pi}\right)^{\frac{1}{2}} \left\{ \sum\limits_{m=1}^{n-1} \frac{(m-1)!}{(n-1)!} \cos\left(\frac{1}{2}\pi n - \frac{1}{2}\pi m - by\right)(a+b)^{-m}(-y)^{n-m-1} \right.$ $\left. - \frac{(-y)^{n-1}}{(n-1)!}\left[\cos\left(ay + \frac{1}{2}\pi n\right) Ci(ay+by) + \sin\left(ay + \frac{1}{2}\pi n\right) si(ay+by)\right]\right\}$ for $n = 1$, $\sum\limits_{m=1}^{n-1}(\,) = 0$

$0 \quad 0 < x < b$ $(a+x)^{-\frac{1}{2}} \quad x > b$	$y^{-\frac{1}{2}}[\cos(ay) - \sin(ay) + 2C(ay+by)\sin(ay) - 2S(ay+by)\cos(ay)]$
$(a^2+x^2)^{-1}$	$(2\pi)^{-\frac{1}{2}} a^{-1}\left[e^{-ay}\overline{Ei}(ay) - e^{ay}Ei(-ay)\right]$
$(a^2+x^2)^{-\frac{1}{2}}\left[(a^2+x^2)^{\frac{1}{2}} - a\right]^{-\frac{1}{2}}$	$\left(\frac{\pi}{a}\right)^{\frac{1}{2}} \mathrm{Erf}\left[(ay)^{\frac{1}{2}}\right]$
$(a^2+x^2)^{-\frac{1}{2}}\left[(a^2+x^2)^{\frac{1}{2}} + a\right]^{-\frac{1}{2}}$	$-i\left(\frac{\pi}{a}\right)^{\frac{1}{2}} \mathrm{Erf}\left[i(ay)^{\frac{1}{2}}\right] \mathrm{Erfc}\left[(ay)^{\frac{1}{2}}\right]$
$(a^2+x^2)^{-\frac{1}{2}}\left[(a^2+x^2)^{\frac{1}{2}} - a\right]^{\frac{1}{2}}$	$y^{-\frac{1}{2}} e^{-ay}$
$x^{2m+1}(a^2+x^2)^{-n-\frac{1}{2}}$ $-1 \leq m < n$ $m, n = 0, 1, \ldots$	$(-1)^{m+1}\, 2^{\frac{1}{2}-n} a^{-n}\left[\Gamma\left(\frac{1}{2}+n\right)\right]^{-1} \frac{d^{2m+1}}{dy^{2m+1}}[y^n K_n(ay)]$
$x^{2m-1}(x^{2n}+a^{2n})^{-1}$ $m \leq n,$ $n, m = 1, 2, 3, \ldots$	$-\left(\frac{1}{2}\pi\right)^{\frac{1}{2}} a^{2m-2n} n^{-1} \sum_{k=1}^{n} e^{-ay\sin\left[\left(k-\frac{1}{2}\right)\pi n^{-1}\right]}$ $\times \cos\left[n^{-1}(2\pi mk - \pi m) + ay\cos\left(k-\frac{1}{2}\right)\pi n^{-1}\right]$

$f(x)$	$g_s(y) = \left(\frac{2}{\pi}\right)^{\frac{1}{2}} \int\limits_0^\infty f(x) \sin(xy)\, dx$
$x^{-\nu} \qquad 0 < Re\, \nu < 2$	$\left(\frac{2}{\pi}\right)^{\frac{1}{2}} \cos\left(\frac{1}{2}\pi\nu\right) \Gamma(1-\nu)\, y^{\nu-1}$
$(a^2+x^2)^{-\nu-\frac{1}{2}}$ $Re\, \nu > -\frac{1}{2}$	$2^{-\nu-\frac{1}{2}} \Gamma\left(\frac{1}{2}-\nu\right) \left(\frac{y}{a}\right)^{\nu} [I_\nu(ay) - \mathbf{L}_{-\nu}(ay)]$
$(a^2-x^2)^{\nu-\frac{1}{2}} \qquad 0 < x < a$ $0 \qquad x > a$ $Re\, \nu > -\frac{1}{2}$	$2^{\nu-\frac{1}{2}} \left(\frac{y}{a}\right)^{-\nu} \Gamma\left(\frac{1}{2}+\nu\right) \mathbf{H}_\nu(ay)$
$0 \qquad 0 < x < a$ $(x^2-a^2)^{-\nu-\frac{1}{2}} \qquad x > a$ $-\frac{1}{2} < Re\, \nu < \frac{1}{2}$	$2^{-\nu-\frac{1}{2}} \Gamma\left(\frac{1}{2}-\nu\right) \left(\frac{y}{a}\right)^{\nu} J_\nu(ay)$
$(2ax+x^2)^{-\nu-\frac{1}{2}}$ $\frac{3}{2} > Re\, \nu > -\frac{1}{2}$	$2^{-\nu-\frac{1}{2}} \Gamma\left(\frac{1}{2}-\nu\right) \left(\frac{y}{a}\right)^{\nu} [J_\nu(ay) \cos(ay) + Y_\nu(ay) \sin(ay)]$

$(2ax - x^2)^{\nu - \frac{1}{2}}, \quad 0 < x < 2a$ $0 \quad x > 2a$ $Re\, \nu > -\frac{1}{2}$	$2^{\nu + \frac{1}{2}} \Gamma\left(\frac{1}{2} + \nu\right) \left(\frac{y}{a}\right)^{-\nu} \sin(ay)\, J_\nu(ay)$
$0 \quad 0 < x < 2a$ $(x^2 - 2ax)^{-\nu - \frac{1}{2}} \quad x > 2a$ $-\frac{1}{2} < Re\, \nu < \frac{1}{2}$	$2^{-\nu - \frac{1}{2}} \Gamma\left(\frac{1}{2} - \nu\right) \left(\frac{y}{a}\right)^{\nu} [J_\nu(ay) \cos(ay) - Y_\nu(ay) \sin(ay)]$
$(a^2 + x^2)^{-\frac{1}{2}} \left\{ \left[(a^2 + x^2)^{\frac{1}{2}} + x \right]^\nu - \left[(a^2 + x^2)^{\frac{1}{2}} - x \right]^\nu \right\}$ $-1 < Re\, \nu < 1$	$\left(\frac{2}{\pi}\right)^{\frac{1}{2}} 2a^\nu \sin\left(\frac{1}{2} \pi \nu\right) K_\nu(ay)$
e^{-ax}	$\left(\frac{2}{\pi}\right)^{\frac{1}{2}} y (a^2 + y^2)^{-1}$
$x^{\nu - 1} e^{-ax}$ $Re\, \nu > -1$	$\left(\frac{2}{\pi}\right)^{\frac{1}{2}} \Gamma(\nu)\, (a^2 + y^2)^{-\frac{1}{2}\nu} \sin\left[\nu \arctan\left(\frac{y}{a}\right)\right]$
$(e^{ax} + 1)^{-1}$	$(2\pi)^{-\frac{1}{2}} y^{-1} - \left(\frac{1}{2}\pi\right)^{\frac{1}{2}} a^{-1} \operatorname{cosech}\left(\pi \frac{y}{a}\right)$

$f(x)$	$g_s(y) = \left(\frac{2}{\pi}\right)^{\frac{1}{2}} \int\limits_0^\infty f(x) \sin(xy)\, dx$
$(e^{ax} - e^{bx})^{-1}$	$i(2\pi)^{-\frac{1}{2}} (a-b)^{-1} \left[\psi\left(\frac{a-iy}{a-b}\right) - \psi\left(\frac{a+iy}{a-b}\right)\right]$
$e^{-ax}(1-e^{-bx})^{\nu-1}$ $Re\, \nu > -1$	$i b^{-1} (2\pi)^{-\frac{1}{2}} \left[B\left(\nu, \frac{a+iy}{b}\right) - B\left(\nu, \frac{a-iy}{b}\right)\right]$
e^{-ax^2}	$-i(2a)^{-\frac{1}{2}} e^{-\frac{1}{4}y^2/a} \operatorname{Erf}\left(\frac{1}{2} i y a^{-\frac{1}{2}}\right)$
$x^{-1} e^{-ax^2}$	$\left(\frac{1}{2}\pi\right)^{\frac{1}{2}} \operatorname{Erf}\left(\frac{1}{2} y a^{-\frac{1}{2}}\right)$
$x^{2n+1} e^{-ax^2}$ $n = 0, 1, 2, \ldots$	$(-1)^n (2a)^{-n-1} e^{-\frac{1}{4}y^2/a} He_{2n+1}\left[(2a)^{-\frac{1}{2}} y\right]$
$x^{\nu-1} e^{-ax^2}$ $Re\, \nu > 0$	$(2\pi)^{-\frac{1}{2}} a^{-\frac{1}{2}\nu-\frac{1}{2}} \Gamma\left(\frac{1}{2}+\frac{1}{2}\nu\right) y\, {}_1F_1\left(\frac{1}{2}+\frac{1}{2}\nu; \frac{3}{2}; -\frac{\frac{1}{4}y^2}{a}\right)$
$x^{\nu-1} e^{-ax-bx^2}$ $Re\, \nu > 0$	$i(2\pi)^{-\frac{1}{2}} \Gamma(\nu) (2b)^{-\frac{1}{2}\nu} e^{\frac{a^2-y^2}{8b}} \left\{e^{i\frac{ay}{4b}} D_{-\nu}\left[(2b)^{-\frac{1}{2}}(a+iy)\right] - e^{-i\frac{ay}{4b}} D_{-\nu}\left[(2b)^{-\frac{1}{2}}(a-iy)\right]\right\}$

$f(x)$	$F_s(y)$
$x^{\nu-1}e^{-bx^{-1}-ax}$	$i\left(\frac{2}{\pi}\right)^{\frac{1}{2}} b^{\frac{1}{2}\nu}\left\{(a+iy)^{-\frac{1}{2}\nu} K_\nu\left[2b^{\frac{1}{2}}(a+iy)^{\frac{1}{2}}\right] - (a-iy)^{-\frac{1}{2}\nu} K_\nu\left[2b^{\frac{1}{2}}(a-iy)^{\frac{1}{2}}\right]\right\}$
$\log x \quad 0<x<1$ $0 \quad x>1$	$\left(\frac{2}{\pi}\right)^{\frac{1}{2}} y^{-1}[Ci(y)-\gamma-\log y]$
$(a^2+x^2)^{-\frac{1}{2}}\log\left[x+(a^2+x^2)^{\frac{1}{2}}\right]$	$\left(\frac{1}{2}\pi\right)^{\frac{1}{2}}\{K_0(ay)+\log a\,[I_0(ay)-\mathbf{L}_0(ay)]\}$
$x^{-1}\log(1+a^2x^2)$	$-(2\pi)^{\frac{1}{2}} Ei\left(-\frac{y}{a}\right)$
$(a^2+x^2)^{-\frac{1}{2}}\log\left[\frac{(a^2+x^2)^{\frac{1}{2}}+x}{a}\right]$	$\left(\frac{1}{2}\pi\right)^{\frac{1}{2}} K_0(ay)$
$x^{-1}\sin(ax)$	$(2\pi)^{-\frac{1}{2}}\log\left\|\frac{y+a}{y-a}\right\|$
$\left[\frac{\sin(ax)}{x}\right]^{2m}$ $m=1,2,3,\ldots$	$(-1)^m\left(\frac{2}{\pi}\right)^{\frac{1}{2}} 2m\, 2^{-2m}\left\{(m!)^{-2} y\log\left(\frac{y}{a}\right)+\sum_{n=1}^{m}(-1)^n[(m-n)!(m+n)!]^{-1} \times\left[(y-2an)^{2m-1}\log\left\|\frac{y}{a}-2n\right\|+(y+2an)^{2m-1}\log\left(\frac{y}{a}+2n\right)\right]\right\}$

$f(x)$	$g_s(y) = \left(\frac{2}{\pi}\right)^{\frac{1}{2}} \int_0^\infty f(x) \sin(xy)\, dx$
$\left[\frac{\sin(ax)}{x}\right]^{2m+1}$ $m = 0, 1, 2, \ldots$	$(-1)^m \left(\frac{2}{\pi}\right)^{\frac{1}{2}} 2^{-2m-1}(2m+1) \sum_{n=0}^{m} (-1)^n [(m+1+n)!\,(m-n)!]^{-1}$ $\times \left\{[(2n+1)a+y]^{2m} \log\left(\frac{y}{a}+(2n+1)\right)\right.$ $\left.-[(2n+1)a-y]^{2m} \log\left\lvert\frac{y}{a}-(2n+1)\right\rvert\right\}$
$[\sin(\pi x)]^{\nu-1}, \quad 0 < x < 1$ $0 \quad x > 1$ $Re\, \nu > 0$	$\left(\frac{2}{\pi}\right)^{\frac{1}{2}} 2^{1-\nu} \sin\left(\frac{1}{2}y\right) \Gamma(\nu) \left[\Gamma\left(\frac{1}{2}+\frac{1}{2}\nu+\frac{y}{2\pi}\right) \Gamma\left(\frac{1}{2}+\frac{1}{2}\nu-\frac{y}{2\pi}\right)\right]^{-1}$
$\sin(ax^2)$	$a^{-\frac{1}{2}} \left[\cos\left(\frac{y^2}{4a}\right) C\left(\frac{y^2}{4a}\right) + \sin\left(\frac{y^2}{4a}\right) S\left(\frac{y^2}{4a}\right)\right]$
$\cos(ax^2)$	$a^{-\frac{1}{2}} \left[\sin\left(\frac{y^2}{4a}\right) C\left(\frac{y^2}{4a}\right) - \cos\left(\frac{y^2}{4a}\right) S\left(\frac{y^2}{4a}\right)\right]$
$x^{-1} \sin(ax^2)$	$\left(\frac{1}{2}\pi\right)^{\frac{1}{2}} \left[C\left(\frac{y^2}{4a}\right) - S\left(\frac{y^2}{4a}\right)\right]$
$x^{-1} \cos(ax^2)$	$\left(\frac{1}{2}\pi\right)^{\frac{1}{2}} \left[C\left(\frac{y^2}{4a}\right) + S\left(\frac{y^2}{4a}\right)\right]$

$\sin(a^3x^3)$	$\left(\frac{1}{2}\pi\right)^{\frac{1}{2}}(3a)^{-\frac{3}{2}}y^{\frac{1}{2}}\left[J_{\frac{1}{3}}(z)+J_{-\frac{1}{3}}(z)-\pi^{-1}3^{\frac{1}{2}}K_{\frac{1}{3}}(z)\right]$ $z=2\left(\frac{y}{3a}\right)^{\frac{3}{2}}$
$\cos(a^3x^3)$	$\frac{(2\pi)^{\frac{1}{2}}}{18a}\left(\frac{y}{a}\right)^{\frac{1}{2}}\left[I_{-\frac{1}{3}}(z)+I_{\frac{1}{3}}(z)-J_{-\frac{1}{3}}(z)+J_{\frac{1}{3}}(z)\right.$ $\left.+2\boldsymbol{J}_{\frac{1}{3}}(z)-2\boldsymbol{J}_{-\frac{1}{3}}(z)+2i\boldsymbol{J}_{\frac{1}{3}}(iz)-2i\boldsymbol{J}_{-\frac{1}{3}}(iz)\right]$ $z=2\left(\frac{y}{3a}\right)^{\frac{3}{2}}$
$x^{\nu-1}\sin(ax^{-1})$ $-2<Re\,\nu<2$	$\left(\frac{1}{2}\pi\right)^{\frac{1}{2}}\left(\frac{a}{y}\right)^{\frac{1}{2}\nu}\left\{\sin\left(\frac{1}{2}\pi\nu\right)J_\nu\left[2(ay)^{\frac{1}{2}}\right]+\cos\left(\frac{1}{2}\pi\nu\right)Y_\nu\left[2(ay)^{\frac{1}{2}}\right]\right.$ $\left.+2\pi^{-1}\cos\left(\frac{1}{2}\pi\nu\right)K_\nu\left[2(ay)^{\frac{1}{2}}\right]\right\}$
$x^{\nu-1}\cos(ax^{-1})$ $-2<Re\,\nu<2$	$\left(\frac{1}{2}\pi\right)^{\frac{1}{2}}\left(\frac{a}{y}\right)^{\frac{1}{2}\nu}\left\{\cos\left(\frac{1}{2}\pi\nu\right)J_\nu\left[2(ay)^{\frac{1}{2}}\right]-\sin\left(\frac{1}{2}\pi\nu\right)Y_\nu\left[2(ay)^{\frac{1}{2}}\right]\right.$ $\left.+2\pi^{-1}\sin\left(\frac{1}{2}\pi\nu\right)K_\nu\left[2(ay)^{\frac{1}{2}}\right]\right\}$

$f(x)$	$g_s(y) = \left(\frac{2}{\pi}\right)^{\frac{1}{2}} \int_0^\infty f(x) \sin(xy)\, dx$
$0 \qquad 0 < x < a$ $(x^2 - a^2)^{-\frac{1}{2}} \cos\left[b(x^2 - a^2)^{\frac{1}{2}}\right] \qquad x > a$	$0 \qquad 0 < y < b$ $\left(\frac{1}{2}\pi\right)^{\frac{1}{2}} J_0\left[a(y^2 - b^2)^{\frac{1}{2}}\right] \qquad y > b$
$\arcsin x \qquad 0 < x < 1$ $0 \qquad x > 1$	$\left(\frac{1}{2}\pi\right)^{\frac{1}{2}} y^{-1}\left[J_0(y) - \cos y\right]$
$\operatorname{sech}(ax)$	$-\left(\frac{1}{2}\pi\right)^{\frac{1}{2}} a^{-1}\left\{\tanh\left(\frac{\frac{1}{2}\pi y}{a}\right) - i\pi^{-1}\left[\psi\left(\frac{1}{4} + i\frac{y}{4a}\right) - \psi\left(\frac{1}{4} - i\frac{y}{4a}\right)\right]\right\}$
$\operatorname{cosech}(ax)$	$\left(\frac{1}{2}\pi\right)^{\frac{1}{2}} a^{-1} \tanh\left(\frac{\frac{1}{2}\pi y}{a}\right)$
$0 \qquad 0 < x < a$ $(\cosh x - \cosh a)^{-\mu} \qquad x > a$ $0 < Re\,\mu < 1$	$(2\pi)^{-\frac{1}{2}} \Gamma(1-\mu)(\sinh a)^{\frac{1}{2}-\mu} \sinh(\pi y)$ $\times \Gamma(\mu + iy)\, \Gamma(\mu - iy)\, \mathfrak{P}^{\frac{1}{2}-\mu}_{-\frac{1}{2}+iy}(\cosh a)$
$x^{-1} - \operatorname{cosech} x$	$(2\pi)^{\frac{1}{2}}(1 + e^{\pi y})^{-1}$

$\cosh(ax)\operatorname{cosech}(bx)$ $a < b$	$\left(\frac{1}{2}\pi\right)^{\frac{1}{2}} b^{-1}\sinh\left(\frac{\pi y}{b}\right)\left[\cosh\left(\frac{\pi y}{b}\right)+\cos\left(\frac{\pi a}{b}\right)\right]^{-1}$
$\sinh(ax)\operatorname{sech}(bx)$ $a < b$	$(2\pi)^{\frac{1}{2}} b^{-1}\sin\left(\frac{\frac{1}{2}\pi a}{b}\right)\sinh\left(\frac{\frac{1}{2}\pi y}{b}\right)\left[\cosh\left(\frac{\pi y}{b}\right)+\cos\left(\frac{\pi a}{b}\right)\right]^{-1}$
$[\sinh(ax)]^{-\nu}$ $0 < Re\,\nu < 2$	$2^{\nu}(2\pi)^{\frac{1}{2}} a^{-1}\cos\left(\frac{1}{2}\pi\nu\right)\Gamma(1-\nu)\left[\Gamma\left(1-\frac{1}{2}\nu+\frac{iy}{2a}\right)\Gamma\left(1-\frac{1}{2}\nu-\frac{iy}{2a}\right)\right]^{-1}$ $\sinh\left(\frac{\frac{1}{2}\pi y}{a}\right)\left[\cosh\left(\frac{\frac{1}{2}\pi y}{a}\right)-\cos(\pi\nu)\right]^{-1}$
$e^{-a\sinh x}$	$\left(\frac{2}{\pi}\right)^{\frac{1}{2}} y\,S_{-1,iy}(a)=\left(\frac{1}{2}\pi\right)^{\frac{1}{2}}\operatorname{cosech}(\pi y)\,[\boldsymbol{J}_{iy}(a)+\boldsymbol{J}_{-iy}(a)-J_{iy}(a)-J_{-iy}(a)]$
$\sin(a\sinh x)$	$\left(\frac{2}{\pi}\right)^{\frac{1}{2}}\sinh\left(\frac{1}{2}\pi y\right)K_{iy}(a)$
$(1-x^2)^{-\frac{1}{2}}\cos\left[a(1-x^2)^{\frac{1}{2}}\right]T_{2n+1}(x)$ $\quad 0 < x < 1$ 0 $\quad x > 1$	$(-1)^n\left(\frac{1}{2}\pi\right)^{\frac{1}{2}} T_{2n+1}\left[y(a^2+y^2)^{-\frac{1}{2}}\right]J_{2n+1}\left[(a^2+y^2)^{\frac{1}{2}}\right]$

$f(x)$	$g_s(y) = \left(\frac{2}{\pi}\right)^{\frac{1}{2}} \int_0^\infty f(x) \sin(xy)\, dx$
$(a^2 - x^2)^{\nu - \frac{1}{2}} C_{2n+1}^{\nu}\left(\frac{x}{a}\right)$ $\quad 0 < x < a$ $0 \quad x > a$	$(-1)^n (2\pi)^{\frac{1}{2}} a^{\nu} \frac{\Gamma(2n + 2\nu + 1)}{(2n+1)!\,\Gamma(\nu)} (2y)^{-\nu} J_{\nu+2n+1}(ay)$
$e^{-\frac{1}{2}x^2/a} He_{2n+1}\left[x(a^{-1} - 1)^{-\frac{1}{2}}\right]$	$(-1)^n a^{n+1} (1 - a)^{-n-\frac{1}{2}} e^{-\frac{1}{2}ay^2} He_{2n+1}(y)$
$\operatorname{Erfc}(ax)$	$\left(\frac{2}{\pi}\right)^{\frac{1}{2}} y^{-1}\left(1 - e^{-\frac{1}{4}y^2/a^2}\right)$
$e^{a^2x^2} \operatorname{Erfc}(ax)$	$2^{-\frac{1}{2}} a^{-1} e^{\frac{1}{4}y^2/a^2} \operatorname{Erfc}\left(\frac{\frac{1}{2}y}{a}\right)$
$(1 - x^2)^{-\frac{1}{2}\mu} P_\nu^\mu(x) \quad 0 < x < 1$ $0 \quad x > 1$	$2^{\mu - \frac{3}{2}} \left[\Gamma\left(\frac{3 - \mu - \nu}{2}\right) \Gamma\left(\frac{4 - \mu + \nu}{2}\right)\right]^{-1}$ $(1 - \mu - \nu)(2 + \nu - \mu)\, y^{\mu - \frac{1}{2}} s_{\frac{1}{2} - \mu, \frac{1}{2} + \nu}(y)$
$0 \quad 0 < x < 1$ $(x^2 - 1)^{-\frac{1}{2}\mu} \mathfrak{P}_\nu^\mu(x), \quad x > 1$ $Re\,\mu > Re\,\nu > -Re\,(1 + \mu)$	$y^{\mu - \frac{1}{2}} \left[\sin\left(\frac{1}{2}\pi\mu - \frac{1}{2}\pi\nu\right) J_{\nu + \frac{1}{2}}(y) - \cos\left(\frac{1}{2}\pi\mu - \frac{1}{2}\pi\nu\right) Y_{\nu + \frac{1}{2}}(y)\right]$

$(x^2-1)^{\frac{1}{2}\mu}\,\mathfrak{P}_\nu^\mu(x)$ $Re\,\mu<\frac{3}{2},\qquad Re\,(\mu+\nu)<1$	$\left[\Gamma\left(\frac{1-\mu-\nu}{2}\right)\Gamma\left(1-\frac{\mu-\nu}{2}\right)\right]^{-1}\left(\frac{1}{2}y\right)^{-\mu-\frac{1}{2}}S_{\mu+\frac{1}{2},\nu+\frac{1}{2}}(y)$
$0\qquad 0<x<a+b$ $\mathfrak{P}_\nu\left(\frac{x^2-a^2-b^2}{2ab}\right),\qquad x>a+b$ $-1<Re\,\nu<0$	$-\left(\frac{1}{2}\pi ab\right)^{\frac{1}{2}}\left[J_{\nu+\frac{1}{2}}(ay)\,Y_{-\nu-\frac{1}{2}}(ay)+Y_{\nu+\frac{1}{2}}(ay)\,J_{-\nu-\frac{1}{2}}(ay)\right]$
$\sinh(\pi x)\,\Gamma\left(\frac{1}{2}-\mu+ix\right)$ $\times\Gamma\left(\frac{1}{2}-\mu-ix\right)\mathfrak{P}^\mu_{-\frac{1}{2}+ix}(\cosh a)$ $Re\,\mu>-\frac{1}{2}$	$\pi\left[\Gamma\left(\frac{1}{2}+\mu\right)\right]^{-1}(\sinh a)^{-\mu}(\cosh y-\cosh a)^{\mu-\frac{1}{2}}\qquad y>a$ $0\qquad y<a$
$J_{2n+1}(ax)$ $n=0,1,2,\ldots$	$(-1)^n\left(\frac{2}{\pi}\right)^{\frac{1}{2}}(a^2-y^2)^{-\frac{1}{2}}T_{2n+1}\left(\frac{y}{a}\right),\qquad y<a$ $0\qquad y>a$
$J_\nu(ax),\qquad Re\,\nu>-2$	$\left(\frac{2}{\pi}\right)^{\frac{1}{2}}(a^2-y^2)^{-\frac{1}{2}}\sin\left[\nu\arcsin\left(\frac{y}{a}\right)\right],\qquad y<a$ $\left(\frac{2}{\pi}\right)^{\frac{1}{2}}\cos\left(\frac{1}{2}\pi\nu\right)a^\nu(y^2-a^2)^{-\frac{1}{2}}\left[y+(y^2-a^2)^{\frac{1}{2}}\right]^{-\nu}\qquad y>a$

$f(x)$	$g_s(y) = \left(\frac{2}{\pi}\right)^{\frac{1}{2}} \int_0^\infty f(x) \sin(xy)\, dx$
$x^\nu J_\nu(ax)$ $-1 < Re\,\nu < \frac{1}{2}$	$0 \qquad y < a$ $2^{\frac{1}{2}} \left[\Gamma\left(\frac{1}{2} - \nu\right)\right]^{-1} (2a)^\nu (y^2 - a^2)^{-\nu - \frac{1}{2}} \qquad y > a$
$x^{-\nu} J_{\nu+2n+1}(ax)$ $n = 0, 1, 2, \ldots$ $Re\,\nu > -\frac{1}{2}$	$(-1)^n \left(\frac{2}{\pi}\right)^{\frac{1}{2}} 2^{\nu-1} a^{-\nu} (2n+1)!\,\Gamma(\nu)$ $[\Gamma(2n + 2\nu + 1)]^{-1} (a^2 - y^2)^{\nu - \frac{1}{2}} C^\nu_{2n+1}\left(\frac{y}{a}\right) \qquad y < a$ $0 \qquad y > a$
$x^{\nu+2n}(b^2 + x^2)^{-1} J_\nu(ax)$ $Re\,(\nu + n) > -1, \quad Re\,(\nu + 2n) < \frac{5}{2}$ $n = 0, \pm 1, \pm 2, \ldots$	$(-1)^n \left(\frac{2}{\pi}\right)^{\frac{1}{2}} b^{\nu+2n-1} \sinh(by)\, K_\nu(ab) \qquad y < a$
$x^{2n+1-\nu}(b^2 + x^2)^{-1} J_\nu(ax)$ $Re\,\nu > 2n - \frac{3}{2}$ $n = -1, 0, 1, \ldots$	$(-1)^n \left(\frac{1}{2}\pi\right)^{\frac{1}{2}} b^{2n-\nu} e^{-by} I_\nu(ab), \qquad y > a$
$J_\nu(ax)\, J_\nu(bx), \quad b \geq a$ $Re\,\nu > -1$	$0 \qquad 0 < y < b - a$ $(2\pi ab)^{-\frac{1}{2}} P_{\nu-\frac{1}{2}}\left(\frac{a^2 + b^2 - y^2}{2ab}\right) \qquad b - a < y < b + a$ $\left(\frac{2}{\pi}\right)^{\frac{1}{2}} \pi^{-1} (ab)^{-\frac{1}{2}} \cos(\pi\nu)\, \mathfrak{Q}_{\nu-\frac{1}{2}}\left(\frac{y^2 - a^2 - b^2}{2ab}\right) \qquad y > a + b$

$Y_0(ax)$	$\left(\frac{2}{\pi}\right)^{\frac{3}{2}}(a^2-y^2)^{-\frac{1}{2}}\arcsin\left(\frac{y}{a}\right)$ $y<a$ $\left(\frac{2}{\pi}\right)^{\frac{3}{2}}(y^2-a^2)^{-\frac{1}{2}}\log\left[\frac{y}{a}-\left(\frac{y^2}{a^2}-1\right)^{\frac{1}{2}}\right]$ $y>a$
0 $0<x<a$ $J_0\left[b(x^2-a^2)^{\frac{1}{2}}\right]$ $x>a$	0 $y<b$ $\left(\frac{2}{\pi}\right)^{\frac{1}{2}}(y^2-b^2)^{-\frac{1}{2}}\cos\left[a(y^2-b^2)^{\frac{1}{2}}\right]$ $y>b$
0 $0<x<a$ $(x^2-a^2)^{-\frac{1}{2}}J_\nu\left[b(x^2-a^2)^{\frac{1}{2}}\right]$ $x>a$ $Re\,\nu>-1$	$\left(\frac{1}{2}\pi\right)^{\frac{1}{2}}J_{\frac{1}{2}\nu}\left\{\frac{1}{2}a\left[y-(y^2-b^2)^{\frac{1}{2}}\right]\right\}J_{-\frac{1}{2}\nu}\left\{\frac{1}{2}a\left[y+(y^2-b^2)^{\frac{1}{2}}\right]\right\}$ $y>b$
0 $0<x<a$ $(x^2-a^2)^{\frac{1}{2}\nu}J_\nu\left[b(x^2-a^2)^{\frac{1}{2}}\right]$ $x>a$ $-1<Re\,\nu<\frac{1}{2}$	0 $y<b$ $a^{\frac{1}{2}+\nu}b^\nu(y^2-b^2)^{-\frac{1}{2}\nu-\frac{1}{4}}J_{-\nu-\frac{1}{2}}\left[a(y^2-b^2)^{\frac{1}{2}}\right],$ $y>b$
$J_\nu(a\sin x)$ $0<x<\pi$ 0 $x>\pi$ $Re\,\nu>-2$	$(2\pi)^{\frac{1}{2}}\sin\left(\frac{1}{2}\pi y\right)J_{\frac{1}{2}\nu-\frac{1}{2}y}\left(\frac{1}{2}a\right)J_{\frac{1}{2}\nu+\frac{1}{2}y}\left(\frac{1}{2}a\right)$

$f(x)$	$g_s(y) = \left(\frac{2}{\pi}\right)^{\frac{1}{2}} \int_0^\infty f(x) \sin(xy)\, dx$
$K_0(ax)$	$\left(\frac{2}{\pi}\right)^{\frac{1}{2}} (a^2+y^2)^{-\frac{1}{2}} \log\left[\frac{y}{a} + \left(1+\frac{y^2}{a^2}\right)^{\frac{1}{2}}\right]$
$K_\nu(ax)$ $-2 < Re\,\nu < 2$	$\frac{1}{4}(2\pi)^{\frac{1}{2}} a^{-\nu} \operatorname{cosec}\left(\frac{1}{2}\pi\nu\right)(a^2+y^2)^{-\frac{1}{2}}$ $\times\left\{\left[(y^2+a^2)^{\frac{1}{2}}+y\right]^\nu - \left[(y^2+a^2)^{\frac{1}{2}}-y\right]^\nu\right\}$
$x^{\nu+1}K_\nu(ax)$ $Re\,\nu > -\frac{3}{2}$	$2^{\nu+\frac{1}{2}} a^\nu \Gamma\left(\frac{3}{2}+\nu\right) y\,(y^2+a^2)^{-\nu-\frac{3}{2}}$
$[I_\nu(ax) + I_{-\nu}(ax)]\,K_\nu(ax)$	$\left(\frac{1}{2}\pi\right)^{\frac{1}{2}} a^{-1}\, \mathfrak{P}_{\nu-\frac{1}{2}}\left(1+\frac{\frac{1}{2}y^2}{a^2}\right)$
$0 \qquad 0 < x < a$ $(x^2-a^2)^{-\frac{1}{2}} K_\nu\left[b(x^2-a^2)^{\frac{1}{2}}\right] \quad x > a$ $-1 < Re\,\nu < 1$	$\frac{1}{2}\left(\frac{\pi}{2}\right)^{\frac{3}{2}} \sec\left(\frac{1}{2}\pi\nu\right)\left[J_{\frac{1}{2}\nu}(z_2)\,Y_{\frac{1}{2}\nu}(z_1) - J_{\frac{1}{2}\nu}(z_1)\,Y_{\frac{1}{2}\nu}(z_2)\right]$ $z_{\frac{1}{2}} = \frac{1}{2}a\left[(b^2+y^2)^{\frac{1}{2}} \pm y\right]$
$x^{-\nu}\,\boldsymbol{H}_\nu(ax)$	$2^{\frac{1}{2}-\nu} a^{-\nu}\left[\Gamma\left(\frac{1}{2}+\nu\right)\right]^{-1}(a^2-y^2)^{\nu-\frac{1}{2}}, \quad y < a$ $0 \quad y > a$

Examples for the exponential Fourier transform

$f(x)$	$g(y) = \int_{-\infty}^{\infty} f(x)\, e^{ixy}\, dx$
$A \quad a \le x \le b$ $0 \quad$ otherwise	$i A y^{-1} (e^{iay} - e^{iby})$
$x^n \quad 0 \le x \le b$ $0 \quad$ otherwise $n = 1, 2, 3, \ldots$	$n!(-iy)^{-n-1} - e^{iby} \sum_{m=0}^{n} \frac{n!}{m!} (-iy)^{m-n-1} b^m$
$x^\nu \quad 0 < x \le b$ $0 \quad$ otherwise $Re\, \nu > -1$	$i e^{\frac{1}{2} i\pi\nu} y^{-\nu-1} \gamma(\nu + 1, -iby)$
$x^\nu \quad x \ge b$ $0 \quad$ otherwise $Re\, \nu < 0$	$-i e^{\frac{1}{2} i\pi\nu} y^{-\nu-1} \Gamma(\nu + 1, -iby)$
$(x - b)^\nu \quad x > b$ $0 \quad$ otherwise $-1 < Re\, \nu < 0$	$i e^{\frac{1}{2} i\pi\nu} \Gamma(1 + \nu)\, y^{-\nu-1} e^{iby}$

$(a+ix)^{-\nu}$ $\quad Re\,\nu>0$	$2\pi[\Gamma(\nu)]^{-1}y^{\nu-1}e^{-ay}$ $\quad y>0$ 0 $\quad y<0$
$(a-ix)^{-\nu}$ $\quad Re\,\nu>0$	0 $\quad y>0$ $2\pi[\Gamma(\nu)]^{-1}y^{\nu-1}e^{-ay}$ $\quad y<0$
$(a+ix)^{-\nu}(b+ix)^{-1}$ $Re\,\nu>-1$	$2\pi[\Gamma(\nu)]^{-1}(a-b)^{-\nu}e^{-by}\gamma(\nu,ay-by)$ $\quad y>0$ 0 $\quad y<0$
$(a+ix)^{-\nu}(b-ix)^{-1}$ $Re\,\nu>-1$	$2\pi[\Gamma(\nu)]^{-1}(a+b)^{-\nu}e^{by}\Gamma(\nu,ay+by)$ $\quad y>0$ $2\pi(a+b)^{-\nu}e^{by}$ $\quad y<0$
$(a-ix)^{-\nu}(b+ix)^{-1}$ $Re\,\nu>-1$	$2\pi[\Gamma(\nu)]^{-1}(a+b)^{-\nu}e^{-by}\Gamma(\nu,-ay-by)$ $\quad y<0$ $2\pi(a+b)^{-\nu}e^{-by}$ $\quad y>0$
$(a-ix)^{-\nu}(b-ix)^{-1}$ $Re\,\nu>-1$	0 $\quad y>0$ $2\pi[\Gamma(\nu)]^{-1}(a-b)^{-\nu}e^{by}\gamma(\nu,by-ay)$ $\quad y<0$
$(ix)^{-\nu}(a^2+x^2)^{-1}$ $-2<Re\,\nu<1$ $\arg(ix)=\pm\frac{1}{2}\pi$ for $x\gtrless 0$	$\pi[\Gamma(\nu)]^{-1}a^{-\nu-1}e^{ay}\Gamma(\nu,ay)$ $+\pi[\Gamma(1+\nu)]^{-1}a^{-1}y^{\nu}e^{-ay}{}_1F_1(\nu;1+\nu;ay)$ $\quad y>0$ $\pi a^{-\nu-1}e^{ay}$ $\quad y<0$
$(-ix)^{-\nu}(a^2+x^2)^{-1}$ $-2<Re\,\nu<1$ $\arg(-ix)=\mp\frac{1}{2}\pi$ for $x\gtrless 0$	$\pi a^{-\nu-1}e^{-ay}$ $\quad y>0$ $\pi[\Gamma(\nu)]^{-1}a^{-\nu-1}e^{-ay}\Gamma(\nu,-ay)$ $+\pi[\Gamma(1+\nu)]^{-1}a^{-1}(-y)^{\nu}e^{ay}{}_1F_1(\nu;1+\nu;-ay)$ $\quad y>0$

$f(x)$	$g(y) = \int_{-\infty}^{\infty} f(x)\, e^{ixy}\, dx$	
$(a + ix)^{-\nu} (b^2 + x^2)^{-1}$ $Re\, \nu > -2$	$\pi b^{-1} (a + b)^{-\nu} e^{by}$	$y < 0$
	$\pi b^{-1} [\Gamma(\nu)]^{-1} [(a - b)^{-\nu} e^{-by} \gamma(\nu, by - ay)$ $+ (a + b)^{-\nu} e^{-by} \Gamma(\nu, -ay - by)]$	$y > 0$
$(a - ix)^{-\nu} (b^2 + x^2)^{-1}$ $Re\, \nu > -2$	$\pi b^{-1} (a + b)^{-\nu} e^{-by}$	$y > 0$
	$\pi b^{-1} [\Gamma(\nu)]^{-1} [(a - b)^{-\nu} e^{by} \gamma(\nu, by - ay)$ $+ (a + b)^{-\nu} e^{-by} \Gamma(\nu, -ay - by)]$	$y < 0$
$[a^2 + (x \pm b)^2]^{-\nu}$ $Re\, \nu > 0$	$2 e^{\mp iby} \pi^{\frac{1}{2}} [\Gamma(\nu)]^{-1} \left(\frac{\lvert y \rvert}{2a}\right)^{\nu - \frac{1}{2}} K_{\nu - \frac{1}{2}}(a \lvert y \rvert)$	
$(a + ix)^{-\mu} (b + ix)^{-\nu}$ $Re\, (\mu + \nu) > 0$	$2\pi [\Gamma(\mu + \nu)]^{-1} e^{-ay} y^{\mu+\nu-1} {}_1F_1(\nu; \nu + \mu; ay - by)$	$y > 0$
	0	$y < 0$
$(a - ix)^{-\mu} (b - ix)^{-\nu}$ $Re\, (\mu + \nu) > 0$	0	$y > 0$
	$2\pi [\Gamma(\mu + \nu)]^{-1} e^{ay} (-y)^{\mu+\nu-1} {}_1F_1(\nu; \nu + \mu; by - ay)$	$y < 0$
$(a - ix)^{-\mu} (b + ix)^{-\nu}$ $Re\, (\mu + \nu) > 0$	$2\pi [\Gamma(\nu)]^{-1} (a + b)^{-\frac{1}{2}(\nu+\mu)} y^{\frac{1}{2}(\nu+\mu-1)}$ $\times e^{\frac{1}{2}(b-a)y} W_{\frac{1}{2}(\nu-\mu), \frac{1}{2}(1-\nu-\mu)}(ay + by)$	$y > 0$
	$2\pi [\Gamma(\mu)]^{-1} (a + b)^{-\frac{1}{2}(\nu+\mu)} (-y)^{\frac{1}{2}(\nu+\mu-1)}$ $\times e^{\frac{1}{2}(a-b)y} W_{\frac{1}{2}(\mu-\nu), \frac{1}{2}(1-\nu-\mu)}(-ay - by)$	$y < 0$

$(1-x)^{\nu-1}(1+x)^{\mu-1}$ $\quad -1 < x < 1$ $0 \quad$ otherwise $Re\,(\nu, \mu) > 0$	$2^{\nu+\mu-1} B(\mu, \nu)\, e^{-iy}\, {}_1F_1(\mu; \nu+\mu; 2iy)$
$(a+ix)^{-\nu} e^{-b(a+ix)^{-1}}$ $Re\,\nu > 0$	$2\pi e^{-ay}\left(\frac{y}{b}\right)^{\frac{1}{2}(\nu-1)} J_{\nu-1}\left[2(by)^{\frac{1}{2}}\right] \quad y > 0$ $0 \quad y < 0$
$[\cosh(ax+b)]^{-\nu}$ $Re\,\nu > 0$	$2^{\nu-1} a^{-1} [\Gamma(\nu)]^{-1} e^{iy\frac{b}{a}} \Gamma\left(\frac{1}{2}\nu - i\frac{y}{2a}\right) \Gamma\left(\frac{1}{2}\nu + i\frac{y}{2a}\right)$
$(-ix)^{\nu} e^{-a^2x^2} \quad Re\,\nu > -1$ $\arg(-ix) = \pm\frac{1}{2}\pi$ for $x \gtrless 0$	$\pi^{\frac{1}{2}} 2^{-\frac{1}{2}\nu} a^{-\nu-1} e^{-\frac{1}{2}y^2a^{-2}} \quad D_{\nu}\left(2^{-\frac{1}{2}} a^{-1} y\right)$
$e^{-\lambda x} \log\lvert 1 - e^{-x}\rvert$ $Re\,\lambda > -1$	$\pi(\lambda - iy)^{-1} \cot(\pi\lambda - i\pi y)$
$e^{-\lambda x} \log(1 + e^{-x})$ $Re\,\lambda > -1$	$\pi(\lambda - iy)^{-1} \operatorname{cosec}(\pi\lambda - i\pi y)$
$e^{-\lambda x}(a + e^{-x})^{-\nu} \log(a + e^{-x})$ $Re\,\lambda > 0$	$a^{\lambda-\nu-iy} B(\lambda - iy, \nu - \lambda + iy)\, [\psi(\nu) - \psi(\nu - \lambda + iy) + \log a]$

$f(x)$	$g(y) = \int_{-\infty}^{\infty} f(x)\, e^{ixy}\, dx$
$e^{-a\cosh x - b\sinh x}$	$2\left(\frac{a-b}{a+b}\right)^{\frac{1}{2}iy} K_{iy}\left[(a^2-b^2)^{\frac{1}{2}}\right]$
$P_n(x) \quad -1 < x < 1$ $0 \quad \lvert x\rvert > 1$ $n = 0, 1, 2, \ldots$	$(-i)^n \left(\frac{2\pi}{y}\right)^{\frac{1}{2}} J_{n+\frac{1}{2}}(y)$
$(1-x^2)^{-\frac{1}{2}} T_n(x) \quad -1 < x < 1$ $0 \quad \lvert x\rvert > 1$ $n = 0, 1, 2, \ldots$	$(-i)^n \pi J_n(y)$
$[\Gamma(\nu - x)\, \Gamma(\mu + x)]^{-1}$	$\left[2\cos\left(\frac{1}{2}y\right)\right]^{\nu+\mu-2} e^{-\frac{1}{2}iy(\mu-\nu)} [\Gamma(\nu+\mu-1)]^{-1} \quad \lvert y\rvert < \pi$ $0 \quad \lvert y\rvert > \pi$
$0 \quad \lvert x\rvert > \frac{1}{2}\pi$ $(\cos x)^{\frac{\nu}{2}} (a^2 e^{-ix} + b^2 e^{ix})^{-\frac{\nu}{2}}$ $\times J_\nu\left\{c\,[2(a^2 e^{-ix} + b^2 e^{ix}) \cos x]^{\frac{1}{2}}\right\}$ $\lvert x\rvert < \frac{1}{2}\pi$ $Re\,(\nu + \mu) > -1$	$\pi\, 2^{-\frac{1}{2}\nu} a^{\frac{1}{2}(y-\nu)} b^{-\frac{1}{2}(y+\nu)} J_{\frac{1}{2}(\nu-y)}(ac)\, J_{\frac{1}{2}(\nu+y)}(bc)$

$\operatorname{sech}(ax)\, e^{-b\tanh(ax)}\, J_\nu[c \operatorname{sech}(ax)]$ $Re\, \nu > -1$	$(ab)^{-1}\Gamma\left(\frac{1}{2}+\frac{1}{2}\nu+\frac{1}{2}\,iya^{-1}\right)\Gamma\left(\frac{1}{2}+\frac{1}{2}\nu-\frac{1}{2}\,iya^{-1}\right)$ $\times M_{i\frac{y}{2a},\frac{\nu}{2}}\left\{\frac{1}{2}\left[b+(b^2-c^2)^{\frac{1}{2}}\right]\right\} M_{i\frac{y}{2a},\frac{\nu}{2}}\left\{\frac{1}{2}\left[b-(b^2-c^2)^{\frac{1}{2}}\right]\right\}$
$\left(\frac{ae^x+be^{-x}}{ae^{-x}+be^x}\right)^{\frac{1}{2}\nu} J_\nu\left[(a^2+b^2+2ab\cosh 2x)^{\frac{1}{2}}\right]$	$-\frac{1}{2}\pi\left[J_{\frac{1}{2}(\nu+iy)}(b)\, Y_{\frac{1}{2}(\nu-iy)}(a)+J_{\frac{1}{2}(\nu-iy)}(a)\, Y_{\frac{1}{2}(\nu+iy)}(b)\right]$
$\left(\frac{ae^x+be^{-x}}{ae^{-x}+be^x}\right)^{\frac{1}{2}\nu} Y_\nu\left[(a^2+b^2+2ab\cosh 2x)^{\frac{1}{2}}\right]$	$\frac{1}{2}\pi\left[J_{\frac{1}{2}(\nu+iy)}(b)\, J_{\frac{1}{2}(\nu-iy)}(a)-Y_{\frac{1}{2}(\nu+iy)}(b)\, Y_{\frac{1}{2}(\nu-iy)}(a)\right]$
$\left(\frac{ae^x+be^{-x}}{ae^{-x}+be^x}\right)^{\frac{1}{2}\nu} K_\nu\left[(a^2+b^2+2ab\cosh x)^{\frac{1}{2}}\right]$	$K_{\frac{1}{2}(\nu+iy)}(b)\, K_{\frac{1}{2}(\nu-iy)}(a)$
$J_{\mu+x}(a)\, J_{\nu-x}(a)$ $Re\,(\mu+\nu)>1$	$e^{-\frac{1}{2}iy(\mu-\nu)} J_{\mu+\nu}\left[2a\cos\left(\frac{1}{2}y\right)\right]$
$a^{-\mu-x} J_{\mu+x}(a)\, b^{-\nu+x} J_{\nu-x}(b)$ $Re\,(\mu+\nu)>1$	$\left(2\cos\frac{1}{2}y\right)^{\frac{1}{2}(\mu+\nu)}\left(a^2e^{-i\frac{1}{2}y}+b^2e^{i\frac{1}{2}y}\right)^{-\frac{1}{2}(\nu+\mu)}$ $\times e^{-\frac{1}{2}iy(\mu-\nu)} J_{\mu+\nu}\left\{\left[2\cos\frac{1}{2}y\left(a^2e^{-i\frac{1}{2}y}+b^2e^{i\frac{1}{2}y}\right)\right]^{\frac{1}{2}}\right\}$

Examples for the Laplace transform

If the Laplace transform $f(s)$ of a given function $F(t)$ is known it is often possible to obtain the Laplace transform $f_1(s)$ of a function $F_1(t)$ which is in one way or another obtained from $F(t)$. A few examples are listed in the following table

$F_1(t)$	$f_1(s) = \int_0^\infty F_1(t)\, e^{-st}\, dt$
$e^{-at} F(bt + c)$	$\frac{1}{b} e^{\frac{c}{b}(s+a)} \left[f\left(\frac{s+a}{b}\right) - \int_0^c e^{-\frac{t}{b}(s+a)} F(t)\, dt \right]$
$t^n F(t)$	$(-1)^n f^{(n)}(s)$
$t^{-n} F(t)$	$\int_s^\infty \cdots \int_s^\infty f(s)\, (ds)^n$ n-th repeated integral
$F^{(n)}(t)$	$s^n f(s) - \sum_{m=0}^{n-1} s^{n-m-1} F^{(m)}(0)$
$F(t) = F(t + a)$ (periodic function of period a)	$(1 - e^{-as})^{-1} \int_0^a e^{-st} F(t)\, dt$
$F(t) = -F(t + a)$	$(1 + e^{-as})^{-1} \int_0^a e^{-st} F(t)\, dt$

$0 \quad t < \frac{b}{a}$ $F(at-b) \quad t > \frac{b}{a}$ $a, b > 0$	$a^{-1} e^{-\frac{b}{a}s} f\left(\frac{s}{a}\right)$
$\int\limits_0^t \cdots \int\limits_0^t F(t)\,(dt)^n$	$s^{-n} f(s)$
$t^m F^{(n)}(t) \qquad m \geq n$	$\left(-\frac{d}{ds}\right)^m [s^n f(s)]$
$t^m F^{(n)}(t),\ m < n$	$\left(-\frac{d}{ds}\right)^m [s^n f(s)] + (-1)^{m-1}\left[\frac{(n-1)!}{(n-m-1)!} s^{n-m-1} F(0) + \frac{(n-2)!}{(n-m-2)!} s^{n-m-2} F'(0) + \cdots + m!\, F^{(n-m-1)}(0)\right]$
$\frac{d^n}{dt^n}[t^m F(t)] \qquad m \geq n$	$(-1)^m s^n f^{(m)}(s)$
$\frac{d^n}{dt^n}[t^m F(t)] \qquad m < n$	$(-1)^m s^n f^{(n)}(s) - m!\, s^{n-m-1} F(0) - \frac{(m+1)!}{1!} s^{n-m-2} F'(0) - \cdots - \frac{(n-1)!}{(n-m-1)!} F^{(n-m-1)}(0)$
$\int\limits_0^t t^{-1} F(t)\, dt$	$s^{-1} \int\limits_s^\infty f(s)\, ds$
$\int\limits_t^\infty t^{-1} F(t)\, dt$	$s^{-1} \int\limits_0^s f(s)\, ds$

$F_1(t)$	$f_1(s) = \int\limits_0^\infty F_1(t)\, e^{-st}\, dt$
$\int\limits_0^t F_1^*(\tau)\, F_2^*(t-\tau)\, d\tau$ (convolution theorem)	$f_1^*(s)\, f_2^*(s)$ $\left(f^*_{\substack{1\\2}}(s) = \int\limits_0^\infty F^*_{\substack{1\\2}}(t)\, e^{-st}\, dt\right)$
$F_1^*(t)\, F_2^*(t)$	$\frac{1}{2\pi i} \int\limits_{c-i\infty}^{c+i\infty} f_1^*(z)\, f_2^*(s-z)\, dz$ $f^*_{\substack{1\\2}}(s)$ as above
$F(t^2)$	$\pi^{-\frac{1}{2}} \int\limits_0^\infty e^{-\frac{1}{4}s^2\tau^{-2}}\, f(\tau^2)\, d\tau$
$t^n F(t^2)$	$2^{-\frac{1}{2}n}\, \pi^{-\frac{1}{2}} \int\limits_0^\infty \tau^{n-2} e^{-\frac{1}{4}\tau^2 s^2}\, He_n\left(2^{-\frac{1}{2}} s t\right) f(\tau^{-2})\, d\tau$
$t^\nu F(t^2)$	$(2\pi)^{-\frac{1}{2}} \int\limits_0^\infty \tau^{\nu-2} e^{-\frac{1}{4}s^2\tau^2}\, D_\nu(s\tau)\, f\left(\frac{1}{2}\tau^{-2}\right) d\tau$
$t^{\nu-1} F(t^{-1})$ $Re\, \nu > -1$	$s^{-\frac{1}{2}\nu} \int\limits_0^\infty \tau^{\frac{1}{2}\nu}\, J_\nu\left[2(s\tau)^{\frac{1}{2}}\right] f(\tau)\, d\tau$

$F(a \sinh t),\ a > 0$	$\int_0^\infty J_s(a\tau)\, f(\tau)\, d\tau$
$\int_0^\infty t^{u-1} [\Gamma(u)]^{-1} F(u)\, du$	$f(\log s)$
$\int_0^\infty u^{-\frac{1}{2}} \sin\left[2(ut)^{\frac{1}{2}}\right] F(u)\, du$	$s^{-1}\left(\frac{\pi}{s}\right)^{\frac{1}{2}} f(s^{-1})$
$t^{-\frac{1}{2}} \int_0^\infty \cos\left[2(ut)^{\frac{1}{2}}\right] F(u)\, du$	$\left(\frac{\pi}{s}\right)^{\frac{1}{2}} f(s^{-1})$
$t^{-\frac{1}{2}} \int_0^\infty e^{-\frac{1}{4}t^{-1}u^2} F(u)\, du$	$\left(\frac{\pi}{s}\right)^{\frac{1}{2}} f\left(\frac{1}{s^2}\right)$
$t^{-\frac{1}{2}(n+1)} \int_0^\infty e^{-\frac{1}{4}t^{-1}u^2} He_n\left[u(2t)^{-\frac{1}{2}}\right] F(u)\, du$	$2^{\frac{1}{2}n} \pi^{\frac{1}{2}} s^{\frac{1}{2}(n-1)} f\left(\frac{1}{s^2}\right)$
$t^{-\nu} \int_0^\infty e^{-\frac{1}{8}t^{-1}u^2} D_{2\nu-1}\left[u(2t)^{-\frac{1}{2}}\right] F(u)\, du$	$2^{\nu-\frac{1}{2}} \pi^{\frac{1}{2}} s^{\nu-1} f\left(\frac{1}{s^2}\right)$
$t^{\nu} \int_0^\infty J_{2\nu}\left[2(ut)^{\frac{1}{2}}\right] u^{-\nu} F(u)\, du$	$s^{-2\nu-1} f(s^{-1})$

$F_1(t)$	$f_1(s) = \int_0^\infty F_1(t)\, e^{-st}\, dt$
$\int_0^t J_0\left[(t^2 - u^2)^{\frac{1}{2}}\right] F(u)\, du$	$(s^2+1)^{-\frac{1}{2}} f\left[(1+s^2)^{\frac{1}{2}}\right]$
$\int_0^t \left(\frac{t-u}{au}\right)^{\nu} J_{2\nu}\left[2(atu - au^2)^{\frac{1}{2}}\right] F(u)\, du$	$s^{-2\nu-1} f(s + as^{-1})$
$\int_0^t \left(\frac{t-u}{t+u}\right)^{\nu} J_{2\nu}\left[(t^2 - u^2)^{\frac{1}{2}}\right] F(u)\, du$	$(s^2+1)^{-\frac{1}{2}}\left[(s^2+1)^{\frac{1}{2}} + s\right]^{-2\nu} f\left[(s^2+1)^{\frac{1}{2}}\right]$

$F(t)$	$f(s) = \int_0^\infty F(t)\, e^{-st}\, dt$
$0 \quad 0 < t < b$ $(t+a)^{-1} \quad b < t < c$ $0 \quad t > c$	$e^{as}\,[Ei(-as-cs) - Ei(-as-bs)]$
$0 \quad 0 < t < b$ $(t+a)^{-n} \quad t > b$ $\lvert\arg(a+b)\rvert < \pi, \quad n \geq 2$	$e^{-bs} \sum_{m=1}^{n-1} \frac{(m-1)!}{(n-1)!} (a+b)^{-m} (-s)^{n-m-1} - \frac{(-s)^{n-1}}{(n-1)!} e^{as}\, Ei(-as-bs)$ $Re\, s > 0$
$(t+a)^{-\frac{1}{2}} \quad \lvert\arg a\rvert < \pi$	$\left(\frac{\pi}{s}\right)^{\frac{1}{2}} e^{as}\, \mathrm{Erfc}\left[(as)^{\frac{1}{2}}\right] \quad Re\, s > 0$
$t^{-\frac{1}{2}}(t+a)^{-1} \quad \lvert\arg a\rvert < \pi$	$\pi a^{-\frac{1}{2}} e^{as}\, \mathrm{Erfc}\left[(as)^{\frac{1}{2}}\right] \quad Re\, s \geq 0$
$0 \quad 0 < t < b$ $t^{-1}(t-b)^{-\frac{1}{2}} \quad t > b$	$\pi b^{-\frac{1}{2}}\, \mathrm{Erfc}\left[(bs)^{\frac{1}{2}}\right] \quad Re\, s \geq 0$
$t^{\nu}(t+a)^{-1}$ $\lvert\arg a\rvert < \pi, \quad Re\, \nu > -1$	$\Gamma(\nu+1)\, a^{\nu} e^{as}\, \Gamma(-\nu, as) \quad Re\, s > 0$
$(a^2+t^2)^{\nu-\frac{1}{2}}$	$2^{\nu-1} \pi^{\frac{1}{2}} \Gamma\left(\frac{1}{2}+\nu\right) \left(\frac{a}{s}\right)^{\nu} [\boldsymbol{H}_{\nu}(as) - Y_{\nu}(as)] \quad Re\, s > 0$

$0 \quad 0 < t < b$ $(t^2 - b^2)^{\nu - \frac{1}{2}} \quad t > b$ $Re\, \nu > -\frac{1}{2}$	$\pi^{-\frac{1}{2}} \Gamma\left(\frac{1}{2} + \nu\right) \left(\frac{2b}{s}\right)^{\nu} K_{\nu}(bs) \qquad Re\, s > 0$
$(b^2 - t^2)^{\nu - \frac{1}{2}} \quad 0 < t < b$ $0 \quad t > b$ $Re\, \nu > -\frac{1}{2}$	$\frac{1}{2} \pi^{\frac{1}{2}} \Gamma\left(\frac{1}{2} + \nu\right) \left(\frac{2b}{s}\right)^{\nu} [I_{\nu}(bs) - \mathbf{L}_{\nu}(bs)]$
$0 \quad 0 < t < b$ $(t - b)^{\nu} (b + t)^{-\nu - \frac{1}{2}} \quad t > b$ $Re\, \nu > -1$	$2^{\frac{1}{2} + \nu} \Gamma(1 + \nu)\, s^{-\frac{1}{2}} D_{-2\nu-1}\left[2(bs)^{\frac{1}{2}}\right] \qquad Re\, s > 0$
$0 \quad 0 < t < b$ $(t - b)^{\nu} (t + b)^{-\nu - \frac{3}{2}} \quad t > b$ $Re\, \nu > -1$	$2^{\frac{1}{2} + \nu} \Gamma(1 + \nu)\, b^{-\frac{1}{2}} D_{-2\nu-2}\left[2(bs)^{\frac{1}{2}}\right] \qquad Re\, s \geq 0$
$t^{\nu} (t + a)^{-\nu - \frac{1}{2}}$ $Re\, \nu > -1$ $\lvert \arg a \rvert < \pi$	$2^{\nu + \frac{1}{2}} \Gamma(\nu + 1)\, s^{-\frac{1}{2}} e^{\frac{1}{2} as} D_{-2\nu-1}\left[(2as)^{\frac{1}{2}}\right] \qquad Re\, s > 0$

$F(t)$	$f(s) = \int_0^\infty F(t)\, e^{-st}\, dt$
$t^\nu (t+a)^{-\nu-\frac{3}{2}}$ $Re\, \nu > 0, \ \lvert\arg a\rvert < \pi$	$2^{\nu+1}\Gamma(\nu+1)\, a^{-\frac{1}{2}} e^{\frac{1}{2}as} D_{-2\nu-2}\left[(2as)^{\frac{1}{2}}\right]$ $\quad Re\, s \geq 0$
$\left[(t^2+1)^{\frac{1}{2}} + t\right]^\nu$	$s^{-1} + \pi\nu(s \sin \pi\nu)^{-1}[J_{-\nu}(s) - \boldsymbol{J}_{-\nu}(s)]$ $\quad Re\, s > 0$
$\left[(t^2+1)^{\frac{1}{2}} - t\right]^\nu$	$s^{-1} + \pi\nu(s \sin \pi\nu)^{-1}[J_\nu(s) - \boldsymbol{J}_\nu(s)]$ $\quad Re\, s > 0$
$(t^2+a^2)^{-\frac{1}{2}}\left[(t^2+a^2)^{\frac{1}{2}} + t\right]^\nu$	$-\pi \operatorname{cosec}(\pi\nu)\, a^\nu [\boldsymbol{J}_{-\nu}(as) - J_{-\nu}(as)]$ $\quad Re\, s > 0$
$(t^2+a^2)^{-\frac{1}{2}}\left[(t^2+a^2)^{\frac{1}{2}} - t\right]^\nu$	$\pi \operatorname{cosec}(\pi\nu)\, a^\nu [\boldsymbol{J}_\nu(as) - J_\nu(as)]$ $\quad Re\, s > 0$
$0 \quad 0 < t < b$ $e^{-at^2} \quad t > b$ $Re\, a > 0$	$\frac{1}{2}\left(\frac{\pi}{a}\right)^{\frac{1}{2}} e^{\frac{s^2}{4a}} \operatorname{Erfc}\left(\frac{1}{2} s a^{-\frac{1}{2}} + b a^{\frac{1}{2}}\right)$
$t^{\nu-1} e^{-at^2}$ $Re\, a > 0, \ Re\, \nu > 0$	$(2a)^{-\frac{1}{2}\nu}\, \Gamma(\nu)\, e^{\frac{s^2}{8a}} D_{-\nu}\left[s(2a)^{-\frac{1}{2}}\right]$

$t^{\nu-1} e^{-\frac{a}{t}}$ $Re\, a > 0$	$2 s^{-\frac{1}{2}} K_\nu \left[2 (a s)^{\frac{1}{2}}\right] \qquad Re\, s > 0$
$t^{\nu-1} e^{-a t^{\frac{1}{2}}}$ $Re\, \nu > 0$	$2^{1-\nu} \Gamma(2\nu)\, s^{-\nu} e^{\frac{a^2}{8s}} D_{-2\nu}\left[a (2s)^{-\frac{1}{2}}\right] \qquad Re\, s > 0$
$t^{\nu-1} \log t$ $Re\, \nu > 0$	$\Gamma(\nu)\, s^{-\nu} [\psi(\nu) - \log s],\ Re\, s > 0$
$t^{\nu-1} \sin (a t)$ $Re\, \nu > -1$	$\Gamma(\nu)\, (s^2 + a^2)^{-\frac{1}{2}\nu} \sin\left[\nu \arctan\left(\frac{a}{s}\right)\right] \qquad Re\, s > 0$
$t^{\nu-1} \cos (a t)$ $Re\, \nu > 0$	$\Gamma(\nu)\, (s^2 + a^2)^{-\frac{1}{2}\nu} \cos\left[\nu \arctan\left(\frac{a}{s}\right)\right] \qquad Re\, s > 0$
$\sin^{2n} (a t)$ $n = 1, 2, 3, \ldots$	$\frac{(2n)!\, a^{2n}}{s [s^2 + (2a)^2]\, [s^2 + (4a)^2] \cdots [s^2 + (2 n a)^2]}$ $Re\, s > 2n\, \lvert Im\, a \rvert$
$\sin t^2$	$\left(\frac{1}{2}\pi\right)^{\frac{1}{2}} \left\{\left[\frac{1}{2} - C\left(\frac{1}{4}s^2\right)\right] \cos\left(\frac{1}{4}s^2\right) + \left[\frac{1}{2} - S\left(\frac{1}{4}s^2\right)\right] \sin\left(\frac{1}{4}s^2\right)\right\}$ $Re\, s > 0$
$\cos t^2$	$\left(\frac{1}{2}\pi\right)^{\frac{1}{2}} \left\{\left[\frac{1}{2} - S\left(\frac{1}{4}s^2\right)\right] \cos\left(\frac{1}{4}s^2\right) - \left[\frac{1}{2} - C\left(\frac{1}{4}s^2\right)\right] \sin\left(\frac{1}{4}s^2\right)\right\}$ $Re\, s > 0$

$F(t)$	$f(s) = \int_0^\infty F(t)\, e^{-st}\, dt$
$t^{\nu-1} \sin\left(a t^{\frac{1}{2}}\right)$ $Re\, \nu > -\frac{1}{2}$	$2^{-\nu-\frac{1}{2}} \pi^{\frac{1}{2}} \sec(\pi\nu)\, s^{-\nu} e^{-\frac{a^2}{8s}} \times \left\{D_{2\nu-1}\left[-a(2s)^{-\frac{1}{2}}\right] - D_{2\nu-1}\left[a(2s)^{-\frac{1}{2}}\right]\right\}$ $Re\, s > 0$
$t^{\nu-1} \cos\left(a t^{\frac{1}{2}}\right)$ $Re\, \nu > -1$	$2^{-\nu-\frac{1}{2}} \pi^{\frac{1}{2}} \operatorname{cosec}(\pi\nu)\, s^{-\nu} e^{-\frac{a^2}{8s}} \times \left\{D_{2\nu-1}\left[-a(2s)^{-\frac{1}{2}}\right] + D_{2\nu-1}\left[a(2s)^{-\frac{1}{2}}\right]\right\}$ $Re\, s > 0$
0 $t < b$ $(t^2 - b^2)^{-\frac{1}{2}} \cos\left[a(t^2 - b^2)^{\frac{1}{2}}\right]$ $t > b$	$K_0\left[b(a^2 + s^2)^{\frac{1}{2}}\right]$ $Re\, s > 0$
$\operatorname{sech} t$	$\frac{1}{2}\left[\psi\left(\frac{1}{4}s + \frac{3}{4}\right) - \psi\left(\frac{1}{4}s + \frac{1}{4}\right)\right]$ $Re\, s > -1$
$\operatorname{sech}^2 t$	$\frac{1}{2}s\left[\psi\left(\frac{1}{2} + \frac{1}{4}s\right) - \psi\left(\frac{1}{4}s\right)\right] - 1$ $Re\, s > -2$
$t^{\nu-1} \operatorname{cosech} t$ $Re\, \nu > 1$	$2^{1-\nu} \Gamma(\nu)\, \zeta\left(\nu, \frac{1}{2} + \frac{1}{2}s\right)$ $Re\, s > -1$

$e^{-a \sinh t}$ $Re\, a > 0$	$\pi \operatorname{cosec}(\pi s)\, [\mathbf{J}_s(a) - J_s(a)]$
$\operatorname{Erf}(at)$ $\lvert \arg a \rvert < \frac{1}{4}\pi$	$s^{-1} e^{\frac{s^2}{4a^2}} \operatorname{Erfc}\left(\frac{s}{2a}\right)$ $\quad Re\, s > 0$
$t^{\nu} \operatorname{Erf}\left(a t^{\frac{1}{2}}\right)$ $Re\, \nu > -\frac{3}{2}$	$2\pi^{-\frac{1}{2}} \Gamma\left(\frac{3}{2} + \nu\right) a s^{-\nu - \frac{3}{2}} {}_2F_1\left(\frac{1}{2}, \frac{3}{2} + \nu; \frac{3}{2}; -\frac{a^2}{s}\right)$ $\quad Re\, s > 0$
$\operatorname{Erfc}\left(a t^{-\frac{1}{2}}\right)$	$s^{-1} e^{-2a s^{\frac{1}{2}}}$ $\quad Re\, s > 0$
$\cos(at)\, Ei(-t)$	$-(s^2 + a^2)^{-1} \left\{\frac{1}{2} s \log[a^2 + (s+1)^2] + a \arctan\left(\frac{a}{1+s}\right)\right\}$ $Re\, s > \lvert Im\, a \rvert$
$t^{-\frac{1}{2}\mu}\, \mathfrak{P}_{\nu}^{\mu}\left[(1+t)^{\frac{1}{2}}\right]$ $Re\, \mu < 1$	$2^{\mu} s^{\frac{1}{2}\mu - \frac{5}{4}} e^{\frac{1}{2}s} W_{\frac{1}{2}\mu + \frac{1}{4}, \frac{1}{2}\nu + \frac{1}{4}}(s)$ $\quad Re\, s > 0$
$[t(1+t)]^{-\frac{1}{2}\mu}\, \mathfrak{P}_{\nu}^{\mu}(1+2t)$ $Re\, \mu < 1$	$\pi^{-\frac{1}{2}} s^{\mu - \frac{1}{2}} e^{\frac{1}{2}s} K_{\nu + \frac{1}{2}}\left(\frac{1}{2}s\right)$ $\quad Re\, s > 0$

$F(t)$	$f(s) = \int_0^\infty F(t)\, e^{-st}\, dt$
$t^\mu J_\nu(at)$ $Re\,(\mu+\nu) > -1$	$\Gamma(\mu+\nu+1)\,(s^2+a^2)^{-\frac{1}{2}(\mu+1)}\, P_\mu^{-\nu}\left[s\,(s^2+a^2)^{-\frac{1}{2}}\right]$ $Re\, s > \lvert Im\, a\rvert$
$J_\nu(at)\, J_\nu(bt)$ $Re\,\nu > -\frac{1}{2}$	$\pi^{-1}(ab)^{-\frac{1}{2}}\, \mathfrak{Q}_{\nu-\frac{1}{2}}\left(\frac{s^2+a^2+b^2}{2ab}\right)$ $Re\, s > \lvert Im\, a\rvert + \lvert Im\, b\rvert$
$t^\nu J_\nu(at)$ $Re\,\nu > -\frac{1}{2}$	$2^\nu \pi^{-\frac{1}{2}}\, \Gamma\left(\frac{1}{2}+\nu\right) a^\nu (s^2+a^2)^{-\nu-\frac{1}{2}}$ $Re\, s > \lvert Im\, a\rvert$
$J_\nu\left(a t^{\frac{1}{2}}\right)$	$\frac{1}{4}\, a\, \pi^{\frac{1}{2}} s^{-\frac{3}{2}} e^{-\frac{a^2}{8s}} \left[I_{\frac{1}{2}\nu-\frac{1}{2}}\left(\frac{a^2}{8s}\right) - I_{\frac{1}{2}\nu+\frac{1}{2}}\left(\frac{a^2}{8s}\right)\right]$ $\quad Re\, s > 0$
$t^{\frac{1}{2}\nu}\, J_\nu\left(a t^{\frac{1}{2}}\right)$	$\left(\frac{1}{2}a\right)^\nu s^{-\nu-1} e^{-\frac{a^2}{4s}}$ $\quad Re\, s > 0$
$J_\nu\left(a t^{\frac{1}{2}}\right) J_\nu\left(b t^{\frac{1}{2}}\right)$ $Re\,\nu > -\frac{1}{2}$	$s^{-1} e^{-\frac{1}{4}(a^2+b^2)s^{-1}}\, I_\nu\left(\frac{ab}{2s}\right)$ $\quad Re\, s > 0$

$Y_0(at)$	$-2\pi^{-1}(s^2+a^2)^{-\frac{1}{2}}\log\left[\frac{s+(s^2+a^2)^{\frac{1}{2}}}{a}\right]$ $\quad Re\, s > \lvert Im\, a \rvert$
$Y_\nu(at)$	$a^\nu \cot(\nu\pi)\,(s^2+a^2)^{-\frac{1}{2}}\left[s+(s^2+a^2)^{\frac{1}{2}}\right]^{-\nu}$ $- a^{-\nu}\operatorname{cosec}(\pi\nu)\,(s^2+a^2)^{-\frac{1}{2}}\left[s+(s^2+a^2)^{\frac{1}{2}}\right]^{\nu}$ $Re\, s > \lvert Im\, a \rvert$
$t^\mu Y_\nu(at)$	$(s^2+a^2)^{-\frac{1}{2}\mu-\frac{1}{2}}\left\{\Gamma(\mu+\nu+1)\cot(\pi\nu)\,P_\mu^{-\nu}\left[s(a^2+s^2)^{-\frac{1}{2}}\right]\right.$ $\left.-\Gamma(\mu-\nu+1)\operatorname{cosec}(\pi\nu)\,P_\mu^{\nu}\left[s(a^2+s^2)^{-\frac{1}{2}}\right]\right\}$ $Re\, s > \lvert Im\, a \rvert$
$J_\nu(at^2)$ $Re\,\nu > -1$	$(2a\pi)^{-\frac{1}{2}}\,\Gamma\left(\frac{1}{2}+\nu\right) D_{-\nu-\frac{1}{2}}\left[s(2ai)^{-\frac{1}{2}}\right] D_{-\nu-\frac{1}{2}}\left[s(-2ai)^{-\frac{1}{2}}\right]$ $Re\, s > 0$
$t^{-1} J_\nu(t^{-1})$	$2 J_\nu\left[(2s)^{\frac{1}{2}}\right] K_\nu\left[(2s)^{\frac{1}{2}}\right],$ $\quad Re\, s > 0$
$0 \quad 0 < t < b$ $J_0\left[a(t^2-b^2)^{\frac{1}{2}}\right] \quad t > b$	$(s^2+a^2)^{-\frac{1}{2}} e^{-b(s^2+a^2)^{\frac{1}{2}}}$ $Re\, s > \lvert Im\, a \rvert$

$F(t)$	$f(s) = \int_0^\infty F(t)\, e^{-st}\, dt$
$0 \qquad 0 < t < b$ $(t^2 - b^2)^{-\frac{1}{2}} J_\nu\left[a(t^2 - b^2)^{\frac{1}{2}}\right]$ $Re\, \nu > -1 \qquad t > b$	$I_{\frac{1}{2}\nu}\left\{\frac{1}{2} b\left[(s^2 + a^2)^{\frac{1}{2}} - s\right]\right\} K_{\frac{1}{2}\nu}\left\{\frac{1}{2} b\left[(s^2 + a^2)^{\frac{1}{2}} + s\right]\right\}$ $Re\, s > \lvert Im\, a\rvert$
$(t^2 + bt)^{\frac{1}{2}\nu} J_\nu\left[a(t^2 + bt)^{\frac{1}{2}}\right]$ $Re\, \nu > -1$ $\lvert \arg b\rvert < \pi$	$\pi^{-\frac{1}{2}}\left(\frac{1}{2} a\right)^\nu b^{\nu + \frac{1}{2}} (s^2 + a^2)^{-\frac{1}{2}\nu - \frac{1}{4}} e^{\frac{1}{2} bs} K_{\nu + \frac{1}{2}}\left[\frac{1}{2} b (s^2 + a^2)^{\frac{1}{2}}\right]$ $Re\, s > \lvert Im\, a\rvert$
$J_\nu(a \sinh t)$ $Re\, \nu > -1$	$I_{\frac{1}{2}\nu + \frac{1}{2}s}\left(\frac{1}{2} a\right) K_{\frac{1}{2}\nu - \frac{1}{2}s}\left(\frac{1}{2} a\right)$ $Re\, s > -\frac{1}{2}$
$t^{-1} Y_\nu(t^{-1})$	$2 Y_\nu\left[(2s)^{\frac{1}{2}}\right] K_\nu\left[(2s)^{\frac{1}{2}}\right]$ $Re\, s > 0$
$I_\nu(at)$	$a^\nu (s^2 - a^2)^{-\frac{1}{2}}\left[s + (s^2 - a^2)^{\frac{1}{2}}\right]^{-\nu}$ $Re\, s > \lvert Re\, a\rvert$

$t^\nu I_\nu(at)$ $Re\,\nu > -\frac{1}{2}$	$2^\nu \pi^{-\frac{1}{2}} \Gamma\left(\frac{1}{2}+\nu\right) a^\nu (s^2-a^2)^{-\nu-\frac{1}{2}}$ $Re\,s > \lvert Re\,a \rvert$
$t^\mu I_\nu(at)$ $Re\,(\mu+\nu) > -1$	$\Gamma(\mu+\nu+1)\,(s^2-a^2)^{-\frac{1}{2}\mu-\frac{1}{2}}\,\mathfrak{P}_\mu^{-\nu}\left[s(s^2-a^2)^{-\frac{1}{2}}\right]$ $Re\,s > \lvert Re\,a \rvert$
$I_\nu(at)\,I_\nu(bt)$ $Re\,\nu > -\frac{1}{2}$	$\pi^{-1}(ab)^{-\frac{1}{2}}\,\mathfrak{Q}_{\nu-\frac{1}{2}}\left(\frac{s^2-a^2-b^2}{2ab}\right)$ $Re\,s > \lvert Re\,(a+b) \rvert$
$K_0(at)$	$(s^2-a^2)^{-\frac{1}{2}} \log\left[\frac{s+(s^2-a^2)^{\frac{1}{2}}}{a}\right]$ $Re\,s > -Re\,a$
$K_\nu(at)$ $-1 < Re\,\nu < 1$	$\frac{1}{2}\pi\,\mathrm{cosec}\,(\pi\nu)\,(s^2-a^2)^{-\frac{1}{2}}\left\{a^{-\nu}\left[s+(s^2-a^2)^{\frac{1}{2}}\right]^\nu - a^\nu\left[s+(s^2-a^2)^{\frac{1}{2}}\right]^{-\nu}\right\}$ $Re\,s > -Re\,a$
$t^\mu K_\nu(at)$ $Re\,(\mu \pm \nu) > -1$	$\left(\frac{\pi}{2a}\right)^{\frac{1}{2}} \Gamma(\mu+\nu+1)\,\Gamma(\mu-\nu+1)\,(a^2-s^2)^{-\frac{1}{2}\mu-\frac{1}{4}}\,P_{\nu-\frac{1}{2}}^{-\mu-\frac{1}{2}}\left(\frac{s}{a}\right)$ $-a < s < a$

$F(t)$	$f(s) = \int_0^\infty F(t)\, e^{-st}\, dt$
$t^\mu K_\nu(at)$ $Re\,(\mu \pm \nu) > -1$	$\left(\frac{\pi}{2a}\right)^{\frac{1}{2}} \Gamma(\mu+\nu+1)\, \Gamma(\mu-\nu+1)\, (s^2-a^2)^{-\frac{1}{2}\mu-\frac{1}{4}}\, \mathfrak{P}_{\nu-\frac{1}{2}}^{-\mu-\frac{1}{2}}\left(\frac{s}{a}\right)$ $Re\, s > -Re\, a$
$t^{\frac{1}{2}\nu} K_\nu\left(a t^{\frac{1}{2}}\right)$ $Re\,\nu > -1$	$\frac{1}{2}\left(\frac{1}{2}a\right)^\nu \Gamma(\nu+1)\, s^{-\nu-1} e^{\frac{a^2}{4s}}\, \Gamma\left(-\nu, \frac{a^2}{4s}\right)$ $Re\, s > 0$
$t^\mu K_\nu\left(a t^{\frac{1}{2}}\right)$ $Re\left(\mu \pm \frac{1}{2}\nu\right) > -1$	$\Gamma\left(1+\mu+\frac{1}{2}\nu\right) \Gamma\left(1+\mu-\frac{1}{2}\nu\right) a^{-1} s^{-\mu-\frac{1}{2}} e^{\frac{a^2}{4s}} W_{-\mu-\frac{1}{2},\nu}\left(\frac{a^2}{4s}\right)$ $Re\, s > 0$
$0 \qquad 0 < t < b$ $I_0\left[a(t^2-b^2)^{\frac{1}{2}}\right] \qquad t > b$	$(s^2-a^2)^{-\frac{1}{2}} e^{-b(s^2-a^2)^{\frac{1}{2}}} \qquad Re\, s > \lvert Re\, a \rvert$
$0 \qquad 0 < t < b$ $(t^2-b^2)^{\frac{1}{2}\nu} I_\nu\left[a(t^2-b^2)^{\frac{1}{2}}\right] \qquad t > b$ $Re\,\nu > -1$	$\left(\frac{\pi}{2b}\right)^{-\frac{1}{2}} (ab)^\nu (s^2-a^2)^{-\frac{1}{2}\nu-\frac{1}{4}} K_{\nu+\frac{1}{2}}\left[b(s^2-a^2)^{\frac{1}{2}}\right]$ $Re\, s > \lvert Re\, a \rvert$

$(t^2+bt)^{\frac{1}{2}\nu} I_\nu\left[a(t^2+bt)^{\frac{1}{2}}\right]$ $Re\,\nu > -1$ $\lvert\arg b\rvert < \pi$	$\left(\frac{\pi s}{b}\right)^{-\frac{1}{2}}\left(\frac{1}{2}ab\right)^{\nu} e^{\frac{1}{2}bs} K_{\nu+\frac{1}{2}}\left(\frac{1}{2}bs\right) \qquad Re\,s > \lvert Re\,a\rvert$
$t^{-1}e^{-\frac{a+b}{t}} K_\nu\left(\frac{a-b}{t}\right)$ $Re\,a > Re\,b > 0$	$2K_\nu\left[(2as)^{\frac{1}{2}}+(2bs)^{\frac{1}{2}}\right] K_\nu\left[(2as)^{\frac{1}{2}}-(2bs)^{\frac{1}{2}}\right]$ $Re\,s > 0$
${}_1F_1(a;c;t)$	$\frac{1}{s}F\left(a,1;c;\frac{1}{s}\right) \qquad Re\,s > 1$
$t^{\nu-1}M_{\varkappa,\mu}(at)$ $Re\,(\mu+\nu) > -\frac{1}{2}$	$a^{\mu+\frac{1}{2}}\Gamma\left(\mu+\nu+\frac{1}{2}\right)\left(s+\frac{1}{2}a\right)^{-\mu-\nu-\frac{1}{2}}$ $\times F\left(\mu+\nu+\frac{1}{2}, \mu-\varkappa+\frac{1}{2}; 1+2\mu; \frac{2a}{2s+a}\right)$ $Re\,s > \frac{1}{2}\lvert Re\,a\rvert$
$t^{\nu-1}W_{\varkappa,\mu}(at)$ $Re\,(\nu\pm\mu) > -\frac{1}{2}$	$\frac{\Gamma\left(\mu+\nu+\frac{1}{2}\right)}{\Gamma(\nu-\varkappa+1)}\Gamma\left(\nu-\mu+\frac{1}{2}\right)\left(s+\frac{1}{2}a\right)^{-\nu-\mu-\frac{1}{2}}$ $\times a^{\frac{1}{2}+\mu}F\left(\mu+\nu+\frac{1}{2}, \mu-\varkappa+\frac{1}{2}; \nu-\varkappa+1; \frac{2s-a}{2s+a}\right)$ $Re\left(s+\frac{1}{2}a\right) > 0$

$F(t)$	$f(s) = \int_0^\infty F(t)\, e^{-st}\, dt$
$\vartheta_1(v, i\pi t)$	$-s^{-\frac{1}{2}} \sinh\left(2v s^{\frac{1}{2}}\right) \operatorname{sech} s^{\frac{1}{2}}$ $\quad -\frac{1}{2} \leq v \leq \frac{1}{2}$ $Re\, s > 0$
$\vartheta_2(v, i\pi t)$	$-s^{-\frac{1}{2}} \sinh\left[(2v-1)\, s^{\frac{1}{2}}\right] \operatorname{sech} s^{\frac{1}{2}}$ $\quad 0 \leq v \leq 1$ $Re\, s > 0$
$\vartheta_3(v, i\pi t)$	$s^{-\frac{1}{2}} \cosh\left[(2v-1)\, s^{\frac{1}{2}}\right] \operatorname{cosech} s^{\frac{1}{2}}$ $\quad 0 \leq v \leq 1$ $Re\, s > 0$
$\vartheta_4(v, i\pi t)$	$s^{-\frac{1}{2}} \cosh\left(2v s^{\frac{1}{2}}\right) \operatorname{cosech} s^{\frac{1}{2}}$ $\quad -\frac{1}{2} \leq v \leq \frac{1}{2}$ $Re\, s > 0$

Examples for the Mellin transform

$f(x)$	$g(s) = \int_0^\infty f(x)\, x^{s-1}\, dx$
$(\alpha + x^n)^{-1}$ $n = 1, 2, 3, \ldots$ $\lvert\arg \alpha\rvert < \pi$	$\pi n^{-1} \operatorname{cosec}\left(\frac{\pi s}{n}\right) \alpha^{\frac{s}{n}-1}$ $\qquad 0 < Re\, s < n$
$(1 + x)^\nu (1 + \alpha x)^\mu$ $\lvert\arg \alpha\rvert < \pi$	$B(s, -\mu - \nu - s)\, F(-\mu, s; -\mu - \nu; 1 - \alpha)$ $0 < Re\, s < -Re\,(\mu + \nu)$
$(1 + x^2)^{-\frac{1}{2}} \left[\alpha + (1 + x^2)^{\frac{1}{2}}\right]^\nu$ $\lvert\arg \alpha\rvert < \pi$	$2^{\frac{1}{2}s-1} (\alpha^2 - 1)^{\frac{1}{2}\nu + \frac{1}{4}s}\, \Gamma\left(\frac{1}{2}s\right) \Gamma(1 - \nu - s)\, \mathfrak{P}^{\nu + \frac{1}{2}s}_{\frac{1}{2}s-1}(\alpha)$ $0 < Re\, s < 1 - Re\, \nu$
$(1 - x)^\nu (1 + \alpha x)^\mu \qquad 0 < x < 1$ $0 \qquad x > 1$ $Re\, \nu > -1, \ \lvert\arg \alpha\rvert < \pi$	$B(\nu + 1, s)\, F(-\mu, s; \nu + \mu + 1; -\alpha)$ $Re\, s > 0$
$(\alpha^2 + x^2)^{-\frac{1}{2}} \left[(\alpha^2 + x^2)^{\frac{1}{2}} + x\right]^\nu$ $Re\, \alpha > 0$	$2^{-s} \alpha^{\nu+s-1} B\left(s, \frac{1}{2} - \frac{1}{2}s - \frac{1}{2}\nu\right)$ $0 < Re\, s < 1 - Re\, \nu$

$f(x)$	$F(s)$
$(1+\alpha x^h)^{-\nu}$ $h>0,\ \lvert\arg\alpha\rvert<\pi$	$h^{-1}\alpha^{-\frac{s}{h}}B\left(\frac{s}{h},\nu-\frac{s}{h}\right)\qquad 0<Re\,s<h\,Re\,\nu$
$(1-x^h)^{\nu-1}\quad 0<x<1$ $0\quad x>1$ $h>0,\ Re\,\nu>0$	$h^{-1}B\left(\nu,\frac{s}{h}\right)$
$\log(1+\alpha x)$ $\lvert\arg\alpha\rvert<\pi$	$\pi s^{-1}\alpha^{-s}\operatorname{cosec}(\pi s),\quad -1<Re\,s<0$
$\sin[a(x-b^2x^{-1})]$ $a,b>0$	$2b^sK_s(2ab)\sin\left(\frac{1}{2}\pi s\right),\quad -1<Re\,s<1$
$\arctan x$	$-\frac{1}{2}\pi s^{-1}\sec\left(\frac{1}{2}\pi s\right)\qquad -1<Re\,s<0$
$\operatorname{arccot} x$	$\frac{1}{2}\pi s^{-1}\sec\left(\frac{1}{2}\pi s\right)\qquad 0<Re\,s<1$
$\operatorname{cosech}(\alpha x)\qquad Re\,\alpha>0$	$\alpha^{-s}\,2\,(1-2^{-s})\,\Gamma(s)\,\zeta(s)\qquad Re\,s>1$
$\operatorname{sech}^2(\alpha x)\qquad Re\,\alpha>0$	$4\alpha^{-s}(1-2^{2-s})\,\Gamma(s)\,2^{-s}\zeta(s-1)\qquad Re\,s>0$
$\operatorname{cosech}^2(\alpha x)\qquad Re\,\alpha>0$	$4\alpha^{-s}\Gamma(s)\,2^{-s}\zeta(s-1),\qquad Re\,s>2$
$(x^2+b^2)^{-\frac{1}{2}\nu}J_\nu\left[a(x^2+b^2)^{\frac{1}{2}}\right]$	$2^{\frac{1}{2}s-1}a^{-\frac{1}{2}s}b^{\frac{1}{2}s-\nu}\Gamma\left(\frac{1}{2}s\right)J_{\nu-\frac{1}{2}}(ab)\qquad 0<Re\,s<\frac{3}{2}+Re\,\nu$

$f(x)$	$g(s) = \int_0^\infty f(x)\, x^{s-1}\, dx$
$(a^2 - x^2)^{\frac{1}{2}\nu} J_\nu\left[a(b^2 - x^2)^{\frac{1}{2}}\right]$ $\quad 0 < x < a$ $0 \quad x > a$ $Re\, \nu > -1$	$2^{\frac{1}{2}s-1}\Gamma\left(\frac{1}{2}s\right) b^{-\frac{1}{2}s} a^{\nu+\frac{1}{2}s} J_{\nu+\frac{1}{2}s}(ab)$ $Re\, s > 0$
$(a^2 - x^2)^{-\frac{1}{2}\nu} J_\nu\left[b(a^2 - x^2)^{\frac{1}{2}}\right]$ $\quad 0 < x < a$ $0 \quad x > a$	$2^{1-\nu}[\Gamma(\nu)]^{-1} a^{\frac{1}{2}s-\nu} b^{-\frac{1}{2}\nu} S_{\nu-1+\frac{1}{2}s,\frac{1}{2}s-\nu}(ab)$ $Re\, s > 0$
$K_\nu(\alpha x)$	$\alpha^{-s}\, 2^{s-2}\Gamma\left(\frac{1}{2}s - \frac{1}{2}\nu\right)\Gamma\left(\frac{1}{2}s + \frac{1}{2}\nu\right), \qquad Re\, s > \lvert Re\, \nu\rvert$
$(\beta^2 + x^2)^{-\frac{1}{2}\nu} K_\nu\left[\alpha(\beta^2 + x^2)^{\frac{1}{2}}\right]$ $Re(\alpha, \beta) > 0$	$\alpha^{-\frac{1}{2}s}\, 2^{\frac{1}{2}s-1} \beta^{\frac{1}{2}s-\nu}\Gamma\left(\frac{1}{2}s\right) K_{\nu-\frac{1}{2}s}(\alpha\beta)$ $Re\, s > 0$

Examples for the Hankel transform

$f(x)$	$g(y) = \int_0^\infty f(x)\,(xy)^{\frac{1}{2}} J_\nu(xy)\,dx$
$x^{-\frac{1}{2}}(a^2+x^2)^{-\frac{1}{2}}$ $Re\, a > 0,\ Re\, \nu > -1$	$y^{\frac{1}{2}} I_{\frac{1}{2}\nu}\left(\frac{1}{2}ay\right) K_{\frac{1}{2}\nu}\left(\frac{1}{2}ay\right)$
$x^{-\nu-\frac{1}{2}}(a^2+x^2)^{-\nu-\frac{1}{2}}$ $Re\, a > 0,\ Re\, \nu > -\frac{1}{2}$	$\pi^{\frac{1}{2}}\, 2^{-\nu} a^{-2\nu}\left[\Gamma\left(\frac{1}{2}+\nu\right)\right]^{-1} I_\nu\left(\frac{1}{2}ay\right) K_\nu\left(\frac{1}{2}ay\right)$
$x^{-\frac{1}{2}}(a^2-x^2)^{-\frac{1}{2}}$ $\quad 0 < x < a$ 0 $\quad x > a$ $Re\, \nu > -1$	$\frac{1}{2}\pi y^{\frac{1}{2}}\left[J_{\frac{1}{2}\nu}\left(\frac{1}{2}ay\right)\right]^2$
0 $\quad 0 < x < a$ $x^{-\frac{1}{2}}(x^2-a^2)^{-\frac{1}{2}}$ $\quad x > a$	$-\frac{1}{2}\pi y^{\frac{1}{2}} J_{\frac{1}{2}\nu}\left(\frac{1}{2}ay\right) Y_{\frac{1}{2}\nu}\left(\frac{1}{2}ay\right)$
0 $\quad 0 < x < a$ $x^{-\nu-\frac{1}{2}}(x^2-a^2)^{-\nu-\frac{1}{2}}$ $\quad x > a$ $-\frac{1}{2} < Re\, \nu < \frac{1}{2}$	$-2^{-\nu-1} a^{-2\nu}\Gamma\left(\frac{1}{2}-\nu\right)\pi^{\frac{1}{2}} y^{\nu+\frac{1}{2}} J_\nu\left(\frac{1}{2}ay\right) Y_\nu\left(\frac{1}{2}ay\right)$

$x^{-\frac{1}{2}}(a^2+x^2)^{-\frac{1}{2}}\left[(a^2+x^2)^{\frac{1}{2}}+x\right]^{\nu-1}$ $Re\, a > 0,\quad -1 < Re\, \nu < \frac{5}{2}$	$2\pi^{-\frac{1}{2}} a^{\nu-\frac{3}{2}} \sinh\left(\frac{1}{2}ay\right) K_{\nu-\frac{1}{2}}\left(\frac{1}{2}ay\right)$
$x^{-\mu-\frac{1}{2}}(a^2+x^2)^{-\frac{1}{2}}\left[(x^2+a^2)^{\frac{1}{2}}+a\right]^{\mu}$ $Re\, a > 0,\quad Re\,(\nu-\mu) > -1$	$(a^2y)^{-\frac{1}{2}} \frac{\Gamma\left(\frac{1}{2}+\frac{1}{2}\nu-\frac{1}{2}\mu\right)}{\Gamma(\nu+1)} W_{\frac{1}{2}\mu,\frac{1}{2}\nu}(ay)\, M_{-\frac{1}{2}\mu,\frac{1}{2}\nu}(ay)$
$x^{-\frac{1}{2}} e^{-ax^2}$ $Re\, a > 0,\quad Re\, \nu > -1$	$\frac{1}{2}\left(\frac{\pi y}{a}\right)^{\frac{1}{2}} e^{-\frac{y^2}{8a}} I_{\frac{1}{2}\nu}\left(\frac{y^2}{8a}\right)$
$x^{-\frac{1}{2}}(b^2+x^2)^{-\frac{1}{2}} e^{-a(b^2+x^2)^{\frac{1}{2}}}$ $Re\, a > 0,\quad Re\, b > 0$ $Re\, \nu > -1$	$y^{\frac{1}{2}} I_{\frac{1}{2}\nu}\left\{\frac{1}{2}b\left[(a^2+y^2)^{\frac{1}{2}}-a\right]\right\} K_{\frac{1}{2}\nu}\left\{\frac{1}{2}b\left[(a^2+y^2)^{\frac{1}{2}}+a\right]\right\}$
$x^{\mu-\frac{1}{2}}(b^2+x^2)^{-\frac{1}{2}}\left[(b^2+x^2)^{\frac{1}{2}}+b\right]^{-\mu}$ $\times\, e^{-a(b^2+x^2)^{\frac{1}{2}}}$ $Re\, a > 0,\quad Re\, b > 0$ $Re\,(\nu+\mu) > -1$	$\frac{\Gamma\left(\frac{1}{2}\nu+\frac{1}{2}\mu+\frac{1}{2}\right)}{b\,\Gamma(\nu+1)}$ $\times\, y^{-\frac{1}{2}} M_{\frac{1}{2}\mu,\frac{1}{2}\nu}\left\{b\left[(a^2+y^2)^{\frac{1}{2}}-a\right]\right\} W_{-\frac{1}{2}\mu,\frac{1}{2}\nu}\left\{b\left[(a^2+y^2)^{\frac{1}{2}}+a\right]\right\}$
$x^{-\frac{1}{2}}(b^2+x^2)^{-\frac{1}{2}} \sin\left[a(b^2+x^2)^{\frac{1}{2}}\right]$ $a > 0,\quad Re\, b > 0\quad Re\, \nu > -1$	$\frac{1}{2}\pi y^{\frac{1}{2}} J_{\frac{1}{2}\nu}\left\{\frac{1}{2}b\left[a-(a^2-y^2)^{\frac{1}{2}}\right]\right\} J_{\frac{1}{2}\nu}\left\{\frac{1}{2}b\left[a+(a^2-y^2)^{\frac{1}{2}}\right]\right\}\quad 0 < y < a$

$f(x)$	$g(y) = \int_0^\infty f(x)\,(xy)^{\frac{1}{2}} J_\nu(xy)\,dx$
$x^{-\frac{1}{2}}(b^2 + x^2)^{-\frac{1}{2}} \cos\left[a(b^2 + x^2)^{\frac{1}{2}}\right]$ $a > 0,\ Re\, b > 0,\ Re\, \nu > -1$	$-\frac{1}{2}\pi y^{\frac{1}{2}} J_{\frac{1}{2}\nu}\left\{\frac{1}{2} b\left[a - (a^2 - y^2)^{\frac{1}{2}}\right]\right\} Y_{-\frac{1}{2}\nu}\left\{\frac{1}{2} b\left[a + (a^2 - y^2)^{\frac{1}{2}}\right]\right\}$ $0 < y < a$
$x^{-\frac{1}{2}}(a^2 - x^2)^{-\frac{1}{2}} \cos\left[b(a^2 - x^2)^{\frac{1}{2}}\right]$ $\quad 0 < x < a$ $0 \quad x > a$ $Re\, \nu - > 1$	$\frac{1}{2}\pi y^{\frac{1}{2}} J_{\frac{1}{2}\nu}\left\{\frac{1}{2} a\left[(b^2 + y^2)^{\frac{1}{2}} - b\right]\right\} J_{\frac{1}{2}\nu}\left\{\frac{1}{2} a\left[(b^2 + y^2)^{\frac{1}{2}} + b\right]\right\}$
$x^{-\frac{1}{2}}[J_\mu(ax)]^2$ $Re\,(\nu + 2\mu) > -1$	$y^{-\frac{1}{2}} \frac{\Gamma\left(\frac{1}{2} + \frac{1}{2}\nu + \mu\right)}{\Gamma\left(\frac{1}{2} + \frac{1}{2}\nu - \mu\right)} \left\{P^{-\mu}_{-\frac{1}{2}+\frac{1}{2}\nu}\left[(1 - 4a^2y^{-2})^{\frac{1}{2}}\right]\right\}^2$ $\quad y > 2a$
$x^{-\frac{1}{2}} J_\mu(ax)\, J_{-\mu}(ax)$ $Re\, \nu > -1$	$y^{-\frac{1}{2}} P^{\mu}_{\frac{1}{2}\nu-\frac{1}{2}}\left[(1 - 4a^2y^{-2})^{\frac{1}{2}}\right] P^{-\mu}_{\frac{1}{2}\nu-\frac{1}{2}}\left[(1 - 4a^2y^{-2})^{\frac{1}{2}}\right]$ $\quad y > 2a$
$x^{-\frac{1}{2}} J_\nu(ax^{-1})$ $a > 0, \quad Re\, \nu > -\frac{1}{2}$	$\left(\frac{y}{a}\right)^{\frac{1}{2}} J_{2\nu}\left[2(av)^{\frac{1}{2}}\right]$

$x^{-\frac{1}{2}} e^{-bx} J_{2\nu}\left(2ax^{\frac{1}{2}}\right)$ $Re\, b > 0, \quad Re\, \nu > -\frac{1}{2}$	$y^{\frac{1}{2}} (b^2 + y^2)^{-\frac{1}{2}} \exp\, [-a^2 b (b^2 + y^2)^{-1}]\, J_\nu [a^2 y (b^2 + y^2)^{-1}]$
$x^{-\frac{1}{2}} Y_\nu (a x^{-1})$ $a > 0, -\frac{1}{2} < Re\, \nu < \frac{1}{2}$	$-2\pi^{-1} y^{-\frac{1}{2}} \left[K_{2\nu}\left(2a^{\frac{1}{2}} y^{\frac{1}{2}}\right) - \frac{1}{2}\pi Y_{2\nu}\left(2a^{\frac{1}{2}} y^{\frac{1}{2}}\right)\right]$
$x^{-\frac{1}{2}} Y_{2\nu}\left(2ax^{\frac{1}{2}}\right)$ $a > 0, \qquad Re\, \nu > -\frac{1}{2}$	$2 \sec(\pi\nu)\, y^{-\frac{1}{2}} \left[\frac{1}{2}\cos(\pi\nu)\, Y_\nu(a^2 y^{-1}) - Y_{-\nu}(a^2 y^{-1}) + \boldsymbol{H}_{-\nu}(a^2 y^{-1})\right]$
$x^{-\frac{1}{2}} I_\mu(ax)\, K_\mu(ax)$ $Re\, a > 0, \quad Re\, \nu > -1$ $Re\, (\nu + 2\mu) > -1$	$e^{i\pi\mu} \Gamma\left(\frac{1}{2}\nu + \frac{1}{2} + \mu\right) \left[\Gamma\left(\frac{1}{2}\nu + \frac{1}{2} - \mu\right)\right]^{-1}$ $\times\, \mathfrak{P}^{-\mu}_{\frac{1}{2}\nu - \frac{1}{2}} \left[(1 + 4a^2 y^{-2})^{\frac{1}{2}}\right] \mathfrak{Q}^{-\mu}_{\frac{1}{2}\nu - \frac{1}{2}} \left[(1 + 4a^2 y^{-2})^{\frac{1}{2}}\right]$
$x^{-\frac{1}{2}} [K_\mu(ax)]^2$ $Re\, a > 0, \quad Re\, (\nu \pm 2\mu) > -1$	$\frac{1}{2}\pi a^{-1} \Gamma\left(\frac{1}{2} + \frac{1}{2}\nu + \mu\right) \Gamma\left(\frac{1}{2} + \frac{1}{2}\nu - \mu\right)$ $\times \left\{\mathfrak{P}^{-\frac{1}{2}\nu}_{\mu - \frac{1}{2}} \left[\left(1 + \frac{1}{4} y^2 a^{-2}\right)^{\frac{1}{2}}\right]\right\}^2$

$f(x)$	$g(y) = \int_0^\infty f(x)\,(xy)^{\frac{1}{2}}\,J_\nu(xy)\,dx$
$x^{\nu+\frac{1}{2}} K_\mu(ax)\,K_\mu(bx)$ $Re\, a > 0, \quad Re\, b > 0$ $Re\, \nu > -1, \quad Re\,(\nu \pm \mu) > -1$	$\frac{1}{2}\left(\frac{1}{2}\pi y\right)^{\frac{1}{2}} \Gamma(\nu+\mu+1)\,\Gamma(\nu-\mu+1)\,(ab)^{-\nu-1}$ $(z^2-1)^{-\frac{1}{2}\left(\nu+\frac{1}{2}\right)} \mathfrak{P}_{\mu-\frac{1}{2}}^{-\frac{1}{2}-\nu}(z), \qquad 2abz = y^2+b^2+a^2$
$x^{-\frac{1}{2}} J_\mu\left(a x^{\frac{1}{2}}\right) K_\mu\left(a x^{\frac{1}{2}}\right)$ $\lvert\arg a\rvert < \frac{1}{4}\pi$ $Re\, \nu > -1, \quad Re\,(\nu+\mu) > -1$	$\Gamma\left(\frac{1}{2}+\frac{1}{2}\mu+\frac{1}{2}\nu\right)[\Gamma(1+\mu)]^{-1}\,a^{-2}y^{\frac{1}{2}}$ $W_{-\frac{1}{2}\nu,\frac{1}{2}\mu}\left(\frac{1}{2}a^2y^{-1}\right) M_{\frac{1}{2}\nu,\frac{1}{2}\mu}\left(\frac{1}{2}a^2y^{-1}\right)$
$x^{\nu+\frac{1}{2}}(b^2+x^2)^{-\frac{1}{2}\mu} K_\mu\left[a(b^2+x^2)^{\frac{1}{2}}\right]$ $Re\, a > 0, \quad Re\, b > 0$ $Re\, \nu > -1$	$a^{-\mu} b^{\nu+1-\mu} y^{\nu+\frac{1}{2}} (a^2+y^2)^{\frac{1}{2}(\mu-\nu-1)} K_{\mu-\nu-1}\left[b(a^2+y^2)^{\frac{1}{2}}\right]$

Examples for the Lebedev, Mehler and generalised Mehler transform

$f(x)$	$g(y) = \int_0^\infty f(x)\, K_{ix}(y)\, dx$
x^{2n}	$(-1)^n \frac{1}{2} \pi \left[\frac{d^{2n}}{dz^{2n}} e^{-y\cosh z} \right]_{z=0}$
$\cos(ax)\cosh(bx)$ $b \leq \frac{\pi}{2}$	$\frac{1}{2} \pi \cos(y \sin b \sinh a)\, e^{-y\cos b\cosh a}$
$\sin(ax)\sinh(bx)$ $b \leq \frac{\pi}{2}$	$\frac{1}{2} \pi \sin(y \sin b \sinh a)\, e^{-y\cos b\cosh a}$
$\cosh(ax)\cosh(bx)$ $a + b \leq \frac{1}{2}\pi$	$\frac{1}{2} \pi \cosh(y \sin a \sin b)\, e^{-y\cos a\cos b}$
$\sinh(ax)\sinh(bx)$ $a + b \leq \frac{1}{2}\pi$	$\frac{1}{2} \pi \sinh(y \sin a \sin b)\, e^{-y\cos a\cos b}$
$\cosh(ax)\, K_{ix}(b)$ $a \leq \pi$	$\frac{1}{2} \pi K_0 \left[(b^2 + y^2 + 2by\cos a)^{\frac{1}{2}} \right]$

$x(c^2+x^2)^{-1}\sinh(\pi x)\,K_{ix}(a)$	$\frac{1}{2}\pi^2 I_c(y)\,K_c(a)$	$y<a$	
	$\frac{1}{2}\pi^2 I_c(a)\,K_c(y)$	$y>a$	$\lvert\arg c\rvert<\pi$
$x\tanh(\pi x)\,P_{-\frac{1}{2}+ix}(z)\,K_{ix}(a)$	$\frac{1}{2}\pi(ay)^{\frac{1}{2}}(a^2+y^2+2azy)^{-\frac{1}{2}}e^{-(a^2+y^2+2azy)^{\frac{1}{2}}}$		

$f(x)$	$g(y) = \int_0^\infty f(x)\, \mathfrak{P}_{-\frac{1}{2}+ix}(y)\, dx$
$x^{-1} \tanh(\pi x)$	$2\left[y + (y^2-1)^{\frac{1}{2}}\right]^{-\frac{1}{2}} K\left\{\left[y + (y^2-1)^{\frac{1}{2}}\right]^{-1}\right\}$
$x(a^2+x^2)^{-1} \tanh(\pi x)$	$\mathfrak{Q}_{a-\frac{1}{2}}(y)$
$\sin(ax) \tanh(\pi x)$	$(2\cosh a - 2y)^{-\frac{1}{2}}$ $\quad y < \cosh a$ 0 $\quad y > \cosh a$
$x \tanh(\pi x)\, K_{ix}(a)$	$\left(\frac{1}{2} a\pi\right)^{\frac{1}{2}} e^{-ay}$
$x \tanh(\pi x)\, K_{i2x}(a)$	$\frac{1}{4} a K_0\left[a\left(\frac{1}{2}+\frac{1}{2}y\right)^{\frac{1}{2}}\right]$
$x \operatorname{sech}(\pi x) \tanh(\pi x)$ $K_{ix}(a)$	$-\left(\frac{2\pi}{a}\right)^{-\frac{1}{2}} e^{ay}\, Ei(-ay-a)$
$x \tanh(\pi x)$ $\times\, [I_{ix}(a)\, K_{ix}(b) - K_{ix}(a)\, I_{ix}(b)]$	$(ab)^{\frac{1}{2}} (a^2+b^2-2aby)^{-\frac{1}{2}} \cosh\left[(a^2+b^2-2aby)^{\frac{1}{2}}\right]$ $\quad y < \frac{1}{2}\frac{a}{b} + \frac{1}{2}\frac{b}{a}$ 0 $\quad$ otherwise
$x \tanh(\pi x)\, K_{ix}(a)\, K_{ix}(b)$	$\frac{1}{2}\pi (ab)^{\frac{1}{2}} (a^2+b^2+2aby)^{-\frac{1}{2}} e^{-(a^2+b^2+2aby)^{\frac{1}{2}}}$

$f(x)$	$g(y) = \int_0^\infty f(x)\, \mathfrak{P}^{\mu}_{-\frac{1}{2}+ix}(y)\, dx$
$x \sinh\left(\frac{1}{2}\pi x\right)$ $\Gamma\left(\frac{1}{2} - k + ix\right) \Gamma\left(\frac{1}{2} - k - ix\right) K_{i\frac{1}{2}x}(a)$	$\pi\, 2^{\frac{1}{2}-k}\, a^{\frac{1}{4}-\frac{1}{2}k}\, \Gamma\left(\frac{3}{2} - k\right) (y^2 - 1)^{-\frac{1}{2}k} e^{ay^2 - a} D_{\mu-\frac{3}{2}}\left(2ya^{\frac{1}{2}}\right)$ $Re\, \mu \leq \frac{1}{2}$
$x \sinh(\pi x)\, K_{ix}(a)$ $K_{ix}(b)$	$0 \qquad y < \tau$ $2^{-\mu-2} \pi^{\frac{3}{2}} c^{\frac{1}{2}+\mu} (y^2 - 1)^{\frac{1}{2}\mu} (y - \tau)^{-\frac{1}{2}\mu-\frac{1}{4}} J_{-\mu-\frac{1}{2}}\left[c(y - \tau)^{\frac{1}{2}}\right] \qquad y > \tau$ $c = (2ab)^{\frac{1}{2}}, \quad \tau = \frac{1}{2}\left(\frac{a}{b} + \frac{b}{a}\right)$
$x \sinh(\pi x)$ $\Gamma\left(\frac{1}{2} - k + ix\right) \Gamma\left(\frac{1}{2} - k - ix\right)$ $K_{ix}(a)\, K_{ix}(b)$	$2^{-\frac{1}{2}} \pi^{\frac{3}{2}} (ab)^{\frac{1}{2}-\mu} (y^2 - 1)^{-\frac{1}{2}\mu} (a^2 + b^2 + 2aby)^{\frac{1}{2}\mu-\frac{1}{4}}$ $\times K_{\frac{1}{2}-\mu}\left[(a^2 + b^2 + 2aby)^{\frac{1}{2}}\right]$ $Re\, \mu \leq \frac{1}{2}$

Example for the Gauss transform

$f(x)$	$g(y) = \int\limits_{-\infty}^{\infty} e^{-h(x-y)^2} f(x)\,dx$
e^{-ax^2}	$\left(\frac{\pi}{a+h}\right)^{\frac{1}{2}} \exp\left(\frac{-ah}{a+h} y^2\right)$

11.1 Several examples of solutions of integral equations of the first kind

The integral transforms considered before are special cases of integral equations of the first kind with singular kernel, where the singularity is due to the fact that the interval of integration extends into infinity. Similar examples involving a finite interval of integration are the following ones (the integrals occuring here are defined as "Cauchy principal value"). The general form considered is

$$f(s) = \int\limits_a^b K(s,t)\, y(t)\, dt.$$

The inversion formulas listed below provide each square integrable function $f(s)$ with a square integrable solution $y(t)$.

1. $$(2\pi)^{-1} \int\limits_{-\pi}^{\pi} \left[1 + \cot\left(\frac{1}{2}s - \frac{1}{2}t\right)\right] y(t)\, dt = f(s),$$

$$y(t) = (2\pi)^{-1} \int\limits_{-\pi}^{\pi} \left[1 + \cot\left(\frac{1}{2}s - \frac{1}{2}t\right)\right] f(s)\, ds$$

(Hilbert's inversion formula),

2. $$(2\pi)^{-1} \int\limits_{-a}^{a} (s-t)^{-1} y(t)\, dt = f(s),$$

$$y(t) = \pi^{-1} (a^2 - t^2)^{-\frac{1}{2}} \left[A - \int\limits_{-a}^{a} f(s)\,(t-s)\,(a^2 - s^2)^{\frac{1}{2}}\, ds\right],$$

A arbitrary,

3. $$\pi^{-1} \int\limits_0^{\pi} \sin t\,(\cos t - \cos s)^{-1} y(t)\, dt = f(s),$$

$$y(t) = \pi^{-1} \int\limits_0^{\pi} \sin t\,(\cos t - \cos s)^{-1} f(s)\, ds$$

$$\text{probided} \int\limits_0^{\pi} f(s)\, ds = 0,$$

4.
$$\pi^{-1}\int_0^{\pi} \sin s\,(\cos t - \cos s)^{-1}\, y(t)\, dt = f(s),$$

$$y(t) = \pi^{-1}\int_0^{\pi} y(t)\, dt - \pi^{-1}\int_0^{\pi} \sin s\,(\cos s - \cos t)^{-1} f(s)\, ds,$$

$$\int_0^{\pi} y(t)\, dt \quad \text{arbitrary},$$

5.
$$\int_0^1 \log|s-t|\, y(t)\, dt = f(s),$$

$$y(t) = -\tfrac{1}{2}\pi^{-2}\left\{[t(1-t)]^{\frac{1}{2}}\log 2\right\}^{-1}\int_0^1 [s(1-s)]^{-\frac{1}{2}} f(s)\, ds$$

$$+\,\pi^{-2}\,[t(1-t)]^{-\frac{1}{2}}\int_0^1 (s-t)^{-1}\left[s(1-s)^{\frac{1}{2}}\right] f'(s)\, ds,$$

6.
$$\pi^{-1}\int_0^{\pi} \log\left[\frac{1-\cos(s+t)}{1-\cos(s-t)}\right] y(t)\, dt = f(s),$$

$$y(t) = \pi^{-1}\int_0^{\pi} \sin t\,(\cos s - \cos t)^{-1} f(s)\, ds,$$

7.
$$\pi^{-1}\int_0^{\pi} \log|\cos s - \cos t|\, y(t)\, dt = f(s),$$

$$y(t) = -(\pi\sin t)^{-1}\int_0^{\pi} \sin s\,(\cos s - \cos t)^{-1} f'(s)\, ds$$

$$-\,(\pi\log 2)^{-1}\int_0^{\pi} f(s)\, ds,$$

8.
$$\pi^{-1}\int_0^{\infty} [(s+t)^{-1} - (s-t)^{-1}]\, y(t)\, dt = f(s),$$

$$y(t) = \pi^{-1}\int_0^{\infty} [(s+t)^{-1} - (s-t)^{-1}]\, f(s)\, ds,$$

9.
$$\int_0^{s} (s-t)^{-\alpha}\, y(t)\, dt = f(s), \qquad 0<\alpha<1,$$

$$y(t) = \pi^{-1}\sin(\alpha\pi)\,\frac{d}{dt}\left[\int_0^{t} f(t)\,(t-s)^{\alpha-1}\, ds\right],$$

Literature

BOCHNER, S.: Lectures on Fourier integrals. Princeton 1959.

DOETSCH, G.: Handbuch der Theorie der Laplace-Transformation. Basel: Birkhauser 1950—1956.

HIRSCHMANN, I. I., and D. V. WIDDER: The convolution transform. Princeton 1955.

SNEDDON, I. N.: Fourier Transforms. New York: MacGraw-Hill 1951.

TITCHMARSH, E. C.: Introduction to the theory of Fourier integrals. Oxford 1948.

Literature concerning tables:

ERDÉLYI, A.: Tables of integral transforms. 2 vols. New York: McGraw-Hill 1954.

OBERHETTINGER, F.: Tabellen zur Fourier-Transformation. Berlin/Göttingen/Heidelberg: Springer 1957.

OBERHETTINGER, F.: Tables of Laplace and Mellin transforms. Berlin/Heidelberg/New York: Springer (To be published).

OBERHETTINGER, F., and T. P. HIGGINS: Tables of Lebedev, Mehler, and generalized Mehler transforms. Report Boeing Scientific Research Laboratories, Seattle, 1961.

Appendix to Chapter XI

Representations of some elementary functions in the form of Fourier series, partial fractions and infinite products

$$\sum_1^\infty n^{-1} \sin(nx) = \frac{1}{2}\pi - \frac{1}{2}x, \qquad 0 < x < 2\pi,$$

$$\sum_1^\infty (-1)^n n^{-1} \sin(nx) = -\frac{1}{2}x, \qquad -\pi < x < \pi,$$

$$\sum_0^\infty (2n+1)^{-1} \sin[(2n+1)x] = \begin{cases} \frac{1}{4}\pi, & 0 < x < \pi, \\ 0, & x = 0, \\ -\frac{1}{4}\pi, & \pi < x < 2\pi, \end{cases}$$

$$\sum_0^\infty (-1)^n (2n+1)^{-1} \sin[(2n+1)x]$$

$$= \begin{cases} \frac{1}{2}\log\cot\left(\frac{1}{4}\pi - \frac{1}{2}x\right), & -\frac{\pi}{2} < x < \frac{\pi}{2}, \\ \frac{1}{2}\log\cot\left(\frac{1}{2}x - \frac{1}{4}\pi\right), & \frac{\pi}{2} < x < \frac{3\pi}{2}, \end{cases}$$

$$\sum_1^\infty n^{-1} \cos(nx) = -\log\left(2\sin\frac{x}{2}\right), \qquad 0 < x < 2\pi,$$

$$\sum_1^\infty (-1)^n n^{-1} \cos(nx) = -\log\left(2\cos\frac{x}{2}\right), \qquad -\pi < x < \pi,$$

$$\sum_0^\infty (2n+1)^{-1} \cos[(2n+1)x] = \frac{1}{2}\log\left(\cot\frac{x}{2}\right), \qquad 0 < x < \pi,$$

$$\sum_1^\infty n^{-2}\cos(nx) = \left(\frac{1}{2}\pi - \frac{1}{2}x\right)^2 - \frac{\pi^2}{12}, \qquad 0 \le x \le 2\pi,$$

$$\sum_1^\infty (-1)^n n^{-2}\cos(nx) = \frac{1}{4}x^2 - \frac{\pi^2}{12}, \qquad -\pi \le x \le \pi,$$

$$\sum_0^\infty (2n+1)^{-2}\cos[(2n+1)x] = \frac{\pi}{4}\left(\frac{\pi}{2} - |x|\right), \qquad -\pi \le x \le \pi,$$

$$\sum_0^\infty (-1)^n (2n+1)^{-2}\sin[(2n+1)x] = \begin{cases} \dfrac{\pi}{4}x, & -\dfrac{\pi}{2} \le x \le \dfrac{\pi}{2}, \\ \dfrac{\pi}{4}(\pi - x), & \dfrac{\pi}{2} \le x \le 3\dfrac{\pi}{2}, \end{cases}$$

$$\sum_1^\infty (-1)^n [n(n+1)]^{-1}\cos[(n+1)x] = \cos x - \frac{1}{2}x\sin x - (1+\cos x)\log\left|2\cos\frac{x}{2}\right|,$$

$$\sum_1^\infty (-1)^n [n(n+1)]^{-1}\sin[(n+1)x] = \sin x - \frac{1}{2}x(1+\cos x) - \sin x\log\left|2\cos\frac{x}{2}\right|,$$

$$\sum_1^\infty (4n^2-1)^{-1}[1-\cos(2nx)] = \frac{1}{4}\pi\,|\sin x|,$$

$$\sum_{n=1}^\infty n^{-2r}\cos(2\pi nx) = (-1)^{r+1}2^{2r-1}\pi^{2r}\frac{B_{2r}(x)}{(2r)!},$$

$$\sum_{n=1}^\infty n^{-2r-1}\sin(2\pi nx) = (-1)^{r+1}2^{2r}\pi^{2r+1}\frac{B_{2r+1}(x)}{(2r+1)!},$$

$$r = 1, 2, 3, \ldots, \qquad 0 \le x \le 1.$$

The $B_r(x)$ are the Bernoulli polynomials (see chap. I)

$$\sum^\infty t^n\cos(nx) = (1 - t\cos x)(1 - 2t\cos x + t^2)^{-1}, \qquad |t| < 1,$$

$$\sum^\infty \varepsilon_n t^n\cos(nx) = (1 - t^2)(1 - 2t\cos x + t^2)^{-1}, \qquad |t| < 1,$$

$$\sum^\infty t^n\sin(nx) = t\sin x\,(1 - 2t\cos x + t^2)^{-1}, \qquad |t| < 1,$$

$$\sum^\infty n^{-1}t^n\cos(nx) = -\frac{1}{2}\log(1 - 2t\cos x + t^2), \qquad |t| \le 1,$$

$$\sum^\infty n^{-1}t^n\sin(nx) = \arctan\left(\frac{t\sin x}{1 - t\cos x}\right), \qquad |t| \le 1,$$

$$\sum^\infty \varepsilon_n e^{-nt}\cos(nx) = \sinh t\,(\cosh t - \cos x)^{-1}, \qquad t > 0,$$

$$\sum_1^\infty e^{-nt}\sin(nx) = \frac{1}{2}\sin x\,(\cosh t - \cos x)^{-1}, \qquad t > 0,$$

$$\sum_1^\infty n^{-1}e^{-nt}\cos(nx) = \frac{1}{2}t - \frac{1}{2}\log(2\cosh t - 2\cos x), \qquad t > 0,$$

$$\sum_1^\infty n^{-1}e^{-nt}\sin(nx) = \arctan\left[\coth\left(\frac{t}{2}\right)\tan\left(\frac{x}{2}\right)\right] - \frac{1}{2}x, \qquad t > 0,$$

$$\sum_1^\infty n^{-1}\sin(nx)\sin(ny)\,e^{-2nt} = \frac{1}{4}\log\left[\frac{\sin^2\left(\frac{1}{2}x + \frac{1}{2}y\right) + \sinh^2 t}{\sin^2\left(\frac{1}{2}x - \frac{1}{2}y\right) + \sinh^2 t}\right]$$

$$= \frac{1}{4}\log\left[\frac{\cosh(2t) - \cos(x+y)}{\cosh(2t) - \cos(x-y)}\right], \qquad t > 0,$$

$$(1 - 2t\cos x + t^2)^{-\nu} = \sum_0^\infty \varepsilon_n A_n \cos(nx),$$

$$A_n = \frac{t^n}{n!}\,\frac{\Gamma(n+\nu)}{\Gamma(\nu)}\cdot F(\nu, n+\nu; n+1; t^2), \qquad t > 0,$$

$$\sum_{-\infty}^\infty e^{-|x+nd|} = \frac{2}{d}\sum_0^\infty \frac{\varepsilon_n\cos\left(2\pi n\frac{x}{d}\right)}{1 + \left(2\pi\frac{n}{d}\right)^2} = \frac{\cosh\left(\frac{1}{2}d - x\right)}{\sinh\left(\frac{1}{2}d\right)}, \qquad 0 \le x \le d,$$

$$\sum_{-\infty}^\infty e^{-(x+nd)^2} = \pi^{\frac{1}{2}}d^{-1}\sum_{-\infty}^\infty e^{-n^2\frac{\pi^2}{d^2}}\cos\left(2\pi n\frac{x}{d}\right) = \pi^{\frac{1}{2}}d^{-1}\vartheta_3\left(\frac{x}{d}, i\frac{\pi}{d^2}\right),$$

$$\sum_{-\infty}^\infty \frac{1}{\cosh(x+nd)} = \frac{\pi}{d}\sum_0^\infty \varepsilon_n \frac{\cos\left(2\pi n\frac{x}{d}\right)}{\cosh\left(n\frac{\pi^2}{d}\right)} = \frac{2}{d}K\,dn\left(2K\frac{x}{d}\right)$$

with $\frac{K'}{K} = \frac{\pi}{d}$.

Here ϑ_3 means the elliptic theta function and $\mathrm{dn}\,u$ one of the Jacobi elliptic functions (see chap. X)

$$\sum_{-\infty}^\infty \frac{e^{i(nd-z)}}{nd - z} = \begin{cases} \dfrac{-\pi e^{-i(2r+1)\pi\frac{z}{d}}}{d\sin\left(\pi\frac{z}{d}\right)}, & 2r\pi < d < 2(r+1)\pi, \\[2ex] \dfrac{-\pi e^{-i2r\pi\frac{z}{d}}}{d\tan\left(\pi\frac{z}{d}\right)}, & d = 2r\pi. \end{cases}$$

Poisson's summation formula

$$\sum_{n=-\infty}^\infty e^{ina}f(x+nb) = \frac{1}{b}\sum_{m=-\infty}^\infty F\left(\frac{2\pi m + a}{b}\right)e^{-i\frac{x}{b}(2\pi m + a)}$$

with

$$F(t) = \int_{-\infty}^{\infty} f(y)\, e^{ity}\, dy,$$

$$\frac{\sinh(\lambda\delta)}{\sinh(\lambda\alpha)} = -2\pi \sum_{n=1}^{\infty} \frac{(-1)^n\, n \sin\left(n\pi\frac{\delta}{\alpha}\right)}{n^2\pi^2 + \lambda^2\alpha^2}, \qquad -\alpha < \delta < \alpha,$$

$$\frac{\cosh(\lambda\delta)}{\sinh(\lambda\alpha)} = \lambda\alpha \sum_{n=0}^{\infty} \frac{(-1)^n\, \varepsilon_n \cos\left(n\pi\frac{\delta}{\alpha}\right)}{n^2\pi^2 + \lambda^2\alpha^2}, \qquad -\alpha \le \delta \le \alpha,$$

$$\frac{\sinh(\lambda\delta)}{\cosh(\lambda\alpha)} = 2\lambda\alpha \sum_{n=0}^{\infty} \frac{(-1)^n \sin\left[\left(n+\frac{1}{2}\right)\pi\frac{\delta}{\alpha}\right]}{\left(n+\frac{1}{2}\right)^2\pi^2 + \lambda^2\alpha^2}, \qquad -\alpha \le \delta \le \alpha,$$

$$\frac{\cosh(\lambda\delta)}{\cosh(\lambda\alpha)} = \pi \sum_{n=0}^{\infty} \frac{(-1)^n (2n+1) \cos\left[\left(n+\frac{1}{2}\right)\pi\frac{\delta}{\alpha}\right]}{\left(n+\frac{1}{2}\right)^2\pi^2 + \lambda^2\alpha^2}, \qquad -\alpha < \delta < \alpha.$$

The corresponding relations for the quotient of two trigonometric functions can be obtained from the formulas above upon replacing α by $i\alpha$ and δ by $i\delta$.

Some finite trigonometric sums

$$\sum_{n=1}^{N} e^{inx} = e^{i(N+1)\frac{x}{2}} \frac{\sin\left(N\frac{x}{2}\right)}{\sin\left(\frac{x}{2}\right)}$$

$$\sum_{n=0}^{N} e^{i(2n+1)x} = e^{i(N+1)x} \frac{\sin(N+1)\,x}{\sin x},$$

$$\sum_{n=1}^{N} \cos(nx) = \frac{1}{2}\left\{\frac{\sin\left[(2N+1)\frac{x}{2}\right]}{\sin\left(\frac{x}{2}\right)} - 1\right\},$$

$$\sum_{n=1}^{N} \sin(nx) = \frac{1}{2}\, \frac{\cos\left(\frac{x}{2}\right) - \cos\left[(2N+1)\frac{x}{2}\right]}{\sin\left(\frac{x}{2}\right)},$$

$$\sum_{n=1}^{N} (-1)^n \cos(nx) = \frac{1}{2}\left\{\frac{(-1)^N \cos\left[(2N+1)\frac{x}{2}\right]}{\cos\left(\frac{x}{2}\right)} - 1\right\},$$

$$\sum_{n=1}^{N} (-1)^n \sin(nx) = \frac{1}{2}\left\{\frac{(-1)^N \sin\left[(2N+1)\frac{x}{2}\right] - \sin\frac{x}{2}}{\cos\left(\frac{x}{2}\right)}\right\},$$

$$\left(\frac{\sin nx}{\sin x}\right)^2 = n + 2(n-1)\cos(2x) + 2(n-2)\cos(4x)$$
$$+ \cdots + 2\cos[2(n-1)x]$$
$$= \sum_{l=0}^{n-1} \frac{\sin[(2l+1)x]}{\sin x}$$

$$\sin^{2n} x = 2^{-2n}(2n)! \sum_{l=0}^{n} \frac{(-1)^l \varepsilon_l \cos(2lx)}{(n+l)!\,(n-l)!},$$

$$\sin^{2n+1} x = 2^{-2n}(2n+1)! \sum_{l=0}^{n} \frac{(-1)^l \sin[(2l+1)x]}{(n+1+l)!\,(n-l)!},$$

$$\cos^{2n} x = 2^{-2n}(2n)! \sum_{l=0}^{n} \frac{\varepsilon_l \cos(2lx)}{(n+l)!\,(n-l)!},$$

$$\cos^{2n+1} x = 2^{-2n}(2n+1)! \sum_{l=0}^{n} \frac{\cos[(2l+1)x]}{(n+1+l)!\,(n-l)!},$$

$$\frac{\sin(2nx)}{\sin(2x)} = 2 \sum_{l=0}^{n-1} \frac{(-1)^l (l+1)(n+l)!}{(n-l-1)!\,(2l+2)!} (2\sin x)^{2l},$$

$$= 2(-1)^{n+1} \sum_{l=0}^{n-1} \frac{(-1)^l (l+1)(n+l)!}{(n-l-1)!\,(2l+2)!} (2\cos x)^{2l},$$

$$\cos(2nx) = (-1)^n n \sum_{l=0}^{n} \frac{(-1)^l (n+l-1)!}{(n-l)!\,(2l)!} (2\cos x)^{2l}$$

$$= n \sum_{l=0}^{n} \frac{(-1)^l (n+l-1)!}{(n-l)!\,(2l)!} (2\sin x)^{2l},$$

$$\frac{\sin[(2n+1)x]}{\sin x} = 2(2n+1) \sum_{=0}^{n} \frac{(-1)^l (l+1)(n+l)!}{(n-l)!\,(2l+2)!} (2\sin x)^{2l}$$

$$= (-1)^n \sum_{l=0}^{n} \frac{(-1)^l (n+l)!}{(n-l)!\,(2l)!} (2\cos x)^{2l},$$

$$\frac{\cos[(2n+1)x]}{\cos x} = (-1)^n 2(2n+1) \sum_{l=0}^{n} \frac{(-1)^l (l+1)(n+l)!}{(n-l)!\,(2l+2)!} (2\cos x)^{2l}$$

$$= \sum_{l=0}^{n} \frac{(-1)^l (n+l)!}{(n-l)!\,(2l)!} (2\cos x)^{2l}.$$

The corresponding relations for hyperbolic functions can be obtained upon replacing x by ix in the formulas above

$$(2a)^{-1} + \frac{1}{2}\pi(ab)^{-\frac{1}{2}} \coth\left[\pi\left(\frac{a}{b}\right)^{\frac{1}{2}}\right] = \sum_{0}^{\infty}(a+bn^2)^{-1},$$

$$\frac{\sinh\left(\frac{2\pi a}{b}\right)}{\cosh\left(\frac{2\pi a}{b}\right) - \cos\left(\frac{2\pi z}{b}\right)} = \pi^{-1} a b \sum_{-\infty}^{\infty} [(z + nb)^2 + a^2]^{-1},$$

$$\psi\left(\frac{1}{2} z + \frac{1}{2}\right) - \psi\left(\frac{1}{2} z\right) = 2 \sum_{0}^{\infty} (-1)^n (z + n)^{-1},$$

$$m\, 2^{1-m} = \prod_{r=1}^{m-1} \sin\left(\pi \frac{r}{m}\right),$$

$$z^{-1} \sin z = \prod_{n=1}^{\infty} \cos(2^{-n} z), \qquad |z| < 1,$$

$$e^{az} - e^{bz} = z(a - b)\, e^{\frac{1}{2}(a+b)z} \prod_{n=1}^{\infty} \left[1 + \frac{(az - bz)^2}{4n^2\pi^2}\right],$$

$$\frac{\sin[\pi(z + a)]}{\sin(\pi a)} = a^{-1}(z + a) \prod_{n=1}^{\infty} \left(1 - \frac{z}{n - a}\right)\left(1 + \frac{z}{n + a}\right),$$

$$1 - \frac{\sin^2(\pi z)}{\sin^2(\pi a)} = \prod_{n=-\infty}^{\infty} \left[1 - \left(\frac{z}{n - a}\right)^2\right],$$

$$\frac{\cosh z - \cos a}{1 - \cos a} = \prod_{n=-\infty}^{\infty} \left[1 + \left(\frac{z}{2n\pi + a}\right)^2\right].$$

For further formulas see chap. I.

Chapter XII

XII. Transformation of systems of coordinates

12.1 General transformation and special cases

12.1.1 General transformation

In many problems of mathematical physics it is better to use an orthogonal coordinate system u, v, w instead of the cartesian coordinates x, y, z. The choice of a particular coordinate system may be motivated by physical reason and can result in a considerably simplified analysis of the problem.

Let u, v, w be an arbitrary orthogonal coordinate system related to the cartesian coordinate system x, y, z by the equations

$$x = x(u, v, w), \quad y = y(u, v, w), \quad z = z(u, v, w),$$

where it is assumed that the surfaces

$$u = a_1 (\text{const.}), \ v = b_1 (\text{const.}), \ w = c_1 (\text{const.})$$

are mutually orthogonal. In terms of the new coordinates u, v, w the element of arclength ds, and the element of volume $d\tau$ are expressed by the following relations:

$$\begin{aligned} (ds)^2 &= (dx)^2 + (dy)^2 + (dz)^2 \\ &= \frac{1}{U^2} (du)^2 + \frac{1}{V^2} (dv)^2 + \frac{1}{W^2} (dw)^2, \\ d\tau &= \frac{1}{UVW} du \, dv \, dw \end{aligned}$$

and the new elements of area

$$\frac{du}{U} \frac{dv}{V}, \quad \frac{dv}{V} \frac{dw}{W}, \quad \frac{dw}{W} \frac{dx}{U},$$

where

$$\begin{aligned} \frac{1}{U^2} &= \left(\frac{\partial x}{\partial u}\right)^2 + \left(\frac{\partial y}{\partial u}\right)^2 + \left(\frac{\partial z}{\partial u}\right)^2, \\ \frac{1}{V^2} &= \left(\frac{\partial x}{\partial v}\right)^2 + \left(\frac{\partial y}{\partial v}\right)^2 + \left(\frac{\partial z}{\partial v}\right)^2, \\ \frac{1}{W^2} &= \left(\frac{\partial x}{\partial w}\right)^2 + \left(\frac{\partial y}{\partial w}\right)^2 + \left(\frac{\partial z}{\partial w}\right)^2. \end{aligned}$$

It may be observed that, in general,

$$dx \, dy \neq \frac{du}{U} \frac{dv}{V}$$

and similarly for other elements of area. If the direction cosines of the normals to the surfaces $u =$ const., $v =$ const., $w =$ const. be given by (l_1, m_1, n_1), (l_2, m_2, n_2) and (l_3, m_3, n_3) respectively, then

$$\begin{aligned} l_1 &= \frac{1}{U} \frac{\partial u}{\partial x}, & m_1 &= \frac{1}{U} \frac{\partial u}{\partial y}, & n_1 &= \frac{1}{U} \frac{\partial u}{\partial z}, \\ &= U \frac{\partial x}{\partial u}, & &= U \frac{\partial y}{\partial u}, & &= U \frac{\partial z}{\partial u}, \\ l_2 &= \frac{1}{V} \frac{\partial v}{\partial x}, & m_2 &= \frac{1}{V} \frac{\partial v}{\partial y}, & n_2 &= \frac{1}{V} \frac{\partial v}{\partial z}, \\ &= V \frac{\partial x}{\partial v}, & &= V \frac{\partial y}{\partial v}, & &= V \frac{\partial z}{\partial v}, \\ l_3 &= \frac{1}{W} \frac{\partial w}{\partial x}, & m_3 &= \frac{1}{W} \frac{\partial w}{\partial y}, & n_3 &= \frac{1}{W} \frac{\partial w}{\partial z}, \\ &= W \frac{\partial x}{\partial w}, & &= W \frac{\partial y}{\partial w}, & &= W \frac{\partial z}{\partial w}. \end{aligned}$$

Let Φ and $\vec{F} = (F_1, F_2, F_3)$ be a scaler and a vector function respectively.

The quantities $\nabla\Phi$ (grad Φ), $\nabla^2\Phi = \nabla \cdot \nabla\Phi$ (Laplacian Φ), $\nabla \cdot \vec{F}$ (div $\vec{F}$) and $\nabla \times \vec{F}$ (curl $\vec{F}$) are represented, in terms of the new coordinates u, v, w, by

$$\nabla\Phi = \operatorname{grad}\Phi = \left(U\frac{\partial\Phi}{\partial u}, V\frac{\partial\Phi}{\partial v}, W\frac{\partial\Phi}{\partial w}\right),$$

$$\nabla \cdot \vec{F} = \operatorname{div}\vec{F} = UVW\left[\frac{\partial}{\partial u}\left(\frac{F_1}{VW}\right) + \frac{\partial}{\partial v}\left(\frac{F_2}{WU}\right) + \frac{\partial}{\partial w}\left(\frac{F_3}{UV}\right)\right],$$

$$\nabla^2\Phi = \operatorname{div}(\operatorname{grad}\Phi)$$
$$= UVW\left\{\frac{\partial}{\partial u}\left(\frac{U}{VW}\frac{\partial\Phi}{\partial u}\right) + \frac{\partial}{\partial v}\left(\frac{V}{UW}\frac{\partial\Phi}{\partial v}\right) + \frac{\partial}{\partial w}\left(\frac{W}{UV}\frac{\partial\Phi}{\partial w}\right)\right\},$$

$$\nabla \times \vec{F} = \operatorname{curl}\vec{F}$$
$$= \left(VW\left[\frac{\partial}{\partial v}\left(\frac{F_3}{W}\right) - \frac{\partial}{\partial w}\left(\frac{F_2}{V}\right)\right], UW\left[\frac{\partial}{\partial w}\left(\frac{F_1}{U}\right) - \frac{\partial}{\partial u}\left(\frac{F_3}{W}\right)\right],\right.$$
$$\left. UV\left[\frac{\partial}{\partial u}\left(\frac{F_2}{V}\right) - \frac{\partial}{\partial v}\left(\frac{F_1}{U}\right)\right]\right).$$

where F_1, F_2, F_3 are the components of $\vec{F}$ in the u, v, w system.
Define

$$(\nabla\times\vec{F})_u = VW\left[\frac{\partial}{\partial v}\left(\frac{F_3}{W}\right) - \frac{\partial}{\partial w}\left(\frac{F_2}{V}\right)\right],$$
$$(\nabla\times\vec{F})_v = UW\left[\frac{\partial}{\partial w}\left(\frac{F_1}{U}\right) - \frac{\partial}{\partial u}\left(\frac{F_3}{W}\right)\right],$$
$$(\nabla\times\vec{F})_w = UV\left[\frac{\partial}{\partial u}\left(\frac{F_2}{V}\right) - \frac{\partial}{\partial v}\left(\frac{F_1}{U}\right)\right].$$

The following relations express the transformation of the vector components

$$F_1 = U\left(F_x\frac{\partial x}{\partial u} + F_y\frac{\partial y}{\partial u} + F_z\frac{\partial z}{\partial u}\right),$$
$$F_2 = V\left(F_x\frac{\partial x}{\partial v} + F_y\frac{\partial y}{\partial v} + F_z\frac{\partial z}{\partial v}\right),$$
$$F_3 = W\left(F_x\frac{\partial x}{\partial w} + F_y\frac{\partial y}{\partial w} + F_z\frac{\partial z}{\partial w}\right),$$

where F_x, F_y, F_z are the components of $\vec{F}$ in the Cartesian coordinates.

12.1.2 Special cases

The results of the preceeding section for some commonly used coordinate systems are given by:

Spherical coordinates (r, φ, θ)

$$x = r \sin\theta \cos\varphi, \qquad u = r,\ v = \theta,\ w = \varphi,$$

$$y = r \sin\theta \sin\varphi,$$

$$z = r\cos\theta, \qquad U = 1,\ V = \frac{1}{r},\ W = \frac{1}{r\sin\theta}.$$

The surfaces $r =$ constant are spheres and the surfaces $\varphi =$ const. are planes through z-axis. The surfaces $\theta =$ const. are right circular cones with vertex at the origin and z-axis as the axis of symmetry

$$(ds)^2 = (dr)^2 + (r\,d\theta)^2 + (r\sin\theta\,d\varphi)^2,$$

$$\nabla\Phi = \left(\frac{\partial\Phi}{\partial r}, \frac{1}{r}\frac{\partial\Phi}{\partial\theta}, \frac{1}{r\sin\theta}\frac{\partial\Phi}{\partial\varphi}\right),$$

$$\nabla\cdot\vec{F} = \frac{1}{r^2}\frac{\partial}{\partial r}(r^2 F_1) + \frac{1}{r\sin\theta}\frac{\partial}{\partial\theta}(\sin\theta\,F_2) + \frac{1}{r\sin\theta}\frac{\partial F_3}{\partial\varphi},$$

$$\nabla^2\Phi = \frac{1}{r^2}\frac{\partial}{\partial r}\left(r^2\frac{\partial\Phi}{\partial r}\right) + \frac{1}{r^2\sin\theta}\frac{\partial}{\partial\theta}\left(\sin\theta\frac{\partial\Phi}{\partial\theta}\right) + \frac{1}{r^2\sin^2\theta}\frac{\partial^2\Phi}{\partial\varphi^2},$$

$$(\nabla\times\vec{F})_r = \frac{1}{r\sin\theta}\left[\frac{\partial}{\partial\theta}(\sin\theta\,F_3) - \frac{\partial F_2}{\partial\varphi}\right],$$

$$(\nabla\times\vec{F})_\theta = \frac{1}{r\sin\theta}\left[\frac{\partial F_1}{\partial\varphi} - \sin\theta\frac{\partial}{\partial r}(rF_3)\right],$$

$$(\nabla\times\vec{F})_\varphi = \frac{1}{r}\left[\frac{\partial}{\partial r}(rF_2) - \frac{\partial F_1}{\partial\varphi}\right],$$

$$F_1 = F_x\sin\theta\cos\varphi + F_y\sin\theta\sin\varphi + F_z\cos\theta,$$

$$F_2 = F_x\cos\theta\cos\varphi + F_y\cos\theta\sin\varphi - F_z\sin\theta,$$

$$F_3 = -F_x\sin\varphi + F_y\cos\varphi.$$

Cylindrical coordinates (ϱ, φ, z)

$$x = \varrho\cos\varphi, \qquad u = \varrho,\ v = \varphi,\ w = z,$$

$$y = \varrho\sin\varphi,$$

$$z = z, \qquad U = 1,\ V = \frac{1}{\varrho},\ W = 1.$$

The surfaces $\varrho =$ constant are right circular cylinders with z-axis as the axis of symmetry, and the surfaces $\varphi =$ const. are planes through z-axis

$$ds^2 = d\varrho^2 + \varrho^2\,d\varphi^2 + dz^2,$$

$$\nabla\Phi = \left(\frac{\partial\Phi}{\partial\varrho}, \frac{1}{\varrho}\frac{\partial\Phi}{\partial\varphi}, \frac{\partial\Phi}{\partial z}\right),$$

$$\operatorname{div}\vec{F} = \nabla\cdot\vec{F} = \frac{1}{\varrho}\frac{\partial}{\partial\varrho}(\varrho F_1) + \frac{1}{\varrho}\frac{\partial F_2}{\partial\varphi} + \frac{\partial F_3}{\partial z},$$

$$\nabla^2\Phi = \frac{\partial^2\Phi}{\partial\varrho^2} + \frac{1}{\varrho}\frac{\partial\Phi}{\partial\varrho} + \frac{1}{\varrho^2}\frac{\partial^2\Phi}{\partial\varphi^2} + \frac{\partial^2\Phi}{\partial z^2},$$

$$(\nabla\times\vec{F})_\varrho = \frac{1}{\varrho}\frac{\partial F_3}{\partial\varphi} - \frac{\partial F_2}{\partial z},$$

$$(\nabla\times\vec{F})_\varphi = \frac{\partial F_1}{\partial z} - \frac{\partial F_3}{\partial z},$$

$$(\nabla\times\vec{F})_z = \frac{1}{\varrho}\left[\frac{\partial}{\partial\varrho}(\varrho F_2) - \frac{\partial F_1}{\partial\varphi}\right],$$

$$F_1 = F_x\cos\varphi + F_y\sin\varphi,$$

$$F_2 = -F_x\sin\varphi + F_y\cos\varphi,$$

$$F_3 = F_z.$$

Parabolic coordinates (ξ, η, φ)

$$x = \xi\eta\cos\varphi, \qquad u = \xi,\ v = \eta,\ w = \varphi,$$

$$y = \xi\eta\sin\varphi,$$

$$z = \frac{1}{2}(\xi^2 - \eta^2), \quad U = V = (\xi^2 + \eta^2)^{-\frac{1}{2}},\ W = (\xi\eta)^{-1}.$$

The surfaces $\xi =$ const., $\eta =$ const. are paraboloids of revolution about z-axis and $\varphi =$ const. are planes through z-axis

$$ds^2 = (\xi^2 + \eta^2)(d\xi^2 + d\eta^2) + \xi^2\eta^2\, d\varphi^2,$$

$$\nabla\Phi = \left(\frac{1}{\sqrt{\xi^2+\eta^2}}\frac{\partial\Phi}{\partial\xi},\ \frac{1}{\sqrt{\xi^2+\eta^2}}\frac{\partial\Phi}{\partial\eta},\ \frac{1}{\xi\eta}\frac{\partial\Phi}{\partial\varphi}\right),$$

$$\nabla\cdot\vec{F} = \frac{1}{\sqrt{\xi^2+\eta^2}}\left\{\frac{\xi F_1 + \eta F_2}{\xi^2+\eta^2} + \frac{1}{\xi}\frac{\partial}{\partial\xi}(\xi F_1) + \frac{1}{\eta}\frac{\partial}{\partial\eta}(\eta F_2) + \frac{\sqrt{\xi^2+\eta^2}}{\xi\eta}\frac{\partial F_3}{\partial\varphi}\right\},$$

$$\nabla^2\Phi = \frac{1}{\xi^2+\eta^2}\left\{\frac{1}{\xi}\frac{\partial}{\partial\xi}\left(\xi\frac{\partial\Phi}{\partial\xi}\right) + \frac{1}{\eta}\frac{\partial}{\partial\eta}\left(\eta\frac{\partial\Phi}{\partial\eta}\right) + \left(\frac{1}{\xi^2} + \frac{1}{\eta^2}\right)\frac{\partial^2\Phi}{\partial\varphi^2}\right\},$$

$$(\nabla\times\vec{F})_\xi = \frac{1}{\xi\eta\sqrt{\xi^2+\eta^2}}\left\{\xi\frac{\partial}{\partial\eta}(\eta F_3) - \sqrt{\xi^2+\eta^2}\frac{\partial F_2}{\partial\varphi}\right\},$$

$$(\nabla\times\vec{F})_\eta = \frac{1}{\xi\eta\sqrt{\xi^2+\eta^2}}\left\{\sqrt{\xi^2+\eta^2}\frac{\partial F_1}{\partial\varphi} - \eta\frac{\partial}{\partial\xi}(\xi F_3)\right\},$$

$$(\nabla\times\vec{F})_\varphi = -\frac{1}{\sqrt{\xi^2+\eta^2}}\left\{\frac{\eta F_1 - \xi F_2}{\xi^2+\eta^2} - \frac{\partial F_2}{\partial\xi} + \frac{\partial F_1}{\partial\eta}\right\},$$

$$F_1 = \frac{1}{\sqrt{\xi^2+\eta^2}}\{F_x\eta\cos\varphi + F_y\eta\sin\varphi + \xi F_z\},$$

$$F_2 = \frac{1}{\sqrt{\xi^2+\eta^2}}\{F_x\xi\cos\varphi + F_y\xi\sin\varphi - \eta F_z\},$$

$$F_3 = -F_x\sin\varphi + F_y\cos\varphi.$$

Parabolic cylinder coordinates (ξ, η, z)

$$x = \xi\eta,$$

$$y = \frac{1}{2}(\xi^2 - \eta^2), \qquad u = \xi,\ v = \eta,\ w = z,$$

$$z = z, \qquad U = V = (\xi^2 + \eta^2)^{-\frac{1}{2}},\ W = 1.$$

The surfaces $\xi =$ const. and $\eta =$ const. are parabolic cylinders with generators parallel to z-axis

$$ds^2 = (\xi^2 + \eta^2)(d\xi^2 + d\eta^2) + dz^2,$$

$$\nabla\Phi = \left(\frac{1}{\sqrt{\xi^2 + \eta^2}}\frac{\partial\Phi}{\partial\xi}, \frac{1}{\sqrt{\xi^2 + \eta^2}}\frac{\partial\Phi}{\partial\eta}, \frac{\partial\Phi}{\partial z}\right),$$

$$\nabla\cdot\vec{F} = \frac{1}{\xi^2 + \eta^2}\left\{\frac{\partial}{\partial\xi}\left(\sqrt{\xi^2 + \eta^2}\,F_1\right) + \frac{\partial}{\partial\eta}\left(\sqrt{\xi^2 + \eta^2}\,F_2\right) + (\xi^2 + \eta^2)\frac{\partial F_3}{\partial z}\right\},$$

$$\nabla^2\Phi = \frac{1}{\xi^2 + \eta^2}\left(\frac{\partial^2\Phi}{\partial\xi^2} + \frac{\partial^2\Phi}{\partial\eta^2}\right) + \frac{\partial^2\Phi}{\partial z^2},$$

$$(\nabla\times\vec{F})_\xi = (\xi^2 + \eta^2)^{-\frac{1}{2}}\left[\frac{\partial F_3}{\partial\eta} - (\xi^2 + \eta^2)^{\frac{1}{2}}\frac{\partial F_2}{\partial z}\right],$$

$$(\nabla\times\vec{F})_\eta = (\xi^2 + \eta^2)^{-\frac{1}{2}}\left[(\xi^2 + \eta^2)^{\frac{1}{2}}\frac{\partial F_1}{\partial z} - \frac{\partial F_3}{\partial\xi}\right],$$

$$(\nabla\times F)_z = \frac{1}{\xi^2 + \eta^2}\left[\frac{\partial}{\partial\xi}\left(\sqrt{\xi^2 + \eta^2}\,F_2\right) - \frac{\partial}{\partial\eta}\left(\sqrt{\xi^2 + \eta^2}\,F_1\right)\right],$$

$$F_1 = (\xi^2 + \eta^2)^{-\frac{1}{2}}[\eta F_x + \xi F_y],$$

$$F_2 = (\xi^2 + \eta^2)^{-\frac{1}{2}}[\xi F_x - \eta F_y],$$

$$F_3 = F_z.$$

Elliptic cylinder coordinates (ξ, η, z)

$$x = c\cosh\xi\cos\eta, \qquad u = \xi,\ v = \eta,\ w = z,$$

$$y = c\sinh\xi\sin\eta,$$

$$z = z, \qquad U = V = \frac{1}{c\tau},\ W = 1,$$

where

$$\tau = (\sinh^2\xi + \sin^2\eta)^{\frac{1}{2}}.$$

The domain of the new variables is given by:

$$0 \leq \xi < \infty,$$
$$0 \leq \eta \leq 2\pi,$$
$$-\infty < z < \infty.$$

The surfaces $\xi = \xi_0$ (const.) are elliptic cylinders with semi-axes $c \cosh \xi_0$, $c \sinh \xi_0$ and given by the equation

$$\frac{x^2}{c^2 \cosh^2 \xi_0} + \frac{y^2}{c^2 \sinh^2 \xi_0} = 1.$$

The generators are parallel to z-axis. The surfaces $\eta = \eta_0$ (const.) are hyperbolic cylinders with generators parallel to z-axis and given by

$$\frac{x^2}{c^2 \cos^2 \eta_0} - \frac{y^2}{c^2 \sin^2 \eta_0} = 1,$$

$$ds^2 = c^2 \tau^2 (d\xi^2 + d\eta^2) + dz^2,$$

$$\nabla \Phi = \left(\frac{1}{c\tau} \frac{\partial \Phi}{\partial \xi}, \frac{1}{c\tau} \frac{\partial \Phi}{\partial \eta}, \frac{\partial \Phi}{\partial z}\right),$$

$$\nabla \cdot \vec{F} = \frac{1}{c\tau^2} \left\{\frac{\partial}{\partial \xi} (\tau F_1) + \frac{\partial}{\partial \eta} (\tau F_2) + c\tau^2 \frac{\partial F_3}{\partial z}\right\},$$

$$\nabla^2 \Phi = \frac{1}{c^2 \tau^2} \left(\frac{\partial^2 \Phi}{\partial \xi^2} + \frac{\partial^2 \Phi}{\partial \eta^2}\right) + \frac{\partial^2 \Phi}{\partial z^2},$$

$$(\nabla \times \vec{F})_\xi = \frac{1}{c\tau} \left(\frac{\partial F_3}{\partial \eta} - c\tau \frac{\partial F_2}{\partial z}\right),$$

$$(\nabla \times \vec{F})_\eta = \frac{1}{c\tau} \left(c\tau \frac{\partial F_1}{\partial z} - \frac{\partial F_3}{\partial \xi}\right),$$

$$(\nabla \times \vec{F})_z = \frac{1}{c\tau} \left(\frac{\partial}{\partial \xi} (\tau F_2) - \frac{\partial}{\partial \eta} (\tau F_1)\right),$$

$$F_1 = \frac{1}{\tau} [F_x \sinh \xi \cos \eta + F_y \cosh \xi \sin \eta],$$

$$F_2 = \frac{1}{\tau} [-F_x \cosh \xi \sin \eta + F_y \sinh \xi \cos \eta],$$

$$F_3 = F_z.$$

Elliptic coordinates (ξ, η, φ) (*Prolate ellipsoidal*)

$$x = c\sqrt{(1 - \eta^2)(\xi^2 - 1)} \cos \varphi,$$
$$y = c\sqrt{(1 - \eta^2)(\xi^2 - 1)} \sin \varphi, \qquad u = \xi,\ v = \eta,\ w = \varphi,$$
$$z = c\xi\eta,$$

$$U = \frac{1}{c} (\xi^2 - 1)^{\frac{1}{2}} (\xi^2 - \eta^2)^{-\frac{1}{2}}, \qquad V = \frac{1}{c} (1 - \eta^2)^{\frac{1}{2}} (\xi^2 - \eta^2)^{-\frac{1}{2}},$$

$$W = \frac{1}{c} (1 - \eta^2)^{-\frac{1}{2}} (\xi^2 - 1)^{-\frac{1}{2}}.$$

The domain of the variables ξ, η φ, is given by:

$$1 \leq \xi < \infty,$$
$$-1 \leq \eta \leq 1,$$
$$0 \leq \varphi \leq 2\pi.$$

The surfaces $\xi = \xi_0$ (const.) are ellipsoids of revolution around the z-axis and given by:

$$\frac{x^2 + y^2}{c^2(\xi_0^2 - 1)} + \frac{z^2}{c^2 \xi_0^2} = 1.$$

The surfaces $\eta = \eta_0$ (const.) are hyperboloids of two sheets and given by

$$\frac{x^2 + y^2}{c^2(1 - \eta_0^2)} - \frac{z^2}{c^2 \eta_0^2} = 1.$$

The surfaces $\varphi = \varphi_0$ (const.) are planes through z-axis and given by:

$$y = x \tan \varphi_0,$$

$$ds^2 = c^2(\xi^2 - \eta^2)\left(\frac{1}{\xi^2 - 1} d\xi^2 + \frac{1}{1 - \eta^2} d\eta^2\right) + c^2(\xi^2 - 1)(1 - \eta^2)\, d\varphi^2,$$

$$\nabla\Phi = \left(\frac{1}{c}\sqrt{\frac{\xi^2 - 1}{\xi^2 - \eta^2}}\frac{\partial\Phi}{\partial\xi}, \frac{1}{c}\sqrt{\frac{1 - \eta^2}{\xi^2 - \eta^2}}\frac{\partial\Phi}{\partial\eta}, \frac{1}{c\sqrt{(\xi^2 - 1)(1 - \eta^2)}}\frac{\partial\Phi}{\partial\varphi}\right),$$

$$\nabla \cdot \vec{F} = \frac{1}{c(\xi^2 - \eta^2)}\left\{\frac{\partial}{\partial\xi}\left(\sqrt{(\xi^2 - \eta^2)(\xi^2 - 1)}\, F_1\right)\right.$$
$$\left. + \frac{\partial}{\partial\eta}\left(\sqrt{(\xi^2 - \eta^2)(1 - \eta^2)}\, F_2\right) + \frac{\xi^2 - \eta^2}{\sqrt{(\xi^2 - 1)(1 - \eta^2)}}\frac{\partial F_3}{\partial\varphi}\right\},$$

$$\nabla^2\Phi = \frac{1}{c^2(\xi^2 - \eta^2)}\left\{\frac{\partial}{\partial\xi}\left[(\xi^2 - 1)\frac{\partial\Phi}{\partial\xi}\right] + \frac{\partial}{\partial\eta}\left[(1 - \eta^2)\frac{\partial\Phi}{\partial\eta}\right]\right.$$
$$\left. + \frac{\xi^2 - \eta^2}{(\xi^2 - 1)(1 - \eta^2)}\frac{\partial^2\Phi}{\partial\varphi^2}\right\},$$

$$(\nabla \times \vec{F})_\xi = \frac{1}{c\sqrt{(\xi^2 - \eta^2)(\xi^2 - 1)}}$$
$$\times \left\{\frac{\partial}{\partial\eta}\left[\sqrt{(\xi^2 - 1)(1 - \eta^2)}\, F_3\right] - (\xi^2 - \eta^2)^{\frac{1}{2}}(1 - \eta^2)^{-\frac{1}{2}}\frac{\partial F_2}{\partial\varphi}\right\},$$

$$(\nabla \times \vec{F})_\eta = \frac{1}{c\sqrt{(\xi^2 - \eta^2)(1 - \eta^2)}}$$
$$\times \left\{(\xi^2 - \eta^2)^{\frac{1}{2}}(\xi^2 - 1)^{-\frac{1}{2}}\frac{\partial F_1}{\partial\varphi} - \frac{\partial}{\partial\xi}\left[\sqrt{(\xi^2 - 1)(1 - \eta^2}\, F_3\right]\right\},$$

$$(\nabla \times \vec{F})_\varphi = \frac{\sqrt{(\xi^2 - 1)(1 - \eta^2)}}{c(\xi^2 - \eta^2)}\left\{\frac{\partial}{\partial\xi}\left(\sqrt{\frac{\xi^2 - \eta^2}{1 - \eta^2}}\, F_2\right) - \frac{\partial}{\partial\eta}\left(\sqrt{\frac{\xi^2 - \eta^2}{\xi^2 - 1}}\, F_1\right)\right\},$$

$$F_1 = (\xi^2 - \eta^2)^{-\frac{1}{2}}$$
$$\times \left\{F_x \xi\sqrt{1 - \eta^2}\cos\varphi + F_y \xi\sqrt{1 - \eta^2}\sin\varphi + F_z \eta\sqrt{\xi^2 - 1}\right\},$$

$$F_2 = (\xi^2 - \eta^2)^{-\frac{1}{2}}$$
$$\times \left\{-F_x \eta \sqrt{\xi^2 - 1} \cos\varphi - F_y \eta \sqrt{\xi^2 - 1} \sin\varphi + F_z \xi \sqrt{1 - \eta^2}\right\},$$
$$F_3 = -F_x \sin\varphi + F_y \cos\varphi.$$

Elliptic coordinates (ξ, η, φ) (*Oblate ellipsoidal*)

$$x = c\sqrt{(1 + \xi^2)(1 - \eta^2)} \cos\varphi,$$
$$y = c\sqrt{(1 + \xi^2)(1 - \eta^2)} \sin\varphi, \qquad u = \xi,\ v = \eta,\ w = \varphi,$$
$$z = c\,\xi\,\eta,$$
$$U = \frac{1}{c}(1 + \xi^2)^{\frac{1}{2}} (\xi^2 + \eta^2)^{-\frac{1}{2}}, \qquad V = \frac{1}{c}(1 - \eta^2)^{\frac{1}{2}} (\xi^2 + \eta^2)^{-\frac{1}{2}},$$
$$W = \frac{1}{c}(1 + \xi^2)^{-\frac{1}{2}} (1 - \eta^2)^{-\frac{1}{2}}.$$

The domain of the variables ξ, η, φ is given by:

$$0 \leq \xi < \infty,$$
$$-1 \leq \eta \leq 1,$$
$$0 \leq \varphi \leq 2\pi.$$

The surfaces $\xi = \xi_0$ (const.) are ellipsoids of revolution about z-axis and given by:

$$\frac{x^2 + y^2}{c^2(1 + \xi_0^2)} + \frac{z^2}{c^2 \xi_0^2} = 1.$$

The surfaces $\eta = \eta_0$ (const.) are hyperboloids and given by

$$\frac{x^2 + y^2}{1 - \eta_0^2} - \frac{z^2}{\eta_0^2} = c^2.$$

The surfaces $\varphi = \varphi_0$ (const.) are the planes through z-axis and given by

$$y = x \tan\varphi_0,$$

$$ds^2 = c^2(\xi^2 + \eta^2)\left(\frac{1}{1 + \xi^2} d\xi^2 + \frac{1}{1 - \eta^2} d\eta^2\right) + c^2(1 + \xi^2)(1 - \eta^2)\, d\varphi^2,$$

$$\nabla\Phi = \left(\frac{1}{c}\sqrt{\frac{1 + \xi^2}{\xi^2 + \eta^2}} \frac{\partial\Phi}{\partial\xi}, \frac{1}{c}\sqrt{\frac{1 - \eta^2}{\xi^2 + \eta^2}} \frac{\partial\Phi}{\partial\eta}, \frac{1}{c\sqrt{(1 + \xi^2)(1 - \eta^2)}} \frac{\partial\Phi}{\partial\varphi}\right),$$

$$\nabla \cdot \vec{F} = \frac{1}{c(\xi^2 + \eta^2)} \left\{\frac{\partial}{\partial\xi}\left(\sqrt{(1 + \xi^2)(\xi^2 + \eta^2)}\, F_1\right)\right.$$
$$\left. + \frac{\partial}{\partial\eta}\left(\sqrt{(1 - \eta^2)(\xi^2 + \eta^2)}\, F_2\right) \frac{\xi^2 - \eta^2}{\sqrt{(1 + \xi^2)(1 - \eta^2)}} \frac{\partial F_3}{\partial\varphi}\right\},$$

$$\nabla^2\Phi = \frac{1}{c^2(\xi^2+\eta^2)}\left\{\frac{\partial}{\partial\xi}\left[(1+\xi^2)\frac{\partial\Phi}{\partial\xi}\right] + \frac{\partial}{\partial\eta}\left[(1-\eta^2)\frac{\partial\Phi}{\partial\eta}\right] + \frac{\xi^2+\eta^2}{(1-\xi^2)(1-\eta^2)}\frac{\partial^2\Phi}{\partial\varphi^2}\right\},$$

$$(\nabla\times\vec{F})_\xi = \frac{1}{c}\left[(1+\xi^2)(\xi^2+\eta^2)\right]^{-\frac{1}{2}} \times\left\{\frac{\partial}{\partial\eta}\left[\sqrt{(1+\xi^2)(1-\eta^2)}\,F_3\right] - \sqrt{\frac{\xi^2+\eta^2}{1-\eta^2}}\frac{\partial}{\partial\varphi}[F_2]\right\},$$

$$(\nabla\times\vec{F})_\eta = \frac{1}{c}(1-\eta^2)^{-\frac{1}{2}}(\xi^2+\eta^2)^{-\frac{1}{2}} \times\left\{\sqrt{\frac{\xi^2+\eta^2}{1+\xi^2}}\frac{\partial F_1}{\partial\varphi} - \frac{\partial}{\partial\xi}\left[\sqrt{(1+\xi^2)(1-\eta^2)}\,F_3\right]\right\},$$

$$(\nabla\times\vec{F})_\varphi = \frac{\sqrt{(1+\xi^2)(1-\eta^2)}}{c(\xi^2+\eta^2)} \times\left\{(1-\eta^2)^{-\frac{1}{2}}\frac{\partial}{\partial\xi}\left(\sqrt{\xi^2+\eta^2}\,F_2\right) - (1+\xi^2)^{-\frac{1}{2}}\frac{\partial}{\partial\eta}\left(\sqrt{\xi^2+\eta^2}\,F_1\right)\right\},$$

$$F_1 = (\xi^2+\eta^2)^{-\frac{1}{2}} \times\left\{\xi\sqrt{1-\eta^2}\,(F_x\cos\varphi + F_y\sin\varphi) + F_z\,\eta\sqrt{1+\xi^2}\right\},$$

$$F_2 = (\xi^2+\eta^2)^{-\frac{1}{2}} \times\left\{-\eta\sqrt{1+\xi^2}\,(F_x\cos\varphi + F_y\sin\varphi) + F_z\,\xi\sqrt{1-\eta^2}\right\},$$

$$F_3 = -F_x\sin\varphi + F_y\cos\varphi.$$

Torus coordinates (ξ, η, φ)

$$x = \frac{c\sinh\xi\cos\varphi}{\cosh\xi-\cos\eta}, \qquad u=\xi,\ v=\eta,\ w=\varphi,$$

$$y = \frac{c\sinh\xi\sin\varphi}{\cosh\xi-\cos\eta}, \qquad U=V=\frac{1}{c}(\cosh\xi-\cos\eta),$$

$$z = \frac{c\sin\eta}{\cosh\xi-\cos\eta}, \qquad W=\frac{\cosh\xi-\cos\eta}{c\sinh\xi}.$$

The domain of the variables ξ, η, φ is given by

$$0 \le \xi < \infty,$$
$$0 \le \eta \le 2\pi,$$
$$0 \le \varphi \le 2\pi.$$

The surfaces $\xi = \xi_0$ (const.) and $\eta = \eta_0$ (const.) are given by the equations

$$\left(\sqrt{x^2+y^2} - c\coth\xi_0\right)^2 + z^2 = c^2\,\mathrm{cosech}^2\,\xi_0$$

and

$$x^2 + y^2 + (z - c \cot \eta_0)^2 = c^2 \operatorname{cosec}^2 \eta_0$$

respectively. The surfaces $\varphi = \varphi_0$ are planes through z-axis and given by

$$y = x \tan \varphi_0,$$

$$ds^2 = \frac{c^2}{(\cosh \xi - \cos \eta)^2} (d\xi^2 + d\eta^2) + \frac{c^2 \sinh^2 \xi}{(\cosh \xi - \cos \eta)^2} d\varphi^2,$$

$$\nabla \Phi = \left(\frac{\cosh \xi - \cos \eta}{c} \frac{\partial \Phi}{\partial \xi}, \frac{\cosh \xi - \cos \eta}{c} \frac{\partial \Phi}{\partial \eta}, \frac{\cosh \xi - \cos \eta}{c \sinh \xi} \frac{\partial \Phi}{\partial \varphi} \right),$$

$$\nabla \cdot \vec{F} = \frac{(\cosh \xi - \cos \eta)^3}{c \sinh \xi} \times \left\{ \frac{\partial}{\partial \xi} \left(\frac{1}{\tau} F_1 \right) + \frac{\partial}{\partial \eta} \left(\frac{1}{\tau} F_2 \right) + \frac{1}{(\cosh \xi - \cos \eta)^2} \frac{\partial F_3}{\partial \varphi} \right\},$$

where

$$\tau = \frac{(\cosh \xi - \cos \eta)^2}{\sinh \xi},$$

$$\nabla^2 \Phi = \frac{(\cosh \xi - \cos \eta)^3}{c^2 \sinh \xi} \left\{ \frac{\partial}{\partial \xi} \left(\frac{\sinh \xi}{\cosh \xi - \cos \eta} \frac{\partial \Phi}{\partial \xi} \right) + \frac{\partial}{\partial \eta} \left(\frac{\sinh \xi}{\cosh \xi - \cos \eta} \frac{\partial \Phi}{\partial \eta} \right) + \frac{1}{\sinh \xi (\cosh \xi - \cos \eta)} \frac{\partial^2 \Phi}{\partial \varphi^2} \right\},$$

$$(\nabla \times \vec{F})_\xi = \frac{(\cosh \xi - \cos \eta)^2}{c \sinh \xi} \left\{ \frac{\partial}{\partial \eta} \left(\frac{\sinh \xi}{\cosh \xi - \cos \eta} F_3 \right) - \frac{1}{\cosh \xi - \cos \eta} \frac{\partial F_2}{\partial \varphi} \right\},$$

$$(\nabla \times \vec{F})_\eta = \frac{(\cosh \xi - \cos \eta)^2}{c \sinh \xi} \left\{ \frac{1}{\cosh \xi - \cos \eta} \frac{\partial F_1}{\partial \varphi} - \frac{\partial}{\partial \xi} \left(\frac{\sinh \xi}{\cosh \xi - \cos \eta} F_3 \right) \right\},$$

$$(\nabla \times \vec{F})_\varphi = \frac{(\cosh \xi - \cos \eta)^2}{c} \left\{ \frac{\partial}{\partial \xi} \left(\frac{F_2}{\cosh \xi - \cos \eta} \right) - \frac{\partial}{\partial \eta} \left(\frac{F_1}{\cosh \xi - \cos \eta} \right) \right\},$$

$$F_1 = F_x \frac{1 - \cosh \xi \cos \eta}{\cosh \xi - \cos \eta} \cos \varphi + F_y \frac{1 - \cosh \xi \cos \eta}{\cosh \xi - \cos \eta} \sin \varphi - F_z \frac{\sinh \xi \sin \eta}{\cosh \xi - \cos \eta},$$

$$F_2 = - F_x \frac{\sinh \xi \sin \eta}{\cosh \xi - \cos \eta} \cos \varphi - F_y \frac{\sinh \xi \sin \eta}{\cosh \xi - \cos \eta} \sin \varphi + F_z \frac{\cosh \xi \cos \eta - 1}{\cosh \xi - \cos \eta},$$

$$F_3 = - F_x \sin \varphi + F_y \cos \varphi.$$

Elliptic coordinates λ, μ, ν (ellipsoid of three axes)

$$x^2 = \frac{(a^2 + \lambda)(a^2 + \mu)(a^2 + \nu)}{(a^2 - b^2)(a^2 - c^2)}, \quad u = \lambda, \ v = \mu, \ w = \nu,$$

$$y^2 = \frac{(b^2 + \lambda)(b^2 + \mu)(b^2 + \nu)}{(b^2 - a^2)(b^2 - c^2)}, \quad a > b > c > 0,$$

$$z^2 = \frac{(c^2 + \lambda)(c^2 + \mu)(c^2 + \nu)}{(c^2 - a^2)(c^2 - b^2)},$$

$$U = \frac{2\sqrt{f(\lambda)}}{\sqrt{(\lambda-\mu)(\lambda-\nu)}}, \qquad V = \frac{2\sqrt{f(\mu)}}{\sqrt{(\mu-\lambda)(\mu-\nu)}},$$

$$W = \frac{2\sqrt{f(\nu)}}{\sqrt{(\nu-\lambda)(\nu-\mu)}},$$

where $f(\lambda) = (a^2+\lambda)(b^2+\lambda)(c^2+\lambda)$; $f(\mu)$ and $f(\nu)$ are similarly defined.

The variables λ, μ, ν are the three solutions of the cubic equation in ϱ

$$\frac{x^2}{a^2+\varrho} + \frac{y^2}{b^2+\varrho} + \frac{z^2}{c^2+\varrho} = 1$$

for given x, y, z such that

$$\lambda > \mu > \nu.$$

The domain of the variables λ, μ, ν is given by:

$$-c^2 < \lambda < \infty,$$
$$-b^2 < \mu < -c^2,$$
$$-a^2 < \nu < -b^2.$$

The surfaces $\lambda = \lambda_0$ (const.) are ellipsoids given by

$$\frac{x^2}{a^2+\lambda_0} + \frac{y^2}{b^2+\lambda_0} + \frac{z^2}{c^2+\lambda_0} = 1.$$

The surfaces $\mu = \mu_0$ (const.) are hyperboloids of one sheet and given by

$$\frac{x^2}{a^2+\mu_0} + \frac{y^2}{b^2+\mu_0} - \frac{z^2}{-(c^2+\mu_0)} = 1.$$

The surfaces $\nu = \nu_0$ are hyperboloids of two sheets and given by

$$\frac{x^2}{a^2+\nu_0} - \frac{y^2}{-b^2-\nu_0} - \frac{z^2}{-c^2-\nu_0} = 1,$$

$$ds^2 = \frac{(\lambda-\mu)(\lambda-\nu)}{4f(\lambda)}\,d\lambda^2 + \frac{(\mu-\lambda)(\mu-\nu)}{4f(\mu)}\,d\mu^2 + \frac{(\nu-\lambda)(\nu-\mu)}{4f(\nu)}\,d\nu^2,$$

$$\nabla\Phi = \left(\frac{2\sqrt{f(\lambda)}}{\sqrt{(\lambda-\mu)(\lambda-\nu)}}\frac{\partial\Phi}{\partial\lambda}, \frac{2\sqrt{f(\mu)}}{\sqrt{(\mu-\lambda)(\mu-\nu)}}\frac{\partial\Phi}{\partial\mu}, \frac{2\sqrt{f(\nu)}}{\sqrt{(\nu-\lambda)(\nu-\mu)}}\frac{\partial\Phi}{\partial\nu}\right),$$

$$\nabla^2\Phi = \frac{4\sqrt{f(\lambda)}}{(\lambda-\mu)(\lambda-\nu)}\frac{\partial}{\partial\lambda}\left(\sqrt{f(\lambda)}\frac{\partial\Phi}{\partial\lambda}\right) + \frac{4\sqrt{f(\mu)}}{(\mu-\lambda)(\mu-\nu)}\frac{\partial}{\partial\mu}\left(\sqrt{f(\mu)}\frac{\partial\Phi}{\partial\mu}\right)$$
$$+ \frac{4\sqrt{f(\nu)}}{(\nu-\lambda)(\nu-\mu)}\frac{\partial}{\partial\nu}\left(\sqrt{f(\nu)}\frac{\partial\Phi}{\partial\nu}\right),$$

which can be written as

$$\nabla^2\Phi = \frac{4}{(\lambda-\mu)(\mu-\nu)(\nu-\lambda)}\left\{(\nu-\mu)\frac{\partial^2\Phi}{\partial\alpha^2} + (\lambda-\nu)\frac{\partial^2\Phi}{\partial\beta^2} + (\mu-\lambda)\frac{\partial^2\Phi}{\partial\theta^2}\right\},$$

where

$$d\alpha = \frac{d\lambda}{\sqrt{f(\lambda)}}, \; d\beta = \frac{d\mu}{\sqrt{f(\mu)}}, \; d\theta = \frac{d\nu}{\sqrt{f(\nu)}}.$$

The relations defining α, β, θ give

$$\lambda = 4p(\alpha) - \frac{a^2 + b^2 + c^2}{3},$$

$$\mu = 4p(\beta) - \frac{a^2 + b^2 + c^2}{3},$$

$$\nu = 4p(\theta) - \frac{a^2 + b^2 + c^2}{3}.$$

Thus

$$\nabla^2 \Phi = \frac{[p(\theta) - p(\beta)] \frac{\partial^2 \Phi}{\partial \alpha^2} + [p(\alpha) - p(\theta)] \frac{\partial^2 \Phi}{\partial \beta^2} + [p(\beta) - p(\alpha)] \frac{\partial^2 \Phi}{\partial \theta^2}}{4 [p(\alpha) - p(\beta)] [p(\beta) - p(\theta)] [p(\theta) - p(\alpha)]}.$$

Bipolar coordinates (ξ, η, z)

$$x = \frac{c \sinh \xi}{\cosh \xi - \cos \eta}, \quad u = \xi, \ v = \eta, \ w = z,$$

$$y = \frac{c \sin \eta}{\cosh \xi - \cos \eta},$$

$$z = z,$$

$$U = V = \frac{1}{c} (\cosh \xi - \cos \eta), \ W = 1.$$

The domain of the variables ξ, η, z is given by

$$-\infty < \xi < \infty,$$

$$0 \leq \eta \leq 2\pi,$$

$$-\infty < z < \infty.$$

The surfaces $\xi = \xi_0$ (const.) are circular cylinders with generators parallel to z-axis and given by

$$(x - c \coth \xi_0)^2 + y^2 = \frac{c^2}{\sinh^2 \xi_0}.$$

The surfaces $\eta = \eta_0$ (const.) are also circular cylinders given by

$$x^2 + (y - c \cot \eta_0)^2 = c^2 \operatorname{cosec}^2 \eta_0,$$

$$ds^2 = \frac{c^2}{(\cosh \xi - \cos \eta)^2} (d\xi^2 + d\eta^2) + dz^2,$$

$$\nabla \Phi = \left(\frac{\cosh \xi - \cos \eta}{c} \frac{\partial \Phi}{\partial \xi}, \frac{\cosh \xi - \cos \eta}{c} \frac{\partial \Phi}{\partial \eta}, \frac{\partial \Phi}{\partial z} \right),$$

$$\nabla \cdot \vec{F} = \frac{(\cosh \xi - \cos \eta)^2}{c^2} \left\{ \frac{\partial}{\partial \xi} \left(\frac{c F_1}{\cosh \xi - \cos \eta} \right) + c \frac{\partial}{\partial \eta} \left(\frac{F_2}{\cosh \xi - \cos \eta} \right) + \frac{c^2}{(\cosh \xi - \cos \eta)^2} \frac{\partial F_3}{\partial z} \right\},$$

$$\nabla^2 \Phi = \frac{c^2}{(\cosh \xi - \cos \eta)^2} \left(\frac{\partial^2 \Phi}{\partial \xi^2} + \frac{\partial^2 \Phi}{\partial \eta^2} \right) + \frac{\partial^2 \Phi}{\partial z^2},$$

$$(\nabla\times\vec{F})_\xi = \frac{\cosh\xi-\cos\eta}{c}\frac{\partial F_3}{\partial\eta} - \frac{\partial F_2}{\partial z},$$

$$(\nabla\times\vec{F})_\eta = \frac{\partial F_1}{\partial z} - \frac{\cosh\xi-\cos\eta}{c}\frac{\partial F_3}{\partial\xi},$$

$$(\nabla\times\vec{F})_z = \frac{(\cosh\xi-\cos\eta)^2}{c}\left\{\frac{\partial}{\partial\xi}\left(\frac{F_2}{\cosh\xi-\cos\eta}\right) - \frac{\partial}{\partial\eta}\left(\frac{F_1}{\cosh\xi-\cos\eta}\right)\right\},$$

$$F_1 = F_x\frac{1-\cosh\xi\cos\eta}{\cosh\xi-\cos\eta} - F_y\frac{\sinh\xi\sin\eta}{\cosh\xi-\cos\eta},$$

$$F_2 = -F_x\frac{\sinh\xi\sin\eta}{\cosh\xi-\cos\eta} + F_y\frac{\cosh\xi\cos\eta-1}{\cosh\xi-\cos\eta},$$

$$F_3 = F_z.$$

12.2 Examples of separation of variables

An important application of the results is the solution of differential equations by the method of separation of variables*. Here the discussion is limited to the following differential equations:

1. Wave equation

$$\nabla^2\psi = \frac{1}{c^2}\frac{\partial^2\psi}{\partial t^2}.$$

2. Equation of heat conduction

$$\nabla^2\psi = \frac{1}{a^2}\frac{\partial\psi}{\partial t}.$$

The method of separation of variable assumes that $\psi(x, y, z, t)$ can be expressed as a product of two functions $f(x, y, z)$ and $\Phi(t)$, where f depends only on the space coordinates xyz and Φ only on time t. The wave equation then becomes

$$\frac{1}{f}\nabla^2 f = \frac{1}{c^2}\frac{\Phi''}{\Phi},$$

which is equivalent to the two differential equations

$$\nabla^2 f + k^2 f = 0$$

and

$$\frac{d^2\Phi}{dt^2} + c^2k^2\Phi = 0,$$

where k is an arbitrary constant.

Clearly

$$\Phi(t) = e^{\pm ickt} = e^{\pm i\omega t},$$

* Such a solution of the differential equation is also called the Bernoulli trial solution.

where the frequency ω, the wave number k and the propagation speed c are related by the equation

$$\omega = kc.$$

The differential equation $\nabla^2 f + k^2 f = 0$ is Called the time independent wave equation. A similar analysis for the heat equation gives

$$\Phi(t) = e^{-k^2 a^2 t}$$

and

$$\nabla^2 f + k^2 f = 0.$$

To obtain a solution of the time independent wave equation in cartesian coordinates by the method of separation of variables, it is assumed that

$$f(x, y, z) = f_1(x)\, f_2(y)\, f_3(z).$$

The wave equation now becomes

$$\frac{1}{f_1}\frac{d^2 f_1}{dx^2} + \frac{1}{f_2}\frac{d^2 f_2}{dy^2} + \frac{1}{f_3}\frac{d^2 f_3}{dz^2} + k^2 = 0,$$

which gives

$$f_1(x) = e^{\pm i\alpha x}, \quad f_2(y) = e^{\pm i\beta y}, \quad f_3(z) = e^{\pm i\sqrt{k^2-\alpha^2-\beta^2}\, z},$$

where α, β are arbitrary (separation) constants. If one introduces arbitrary orthogonal curvilinear coordinates u, v, w in place of the cartesian coordinates xyz, the function f in the time independent wave equation becomes a function of u, v, w and the wave equation becomes

$$\nabla^2 f(u, v, w) + k^2 f(u, v, w) = 0$$

in which the independent variables are u, v, w. Under the assumption $f(u, v, w) = f_1(u)\, f_2(v)\, f_3(w)$, (i.e. provided the desired function $f(u, v, w)$ may be expressed as a product of three functions $f_1(u), f_2(v), f_3(w)$ each one of which depends on only one of the three variables u, v, w), are stated for several special coordinate systems the differential equations that must be satisfied by $f_1(u), f_2(v), f_3(w)$ and the linearly independent solutions of each one of those three differential equations.

Cylindrical coordinates (ϱ, φ, z)

$$u = \varrho, \; v = \varphi, \; w = z,$$

$$\nabla^2 f + k^2 f \equiv \frac{\partial^2 f}{\partial \varrho^2} + \frac{1}{\varrho}\frac{\partial f}{\partial \varrho} + \frac{1}{\varrho^2}\frac{\partial^2 f}{\partial \varphi^2} + \frac{\partial^2 f}{\partial z^2} + k^2 f = 0,$$

$$f \equiv f_1(\varrho)\, f_2(\varphi)\, f_3(z),$$

$$\frac{d^2 f_1}{d\varrho^2} + \frac{1}{\varrho}\frac{d f_1}{d\varrho} + \left(k^2 - \alpha^2 - \frac{\mu^2}{\varrho^2}\right) f_1 = 0,$$

$$f_1(\varrho) = C_\mu\left(\varrho\sqrt{k^2-\alpha^2}\right),$$
$$\frac{d^2 f_2}{d\varphi^2} = -\mu^2 f_2, \qquad f_2(\varphi) = e^{\pm i\mu\varphi},$$
$$\frac{d^2 f_3}{dz^2} = -\alpha^2 f_3, \qquad f_3(z) = e^{\pm i\alpha z},$$

where α, μ are arbitrary constants and $C_\mu(z)$ are the cylinder functions (see chap. III).

The plane wave and the spherical wave may be expressed in terms of particular solutions of the wave equation in cylindrical coordinates as follows:

Plane wave:

$$e^{ikx} = e^{ik\varrho\cos\varphi} = \sum_{n=0}^{\infty} i^n \varepsilon_n J_n(k\varrho)\cos(n\varphi),$$
$$\varepsilon_0 = 1,\ \varepsilon_n = 2 \text{ for } n \geq 1.$$

Spherical wave:

$$\frac{e^{ikR}}{R} = \frac{i}{2}\sum_{n=0}^{\infty}\varepsilon_n \cos n(\varphi-\varphi_0)\int_{\alpha=-\infty}^{\infty} J_n\left(\varrho\sqrt{k^2-\alpha^2}\right) H_n^{(1)}\left(\varrho_0\sqrt{k^2-\alpha^2}\right) \times e^{-i\alpha|z-z_0|}\,d\alpha,$$
$$\varrho < \varrho_0,$$
$$= \frac{i}{2}\sum_{n=0}^{\infty}\varepsilon_n \cos n(\varphi-\varphi_0)\int_{-\infty}^{\infty} J_n\left(\varrho_0\sqrt{k^2-\alpha^2}\right) H_n^{(1)}\left(\varrho\sqrt{k^2-\alpha^2}\right) \times e^{-i\alpha|z-z_0|}\,d\alpha,$$
$$\varrho > \varrho_0,$$
$$= \sum_{n=0}^{\infty}\varepsilon_n \cos n(\varphi-\varphi_0)\int_{\alpha=0}^{\infty} J_n(\alpha\varrho)\, J_n(\alpha\varrho_0)\, e^{-|z-z_0|\sqrt{\alpha^2-k^2}}\frac{\alpha\,d\alpha}{\sqrt{\alpha^2-k^2}},$$

where

$$R = [\varrho^2 + \varrho_0^2 - 2\varrho\varrho_0\cos(\varphi-\varphi_0) + (z-z_0)^2]^{\frac{1}{2}}.$$

Spherical coordinates (r, θ, φ)

$$u = r,\ v = \theta,\ w = \varphi,$$
$$\nabla^2 f + k^2 f \equiv \frac{1}{r^2}\frac{\partial}{\partial r}\left(r^2\frac{\partial f}{\partial r}\right) + \frac{1}{r^2\sin\theta}\frac{\partial}{\partial\theta}\left(\sin\theta\frac{\partial f}{\partial\theta}\right) + \frac{1}{r^2\sin^2\theta}\frac{\partial^2 f}{\partial\varphi^2} + k^2 f = 0,$$
$$f(r,\theta,\varphi) = f_1(r)\, f_2(\theta)\, f_3(\varphi),$$
$$\frac{d^2 f_1}{dr^2} + \frac{2}{r}\frac{df_1}{dr} + \left(k^2 - \frac{\nu(\nu+1)}{r^2}\right) f_1 = 0,$$

$$f_1(r) = r^{-\frac{1}{2}} C_{\nu+\frac{1}{2}}(kr),$$

$$\sin^2\theta \frac{d^2 f_2}{d\theta^2} + \sin\theta\cos\theta \frac{d f_2}{d\theta} + [\nu(\nu+1)\sin^2\theta - \mu^2] f_2 = 0,$$

$$f_2(\theta) = P_\nu^\mu(\cos\theta),$$

$$\frac{d^2 f_3}{d\varphi^2} + \mu^2 f_3 = 0,$$

$$f_3(\varphi) = e^{\pm i\mu\varphi}.$$

Representation of the plane and spherical waves by particular solutions of the wave equation in spherical coordinates.

Plane waves:

$$e^{ikz} = e^{ikr\cos\theta} = \sqrt{\frac{\pi}{2k}} \sum_{n=0}^{\infty} i^n (2n+1) \frac{1}{\sqrt{r}} J_{n+\frac{1}{2}}(kr) P_n(\cos\theta).$$

Spherical wave:

$$\frac{e^{ikR}}{R} = i\frac{\pi}{2} \sum_{n=0}^{\infty} \frac{m^2}{\sqrt{r r_0}} J_{n+\frac{1}{2}}(kr) H^{(1)}_{n+\frac{1}{2}}(kr_0) P_n(\cos\gamma), \quad r < r_0,$$

$$= i\frac{\pi}{2} \sum_{n=0}^{\infty} \frac{m^2}{\sqrt{r r_0}} J_{n+\frac{1}{2}}(kr_0) H^{(1)}_{n+\frac{1}{2}}(kr) P_n(\cos\gamma), \quad r > r_0,$$

where

$$m = 2n + 1;$$

$$R = (r^2 + r_0^2 - 2rr\cos\gamma)^{\frac{1}{2}}; \ \cos\gamma = \cos\theta\cos\theta_0 + \sin\theta\sin\theta_0\cos(\varphi - \varphi_0),$$

$$P_n(\cos\gamma) = \sum_{l=0}^{n} \varepsilon_l \frac{(n-l)!}{(n+l)!} P_n^l(\cos\theta) P_n^l(\cos\theta_0) \cos l(\varphi - \varphi_0),$$

$$\varepsilon_0 = 1, \varepsilon_n = 2 \ \text{for} \ n \geq 1.$$

Parabolic coordinates (ξ, η, φ)

$$u = \xi, \ v = \eta, \ w = \varphi.$$

The wave equation $\nabla^2 f + k^2 f = 0$ becomes

$$\frac{\partial^2 f}{\partial\xi^2} + \frac{1}{\xi}\frac{\partial f}{\partial\xi} + \frac{\partial^2 f}{\partial\eta^2} + \frac{1}{\eta}\frac{\partial f}{\partial\eta} + \frac{(\xi^2+\eta^2)}{\xi^2\eta^2}\frac{\partial^2 f}{\partial\varphi^2} + k^2(\xi^2+\eta^2) f = 0,$$

$$f(\xi, \eta, \varphi) = f_1(\xi) f_2(\eta) f_3(\varphi),$$

$$\frac{d^2 f_1}{d\xi^2} + \frac{1}{\xi}\frac{d f_1}{d\xi} + \left(k^2\xi^2 - \frac{\mu^2}{\xi^2} + \lambda\right) f_1 = 0,$$

$$f_1(\xi) = \xi^\mu e^{\pm i\frac{k}{2}\xi^2} {}_1F_1\left(\frac{1+\mu}{2} - \frac{i\lambda}{4k}; 1+\mu; \mp ik\xi^2\right)$$

or

$$f_1(\xi) = \frac{1}{\xi} W_{\frac{\lambda}{4ik}, \frac{\mu}{2}} (ik\xi^2),$$

$$\frac{d^2 f_2}{d\eta^2} + \frac{1}{\eta}\frac{df_2}{d\eta} + \left(k^2\eta - \frac{\mu^2}{\eta^2} - \lambda\right) f_2 = 0,$$

$$f_2(\eta) = \eta^\mu e^{\pm i \frac{k}{2}\eta^2} {}_1F_1\left(\frac{1+\mu}{2} + \frac{i\lambda}{4k}; 1+\mu; \mp ik\eta^2\right)$$

or

$$f_2(\eta) = \eta^{-1} W_{\frac{i\lambda}{4k}, \frac{\mu}{2}} (ik\eta^2),$$

$$\frac{d^2 f_3}{d\varphi^2} + \mu^2 f_3 = 0, \; f_3(\varphi) = e^{\pm i\mu\varphi}.$$

For Laplace's equation, i.e. $\nabla^2 f = 0$

$$f_1(\xi) = C_\mu(\sqrt{\lambda}\,\xi),$$
$$f_2(\eta) = C_\mu(i\sqrt{\lambda}\,\xi),$$
$$f_3(\varphi) = e^{\pm i\mu\varphi},$$

where $C_\mu(z)$ are the cylinder functions.

Parabolic cylinder coordinates (ξ, η, z)

$$u = \xi, \; v = \eta, \; w = z.$$

The wave equation $\nabla^2 f + k^2 f = 0$ becomes

$$\frac{\partial^2 f}{\partial \xi^2} + \frac{\partial^2 f}{\partial \eta^2} + (\xi^2 + \eta^2)\frac{\partial^2 f}{\partial z^2} + k^2(\xi^2 + \eta^2) f = 0,$$

$$f(\xi, \eta, z) = f_1(\xi)\, f_2(\eta)\, f_3(z),$$

$$\frac{d^2 f_1}{d\xi^2} + (\lambda + l^2\xi^2) f_1 = 0; \;\; l^2 = k^2 - \alpha^2,$$

$$f_1(\xi) = D_{-\frac{l+i\lambda}{2l}} [\pm \xi \sqrt{l}\,(1+i)],$$

$$\frac{d^2 f_2}{d\eta^2} + (l^2\eta^2 - \lambda) f_2 = 0; \;\; l^2 = k^2 - \alpha^2;$$

$$f_2(\eta) = D_{-\frac{l-i\lambda}{2l}} [\pm \eta \sqrt{l}\,(1+i)],$$

$$\frac{d^2 f_3}{dz^2} + \alpha^2 f_3 = 0; \;\; f_3(z) = e^{\pm i\alpha z}.$$

Representation of a cylindrical wave:

$$H_0^{(2)}(k\varrho) = H_0^{(2)}\left(k\,\frac{\xi^2+\eta^2}{2}\right)$$
$$= \frac{1}{\pi^2\sqrt{2}} \int_{\sigma-i\infty}^{\sigma+i\infty} D_\nu\left[(1+i)\sqrt{k}\,\xi\right] D_{-\nu-1}\left[(1+i)\sqrt{k}\,\eta\right]$$
$$\times \Gamma\left(-\frac{\nu}{2}\right)\Gamma\left(\frac{1+\nu}{2}\right) d\nu,$$
$$\nu = \sigma + i\tau;\quad -1 < \sigma < 0;\quad Re\,(ik\xi^2) \geq 0;\quad Re\,(ik\eta^2) \geq 0.$$

More generally the partial differential equation $\nabla^2 f + \lambda(z)\, f = 0$ in parabolic cylindrical coordinates becomes

$$(\xi^2+\eta^2)^{-1}\left(\frac{\partial^2 f}{\partial\xi^2} + \frac{\partial^2 f}{\partial\eta^2}\right) + \frac{\partial^2 f}{\partial z^2} + \lambda(z)\, f = 0,$$
$$f(\xi,\eta,z) = f_1(\xi)\, f_2(\eta)\, f_3(z),$$
$$\frac{d^2 f_1}{d\xi^2} + (l^2\xi^2 + \lambda_0)\, f_1 = 0,$$
$$\frac{d^2 f_2}{d\eta^2} + (l^2\eta^2 - \lambda_0)\, f_2 = 0,$$
$$\frac{d^2 f_3}{dz^2} + [\lambda(z) - l^2]\, f_3 = 0,$$

where l, λ_0 are arbitrary constants. The solutions of the differential equations for f_1 and f_2 are again Parabolic cylinder functions.

Elliptic cylinder coordinates (ξ, η, z)

$$u = \xi,\quad v = \eta,\quad w = z.$$

The wave equation $\nabla^2 f + k^2 f = 0$ becomes

$$\frac{\partial^2 f}{\partial\xi^2} + \frac{\partial^2 f}{\partial\eta^2} + \frac{c^2}{2}(\cosh 2\xi - \cos 2\eta)\frac{\partial^2 f}{\partial z^2} + \frac{k^2 c^2}{2}(\cosh 2\xi - \cos 2\eta)\, f = 0,$$
$$f(\xi,\eta,z) = f_1(\xi)\, f_2(\eta)\, f_3(z),$$
$$\frac{d^2 f_1}{d\xi^2} + \left[-\lambda + \frac{c^2}{2}(k^2-\alpha^2)\cosh 2\xi\right] f_1 = 0,$$
$$\frac{d^2 f_2}{d\eta^2} + \left[\lambda - \frac{c^2}{2}(k^2-\alpha^2)\cos 2\eta\right] f_2 = 0,$$
$$\frac{d^2 f_3}{dz^2} + \alpha^2 f_3 = 0;\quad f_3(z) = e^{\pm i\alpha z}.$$

The differential equations for f_1 and f_2 are called the Mathieu differential equations and their solutions, the Mathieu functions. The differential equation for f_2 can be obtained from the one for f_1 by putting $-i\eta$ inplace of ξ. The corresponding statement for the solutions of the two differential equations is also true. For more on Mathieu functions see McLachlan (1947), Erdélyi (1955) and Strutt (1932).

Elliptic coordinates (ξ, η, φ) (*Prolate ellipsiodal*)

$$u = \xi, \; v = \eta, \; w = \varphi.$$

The wave equation $\nabla^2 f + k^2 f = 0$ becomes

$$\frac{\partial}{\partial \xi}\left[(\xi^2 - 1)\frac{\partial f}{\partial \xi}\right] + \frac{\partial}{\partial \eta}\left[(1 - \eta^2)\frac{\partial f}{\partial \eta}\right]\left(\frac{1}{\xi^2 - 1} + \frac{1}{1 - \eta^2}\right)\frac{\partial^2 f}{\partial \varphi^2} + k^2 c^2 (\xi^2 - \eta^2) f = 0,$$

$$f(\xi, \eta, \varphi) = f_1(\xi)\, f_2(\eta)\, f_3(\varphi),$$

$$\frac{d}{d\xi}\left[(1 - \xi^2)\frac{df_1}{d\xi}\right] + \left(\lambda - k^2 c^2 \xi^2 - \frac{\mu^2}{1 - \xi^2}\right) f_1 = 0,$$

$$\frac{d}{d\eta}\left[(1 - \eta^2)\frac{df_2}{d\eta}\right] + \left(\lambda - k^2 c^2 \eta^2 - \frac{\mu^2}{1 - \eta^2}\right) f_2 = 0,$$

$$\frac{d^2 f_3}{d\varphi^2} + \mu^2 f_3 = 0; \; f_3(\varphi) = e^{\pm i\mu\varphi}.$$

For Laplace's equation, i.e. $k = 0$, and $\lambda = \nu(\nu + 1)$ the solutions of the differential equations for f_1 and f_2 are the Legendre functions (see chap. IV). For $k \neq 0$ the differential equations for f_1 and f_2 are the so-called Lamé differential equations and their solutions, the Lamé wave functions. For further information see ERDÉLYI (1955) and STRUTT (1932).

Elliptic coordinates (ξ, η, φ) (*Oblate ellipsoidal*)

$$u = \xi, \; v = \eta, \; w = \varphi.$$

The wave equation $\nabla^2 f + k^2 f = 0$ becomes

$$\frac{\partial}{\partial \xi}\left[(1 + \xi^2)\frac{\partial f}{\partial \xi}\right] + \frac{\partial}{\partial \eta}\left[(1 - \eta^2)\frac{\partial f}{\partial \eta}\right] + \left(\frac{1}{1 - \eta^2} - \frac{1}{1 + \xi^2}\right)\frac{\partial^2 f}{\partial \varphi^2} + k^2 c^2 (\xi^2 + \eta^2) f = 0,$$

$$f(\xi, \eta, \varphi) = f_1(\xi)\, f_2(\eta)\, f_3(\varphi),$$

$$\frac{d}{d\xi}\left[(1 + \xi^2)\frac{df_1}{d\xi}\right] + \left(-\lambda + k^2 c^2 \xi^2 + \frac{\mu^2}{1 + \xi^2}\right) f_1 = 0,$$

$$\frac{d}{d\eta}\left[(1 - \eta^2)\frac{df_2}{d\eta}\right] + \left(\lambda + k^2 c^2 \eta^2 - \frac{\mu^2}{1 - \eta^2}\right) f_2 = 0,$$

$$\frac{d^2 f_3}{d\varphi^2} + \mu^2 f_3 = 0; \; f_3(\varphi) = e^{\pm i\mu\varphi}.$$

The differential equations for f_1 and f_2 are again the Lamé differential equations. The differential equation for f_2 may be obtained from that f_1 by putting $-i\eta$ inplace of ξ.

Torus coordinates (ξ, η, φ)

$$u = \xi, \; v = \eta, \; w = \varphi.$$

The wave equation $\nabla^2 f + k^2 f = 0$ becomes

$$\frac{\partial}{\partial \xi}\left(\frac{\sinh \xi}{\cosh \xi - \cos \eta} \frac{\partial f}{\partial \xi}\right) + \frac{\partial}{\partial \eta}\left(\frac{\sinh \xi}{\cosh \xi - \cos \eta} \frac{\partial f}{\partial \eta}\right) + \frac{1}{\sinh \xi (\cosh \xi - \cos \eta)} \frac{\partial^2 f}{\partial \varphi^2} + \frac{k^2 c^2 \sinh \xi}{(\cosh \xi - \cos \eta)^3} f = 0.$$

This differential equation is not separable for $k \neq 0$. For $k = 0$ (Laplace's equation) one has

$$\frac{\partial}{\partial \xi}\left(\frac{\sinh \xi}{\cosh \xi - \cos \eta} \frac{\partial f}{\partial \xi}\right) + \frac{\partial}{\partial \eta}\left(\frac{\sinh \xi}{\cosh \xi - \cos \eta} \frac{\partial f}{\partial \eta}\right) + \frac{1}{\sinh \xi (\cosh \xi - \cos \eta)} \frac{\partial^2 f}{\partial \varphi^2} = 0.$$

Making the change of variables

$$\cosh \xi = s \text{ and } f(\xi, \eta, \varphi) = \sqrt{\cosh \xi - \cos \eta}\, g(s, \eta, \varphi)$$

the Laplace's equation becomes

$$\frac{\partial}{\partial s}\left[(s^2 - 1) \frac{\partial g}{\partial s}\right] + \frac{\partial^2 g}{\partial \eta^2} + \frac{1}{s^2 - 1} \frac{\partial^2 g}{\partial \varphi^2} + \frac{1}{4} g = 0,$$

$$g(s, \eta, \varphi) = f_1(s)\, f_2(\eta)\, f_3(\varphi),$$

$$(1 - s^2) \frac{d^2 f_1}{ds^2} - 2s \frac{d f_1}{ds} + \left[\left(\nu^2 - \frac{1}{4}\right) - \frac{\mu^2}{1 - s^2}\right] f_1 = 0,$$

which is Legendre's differential equation and has its solutions the Legendre functions (see chap. IV)

$$\frac{d^2 f_2}{d\eta^2} + \nu^2 f_2 = 0; \; f_2(\eta) = e^{\pm i \nu \eta},$$

$$\frac{d^2 f_3}{d\varphi^2} + \mu^2 f_3 = 0; \; f_3(\varphi) = e^{\pm i \mu \varphi}.$$

Literature

ERDÉLYI, A.: [1] Higher transcendental functions, Vol. 2. New York: McGraw-Hill 1953.

— Vol. 3 (1955).

McLACHLAN, N. W.: Theory and application of Mathieu functions. Oxford: Oxford Univ. Press. 1947.

MORSE, P. M., and H. FESHBACH: Methods of theoretical physics, part I. New York: Mc-Graw-Hill 1953.

STRUTT, M. J. O.: Ergebnisse der Mathematik und ihrer Grenzgebiete, Vol. 1, no. 3. Berlin: Springer 1932.

List of special symbols

Let

z	$= x + iy$ be a complex number. Then
$Re\, z$	$= x$, Real part of z.
$Im\, z$	$= y$, Imaginary part of z.
$\bar{z}$	$= x - iy$, complex conjugate of z.
$\lvert z\rvert$	$= (x^2 + y^2)^{\frac{1}{2}}$, Absolute value of z.
$\arg z$	= argument of z.
$\log z$, $\ln z$	= Principal value of the natural logarithm of z; $\log z = \log \lvert z\rvert + i \arg z$ with $-\pi < \arg z < \pi$; if z is real and negative, it is always specified whether $\arg z = \pi$ or $-\pi$ is to be taken.
z^α	$= \exp(\alpha \log z)$.
$[x]$	= largest integer less than or equal to the real number x.
$\operatorname{sgn} x$	$= \begin{cases} 1, & x > 0 \\ 0, & x = 0 \\ -1, & x < 0 \end{cases}$ sign of the real number x.
$f(a + 0)$	$= \lim_{\varepsilon\to 0} f(a + \varepsilon)$, $\varepsilon > 0$; the limiting value of $f(x)$ as x approaches a through values of $x > a$.
$f(a - 0)$	$= \lim_{\varepsilon\to 0} f(a - \varepsilon)$, $\varepsilon > 0$; the limiting value of $f(x)$ as x approaches a through values of $x < a$.
$f(x \pm 0i)$	$= \lim_{\varepsilon\to 0} f(x \pm i\varepsilon)$, $\varepsilon > 0$.
n	= means a non-negative integer unless stated otherwise.
$n!$	$= 1, 2, 3, \ldots, n$.
$0!$	$= 1$.
$(a)_n$	$= a(a + 1)(a + 2) \cdots (a + n - 1)$ for $n = 1, 2, 3, \ldots$
$(a)_0$	$= 1$.
$\binom{\lambda}{n}$	$= \dfrac{\lambda(\lambda - 1)(\lambda - 2) \cdots (\lambda - n + 1)}{n!} = (-1)^n \dfrac{(-\lambda)_n}{n!}$.
D_x^α	= Fractional derivative operator.
ε^n	$= \begin{cases} 1 & n = 0 \\ 2 & n = 1, 2, 3, \ldots, \end{cases}$ Neumann symbol.
(λ, n)	$= 2^{-2n}\, n! \prod_{k=1}^{n} [4\lambda^2 - (2k - 1)^2]$, $= \dfrac{\left(\lambda + \frac{1}{2}\right)_n \left(\lambda - n + \frac{1}{2}\right)_n}{n!}$, $n = 1, 2, 3, \ldots$
$(\lambda, 0)$	$= 1$.
$\gamma_{\nu,m}$	$= m^{\text{th}}$ +ve root of $J_\nu(x) = 0$.
$\lambda_{\nu,m}$	$= m^{\text{th}}$ +ve root of $x J'_\nu(x) + a J_\nu(x) = 0$.
Σ'	= restricted sum.
$o[f(x)]$	= order of magnitude notation; if $\lim_{x\to x_0} \dfrac{g(x)}{f(x)} = 0$ in some neighborhood of $x = x_0$ we write $g(x) = o[f(x)]$. In general the point x_0 is the point at ∞.

$O[f(z)]$ = Order of magnitude of $f(z)$. When z approaches a limiting value z_0 (usually z_0 equals ∞; the limiting value z_0 is always evident from the context), we write $g(z) = O[f(z)]$, if there exists a real non-negative constant M, such that in a sufficiently small neighborhood of $z = z_0$ everywhere $|g(z)| \leq M\,|f(z)|$.

$\gg$ = "large compared to"
$\ll$ = "small compared to"
expressions used to indicate the applicability of approximation formulas.

$\sim$ = "Approximately equal" in formulas without explicit estimation of the error, used primarily when giving the first term of an asymptotic (semi-convergent) expansion for a function.

$\approx$ = "Asymptotically equal." The symbol is used when giving a semiconvergent expansion for a function.

List of functions

Symbol	Name of the function	Chapter and section
$A_{n,\nu}(z)$	—	3.11.2
$Ai(z)$	Airy function	3.4
$\alpha_n(x)$	—	9.2.1
$B(x, y)$	Beta function	1.1
$B(x, y, \alpha)$	Incomplete beta function	9.2.5
B_n	Bernoulli numbers	1.5.1
B_n^m	Generalized Bernoulli numbers	1.5.1
$B_n(x)$	Bernoulli polynomials	1.5.1
$B_n^m(x)$	Generalized Bernoulli polynomials	1.5.1
$B_{n;\mu,\nu}(z)$	—	3.11.2
$\beta_n(x)$	—	9.2.1
$Bi(z)$	Airy function	3.4
$\mathrm{bei}_\nu(z)$, $\mathrm{bei}(z)$ $\mathrm{ber}_\nu(z)$, $\mathrm{ber}(z)$	Kelvin functions	3.16.1
$C(x)$ $C_1(x)$ $C_2(x)$	Fresnel cosine integral	9.2.4
$C_\nu(z)$	Bessel, Neumann or Hankel function	3.10.2
$C_n^{(\lambda)}(x)$	Gegenbauer or Ultraspherical polynomials	5.3
$Ci(x)$	cosine integral	9.2.2
$Chi(x)$	Hyperbolic cosine integral	9.2.2
$\mathrm{cn}(z, k)$, $\mathrm{dn}(z, k)$ $\mathrm{sn}(z, k)$	Jacobi elliptic functions	10.3
$\mathrm{cs}\, z = \frac{\mathrm{cn}\, z}{\mathrm{sn}\, z}$ $\mathrm{cd}\, z = \frac{\mathrm{cn}\, z}{\mathrm{dn}\, z}$	Jacobi elliptic functions	10.3
$\mathrm{dc}\, z = \frac{\mathrm{dn}\, z}{\mathrm{cn}\, z}$ $\mathrm{dn}\, z$ $\mathrm{ds}\, z = \frac{\mathrm{dn}\, z}{\mathrm{sn}\, z}$	Jacobi elliptic functions	10.3
$D_\nu(z)$	Parabolic cylinder functions	8.1
e_1, e_2, e_3	—	10.5
$E(k)$	Complete elliptic integral of the 2nd kind	10.1

Symbol	Name of the function	Chapter and section
$E(k, \varphi)$	Incomplete elliptic integral of the 2nd kind	10.1
$E(\alpha, \beta::z)$	MacRobert's E-function	6.7.2
$\mathrm{Erf}(z)$	Error function	9.2.3
$\mathrm{Erfc}(z)$	Complementary error function	9.2.3
$\{1\}^n \mathrm{Erfc}(z)$	Repeated integrals of error functions	9.2.3
$E^*(x)$, $Ei(x)$, $E_1(x)$	Exponential integrals	9.2.1
E_n	Euler numbers	1.5.2
$E_n(x)$	Euler polynomials	1.5.2
E_n^m	Generalized Euler numbers	1.5.2
$E_n^m(x)$	Generalized Euler polynomials	1.5.2
$\mathsf{E}_n(z)$	Weber's function	3.10.4
$F(k, \varphi)$	Elliptic integral of the 1st kind	10.1
$F_1(\alpha, \beta, z)$	Meixner's function	6.7.2
$F_n(z)$	—	2.10
$F_k(\eta)$	Fermi-Dirac function	1.6
$F(a, b; c; z)$, ${}_2F_1(a, b; c; z)$	Gauss hypergeometric series	2.1.1
${}_1F_1(a; c; z)$	Confluent hypergeometric function or Kummer's function	6.1
${}_pF_q$	Generalized hypergeometric series	2.9
$\Gamma(z)$	Gamma function	1.1
γ	Euler's constant	
$\Gamma(a, x)$, $\gamma(a, x)$, $\gamma^*(a, x)$	Incomplete gamma function	9.1.1
$G_n(p, q, x)$	—	5.2.1
g_2, g_3	—	10.5
$H_n\ x)$, $He_n(x)$	Hermite polynomials	5.6.1
$H_n(\xi, \alpha, z)$	—	2.10
$\mathsf{H}_\nu(z)$	Struve's function	3.10.3
$H_\nu^{(1)}(z)$, $H_\nu^{(2)}(z)$	Hankel functions	3.1

Symbol	Name of the function	Chapter and section
$I(x, y, \alpha)$	Normalized incomplete beta function	9.2.5
$I_\nu(z)$	Modified Bessel function	3.1
$I_{\mu,\nu}$	Hardy's integral	3.8.5
$\{1\}^n \operatorname{Erfc}(z)$	Repeated integrals of the error function	9.2.3
$J_\nu(z)$	Bessel functions of the 1st kind	3.1
$\mathsf{J}_\nu(z)$	Anger's function	3.10.4
$J_n^{u,v}(z)$	—	2.10
k	Modulus of Jacobi elliptic functions and integrals	10.1
$K(k)$	Complete elliptic integral of the 1st kind	10.1
$K_\nu(z)$	Modified Bessel function	3.1
$k_\nu(z)$	Bateman's function	6.7.2
$\operatorname{kei}_\nu(z)$, $\operatorname{kei}(z)$ $\operatorname{ker}_\nu(z)$, $\operatorname{ker}(z)$	Kelvin functions	3.16.1
$L_n(x)$	Laguerre polynomials	5.5.1
$L_n^{(\alpha)}(x)$	Generalized Laguerre polynomials	5.5.1
$L_\nu^{(\alpha)}(z)$	Laguerre functions	6 (App.)
$\mathsf{L}_\nu(z)$	Modified Struve function	3.10.3
$li(x)$	Logarithmic-integral function	9.2.1
$M_{\varkappa,\mu}(z)$	Whittaker functions	7.1
$\mathcal{M}_{\varkappa,\mu}(z)$	—	7.1.1
$N_{\varkappa,\mu}(z)$	Confluent hypergeometric functions	7.1.1
$\operatorname{nc} z = \frac{1}{\operatorname{cn} z}$ $\operatorname{nd} z = \frac{1}{\operatorname{dn} z}$ $\operatorname{ns} z = \frac{1}{\operatorname{sn} z}$	Jacobi elliptic functions	10.3
$O_n(z)$	Neumann polynomials	3.11.2
$\Omega_n(z)$	Neumann polynomials	3.11.2
$\wp(z)$	Weierstrass elliptic function	10.5
$\psi(z)$	Euler psi-function	1.2
$\psi_n(z)$	Spherical Bessel functions	3.3
$\Phi(z, s, \alpha)$	Lerch's transcendent	1.6
$\Pi(\varphi, n, k)$	Elliptic integral of the 3rd kind	10.1

Symbol	Name of the function	Chapter and section
$P\begin{Bmatrix} a & b & c & \\ \alpha & \beta & \gamma & z \\ \alpha' & \beta' & \gamma' & \end{Bmatrix}$	Riemann's differential equation	2.7
$p_n(z)$	Poisson-Charlier polynomials	6.7.2
$P_n(x)$	Legendre polynomials	5.4
$P_\nu^\mu(x)$	Associated Legendre functions of the 1st kind	4.1.1
$\mathfrak{P}_\nu(z)$	Associated Legendre functions of the 1st kind	4.4.2
$\mathfrak{P}_\nu^\mu(z)$	Associated Legendre functions of the 1st kind	4.1.1
$P_n^{(\alpha,\beta)}(x)$	Jacobi polynomials	5.2.1
$Q_n(x)$	Legendre function of the 2nd kind	5.4.2
$Q_\nu^\mu(x)$	Associated Legendre function of the 2nd kind	4.1.1
$\mathfrak{Q}_\nu(z)$	Associated Legendre function of the 2nd kind	4.4.2
$\mathfrak{Q}_\nu^\mu(z)$	Associated Legendre function of the 2nd kind	4.4.2
$Q_n^{(\alpha,\beta)}(x)$	Jacobi function of the 2nd kind	5.2.2
$R_{m,\nu}(z)$	Lommel polynomials	3.11.1
$S(z)$, $S_1(z)$, $S_2(z)$	Fresnel sine integrals	9.2.4
$si(x)$, $Si(x)$	Sine integrals	9.2.2
$Shi(x)$	Hyperbolic sine integral	9.2.2
$S_n(z)$	Schlaefli's polynomials	3.11.3
$S_{\nu,\mu}(z)$	Lommel's functions	3.10.1
$\sigma(z)$	Weierstrass sigma function	10.5
$\sigma_\alpha(z)$	Weierstrass sigma function	10.5
$\mathrm{sc}\, z = \frac{\mathrm{sn}\, z}{\mathrm{cn}\, z}$, $\mathrm{sd}\, z = \frac{\mathrm{sn}\, z}{\mathrm{dn}\, z}$, $\mathrm{sn}\, z$	Jacobi elliptic functions	10.3

Symbol	Name of the function	Chapter and section
$T(m, n, x)$	Toronto functions	6.7.2
$T_n(x)$	Chebyshev (Tchebicheff) polynomials of the 1st kind	5.7.1
$T_n^*(x)$	Shifted Chebyshev polynomials of the 1st kind	5.7.1
$T_n(t, x)$	—	3.4
$\vartheta_i(z, t)$, $i = 0, 1, 2, 3, 4$; $\vartheta_i(z)$, $i = 1, 2, 3, 4$	Elliptic theta functions	10.2
$U_n(x)$	Chebyshev polynomials of the 2nd kind	5.7.1
$U_n^*(x)$	Shifted Chebyshev polynomials of the 2nd kind	5.7.1 6.1.1
$U(a, b, z)$	Confluent hypergeometric function	6.1.1
$W_{\varkappa,\mu}(z)$	Whittaker function	7
$Y_\nu(z)$	Neumann function	3.1
$\mathrm{zn}(z, k)$	Jacobi zeta function	10.4
$Z_n(z)$	—	2.10
$\zeta(z)$	Riemann zeta function	1.3
$\zeta(z, \alpha)$; $\zeta^*(z, \alpha)$	Generalised zeta function	1.4
$\zeta_n(z)$	Spherical Bessel functions	3.3
$\zeta(z)$	Weierstrass zeta function	10.5

Index